Contents

Color Vision

Color Vision: From Genes to Perception documents the current state of understanding about primate color vision in 20 review articles written by 35 leading international experts. The articles range from genes – the molecular genetics of the human cone photopigment genes – to perception – the color processing of complex scenes. Detailed overviews of such basic topics as cone spectral sensitivity and color processing in the retina and cortex are included. Introductions are given to important and innovative technologies such as molecular genetics, anatomical staining, visual psychophysics, intracellular and extracellular physiological recordings, and functional magnetic resonance imaging.

Color Vision is intended for graduate students and research specialists. By bringing together scientists from different disciplines, the book will clarify issues of general interest for the expert and nonexpert alike.

Karl Genenfurtner is Professor of Biological Psychology at Institut für Psychologie, Otto-von-Guericke-Universität Magdeburg, Germany.

Lindsay Sharpe is professor of experimental ophthalmology at Eberhard-Karls-Universität, Tübingen, Germany.

Color Vision

From Genes to Perception

Edited by

Karl R. Gegenfurtner

Institut für Psychologie,
Otto-von-Guericke-
Universität Magdeburg

Lindsay T. Sharpe

Eberhard-Karls-Universität, Tübingen

PUBLISHED BY THE PRESS SYNDICATE OF THE UNIVERSITY OF CAMBRIDGE
The Pitt Building, Trumpington Street, Cambridge, United Kingdom

CAMBRIDGE UNIVERSITY PRESS
The Edinburgh Building, Cambridge CB2 2RU, UK
40 West 20th Street, New York, NY 10011-4211, USA
10 Stamford Road, Oakleigh, Melbourne 3166, Australia
Ruiz de Alarcón 13, 28014 Madrid, Spain
Dock House, The Waterfront, Cape Town 8001, South Africa

http://www.cambridge.org

First published 1999
First paperback edition 2001
Reprinted 2001

Printed in the United Kingdom at the University Press, Cambridge

Typeset in Times Roman in FrameMaker [au]

A catalog record for this book is available from the British Library

Library of Congress Cataloging in Publication data is available

ISBN 0 521 59053 1 hardback
ISBN 0 521 00439 X paperback

Contributors

Heidi A. Baseler (heidi@white.stanford.edu) Stanford University, Department of Psychology, Building 420, Stanford, CA 94305-2130, USA

Geoffrey M. Boynton (boynton@salk.edu) Salk Institute Biological Studies, PO Box 85800, San Diego, CA 92186-5800, USA

David Calkins (calkins@cvs.rochester.edu) Department of Ophthalmology, University of Rochester Medical Center, Room 3-3750, 601 Elmwood, Rochester, NY 14642, USA

Michael D'Zmura (mdzmura@uci.edu) Department of Cognitive Sciences, University of California Irvine, Irvine, CA 92717, USA

Dennis M. Dacey (dmd@u.washington.edu) Biological Structure HSB G-514, University of Washington, Seattle, WA 98195-7420, USA

Stephen A. Engel (engel@psych.ucla.edu) Department of Psychology, University of California Los Angeles, Franz Hall Box 951563, Los Angeles, CA 90095-1563, USA

Rhea T. Eskew Jr. (eskew@neu.edu) Department of Psychology, 125-NI, Northeastern University, Boston, MA 02115, USA

Karl R. Gegenfurtner (karl@kyb.tuebingen.mpg.de) Max-Planck-Institut für biologische Kybernetik, Spemannstr. 38, 72076 Tübingen, Germany

Franco Giulianini (fgiulian@bu.edu) Biomedical Engineering, Boston University, 44 Cummington Street, Boston, MA 02115, USA

Michael J. Hawken (mjh@cns.nyu.edu) New York University, Center for Neural Science, 4 Washington Place, 8th floor, New York, NY 10003, USA

Herbert Jägle (herbert.jaegle@uni-tuebingen.de) Universitäts-Augenklinik, Forschungsstelle für Experimentelle Ophthalmologie, Röntgenweg 11, 72076 Tübingen, Germany

Daniel C. Kiper (kiper@ini.phys.ethz.ch) Institute of Neuroinformatics, UNIZH/ETHZ, Winterthurerstr. 190, 8057 Zürich, Switzerland

John Krauskopf (jkr@cns.nyu.edu) New York University, Center for Neural Science, 4 Washington Place, 8th floor, New York, NY 10003, USA

Jan Kremers (jan.kremers@uni-tuebingen.de) Universitäts-Augenklinik, Forschungsstelle für Experimentelle Ophthalmologie, Röntgenweg 11, 72076 Tübingen, Germany

Trevor Lamb (tdl1@cam.ac.uk) Physiological Laboratory, Cambridge University, Downing Street, Cambridge CB2 3EG, England

Barry B. Lee (blee@gwdg.de) Abteilung Neurobiologie, Max-Planck-Institut für Biophysikalische Chemie, Am Faßberg, 37077 Göttingen, Germany

Peter Lennie (pl@cns.nyu.edu) New York University, Center for Neural Science, 4 Washington Place, 8th floor, New York, NY 10003, USA

Jonathan B. Levitt (jbl@sci.ccny.cuny.edu) City College of the City University of New York,

Department of Biology, 138 Street & Convent Avenue, New York, NY 10031, USA

Laurence T. Maloney (ltm@cns.nyu.edu) New York University, Department of Psychology, 6 Washington Place, New York, NY 10003, USA

James S. McLellan (mclellan@vision.eri.harvard.edu) Schepens Eye Research Institute, 20 Staniford Street, Boston, MA 02114, USA

Jeremy Nathans (jnathans@bs.jhmi.edu) Howard Hughes Medical Institute Research Laboratories, The Johns Hopkins University School of Medicine, 725 N. Wolfe Street, Room 805 PCTB, Baltimore, MD 21205, USA

Allen B. Poirson (poirson@accuimage.com) AccuImage Diagnostics Corp., 400 Oyster Point Blvd., Ste. 114, So. San Francisco, CA 94080, USA

Austin Roorda (aroorda@popmail.opt.uh.edu) University of Houston, College of Optometry, 4800 Calhoun Road, Houston, TX 77204-6052, USA

Julie L. Schnapf (schn@phy.ucsf.edu) Departments of Physiology and Ophthalmology, University of California, San Francisco, CA 94143-0730, USA

David M. Schneeweis (schnee@skivs.ski.org) Smith-Kettlewell Eye Research Institute, 2232 Webster Street, San Francisco, CA 94115, USA

Robert M. Shapley (shapley@cns.nyu.edu) New York University, Center for Neural Science, 4 Washington Place, 8th floor, New York, NY 10003, USA

Lindsay T. Sharpe (lindsay.sharpe@uni-tuebingen.de) Universitäts-Augenklinik, Forschungsstelle für Experimentelle Ophthalmologie, Röntgenweg 11, 72076 Tübingen, Germany

Luiz Carlos Silveira (luiz@marajo.secom.ufpa.br) Dept Fysiologia, Universidade federal do Para, Belem, Brazil

Benjamin Singer (bens@cvs.rochester.edu) Center for Visual Science, University of Rochester, 262 Meliora Hall, Rochester, NY 14627, USA

Andrew Stockman (astockman@ucsd.edu) Department of Psychology, University of California San Diego, 9500 Gilman Drive, La Jolla, CA 92093-0109, USA

Brian A. Wandell (wandell@stanford.edu) Stanford University, Department of Psychology, Building 420, Stanford, CA 94305-2130, USA

Heinz Wässle (waessle@mpih-frankfurt.mpg.de) Abt. Neuroanatomie, Max-Planck-Institut für Hirnforschung, Deutschordenstraße 46, Frankfurt am Main 60528, Germany

David R. Williams (david@cvs.rochester.edu) University of Rochester, Center for Visual Science, Box 270270, Rochester, NY 14627-0270, USA

Elizabeth S. Yamada (esyamada@ufpa.br) Dept. Fysiologia, Universidade federal do Para, Belem, Brazil

Qasim Zaidi (qz@cns.nyu.edu) State University of New York, College of Optometry, 100 East 24th Street, New York, NY 10010, USA

Foreword by Brian B. Boycott

This book arises from a meeting organized by the two editors in Tübingen (5–7 September 1996). It was a happy meeting with good interaction between the participants. The editors have continued their hard work by supervising most of the contributors into writing comprehensive, well-balanced articles. In my opinion, a successful effort has been made to make the papers accessible to a wider readership than is usual for the published offerings of a specialist gathering. Given the large array of techniques that today have to be mastered to do a worthwhile piece of neurobiological research, all of us, especially graduate students, are increasingly unable to find time to reach out beyond the constricted horizons of our own specialities, in short, to take a broader view. It was bold to organize a meeting spanning the genetic determination of cone photopigments in primates, their electrophysiology and evolution, on through retinal circuitry to cerebral cortical processing of chromatic signals, the interaction between color and motion in the primate visual system and to end with papers on the perception of color. The result is a welcome gathering together of diverse approaches being made to understand the neural mechanisms of color vision.

There is a tension between the increasing technical effort and specialization required in modern neurobiological research and the achievement of sufficient general understanding to enunciate basic general rules of neural functioning. About twenty-five years ago, H. B. Barlow (1972, *Perception*, Vol. 1, pp. 371–394) sought a "neuron doctrine for perceptual psychology." He reviewed the then available literature seeking basic formulations (dogmas) that might have the power to do for neurobiology what Crick and Watson's formulation, "DNA codes protein," has done for molecular biology. It is no criticism of the present volume to say that it would be interesting to use it to examine to what extent modern understanding of the mechanisms of color vision fits into or modifies those dogmas. Are there compelling reasons to reformulate them? To what extent does an understanding of other sensory mechanisms affect generalizations in color vision, and vice versa? For example, suppose that we now knew the neural networking for the perception of moving colored buses; would that predict the mechanisms for an equivalent auditory perception? That is a task for another place and another time, but it is worth reminding readers, especially because they are not addressed frequently, that the issues of the basic principles that Barlow raised will not go away.

While making these, perhaps rather obvious, remarks, it is also worthwhile to remind ourselves of a current great inhibition to making reductionist generalizations about central nervous systems. We are all inclined to forget, or pay only lip-service to, the fact that for most parts of the brain we do not know how many types of nerve cells we have to deal with and nothing, therefore, about what they may do to our current interpretations. For example, it is only recently that we have come to know that amacrine cells in the inner nuclear layers of monkey and rabbit retinae comprise between 28 and 35% of the population of cells. Furthermore, in rabbit retina these have been shown to constitute some 26 different morphological types of interneuron. Only for four of these types (the AII, the starburst, the indoleamine-accumulating and the dopaminergic cells) do we have evidence of what these cells do physiologically (M. A. MacNeil & R. H. Masland, 1998, *Neuron*, Vol. 20, pp. 971–982). The primate amacrine cell system is certainly equally complex. Furthermore, as those authors point out, the work of J. Lund and colleagues (e.g., Lund & Wu, 1997, J. Comp. Neurol., Vol. 384, pp 109-126) reveals some 50 anatomical types of local circuit neurons in the striate cortex about which we know nothing physiologically and which are hardly yet incorporated into functional models.

The need for knowledge of what structures are actually in the areas of the brain that we analyze physiologically and perceptually; the need to look for the basic rules of the operation of the neural nets that we study; both seem to me to need emphasis in a foreword to a book of papers that I hope will speed us towards an even more fundamental and general understanding of the problems that they address.

Brian B. Boycott
December 8 1998

Acknowledgments

There have been monumental advances in our knowledge of color vision during the last 100 years. The spectral sensitivities of the cone photoreceptors are now known in great detail down to their molecular genetic foundations. The characteristics of subsequent color opponent processes have been well studied, even though it is still not known how exactly they are implemented by the retinal circuitry. Finally, big advancements have been made in our understanding of the cortical mechanisms of color vision. This book tries to offer a snapshot of our current understanding about the visual neuroscience of color vision. We hope that the reader gets the impression that it is a vibrant and active field of research.

Although we do cover many different aspects of color vision in this book, the list of topics that are missing – owing to limited space – is just as long. In particular, we regret that we could not include chapters on the development of color vision, effects of aging, and, most notably, the current state of understanding on color appearance.

We are grateful to all our colleagues who contributed to this book, and at the same time, we would like to apologize to them for our persistent nagging about the delivery of their chapters. We are grateful to our editors at Cambridge University Press: Robin Smith (now at Springer) who got us started and to Michael Penn and Cathy Felgar, who brought it to an end.

This book is based on a workshop organized by the editors in 1996. We are grateful to the Max-Planck-Society and the *f*ortüne-program of Tübingen University. Without their generous financial support the workshop would not have been possible. We are particularly grateful to Heinrich Bülthoff and Eberhart Zrenner for their encouragement. We would also like to thank our friends and families for support, encouragement, and food.

Finally, we recall what Ragnar Granit warned William Rushton long ago: "Colour is the femme fatale of vision. When once seduced, you will never be a free man again." Truly, many a dangerous temptation comes arrayed in fine bright colors.

Karl R. Gegenfurtner and Lindsay T. Sharpe
Tübingen, May 1999

Part I: Photoreceptors

1

Opsin genes, cone photopigments, color vision, and color blindness

Lindsay T. Sharpe, Andrew Stockman, Herbert Jägle, and Jeremy Nathans

In this chapter, we introduce the molecular structure of the genes encoding the human cone photopigments and their expression in photoreceptor cells. We also consider the consequences that alterations in those genes have on the spectral sensitivity of the photopigments, the cone photoreceptor mosaic, and the perceptual worlds of the color normal and color blind individuals who possess them. Throughout, we highlight areas in which our knowledge is still incomplete.

Trichromacy. Human color vision is trichromatic; this has three consequences. First, as was recognized in the eighteenth century (e.g., Le Blon, 1722; see Birren, 1963, 1980), but only formally postulated (Grassman, 1853) and verified (Maxwell, 1855, 1860) in the nineteenth century, the number of independent variables in color vision is three. That is, all colors can be matched by just three parameters: either by the three primaries of additive light mixture – typically, violet, green, and red – or by the three primaries of subtractive pigment mixture – typically, cyan, yellow, and magenta.

Second, as intimated by Palmer (1777, 1786; see also Voigt, 1781; Walls, 1956; Mollon, 1997), definitively stated by Young (1802, 1807), and revived by Helmholtz (1852), trichromacy is not a physical property of light but a physiological limitation of the eye: All color perceptions are determined by just three physiological response systems.

Third, as pointed out by Maxwell (1855) and applied by König and Dieterici (1886), a linear transform must exist between the tristimulus color matching properties of the eye, as established by the three primaries of additive light mixture, and the spectral sensitivities of the three physiological systems mediating the matches (see Chapter 2).

The three physiological response systems are universally acknowledged to be the three types of retinal photoreceptor cell, each containing a different photopigment: the short (S)-, middle (M)-, and long (L)-wave sensitive cones.[1] These have distinct, spectral sensitivities (Fig. 1.1A) or absorption spectra (Fig. 1.1B), which define the probability of photon capture as a function of wavelength. The absorbance spectra of the S-, M-, and L-cone photopigments overlap considerably, but have their wavelengths of maximum absorbance (λ_{max}) in different parts of the visible spectrum: ca. 420, 530, and 558 nm, respectively. When estimated in vivo, the λ_{max}'s are shifted to longer wavelengths (ca. 440, 545, and 565 nm, respectively) by the transmission properties of the intervening ocular media: the yellowish crystalline lens and the macular pigment of the eye (see Chapter 2).

The individual cone photopigments are blind to the wavelength of capture; they signal only the rate at

[1]The fourth type of photoreceptor cell, the rods, contain rhodopsin as their photopigment. They are by far the most prevalent in the human retina, constituting more than 95% of all photoreceptor cells. However, they do not contribute to color vision, except under limited, twilight conditions (see section on rod monochromacy). Under most daylight conditions, where we enjoy color vision, the rod photoreceptor response is saturated by excessive light stimulation.

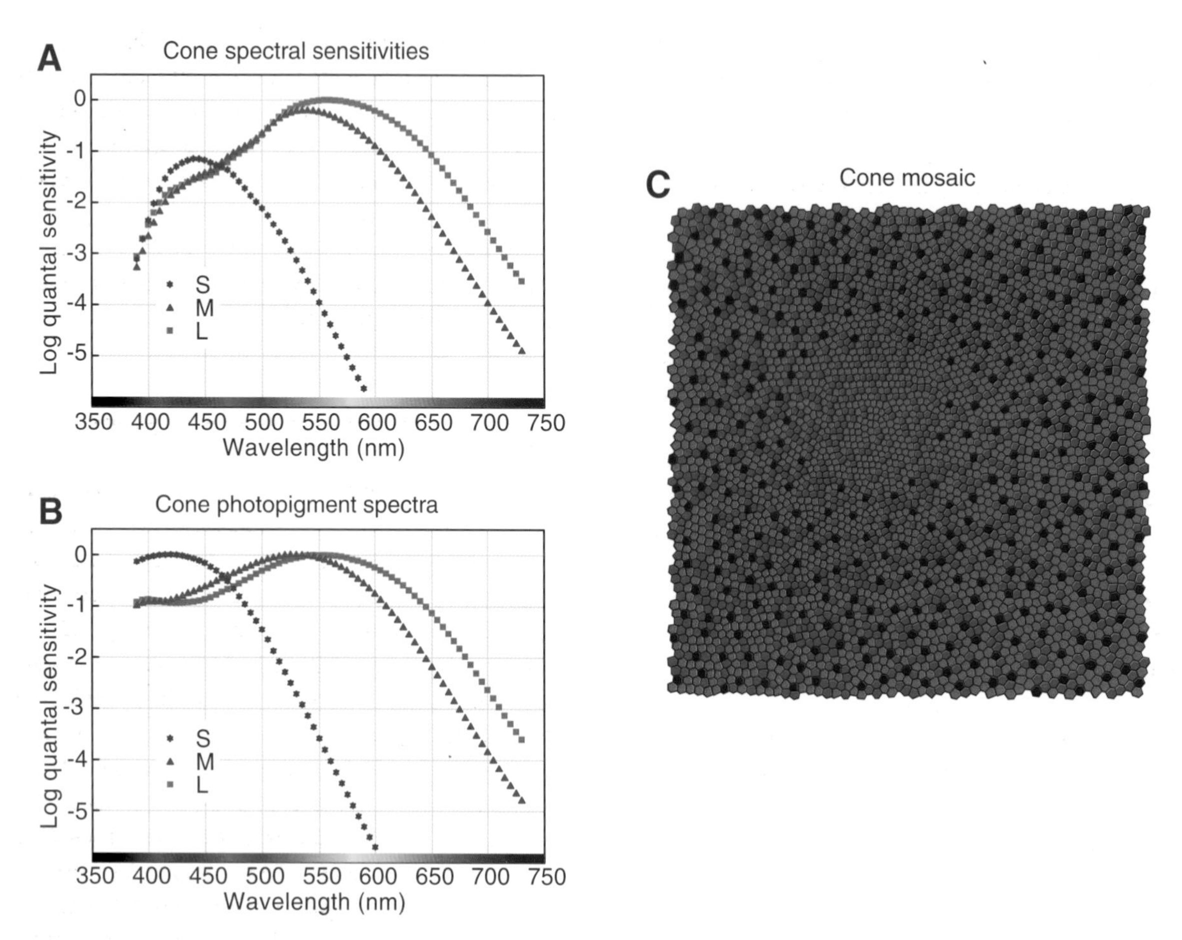

Figure 1.1: Cone spectral sensitivities and their representations in the photoreceptor mosaic. (A) Estimates of the light absorbing properties of the L- M-, and S-cones, measured at the cornea, as a function of wavelength (see Chapter 2, Table 2.1, for values). The heights of the curves have been adjusted according to the assumption that the relative cone sensitivities depend on the relative numbers of the different cone types; namely, that 7% of all cones contain the S-cone pigment and that, of the remaining 93%, those containing the L-cone pigment are 1.5 times more frequent than those containing the M-cone pigment (see Chapter 2). (B) The cone pigment absorption spectra. These were determined from the cone spectral sensitivity functions, by correcting the latter for the filtering of the ocular media and the macular pigment and for the self-screening of the pigment in the outer segment (see Chapter 2, Table 2.1, for values). (C) The cone mosaic of the rod-free inner fovea of an adult human retina at the level of the inner segment (tangential section). Superior is at the top and nasal to the left. The region is ca. 1 deg of visual angle in diameter (ca. 300 μm). The center coordinates of the cone cross sections shown were obtained from the retina of a 35-year-old male (Curcio & Sloan, 1992). The outer dimensions of the cone cross sections have been defined mathematically by Voronoi regions and computer-colored according to the following assumptions: (1) only three cone opsin genes, those encoding the S-, M-, and L-cone pigments are expressed; (2) the inner roughly circular area (ca. 100 μm or 0.34 deg in diameter), displaced slightly to the upper left quadrant of the mosaic, is free of S-cones (Curcio et al., 1991); (3) S-cone numbers in the rest of the retina do not exceed 7% and are semiregularly distributed (Curcio et al., 1991); and (4) there are approximately 1.5 times as many L- as M-cones in this region of the retina and they are randomly distributed (see Chapter 5). The diameters of the cross sections in the center are slightly smaller than those at the outer edge to allow for close packing.

which photons are caught (cf. Rushton, 1972). Lights of different spectral distributions, therefore, will appear identical, if they produce the same absorptions in the three cone photopigments, and different, if they do not (see Chapter 2). Thus color vision – the ability to discriminate on the basis of wavelength – requires comparisons of photon absorptions in different photopigments. And, accordingly, trichromatic color vision requires three such independent comparisons. Merely summing the absorptions in the three cone photopigments at some later neural stage will permit brightness or contrast discrimination, but not color vision.

The loss of one of the cone photopigments, as occurs in certain congenital disorders, reduces (photopic) color vision to two dimensions or dichromacy. The loss of two further reduces it to one dimension or monochromacy. And, the loss of all three completely extinguishes it. Vision, then, is purely scotopic and limited to the rods.

Cone pigments and visual pathways. In man and the higher primates, the primary visual or retino-geniculostriate pathway has evolved into three postreceptoral neuronal systems for transmitting the cone signals that arise from the photopigment absorptions (see Chapter 11). These have been characterized as: (i) a luminance subsystem, which mainly carries information about luminance contrast by summing the relative rates of quantum catch in the M- and L-cones (and is sensitive to high spatial and temporal frequencies); (ii) a yellow-blue color subsystem, which mainly carries information about color contrast by comparing the relative rate of quantum catch in the S-cones with those in the M- and L-cones; and (iii) a red-green color subsystem, which carries information about color contrast by comparing the relative rates of quantum catch in the M- and L-cones. Roughly, it can be said that the three subsystems allow three kinds of discriminations: light from dark, yellow from blue, and red from green.

The dimensionality of the color information transmitted by these postreceptoral subsystems is, in the first instance, limited by the number of available cone photopigments. If one or more of the three normal cone photopigments is absent, then the dimensionality is correspondingly reduced from trichromacy to dichromacy or monochromacy. On the other hand, if an extra, fourth cone photopigment is present, as occurs in certain heterozygotic carriers of color blindness, full four-dimensional or tetrachromatic color vision does not seem readily possible (see page 38). The limitation may be the inability of the postreceptoral subsystems to convey more than three independent color signals (see Chapter 6).

Molecular genetics of the opsin genes

The spectral sensitivity of the cone photopigments is intimately related to the structure of the cone pigment molecules. These are concentrated in the photoreceptor outer segment, a specialized cilium containing the phototransduction machinery (see Fig. 1.2 and Chapter 3). Each pigment molecule consists of a transmembrane opsin (or apoprotein) covalently linked to the same, small conjugated chromophore (11-*cis*-retinal), which is an aldehyde derivative of vitamin A.

All opsins are heptahelical proteins, composed of seven transmembrane helices that are linked together by intra- and extracellular loops. Structural work on the opsin of the rod pigment rhodopsin (Unger & Schertler, 1995), about which we have the most information, indicates that the membrane-embedded helices form a barrel around a central retinal binding pocket (see Fig. 1.2). The binding site of the chromophore in both the cone and rod opsins is located in helix 7, a region that has been relatively conserved during the process of divergent evolutionary change (see the following section).

Photon absorption by the pigment molecules initiates visual excitation by causing an 11-*cis* to all-*trans* isomerization of the chromophore, which activates a transducin G-protein (see Chapter 3). Opsins absorb maximally in the ultraviolet region of the spectrum below 300 nm, whereas retinal absorbs maximally at about 380 nm (Knowles & Dartnall, 1977). It is only by binding together that a broad absorbance band (known as the α-band) in the visible spectrum is created. The λ_{max} of the α-band depends on the geneti-

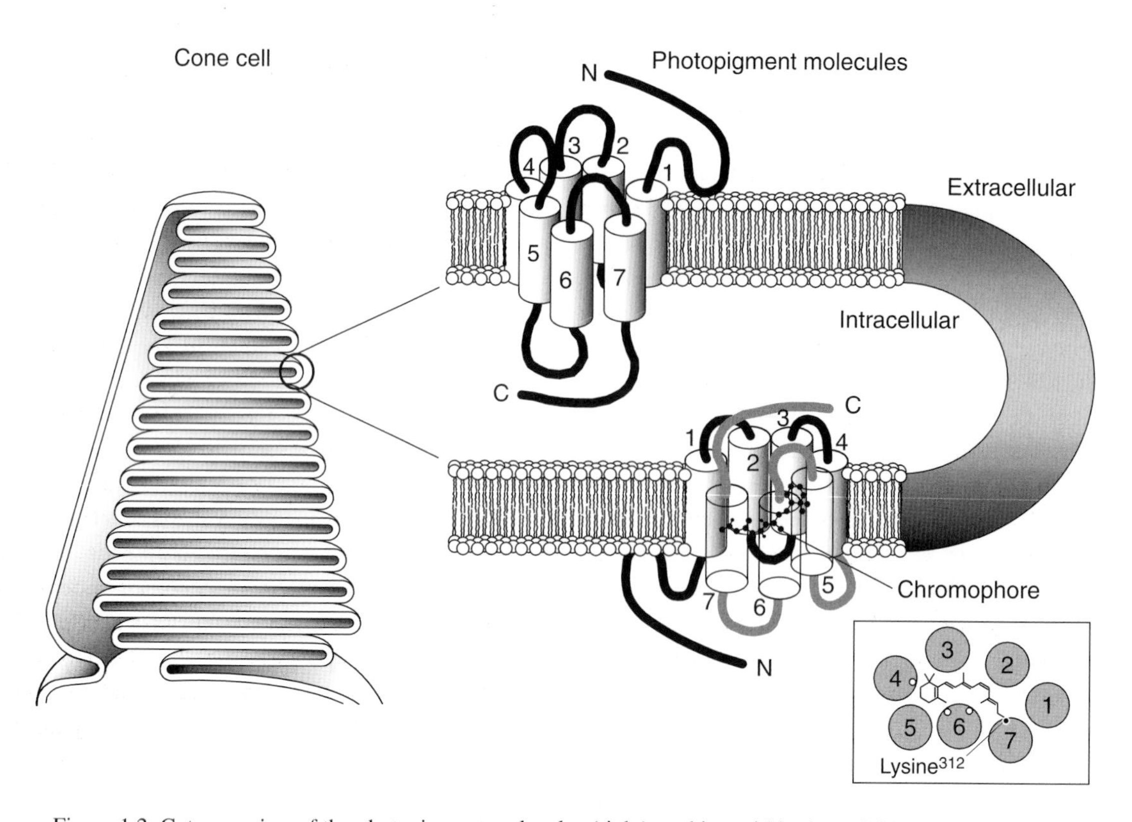

Figure 1.2: Cutaway view of the photopigment molecules (right) packing within the enfolded membrane discs in the outer segments of the cone photoreceptor cells (left). Each molecule consists of a transmembrane opsin bound to a chromophore, 11-*cis*-retinal. The opsin or protein portion of the molecule is a chain of amino acids, running from an amino-terminal end (N), exposed on the external aqueous surface of the membrane discs, to a carboxyl terminal region (C), exposed on the internal aqueous surface of the discs. The chain has seven coils, termed α-helices, spanning the membrane (Hargrave et al., 1984). Linked together by loops in the rest of the chain, the α-helices encircle the chromophore (right, lower cutaway view). The loops are distinguished by whether they occur in the luminal (extracellular) or cytoplasmic (intracellular) face of the cell. The view is from the intracellular surface. It indicates the approximate position of the α-helices and of the three amino acids (open circles) that are believed to have the major influence on the λ_{max} of the pigment (see Fig. 1.5B). The tail of the chromophore is attached by a protonated Schiff base to a charged lysine amino acid residue lying at nucleotide position 312 in the chain of the L- and M-cone opsins (filled circle, see Fig. 1.5B) and at position 293 in the chain of the S-cone opsin (corresponding to position 296 in rhodopsin). Features critical to the function of the opsin are well conserved in all known mammalian species, with the interhelical loops being, on average, more conserved than the transmembrane helical regions.

cally determined amino acid sequence of the opsin and the relationship that the opsin establishes with the chromophore. A second, but lower, absorbance band, known as the β-band, may also be present, which is due to the *cis*-band of the chromophore. The upturn of the L- and M-cone photopigment spectra at very short wavelengths has been interpreted as indicating the presence of such a β-band (see Fig. 1.1B).

The different opsins of the S-, M-, and L-cone photopigments and of the rod photopigment are encoded by four separate genes. These have been formally identified by the HUGO/GDB (genome data base) Nomenclature Committee as the BCP (blue cone pigment), GCP (green cone pigment), RCP (red cone pigment), and RHO (rhodopsin) genes. Visual psychophysicists, however, often refer to them as the S-, M-, L-cone and rod pigment genes.

The genes encoding the S-cone and rod pigments reside alone as single copies. The former is found on the long or q-arm of chromosome 7 (Nathans, Thomas, & Hogness, 1986) within a cytogenetic location between 7q31.3 and 7q32 (Nathans, Thomas, & Hogness, 1986; Fitzgibbon et al., 1994) and the latter on the q-arm of chromosome 3, between 3q21.3 and 3q24 (Nathans, Thomas, & Hogness, 1986). In contrast, the genes encoding the L- and M-cone pigments reside on the q-arm of the X-chromosome at location Xq28 (Nathans, Thomas, & Hogness, 1986) within a head-to-tail tandem array (Vollrath, Nathans, & Davis, 1988; Feil et al., 1990), which may contain as many as six gene copies. In general, the L-cone pigment gene is present in only a single copy and precedes the multiple M-cone pigment genes in the array (see the following section). In addition, the array contains the five nested exons of a complete gene (termed TEX28), the first exon of which is located ca. 700 base pairs downstream of the end of the visual pigment gene cluster (see the section on visual pigment gene structure). Extra, truncated (lacking exon 1) nonfunctional copies may be interdigitated between the opsin genes, filling up most of the intervening area. The TEX28 gene is expressed in testes but not in the cone photoreceptor cells (Hanna, Platts, & Kirkness, 1997). It is transcribed in the orientation opposite to the cone opsin genes (see mRNA transcription).

Pigment gene evolution. The reason for the separate chromosomal locations of the opsin genes is unknown, and their evolutionary development is subject to speculation based on comparisons of their sequence homologies (see the section on opsin gene sequence homologies). One plausible alternative, although by no means the only one, is that the three cone opsin genes and human trichromacy evolved in the following steps (see also Chapters 6 and 7; Goldsmith, 1991):

1. The emergence of a primordial opsin gene on the X-chromosome that encodes a pigment with its λ_{max} in the region conferring the greatest sensitivity to the quantum-intensity–based spectral distribution of sunlight (Dartnall, 1962, but see Lynch & Soffer, 1999) and reflectance of green plants (Lythgoe, 1972). This system formed the basis of the contrast or luminance subsystem of vision, which has a λ_{max} near 555 nm.

2. The emergence of a second opsin gene, about 500 million years ago (Nathans, 1987; Chiu et al., 1994; Hisatomi et al., 1994), located on chromosome 7. Through the accumulation of DNA sequence changes (see the section on opsin gene sequence homologies), it encoded a pigment with its λ_{max} placed at short wavelengths (the S-cone pigment) and expressed it in a different subset of (anatomically distinct) cones from that in which the primordial opsin gene was expressed. The subsequent development of second-order neurons (the yellow-blue opponent color subsystem), which is sensitive to differences in the excitations of the two sets of cones, enabled the discrimination of many forms of natural green vegetation, differing mainly in their reflectances of short-wave light (Hendley & Hecht, 1949; Mollon, 1996).

3. The emergence of a third opsin gene, about 30 to 40 million years ago (Nathans, 1987; Yokoyama & Yokoyama, 1989; Yokoyama, Starmer, & Yokoyama, 1993), as a result of the duplication of the ancestral opsin gene on the X-chromosome. The event copied the transcription unit, but not the locus control region (see the following section). According to one view, the accumulation of DNA sequence changes in the duplicated genes resulted in them encoding distinct M- and L-cone pigments and being expressed in different subsets of (anatomically similar) cones. An alternative view is that the duplication event resulted from unequal crossing over (involving Alu[2] repeats) between two alleles of the ancestral gene that had different spectral sensitivities, so that trichromatic color

[2]An Alu element is a dispersed repetitive DNA sequence that is about 300 bp in length. The name derives from the restriction endonuclease Alu I that cleaves it. The sequence occurs in about 300,000 copies in the human genome and is believed to have no coding value. An Alu repeat element at the site of insertion of the duplicated opsin gene may have been important in promoting crossing over within the array.

vision did not have to await the accumulation of mutations. Initially, changes in the relative excitations of the two pigments caused by changes in wavelength were undifferentiated, but the subsequent recruitment of existing second-order neurons or the development of new ones (the red-green opponent color subsystem) enabled discriminations in the yellow-green to orange-red spectral region. This duplication event may have occurred in our arboreal ancestors, after the divergence of the Old- and New-World monkeys (see Chapters 6 and 7), as an adaptation to frugivory, assisting the detection of fruit amid foilage (Mollon, 1989, 1991; Osorio & Vorobyev, 1996).

This story is necessarily complicated by the coevolution of the rhodopsin gene, which is similar in structure and sequence to the cone opsin genes. It appears to have derived from the S-cone opsin gene, after the divergence of the latter from the common ancestral gene (Okano et al., 1992). The tight clustering of the λ_{max}'s of almost all vertebrate rhodopsins near 500 nm – the human rod spectral sensitivity measured in vivo peaks at 507 nm and the absorbance spectrum at 493 nm – has so far eluded easy explanation (Goldsmith, 1991). It does not directly correspond to the λ_{max} of starlight, moonlight, or twilight (Lythgoe, 1972).

Visual pigment gene structure. Structurally, each visual pigment gene is a large deoxyribonucleic acid (DNA) molecule, consisting of thousands of subunits – nucleotides – linked together. These are the nucleotide base sequences. Each comprises a nitrogenous base (adenine, guanine, thymine, or cytosine), a phosphate molecule, and a sugar molecule (deoxyribose). Owing to the double-helical structure of DNA, the nucleotide in one DNA strand has a complementary nucleotide in the opposite strand. The two are held together, in nucleotide or base pairs (bp), by weak hydrogen bonds. Adenine (A) conjoins with thymine (T) and guanine (G) with cytosine (C); no other combinations are possible. The base sequences can be divided into promoter, noncoding (intron), and coding (exon) sequences (see Fig. 1.3A).

(i) Transcription unit: The term "transcription unit" is often used to refer to the exons and the intervening introns to indicate the region that is actually synthesized into messenger ribonucleic acid (mRNA) before being translated into the opsin. The transcription region begins at the start or cap site at the 5' (upstream or head) end of the gene. It is followed by a short leader sequence – 6 bp long in the S-cone pigment gene and 40 bp long in the M- and L-cone pigment genes – that is not translated into the opsin. Downstream of the leader sequence is the start codon, a trinucleotide sequence, ATG, which specifies the initiation of opsin translation. It is paired with a stop codon, TGA, at the 3' (downstream or tail) end, which specifies the termination of opsin translation. In the transcription region, the stop codon is followed by untranslated tail sequences. These include a signal – the polyadenylation or poly (A) site – for the addition of a string of adenosine residues. The exact function of the residues is unknown, but there is evidence that mRNA degradation occurs from the 3' end and that the poly (A) tail together with the poly (A)–specific RNA-binding proteins increases the half-life of the mRNA during translation (see the section on opsin translation).

(ii) Promoters: Promoters are specific regulatory sequences or boxes upstream of the transcription start site. They bind the enzyme (RNA polymerase) that catalyzes the synthesis of the RNA chain, a reaction that is referred to as transcription (see Fig. 1.3). The first promoter sequence, the TATA regulatory box, is ca. 25 bp upstream of the transcription start site. It is involved in binding RNA polymerase via a TATA binding protein. Another promoter sequence, the CCAAT box, is ca. 70–90 bp upstream of the transcription start site. The promoter sequences also interact with transcriptionally active sequences contained in the upstream locus control region (LCR) to regulate the rate of DNA transcription into RNA and hence the amount of opsin gene expression.

(iii) Introns: The intron sequences are silent or noncoding sequences usually believed to have no apparent function (but see the section on intergenic recombination). The possibility that they contain regulatory sequences involved in gene expression, however, cannot be ruled out. They are delimited by recognition

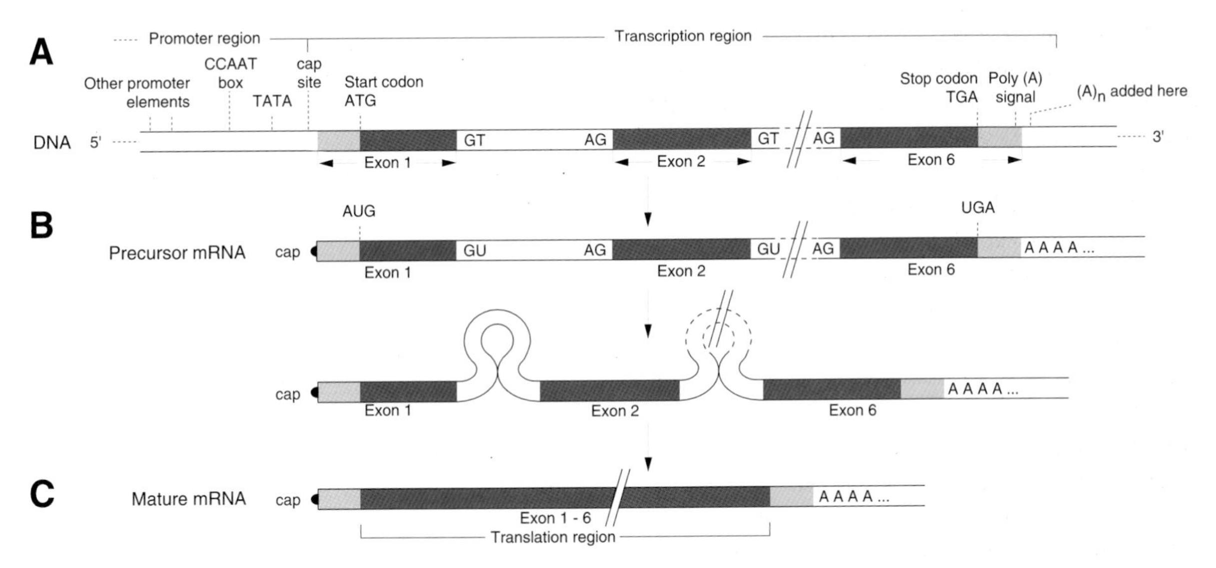

Figure 1.3: Schematic representation of an opsin gene and its transcription into mRNA. (A) The structure of an L-cone opsin gene (DNA), indicating the promoter region, and the untranslated regions at the 5' and 3' ends (gray) the coding regions (black) and the intervening introns (white). (B) and (C) The transcription of the DNA into precursor mRNA and the splicing out of the introns to form mature mRNA. The structure of the mRNA is similar to that of the DNA, except that the sugar molecule is ribose instead of deoxyribose, uracil (U) replaces thymine (T), and the molecule is single-stranded. mRNA is capped or blocked with 7-methylguanosine at its 5' end. There follows a short untranslated region and then the start codon, ATG (AUG), which specifies the initiation of translation. A stop codon, TGA (UGA), indicates the termination of the translated region. Capping and tailing of precursor mRNA precede the splicing out of introns.

sites, which are necessary for identifying and splicing them out from the mRNA precursor (see Fig. 1.3). Introns typically begin with the dinucleotide GT (the splice donor; GU in the precursor mRNA) and end with the dinucleotide AG (the splice acceptor).

(iv) Exons: The opsin-coding sequences are divided into exons, which are separated by the introns and numbered according to their proximity to the 5' end (see Fig. 1.3A). Within the exons, the nucleotide sequences are grouped into triplets – the 3-base sequences or codons – each of which specifies a constituent amino acid (monomer) of the polypeptide chain of the visual pigment opsin (see Table 1.1). There are 64 possible codons (the possible combinations of the four nucleotides), but only 20 unmodified amino acids in the opsin. Thus each amino acid may have more than one codon. Many of the different codons for single amino acids differ only in the third nucleotide of the 3-base sequence.

(v) Gene length: The S-cone pigment gene comprises 5 exons (1,044 bp of which are protein coding) and 4 introns (total length: 2,200 bp). The length of the gene from its mRNA start or cap site (nucleotide base sequence 403) to its poly (A) site (nucleotide base sequence 1,510) is 3,308 bp. (The extra base pairs occur because the exons include 5' and 3' untranslated, nonprotein-coding, sequences that also end up in the mature mRNA; see Fig. 1.3). The M- and L-cone pigment genes each comprise six exons (1,092 bp of which are protein coding) and five introns (total length: 12,036 bp and 14,000 bp, respectively, for the M- and L-cone pigment genes). The length of the M-cone pigment gene is 13,300 bp and that of the L-cone pigment gene is 15,200 bp.

A small, extra exon, encoding only 38 amino acid residues (114 bp), is found at the beginning of the L- and M-cone pigment genes. It may have been added at some point during evolution to the primordial visual pigment gene to facilitate transcription of more than one gene copy in the tandem array.

mRNA transcription and opsin translation. The base sequences in the DNA are transcribed into RNA, which is subsequently translated to produce the opsin. The primary product in transcription – the mRNA precursor, often called pre-mRNA – contains all of the base sequences, those defining the introns as well as the exons (see Fig. 1.3B). It is blocked or capped with 7-methylguanosine at its 5' end and tailed by a string of adenosine residues at its 3' end. The capping and tailing are believed to permit the export of mRNA from the cell nucleus.

During processing in the cell nucleus, the introns are spliced out, so that the final product in transcription – the mature mRNA (Fig. 1.3C) – only contains the exon sequences. The mature mRNA is exported to the cytoplasm of the photoreceptor cells, where it serves as a template for the synthesis of the opsin from its constituent amino acid residues (see Fig. 1.4). The translation process is complex, involving several stages and a family of transfer RNAs (tRNA), the role of which is to bond with amino acids and transfer them to the ribosome (the site of protein synthesis). The amino acids are assembled sequentially in the growing polypeptide chain of the opsin, from the amino-terminal end to the carboxyl-terminal end, according to the order of codons carried by the mRNA. In the chain, the amino acids are linked by the carboxyl group (COOH) of one amino acid and the amino group (NH_2) of another. Hence, opsins can be identified by their NH_2 (or N) and COOH (or C) ends.

DNA codons	Amino acid	Abbreviations		Class
GCA, GCC, GCG, GCT	Alanine	ala	A	(1)
AGA, AGG, CGA, CGC, CGG, CGT	Arginine +	arg	R	(4)
GAC, GAT	Aspartic acid -	asp	D	(3)
AAC, AAT	Asparagine	asn	N	(2)
TGC, TGT	Cysteine	cys	C	(1, 2)
GAA, GAG	Glutamic acid -	glu	E	(3)
CAA, CAG	Glutamine	gln	Q	(2)
GCA, GGC, GGG, GGT	Glycine	gly	G	(1)
CAC, CAT	Histidine (+)	his	H	(4, 6)
ATA, ATC, ATT	Isoleucine	ile	I	(1)
TTA, TTG, CTA, CTC, CTG, CTT	Leucine	leu	L	(1)
AAA, AAG	Lysine +	lys	K	(4)
ATG	Methionine (Start)	met	M	(1)
TTC, TTT	Phenylalanine	phe	F	(1, 5)
CCA, CCC, CCG, CCT	Proline	pro	P	(1)
AGC, AGT, TCA, TCC, TCG, TCT	Serine	ser	S	(2)
ACA, ACC, ACG, ACT	Threonine	thr	T	(2)
TGC	Tryptophan	trp	W	(1, 5, 6)
TAC, TAT	Tyrosine	tyr	Y	(2, 5)
GTA, GTC, GTG, GTT	Valine	val	V	(1)
TAA, TAG, TGA	(Stop)			

Table 1.1: The genetic code and the chemical properites of amino acids. The four types of nucleotides forming the deoxyribonucleic acid (DNA) codons are adenylic (A), guanylic (G), cytidylic (C), and thymidylic (T) acid. In ribonucleic acid (RNA) codons, uridylic (U) replaces thymidylic (T) acid. Amino acids belonging to the same class (1–6) are considered homologous and their substitution conservative. (+ or - indicates those amino acids most likely to be positively or negatively charged. The charge of histidine (+) depends on the local environment, and is, therefore, indicated in parentheses.)

Opsin gene sequence homologies. The nature of the pigment defined by an opsin gene depends on the nucleotide sequences of its exons, which are grouped into triplets (codons) encoding amino acid residues and numbered sequentially beginning with the first codon.

The S-cone pigment gene comprises 348 codons divided over its 5 exons, while the M- and L-cone pigment genes comprise 364 codons divided over their 6 exons. The S-cone pigment gene shows only 43 ± 1%

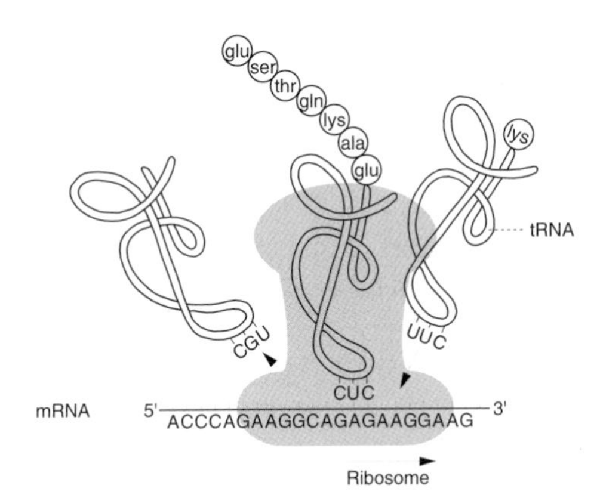

Figure 1.4: The mechanism of translation of the mRNA (from exon 5) into the L-cone opsin. Translation involves tRNA, each of which has a recognition site or triplet nucleotide sequence (anticodon) that is complementary to the triplet nucleotide coding sequence (codon) of mRNA. For example, the tRNA at the right carries lysine and its (RNA) anticodon CUU recognizes the lysine (RNA) codon AAG. Amino acids attach serially to the growing polypeptide chain and the respective tRNAs are jettisoned (as shown at left). The process is mediated by ribosomes (one is shown in outline) moving along the mRNA in a 5' to 3' direction.

amino acid identity with either the M- or L-cone pigment gene (see Fig. 1.5), which is about the same amount of homology with the rod pigment gene (41 ± 1%). In contrast, the M- and L-cone pigment genes show 96% mutual identity for the 6 exons (they are 98% identical at the DNA sequence level if the introns and 3'-flanking sequences are included). From the sequence homologies, it is possible to estimate the evolutionary divergence of the genes: The greater the identity, the more recent the divergence (see the section on pigment gene evolution). A curiosity is that the noncoding intron sequences of M- and L-cone pigment genes are more homologous than the coding exon sequences, even though the former should be freer to diverge than the latter during the course of evolution (see Mollon, 1997).

(i) L- and M-cone exon sequences: There are only 15 codon differences between the L- and M-cone pigment genes (Fig. 1.5B). They are confined to exons 2–5, which encode the seven membrane-embedded α-helices that together form the chromophore binding pocket (see Fig. 1.2). Three sites are in exon 2 (at codons 65, 111, and 116; for numbering system, see Nathans, Thomas, & Hogness, 1986); two in exon 3 (at codons 153 and 180), three in exon 4 (at codons 230, 233, and 236), and seven in exon 5 (at codons 274, 275, 277, 279, 285, 298, and 309). Six of these differences involve conservative substitutions of hydrophobic residues, which do not influence the interaction of the opsin with the chromophore. Of the remaining nine, one of the sites (codon 116) lies in the first extracellular loop of the molecule; it is therefore unlikely to be involved in a direct interaction with the chromophore. Seven sites, however, lie in the transmembrane helices and may contact the chromophore. These are codons 65, 180, 230, 233, 277, 285, and 309. They involve the substitution of an amino acid residue that lacks a hydroxyl group (a nonpolar or uncharged amino acid) by one that carries a hydroxyl group (a polar or charged amino acid). On theoretical grounds, amino acid substitutions that change the number or locations of polar side chains (e.g., those carrying an hydroxyl group) in the retinal-binding pocket could alter the spectral tuning of the 11-*cis*-retinal chromophore by readjusting its three-dimensional packing or by changing the electrical properties of its immediate environment (Kropf & Hubbard, 1958; Mathies & Stryer, 1976; Hays et al., 1980).

The effect of replacing polar amino acids by nonpolar ones is supported by the results of site-directed mutagenesis experiments in bovine rhodopsin (Chan, Lee, & Sakmar, 1992) and in human hybrid cone pigments (Merbs & Nathans, 1992b; Asenjo, Rim, & Oprian, 1994), as well as by comparisons between electroretinographic (ERG) measurements of primate cone pigment spectral sensitivities and corresponding amino acid sequences (Neitz, Neitz, & Jacobs, 1991). The largest shifts in λ_{max} are produced by substituting alanine for threonine at codon 285 (ca. -14 nm), phenylalanine for tyrosine at position 277 (ca. -7 nm), and alanine for serine at position 180 (ca. -4 nm) (Merbs & Nathans, 1992a). By contrast, substitutions at positions 65, 230, 233, and 309 produce shifts of approximately 1 nm or less at λ_{max} (Merbs & Nathans, 1993).

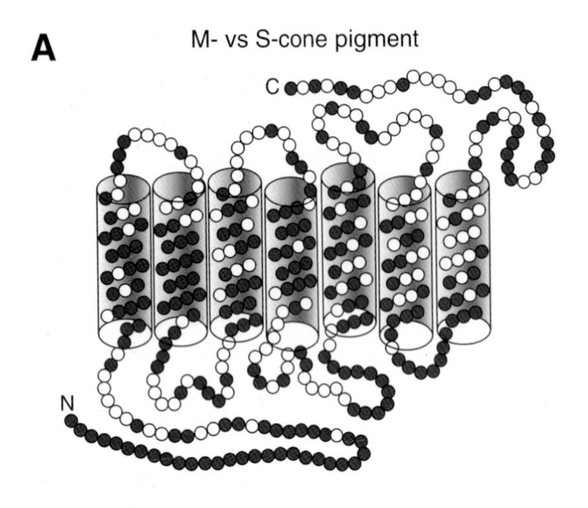

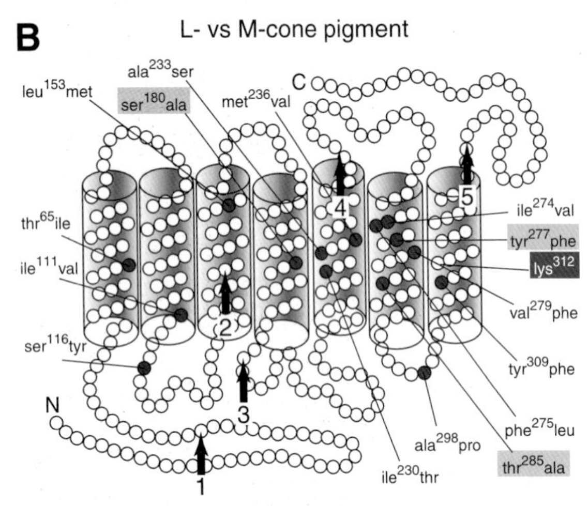

Figure 1.5: Pairwise comparisons of human visual pigment molecules showing amino-acid identities (open circles) and differences (filled circles) (after Nathans, Thomas, & Hogness, 1986). In each representation, the seven α-helices are arranged in a line. When intramembrane regions are optimally aligned, the amino-proximal tails (extracellular face) of the M- (or L-) cone pigments are 16 amino acids longer than for the S-cone pigment. The alignment can be improved by inserting into the M- (or L-) cone pigment sequences gaps of two amino acids and of one amino acid, respectively, at positions 4 residues and 29 residues from the carboxyl terminus. (A) Identity between the M- and S-cone pigments. (B) Identity between the L- and M-cone pigments. The location of lysine^{312}, the site of covalent attachment of 11-*cis* retinal, and the 15 amino acid substitutions are indicated. The start of each of the 5 intron positions are indicated by numbered vertical arrows. The subsitutions at codons 180, 277, and 285 (highlighted) are believed to contribute the majority of the spectral difference between the M- and L-cone pigments.

Such single amino acid substitution data, however, do not explain the nonadditive shifts in λ_{max} that are observed when more than one hydroxyl group is simultaneously substituted (Merbs & Nathans, 1992b; Asenjo, Rim, & Oprian, 1994), nor do they take into account the influence of multiple aliphatic amino acid differences on side-chain packing (Merbs & Nathans, 1993).

(ii) L- and M-cone 5' and intron sequences: L- and M-cone pigment genes differ not only in their coding sequences, but also in their 5' and intron sequences (Nathans, Thomas, & Hogness, 1986; Vollrath, Nathans, & Davis, 1988). In the vast majority of color normal arrays sequenced in Caucasian males, there is only one L-cone pigment gene, which is longer than all of the other gene copies (see the section on the arrangement of the gene array). It is located at the 5' (upstream) end of the array and abuts single copy DNA sequences, which are not found in front of the other, downstream genes. The length difference arises because its intron 1 *typically* contains 1,612 bp extra sequences (comprising 1,284 bp of three Alu elements and 328 bp of intervening unique-sequence DNA), which are also not found in the downstream genes (cf. the L- and M-cone pigment genes in Fig. 1.7A).

Although the extra intron sequences are found in >99% of Caucasian males, in ca. 45, 35, and 2.5 of African, Afro-American, and Japanese males, respectively, the most proximal gene in the array lacks the extra sequences in intron 1 and is the same size as the downstream M-cone pigment genes (Jørgensen, Deeb, & Motulsky, 1990; Meagher, Jørgensen, & Deeb, 1996). A reason for this may be that it contains inserted (exon 2) M-cone pigment-specific sequences (see the next section).

Normal and hybrid pigment genes. The S-cone opsin gene sequence seems to be nearly invariant in the human population. In contrast, the M- and L-cone opsin genes are diversiform, owing to hybrid variants and shared polymorphisms (see Fig. 1.6).

Hybrids are fusion genes containing the coding sequences of both L- and M-cone pigment genes. They are produced by intragenic crossing over: the breaking during meiosis of one maternal and one paternal chromosome at the opsin gene locus; the exchange of the corresponding sections of nucleotide sequences; and the rejoining of the chromosomes (see the section on intragenic recombination and Fig. 1.15C).

Intragenic crossing over between the M- and L-cone pigment genes is much more likely to occur within intron sequences than within exon sequences, owing to the approximately tenfold greater size of the introns compared with the exons and the paradoxically greater DNA sequence similarity of the M- and L-cone pigment gene introns compared with the exons (Shyue et al., 1995). Thus, in general, hybrid genes contain some number of contiguous exons from one end of an L-cone pigment gene joined to the remaining exons from the other end of an M-cone pigment gene. Those beginning with L-cone exon sequences are known as 5'L-3'M (or 5'red-3'green) hybrid or fusion genes, and those beginning with M-cone exon sequences as 5'M-3'L (or 5'green-3'red) hybrid genes. The 5'L-3'M hybrid genes encode M or M-like anomalous pigments; whereas the 5'M-3'L hybrid genes encode L or L-like pigments. Therefore, a convenient shorthand terminology for referring to normal and hybrid genes is to identify their exon sequences as being M-cone or L-cone pigment-specific (see Fig. 1.6). Two factors, however, complicate this simple picture.

First, exon 3 is more variable than exons 2, 4, or 5 in its amino acid residues (Winderickx, Battisti, et al., 1993; Sharpe et al., 1998), owing to the existence of several shared genetic polymorphisms between the M- and L-cone pigment genes. Genetic polymorphisms (or dimorphisms, if confined to two forms) are allelic variants of a gene occurring with a frequency greater than 1%. Most of the polymorphisms in exon 3 are confined to a dimorphic subsitution of a single nucleotide sequence. These alter the encoded amino acid without apparently affecting the properties of the photopigment. However, one – the substitution of a serine for alanine residue at codon 180 (the only one involving the substitution of a hydroxyl group) – produces a phenotypic variation. It causes a slight red shift (see below). Current estimates in normal observers suggest that the polymorphism is not equally distributed (see

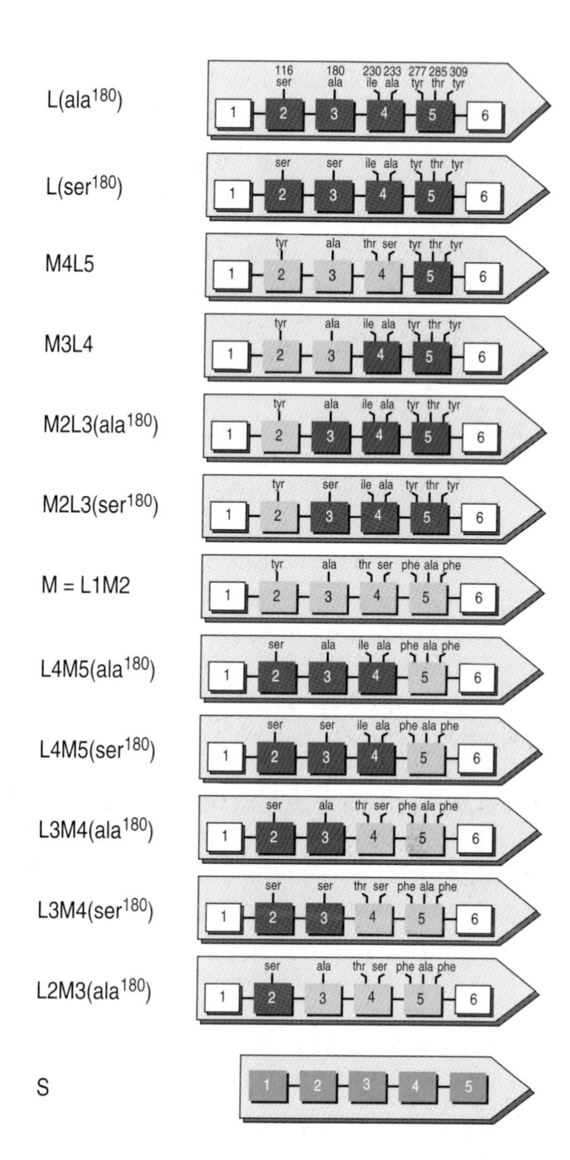

Figure 1.6: Exon arrangement of the S-, M-, L-, 5'L-3'M-hybrid, and 5'M-3'L-hybrid pigment genes. The S-cone pigment gene has one fewer exon, missing from its 5' end, than the X-linked pigment genes. There are no sequence differences between X-linked genes in exons 1 and 6. The 7 amino acid residues indicated above exons 2 to 5 are those responsible for the spectra shift between the normal and anomalous pigments. Dark gray indicates an L-cone pigment gene-specific sequence; light gray, an M-cone pigment gene-specific sequence.

Table 1.2): Among human L-cone pigment genes approximately 56.3% have serine and 43.7% have alanine at position 180, whereas, among M-cone pigment genes approximately 6% have serine and 94% have alanine (Winderickx et al., 1992b; Winderickx, Battisti, et al., 1993; Neitz & Neitz, 1998; Sharpe et al., 1998; Schmidt et al., 1999). However, large variability may occur between groups of different ethnic origin. In one report, 80% of African (N = 56), 84% of Japanese (N = 49), and 62% of Caucasian (N = 49) males had serine at codon 180 (Deeb & Motulsky, 1998). Therefore, it is useful to designate an M- or L-cone pigment gene, by an abbreviation that reflects the identity of the polymorphic residue at position 180 in exon 3.

A further complication is that frequently pigment gene sequences reveal an M-cone pigment gene exon 2 embedded within an L-cone pigment gene or a 5'L-3'M hybrid gene (Sharpe et al., 1998), indicating a complicated history of recombination events, and, therefore, making it pertinent to designate whether exon 2 is derived from an M- or L-cone pigment gene.

Thus, each normal or hybrid gene is more properly referred to by an abbreviation that reflects not only the origin of its various exons, but also the identity of the polymorphic residue at position 180. For example, L-cone pigment genes can be designated L(ala^{180}) or L(ser^{180}) to indicate the presence of alanine or serine, respectively, at position 180. L4M5(ala^{180}) is a hybrid pigment encoded by a gene in which exons 1–4 are derived from an L-cone pigment gene, exons 5 and 6 are derived from an M-cone pigment gene (i.e., the point of crossing over is in intron 4), and position 180 is occupied by alanine. L(M2;ser^{180}) is an L-cone pigment gene in which exon 2 is derived from an M-cone pigment gene and position 180 is occupied by serine. The fact that sequence differences between L- and M-cone pigment genes are confined to exons 2–5 implies that an L1M2 hybrid gene encodes a de facto M-cone pigment and, likewise, that an M1L2 hybrid gene encodes a de facto L-cone pigment.

Study	Population	Polymorphic residue	Allele frequency
L-cone opsin genes			
Winderickx, Battisti, et al. (1993)	74 normals 35 deutans	ala^{180} ser^{180}	0.440 0.560
Sjoberg et al. (1997), see Neitz & Neitz (1998)	130 normals	ala^{180} ser^{180}	0.485 0.515
Sharpe et al. (1998)	27 deuteranopes	ala^{180} ser^{180}	0.259 0.741
Schmidt et al. (1999)	36 normals 2 deuteranopes	ala^{180} ser^{180}	0.395 0.605
Mean	304	ala^{180} ser^{180}	0.437 0.563
M-cone opsin genes			
Winderickx, Battisti, et al. (1993)	52 normals† 12 deutan 8 protans	ala^{180} ser^{180}	0.959 0.041
Sjoberg et al. (1997), see Neitz & Neitz (1998)	130 normals	ala^{180} ser^{180}	0.930 0.070
Mean		ala^{180} ser^{180}	0.940 0.060

Table 1.2: Polymorphisms in the L- and M-cone opsin genes at codon 180 in human males. For descriptions of deutans, protans, and deuteranopes, see section on color blindness. (†Selected from a larger population of 72; only those were included who had one type of M-cone opsin gene or two types differing by only a single polymorphism.)

Protein sequence variation and spectral sensitivity. Several in vitro and in vivo techniques have been applied to studying the variation in normal and, less frequently, hybrid pigment spectral sensitivities (for a review, see Stockman et al., 1999). A partial list, emphasizing studies that have investigated the hybrid pigments, is given in Table 1.3.

The in vitro measurements include ERG, single photoreceptor suction electrode action spectra (see Chapter 4), microspectrophotometry (MSP), and photobleaching difference absorption spectra measurements of recombinant cone pigments produced in tissue culture cells. The in vivo measurements include reflection densitometry, linear transforms of psycho-

Genotype	In vivo	In vitro	
	Sharpe et al. (1998)	Merbs & Nathans (1992a, 1992b)	Asenjo, Rim, & Oprian (1994)
S	418.9 ± 1.5†	426.3 ± 1.0	424.0*
M(ala[180]) = L1M2(ala[180])	527.8 ± 1.1	529.7 ± 2.0	532 ± 1.0
L2M3 (ala[180])	528.5 ± 0.7	529.5 ± 2.6	532 ± 1.0
L3M4 (ser[180])	531.5 ± 0.8	533.3 ± 1.0	534 ± 1.0
L4M5 (ala[180])	535.4	531.6 ± 1.8	—
L4M5 (ser[180])	534.2	536.0 ± 1.4	538 ± 1.0
M2L3 (ala[180])	—	549.6 ± 0.9	—
M2L3 (ser[180])	—	553.0 ± 1.4	559 ± 1.0
M3L4	—	548.8 ± 1.3	555 ± 1.0
M4L5	—	544.8 ± 1.8	551 ± 1.0
L (ala[180])	557.9 ± 0.4	552.4 ± 1.1	556 ± 1.0
L (M2, ala[180])	556.9	—	—
L (M2, ser[180])	558.5	—	—
L (ser[180])	560.3 ± 0.3	556.7 ± 2.1	563 ± 1.0

Table 1.3: Absorbance spectrum peaks (λ_{max} ± SD) of the human normal and hybrid cone pigments. (†Value from Stockman et al., 1999; *value from Oprian et al., 1991.)

physical color matching functions (CMFs), and spectral sensitivity measurements in normal and color-deficient observers of known genotype under conditions chosen to isolate preferentially a single cone pigment (see Chapter 2 for a review of measurements of the normal cone absorption spectra).

(i) S-cone pigment: The λ_{max} (± the standard deviation) of the human S-cone pigment, measured at the retina (see legend to Fig. 1.1), has been placed at: (i) 419.0 ± 3.6 nm by in vitro MSP of human cones (Dartnall, Bowmaker, & Mollon, 1983); (ii) 424 (Oprian et al., 1991) or 426 nm (Merbs & Nathans, 1992a) by in vitro absorption spectroscopy of recombinant cone pigments; (iii) 419.0 and 419.7 nm by in vivo central and peripheral spectral sensitivity measurements, respectively, in normal and blue-cone monochromat (see page 41) observers (Stockman, Sharpe, & Fach, 1999); and (iv) 420.8 nm by transforms of the Stiles and Burch 10-deg CMFs (Stockman, Sharpe, & Fach, 1999). Determination of the λ_{max} of the S-cone pigment in solution is complicated by short-wavelength–absorbing bleaching products that partially overlap the pigment absorbance and must be subtracted from it. On the other hand, the in vivo determinations are complicated by several factors, including individual differences in the absorption of the lens and macular pigment (see Chapter 2). Some variability in the λ_{max} of the S-cone pigment has been suggested (Stockman, Sharpe, & Fach, 1999; see p. 44).

(ii) L- and M-cone pigments: The λ_{max}'s of the normal M-, L(ala[180])-, and L(ser[180])-cone pigments have been placed, respectively, at: (i) 530.8 ± 3.5, 554.2 ± 2.3, and 563.2 ± 3.1 nm (558.4 ± 5.2 nm for the mixed L-cone pigments) by MSP[3] of human cones (Dartnall, Bowmaker, & Mollon, 1983); (ii) 529.7 ± 2.0, 552.4 ± 1.1, and 556.7 ± 2.1 nm (Merbs & Nathans, 1992a) or 532 ± 1.0, 556 ± 1.0, and 563 ± 1.0 nm (Asenjo, Rim, & Oprian, 1994) by in vitro spectroscopy of recombinant cone pigments; (iii) 530 and 560 nm (mixed L-cone pigments; Schnapf, Kraft, & Baylor, 1987) or 531 (Kraft, private communication), 559.2, and 563.4 nm (Kraft, Neitz, & Neitz, 1998) by suction electrode data; (iv) 528.6 ± 0.5, 557.5 ± 0.4, and 560.2 ± 0.3 nm by foveal spectral sensitivity measurements in dichromat observers (Sharpe et al., 1998, 1999; Stockman & Sharpe, 2000a); and (v) 530.6 and 559.1 nm (mixed L-cone pigments) by transforms of the Stiles and Burch 10-deg CMFs (Stockman & Sharpe, 2000a).

The in vivo spectral sensitivity measurements obtained from dichromats (Sharpe et al., 1998) show a mean separation of ca. 2.7 nm between the L(ala[180])- and L(ser[180])-cone pigments. This value is somewhat less than that which has been obtained from site-directed mutagenesis experiments and other tech-

[3]These values are based on ad hoc subgrouping according to whether the individual λ_{max} lies above or below the group mean and not according to genotype.

niques. However, all of the previous reported values, whether based on inferences from Rayleigh matches (Winderickx et al., 1992b; Sanocki et al., 1993; Sanocki, Shevell, & Winderickx, 1994; He & Shevell, 1994), spectral sensitivities (Eisner & MacLeod, 1981), cloned pigment (Merbs & Nathans, 1992a, 1992b; Asenjo, Rim, & Oprian, 1994), or on the ERG (Neitz, Neitz, & Jacobs, 1995), have shown that serine-containing pigments are red-shifted with respect to alanine-containing pigments. The in vivo estimates accord with other psychophysical measures of the variability of the L-cone λ_{max} in the normal population, based on the analysis of color-matching data by Stiles and Burch (1959), which preclude shifts greater than 3.0 nm (Neitz & Jacobs, 1989, 1990; Webster & MacLeod, 1988; Webster, 1992).

As of yet, no reliable in vivo spectral sensitivity data exist for the two polymorphic variants of the M-cone pigment. However, in vivo comparisons between subjects with a L2M3(ala^{180}) pigment and those with a L3M4(ser^{180}) pigment, for whom the only important amino acid difference is at position 180, show a mean shift of 3.0 nm (see Table 1.3). Further, the spectral shift between the λ_{max}'s of the M(ala^{180})- and M(ser^{180})-cone pigments has been estimated provisionally (awaiting in vivo confirmation) at: (i) 5.9 nm by ad hoc analysis of MSP data (Dartnall, Bowmaker, & Mollon, 1983); and (ii) 4.3 to 4.4 nm (Merbs & Nathans, 1992b) and 2 nm (Asenjo, Rim, & Oprian, 1994) by in vitro spectroscopy of recombinant normal and 5'M-3'L hybrid cone pigments.

(iii) Hybrid cone pigments: Estimates of the λ_{max}'s of the hybrid pigments encoded by 5'L-3'M and 5'M-3'L hybrid genes, which presumably underlie anomalous trichromacy (see color blindness), are summarized in Table 1.3. Differences between the in vivo and in vitro estimates probably reflect the limitations of the measuring techniques (see Chapter 2). For instance, the absorption measurements of visual pigment in vitro are accurate within only about 0.5 to 1.0 log unit of the λ_{max} and thus encompass only a limited range of wavelengths. Additionally, they will differ from the in vivo measurements because they do not account for waveguiding in the photoreceptor.

Nonetheless, the in vivo and in vitro data support one another in indicating that both 5'L-3'M and 5'M-3'L hybrid genes encode a range of pigments with spectral sensitivities, which, in every case so far examined, lie between those of the normal L- and M-cone pigments. The data further indicate that the spectral sensitivity of the hybrid pigment depends on the position of the crossing-over and on the identity of the polymorphic amino acids at position 180. For each exon, the set of amino acids normally associated with the L- or M-cone pigments produce, respectively, spectral shifts to longer or shorter wavelengths, thus producing a monotonic relationship between the λ_{max} and the fraction of the hybrid pigment derived from the L and M parental pigments (see Sharpe et al., 1998).

The primary determinants of the spectral shift are located in exon 5, as seen by the clustering of the λ_{max} of all of the pigments encoded by 5'L-3'M genes within 8 nm of the maxima of the normal M pigments. Further, a comparison of the in vivo measured λ_{max}'s in Table 1.3 indicates that the L/M sequence differences in exon 5 – principally the residues at 277 and 285 – result in spectral shifts of 15–25 nm, the exact value depending on sequences in exons 2–4. The in vivo measured data further suggest that substitutions at the sites confined to exons 2–4 produce much smaller spectral shifts: Exon 2 contributes at most 0–2.0 nm; exon 3, 1.0–4.0 nm; and exon 4, 2.5–4.0 nm. These results are in approximate agreement with the in vitro results (Merbs & Nathans, 1992b; Asenjo, Rim, & Oprian, 1994) and with inferences based on a comparison of primate visual pigment gene sequences and cone spectral sensitivity curves (Neitz, Neitz, & Jacobs, 1991; Ibbotson et al., 1992; Williams et al., 1992).

(iv) Sequence variation and opsin viability: Amino acid substitutions may have other consequences than shifting the spectral sensitivity of the X-chromosome–linked opsins. They also could alter the quantum efficiency or the optical density of the pigment. As Williams et al. (1992) point out, in vitro expression studies have noted that some hybrid pigments may be unstable or of reduced optical density (Merbs & Nathans, 1992b; Asenjo, Rim, & Oprian, 1994). Further, both

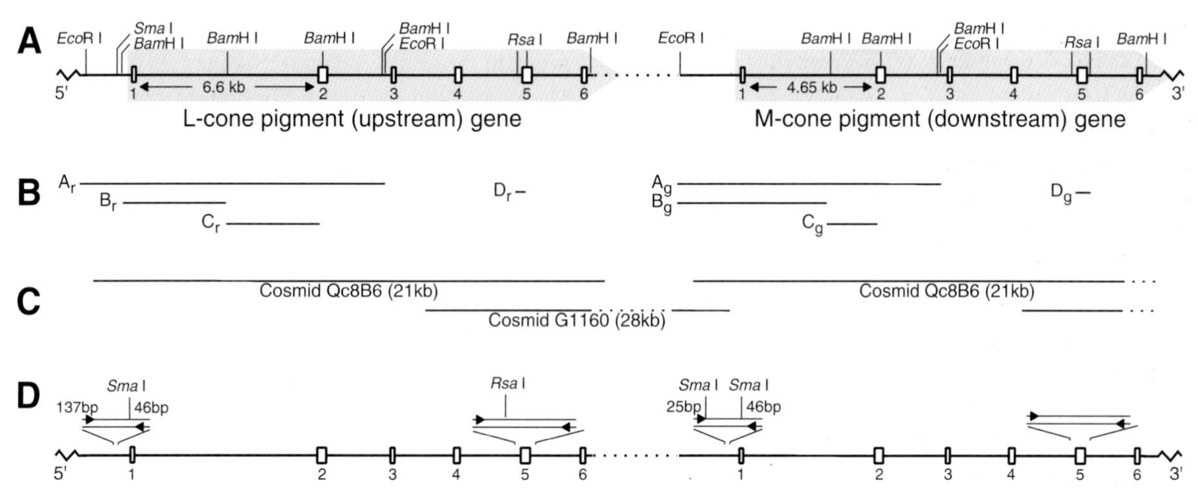

Figure 1.7: A schematic representation of the restriction enzymes and probes used to quantitate or directly visualize opsin genes in the X-chromosome linked visual pigment gene array. (A) The recognition or cleavage sites of the restriction enzymes, *Bam*H I, *Eco*R I, *Hind* III, *Rsa* I, and *Sma* I, used to define fragment sequences. The exons and intervening introns of a single L-cone pigment gene (upstream) and a single M-cone pigment (downstream) gene are indicated. (B) The L-cone pigment gene restriction fragments, A_r, B_r, C_r, and D_r, and the M-cone pigment gene restriction fragments, A_g, B_g, C_g, and D_g used in the conventional gel electrophoresis/Southern blot hybridization method to estimate the ratios and numbers of L- and M-cone pigment genes in the array (Nathans, Thomas, & Hogness, 1986). (C) Positions of the hybridization pattern of the cosmids, Qc8B6 (Gene Bank Accession number Z68193), and G1160 (Accession number Z46936), used to directly visualize the opsin genes by in situ hybridization. (D) Primer sequences from a 183-bp fragment containing promoter sequences (ca. 50 bp upstream exon 1) and from a ca. 300-bp fragment containing the sequences of exon 5, used by the endlabeled PCR-product method to count the number of genes and the ratio between L-cone and M-cone pigment genes, respectively, in the opsin gene array (Neitz & Neitz, 1995).

psychophysical (Miller, 1972; Smith & Pokorny, 1973; Knau & Sharpe, 1998) and retinal densitometric (Berendschot, van de Kraats, & van Norren, 1996) data suggest that the L- and M-cone pigments, as estimated in deuteranopes and protanopes, respectively (see section on color blindness), differ in optical density.

It is unlikely that other differences between the opsin genes, in particular the extra sequence in intron 1 of the L-cone opsin gene, could contribute to levels of expression and optical density. Although some genes have enhancer sequences in their introns or even 3' of the transcription unit, the available transgenic data for M- and L-cone opsin genes indicate that cone-specific expression is only regulated by the promoter and LCR (however, transgenic expression in the absence of introns does not preclude a role in level of expression). Moreover, most of the extra sequence in intron 1 of the L-cone opsin gene is made up of Alu repeat elements.

The size of the opsin gene array. The opsin gene array on the X-chromosome varies in size, typically containing more than two opsin genes (Nathans, Thomas, & Hogness, 1986). Its variability in the normal population is a subject of controversy; the differences between investigators have been used to challenge the Young–Helmholtz trichromatic theory of color vision as well as current models about the evolution of the human photopigments (Neitz & Neitz, 1995). At the heart of the controversy are the different quantitation and direct techniques (see Figs 1.7 and 1.8) used to assess the total copy number and ratio of L- and M-cone pigment genes within the array (see Table 1.4).

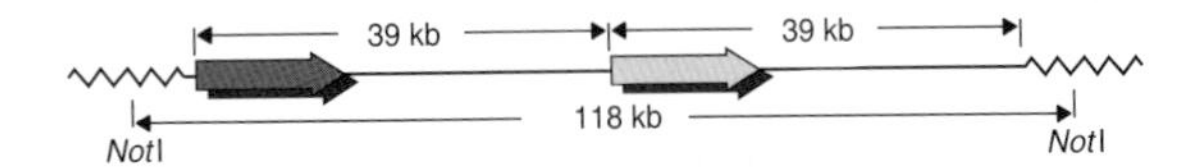

Figure 1.8: The *Not* I fragment that carries the entire opsin gene array, used to determine the total number of genes (Vollrath, Nathans, & Davis, 1988; Macke & Nathans, 1997). It comprises: (i) 40 kb of nonrepeated single-copy flanking DNA; and (ii) repeating units of ca. 39-kb length, each unit consisting of one complete opsin gene (13.2–15.2 kb) and the 24-kb highly conserved flanking region. Thus, the size of the *Not* I fragment will vary in steps of ca. 39 kb; which can be resolved by pulsed field gel electrophoresis. The total gene number will be equal to: (the size of the *Not* I fragment - 40 kb)/39 kb. In the example shown, the 118-kb-long *Not* I fragment contains two genes.

(i) Techniques: The techniques rely on bacterial restriction enzymes (endonucleases) that cleave the opsin gene at specific base sequences to produce fragments that differ between the M- and L-cone photopigment genes (Nathans, Thomas, & Hogness, 1986). When isolated on a suitable filter gel, the restriction fragments can be electrically separated according to molecular weight by conventional gel electrophoresis and visualized by Southern blotting by hybridizing them with radioactively labeled probe DNA that recognizes specific nucleotide sequences.

Targeted regions of interest are several restriction fragment length polymorphisms (RFLP). These are pairs of small fragments cleaved by the *Eco*R I, *Bam*H I, and *Rsa* I enzymes (Fig. 1.7A). They include A_r/A_g, B_r/B_g, C_r/C_g, and D_r/D_g (Fig. 1.7B), one of which is specific to the L- (subscript r for red) and the other to the M- (subscript g for green) cone pigment gene. Their labeled intensities can be quantitated by autoradiography or phosphor-imaging to provide information about the relative number of L- and M-cone pigment genes in the array and, by inference, about the total number of gene copies (Nathans, Thomas, & Hogness, 1986).

A targeted large fragment, which is cleaved by the *Not* I enzyme, carries the entire opsin gene array (see Fig. 1.8). It can be resolved by pulsed field gel electrophoresis and visualized by Southern blotting to provide direct information about the total number of genes in the array (Vollrath, Nathans, & Davis, 1988; Macke & Nathans, 1997). *Not* I fragments differing in length will separate in an electrophoresis gel according to the number of genes that they contain. The lengths can then be measured in kilobases relative to size standards (e.g., concatamers of bacteriophage λ DNA).

Alternatively, fragments cleaved by the *Sma* I and *Rsa* I enzymes (Fig. 1.7D) can be amplified by using polymerase chain reaction (PCR). Opposite ends of the targeted region of the fragments that differ between the L- and M-cone opsin genes are annealed with primer pairs that recognize specific nucleotide sequences (end-product labeling). The primers are then extended in opposite directions by using a DNA polymerase (an enzyme that catalyzes the synthesis of DNA) to add nucleotide bases to cover the entire targeted area. Repetition of the cycle generates copies of the target DNA between the primers in an exponential manner. Amplified 183-bp fragments from the M- and L-cone gene promoter sequences and ca. 300-bp fragments from exon 5 of the M- and L-cone pigment genes (Fig. 1.7D) can be resolved by gel electrophoresis and quantitated to provide information about the number of genes and the ratio of genes, respectively, in the array (Neitz & Neitz, 1995).

(ii) Results: The techniques involving either quantitation of RFLPs detected by Southern blot hybridization after conventional gel electrophoresis (Nathans, Thomas, & Hogness, 1986; Drummond-Borg, Deeb, & Motulsky, 1989; Schmidt et al., 1999) or quantitation of sequence differences by denaturing electrophoresis that resolves DNA fragments based on nucleotide sequence as well as size (single-strand conformation polymorphism electrophoresis; SSCP) after PCR of M- and L-cone pigment gene promoter sequences (Yamaguchi, Motulsky, & Deeb, 1997) yield smaller copy number estimates than those relying on quantitation of end-labeled restriction products after PCR amplification (Neitz & Neitz, 1995; Neitz, Neitz, & Grishok, 1995; Schmidt et al., 1999). The former studies report an average of three pigment genes, with only a single L-cone pigment copy present and not more than five M-cone pigment copies,

Study	Method	Technique	No.	Mean (± SD)	Range
Nathans, Thomas, & Hogness (1986)	RFLP quantitation	gel electrophoresis/ Southern blotting	18	3.1 ± 0.6	2–4
Drummond-Borg et al. (1989)	RFLP quantitation	gel electrophoresis/ Southern blotting	134	3.2 ± 1.0	2–6
Neitz & Neitz (1995)	RFLP quantitation	endlabeled PCR products	27	4.3 ± 1.9	2–9
Neitz, Neitz, & Grishok (1995)	RFLP quantitation	endlabeled PCR products	26	4.5 ± 1.9	2–9
Yamaguchi et al. (1997)	SSCP quantitation	PCR and SSCP	51	2.9 ± 0.8	2–5
Macke & Nathans (1997)	sizing of the *Not* I fragment	pulsed field gel electrophoresis	67	2.9 ± 0.9	1–5
Wolf et al. (1999)	*in situ* visualization	fiber FISH	8	3.5 ± 1.7†	1–6
Schmidt et al. (1999)	sizing of the *Not* I fragment	pulsed field gel electrophoresis	35*	3.3 ± 1.0	2–6
	RFLP quantitation	gel electrophoresis/ Southern blotting	35*	3.9 ± 1.0	2–6
	RFLP quantitation	endlabeled PCR products	35*	4.3 ± 1.4	2–9

Table 1.4: Typical number of genes in the X-chromosome–linked visual pigment gene arrays of white Caucasian males of unselected or color normal (*) phenotype. Some of the methods overestimate the number of genes (see text). († The population was skewed to include extremes.)

whereas the latter studies (Neitz & Neitz, 1995; Neitz, Neitz, & Grishok, 1995) suggest that nearly 50% of all subjects carry two or more (up to four) L-cone pigment genes, or 5'M-3'L hybrid genes, with some having as many as nine gene copies in total (Table 1.4).

Light on the controversy has been shed, however, by recent developments: (i) the application of direct visualization techniques, including pulsed field gel electrophoretic sizing of *Not* I fragments (Macke & Nathans, 1997; Schmidt et al., 1999; see Fig. 1.8) and the fiber FISH (fluorescent in situ hybridization; Parra & Windle, 1993) protocol (Wolf et al., 1999; see Fig. 1.7C); and (ii) comparisons between the various methods in the same population of individuals (Wolf et al., 1999; Schmidt et al., 1999). The two direct procedures, which agree exactly in their results on the same individuals (Wolf et al., 1999), demonstrate that on average a typical array contains three pigment genes. Further, they suggest that reports of frequent occurrences of larger arrays, including those with two or more (up to four) L-cone-pigment genes (Neitz & Neitz, 1995; Neitz, Neitz, & Grishok, 1995), may reflect technical artefacts that are inherent in PCR methods (Macke & Nathans, 1997; Yamaguchi et al., 1997; Schmidt et al., 1999) and ambiguities arising from an inability to distinguish 5'M-3'L hybrid genes from L-cone pigment genes.

Examples of visual pigment gene arrays differing widely in gene number are shown in Fig. 1.9. It presents digitized images of single DNA fibers that have been subjected to dual-color FISH, using the cosmids[4] Qc8B6 and G1160 as probes (see Fig. 1.7C; Wolf et al., 1999). In the images, identification by gene or exon type is not possible because each gene is pseudocolored red and each intergenic region – the ~25-kb region at the downstream or 3' end of each gene – is pseudocolored green. The first (upper) fiber is from a deuteranope (see color blindness). His hybridized fibers exhibit one red and one green signal, indicating the occurrence of a single gene copy. The second, third, and fourth fibers are from trichromats, whose hybridized fibers display two, three, and six gene cop-

[4]Cosmids are artificially constructed cloning vectors containing the cos site of bacteriophage λ. They permit cloning of larger DNA segments than can be introduced into bacterial hosts in conventional plasmid vectors.

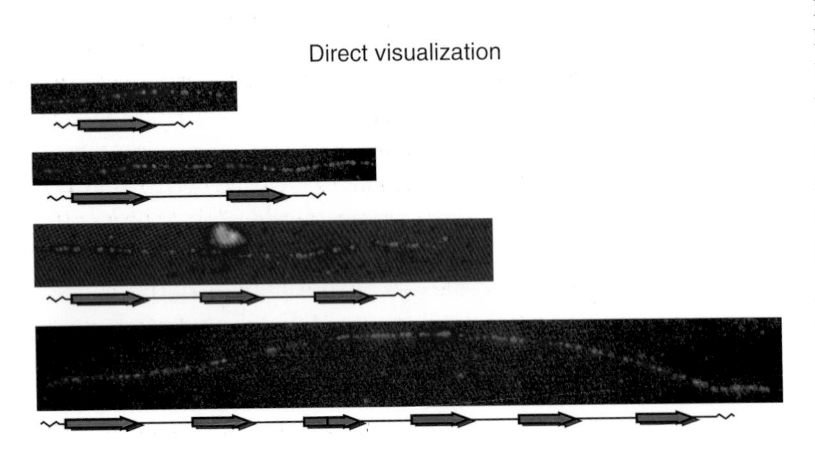

Figure 1.9: Opsin gene array. Single hybridized DNA fibers of four individuals with differently sized opsin gene arrays, containing one, two, three, and six genes (top to bottom). The genes appear red; the intergenic regions, green. The magnification factor is reduced for the six-gene array. For clarity, the location of the genes (arrows) and intergenic regions (straight lines) in each array are schematically depicted below the hybridized images. The color coding has been altered so that red now refers to the L-cone pigment gene, green to the M-cone pigment gene, and red-green to a 5'M-3'L hybrid gene.

ies, respectively. In each array, the upstream (5') end can be clearly identified because of the characteristic greater length of the most proximal gene.

The arrangement of the gene array. Precise sizing of the number of genes in the array is the important first step in determining the composition of the array. The next steps are to determine the L- to M-cone pigment gene ratio and order of the array. Owing to the high (96%) sequence homology of the L- and M-pigment genes, it is not yet possible to do either of these by direct procedures.

(i) Techniques: Information about the L- to M-cone pigment gene ratio can be obtained by RFLP quantitation of the relative band intensities of fragment pairs. However, the interpretation of these procedures is complicated by the difficulty in distinguishing normal from hybrid genes on the basis of limited restriction fragment pairs (it often requires looking at the D_r/D_g fragments, which encompass exon 5; see Fig. 1.7B) and by potential artefacts such as the background level of radioactivity in the gels and the separation of peaks for the fragment pairs. No procedure is currently available for strictly determining the order of genes in the array, regardless of the array size. A prerequisite for nucleotide sequencing would be the stable propagation in *E. coli* or yeast of large cloned segments with multiple pigment genes. This has not yet been demonstrated. However, very recently Hayashi et al. (1999), using long-range PCR amplification of a 27.4 kb opsin gene fragment, have been able for the first time to completely define the order of gene types in a three gene array (see section on deuteranomaly). This feat is achieved by employing standard techniques to define the most 5' (upstream) gene and also to define the types of genes present without regard to their order in the array; and by employing the long-range PCR to define the most 3' (downstream) gene in the array.

(ii) Results: Generally, the RFLP quantitation methods support the interpretation that there is only a single L-cone pigment gene in the array, occupying the most proximal position, followed by one or more M- or 5'M-3'L hybrid pigment genes (see Yamaguchi et al., 1997; Schmidt et al., 1999). In the Caucasian male population, the range appears to be one to five M-cone pigment gene copies, with a mean of two (Macke & Nathans, 1997; Schmidt et al., 1999). In non-Caucasian populations, both the range and mean are smaller: About one-half of Japanese (48.5%) and Afro-American (42%) males have a single downstream M-cone pigment gene as opposed to about one-fifth (22%) of Caucasian (Jørgensen et al., 1990; Deeb et al., 1992).

The reason for multiple M-cone pigment genes in the array is unclear. It has been speculated that variations in the L- to M-cone ratio in the photoreceptor mosaic (see Fig. 1.1C and Chapter 6) may be related to the number of M-cone pigment genes in the array: The higher the number of M-cone pigment genes, the greater the number of M-cones. However, Nathans et al. (1992) failed to find any correlation between the relative sensitivities to red and green spectral lights and the number of M-cone pigment genes.

Examples of the variation in pigment gene arrangment are shown in Fig. 1.9 below the hybridized images. The inferred arrangments are based on the FISH results combined with direct sequencing or indirect RFLP quantitation. In the first array, sequencing of exons 2 to 5 establishes that the single gene has only the L-cone pigment-specific sequences (Sharpe et al., 1998). The presence of a single L-cone pigment gene accords with the phenotype, deuteranopia (see page 27). In the second (two gene copies) and third (three gene copies) arrays, RFLP quantitation (based on the A, B, and C fragment pairs) establishes a 1:1 and 1:2 L- to M-cone pigment gene ratio, respectively (Schmidt et al., 1999). Further, the phenotype in both is normal. Therefore, it is reasonable to infer that in both a single L-cone pigment gene is followed by normal M-cone pigment genes. In the fourth (six gene copies) array, RFLP quantitation (based on A and C fragment pairs) establishes a 1:3 L- to M-cone pigment gene ratio; and examination of the B and D fragment pairs reveals a 5'M-3'L hybrid gene (Schmidt et al., 1999). Thus, an L-cone, a 5'M-3'L hybrid, and four M-cone pigment genes are probably present. Because the phenotype is normal, the hybrid gene has been placed at an arbitrary position downstream from a normal M-cone pigment gene (its actual position cannot yet be determined). The placement is consistent with models of gene expression in the array that are described below.

Gene expression in the array. Which opsin genes in the array are actually expressed in the cone photoreceptor cells, quite apart from the number of genes available in the array, can be determined by assessing the ratio of gene mRNA transcripts in extracts of retina. This involves reverse-transcribing cellular RNA into cDNA and then amplifying and quantitating different sequence variants.

mRNA analysis, however, is limited not only by problems inherent to quantitation, but also by two other factors. First, the method can only differentiate whether genes of different types are expressed. That is, if there are five M-cone pigment genes in the array, all with the same sequences, it is not possible to say whether only one or all five are expressed. Second, the method cannot take into account variations in the levels of expression of the gene among individual cone photoreceptor cells. For instance, a difference in the ratio of mRNA transcripts may reflect differences in the number of cones containing L-pigment as compared with the number containing M-pigment, or it may reflect differences in the amount of pigment contained in L-cones as compared with the amount contained in M-cones (e.g., a difference in photopigment optical density; see Chapter 2).

Two issues have been investigated by mRNA transcript analysis: (i) selective expression of the visual pigment genes (Are all genes in complex arrays expressed?) and (ii) differential transcription of the expressed genes (Are L-cone pigment genes more frequently expressed than M-cone pigment genes in the cone photoreceptors?).

(i) Selective expression: Winderickx et al. (1992a) detected in male donor eyes only two retinal mRNA transcripts: one coding for an L-cone pigment and the other for an M-cone pigment. In those donors who had two or more M-cone (or 5'M-3'L hybrid) pigment genes, only one allele was represented in the retinal mRNA. Yamaguchi, Motulsky, and Deeb (1997) and Hayashi et al. (1999) subsequently confirmed and extended these findings. In donors of unknown phenotype with a 5'M-3'L hybrid gene in addition to normal L- and M-cone pigment genes in their array, either (i) the normal L- and M-cone pigment genes were expressed, but not the hybrid gene, or (ii) the normal L-cone and the hybrid pigment genes were expressed, but not the normal M-cone pigment gene. What is decisive is the position of the M and 5'M-3'L hybrid pigment genes in the array. None of these studies found evidence of the presence or expression of more than one L-cone pigment gene (see deuteranomaly). On the other hand, Sjoberg, Neitz, Balding, and Neitz (1998) reported that about 10% of men express more than one L-cone pigment gene. It is unclear, however,

whether the authors, when they refer to extra L-cone pigment genes, are describing L-cone pigment genes or 5'M-3'L hybrid pigment genes that express an L-cone–like pigment. Regardless, they reject the hypothesis that only two opsin genes from one X-chromosome array can be expressed.

(ii) Differential transcription: Yamaguchi, Motulsky, and Deeb (1997) found, in extracts of whole retina, that the ratio of expressed L- to M-cone opsin retinal mRNA varies widely (from unity to ten times greater L-cone opsin expression, with a mode of four) and is not correlated with the ratio of L- to M-cone opsin genes (Yamaguchi et al., 1997). Hagstrom, Neitz, and Neitz (1997), looking at 6-mm-diameter patches of retina (corresponding to about 20 deg of visual angle), reported that the average ratio of L- to M-cone opsin mRNA in patches centered on the fovea was roughly 1.5:1.0, whereas in patches centered at 12 mm (ca. 41 deg) eccentricity it increased to 3.0:1.0 (see also Hagstrom et al., 1998). There were, however, large individual differences among eyes examined: The L- to M-cone opsin mRNA ratios in the fovea patches differed by a factor of greater than 3.

A favoring of L- over M-cone pigment expression, in both the fovea and retinal periphery, is supported by other indirect evidence, including psychophysical spectral sensitivity measurements (e.g., DeVries, 1948a; Brindley, 1954a; Vos & Walraven, 1971; Kelly, 1974; Walraven, 1974; Smith & Pokorny, 1975; Cicerone & Nerger, 1989; Vimal et al., 1989, 1991; Pokorny, Smith, & Wesner, 1991; Wesner et al., 1991; Cicerone et al., 1994), retinal densitometry (Rushton & Baker, 1964), flicker electroretinography (Shapley & Brodie, 1993; Usui et al., 1998), and MSP of human cones (Bowmaker & Dartnall, 1980; Dartnall, Bowmaker, & Mollon, 1983). Taken together, the mean ratios yielded by these methods suggest that there are roughly twice as many L- as M-cones in the central fovea. However, it should be pointed out that individual ratios that are estimated by such methods are highly variable between observers (ranging from 0.33:1 to 10:1) and further that each method, other than MSP, has serious problems of interpretation (see Chapters 2 and 4).

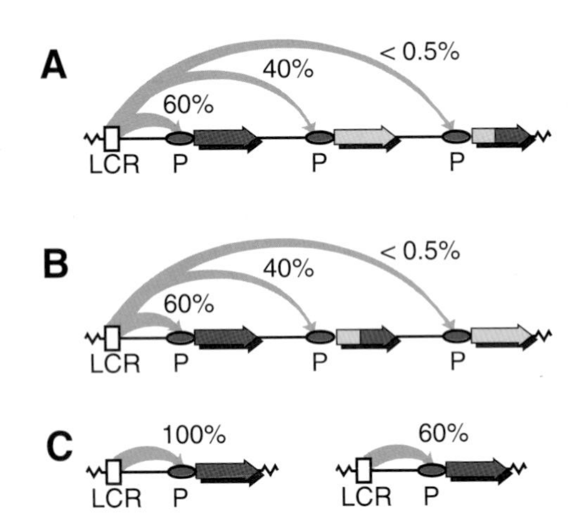

Figure 1.10: Model of opsin gene expression (after Nathans et al., 1989; Winderickx et al., 1992a). (A) Gene expression in an opsin gene array in which a 5'M-3'L hybrid gene (encoding an anomalous L-cone–like pigment) occupies a distal position relative to the normal L- and M-cone pigment genes. An individual with such an array would test as color normal. The expression ratios shown are based on the presumed 1.5 L- to M-cone ratio in the foveal photoreceptor mosaic (see Fig. 1.1C and Chapter 5). (B) Gene expression in an opsin gene array in which the 5'M-3'L hybrid gene occupies a proximal position relative to the normal M-cone pigment gene. An individual with such an array would test as deuteranomalous or deuteranopic, depending on the fusion point of the hybrid gene. This assumes that if the third gene in the array is expressed (< 0.5%), it is not expressed in sufficient amounts to predominate over the upstream genes to enable normal color vision. (C) Two possibilities for gene expression in the reduced, single-gene array of a deuteranope (see section on color blindness). The alternatives are related to the missing cone and replacement cone models in dichromats. (LCR = locus control region; P = promoter region.)

(iii) A model: A possible model of gene expression that incorporates selective expression and differential transcription is shown in Fig. 1.10 (based on Nathans et al., 1989; Winderickx et al., 1992a). The presence of a locus control region (LCR, see blue-cone monochromacy) located between 3.1 and 3.7 kb upstream (5') of the transcription inititation (cap) site of the most proximal gene in the array is known to be required for cone photoreceptor-specific expression (Nathans et al., 1989; Wang et al., 1992; Nathans et al., 1993). LCRs

and other upstream regulatory elements are hypothesized to exert their effects on promoters by a looping mechanism that brings the upstream elements into the vicinity of the transcription initiation site through protein–protein interactions (Knight et al., 1991; Mastrangelo et al., 1991).

Only if the LCR forms a stable transcriptionally active complex with an opsin gene promoter is the gene product expressed in a cone photoreceptor cell. Expression is assumed to be regulated according to a falling gradient: Transcriptionally active complexes between the LCR and the promoters of the most proximal pigment genes are considered more likely than those between the LCR and the promoters of more distal pigment genes. In general, only the two most proximal pigment genes of an array will be expressed in a significant number of cone photoreceptors (for a confirmation of this hypothesis in three-arrays, see page 37), and the first gene in the array (usually an L-cone opsin gene) will be expressed in a higher percentage than the second (usually an M-cone opsin gene).

In accord with the idea that more distal genes are disadvantaged, Shaaban and Deeb (1998) reported that the M-cone pigment gene promoter is two to four times more active than that of the L-cone pigment. Thus, a stronger intrinsic promoter activity may have evolved to offset the distance handicap in competing with the L-cone pigment promoter for coupling to the LCR. However, if LCR complex formation is presumed to be an all-or-none mechanism, promoter strength would not be relevant.

(iv) L- versus M-cone pigment choice: The model allows for the mutually exclusive expression of L- and M-cone pigment genes in single-cone photoreceptor cells, but does not address the question of how, during development, a photoreceptor cell chooses which visual pigment to produce. It simply assumes that L- and M-cones are merely distinguished by the pigment that they contain, and not by other membrane and/or transcriptional factors as well. A mutually exclusive stochastic process that gives preference to a single gene, combined with male hemizygosity and female X-chromosome inactivation (see heterozygotic carriers of protan and deutan defects), allows only one L- and M-cone pigment gene locus to be expressed per cell. The choice is also preserved in all descendant cells, obviating any requirement for coordinating or suppressing promoter–LCR interactions at a second locus (Wang et al., 1992). In short, this "stochastic" model predicts that all the machinery needed for L- versus M-cone segregation is inherent in the gene array, not only in humans but in all mammals who share identical transcriptional regulatory proteins.

An alternative model is that L- versus M-cone specificity in cell to cell contact and information transfer is predetermined during development. This "standard" model assumes that a set of regulatory molecules, including transcriptional factors, differs between L- and M-cone photoreceptor cells and orchestrates the choice of L- and M-cone pigment gene interaction with the LCR as well as the production of any additional proteins that differ between the two cell types. Shared transcriptional factors could also determine the matching up of L- and M-cones with their postreceptoral neurons. Some indirect evidence is consistent with regulatory molecules playing a crucial role in the differentiation of cone photoreceptor cells (Furukawa et al., 1997) and bipolar neurons (Chiu & Nathans, 1994). This model predicts that only primates with trichromatic color vision, among mammals, have evolved the requisite transcriptional regulators to recognize L- and M-cone opsin genes.

To distinguish between the competing models, Wang et al. (1999) investigated whether a mouse, which normally possesses only a single X-linked opsin gene, could support mutually exclusive expression of the human L- and M-cone opsin genes when these are integrated into its genome. They generated transgenic mice carrying a single copy of a minimal human X-chromosome opsin gene array in which the L- and M-cone opsin gene transcription units and 3' intergenic sequences were replaced by alkaline phosphatase (AP) and β-galactosidase (β-gal) reporters, respectively, and then determined the pattern of expression in the cone photoreceptor cells: 63% of expressing cones had AP activity, 10% had β-gal activity, and 27% had activity for both reporters. Thus a large fraction of cone photoreceptors in the mouse retina can efficiently and selectively express either of the reporter transgenes. The sixfold higher average frequency of the cones expressing AP (the L-cone opsin gene reporter) over β-gal (the M-cone opsin gene reporter) may have arisen from the former's greater proximity to the LCR or from the inherently greater efficiency of activation of the L-cone promoter. The doubly labeled cones could have arisen from simultaneous expression of both reporters or from an occasional switch between expression of one reporter and the other (the phenomenon deserves closer inspection).

The mixed expression aside, these findings tend to support the stochastic model and thereby a simple evolutionary path for the development of trichromacy after visual pigment gene duplication (see pigment gene evolution). They imply that the differentiation of L- and M-cone signals occurs opportunistically at more distal stages of visual processing.

Color blindness

Normal color vision can vary significantly among males and females. A large part of the diversity is due to individual differences in the optical densities of the photopigments and the lens and macular screening pigments (see Chapter 2). A substantial part is also due to polymorphisms in the normal L- and M-cone pigment genes (see page 13), which constitute a hypervariable locus in the human genome (see Neitz & Neitz, 1998).

Trichromacy, however, is not enjoyed by all. Many people are partially color blind or color deficient, confusing colors that full trichromats – regardless of their individual differences in color vision – distinguish easily. A very few are completely color blind, unable to discriminate color at all.

Even among color normals, trichromacy fails under several conditions. A complete collapse is associated with the transition from cone-mediated day vision to rod-mediated night vision: As night falls, colors pale and merge into darker and lighter shades of gray. And a partial reduction to a form of dichromacy – tritanopia or the loss of S-cone function (see page 44) – is associated with small, brief targets: (i) viewed centrally (foveal tritanopia), where S-cones are missing;[5] (ii) viewed in the periphery (small field tritanopia), where S-cones are sparse (see Figs 1.1C and 1.17B); and (iii) viewed immediately after the extinction of strong, yellow, adapting fields (transient tritanopia), which is assumed to polarize the yellow-blue color subsystem (Stiles, 1949; Mollon & Polden, 1977a). Trichromacy also fails, to a limited extent, under several extreme conditions of chromatic adaptation and photopigment bleaching.

Inherited color vision defects. The most common forms of color blindness (a.k.a. color vision deficiency or daltonism) are inherited. They arise from alterations in the genes encoding the opsin molecules, an explanation remarkably guessed at by George Palmer (Voigt, 1781; Mollon, 1997). Either genes are lost (due to intergenic nonhomologous recombination), rendered nonfunctional (due to missense or nonsense mutations or coding sequence deletions), or altered (due to intragenic recombination between genes of different types or possibly point mutations).

Phenotypically, the results of the gene alterations are either: (i) anomalous trichromacy (when one of the three cone pigments is altered in its spectral sensitivity but trichromacy is not fully impaired); (ii) dichromacy (when one of the cone pigments is missing and color vision is reduced to two dimensions); or (iii) mono-

[5]S-cones are completely missing in a central retinal disc, which is about 100 μm (ca. 0.34 deg) in diameter (Curcio et al., 1991). Their absence, confirmed by labeling with anti-blue opsin in whole-mounted retinae (Curcio et al., 1991), accords with much prior psychophysical evidence that the center of the human fovea is tritanopic for very small objects (König, 1894; Wilmer, 1944; Wilmer & Wright, 1945; Wald, 1967; Williams, MacLeod, & Hayhoe, 1981a, 1981b; Castano & Sperling, 1982). An estimate of the blue-blind region obtained from psychophysics, which takes into account the blur introduced by ocular optics and eye movements, is 0.42 deg (ca. 118 μm) in diameter (Williams, MacLeod, & Hayhoe, 1981a,b). The exclusion of S-cones from the central foveola is usually attributed to the need to counteract the deleterious effects of light scattering and axial chromatic aberration on spatial resolution, which cause blurring and/or defocus particularly at short wavelengths (see also Chapter 5).

chromacy (when two or all three of the cone pigments are missing and color and lightness vision is reduced to one dimension). Under special conditions, color vision may be in some respects tetrachromatic, when an extra cone pigment is present, owing to X-chromosome inactivation (see page 38).

Those inherited alterations affecting a single cone pigment are referred to by the generic names protan (from the Greek *protos*: first + *an*: not), deutan (*deuteros*: second), and tritan (*tritos*: third), to distinguish disorders in the L-, M-, and S-cone pigments, respectively. The ordering is related to the history of discovery and to early assumptions about the nature of the disorders (von Kries & Nagel, 1896; Boring, 1942; Judd, 1943). The suffix "-anomaly", when appended to the generic names, indicates a deviation or abnormality in the function of the L- (protanomaly), M- (deuteranomaly), or S- (tritanomaly) cone pigments. Likewise, the suffix "-anopia" indicates the absence of function of the L- (protanopia), M- (deuteranopia), or S- (tritanopia) cone pigments.

Those inherited alterations resulting in the loss of two cone pigments are referred to as "cone monochromacies" (Pitt, 1944; Weale, 1953). They include blue- or S-cone monochromacy (affecting both the L- and M-cone opsin genes), green or M-cone monochromacy (affecting the S- and L-cone opsin genes), and red or L-cone monochromacy (affecting the S- and M-cone opsin genes).

Other inherited forms of color blindness arise from mutations in genes not encoding the cone opsins, but rather components of cone structure and function. Those resulting in the loss of function of all three cone types are referred to as "complete achromatopsia" or "rod monochromacy."

Acquired color vision defects. The less common forms of color blindness arise from factors other than inherited alterations in the opsin genes. For instance, cerebral achromatopsia or dyschromatopsia (see Chapter 14), a form of total color blindness, can arise adventitiously after brain fever (Boyle, 1688; Mollon et al., 1980), cortical trauma, or cerebral infarction (Critchley, 1965; Meadows, 1974; Damasio et al., 1980; Zeki, 1990; Grüsser & Landes, 1991; Kennard et al., 1995; Rüttiger et al., 1999).

Yet other forms of color blindness may be associated with: (i) disorders of the prereceptoral ocular media; (ii) fundus detachment; (iii) progressive cone dystrophies or degenerations affecting all cone classes with or without involvement of the rods; (iv) macular dystrophies and degenerations; (v) vascular and hematologic diseases; (vi) glaucoma; (vii) hereditary dominant optic atrophy and other optic nerve diseases; (viii) diseases of the central nervous system (e.g., multiple sclerosis) or other organs (e.g., diabetes mellitus); and (ix) toxic agents (e.g., lead, tobacco, alcohol) that affect the retina or the optic tracts (see Birch et al., 1979).

Protan and deutan defects

The most common, hereditary color blindnesses are the loss (protanopia and deuteranopia) and alteration (protanomaly and deuteranomaly) forms of protan (L-cone) and deutan (M-cone; sometimes written as deuteran) defects. Also known as red-green color vision deficiencies, they are associated with disturbances in the X-linked opsin gene array. Their characteristic X-linked recessive pedigree pattern (see Fig. 1.16A) was early remarked upon by Earle (1845) and Horner (1876) and clearly recognized as such by Wilson (1911). Wilson's assignment of the gene(s) responsible for red-green color blindness to the X-chromosome is the event that is often taken to mark the beginning of the mapping of the human genome. That two gene loci, one for protan and one for deutan defects, were involved was subsequently deduced from the characteristics of doubly heterozygous females[6] (Waaler, 1927; Brunner, 1932; Kondo, 1941; Drum-

[6]Males who carry genes on the X-chromosome for protan and deutan color vision defects are hemizygous (the mutant gene is present in one copy). In contrast, females can be heterozygous (when a mutant gene is paired with a normal gene), homozygous (when two mutant genes of the same type are paired), or doubly heterozygous (when two mutant genes of different types are paired).

mond-Borg et al., 1989). Such females were known to carry genes for both types of color blindness but exhibited normal color vision, implying the existence of distinct defective and normal visual pigment genes on each of their X-chromosomes (Vanderdonck & Verriest, 1960; Siniscalco et al., 1964; Drummond-Borg et al., 1989).

Phenotypes. A variety of special color confusion charts (e.g., the Dvorine, Ishihara, and Stilling pseudoisochromatic plates), hue discrimination or arrangement tasks (e.g., the Farnsworth–Munsell 100-Hue test, the Farnsworth Panel D-15, the Lanthony Desaturated D-15), and lantern detection tests (e.g., the Edridge–Green, Holmes–Wright), all of which exploit the color deficits of the color blind, have been designed to screen for protan and deutan defects (for an overview of the available clinical tests, see Lakowski, 1969a, 1969b; Birch et al., 1979; Birch, 1993). For more precise characterizations, spectral appearance data (including unique hue loci for anomalous trichromats and spectral neutral points for dichromats), luminous efficiency, and hue discrimination functions may be determined.

Traditionally, however, observers with protan and deutan defects are most efficiently and elegantly characterized by the nature of their Rayleigh matches (Rayleigh, 1881) on a small viewing field (≤2 deg in diameter) anomaloscope (the best known is the Nagel Type I). In this task, the observer is required to match a spectral yellow (ca. 589-nm) primary light to a juxtaposed mixture of spectral red (ca. 679-nm) and green (ca. 544-nm) primary lights. There are two variables: the intensity of the yellow and the relative mixture of the red and green lights. Ideally, the S-cones and rods are excluded so that the match is determined solely by the relative absorptions in the M- and L-cones.

Most trichromats reproducibly choose a unique match between the red/green mixture ratio and the yellow intensity (see Fig. 1.11). Slight differences between their match midpoints are usually attributed to the normal variability in the λ_{max} of the L- and M-pigments, such as caused by the alanine/serine polymorphism at codon 180 in exon 3. Variation can also be attributed to differences in photopigment optical density. Some studies have reported match points as being bimodally (Neitz & Jacobs, 1986; Winderickx et al., 1992b) or quadrimodally (Neitz & Jacobs, 1990) distributed in male trichromats. However, other studies have found only unimodal distributions in both male and female trichromats (Jordan & Mollon, 1988; Lutze et al., 1990). The differences may be specific to the psychophysical methodology: Unimodality may be associated with the method-of-adjustment and multimodality with the forced-choice method (Piantandia & Gille, 1992).

In contrast, individuals with protan and deutan defects have displaced Rayleigh match midpoints (i.e., the mean value of the red–green ratio required to match the yellow primary falls outside the normal range) and/or extended matching ranges (they accept more than one red–green ratio).

Incidences. Incidences of X-chromosome–linked defects vary between human populations of different racial origin (see Table 1.5). The highest rates are found in Europeans and the Brahmins of India; the lowest in the aborigines in Australia, Brazil, the South Pacific Islands, and North America. The differences between populations are well documented (e.g., Garth, 1933; Cruz-Coke, 1964; Iinuma & Handa, 1976, Fletcher & Voke, 1985) and are usually explained by relaxation in selection pressures, which could arise as hunting and gathering cultures evolve toward industrialized societies (e.g., Post, 1962, 1963; however, see Kalmus, 1965). Other possible biological factors include gene flow, the rise of mutant genes, and the migration and mixture of races. Interestingly, there is no evidence for analogous within-species variations in color vision in the best-investigated Old-World monkey, the macaque (for a review, see Jacobs, 1993). An analysis of behavioral results suggests that if any individual variations in color vision among macaques occur, their frequency must be less than 5% (Jacobs & Harwerth, 1989).

Of relevance perhaps to the population differences is the fact that there tends to be fewer opsin genes in the arrays of non-Europeans. Extra M-cone genes are

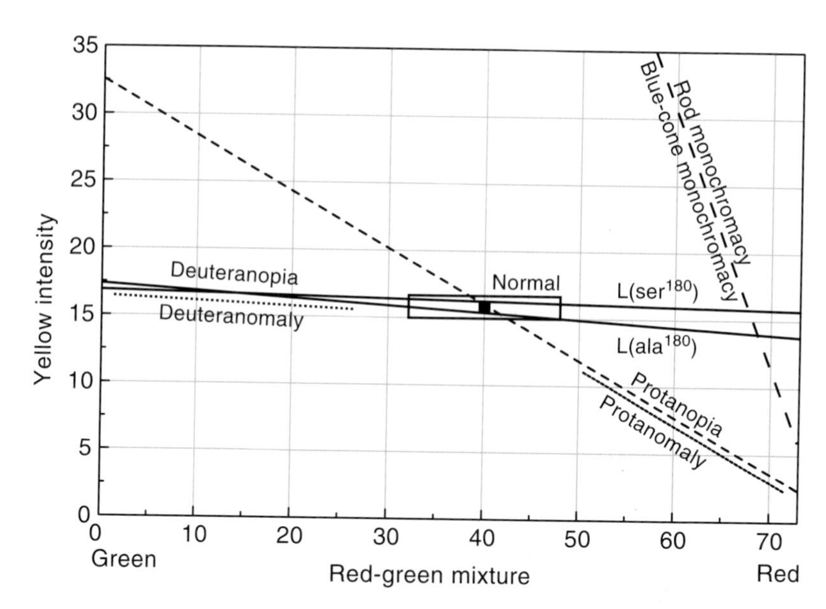

Figure 1.11: Rayleigh matches obtained on a standard Nagel Type I anomaloscope. Characteristic match points or ranges are shown for normal trichromats (the black square indicates the mean, the open rectangle the spread of normal matches), for protanopes with a single L1M2 gene in the X-linked opsin gene array, for deuteranopes with either a single L(ser^{180}) or L(ala^{180}) gene in the array, for an extreme protanomalous trichromat, for an extreme deuteranomalous trichromat, for a blue-cone monochromat, and for a rod monochromat.

believed to increase the opportunities for intragenic and intergenic recombinations between the L- and M-cone pigment genes, which gives rise to phenotypic color vision defects (see unequal crossing over).

The incidences in females are much lower than those in males because the defects are inherited as recessive traits. Males, who have only one X-chromosome, are hemizygous and will always manifest a color defect if they inherit an aberrant gene at one of the first two positions in the gene array. Females, on the other hand, have two X-chromosomes, one inherited from each parent, so they will not usually show a complete manifestation of the typical colour defect unless they are homozygous. Thus, the incidence in Caucasian females should correspond to the sum of the squared frequencies of singly heterozygous females, which should be the same as the frequencies for protan (p_p) and deutan (p_d) defects in hemizygous males: ($p_p^2 + p_d^2$) or 0.39%. (Table 1.6 provides the incidences of protan and deutan defects, divided according to the dichromatic and anomalous trichromatic forms.)

The slightly higher observed figure (0.42%, see Table 1.6, and 0.50%, see Table 1.5) in Caucasian females, and the even larger discrepancy in Asians, is probably explained by the partial manifestation of color blindness in heterozygotic carriers (see page 38) and by problems of testing methodology and inclusion criteria (for a discussion, see Fletcher & Voke, 1985).

Protanopia and deuteranopia

Protanopia and deuteranopia are the dichromatic or loss forms of protan and deutan defects.

Phenotypes. The classical theory was that protanopes and deuteranopes each lack one of the normal pigments, the L- and M-cone pigments, respectively, and contain the others unchanged (König & Dieterici, 1886). Everything else being equal, they were expected to accept the color matches of color normals; their dichromatic color world being a reduced version of the trichromatic one. We now know that the situation is more complex. Considerable variation occurs in both phenotypes and genotypes.

Although some protanopes and deuteranopes are true reduction dichromats, having only one X-chromosome–linked cone photopigment, which is identical to the normal's M- or L-cone pigment, others are not. Some protanopes have only one X-chromosome–

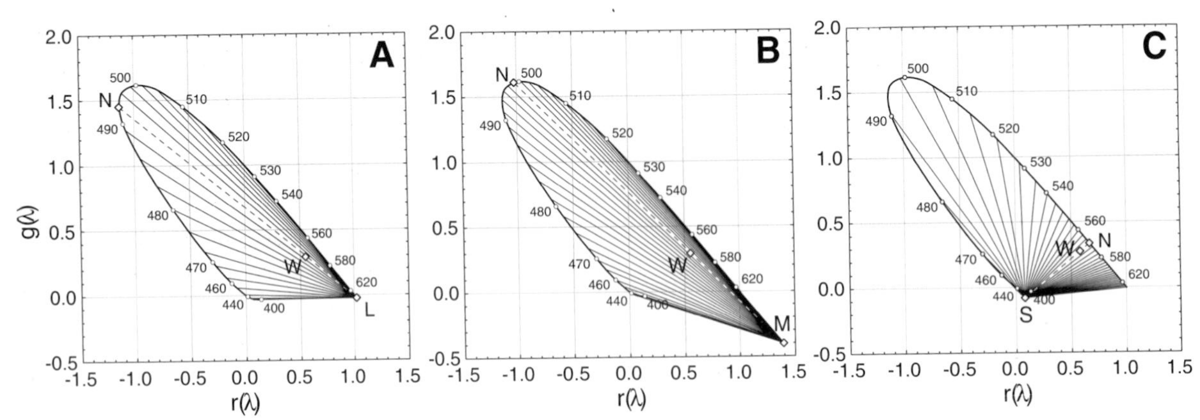

Figure 1.12: Dichromatic lines of constant chromaticity or hue for (A) protanopes, (B) deuteranopes, and (C) tritanopes, drawn in the Stiles and Burch (1955) 2-deg $r(\lambda)$, $g(\lambda)$ chromaticity space (see Chapter 2). The lines have been extrapolated to converge at the confusion or copunctal point, which defines the chromaticity of the missing 2-deg L- (for protanopes), M- (for deuteranopes), or S- (for tritanopes) cone fundamental of Stockman and Sharpe (2000a). The spacing of the lines is based on the spectral wavelength discrimination data of protanopes (Pitt, 1935), deuteranopes (Pitt, 1935), and tritanopes (Fischer et al., 1951; Wright, 1952). The neutral points, N, for the equal-energy white, W, and the corresponding confusion lines (dotted lines) are shown.

linked cone photopigment, a hybrid, that is not identical to the normal M-cone. Among deuteranopes, the situation is complicated by the L(ser^{180}) and L(ala^{180}) polymorphism. This diversity explains – to a large extent – why systematic individual differences in color discrimination and Rayleigh matches among protanopes (Alpern & Wake, 1977) and deuteranopes have been observed. And, accordingly, why, in so-called classical confrontation experiments, a given protanope did not always accept the color matches made by a normal trichromat (Mitchell & Rushton, 1971; Rushton et al., 1973).

(i) Confusion colors: By definition, dichromats require only two primaries to match all color stimuli. As a result, they confuse or fail to discriminate colors that are easily distinguished by normal trichromats. As first pointed out by Maxwell (1855) and demonstrated by Helmholtz in his *Physiological Optics* (1867), when the colors confused by dichromats are plotted in a chromaticity diagram, the axes of which may be generated from transformations of standard color matching functions or from representations of cone excitations (see Chapter 2), they lie on a series of straight lines called "confusion loci."

The confusion lines of protanopes and deuteranopes are shown in Fig. 1.12. The line spacing is adjusted to correspond to the number of just noticeable wavelength differences within the equal-energy spectrum (i.e., the wavelength represented by one line is just distinguishable from that represented by its neighbor). For protanopes, only about 21 distinct wavelengths can be discriminated; whereas for deuteranopes, only about 31 can be (slightly lower estimates were obtained by Pitt, 1935). In contrast, the normal discriminates about 150 wavelengths[7] in the spectrum (Wright & Pitt, 1934; Boring, 1942).

(ii) Neutral zone and spectral colors: In both protanopes and deuteranopes, the spectrum is dichromic, consisting of just two pure hues (Herschel, 1845). In contrast, normals see at least seven pure hues: red, orange, yellow, green, cyan, blue, and violet. This property was first remarked upon by John Dalton (1798), a deuteranope (Hunt et al., 1995, see below): "I see only two, or at most three, distinctions [in the solar

[7]The total number of discriminable, nonspectral, hues, of course, is many more. Wool graders at the Gobelin Tapestry works, in the 19th century, were known for being able to distinguish at least 20,000 different hues (Chevreul, 1839).

image]. My yellow comprehends the red, orange, yellow, and green of others; and my blue and purple coincide with theirs" (p. 31).

The two spectral regions for protanopes and deuteranopes – one blue, the other yellow – are separated by a neutral or achromatic zone that is indistinguishable from white and corresponds to where the relative absorptions of the two remaining cone classes are the same as for white (see Fig. 1.13). The midpoint of the zone – the neutral point – which, by definition, falls on the confusion line passing through the physiological white point, is relatively easy to specify. Quantitative measures, however, are obscured by the arbitrariness of the physiological white (see Walls & Heath, 1956), individual variability in the density of the ocular and macular pigmentation (which affects the chromaticity of the physiological white), and the failure to take into account phenotypic differences between dichromats of the same class. For protanopes and deuteranopes, representative neutral point values for a white standard light (of color temperature 6774 K) are 492.3 nm (Walls & Mathews, 1952; Walls & Heath, 1956; Sloan & Habel, 1955) and 498.4 nm (Walls & Mathews, 1952; Walls & Heath, 1956; Massof & Bailey, 1976), respectively.

But what are the spectral hues actually seen by the red-green dichromats above and below the neutral zone? Although Dalton and other dichromats often assign conventional hue names to the shorter (blue) and longer (yellow) wavelength regions of the spectrum, it does not follow that their usage coincides with that of the color normal. In fact, the impossibility of passing from which colors color blind observers confuse to which they actually see was recognized almost as soon as color blindness was discovered (Wilson, 1855; see Judd, 1948). The preferred approach to overcoming the ambiguity has been to study the vision of unilateral color blinds, born with one normal eye and one color-deficient eye (the earliest report is from von Hippel, 1880, 1881). Observations, which must be treated cautiously,[8] suggest that, for both protanopes and deuteranopes, below the neutral zone, the blue at 470 nm, and, above the neutral zone, the yellow at 575 nm have the same hue for the two eyes (Judd, 1948 provides a review of historical cases; see also Sloan & Wallach, 1948; Graham & Hsia, 1958; Graham et al., 1961; MacLeod & Lennie, 1976).

(iii) Luminous efficiency: The two types of dichromats also differ in which part of the equal-energy spectrum appears brightest (see Chapter 2 and Wyszecki & Stiles, 1982a). For the normal, the maximum of the luminosity function is on average near 555 nm, for the protanope it is closer to 540 nm (blue-shifted to the λ_{max} of the M-cone pigment), and for the deuteranope it may be closer to 560 nm (red-shifted to the λ_{max} of the L-cone pigment).

(iv) Color perception: Given that red-green dichromats only see two colors in the spectrum, and assuming that the colors are known, it is possible to simulate for the color normal the extent of the color confusions made by the dichromats. The history of such simulations begins with von Goethe (1810). In his *Farbenlehre*, he included a reproduction of a small water color that he painted to demonstrate how the landscape would appear to those lacking the blue sensation (Akyanoblepsie). Since then, many other simulations have been attempted (Holmgren, 1881; Rayleigh, 1890; Ladd-Franklin, 1932; Evans, 1948; Viénot et al., 1995).

Figure 1.14 presents simulations for the normal observer of how a colorful fruit market is perceived by a protanope (Fig. 1.14B) and a deuteranope (Fig. 1.14C). For both, the normally wide gamut of yellowish-green to red colors is dramatically reduced. Clearly, protanopes are not red-blind (because they lack the L- or "red" cones) and deuteranopes green-blind (because they lack the M- or "green" cones); rather, both are red-green blind. They only distinguish such colors on the basis of saturation and lightness variations. The major difference in the images is that reds appear relatively darker in the protanopic simula-

[8]Such cases are extremely rare, and it is never been fully demonstrated that the eyes described as normal and dichromatic are in fact really so (see Bender et al., 1972; MacLeod & Lennie, 1976). Further, the etiology of such a discrepancy between eyes is uncertain. It could be owing to somatic forward or back mutations or, in heterozygotes, to differential lyonization in the two eyes (see heterozygotic carriers of protan and deutan defects).

Racial group	Male		Female	
	No.	Incidence	No.	Incidence
European descent	250,281	7.40	48,080	0.50
Asian	349,185	4.17	231,208	0.58
Africans	3,874	2.61	1,287	0.54
Australian aborigines	4,455	1.98	3,201	0.03
American Indians	1,548	1.94	1,420	0.63
South Pacific Islanders	608	0.82	—	—

Table 1.5: Racial incidence of red-green color vision deficiencies in males and females. The incidences (in percent) correspond to the total number of color defectives found divided by the total number examined in each racial group. The raw numbers were obtained from 67 studies listed in Waaler (1927), Cox (1961), Post (1962), Waardenburg (1963a), Crone (1968), Iinuma and Handa (1976), and Koliopoulos et al. (1976).

Gender	No.	Protan		Deutan	
		Anomaly	Anopia	Anomaly	Anopia
Male	45,989	1.08	1.01	4.63	1.27
Female	30,711	0.03	0.02	0.36	0.01

Table 1.6: Incidences (in percent) for X-chromosome–linked color vision deficiencies in men and women. The raw numbers were taken from populations in Norway (Waaler, 1927), Switzerland (von Planta, 1928), Germany (Schmidt, 1936), Great Britain (Nelson, 1938), France (François et al., 1957), The Netherlands (Crone, 1968), Greece (Koliopoulos et al., 1976), and Iran (Modarres et al., 1996–1997). Many studies included in Table 1.4 could not be included in this table because they did not separate deficiencies according to phenotype.

tion than in the deuteranopic, owing to the reduced spectral sensitivity of the protanope to long wavelengths. And, protanopes tend to confuse reds, grays, and bluish blue-greens; whereas deuteranopes tend to confuse purples, grays, and greenish blue-greens.

Diagnosis. The Rayleigh match for dichromats, like all color vision tests, exploits the dichromat's color confusions. Because they have only one pigment (or only one functionally distinct pigment), they have one degree of freedom in the Rayleigh equation and are able to fully match the spectral yellow primary to any mixture of the spectral red and green primary lights by merely adjusting the intensity of the yellow, regardless of the red-to-green ratio. Thus, instead of a unique match, they will have a fully extended matching range that encompasses both the red and green primaries. Although the deuteranope will display a normal or near-normal relation in the luminosity of his matches, the protanope will display a luminosity loss for the red primary, requiring less yellow light to match it.

When a regression line is fitted to the matches of either a protanope or deuteranope, its slope and intercept with the yellow intensity axis will depend on the λ_{max} of the underlying photopigment (see Fig. 1.11). In the Nagel Type I instrument, the protanope needs a very dim yellow intensity to match the red primary and a bright intensity to match the green primary (hence a steep slope); whereas the deuteranope uses about the same yellow brightness to match both primaries (hence a flat slope). The intersection of the regression lines of protanopes, who retain normal M-cone pigments, and deuteranopes, who retain normal L-cone pigments, will coincide with the average match of normal trichromats. Regression lines of intermediate slope will be generated by dichromats who have a single X-linked anomalous pigment, such as those encoded by 5'L-3'M hybrid genes. These will not intersect with the average normal match (Jägle, Sharpe, & Nathans, unpublished).

Incidences. Protanopia (1.01%) and deuteranopia (1.28%) are about equally frequent in the European

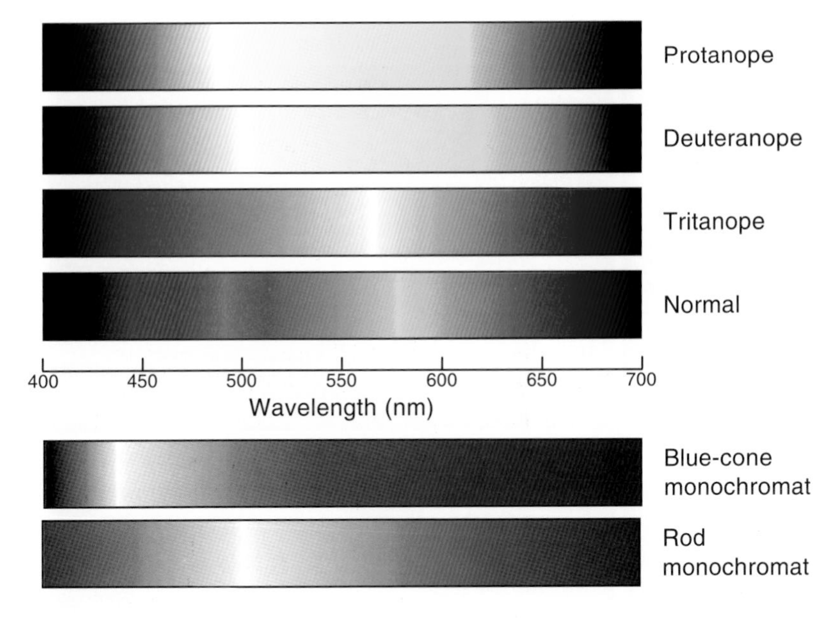

Figure 1.13: The appearance of the visible spectrum for five types of color blindness: protanopia, deuteranopia, tritanopia, blue-cone monochromacy, and rod monochromacy. Neutral points (or zones) are indicated by the white region in the spectrum for the three types of dichromats. These divide the spectrum into two hues. The shorter wavelength hue is perceived below the neutral point, the longer above. The saturation of both increases with distance from the neutral point. The brightest part of the spectrum roughly corresponds to the λ_{max} of the luminosity function (photopically determined, save for the rod monochromat).

(Caucasian) male population (see Table 1.6). This is in marked contrast to protanomaly and deuteranomaly and is worthy of note.

Molecular genetics. The opsin genes may be compromised in three ways to give rise to red-green dichromacy: by point mutations, major sequence deletions, and by unequal crossing over.

(i) Point mutations: In principle, red-green dichromacy can arise from a point mutation[9]: a single-base alteration in which the normal nucleotide has been substituted by an incorrect one. The point mutation could occur either (i) in the coding sequences of the L- or M-cone pigment gene, which vitiates or destabilizes the opsin molecule or impairs its quantum efficiency, or (ii) in the upstream promoter sequences, which leads to failure of expression of the opsin. One such naturally occurring missense mutation has already been identified (Winderickx et al., 1992c; see also blue-cone monochromacy). It is a thymine-to-cytosine substitution at nucleotide 648 in the coding region, resulting in the replacement of cysteine by arginine at codon 203 ($cys^{203}arg$) in the second extracellular loop of the amino acid sequences in exon 4. The substitution disrupts a highly conserved disulfide bond formed with another cysteine at codon 126, in the first extracellular loop. The rupture is thought to damage the three-dimensional structure of the opsin, rendering it nonviable.

So far, the $cys^{203}arg$ mutation has been found in all of the M-cone pigment genes in one individual with deuteranopia or extreme deuteranomaly (Winderickx et al., 1992c) and in the cone pigment genes of multiple blue-cone monochromats (Nathans et al., 1989, 1993). Although fairly common in the population – estimated frequency of 2% by Winderickx et al. (1992c) and 0.5% by Nathans et al. (1993) – it is not always associated with a color vision defect. Its

[9]Point mutations include: (i) synonymous mutations, which do not alter the encoded amino acid residue and, therefore, have no effect on the properties of the opsin; (ii) missense mutations, which alter the encoded amino-acid residue, either with properties similar to (in which case they may be harmless) or different from (in which case they may be disastrous) those of the replaced residue; and (iii) nonsense mutations, which convert an amino-acid–specifying codon to a stop codon, inappropriately signalling the termination of translation and ending the polypeptide chain synthesis of the opsin.

Figure 1.14: A scene from a fruit market as perceived by a normal trichromat (A), a protanope (B), a deuteranope (C), and a tritanope (D). The simulations are based on an algorithm incorporating a colorimetric transformation, which also makes explicit assumptions about the residual sensations experienced by dichromats (see Viénot et al., 1995; Brettel et al., 1997). The transformation replaces the value of each original element by the corresponding value projected onto a reduced color stimulus surface, parallel to the direction of the missing cone fundamental axis in an LMS cone excitation space (see Chapter 2). It is informed by assumptions about the neutral zones of dichromats and the colors perceived by unilateral colorblind observers. The simulations must be treated as approximate, not only because of the limitations of the strictly colorimetric transformation involved and of the printing procedure, but also because the transformed image is viewed by the normal visual system, which will process it differently than the color blind system. (The original scene is reproduced by kind permission of the Minolta Corp.)

expression may depend on the position in the array of the gene in which it occurs (see gene expression above).

(ii) Major sequence deletions: Red-green dichromacy could also arise from major deletions in the coding sequences or promoter region of one of the opsin genes. A case in which a rearrangement, possibly a deletion, between exons 1 and 4 of the L-cone opsin gene engendered a protan defect with a progressive macular degeneration has been reported (Reichel et al., 1989).

(iii) Unequal crossing over: Dichromacy is most typically the result of unequal homologous crossing over or recombination between the L- and M-cone pigment genes. During meiosis, when the maternal and paternal X-chromosomes in a female are aligning to

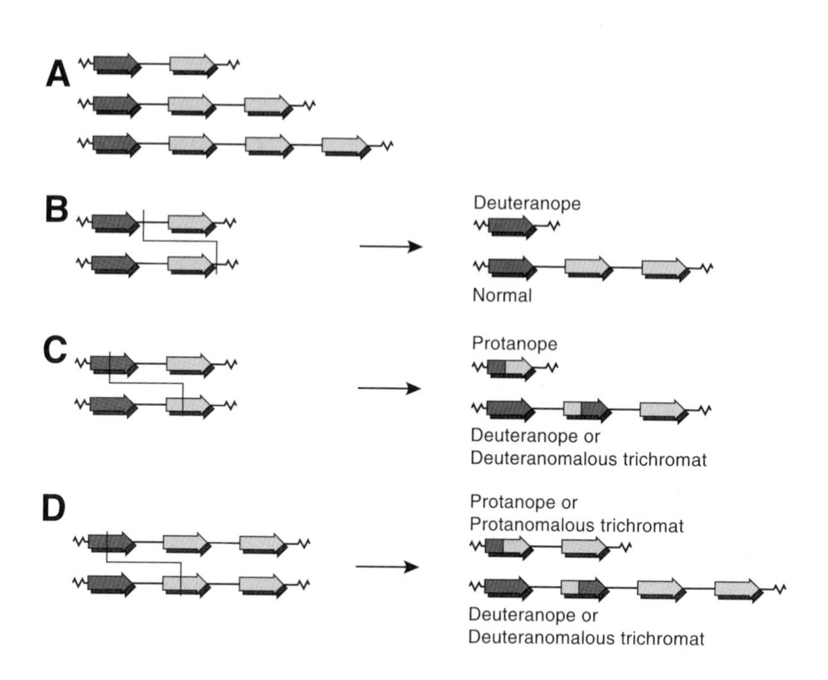

Figure 1.15: Examples of unequal recombination within the tandem array of X-linked genes responsible for the common protan and deutan defects of color vision. Each gene is represented by an arrow: The base corresponds to the 5' end and the tip to the 3' end. Dark gray arrows, L-cone pigment genes; light gray arrows, M-cone pigment genes. Unique flanking DNA is represented by zig-zag lines and homologous intergenic DNA by straight lines. For each recombination event, both products are shown, most of which lead to a color-defective phenotype when inherited in males. (A) Three wild-type (naturally occurring) gene arrays, each containing one L-cone pigment gene and a variable number of M-cone pigment genes. (B) Intergenic recombination events leading to deuteranopia and an enlarged gene array. (C) and (D) Intragenic recombination events leading to protanopia, protanomaly, deuteranopia, or deuteranomaly.

form the X-chromosome of an egg (prophase I), breakage occurs, with an exchange of corresponding sections of DNA sequences and the subsequent rejoining. Normally the exchange is reciprocal, but sometimes, when the chromosomes are misaligned, it is unequal. Such unequal recombination events are facilitated by the juxtaposition of the genes in the head-to-tail tandem array and by their high degree of homology (Nathans, Thomas, & Hogness, 1986; Drummond-Borg et al., 1989).

The mechanisms of recombination inducing dichromacy are summarized in Fig. 1.15 (after Nathans, Thomas, & Hogness, 1986; Nathans, Piantandia, et al., 1986). For simplicity, the common arrangement of the gene array is assumed to be a single L-cone opsin gene followed by one, two, or three M-cone opsin genes (Fig. 1.15A).

Unequal intergenic recombination, in which the crossing-over point occurs in the region between genes, can reduce the gene array to a single L-cone pigment gene (Fig. 1.15B). Males who inherit such an X-chromosome arrangement from their mothers will be single-gene deuteranopes (Nathans, Thomas, & Hogness, 1986; Nathans, Piantanida, et al., 1986; Vollrath et al., 1988; Drummond-Borg et al., 1989). Less than one-half of all deuteranopes display such a pattern (Sharpe et al., 1998). Those who do constitute a true L-cone reduction type of deuteranope (König & Dieterici, 1886), save perhaps for those who carry an M-cone–specific exon 2 embedded within an otherwise L-cone pigment gene, which may slightly change the L-cone photopigment spectrum (see Table 1.3).

Unequal intragenic recombination, in which the crossing-over point typically occurs in the introns between exons, can have several outcomes. It can reduce the array to a single 5'L-3'M hybrid gene (Fig. 1.15C), which will usually be associated with protanopia. Less than one-half of all protanopes display such a pattern (Sharpe et al., 1998); of those, more than one-half possess any one of several M-cone–like hybrid pigments, the spectral sensitivity of which is shifted relative to the normal M-cone pigment (see Table 1.3). Such protanopes can be thought of as anomalous, for their photopigment spectral sensitivities will be nonrepresentative of those of color normals. The remainder, whose array carries a single 5'L-3'M hybrid gene encoding either an L1M2 or an L2M3 pigment, are of the true reduction type, with a spectral sensitivity resembling that of the normal M-cone pigment (Sharpe et al., 1998).

Alternatively, unequal intragenic recombination can replace the normal L-cone pigment gene with a 5'L-3'M hybrid gene (Fig. 1.15D). The array consists of a 5'L-3'M hybrid gene and one or more normal M-cone pigment genes (Deeb et al., 1992; Sharpe et al., 1998). More than one-half of all protanopes display this pattern (Sharpe et al., 1998). Presumably, they express more than one X-linked visual pigment gene. Protanopia is implied only if the fusion in the hybrid gene occurs before exon 3 (because the gene will encode an M-cone pigment). Otherwise, protanomaly is predicted, because the gene will encode an anomalous pigment that differs in spectral sensitivity from the normal M-cone pigment by 2 to 8 nm, depending on which amino acid residues have been substituted (see Table 1.3).

Finally, unequal intragenic recombination can replace the normal M-cone pigment gene with a 5'M-3'L hybrid gene (Deeb et al., 1992; Sharpe et al., 1998). The array consists of a normal L-cone pigment gene and a 5'M-3'L hybrid gene (Figs. 1.15C and 1.15D). More than one-half of all deuteranopes display this pattern (Sharpe et al., 1998). Deuteranopia is only implied if the fusion occurs before exon 2 (because the gene will encode a pigment with a spectral sensitivity similar to that of the normal L-cone pigment). Otherwise, deuteranomaly is predicted, because the gene will encode an anomalous pigment that differs in spectral sensitivity from the normal L-cone pigment by 4 to 12 nm, depending on which amino acid residues have been substituted (see Table 1.3).

The position in the array of the hybrid pigment gene is likely to be critical. Four to eight percent of Caucasian males with normal color vision have, in addition to normal L- and M-cone pigment genes, 5'M-3'L hybrid genes that apparently are not expressed or are insufficiently expressed to disrupt normal color vision (Drummond-Borg et al., 1989; Deeb et al., 1992). Presumably, in these individuals, the hybrid genes occupy a more distal, 3' position in the gene array (see gene expression, above).

(iv) John Dalton's genotype: Historically, John Dalton is the most famous person known to be color blind. In his honor, the term *daltonism* is used in many languages, including English, French, Spanish, and Russian, to refer to color vision disorders.

Believing his own colorblindness to be caused by his eye humors being blue-tinted, he directed that they be examined upon his death. However, a postmortem analysis conducted on 28 July 1844 failed to confirm his hypothesis (Wilson, 1845; Henry, 1854). Fortunately, sufficient eye tissue was preserved after the autopsy to permit, 150 years later, a molecular genetic analysis (Hunt et al., 1995). PCR amplification of opsin gene fragments from the tissue revealed that only a single L-cone opsin sequence was present, consistent with Dalton being a single-gene deuteranope (see Figs. 1.9A and 1.15B). The genotype is contrary to previous interpretations, including that of Thomas Young (1807), who thought Dalton a protanope. But it accords with a close reading of the historical record and with colorimetric analysis of the color confusions and failures of color constancy reported by Dalton himself (Dalton, 1798; Hunt et al., 1995).

These findings represent, to date, the most impressive example of genotyping of a known historical figure, and they provide a satisfying historical continuity to the field of the molecular genetics of color blindness.

Large-field trichromacy. Nagel (1905, 1907) was the first to observe that many red-green dichromats, himself included, are only completely dichromatic for small viewing fields restricted to the central fovea. With larger viewing fields, they become partially trichromatic and are able to make red-green color discriminations. Nagel's observations have been confirmed by many others relying on small- and large-field Rayleigh or neutral point matches (Jaeger & Kroker, 1952; Smith & Pokorny, 1977; Nagy, 1980; Breton & Cowan, 1981) and color naming (Scheibner & Boynton, 1968, Nagy & Boynton, 1979; Montag,

1994). It is now widely believed that the majority of dichromats display such behavior.

In cases in which the dichromats have more than one gene in the opsin gene array, the improvement in color discrimination with large fields might be attributed to genotype. Some dichromats could have two separate genes, one normal and one a hybrid, encoding slightly different pigments, the benefit of which only manifests under large-field conditions. In particular, a small minority of deuteranopes may have a normal M-cone pigment gene, distal to the 5'M-3'L hybrid gene (Deeb et al., 1992). If the downstream, normal gene is expressed (in reduced amounts according to its greater distance from the LCR), it might influence color discrimination under large-field conditions. However, 5'M-3'L hybrid genes may only be expressed and only influence phenotype when they occupy the second position of the opsin gene array (Hayashi et al., 1999). A similar explanation is unlikely to apply to protanopes because the presence of downstream L-cone, or, for that matter, 5'M-3'L hybrid pigment genes, has never been demonstrated in these subjects.

However, these explanations fail to account for another observation: some, but not all, single-gene dichromats also display large-field trichromacy (Nathans et al., 1986; Neitz, Neitz, & Jacobs, 1989; Deeb et al., 1992; Crognale et al., 1999). Moreover, at least one individual, who was missing all of the M-cone pigment genes in the X-chromosome array, behaved as an extreme deuteranomalous observer, even under small-field viewing conditions (Nathans et al., 1986). This suggests that Rayleigh matches are not dependent solely on the number of different X-chromosome–linked photopigments. Presumably other factors contribute to the rejection of red-to-green mixture ratios (see also Smith & Pokorny, 1977; Nagy & Boynton, 1979; Nagy, 1980; Breton & Cowan, 1981; Nagy & Purl, 1987; Crognale et al., 1999). These could include: (i) rod intrusion; (ii) S-cone intrusion; (iii) differences in pigment optical density between central and parafoveal cones; (iv) spatial variation in the macular pigment; (v) changes in cone receptor geometry and orientation with eccentricity (e.g., wave-guiding); and (vi) dissimilarities in the morphology and/or function of M- and L-cone photoreceptors, apart from the photopigment that they contain (e.g., placing an L-cone pigment in an M-cone structure might produce a different spectral sensitivity than placing it in an L-cone structure).

Cone mosaic. Dichromats may have fewer cones than normal or some of their cones may lack pigment: the L-cones may be missing or empty in protanopes and the M-cones may be missing or empty in deuteranopes. On the other hand, they may have a full complement of functioning cones: either the missing pigment is entirely replaced by ones of the available type (the M-cones in protanopes and the L-cones in deuteranopes) or the empty cone structures are filled with photopigment of the available type (M-cone pigment in protanopes and L-cone pigment in deuteranopes). A complication here, as alluded to above, is that we do not yet know whether M- and L-cones differ in ways other than the pigment that they contain in their outer segments. It is unlikely that S-cones could replace the missing cones because their numbers are far too small (although it is unknown what ultimately limits their number). Likewise, it is unlikely that the S-cone pigment could replace the M- or L-cone pigment because the S-cone opsin is specified by a gene on a different chromosome.

Surprisingly, the predictions about effects of changes in the cone array on absolute threshold and luminous efficiency have rarely been tested (Abney, 1913; DeVries, 1948a; Hecht & Hsia, 1948; Hecht, 1949; Wald, 1966; Berendschot et al., 1996; Knau & Sharpe, 1998).

In hindsight of the molecular genetics, it is possible to make specific predictions. Dichromats possessing two (or more) pigment genes in their array should not have any missing or nonfunctioning foveal cones (see Fig. 5.2 for a high-resolution image of the living retina of a multigene protanope, subject MM, whose phenotype has been established by Nathans, personal communication). They should have a normal complement of cones with the remaining pigment (either L or M) plus an extra complement with the de facto pigment encoded by the hybrid gene. This could result in an effective doubling or tripling of the number of L- or M-cones compared with the trichromat eye (assuming a normal ratio of 1.5:1.0; see Fig. 1.1C and Chapter 5).

On the other hand, those dichromats possessing a single L- or 5'L-3'M hybrid pigment gene may or may not have cones that are missing pigment. It would depend on whether all of the non-S-cones are filled with the only available longer wavelength sensitive photopigment.

If the missing pigment is not replaced, then poorer performance on some visual sensitivity and visual acuity tests might be expected (ca. 40% of the cones will be empty). However, an impairment would be difficult to measure because of the limiting effects of the eye's optics and because of the random clustering of the remaining cone photoreceptors (see Fig. 1.1C and Chapter 5) and the constant microsaccadic movements of the eye, which would tend to compensate for any blind areas.

Protanomaly and deuteranomaly

Protanomaly and deuteranomaly are the alteration forms of protan and deutan defects.

Phenotypes. The color vision deficits of anomalous trichromats are usually less severe than those of dichromats, but there is considerable variability among individuals. They can be categorized as simple or extreme, according to their matching behavior on the Rayleigh equation (Franceschetti, 1928). Many simple anomals may be unaware of their color vision deficiency, whereas many extreme anomals may have nearly as poor color discrimination as dichromats.

(i) Color perception: Color terms may be skillfully used by many anomalous trichromats. Unlike dichromats, they do not have a neutral zone and see more than two hues in the spectrum. The saturation of spectral lights may differ from normal (Pokorny & Smith, 1977). But many can assign some hue names, such as blue, cyan, green, yellow, and orange, to unique parts of the spectrum, although the wavelength locations may differ from those of normal trichromats (von Kries, 1919; Rubin, 1961; Hurvich & Jameson, 1964; Smith, Cole, & Isaacs, 1973; Romeskie, 1978). In general, their hue locations are shifted to shorter (protanomals) and longer (deuteranomals) wavelengths. Their spectrum may be conceived as encompassing saturated blues, extremely desaturated bluish-greens and greens, yellows, and relatively desaturated oranges (Pokorny, Smith, & Verriest, 1979).

(ii) Luminous efficiency: Protanomals, like protanopes, have an insensitivity to the far red end of the visible spectrum, with a blue-shifted luminosity function. But deuteranomals, like deuteranopes, may have a normal or slightly red-shifted luminosity function.

Diagnosis. *Rayleigh equation:* The Rayleigh matches of protanomalous and deuteranomalous trichromats are properly characterized by both their displaced match midpoints and their matching ranges. The two measures vary independently (Schmidt, 1955; Hurvich, 1972).

Generally, simple protanomals have a narrow matching range displaced to red (i.e., they require more red than normal in the matching field), with a concomitant luminosity loss for the red primary. Likewise, simple deuteranomals have a narrow matching range displaced to green, with an essentially normal luminosity. Neither tend to accept the matches of normals, and their deviant anomalous ratios are expressed relative to normal ratios by the so-called anomalous quotient (see Pokorny, Smith, & Verriest, 1979).

A very few simple anomalous trichromats accept only one mixture ratio setting (i.e., a unique match), which is outside the normal limits. They are known as minimally affected or *red-green deviants* (Vierling, 1935) and have excellent chromatic discrimination.

On the other hand, extreme anomalous trichromats exhibit a wide matching range, often encompassing the matches of normal trichromats as well as those of anomalous trichromats and sometimes even one (but never both) of the primaries (see Fig. 1.11). The extreme protanomal will match the red-green mixture to the yellow primary except when pure or near pure green is used; the extreme deuteranomal will do the same except when pure or near pure red is used. The midpoint of the matching range is usually displaced from the center of normal matches.

Incidence. Deuteranomaly (4.61%) is more than four times more frequent than protanomaly (1.07%) in Caucasian males (see Table 1.6). Its greater frequency may be related to a peculiarity of intragenic recombination. Deuteranomaly can arise from unequal crossing over in simple arrays, each having only a single M-cone pigment, whereas protanomaly can only arise if one of the arrays contains two or more normal M-cone pigment genes (see Figs. 1.15C & 1.15D).

Molecular genetics. *(i) Crossing over:* Simple and extreme anomalous trichromacy typically occur as the result of unequal intragenic crossing over between the L- and M-cone pigment genes. Either there is a replacement of the normal L- by a 5'L-3'M hybrid pigment gene or the normal M- by a 5'M-3'L hybrid pigment gene. The position of the hybrid gene in the array is presumably critical (see Figs. 1.10A and 1.10B). For instance, protanomaly results only if a 5'L-3'M hybrid gene replaces the normal L-cone pigment gene in the most proximal position (for examples, see Balding et al., 1998). Likewise, deuteranomaly results only if a 5'M-3'L hybrid gene replaces the normal M-cone pigment gene and is expressed preferentially over the downstream normal M-cone pigment genes (for examples, see Hayashi et al., 1999). This would explain the not infrequent presence of 5'M-3'L hybrid genes in individuals with normal color vision (estimated to be 4–8% in Caucasians and <1% in Japanese).

However, it is still debatable in the case of deuteranomalous trichromats with three or more genes in the array, one of which is a normal M-cone opsin gene, whether the pigment with the anomalous spectral sensitivity encoded by the 5'M-3'L hybrid gene simply replaces the normal M-cone pigment, is present in cones as a mixture with the normal M-cone pigment, or is present in a subset of M-cones.

The first of these possibilities is strongly supported by experiments using long-range PCR amplification. Hayashi et al. (1999) studied 10 deutan males (8 deuteranomalous and two deuteranopic) with three opsin genes on the X-chromosome: an L, an M, and a 5'M-3'L hybrid gene. The 5'M-3'L hybrid gene was always at the second position; the first position being occupied by the L-cone opsin gene. Conversely, in two men with L-cone, M-cone, and 5'M-3'L hybrid opsin genes and normal color vision, the 5'M-3'L hybrid gene occupied the third position. When pigment gene mRNA expression was assessed in postmortem retinae of three men with the L-cone, M-cone and 5'M-3'L hybrid genotype, the 5'M-3'L hybrid gene was expressed only when located in the second position. Thus they conclude that the 5'M-3'L gene will only cause deutan defects when it occupies the second position of the opsin gene array. This is consistent with "two gene only" expression at this locus (see section on gene expression and Fig. 1.10B). An alternative hypothesis for the molecular basis of deuteranomaly (Sjoberg, Neitz, Balding, & Neitz, 1998) – namely, that the M-cone opsin genes are mutated and express a nonfunctional pigment – is unlikely because Hayashi et al. (1999) could not find mutations in the promoter or in the coding sequences of the M-cone opsin genes of the deutan subjects that they investigated (Deeb et al., 1992; Winderickx et al.,1992a; Yamaguchi, Motulsky, & Deeb, 1997). However, it is conceivable that distal gene expression is silenced by some elements in the 3' flanking region of the locus.

(ii) Hybrid pigments: Any comprehensive model of anomalous trichromacy must account for the fact that anomalous trichromats differ greatly in the location and range of their Rayleigh matching points (for review, see Hurvich, 1972; Pokorny, Smith, & Verriest, 1979; Mollon, 1997). The finding of a diverse family of hybrid pigments in single-gene protanopes (Sharpe et al., 1998, 1999) supports a model of anomalous trichromacy in which any one of many M- or L-like anomalous pigments can be paired with either of two major polymorphic versions of the more similar normal pigment (Merbs & Nathans, 1992b; Neitz, Neitz, & Kainz, 1996). As the spectral sensitivities of the normal or the anomalous pigment shift, the midpoint of the Rayleigh match will shift, and as the separation between the spectral sensitivities of the normal and anomalous pigments increases or decreases, the better

or poorer will be the subject's chromatic discrimination (see also Shevell et al., 1998).

Inspection of Table 1.3 reveals a wide range of possible anomalous pigments lying between the normal L- and M-cone pigments. Rather than a continuous distribution, there is a clustering of 5'L-3'M hybrid pigments having λ_{max}'s within about 8 nm of the λ_{max} of the M-cone pigment and a clustering of 5'M-3'L hybrid pigments having λ_{max}'s within about 12 nm of the λ_{max} of the L-cone pigment.

Although the λ_{max} of the anomalous pigment may be the decisive factor in determining the severity of anomalous trichromacy, other factors such as variation in the optical density of the expressed hybrid pigment may also have to be taken into account (He & Shevell, 1995). This could come about if the amino acid substitutions alter the stability and/or activity of the expressed pigment as well as its spectral position. As of yet there is no solid evidence that hybrid pigments have different optical densities than normal M- or L-cone pigments (see sequence variability and opsin viability above).

(iii) Alternative theories: The in vivo identification of the hybrid pigments (see Table 1.3) is inconsistent with the theory that all anomalous trichromats share a single anomalous pigment with a spectral sensitivity at a position intermediate between the normal L- and M-cone pigments (DeVries, 1948a; MacLeod & Hayhoe, 1974; Hayhoe & MacLeod, 1976). Nor does it accord with the theory that there is one anomalous pigment common to all deuteranomalous trichromats and another common to all protanomalous trichromats (Rushton et al., 1973; Pokorny, Moreland, & Smith, 1975; Pokorny & Smith, 1977; DeMarco et al., 1992). Finally, it does not support, in its most extreme form, the theory that there are no distinct anomalous pigments but instead clusters of normal L- and M-cone pigments that differ in their peak sensitivities, and that anomalous trichromats draw their pigments from the same rather than different clusters (Alpern & Moeller, 1977; Alpern & Wake, 1977). While this theory fails to account for the existence of bona fide hybrid pigments, it explains the phenotypes of that subset of anomalous trichromats in which the normal and anomalous pigments are distinguished only by the polymorphic alanine versus serine difference at position 180. It is also prescient in recognizing how the anomalous hybrid pigments cluster near the remaining normal pigment.

(iv) Red-green deviants: Molecular genetic explanations of red-green deviants are not yet forthcoming. Very few have ever been identified and none have been investigated by molecular techniques. One possible explanation is that red-green deviants possess an anomalous pigment, produced by a point mutation, whose λ_{max} lies outside the range defined by the normal M- and L-cone pigments (i.e., either blue-shifted relative to the M-cone pigment or red-shifted relative to the L-cone pigment).

Cone Mosaic. The cone mosaic of anomalous trichromats should resemble those of normal trichromats. The only difference being that the pigment in one of the normal cone classes is replaced by an anomalous pigment.

Heterozygotic carriers of protan and deutan defects

About 15% of women inherit an X-chromosome carrying an abnormal opsin gene array from one parent and an X-chromosome carrying a normal opsin gene array from the other (see Fig. 1.16A). As heterozygous carriers for red-green color vision defects, they partially possess that which they fully bequeath to one-half of their sons. Early in their own embryonic development, by a process of dosage compensation known as X-chromosome inactivation or lyonization (Lyon, 1961, 1972), one of the two X-chromosomes in their somatic cells is transcriptionally silenced (see Fig. 1.16B). The process is normally random, with an equal probability that either the maternal or paternal inherited X-chromosome will be inactivated in the cone precursor cells. However, extremes of asymmetrical inactivation in which the same X-chromosome is inactivated in all, or almost all, of the precursor cells have been reported in female monozygotic twins and triplets (Zanen & Meunier, 1958a, 1958b; Pickford, 1967;

Koulischer et al., 1968; Philip et al., 1969; Yokota et al., 1990; Jørgensen et al., 1992), as have differences in the symmetry between the two eyes (Jaeger, 1972; Feig & Ropers, 1978).

Phenotypes. As a result of X-chromosome inactivation, the retinal mosaic of heterozygotic carriers of dichromacy and anomalous trichromacy will be mottled with patches of color-blind and normal areas (cf., Krill & Beutler, 1965; Born et al., 1976; Grützner et al., 1976; Cohn et al., 1989). Thus, they may partly share in the color vision defects of their sons and fathers (Pickford, 1944, 1947, 1949, 1959). But some carriers of anomalous trichromacy and carriers of anomalous protanopia may actually benefit from the presence of the fourth (anomalous) cone pigment in their retinae. They could be tetrachromatic (DeVries, 1948a), enjoying superior color discriminability and an extra dimension of color vision. The more dissimilar the λ_{max} of the hybrid pigment is to those of the normal M- and L-cone pigments, the greater the potential advantage.

A related issue is to determine the psychophysical consequences, (in the heterozygous human female eye) as the result of X-chromosome inactivation, when the two polymorphic variants of the L-cone opsin gene are both expressed in those women who inherit a gene for the serine variant on one X-chromosome and a gene for the alanine variant on the other (Neitz, Kraft, & Neitz, 1998).

(i) Weak tetrachromacy: A weak form of tetrachromacy arises if the heterozygous carriers have four different types of cones, but lack the postreceptoral capacity to transmit four truly independent color signals. This form has been demonstrated in heterozygotes who accept trichromatic color matches but do not exhibit the stability of matches under the chromatic adaptation that is required when only three types of cones are present in the retina (Nagy et al., 1981). The condition may be analogous to rod participation influencing color matching in the normal trichromatic eye (cf. Bongard et al., 1958; Trezona, 1973).

(ii) Strong tetrachromacy: A strong form of tetrachromacy arises if the heterozygous carriers have four different types of cones plus the capacity to transmit four independent cone signals. Such heterozygotes would not accept trichromatic color matches; rather, they would require four variables to match all colors. Up to now, strong tetrachromacy has never been convincingly demonstrated (Jordan & Mollon, 1993), and it is unclear whether the human visual system is sufficiently labile to allow for it during embryonic and postnatal development of the retina and cortex. A model does exist in New-World monkeys, however, demonstrating how heterozygous females can acquire an extra dimension of color vision from the action of X-chromosome inactivation upgrading dichromacy to trichromacy (Mollon, Bowmaker, & Jacobs, 1984; Mollon, 1987; Mollon & Jordan, 1988; Jordan & Mollon, 1993; see Chapter 6).

Diagnosis. Viewing field size is an important factor in detecting color vision defects in heterozygotic carriers (Möller-Ladekarl, 1934; Walls & Mathews, 1952; Krill & Schneidermann, 1964; Ikeda et al., 1972; Verriest, 1972; De Vries-De Mol et al., 1978; Swanson & Fiedelman, 1997; Miyahara et al., 1998). They are clearly disadvantaged relative to normals, when the target is reduced to a point source (Wald et al., 1974; Born et al., 1976; Grützner et al., 1976; Cohn et al., 1989) or when the time allotted for discriminating small color targets is restricted (Verriest, 1972; Cohn et al., 1989; Jordan & Mollon, 1993). This suggests that eye movements may mitigate against retinal mottle by preventing the target from merely falling on dichromatic areas.

(i) Luminous efficiency (the Schmidt sign): In heterochromatic flicker photometry and other measures of relative spectral sensitivity (see Chapter 2), carriers of protanopia (and protan defects in general) exhibit a more protan-like function (i.e., reduced sensitivity to long-wave lights) – known as the Schmidt sign (Schmidt, 1934, 1955; Walls & Mathews, 1952; Crone, 1959; Krill & Beutler, 1965; Adam, 1969; Ikeda et al., 1972; De Vries-De Mol et al., 1978; Miyahara et al., 1998). On the other hand, carriers of deuteranopia (and deutan defects in general) may exhibit a more deutan-like (shifted to longer wavelengths) func-

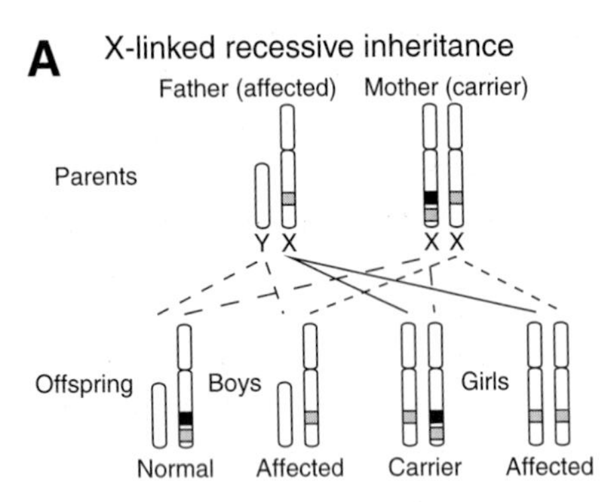

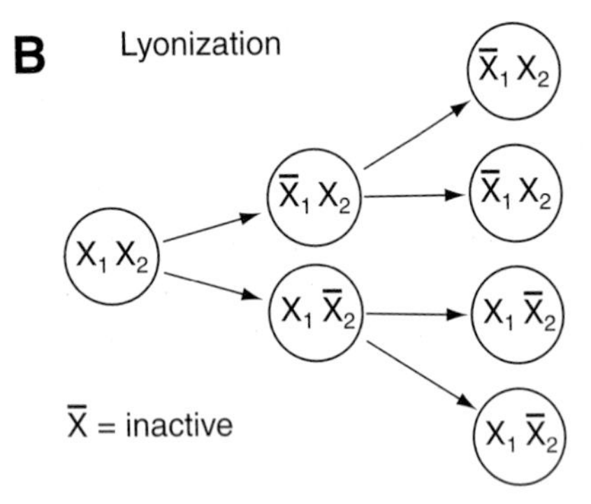

Figure 1.16: Heterozygous defects. (A) X-linked recessive inheritance for protan (or deutan) defects. In the example shown, the father (affected) lacks the L-cone opsin gene on his single X-chromosome, while the mother (carrier) lacks it on one of her X-chromosomes. On average, one-half of the boys will be affected hemizygotes and one-half will be normal; one-half of the girls will be obligate carriers and one-half affected homozygotes. The black and gray squares on the q-arm of the X-chromosome indicate normal L- and M-cone opsin genes, respectively. (B) Schematic representation of X-chromosome inactivation or lyonization. One X-chromosome is normal (X_1), the other is not (X_2). In the early embryonic progenitor cell (left), both X-chromosomes are active. During embryonic development, X-inactivation (middle) is random (50% probability is assumed), so in 50% of the cells X_1 will be inactivated, and in the other 50% X_2 will be inactivated. The descendant lines (right) derived from the original inactivation events retain the same active and inactive chromosomes.

tion (DeVries, 1948a; Crone, 1959; Adam, 1969; Yasuma et al., 1984).

(ii) Rayleigh equation: Both abnormal and normal Rayleigh matches have been reported in heterozygotic carriers (see Jordan & Mollon, 1993; Miyahara et al., 1998). For most carriers of dichromacy, abnormal matches are predicted only when the matching field size is so small that it falls on a dichromatic region or so that one-half of the field falls on a dichromatic region and the other half on a trichromatic region. For carriers of anomalous trichromacy and for dichromat carriers with a hybrid gene, the presence of a fourth anomalous pigment might disturb the Rayleigh match; introducing the possibility of multiple (3) acceptable matchpoints and/or an extended matching range.

Incidences. Approximately 4.5% of Caucausian females are heterozygous carriers of either protanopia or deuteranopia and about 11% are carriers of anomalous trichromacy (see Table 1.6; Feig & Ropers, 1978; Cohn et al., 1989).[10]

Molecular genetics. The molecular genetic causes of heterozygotic red-green color vision defects are the same as for the hemizygotic defects. A method for the molecular genetic detection of female carriers of protan defects has recently been reported (Kainz, Neitz, & Neitz, 1998).

Cone mosaic. *(i) Carriers of dichromacy:* The cone mosaic of heterozygotic carriers of dichromacy will be composed of mottled patches of color-blind (in which either the L- or M-cones are lacking) and normal areas (Krill & Beutler, 1965; Born et al., 1976; Grützner et al., 1976; Cohn et al., 1989). The relative complement of L- and M-cones in the mosaic can be predicted given certain assumptions, such as a 50% inactivation of both X-chromosomes, a 1.5 ratio of expression of the L- and M-cone pigment genes in the normal opsin gene array, and photopigment replacement operating in the defective opsin gene array. The predicted L- to M-cone ratio is 0.43:1.0 for protanopic carriers and 4.0:1.0 for deuteranopic carriers (if it is

[10] The percentage of heterozygotes can be calculated by the equation $2p\ (1 - p)\ (1 - d^2) + 2d\ (1 - d)\ (1 - p^2)$, where p and d are the incidences of protanopia (or protanomaly) and deuteranopia (deuteranomaly), respectively, taken from Table 1.6.

assumed instead that no replacement is operating, the ratios will be 0.6:1.0 and 3.0:1.0, respectively).

(ii) Carriers of anomalous trichromacy: For carriers of protanomaly and deuteranomaly, the mosaic will be a pattern of retinal areas containing normal L- and M-cones alternating with retinal areas containing either L- or M-cones and hybrid pigment cones containing an anomalous pigment. Assuming 50% inactivation of both X-chromosomes and a normal 1.5:1 L- to M-cone ratio, the predicted ratios will be 0.75:0.75:1.0 for the L-, 5'L-3'M hybrid (M-like), and M-cone pigments in carriers of protanomaly and 3.0:1.0:1.0 for the L-, 5'M-3'L hybrid (L-like), and M-cone pigments in carriers of deuteranomaly.

Blue-cone monochromacy

Blue-cone monochromacy is a rare form of monochromacy or total color blindness caused by loss or rearrangement of the X-linked opsin gene array (Nathans et al., 1989, 1993). It is also known as blue-mono-cone-monochromacy (Blackwell & Blackwell, 1957, 1961), S-cone monochromacy, and X-chromosome–linked incomplete achromatopsia. The S-cones are the only photoreceptors, other than the rods, believed to be functioning (Blackwell & Blackwell, 1957, 1961; Alpern, Lee, & Spivey, 1965; Hess et al., 1989; Hess, Mullen, & Zrenner, 1989).

Phenotype. The disorder is characterized by severely reduced visual acuity (Green, 1972; Zrenner, Magnussen, & Lorenz, 1988; Hess et al., 1989; Hess, Mullen, & Zrenner, 1989); a small, central scotoma (corresponding to the blue-blind central foveola; see Fig. 1.1C); eccentric fixation; infantile nystagmus (which diminishes and/or disappears with age); and nearly normal appearing retinal fundi. There also may be associated myopia (Spivey, 1965; François et al., 1966). A cone-rod break (Kohlrausch kink) in dark adaptation curves occurs, denoting the transition from S-cone to rod function.

In some pedigrees with individuals suspected of having blue-cone monochromacy, considerable intrafamilial variation with respect to residual color discrimination has been reported (Smith et al., 1983; Ayyagari et al., 1998). Thus, phenotypes may vary considerably. Indeed, it may be difficult to distinguish between blue-cone monochromats and some types of incomplete rod monochromats (see section on rod monochromacy).

In some cases, residual L-cone function has been reported (Smith et al., 1983); in others, the presence of cones whose outer segments contain rhodopsin instead of a cone photopigment, replacing the L- and M-cones (Pokorny, Smith, & Swartley, 1970; Alpern et al., 1971), has been inferred. The former reports may be evidence for incomplete manifestation of the disorder; the latter have not been confirmed by other investigators (Daw & Enoch, 1973; Hess et al., 1989) and are unlikely on molecular biological grounds.

(i) Heterozygotic manifestation: Minor abnormalities have been reported in some heterozygotic female carriers of blue-cone monochromacy (Krill, 1964; Spivey et al., 1964; Krill, 1969).

(ii) Color perception: In complete blue-cone monochromats, there is no neutral point; the entire spectrum is colorless (see Fig. 1.13). The photopic luminosity function peaks near 440 nm, the λ_{max} of the S-cone spectral sensitivity function (see Fig. 1.1A). Although blue-cone monochromats are usually considered to be totally color blind (the term monochromacy implies that they can match all spectral colors with one variable), they may have residual dichromatic color perception that arises from interactions between the S-cones and rods at mesopic (twilight) levels where both are active (Alpern et al., 1971; Daw & Enoch, 1973; Pokorny, Smith, & Swartley, 1970; Young & Price, 1985; Hess et al., 1989; Reitner et al., 1991).

Diagnosis. *(i) Specific tests:* The disorder is revealed by its obvious clinical signs and by the accumulation of results on conventional color vision tests. Two tests exist, however, that are specifically designed to detect monochromats: the Sloan Achromatopsia test and the François-Verriest-Seki (FVS; see Verriest & Seki, 1965) test plates (see Birch, 1993). To distinguish blue-cone monochromats from rod monochromats, a special four-color plate test (Berson et al., 1983) and a two-color filter test (Zrenner, Magnussen, & Lorenz, 1988) exist. The latter test is based on the observation that the blue-cone monochromat's spatial acuity is improved by viewing through blue cut-off filters (Blackwell & Blackwell, 1961; Zrenner, Magnussen, & Lorenz, 1988).

(ii) The Rayleigh equation: Although not optimumly suited for the task, the phenotype can be identified with the Rayleigh equation, in conjunction with other tests. If the luminance levels are such that the rods are functioning, blue-cone monochromats will be able to match the red primary light and some mixtures of the green and red primary lights to the yellow reference light. But they will be unable to match the green primary light or most red-green mixtures to the yellow reference because of their great sensitivity to the green primary light (i.e., there is too little light in the yellow primary). The slope of the regression line fitted to their limited matches corresponds to the λ_{max} of the rods and should be the same as that of rod monochromats (see Fig. 1.11).

Incidence. Like protan and deutan defects, blue-cone monochromacy is transmitted as an X-linked, recessive trait (Spivey et al., 1964; Spivey, 1965; see Fig. 1.13A). The frequency is estimated at about 1:100,000 (exact incidences are unavailable), and it is not associated with or predicted by combined protanopia and deuteranopia inheritances (see Table 1.6). All cases reported so far, beginning with Blackwell and Blackwell (1957, 1961), have been male (see Sharpe & Nordby, 1990a, for a review), which is not surprising because the frequency in females is estimated to be as low as 1:10,000,000,000.

Molecular genetics. The genetic alterations that give rise to blue-cone monochromacy all involve either a loss or a rearrangement of the L- and M-cone pigment gene cluster. They fall into two classes and are characterized by either one-step or two-step mutational pathways (Nathans, Davenport, et al., 1989, Nathans, Maumenee, et al., 1993).

In the one-step pathway, nonhomologous deletion of genomic DNA encompasses a region between 3.1 and 3.7 kb upstream (5') of the L- and M-cone pigment gene array (see Fig. 1.10). The deletions, which range in size from 0.6 to 55 kb, can extend into the gene array itself, which is otherwise unaltered in structure. These deletions occur in an essential segment of DNA, referred to as the locus control region (LCR). Experiments with reporter constructs[11] in transgenic mice show that the LCR is essential for cone-specific gene expression (Wang et al., 1992)

In the two-step pathway, there are multiple causes. Most typically, in one of the steps, unequal homologous recombination reduces the number of genes in the tandem array to a single gene. In the other step, the remaining gene is inactivated by a point mutation. The temporal order of the steps is unknown, but for the most common point mutation unequal homologous recombination is likely to be the last step.

Three point mutations have so far been identified (Nathans, Davenport, et al., 1989, Nathans, Maumenee, et al., 1993): two missense and one nonsense mutation (see Fig. 1.17A). The most frequent missense mutation is in an extracellular loop. It is a thymine-to-cytosine substitution at nucleotide 1,101 (Nathans, Davenport, et al., 1989), with the result that the essential cysteine residue at codon 203 in exon 4 is replaced by arginine ($cys^{203}arg$), thereby disrupting a highly conserved disulfide bridge. The less frequently identified missense mutation is in a membrane loop: a cytidine-to-thymidine substitution at nucleotide 1,414, resulting in a replacement of the proline residue by leucine at codon 307 ($pro^{307}leu$) in exon 5. The nonsense mutation is also in a membrane loop, and it, too, involves a cytidine-to-thymidine substitution at nucleotide 1,233, resulting in the replacement of arginine by a termination codon (TGA; see Table 1.1) at 247 ($arg^{247}ter$) in exon 4.

In another variety of the two-step pathway, the gene array is not reduced, but a $cys^{203}arg$ missense mutation

[11]These are recombinant DNA constructs in which a gene whose phenotypic expression is easy to monitor is attached to a promoter region of interest and then stably introduced into the mouse genome.

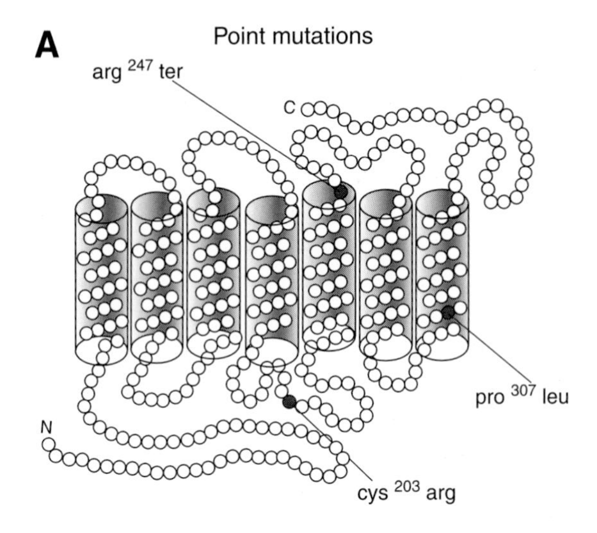

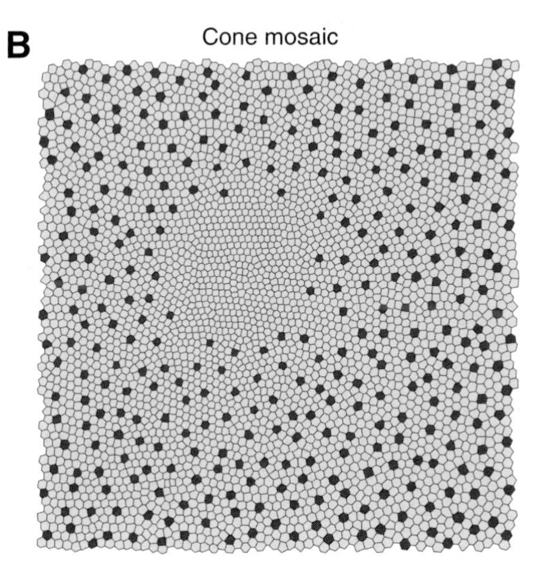

Figure 1.17: Blue-cone monochromacy. (A) Schematic representation of a 5'M-3'L hybrid pigment showing the locations of identified point mutations. The seven α-helices are shown embedded within the membrane. N and C denote amino- and carboxyl-termini, respectively. (B) The cone mosaic of the rod-free inner fovea of an adult human retina at the level of the inner segment; computer altered to simulate the consequences of blue-cone monochromacy. The positions of the S-cones are shown in black; the positions of the M- and L-cones, in white to indicate the lack of a functioning cone pigment. See legend to Fig. 1.1C for other details.

in the M-cone pigment gene is transferred to the L-cone pigment gene by gene conversion or intragenic crossing over (Nathans, Maumenee, et al., 1993; Reyniers et al., 1995).

Cone mosaic. Although no histology has been performed on known blue-cone monochromat donor eyes, knowledge about the distribution of S-cones in normal eyes makes it possible to predict the appearance of their cone mosaic (Fig. 1.17B). The drastic effects – the central scotoma, the paucity of pigment-containing cones – arise because the deletion of the LCR or the loss of function of the pigment genes themselves does not permit replacement of the missing L- and M-cone photopigments by another pigment type. Because the gene for the S-cone pigment resides on another chromosome (7), there is no plausible mechanism for placing the S-cone pigment in morphologically intact, but empty L- and M-cones. The same argument applies to the rod pigment (the opsin gene is located on chromosome 3), largely ruling out, on molecular biological grounds, the possibility for the occurrence of rhodopsin-containing cones.

The lack of replacement of the missing M- and L-cone pigments may have immediate consequences for the morphological integrity of the empty cone photoreceptors. The outer segment is relatively fragile, degenerating in response to many genetic perturbations. But there could also be a delayed onset of degeneration. Interestingly, a slowly progressive macular dystrophy involving extensive peripapillary retinal pigment epithelium regression or thinning has been associated with blue-cone monochromacy (Fleischman & O'Donnell, 1981; Nathans, Davenport, et al., 1989).

Cone monochromacy

Besides blue-cone monochromacy, there are other forms of cone monochromacy, known as complete achromatopsia with normal visual acuity (Pitt, 1944; Weale, 1953). Few cases have ever been described

(Pitt, 1944; Weale, 1953, 1959; Fincham, 1953; Crone, 1956; Gibson, 1962; Ikeda & Ripps, 1966; Alpern, 1974; Vajoczki & Pease, 1997; Ross et al., unpublished), and none is fully accepted as authentic.

Phenotypes. Cone monochromats are conventionally assumed to have either L-cones (in which case they are known as L- or red-cone monochromats) or M-cones (in which case they are known as M- or green-cone monochromats), but not both. Although the S-cones are assumed to be totally absent or inactive, they may be partially functioning, contributing to luminance but not color discrimination (Alpern, 1974; Ross et al., unpublished). Evidence of remnant cone function has led to speculation that the defect may be wholly or partially postreceptoral in origin (Weale, 1953, 1959; Fincham, 1953; Gibson, 1962). Weale (1953), for instance, using retinal densitometry in a careful study of three cases, found evidence for normal M- and L-cone pigments. Unlike blue-cone or rod monochromacy, there is no reduced visual acuity, nystagmus, or light aversion.

(i) Color perception: There is no neutral point, and the entire spectrum appears colorless. Although cone monochromats fail to discriminate colors reliably under conventional test situations, they may be able to teach themselves to identify and to discriminate large and saturated colors, slowly and imperfectly, by secondary signs of color, such as brightness and chromatic aberrations that alter the size, sharpness, and apparent depth of objects according to their hue (Ross et al., unpublished).

Diagnosis. The disorder is revealed by the accumulation of results on conventional color vision tests. Cone monochromats make the same color confusions as X-chromosome–linked dichromats in the red-green range.

The Rayleigh matches of cone monochromats are those of a protanope (M-cone monochromat) or a deuteranope (L-cone monochromat). The Moreland equation or adaptation tests (see page 46) can be used, guardedly, to reveal the amount of S-cone function present.

Incidence. The incidence is extremely rare, estimated at 1:1,000,000 (Weale, 1953) or 1:100,000,000 (Pitt, 1944). The latter estimate is based on assumptions about double dichromacy; that is, tritanopia (see below) combined with either deuteranopia or protanopia (see Table 1.6). Revised estimates of the frequency of tritanopia suggest an upper estimate of ca. 1:10,000,000. However, if cone monochromacy is due, in part or whole, to a post receptoral defect, these estimates are inapplicable.

Molecular genetics. Molecular genetic analysis has only been conducted on a single functional M-cone monochromat (Ross et al., unpublished; Nathans, unpublished). Southern blotting/hybridization after gel electrophoresis revealed that the L-cone pigment gene was missing and replaced by a 5'L-3'M hybrid gene, presumably producing an M-like pigment. None of the known mutations in the S-cone pigment gene (see below) were found. The disorder in the S-cone system may be postreceptoral or associated with sequences adjacent to the S-cone pigment gene that direct its expression.

Cone mosaic. The appearance of the cone mosaic in cone monochromats is predicted to be similar to that of a protanope or deuteranope, with possible replacement of the missing L- or M-cone pigment with the remaining one. If the S-cone pigment is rendered partially or totally inviable, no replacement is deemed likely; so that the S-cones will be missing or present but nonfunctioning (see tritan defects below).

Tritan defects

Tritan defects affect the S-cones. They are often referred to as yellow-blue disorders, but the term blue-green disorder is more accurate (see color perception, page 46). As for protan and deutan defects, congenital tritanopia arises from alterations in the gene encoding the opsin; but, unlike protan and deutan defects, it is autosomal (linked to chromosome 7) in nature.

Phenotypes. Tritan defects affect the ability to discriminate colors in the short- and middle-wave regions of the spectrum. They often go undetected because of their incomplete manifestation (incomplete tritanopia) and because of the nature of the color vision loss involved. From a practical standpoint, even complete tritanopes are not as disadvantaged as many protanomalous and deuteranomalous trichromats because they can distinguish between the environmentally and culturally important red, yellow, and green colors.

On the other hand, the most frequently acquired color vision defects, whether due to aging or to choroidal, pigment epithelial, retinal, or neural disorders, are the Type III acquired blue-yellow defects (see Birch et al., 1979). These are similar, but not identical, to tritan defects. Unlike tritan defects (which are assumed to be stationary), acquired defects are usually progressive and have other related signs, such as associated visual acuity deficits. One acquired disorder – autosomal dominant optic atrophy (DOA) – is commonly described as being predominately tritan in nature (Kjer, 1959) and capable of mimicking congenital tritanopia (Krill et al., 1971; Smith, Cole, & Isaacs, 1973; Went et al., 1974; Miyake et al., 1985). However, it has been linked to chromosome 3q28-qter in many pedigrees (e.g., Eiberg et al., 1994; Bonneau et al., 1995; Lunkes et al., 1995; Jonasdottir et al., 1997; Votruba et al., 1997) and not to the S-cone opsin gene.

(i) Complete and incomplete tritanopia: Tritanopia is the loss form of tritan defects. Like many autosomal dominant disorders, it is complicated by frequent incomplete manifestation (Kalmus, 1955; Henry et al., 1964). The amount of loss of S-cone function ranges from total to minor, but the majority of tritan observers seem to retain some, albeit reduced, S-cone function (Pokorny, Smith, & Went, 1981). Variable penetrance, the degree or frequency with which a gene manifests its effect, has been reported even within families (Sperling, 1960; Cole et al., 1965; Went et al., 1974; Alpern, 1976). However, separate pedigrees have also been documented with complete (Went & Pronk, 1985) and incomplete (Kalmus, 1955; Neuhann et al., 1976) penetrance. Occasionally, tritan defects are observed in combination with deutan defects (van de Merendonk & Went, 1980; van Norren & Went, 1981).

Complete penetrance might only be found in homozygotes and, therefore, would be extremely rare. But it seems more likely that penetrance depends on the location of the mutation or on the influence of other modifier genes.

(ii) Tritanomaly: The classical alteration or anomalous trichromatic form of tritan defects is known as tritanomaly. True cases of tritanomaly, as distinct from partial or incomplete tritanopia, have never been satisfactorily documented. Although the separate existence of tritanopia and tritanomaly, with different modes of inheritance, has been postulated (Engelking, 1925; Oloff, 1935; Kalmus, 1965), it now seems more likely that tritanomaly does not exist, but rather has been mistaken for incomplete tritanopia (Cole et al., 1965) or for acquired disorders such as DOA.

(iii) Tetartan defects: Occasionally, references are found in the literature to a hypothesized second subtype of blue-green disorder, known as tetartan defects (from Greek *tetartos*: fouth). Tetartanopia is the loss and tetartanomaly, the altered form. These are conjectures of the zone (opponent process) theory of Müller (1924), in which tritanopia is considered an outer or retinal (presumably receptoral) defect and tetartanopia, an inner or neural (presumably postreceptoral) defect. No congenital tetartanope or tetartanomalous observer has ever been convincingly demonstrated. However, patients with acquired Type III blue-yellow disorders sometimes exhibit tetartan-like confusions on conventional hue arrangment tests (see Birch et al., 1979). Generally, these disorders involve retinopathies, not disturbances in the S-cone opsin gene.

(iv) Color confusions and neutral point: In the complete tritanope, confused colors should fall along lines radiating from a single tritanopic copunctal point, corresponding to the chromaticity of the S-cones. Calculated tritanopic confusion lines, based on the tritanopic hue discrimination data of Fischer, Bouman, and ten Doesschate (1951) and Wright (1952) are shown in Fig. 1.12C. They indicate that about 44 distinct spectral wavelengths can be discriminated.

The spectrum is divided by a neutral zone (see Fig. 1.10), which occurs near yellow, ca. 569 nm (6500 K; Walls, 1964; Cole et al., 1965). The violet end of the spectrum may also appear colorless (a complementary[12] neutral zone is predicted to occur in this region, although careful searching often does not reveal it (Cole et al., 1965). But the luminosity is essentially normal, without marked loss of sensitivity at the short wavelengths.

Observations in individuals with unilaterally acquired tritan disorders suggest that, below the neutral zone, a blue-green at 485 nm, and, above the neutral zone, a red at 660 nm have the same hue for the normal and color deficient eyes (Fig. 1.13). The best documented study is Alpern, Kitahara, and Krantz (1983); see also Graham and Hsia (1967).

(v) Color perception: The resulting loss of hue discrimination is in the violet, blue, and blue-green portions of the spectrum (see Fig 1.14D). "Yellow" and "blue" do not occur in the color world of the complete tritanope. They confuse blue with green and yellow with violet and light gray, but not yellow with blue, as the descriptive classification "yellow-blue" defect would seem to imply.

(vi) Large-field trichromacy: Pokorny, Smith, and Went (1981) found that the majority of tritan subjects are capable of establishing trichromatic matches for centrally fixated fields of large diameter (8 deg). The explanation is that S-cones are usually present, albeit in reduced numbers. But the interpretation of trichromatic matching in tritan subjects is complicated by several factors.

Diagnosis. The diagnosis of tritan defects is difficult and the condition often eludes detection.

(i) Plate and arrangement tests: Special pseudoisochromatic plate tests have been designed, including the AO (HRR), Tokyo Medical College, Farnsworth or F2 (reproduced as Plate I in Kalmus, 1965; also see Taylor, 1975), Velhagen, Standard (2nd edition), Stilling, and the Lanthony tritan album. Additional information can be gained from examining chromatic discrimination on pigment arrangement tests, such as the Farnsworth–Munsell 100-Hue test or the Farnsworth Panel D-15 (both have a tritan axis). However, none of these tests, alone or in combination, identify a tritan disturbance unequivocally.

(ii) Blue-green equation: Equations similar in concept to the Rayleigh red-green equation, including the Engelking–Trendelenburg equation (Engelking, 1925; Trendelenburg, 1941) and the Moreland blue-green equation (Moreland & Kerr, 1979; Moreland, 1984; Moreland & Roth, 1987), have been designed to detect tritan defects. The Moreland equation, which is now the most frequently employed, involves matching indigo (436 nm) and green (490 nm) mixture primaries to a cyan standard (480 nm plus 580 nm mixed in a fixed ratio). However, many problems limit the diagnostic ability of this test. A central problem is that the equation applies in a trichromatic region of the spectrum, unlike the Rayleigh equation, which applies to a dichromatic region. Further, the primary wavelengths do not fall on a tritanopic confusion axis (see Fig. 1.12C). Thus, the Moreland equation becomes dichromatic in cases of tritanopia, unlike the Rayleigh equation, which becomes monochromatic in cases of protanopia and deuteranopia. Further, the equation is also affected by individual variations in lens and, to a lesser extent, macular pigment density. Thus, it is not a sufficient test for complete or incomplete tritanopia. Even for normal observers, there is a wide distribution of midmatch points and, frequently, an ambiguous diagnosis.

(iii) Adaptation tests: Two adaptation tests have been designed to isolate any residual S-cone responses: the TNO Tritan test (van Norren & Went, 1981) and the Berkeley Color threshold test (Adams et al., 1987). These tests present a flickering blue target, chosen to favor S-cone detection, on a high-luminance, steady, yellow-orange background chosen to repress the M- and L-cones. Their results, however, are also often indecisive.

(iv) ERG techniques: ERG techniques, using chromatic adaptation or silent substitution procedures, have been applied to determining the amount of

[12] Wavelengths that add together to produce white are complementary.

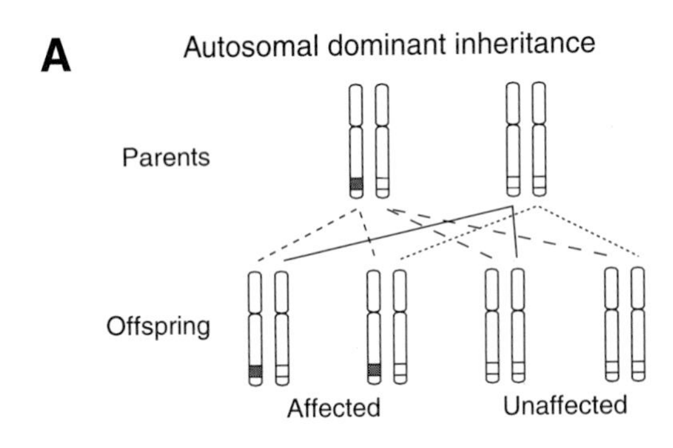

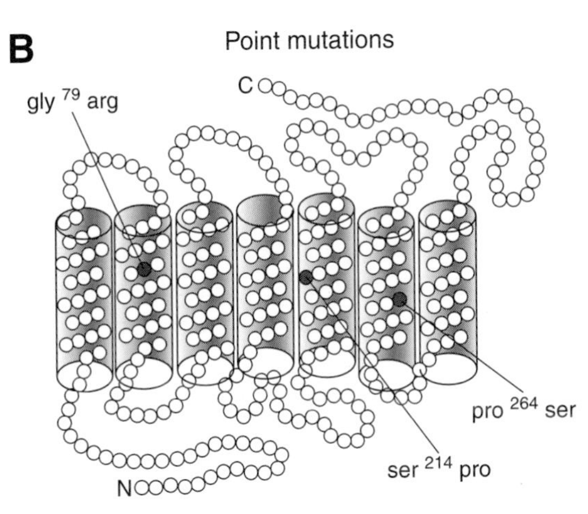

Figure 1.18: Tritan defects. (A) Autosomal dominant inheritance. On average, one-half of the offspring of a heterozygotic parent will inherit the disorder. An open square on the q-arm of chromosome 7 indicates a normal S-cone pigment gene; a filled square, one carrying a missense mutation. (B) Schematic representation of the S-cone pigment showing the locations of three identified amino-acid substitutions.

remaining S-cone function in tritanopes (e.g., Arden et al., 1999). However, isolation of the electrically weak S-cone signal is difficult.

Incidences. Unlike the disorders of the M- and L-cone pigment genes, disorders in the S-cone pigment gene are inherited as autosomal dominant traits (see Fig. 1.18A). Thus their frequencies should be equivalent in males and females, but actual incidences have never been satisfactorily established.

In the United Kingdom the frequency of inherited tritan defects has been estimated as being as low as 1:13,000 to 1:65,000 (Wright, 1952; Kalmus, 1955, 1965), but in The Netherlands it has been estimated as being as high as 1:1,000, (van Heel et al., 1980; van Norren & Went, 1981). The British values pertain only to identified tritanopes and rely on involved calculations, whereas the Dutch samples are too small (480 and 1,023, respectively) and the selection procedures may have been biased.

Molecular genetics. (*i*) *Missense mutations:* Studies of tritan disorders have established that it is associated with at least three different amino-acid subsitutions that cause missense mutations in the gene encoding the S-cone opsin (see Fig. 1.18B). One substitution, involving a G to A substitution at nucleotide 644 in exon 1, leads to the replacement of glycine by arginine at codon 79 (gly^{79}arg; Weitz et al., 1992); another, involving a C to T substitution at nucleotide 1,049 of exon 3, leads to the replacement of serine by proline at codon 214 (ser^{214}pro; Weitz et al., 1992); and a third, involving a T to C substitution at nucleotide 1,199 in exon 4, leads to the replacement of proline by serine at codon 264 (pro^{264}ser; Weitz, Went, & Nathans, 1992). All three substitutions are in the transmembrane domain of the pigment and are believed to give rise to mutant proteins that perturb the structure or stability of the S-cone pigment and thereby actively interfere with the function or viability of the S-cone photoreceptors.

(*ii*) *Incomplete penetrance*: The variable penetrance of the disorder appears to depend, at least in part, on the type of point mutation (Weitz et al., 1992, Weitz, Went, & Nathans, 1992). However, a definitive phenotype–genotype study to examine this question has yet to be performed.

(iii) Polymorphism and tritanomaly: The mechanism that permits the frequent manifestation of anomaly in protan and deutan defects – intragenic crossing over – has no analogy in tritan defects: The S-cone opsin gene resides alone on chromosome 7 and has no neighbor with similar DNA sequences.

Tritanomaly, however, could arise, in principle, from inherited point mutations or polymorphisms that cause shifts in the λ_{max} of the photopigment, but none have been reported so far (see S-cone spectral sensitivity above) and not much variation has been seen in the S-cone pigment gene sequences of the many observers who have been examined. Results from molecular analysis of the S-cone opsin genes in a population of single-gene dichromats and normal subjects revealed only a single common polymorphism that was silent (involving an A to C substitution at nucleotide 775 in exon 2 that leads to the replacement of glycine by alanine at codon 122). No other substitutions were found in the coding sequences and exon–intron junctions (Crognale et al., 1999). In vivo spectral sensitivity (Stockman, Sharpe, & Fach, 1999) and in vitro MSP measurements of the λ_{max} are very limited (see Chapter 2). Moreover, any small variability in λ_{max} would be difficult to disambiguate from individual differences in macular and lens pigmentation in *in vivo* measurements (Stockman et al., 1999) and from the influence of photoproducts produced by photopigment bleaching in *in vitro* experiments. However, some indirect evidence for variation in the λ_{max} of the S-cone pigment, involving factor analysis to disentangle the macular and lens pigment factors, has been reported (Webster & MacLeod, 1988).

Cone mosaic. Expression of the S-cone opsin gene is controlled by adjacent DNA sequences that are located on the same chromosome as the gene itself and that are, therefore, distinct from the sites controlling the expression of the M- and L-cone pigment genes. No LCR seems to be involved. Indirect evidence for the lack of an LCR derives from the finding that, in transgene mice, the ability of a small promoter-proximal sequence to direct expression of the S-cone pigment transgene (Chiu & Nathans, 1994) markedly contrasts with the inactivity of the human L-cone pigment promoter in the absence of an LCR (Wang et al., 1992).

Thus, there is no known mechanism that would allow an M- or L-cone pigment to replace a missing S-cone pigment in an otherwise intact cone structure, if expression of the S-cone pigment gene is invalidated or if the S-cone photoreceptors receive a nonviable gene product. Consequently, empty, nonfunctional, or missing S-cones in tritan defects would be predicted. No histology is presently available.

The lack of replacement should not affect visual resolution, because S-cones are banished from the very central fovea and are too few and widely separated in the periphery to be the limiting factor in visual acuity (see Fig. 1.17B).

Psychophysical and ERG evidence (see Arden et al., 1999) suggests that in some tritanopes the abnormality may lie, in part, in the irregular topographic distribution of the S-cones.

Rod monochromacy

This rare congenital disorder (see Fig. 1.19A) is also referred to as typical, complete achromatopsia, complete achromatopsia with reduced visual acuity, total colorblindness (OMIM 216900), or day blindness (hemeralopia). Its autosomal recessive inheritance distinguishes it from the disorders caused by mutations or structural alterations of the cone opsin genes. However, its manifestation is functionally equivalent to the total loss of all three cone opsin genes. Even if the opsin genes are expressed, they are, nevertheless, never engaged for vision.

Public awareness about the disorder has recently increased, in large part due to the publication of *The Island of the Colorblind* (Sacks, 1997), a description of the Micronesian island of Pingelap and its inhabitants (Brody et al., 1970; Carr et al., 1970) and to the establishment of an Achromatopsia Network on the World Wide Web (http://www.achromat.org/) that is operated for and by achromats (Network facilitator: Frances Futterman, Berkeley, California).

Phenotypes. Rod monochromacy is characterized by photophobia, severely reduced visual acuity (about 20/200), hyperopia, pendular nystagmus, a central scotoma that is associated with the central rod-free foveola (which is often difficult to demonstrate

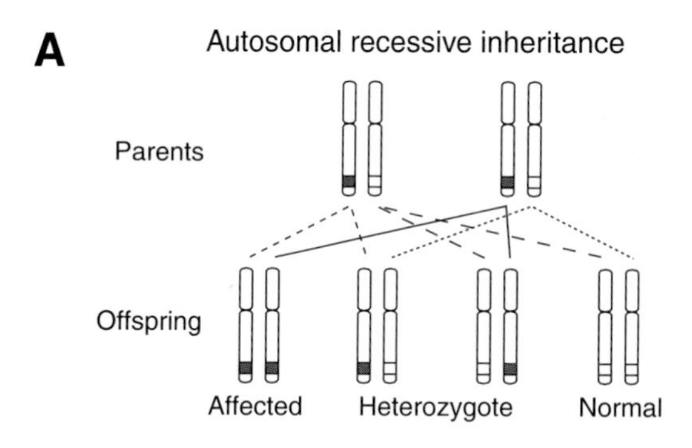

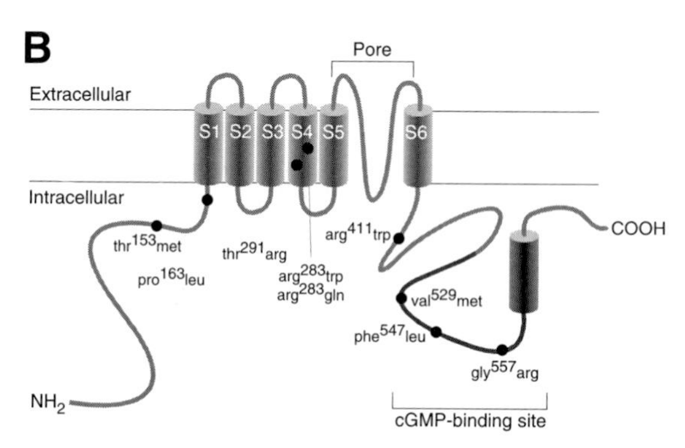

Figure 1.19: Rod monochromacy. (A) Autosomal recessive inheritance for rod monochromacy. On average, one-quarter of the offspring will be affected homozygotes, one-half will be heterozygote carriers, and one-quarter will be normal homozygotes. The light bands indicate a normal gene, the dark bands a gene carrying a mutation for the defect. (B) Putative topology of the human cone cGMP-gated channel α-subunit. The locations of 8 amino-acid substitutions responsible for mutations are indicated by the black dots. The membrane-spanning domains S5 and S6 are thought to line the ion-conducting pore; the cGMP-binding site is located at the intracellular C-terminus (Kohl et al., 1998).

because of the nystagmus), and the complete inability to discriminate between colors (see Waardenburg, 1963b; Sharpe & Nordby, 1990a).

As first proposed by Galezowski (1868), these symptoms can be explained if the cones are deviant, incompletely developed or completely absent, and the visual functions take place wholly in the rods. This interpretation has since been confirmed by extensive psychophysical measurements and electroretinographic recordings (see Sharpe & Nordby, 1990b, for a review). For instance, there is an absence of the Kohlrausch kink in the dark-adaptation curve. The luminosity function peaks at 507 nm (the λ_{max} of the rod or scotopic visual system), the same as for the dark-adapted normal observer (see Fig. 1.13), instead of 555 nm (the λ_{max} of the cone or photopic visual system). And, in the single-flash ERG, there is a pronounced absence or diminuation of the photopic response.

The Pingelap variant (also known as Pingelapese blindness, total colorblindness with myopia, or achromatopsia with myopia) has been thought to be genetically distinct, mainly because of informal reports of the consistent occurrence of severe myopia (Maumenee, unpublished observations). However, these have not always been confirmed (Wassermann, unpublished observations). In fact, although hyperopia is typically reported in cases of rod monochromacy, myopia is also not infrequent. Moreover, the genetic locus associated with the Pingelapese (Maumenee et al., 1998; Winick et al., 1999) has also been associated with families of European origin (Milunsky et al., 1998; see molecular genetics below).

(i) Incomplete rod monochromacy: An associated disorder is atypical, incomplete achromatopsia (a.k.a. incomplete achromatopsia with reduced visual acuity), in which one or more cone type may be partially spared and functioning with the rods. The symptoms are similar but less severe (see Waardenburg, 1963b; Sharpe & Nordby, 1990a).

(ii) Heterozygotic manifestation: Deviation of color vision, both in the red-green and yellow-blue matching ranges, has been reported in the relatives of rod monochromats (Pickford, 1957; Nordström & Polland, 1980). This has led to speculation that the defect has a tendency toward heterozygotic manifestation (Pickford, 1957; Waardenburg, 1963b; Nordström & Polland, 1980). However, other signs, such as diminuition of the photopic response in the flash ERG, have not been reported.

(iii) Color perception: There is no neutral point and no specific confusion colors (see Fig. 1.13). All colors are monochromatic, such as the normal observer per-

ceives at night. But many rod monochromats learn to associate certain colors with objects and to recognize some colors by discerning differences in brightness. For instance, deep reds appear dark and blue-greens bright, owing to the blue-shifted luminosity function (for a personal account, see Nordby, 1990).

Diagnosis. Diagnosis of rod monochromacy is accomplished through the accumulation of abnormal results on a battery of color vision, ERG, and visual acuity tests. No specific axis of color confusion is found on the Farnsworth–Munsell 100-Hue test, but an achromat axis is characteristic on the Panel D-15 test.

A color match can be made over the entire range of Rayleigh matches, but a brightness match is only possible at the red end of most anomaloscopes. The matches, mediated by the rods, resemble those made by blue-cone monochromats (Fig. 1.11).

Incidence. The prevalence of rod monochromacy for Caucasians is often given as 1:33,300 for men and 1:50,000 for women (cf., Judd, 1943), based on Köllner's summary of studies by Göthlin (1924) and Peter (1926); however, estimates vary widely between 1:1,000 and 1:100,000. There is no reason to expect a different prevalence for men and women. A recent, but preliminary, survey in Norway yielded an incidence of 1:50,000 for both sexes (see Sharpe & Nordby, 1990a).

Parental consanguinity is a frequent factor in the disorder. Waardenburg (1963b) estimated a minimum of 28.2% parental consanguinity in Caucasians. The factor is much higher in Japan, owing to the high incidence of consanguineous marriages (see Waardenburg, 1963b, for a review).

On the island of Pingelap, the incidence varies between 4 and 10%. The high incidence is attributed to a founder effect (genetic drift observed in a population founded by a small nonrepresentative sample of a larger population) as the result of a typhoon that completely inundated the island in around 1775. Subsequent starvation, together with isolation, reduced the population to about 20, including the hereditary chief. Pedigrees establish that he was a carrier for the disorder (Brody et al., 1970; Carr et al., 1970).

Molecular genetics. The disorder is heterogeneous with loci assigned to chromosomes 2q11 (Arbour et al., 1997; Wissinger et al., 1998), 8 (Maumenee et al., 1998; Milunsky et al., 1998; Winick et al., 1998), and 14 (Pentao et al., 1992). Phenotype variability may depend on genotype.

Mutations are implied in genes that are common to all three cone types. In those individuals with the locus assigned to chromosome 2, the gene has been identified. The disorder is caused by missense mutations (eight have been identified so far) in the gene (*CNGA3*) encoding the α-subunit of the cone photoreceptor cGMP-gated cation channel (Wissinger et al., 1998; Kohl et al., 1998) – a key component of the phototransduction pathways (Fig. 1.19B; see Chapter 3). The mutant proteins may be either improperly folded within the plasma membrane or inherently unstable or unable to be transported and integrated into the plasma membrane. A lack of membrane targeting has also been demonstrated for mutations in the homologous gene in rod photoreceptors that cause autosomal recessive retinitis pigmentosa (Dryja et al., 1995). An inability to bind cGMP may result in a permanent closure of the channel and elimination of the dark current, a situation comparable to a continuous photoreceptor stimulation. The inappropriate functioning of the phototransduction cascade and the continuous activation of photoreceptors has been speculated to be involved in other stationary retinopathies (Dryja et al., 1993). This finding – the first demonstration of a color vision disorder caused by defects other than mutations in the cone opsin genes – implies a common genetic basis of phototransduction in the three cone types (Kohl et al., 1998).

The genetic locus of the disturbance for the Pingelapese has now been identified on chromosome 8q21-q22 (Winick et al., 1999). It is not unique to the Pingelap; for it has also been reported in families of European origin (Milunsky et al., 1998).

The assignment to chromosome 14 relies on a single case of a very rare disorder, uniparental isodisomy, in which the patient inherited two maternal isochromes 14 (Pentao et al., 1992). In light of other data (Arbour et al., 1997; Wissinger et al., 1998), it seems likely that this association occurred by chance or that only a minor fraction of cases with rod monochromacy are caused by a genetic defect on chromosome 14.

Cone mosaic. Unlike for the other forms of color blindness, some histology has been performed on rod monochromat eyes. However, the results are very discrepant, presumably reflecting the heterogeneity of the disorder (see Sharpe & Nordby, 1990a, for a review). In one histological case study, a 29-year-old woman, cones were scarce and malformed in the fovea but were normally distributed and shaped in the periphery (Larsen, 1921); in another, a 19-year-old male, they were imperfectly shaped and markedly reduced in numbers throughout the entire retina (Harrison et al., 1960); in yet another, a 69-year-old woman, they were normally distributed, although abnormally shaped, in the foveal region and scarce, although less often malformed, in the periphery (Falls et al., 1965); and in a fourth, an 85-year-old man (the best documented clinically and psychophysically), they were completely absent in the fovea, abnormally shaped near the fovea, and severely reduced in number (5–10% of normal values) throughout the entire retina (Glickstein & Heath, 1975).

The lack of unanimity in these anatomical findings may reflect different pathologies: Some of the patients may have suffered from stationary, congenital color blindness, while others may have suffered from an early onset, progressive cone degeneration disease. However, it is conceivable that they may also reflect the nature of the different genetic mutations involved. In some disorders, the mutation may be such that the cones remain intact, whereas in others they may degenerate or never develop.

Interestingly, one of the subjects known to have the missense mutation in the *CNGA3* gene, fixated normally and kept his eyes fully open, even in bright light during the first seven to eight months of life. Only thereafter did he exhibit squinting, nystagmus, and light aversion behavior (Nordby, 1990). In contrast, other rod monochromats are known to already exhibit light avoidance symptoms during the first months of life (see Sharpe & Nordby, 1990a).

Conclusions

The discovery by recombinant DNA techniques of the opsin genes responsible for encoding the cone photopigments (Nathans, Thomas, & Hogness, 1986; Nathans et al., 1986) has significantly advanced our understanding of the initiating stages of color vision. Not only has it revealed how common variants in color vision are produced by alterations in the opsin genes, it also informs speculations about the concatenated development of the cone retinal mosaic and associated neuronal network. The logical consequence of changes at the lowest levels of visual processing are changes at the highest, because whatever alters the function of the cone photopigments and their distribution in the cone mosaic will influence the organization of the retinal subsystems and their representation in the cortex. Genetic variation in the opsin genes implies separate perceptual worlds that differ in more than color.

Acknowledgments

This work was supported by the Deutsche Forschungsgemeinschaft (Bonn) under grants SFB 340 Tp A6 and Sh 23/5-1 and a Hermann-und-Lilly-Schilling-Professur award (LTS), by National Institutes of Health grant EY 10206 (AS), and by the Howard Hughes Medical Institute (JN). We thank Samir Deeb, Frances Futtermann, Karl Gegenfurtner, Jan Kremers, Anne Kurtenbach, Donald MacLeod, Ethan Montag, Hans-Jürgen Schmidt and Bernd Wissinger for comments and Sabine Apitz and Christine Jägle for help and encouragement.

2

Cone spectral sensitivities and color matching

Andrew Stockman and Lindsay T. Sharpe

The eye's optics form an inverted image of the world on the dense layer of light-sensitive photoreceptors that carpet its rear surface. There, the photoreceptors transduce arriving photons into the temporal and spatial patterns of electrical signals that eventually lead to perception. Four types of photoreceptors initiate vision: The rods, more effective at low light levels, provide our nighttime or scotopic vision, while the three classes of cones, more effective at moderate to high light levels, provide our daytime or photopic vision. The three cone types, each with different spectral sensitivity, are the foundations of our trichromatic color vision. They are referred to as long-, middle-, and short-wavelength–sensitive (L, M, and S), according to the relative spectral positions of their peak sensitivities. The alternative nomenclature red, green, and blue (R, G, and B) has fallen into disfavor because the three cones are most sensitive in the yellow-green, green, and violet parts of the spectrum and because the color sensations of pure red, green, and blue depend on the activity of more than one cone type.

A precise knowledge of the L-, M-, and S-cone spectral sensitivities is essential to the understanding and modeling of normal color vision and "reduced" forms of color vision, in which one or more of the cone types is missing. In this chapter, we consider the derivation of the cone spectral sensitivities from sensitivity measurements and from color matching data.

Univariance. Although the probability that a photon is absorbed by a photoreceptor varies by many orders of magnitude with wavelength, its effect, once it is absorbed, is independent of wavelength. A photoreceptor is essentially a sophisticated photon counter, the output of which varies according to the number of photons that it absorbs (e.g., Stiles, 1948; Mitchell & Rushton, 1971). Since a change in photon count could result from a change in wavelength, from a change in intensity, or from both, individual photoreceptors are color blind. The visual system is able to distinguish color from intensity changes only by comparing the outputs of two or three cone types with different spectral sensitivities. The chromatic postreceptoral pathways, which difference signals from different cone types (e.g., L-M and [L + M] - S), are designed to make such comparisons.

Historical background. The search for knowledge of the three cone spectral sensitivities has a long and distinguished history, which can confidently be traced back to the recognition by Young (1802) that trichromacy is a property of physiology rather than physics (see Chapter 1). But it was only after the revival of Young's trichromatic theory by Helmholtz (1852), and the experimental support provided by Maxwell (1855), that the search for the three "fundamental sensations" or "Grundempfindungen" began in earnest. The first plausible estimates of the three cone spectral sensitivities, obtained by König and Dieterici in 1886 from normal and dichromat color matches, are shown as the gray dotted triangles in Figs. 2.7 and 2.9 later. Their derivation depended on the "loss," "reduction," or "König" hypothesis that protanopes, deuteranopes, and tritanopes lack one of the three cone types but retain two that are identical to their counterparts in normals (Maxwell, 1856, 1860).

Since 1886, several estimates of the normal cone spectral sensitivities have been based on the loss hypothesis, notably those by Bouma (1942), Judd (1945, 1949b), and Wyszecki and Stiles (1967). Here we consider the more recent loss estimates by Vos and Walraven (1971) (which were later slightly modified by Walraven, 1974, and Vos, 1978), Smith and Pokorny (1975) (a recent tabulation of which is given in DeMarco, Pokorny, and Smith, 1992), Estévez (1979), Vos, Estévez, and Walraven (1990), and Stockman, MacLeod, and Johnson (1993) (see Figs. 2.7 and 2.9). Parsons (1924), Boring (1942), and Le Grand (1968) can be consulted for more information on earlier cone spectral sensitivity estimates.

Overview. The study of cone spectral sensitivities now encompasses many fields of inquiry, including psychophysics, biophysics, physiology, electrophysiology, anatomy, physics, and molecular genetics, several of which we consider here. Our primary focus, however, is psychophysics, which still provides the most relevant and accurate spectral sensitivity data.

Despite the confident use of "standard" cone spectral sensitivities, there are several areas of uncertainty, not the least of which is the definition of the mean L-, M-, and S-cone spectral sensitivities themselves. Here we review previous estimates and discuss the derivation of a new estimate based on recent data from monochromats and dichromats. The new estimate, like most previous ones, is defined in terms of trichromatic color matching data.

Several factors, in addition to the variability in photopigments (for which there is now a sound genetic basis; see Chapter 1), can cause substantial individual variability in spectral sensitivity. Before reaching the photoreceptor, light must pass through the ocular media, including the pigmented crystalline lens, and, at the fovea, through the macula lutea, which contains macular pigment. The lens and macular pigments both alter spectral sensitivity by absorbing light mainly of short wavelengths, and both vary in density between individuals. Another factor that varies between individuals is the axial optical density of the photopigment in the receptor outer segment. Increases in photopigment optical density result in a flattening of cone spectral sensitivity curves. In this chapter, we examine these factors and the effect that each has on spectral sensitivity. Since macular pigment and photopigment optical density decline with eccentricity, both factors must also be taken into account when standard cone spectral sensitivities, which are typically defined for a centrally viewed 2-deg- (or 10-deg-) diameter target, are applied to nonstandard viewing conditions.

Psychophysical methods measure the sensitivity to light entering the eye at the cornea. In contrast, other methods measure the sensitivity (or absorption) of photopigments or photoreceptors with respect to directly impinging light. To compare photopigment or photoreceptor sensitivities with psychophysical ones, we must factor out the effects of the lens and macular pigments and photopigment optical density. We discuss the necessary adjustments and compare the new spectral sensitivity estimates, so adjusted, with data from isolated photoreceptors.

With so much vision research being carried out under conditions of equal luminance, the relationship between the cone spectral sensitivities and the luminosity function, $V(\lambda)$, has become increasingly important. Unlike cone spectral sensitivity functions, however, the luminosity function changes with adaptation. Consequently, any $V(\lambda)$ function of fixed shape is an incomplete description of luminance. We review previous estimates of the luminosity function and present a new one, which we call $V^*(\lambda)$, that is consistent with the new cone spectral sensitivities. Like the previous estimates, however, the new estimate is appropriate only under a limited range of conditions.

What follows is a necessarily selective discussion of cone spectral sensitivity measurements and their relationship to color matching data and luminance, and of the factors that alter spectral sensitivity. Our ultimate goal is to present a consistent set of L-, M-, and S-cone and $V(\lambda)$ spectral sensitivity functions, photopigment optical density spectra, and lens and macular density spectra that can together be easily applied to predict normal and reduced forms of color vision. We begin with cone spectral sensitivity measurements in normals. (Readers are referred to Chapter 1 for information about the molecular genetics and characteristics of normal and deficient color vision.)

Spectral sensitivity measurements in normals

The three cones types peak in sensitivity in different parts of the spectrum, and their spectral sensitivities overlap extensively (see Figs. 2.2 and 2.12, later). Consequently, spectral sensitivity measurements, in which the threshold for some feature of a target is measured as a function of its wavelength, typically reflect the activity of more than one cone type and often interactions between them. The isolation and measurement of the spectral sensitivity of a single cone type require special procedures to favor the wanted cone type and disfavor the two unwanted ones. Many isolation techniques are based on the two-color threshold technique of Stiles (1939, 1978), so called because the detection threshold for a target or test field of one wavelength is measured on a larger adapting or background field usually of a second wavelength (or mixture of wavelengths). There are two procedures. In the field sensitivity method, a *target* wavelength is chosen to which the cone type to be isolated is relatively sensitive; while in the test sensitivity method, a *background* wavelength is chosen to which it is relatively insensitive.

Field sensitivities. In the field sensitivity method, spectral sensitivity is measured by finding the field radiance that raises the threshold of a fixed-wavelength target by some criterion amount (usually by a factor of ten) as a function of field wavelength. The field sensitivity method was used extensively by Stiles. Through such measurements, and studies of the dependence of target threshold on background radiance for many combinations of target and background wavelength (i.e., threshold versus radiance functions), he identified seven mechanisms, which he referred to as π-mechanisms.

Although it has been variously suggested that the field sensitivities of some of the π mechanisms, such as π_3 (*S*), π_4 or π'_4 (*M*), and π_5 or π'_5 (*L*), might be the spectral sensitivities of single cones (e.g., Stiles, 1959; Pugh & Sigel, 1978; Estévez, 1979; Dartnall, Bowmaker, & Mollon, 1983), it now seems clear that none reflect the spectral sensitivities of isolated cones. For cone isolation to be achieved using the field sensitivity method requires: (i) that the target is detected by a single cone type at all field wavelengths and (ii) that the threshold for the target is raised *solely* by the effect of the field on that same cone type. The second requirement, of adaptive independence (Boynton, Das, & Gardiner, 1966; Mollon, 1982), fails under many, but not all, conditions (e.g., Pugh, 1976; Sigel & Pugh, 1980; Wandell & Pugh, 1980a, 1980b). Whether adaptive independence holds or not, however, the field spectral sensitivities of Stiles's π-mechanisms, with the exception perhaps of π'_4, are inconsistent with the cone spectral sensitivities obtained in dichromats and blue-cone (or S-cone) monochromats in some part or parts of the visible spectrum (see below).

Test sensitivities. In the test sensitivity method, the background field wavelength is fixed at a wavelength that selectively suppresses the sensitivities of two of the three cone types but spares the one of interest. Spectral sensitivity is then determined by measuring the target radiance required to detect some feature of the target as a function of its wavelength. Since the background field wavelength and radiance are held constant in a test sensitivity determination, adaptive independence is not a requirement for the test spectral sensitivity to be a cone spectral sensitivity. All that is necessary is target isolation: A single cone type must mediate detection at all test wavelengths.

There have been several attempts to measure complete cone spectral sensitivities using the test sensitivity method, perhaps the most well known of which are those of Wald (1964). Stiles also made extensive test sensitivity measurements but did not publish many of them until 1964, in a paper accompanying Wald's

(Stiles, 1964). A likely reason for his reluctance to publish test spectral sensitivity data was his recognition, which apparently eluded Wald, of the difficulties involved.

In a test sensitivity determination, cone isolation becomes increasingly difficult as the target wavelength approaches the background wavelength. Since the purpose of the background is to maximally suppress the unwanted cone types relative to the cone type to be isolated, its wavelength is typically one to which the wanted cone type is maximally *insensitive* relative to the unwanted cone types. Consequently, when the target wavelength is the same as the background wavelength (as it must be in any complete spectral sensitivity determination), the target works against cone isolation, since it favors detection by the unwanted cones. When the target and background are the same wavelength, the improvement in isolation achieved by the selective suppression of the unwanted cone types by the background is offset by the insensitivity of the wanted cone type to the target. If the sensitivities of the cone types are independently set in accordance with Weber's Law (i.e., if the target threshold rises in proportion to the background intensity), the two factors cancel each other completely: The background raises the thresholds of the unwanted cones, relative to that of the wanted cone, by the *same* amount that the target lowers them. The cone types are then equally sensitive to the target.

Complete isolation can be achieved with the test sensitivity method, but only if the selective sensitivity losses due to adaptation by the background *exceed* the selective effect of the target (King-Smith & Webb, 1974; Stockman & Mollon, 1986). Adaptation, in other words, must exceed Weber's Law independently for each cone type (see Stockman & Mollon, 1986).

(i) S-cone test sensitivities: In terms of the spectral range over which cone isolation can be achieved in normal subjects, the test sensitivity method is least successful for S-cone isolation. Even with optimal backgrounds of high intensity, S-cone isolation is possible only from short wavelengths to about 540 nm. S-cone isolation is difficult because S-cone-mediated vision is generally less sensitive than vision mediated by the M- or L-cones (e.g., Stiles, 1953). The measurement of S-cone test sensitivities throughout the visible spectrum can be achieved with the use of rare blue-cone monochromat observers (see Blue-cone monochromats, below) who lack functioning M- and L-cones. Nevertheless, S-cone spectral sensitivity data measured in color normals obtained over the range over which S-cone isolation is possible remain important as a means of checking the blue-cone monochromat data for abnormalities, which could be introduced, for example, by their typically eccentric fixation.

Figure 2.1 shows S-cone spectral sensitivities (dotted symbols) measured in five normal observers by Stockman, Sharpe, and Fach (1999). The sensitivities are for the detection of a 1-Hz flicker presented on an intense yellow (580-nm) background field that was there to suppress the M- and L-cones and rods. The normal data are consistent with detection by S-cones and with the blue-cone monochromat data (filled symbols) until about 540 nm, after which the M- and L-cones take over target detection. The suggested S-cone spectral sensitivity is indicated in each case by the continuous line.

(ii) L- and M-cone test sensitivities (steady adaptation): A strategy that can be employed to disadvantage detection mediated by S-cones is the use of targets of high temporal and/or spatial frequencies, to which S-cone vision is relatively insensitive (e.g., Stiles, 1949; Brindley, 1954b; Brindley et al., 1966). The use of moderate- to high-frequency heterochromatic flicker photometry (HFP) to measure spectral sensitivity, in which continuously alternating lights of different wavelengths are matched in intensity to minimize the perception of flicker, is also thought to eliminate contributions from the S-cones (Eisner & MacLeod, 1980; but see Stockman, MacLeod, & DePriest, 1991). With S-cone detection disadvantaged, steady chromatic backgrounds can be used to isolate the L-cones from the M-cones, and vice versa, throughout most, but not all, of the visible spectrum. Eisner and MacLeod (1981) found that chromatic backgrounds produced better M-cone or L-cone isolation than predicted by Weber's Law when spectral sensitivity was measured with a 17-Hz HFP. Nevertheless, isolation remains

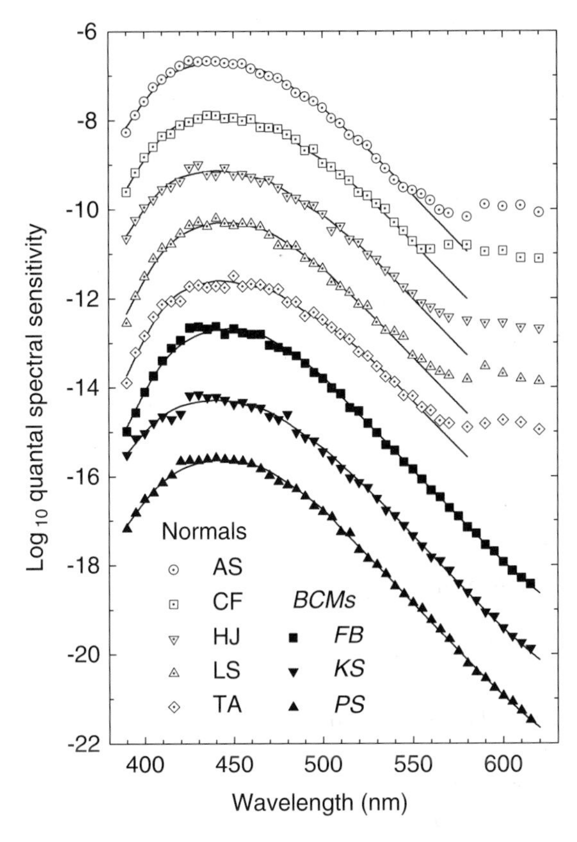

Figure 2.1: Individual 1-Hz spectral sensitivities obtained with central fixation, under S-cone isolation conditions. Each data set, except that for AS, has been displaced vertically for clarity: by –1.2 (CF), –2.0 (HJ), –3.8 (LS), –4.0 (TA), –6.3 (FB), –8.1 (KS), and –9.7 (PS) log units, respectively. Dotted symbols denote observers with normal color vision: AS (circles), CF (squares), HJ (inverted triangles), LS (triangles), and TA (diamonds). Filled symbols denote blue-cone monochromats: FB (squares), KS (inverted triangles), and PS (triangles). The continuous lines drawn through the data are macular- and lens-corrected versions of the Stockman, Sharpe, and Fach (1999) S-cone spectral sensitivities tabulated in the Appendix.

incomplete (Stockman, MacLeod, & Vivien, 1993).

Adaptation with steady fields can exceed Weber's Law by enough to produce M- and L-cone isolation if very small test (3-min-diameter) and background (7-min) fields are used (Stockman & Mollon, 1986). Under such conditions, M- and L-cone adaptation and detection can be monitored separately throughout the visible spectrum; the resulting M- and L-cone test spectral sensitivities agree well with dichromatic spectral sensitivities and with the cone spectral sensitivities tabulated below (see Appendix, Table 2.1). The main drawback of this technique is the need for very small targets, which makes measurements, especially in naïve subjects, challenging.

(iii) L- and M-cone test sensitivities (transient adaptation): Another way of causing adaptation to exceed Weber's Law is to make the adaptation transient. Stockman, MacLeod, and Vivien (1993) found that temporally alternating the adapting field in both color and intensity suppressed the unwanted cone type sufficiently to isolate either the M- or the L-cone types throughout the visible spectrum. They called this method, in which spectral sensitivity is measured with a 17-Hz flickering target immediately after the exchange of two background fields of different colors, the "exchange" method (see also King-Smith & Webb, 1974). M-cone spectral sensitivity was measured immediately following the exchange from a blue (485-nm) to a deep red (678-nm) field, while L-cone spectral sensitivity was measured following the exchange from a deep red to a blue field. The moderately high flicker frequency and the use of an auxiliary steady, violet background ensured that the S-cones did not contribute to flicker detection.

The mean M-cone spectral sensitivity of 11 normals and 2 protanopes (dotted triangles), and the mean L-cone spectral sensitivity of 12 normals and 4 deuteranopes (dotted inverted triangles) from Stockman, MacLeod, and Johnson (1993) can be compared with the dichromat data of Sharpe et al. (1998) in Figs. 2.2 and 2.10.

Spectral sensitivity measurements in monochromats and dichromats

The isolation and measurement of cone spectral sensitivities is most easily achieved in monochromats and dichromats who lack one or two of the three normal cone types. However, the use of such observers to define normal cone spectral sensitivities requires that their color vision is truly a "reduced" form of normal color vision (Maxwell, 1860; König & Dieterici, 1886); that is, that their surviving cones have the same spectral sensitivities as their counterparts in color normal trichromat observers.

We can be more secure in this assumption, since it is now possible to sequence and identify the photopigment genes of normal, dichromat, and monochromat observers (Nathans et al., 1986; Nathans, Thomas, & Hogness, 1986) and so distinguish those individuals who conform, genetically, to the "reduction" hypothesis. Yet, factors other than the photopigment type can affect the corneally measured spectral sensitivities (see Factors that influence spectral sensitivity). Thus, it is important, in those spectral regions in which it is possible, to compare the spectral sensitivities of monochromats and dichromats with those of normals. Blue-cone monochromats (Stockman, Sharpe, & Fach, 1999) and protanopes and deuteranopes (Berendschot et al., 1996) may have narrower foveal cone spectral sensitivities than normals, because the photopigment in their foveal cones is lower in density than that in the foveal cones of normals.

Blue-cone monochromats. Blue-cone monochromats (or S-cone monochromats) were first described by Blackwell and Blackwell (1957; 1961), who concluded that they had rods and S-cones but lacked M- and L-cones. Although two psychophysical studies suggested that blue-cone monochromats might also possess a second cone type containing the rod photopigment (Pokorny, Smith, & Swartley, 1970; Alpern et al., 1971), subsequent studies support the original conclusion of Blackwell and Blackwell (Daw & Enoch, 1973; Hess et al., 1989), as does our knowledge of the molecular biology (see Chapter 1).

Spectral sensitivities in blue-cone monochromats of unknown genotype have been measured several times (e.g., Blackwell & Blackwell, 1961; Grützner, 1964; Alpern, Lee, & Spivey, 1965; Alpern et al., 1971; Daw & Enoch, 1973; Smith et al., 1983; Hess et al., 1989), and are typical of the S-cones. Recently, Stockman, Sharpe, and Fach (1999) measured S-cone spectral sensitivities in three blue-cone monochromats of known genotype. Their results are shown in Fig. 2.1 (filled symbols). The results were obtained in the same way as those for the normal subjects (dotted symbols), except that the flickering target was presented on an orange (620-nm) background of moderate intensity, which was sufficient to saturate their rods.

X-chromosome–linked (red-green) dichromats. A traditional method of estimating the M- and L-cone spectral sensitivities is to use X-chromosome–linked dichromats, or, as they are also known, red-green dichromats: protanopes, who are missing L-cone function, and deuteranopes, who are missing M-cone function. If the experimental conditions are chosen so that the S-cones do not contribute to sensitivity, L- or M-cone spectral sensitivity can, in principle, be measured directly in such observers.

Protanopes and deuteranopes, however, can each differ in both phenotype and genotype. Some may have one gene in the L- and M-cone photopigment gene array while others may have multiple genes (which yield similar photopigments), and some may have normal photopigment genes while others may have hybrid genes (see Chapter 1).

The estimation of normal L- and M-cone spectral sensitivities from dichromat sensitivities requires the use of protanopes and deuteranopes with normal cone photopigments. There are two slightly different normal L-cone photopigments produced by genes with either alanine [L(ala^{180})] or serine [L(ser^{180})] at position 180. A similar polymorphism occurs in the M-cone photopigment, but the serine variant is much less frequent than the alanine. The protanope data shown in Figs. 2.2, 2.9, and 2.10 were obtained from subjects who all had alanine at position 180 of their M-cone opsin genes. Strictly speaking, protanopes have hybrid rather than normal M-cone opsin genes, but because the first exons of the L- and M-cone opsin genes are identical, a hybrid L1M2 gene is equivalent to an M-cone opsin gene. A photopigment that is practically indistinguishable from the M-cone photopigment is produced by the hybrid gene L2M3, its λ_{max} being only 0.2 (Merbs & Nathans, 1992a) or 0.0 nm (Asenjo et al., 1994) different from that of the photopigment

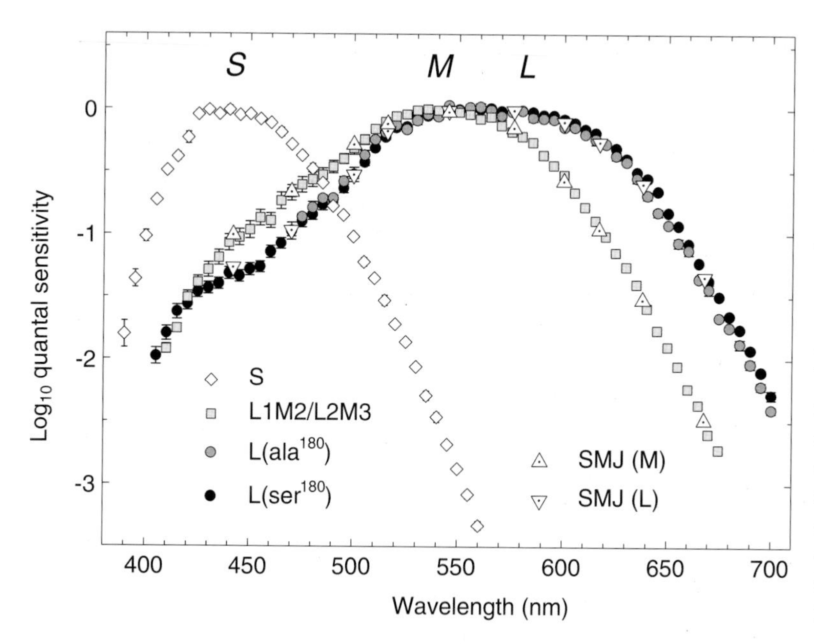

Figure 2.2: Mean spectral sensitivity data. L-cone data from 15 L(ser^{180}) subjects (black circles), 5 L(ala^{180}) subjects (gray circles), and M-cone data from 9 L1M2/L2M3 protanopes (gray squares) measured by Sharpe et al. (1998); and S-cone data from five normals and three blue-cone monochromats (white diamonds) measured by Stockman, Sharpe, and Fach (1999). Also shown are L-cone data from 12 normals and 4 deuteranopes (dotted inverted triangles) and M-cone data for 9 normals and 2 protanopes (dotted triangles) obtained by Stockman, MacLeod, and Johnson (1993).

expressed by the M-cone opsin gene. These values are smaller than the error estimates of the methods used to measure them. Thus, spectral sensitivities from L1M2 and L2M3 protanopes can be reasonably combined.

Dichromats with single photopigment genes in the M- and L-cone pigment gene arrays [L(ala^{180}), L(ser^{180}), L1M2, or L2M3] are especially useful for measuring normal cone spectral sensitivities, because they should possess only a single longer wavelength photopigment. Dichromats with multiple photopigment genes are less useful, unless the multiple genes produce photopigments with the same or nearly the same spectral sensitivities: for example, if an L1M2 or L2M3 gene is paired with an M gene.

With the recent advances in molecular genetics, we can now select protanopes and deuteranopes for spectral sensitivity measurements with the appropriate M- or L-cone photopigment gene(s), as was done in Stockman and Sharpe (2000a) based on the genetic analysis of Sharpe et al. (1998). Some of the protanopes and deuteranopes used in older spectral sensitivity studies (e.g., Pitt, 1935; Hecht, 1949; Hsia & Graham, 1957) may have had hybrid photopigments or multiple longer wavelength photopigments, so that they are unrepresentative of subjects with normal cone spectral sensitivities.

Figure 2.2 shows the mean data obtained by Sharpe et al. from 15 single-gene deuteranopes with an L(ser^{180}) gene (black circles) and from five single-gene deuteranopes with an L(ala^{180}) gene (gray circles). The spectral sensitivity functions for the two groups are separated by ~2.7 nm (Sharpe et al., 1998). Also shown are the data from nine protanopes (gray squares). Of the nine protanopes, three had a single L1M2 gene, three had a single L2M3 gene, one had an L1M2 and an M gene, and two had an L2M3 and an M gene (all genes had alanine at position 180). The mean M- and L-cone data of Stockman, MacLeod, and Johnson (1993) are also shown as the dotted triangles and inverted triangles, respectively. The Stockman, MacLeod, and Johnson data, which are from mainly normals and some dichromats, agree well with the protanope and deuteranope data of Sharpe et al. (1998). Since their group should contain examples of both normal variants of the L-cone photopigment, the mean Stockman, MacLeod, and Johnson L-cone data lie, as expected, between the L(ser^{180}) and L(ala^{180}) means. We will return to the mean spectral sensitivities again

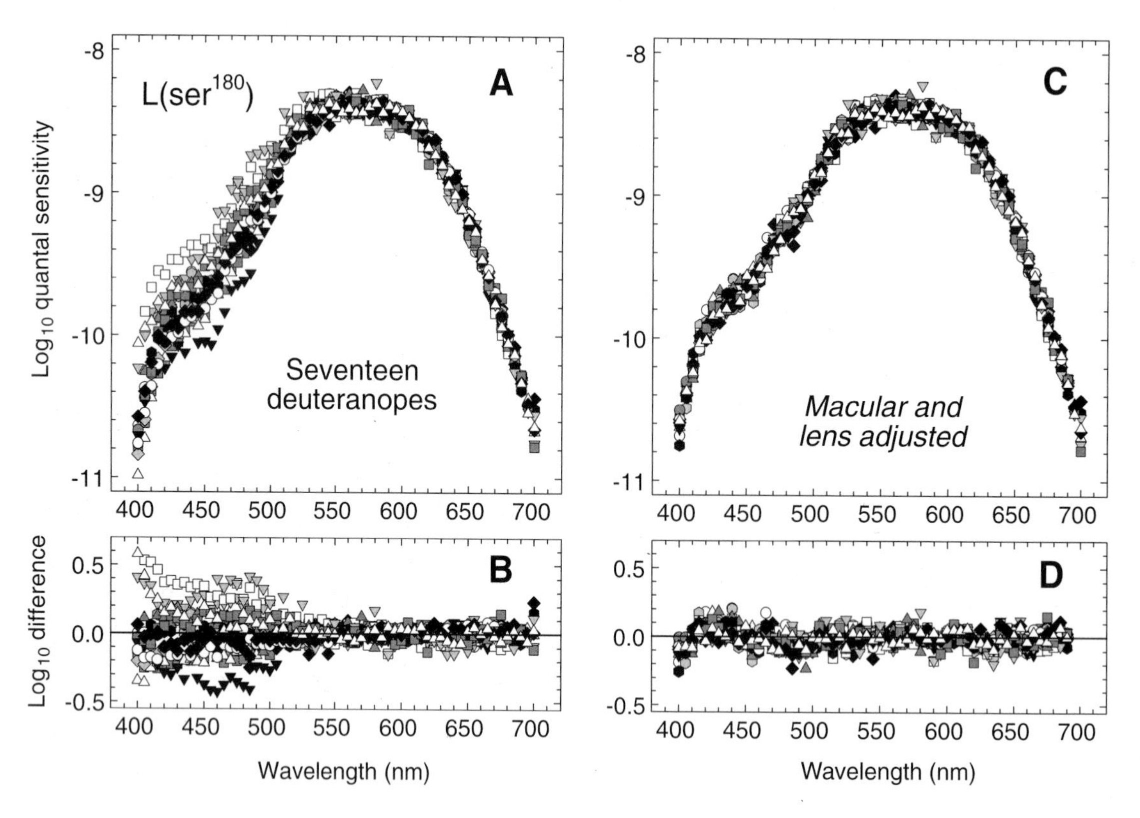

Figure 2.3: Individual differences in macular and lens pigment densities cause individual spectral sensitivity data to appear highly discrepant even if they are determined by the same underlying photopigment. (A) Raw individual L-cone spectral sensitivity data for 17 L(ser^{180}) observers from Sharpe et al. (1998) vertically aligned with the mean at middle and long wavelengths, and (B) differences between each data set and the mean. (C) Same data individually corrected to best-fitting mean macular and lens optical densities and vertically aligned with mean, and (D) differences between each corrected data set and the mean.

later, when we consider their relationship to color matching data.

Factors that influence spectral sensitivity

Individual spectral sensitivity data can appear highly discrepant, even if they depend on the same underlying photopigment. Examples of the range of differences that are found in actual data are shown in Fig. 2.3, which shows the 17 individual L-cone spectral sensitivities for single-gene deuteranopes with L(ser^{180}) (Sharpe et al., 1998). Figure 2.3A shows the raw spectral sensitivity data and Fig. 2.3B their differences from the mean.

The main causes of the individual differences seen in Fig. 2.3 are differences in the densities of the macular and lens pigments. We will consider each factor in turn, and also the effect of differences in the density of the photopigment in the cone outer segment. In each case, two issues are important: first, the changes in spectral sensitivity that are caused by variability in each factor; and, second, the effect of each factor on the mean cone spectral sensitivities and color matching data.

Lens density spectra. The lens pigment absorbs light mainly of short wavelengths. The inset of Fig. 2.4A shows three estimates of the lens density spectrum by van Norren and Vos (1974) (open circles); by

Wyszecki and Stiles (1982a) (filled circles); and the slightly modified van Norren and Vos spectrum proposed by Stockman, Sharpe, and Fach (1999) (continuous line). The lens spectrum given in the Appendix to this chapter is that of Stockman, Sharpe, and Fach (1999) for a small pupil. The tabulated densities are correct for the proposed cone fundamentals that are also given in the Appendix. There is evidence that the shape of the lens density spectrum changes with age (e.g., Pokorny, Smith, & Lutze, 1988; Weale, 1988). When unusually young or old groups of subjects or individuals are employed, such changes should be taken into account.

Because of the way in which it was estimated, the "lens pigment" spectrum tabulated in the Appendix, although dominated by the lens pigment itself, is likely to reflect filtering by any other ocular components or perhaps pigments (e.g., Snodderly et al., 1984; Bowmaker et al., 1991) that intervene between the cornea and the photoreceptors and alter spectral sensitivity. The same is true of other lens pigment density spectra, such as the van Norren and Vos (1974) function.

Lens pigment density differences. Individual differences in the density of the lens pigment can be large. One way of estimating lens density differences between observers is to compare their rod spectral sensitivity functions (or scotopic luminosity functions) measured in a macular-pigment free area of the peripheral retina. By assuming that the differences in spectral sensitivity are due to differences in lens density (see Ruddock, 1965), it is possible to estimate the lens density of each observer relative to other observers.

In the 50 observers measured by Crawford (1949) to obtain the mean standard rod spectral sensitivity function, $V'(\lambda)$, the range of lens densities was approximately ±25% of the mean density (see van Norren & Vos, 1974). Since lens density increases with the age of the observer (e.g., Crawford, 1949; Said & Weale, 1959), and Crawford's subjects were under 30, the variability in the general population will be even larger.

Figure 2.4A shows the changes in S-cone spectral sensitivity that result from changes in lens pigment

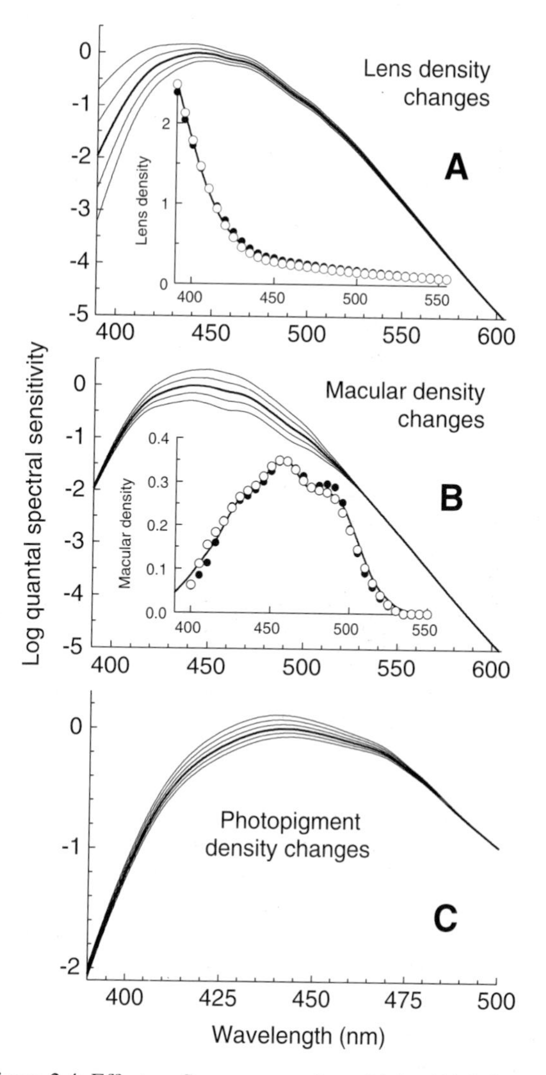

Figure 2.4: Effect on S-cone spectral sensitivity (thick lines) of changes in (A) lens, (B) macular, and (C) photopigment optical densities. (A) From top to bottom, 0.5, 0.75, 1, 1.5, and 2 times the typical lens density. Inset of panel A, lens pigment density spectra of Wyszecki and Stiles (1982a; filled circles), van Norren and Vos (1974; open circles), and the modified version of the van Norren and Vos spectrum proposed by Stockman, Sharpe, and Fach (1999; continuous line, and Appendix). (B) From top to bottom, 0, 0.5, 1, 1.5, and 2 times the typical macular density. Inset of panel B, macular density spectra of Wyszecki and Stiles (1982a; filled circles), Vos (1972; open circles), and one based on Bone, Landrum, and Cains (1992; continuous line, and Appendix). (C) From top to bottom, peak photopigment optical densities of 0.2, 0.3, 0.4 (thick line), 0.5, 0.6, and 0.7. These functions have been normalized at long wavelengths.

density. A typical S-cone spectral sensitivity is indicated by the thickest line, and the effect of varying the lens density in 0.25 steps from one-half the typical density to twice the typical density is indicated by the thinner lines. Changes in lens pigment density variations affect spectral sensitivity mainly at short wavelengths.

Stockman, Sharpe, and Fach (1999) and Sharpe et al. (1998) estimated the lens pigment densities of 40 of their subjects, including those whose data are shown in Figs. 2.1 and 2.3, by measuring rod spectral sensitivities at four wavelengths and comparing the results with the standard rod spectral sensitivity function $V'(\lambda)$. They found that the mean lens densities of their observers was 103.7% of that implied by the $V'(\lambda)$ function with a standard deviation of 16%. The lens density estimates were used to adjust the individual data shown in Fig. 2.3A to the mean lens density value shown in Fig. 2.3C.

Macular density spectrum. The macular pigment also absorbs light mainly of short wavelengths. The inset of Fig. 2.4B shows three estimates of the macular density spectrum by Vos (1972) (open circles); by Wyszecki and Stiles (1982a) (filled circles); and a spectrum (continuous line) based on direct measurements obtained by Bone, Landrum, and Cains (1992). Stockman, Sharpe, and Fach (1999) used the Bone et al. spectrum in their analysis of S-cone spectral sensitivity data, which, in contrast to the Vos (1972) and Wyszecki and Stiles (1982a) spectra, produced plausible estimates of the S-cone photopigment optical density change from central to peripheral retina.

Macular pigment density is typically estimated from the differences between cone spectral sensitivities measured centrally and peripherally, yet both macular pigment density and photopigment optical density vary with eccentricity. Figure 2.5 shows predictions of the L-cone (top panel), M-cone (middle panel), and S-cone (bottom panel) peripheral and central spectral sensitivity differences normalized at long wavelengths. The filled circles show the differences that should be expected if only macular pigment density varies with eccentricity. The lines show the differences that should be expected if, *in addition*, the peak photopigment optical density falls by 0.1 from center to periphery (lowest line) to 0.5 from center to periphery (highest line).

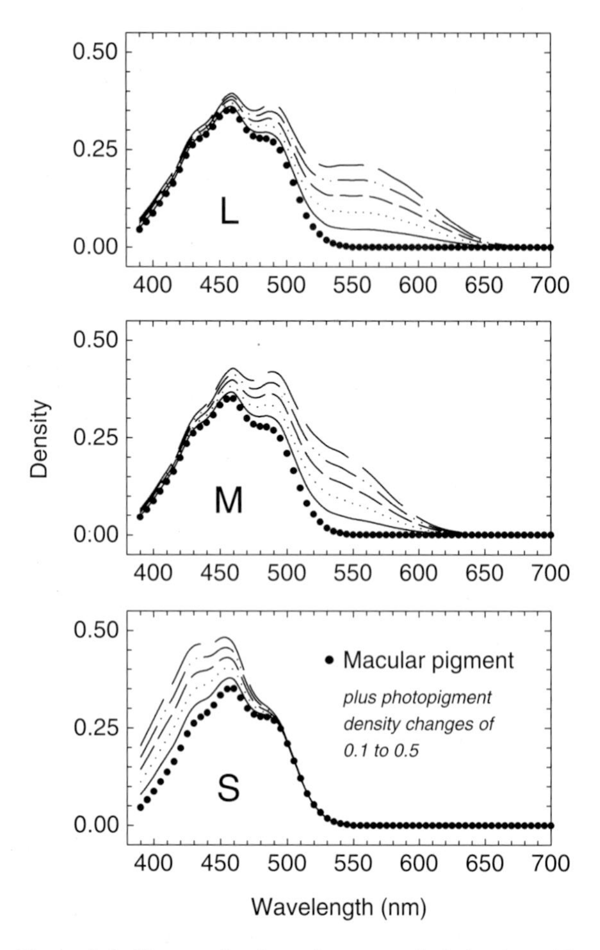

Figure 2.5: Changes in photopigment optical density with eccentricity can substantially distort macular pigment density spectra estimated from peripheral and central spectral sensitivity differences. Predicted differences between peripheral and central spectral sensitivities for a fixed macular pigment spectrum (filled circles, from the Appendix) and peak peripheral and central photopigment optical density differences varying from 0.0 (filled circles) to 0.5 in 0.1 steps for L- (top panel), M- (middle panel), and S- (bottom panel) cone spectral sensitivities (Sharpe et al., 1998).

The potential dangers of ignoring photopigment density changes with eccentricity can be inferred from

Fig. 2.5. For macular pigment density estimates obtained from peripheral and central sensitivity measurements made at a few wavelengths, photopigment density changes could, depending on the cone type isolated, cause a serious overestimation or underestimation of the actual macular pigment density. For macular estimates obtained from peripheral and central measurements made at several wavelengths, the combined effect of the photopigment and macular density changes could be misinterpreted as a novel macular pigment spectrum (e.g., Pease et al., 1987)

Macular pigment density differences. Individual differences in macular pigment density can also be large: In studies using more than ten subjects, macular pigment density has been found to vary from 0.0 to 1.2 at 460 nm (Wald, 1945; Bone & Sparrock, 1971; Pease, Adams, & Nuccio, 1987). Figure 2.4B shows the changes in S-cone spectral sensitivity that result from changes in macular pigment density, assuming the density spectrum tabulated in the Appendix. A typical S-cone spectral sensitivity is shown by the thick line, and the effect of varying the macular pigment density from zero to twice the typical density (in 0.5 steps) is shown by the thinner lines.

The peak macular density most often assumed at 460 nm is the 0.50 value tabulated in Wyszecki and Stiles (1982a). This value, however, is inappropriate for the standard 2-deg target size that is used to define cone spectral sensitivities. Most macular pigment density determinations, including those on which Wyszecki and Stiles based their estimate, were carried out using fields smaller than 2 deg.

Psychophysically, macular pigment density is most often estimated by comparing spectral sensitivities for a centrally presented target with those for the target presented at an eccentricity of 10 deg or more. Given that macular pigment is wholly or largely absent by an eccentricity of 10 deg (e.g., Bone et al., 1988, Table 2 and p. 847), the change in spectral sensitivity in going from periphery to center can provide an estimate of the macular density spectrum (at the few wavelengths usually measured) and its overall density. This type of estimate is, however, complicated by changes in the photopigment density between the central and peripheral measurements (see Fig. 2.5).

Nevertheless, several studies have estimated the macular pigment density using 2-deg fields presented centrally and peripherally and have, for simplicity, ignored changes in photopigment optical density. For M- or L-cone-detected lights, Smith and Pokorny (1975) found a mean peak macular density for their 9 subjects (estimated from their Fig. 3) of about 0.36; Stockman, MacLeod, and Johnson (1993) found a mean value of 0.32 for their 11 subjects; and Sharpe et al. (1998) a value of 0.38 for their 38 observers. For S-cone–detected lights, Stockman, Sharpe, and Fach (1999) found a mean value of 0.26 for 5 observers. The mean peak density from these 2-deg studies is approximately 0.35.

Another difficulty that is often ignored is that the macular pigment density over the central 2 deg is likely to be lower for S-cones than for M- and L-cones, since S-cones, unlike M- and L-cones, are absent at the very center of vision, where the macular density is highest, becoming most common at about 1 deg of visual eccentricity (e.g., Stiles, 1949; Wald, 1967; Williams, MacLeod, & Hayhoe, 1981a).

Figure 2.3C shows again the data for the 17 individual L-cone spectral sensitivity curves for single-gene deuteranopes with L(ser^{180}) measured by Sharpe et al. (1998), but now each curve has been adjusted to the mean lens and macular densities using best-fitting estimates of each individual's macular and lens densities. Much of the variability seen in Fig. 2.3A has been removed by the macular and lens density adjustments. The remaining variability is considered below (see Variability in λ_{max}).

Photopigment optical density. The optical density of the photopigment is related to the axial length of the outer segment in which it resides, the concentration of the photopigment in the outer segment, and the photopigment extinction spectrum (see Knowles & Dartnall, 1977, for further information). Figure 2.4C shows the effects of increasing the peak S-cone photopigment optical density from 0.20 to 0.70 in 0.10 steps. Increases in the photopigment optical density improve sensitivity least near the photopigment λ_{max}. As the wavelength decreases or increases away from the λ_{max}, the sensitivity improvements become larger but reach a constant level at wavelengths far away from the λ_{max}. To emphasize the changes in the shapes of the spectral sensitivity functions, in Fig. 2.4C we have normalized them at longer wavelengths, where the sensitivity improvement is constant with wavelength.

The photopigment optical density can be estimated from the differences between spectral sensitivities or color matches obtained when the concentration of the photopigment is dilute and those obtained when it is in its normal concentration. This can be achieved psychophysically by comparing data obtained under bleached versus unbleached conditions or for obliquely versus axially presented lights. Estimates can also be obtained by microspectrophotometry (MSP) or from retinal densitometry. Most data refer to the M- and L-cones. Comparing central and peripheral spectral sensitivities is less useful, since macular pigment density, as well as photopigment optical density, declines with eccentricity (see Fig. 2.5). The peak photopigment optical densities referred to here are mainly foveal densities.

(i) L- and M-cone photopigment optical densities

(a) Bleaching: In color normals, peak optical density estimates include 0.51 in seven observers (Alpern, 1979); 0.7–0.9 in one observer (Terstiege, 1967); and 0.44 and 0.38, respectively, for the L- and M-cones also in a single observer (Wyszecki & Stiles, 1982b). Two studies have used dichromatic observers. Miller (1972) estimated the peak density to be 0.5–0.6 for the deuteranope and 0.4–0.5 for the protanope, and Smith and Pokorny (1973) found mean peak photopigment densities of 0.4 for four deuteranopes and 0.3 for three protanopes. Burns and Elsner (1993) have obtained mean peak photopigment densities of 0.48 for the L-cones but only 0.27 for the M-cones of six observers.

(b) Oblique presentation: The change in color of monochromatic lights when their incidence on the retina changes from axial to oblique can be accounted for by a self-screening model in which the effective photopigment density is less for oblique incidence (but see Alpern, Kitahara, & Fielder, 1987). Such analyses have yielded higher estimates of photopigment peak density of between 0.69 and 1.0 (Walraven & Bouman, 1960; Enoch & Stiles, 1961), generally for a 1-deg field.

(c) Direct measures: MSP suggests a specific density in the macaque of $0.015 \pm 0.004\ \mu m^{-1}$ for the M-cones and $0.013 \pm 0.002\ \mu m^{-1}$ for the L-cones (Bowmaker et al., 1978). If we assume a foveal cone outer segment length of 35 μm (Polyak, 1941), these values give axial peak photopigment densities of approximately 0.5 (see Bowmaker & Dartnall, 1980). Retinal densitometry gives a value of 0.35 for the M-cones (Rushton, 1963) and 0.41 for the L-cones (King-Smith, 1973a, 1973b). Recently, Berendschot, van de Kraats, and van Norren (1996), also using retinal densitometry, found mean peak photopigment optical densities of 0.57 in ten normal observers, 0.39 in ten protanopes, and 0.42 in seven deuteranopes.

In summary, with the exception of the work of Terstiege (1967), bleaching measurements yield mean peak optical density values in the range 0.3 to 0.6, Stiles–Crawford analyses in the range 0.7 to 1.0, and objective measures in the range 0.35 to 0.57.

Some evidence now suggests that the optical densities in red-green dichromats may be lower than those in color normals (Berendschot et al., 1996). The resulting separation of data obtained in normals from those obtained in red-green dichromats will lead to a higher estimate of the normal photopigment optical densities. Indeed, for the M- and L-cones, a peak value as high as 0.45 or 0.55 seems appropriate. The mainly color normal M-cone data of Stockman, MacLeod, and Johnson (1993), however, agree well at long wavelengths (see Fig. 2.2) with the protanope data of Sharpe et al. (1998), which suggests that the two groups have similar M-cone photopigment optical densities.

Most of the data reviewed above suggest a lower optical density for M- than for L-cones. Other considerations, however, contradict such a difference. Spectral lights of 548 nm (±5 nm standard error) retain the same appearance when directly or obliquely incident on the retina, while longer wavelengths appear redder and shorter ones greener (Stiles, 1937; Enoch & Stiles,

1961; Alpern, Kitahara, & Tamaki, 1983; Walraven, 1993). The self-screening model of the change in color with change in the angle of presentation requires that the M- and L-cone photopigment densities are the same at the invariant wavelength. Since the invariant wavelength roughly bisects the M- and L-cone photopigment λ_{max} wavelengths, their peak photopigment densities must be similar (Stockman, MacLeod, & Johnson, 1993).

(ii) S-cone photopigment optical density: All of the evidence reviewed so far has concerned M- and L-cone photopigment optical densities. Not surprisingly, since lights that strongly bleach the S-cone photopigment may be damaging (see Harwerth & Sperling, 1975), there is a lack of information about S-cone photopigment optical density from bleaching experiments.

Stockman, Sharpe, and Fach (1999) estimated the difference in S-cone photopigment optical density and macular pigment density from the changes in S-cone spectral sensitivity between a centrally viewed 2-deg target and the same target viewed at an eccentricity of 13 deg. In addition to changes in macular pigment density, they found differences in peak photopigment optical densities for five normals of 0.19, 0.20, 0.25, 0.26, and 0.26. For three blue-cone monochromats in their study, however, the changes were only –0.04, – 0.01, and 0.15; and the results were consistent with the blue-cone monochromats having central and peripheral photopigment densities that were as low as those found with eccentric presentation in normals. These differences highlight the potential dangers of using spectral sensitivity data from monochromats and dichromats to estimate normal spectral sensitivities. Before being used to define normal spectral sensitivities, the S-cone spectral sensitivity data from blue-cone monochromats were adjusted to normal photopigment and macular pigment densities (Stockman, Sharpe, & Fach, 1999).

Unfortunately, little evidence exists concerning the absolute optical density of the S-cone photopigment, although inferences can be made from anatomical differences between L- and M-cone and S-cone outer segment lengths. In general, the S-cone outer segments are shorter than the L- or M-cone outer segments at the same retinal location, so that the S-cone optical density should be less than that of the L- or M-cone. Ahnelt (personal communication) suggested that, at the fovea, outer segments of S-cones may be 5% shorter than those of the M- and L-cones; whereas in the periphery, at retinal eccentricities greater than 5 mm (~18 deg of visual angle), they may be shorter by 15–20%. In the single electron micrographs showing outer segments, the histological study of Curcio et al. (1991, Fig. 3) indicates that, at a similar parafoveal location, the outer segment of an S-cone (~4.1 μm) is almost 40% smaller than that of an L/M-cone (7 μm).

The anatomical data suggest that the S-cone photopigment optical density for the central 2 deg must be less than for the L- or M-cone, but the actual density is uncertain. The 5% difference, suggested by the Ahnelt data, may be too small for our purposes, because the S-cones are absent in the central fovea where the L- and M-cones are longest. A photopigment optical density of somewhere between 5 and 20% lower for the S-cones than for the L- and M-cones could be appropriate for the central 2 deg of vision.

Stockman and Sharpe (2000a) and Stockman, Sharpe, and Fach (1999) assumed mean peak photopigment optical densities of 0.50, 0.50, and 0.40 for the L-, M-, and S-cones, respectively, for the central 2 deg of vision; and 0.38, 0.38 and 0.30, respectively, for the central 10 deg of vision. The absolute densities only minimally affect the cone spectral sensitivity calculations. The relative density changes with eccentricity, however, are critical. They were determined by a comparison of 2-deg and 10-deg CMFs and cone fundamentals (see also Stockman, MacLeod, & Johnson, 1993, Fig. 9C).

Variability in λ_{max}. Interest in the variability in photopigment λ_{max} has been revived by the identification of the genes that encode the M- and L-cone photopigments. Estimating the λ_{max} of the M- and L-cones, like their spectral sensitivities, is easier in red-green dichromats. The most extensive data on the variability in the λ_{max} of dichromats come from spectral sensitivity measurements done by Matt Alpern and his associates. Alpern and Pugh (1977) reported L-cone spectral sensitivity curves in eight deuteranopes that varied in λ_{max} over a total range of 7.4 nm, with a standard deviation of about 2.4 nm. Alpern (1987), analyzing the results from Alpern and Wake (1977) and Bastian (1976), estimated the range of λ_{max} in 38 protanopes to be 12.4 nm and that in 38 deuteranopes to be 6.4 nm. These ranges are large, yet the standard deviations of the λ_{max} calculated from Fig. 1 of Alpern (1987) are only 2.3 nm for the protanopes and 1.6 nm for the deuteranopes. Ranges this large would be expected if the red-green dichromats had a mixture of hybrid and normal X-chromosome–linked photopigment genes (see Chapter 1).

From the individual 10-deg color matching data of the 49 color normal observers in the Stiles and Burch (1959) study, MacLeod and Webster (1983), and Webster and MacLeod (1988) estimated the L-cone λ_{max} values to have a standard deviation of 1.5 nm and the M-cone λ_{max} values to have a standard deviation of 0.9 nm. MSP data from the eyes of seven persons, however, suggest a greater variability, with standard deviations in λ_{max} of 3.5 and 5.2 nm, respectively, for 45 human M- and 58 L-cones (Dartnall, Bowmaker, & Mollon, 1983).

Differences in λ_{max} are to be expected between individuals with different photopigment genes, and it is likely that the observers who made up the λ_{max} studies so far described differed in photopigment genotype. The difference in photopigment λ_{max} estimated from the mean L(ser^{180}) and L(ala^{180}) spectral sensitivities shown in Fig. 2.2, for example, is about 2.7 nm (Sharpe et al., 1998). λ_{max} estimates for other genotypes are noted in Chapter 1.

Also of interest in this context is the variability in the measured λ_{max} in observers with the *same* photopigment. The data of Sharpe et al. (1998) are useful here, since spectral sensitivities were measured in 17 single-gene L(ser^{180}) deuteranopes. For the 17 observers, the mean estimate of λ_{max} was 560.14 nm and the standard deviation 1.22 nm (for details, see Sharpe et al., 1998). If these observers had the same photopigment, the variability in λ_{max} must be due to other factors, such as experimental error, subject error, inappropriate lens and macular corrections, differences in photopigment optical density, differences in photoreceptor size, differences in photoreceptor orientation, and so on.

Comparable data for the S-cone λ_{max} comes from the work of Stockman, Sharpe, and Fach (1999). After correcting for lens pigment, macular pigment, and photopigment density differences, they found a mean S-cone photopigment λ_{max} of 418.8 nm for eight observers and a standard deviation of 1.5 nm. The variability is comparable to that found for the L(ser^{180}) group of observers, and, given that the corrections for the macular and lens pigment differences will add variability to the S-cone λ_{max} estimates, it is relatively small.

Color matching and cone spectral sensitivities

The trichromacy of individuals with normal color vision is evident in their ability to match any light to a mixture of three independent "primary" lights. The stimuli used in a typical trichromatic color matching experiment are illustrated in the upper panel of Fig. 2.6. The observer is presented with a half-field illuminated by a "test" light of variable wavelength, λ, and a second half-field illuminated by a mixture of the three primary lights. At each λ the observer adjusts the intensities of the three primary lights, which in this example are 645, 526, and 444 nm, so that the test field is perfectly matched by the mixture of primary lights. The results of a matching experiment carried out by Stiles and Burch (1955) are shown in the lower panel of Fig. 2.6, for equal-energy test lights spanning the visible spectrum. The three functions are the relative intensities of the red, green, and violet primary lights required to match the test light λ. They are referred to as the red, green, and "blue" color matching functions (CMFs), respectively, and written $\bar{r}(\lambda)$, $\bar{g}(\lambda)$, and $\bar{b}(\lambda)$.

Although the CMFs shown in Fig. 2.6 are for primaries of 645, 526, and 444 nm, the data can be linearly transformed to any other set of real primary lights and to imaginary primary lights, such as the X,

Y, and Z primaries favored by the CIE or the L-, M-, and S-cone fundamentals or primaries that underlie all trichromatic color matches. Each transformation is accomplished by multiplying the CMFs by a 3×3 matrix. The goal is to determine the unknown 3×3 matrix that will transform the $\bar{r}(\lambda)$, $\bar{g}(\lambda)$, and $\bar{b}(\lambda)$ CMFs to the three cone spectral sensitivities, $\bar{l}(\lambda)$, $\bar{m}(\lambda)$, and $\bar{s}(\lambda)$ (using a similar notation for the cone spectral sensitivities, or "fundamental" color matching functions, as for the color matching functions).

Color matches are matches at the cone level. When matched, the test and mixture fields appear identical to S-cones, to M-cones, and to L-cones. For matched fields, the following relationships apply:

$$(1)\quad \bar{l}_R\bar{r}(\lambda) + \bar{l}_G\bar{g}(\lambda) + \bar{l}_B\bar{b}(\lambda) = \bar{l}(\lambda),$$

$$\bar{m}_R\bar{r}(\lambda) + \bar{m}_G\bar{g}(\lambda) + \bar{m}_B\bar{b}(\lambda) = \bar{m}(\lambda), \text{ and}$$

$$\bar{s}_R\bar{r}(\lambda) + \bar{s}_G\bar{g}(\lambda) + \bar{s}_B\bar{b}(\lambda) = \bar{s}(\lambda),$$

where $\bar{l}_R$, $\bar{l}_G$, and $\bar{l}_B$ are, respectively, the L-cone sensitivities to the R, G, and B primary lights. Similarly, $\bar{m}_R$, $\bar{m}_G$, and $\bar{m}_B$ are the M-cone sensitivities to the primary lights and $\bar{s}_R$, $\bar{s}_G$, and $\bar{s}_B$ are the S-cone sensitivities. We know $\bar{r}(\lambda)$, $\bar{g}(\lambda)$, and $\bar{b}(\lambda)$, and we assume that for a long-wavelength R primary $\bar{s}_R$ is effectively zero, since the S-cones are insensitive in the red. (The intensity of the spectral light λ, which is also known, is equal in energy units throughout the spectrum and so is discounted from the above equations.)

There are therefore eight unknowns required for the linear transformation:

$$(2)\quad \begin{pmatrix} \bar{l}_R & \bar{l}_G & \bar{l}_B \\ \bar{m}_R & \bar{m}_G & \bar{m}_B \\ 0 & \bar{s}_G & \bar{s}_B \end{pmatrix} \begin{pmatrix} \bar{r}(\lambda) \\ \bar{g}(\lambda) \\ \bar{b}(\lambda) \end{pmatrix} = \begin{pmatrix} \bar{l}(\lambda) \\ \bar{m}(\lambda) \\ \bar{s}(\lambda) \end{pmatrix}.$$

Moreover, since we are often unconcerned about the absolute sizes of $\bar{l}(\lambda)$, $\bar{m}(\lambda)$, and $\bar{s}(\lambda)$, the eight unknowns collapse to just five:

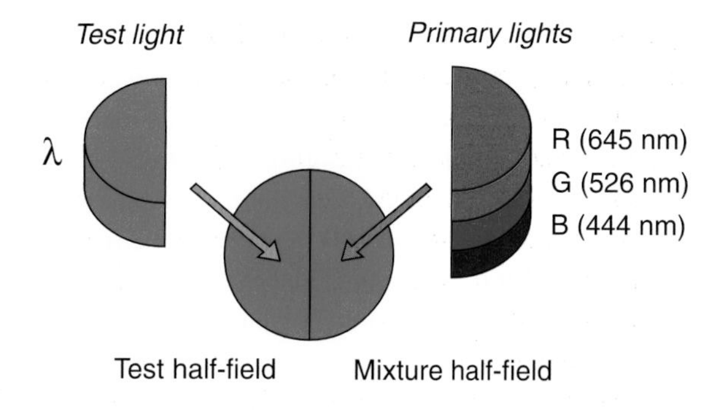

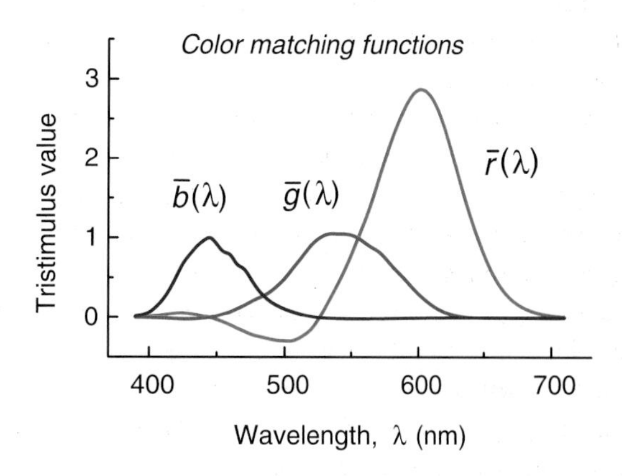

Figure 2.6: A test field of any wavelength (λ) can be matched precisely by a mixture of red (645 nm), green (526 nm), and blue (444 nm) primaries lights. The amounts of each of the three primaries or tristimulus values required to match monochromatic lights spanning the visible spectrum are known as the red, $\bar{r}(\lambda)$, green, $\bar{g}(\lambda)$, and blue, $\bar{b}(\lambda)$, color matching functions (red, green, and blue lines, respectively) shown in the lower panel. The data are from Stiles and Burch (1955). A negative sign means that that primary must be added to the target to complete the match. The matching example shown here is actually impossible, since in the blue-green spectral region the red primary is negative. Consequently, it should be added to the target to complete the match, not as shown.

$$(3)\quad \begin{pmatrix} \bar{l}_R/\bar{l}_B & \bar{l}_G/\bar{l}_B & 1 \\ \bar{m}_R/\bar{m}_B & \bar{m}_G/\bar{m}_B & 1 \\ 0 & \bar{s}_G/\bar{s}_B & 1 \end{pmatrix} \begin{pmatrix} \bar{r}(\lambda) \\ \bar{g}(\lambda) \\ \bar{b}(\lambda) \end{pmatrix} = \begin{pmatrix} k_l\bar{l}(\lambda) \\ k_m\bar{m}(\lambda) \\ k_s\bar{s}(\lambda) \end{pmatrix},$$

where the absolute values of k_l (or $1/\bar{l}_B$), k_m (or

$1/\bar{m}_B$), and k_s (or $1/\bar{s}_B$) remain unknown but are typically chosen to scale three functions in some way: for example, so that $k_l\bar{l}(\lambda)$, $k_m\bar{m}(\lambda)$, and $k_s\bar{s}(\lambda)$ peak at unity. In one formulation (Smith & Pokorny, 1975), $k_l\bar{l}(\lambda) + k_m\bar{m}(\lambda)$ sum to $V(\lambda)$, the luminosity function.

Equations (1) to (3) [and (4) to (6) below] could be for an equal-energy or an equal-quanta spectrum. Since the CMFs are invariably tabulated for test lights of equal energy, we, like previous workers, use an equal-energy spectrum to define the coefficients and calculate the cone spectral sensitivities from the CMFs. We then convert the relative cone spectral sensitivities from energy to quantal sensitivities (by multiplying by λ^{-1}).

The validity of Eqn. (3) depends not only on determining the correct unknowns, but also on the accuracy of the CMFs themselves. There are several CMFs that could be used to derive cone spectral sensitivities. For the central 2-deg of vision, the main candidates are the CIE 1931 functions (CIE, 1932), the Judd (1951) and Vos (1978) corrected version of the CIE 1931 functions, and the Stiles and Burch (1955) functions. Additionally, the 10-deg CMFs of Stiles and Burch (1959), or the 10-deg CIE 1964 CMFs [which are based mainly on the Stiles and Burch (1959) data but also partly on data from Speranskaya (1959), see below] can be corrected to correspond to 2-deg macular and photopigment optical densities.

Color matching data. *(i) CIE 1931 2-deg color matching functions:* The color matching data on which the CIE 1931 2-deg CMFs (CIE, 1932) are based are those of Wright (1928–29) and Guild (1931). Those data, however, are relative color matching data and give only the ratios of the three primaries required to match test lights spanning the visible spectrum. To create color matching functions, however, we also need to know the radiances of the three primaries required for each match. The CIE attempted to reconstruct this information by assuming that a linear combination of the three unknown CMFs must equal the 1924 CIE $V(\lambda)$ function (CIE, 1926) as well as making several other adjustments to the original data (CIE, 1932). Unfortunately, the validity of the $V(\lambda)$ curve used in the reconstruction is highly questionable. The original sources of short-wavelength luminosity data from which the $V(\lambda)$ curve was derived differed by as much as 10 in the violet (Gibson & Tyndall, 1923; CIE, 1926), and, remarkably, the final derivation at short wavelengths was based on the least sensitive (and least plausible) data (see Fig. 2.13A, later).

Unfortunately, the incorrect CIE 1931 CMFs and the 1924 $V(\lambda)$ function [which is also the $\bar{y}(\lambda)$ CMF of the $\bar{x}(\lambda)$, $\bar{y}(\lambda)$, and $\bar{z}(\lambda)$ transformation of the 1931 CMFs] remain international standards in both colorimetry and photometry.

(ii) Judd–Vos modified CIE 2-deg color matching functions: The use of the CIE 1924 $V(\lambda)$ curve to derive the CIE 1931 2-deg CMFs causes a serious underestimation of sensitivity at wavelengths below 460 nm. To overcome this problem, Judd (1951) proposed a revised version of the $V(\lambda)$ function and derived a new set of CMFs [see Wyszecki & Stiles, 1982a, Table 1 (5.5.3)]. Subsequently, Vos made additional corrections to Judd's revision below 410 nm and incorporated the infrared color reversal described by Brindley (1955) to produce the Judd–Vos modified version of the CIE 1931 2-deg CMFs in common usage in color vision research today (Vos, 1978, Table 1). The Judd–Vos modified $V(\lambda)$ function, which is also known as $V_M(\lambda)$, is shown in Fig. 2.13A, later.

The substantial modifications to the CIE 1924 $V(\lambda)$ introduced by Judd are confined mainly to wavelengths below 460 nm, but even above that wavelength [where Judd retained the original CIE 1924 $V(\lambda)$ function] the CIE $V(\lambda)$ function may be incorrect. If the original CIE 1924 luminosity values are too low at and just above 460 nm (as well as at shorter wavelengths, where Judd increased the luminosity values), then the Judd modification creates a "standard" observer whose sensitivity is too low at 460 nm and who could thus be roughly characterized as having artificially high macular pigment density (see Stiles & Burch, 1955, p. 171). Indeed, the Judd modified CIE 2-deg observer does seem to deviate in this way from typical real observers, the Stiles and Burch (1955) 2-deg standard observer, and other relevant data (e.g., Smith, Pokorny, & Zaidi, 1983; Stockman & Sharpe, 2000a).

The validity of both the Judd–Vos modified CIE 2-deg CMFs and the original CIE 1931 CMFs depends on the assumption that $V(\lambda)$ is a linear combination of the CMFs. This assumption was tested experimentally by Sperling (1958), who measured color matches and luminosity functions in the same observers and found deviations from additivity of up to 0.1 $\log_{10}$ unit in the violet, blue, and far-red parts of the spectrum between a flicker-photometric $V(\lambda)$ and the CMFs (see also Stiles & Burch, 1959). This finding suggests that the use of any $V(\lambda)$ function to reconstruct CMFs will result in substantial errors. This problem is compounded in the case of the CIE 1931 functions, because the CIE 1924 $V(\lambda)$ used in their reconstruction was partly determined by side-by-side brightness matches, for which the failures are even greater (Sperling, 1958).

(iii) Stiles and Burch (1955) 2-deg color matching functions: Color matching functions for 2-deg vision can be measured directly instead of being reconstructed using $V(\lambda)$. The Stiles and Burch (1955) 2-deg CMFs are an example of directly measured functions. With characteristic caution, Stiles referred to these 2-deg functions as "pilot" data, yet they are the most extensive set of directly measured color matching data for 2-deg vision available, being averaged from matches made by ten observers. A version of the Stiles and Burch (1955) 2-deg CMFs is tabulated in Wyszecki and Stiles (1982a), Table I (5.5.3). There are some indications, however, that the raw color matching data, after correction for a calibration error noted in Stiles and Burch (1959), should be preferred.

Despite the differences between the Stiles and Burch (1955) pilot 2-deg CMFs and the CIE 1931 2-deg CMFs, the CIE chose not to modify or remeasure their 2-deg functions. However, even in relative terms (i.e., as ratios of primaries), and plotted in a way that eliminates the effects of macular and lens pigment density variations, there are real differences between the CIE 1931 and the Stiles and Burch (1955) 2-deg color matching data in the range between 430 and 490 nm. Within that range, the CIE data repeatedly fall *outside* the range of the individual Stiles and Burch data (see Stiles & Burch, 1955, Fig. 1).

(iv) Stiles and Burch (1959) 10-deg color matching functions: The most comprehensive set of color matching data are the "large-field" 10-deg CMFs of Stiles and Burch (1959). Measured in 49 subjects from 392.2 to 714.3 nm (and in 9 subjects from 714.3 to 824.2 nm), these data are probably the most secure set of existing color matching data and are available as individual as well as mean data. For the matches, the luminance of the matching field was kept high to reduce possible rod intrusion, but nevertheless a small correction for rod intrusion was applied (see also Wyszecki & Stiles, 1982a, p. 140). Like the Stiles and Burch (1955) 2-deg functions, the Stiles and Burch (1959) 10-deg functions represent directly measured CMFs and so do not depend on measures of $V(\lambda)$.

(v) CIE (1964) 10-deg color matching functions: The large-field CIE 1964 CMFs are based mainly on the 10-deg CMFs of Stiles and Burch (1959) and to a lesser extent on the 10-deg CMFs of Speranskaya (1959). While the CIE 1964 CMFs are similar to the 10-deg CMFs of Stiles and Burch (1959), they differ in ways that compromise their use as the basis for cone fundamentals. First, at short wavelengths, the CIE 1964 functions were artificially extended to 360 nm, which is well beyond the short-wavelength limit of the color matches (392 nm) measured by Stiles and Burch. While a straightforward extrapolation could simply be ignored, the CIE chose to accommodate their extension by making small changes to the CMFs in the measured range. Although less than 0.1 $\log_{10}$ unit, the changes conspicuously distort the shape of the cone photopigment spectra derived from CIE 10-deg CMFs at short wavelengths. Second, large adjustments were made to the blue CMF above 520 nm. These changes mean that the CIE 1964 10-deg CMFs cannot be used to derive the S-cone fundamental by finding the ratio of $\bar{b}(\lambda)$ to $\bar{g}(\lambda)$ at middle and long wavelengths (which is possible for the original Stiles and Burch 10-deg functions), and, furthermore, that the CIE 1964 10-deg CMFs cannot be used to define the S-cone fundamental above 520 nm.

(vi) Conclusions: Previous estimates of the cone spectral sensitivities are linear transformations of the Judd modified or Judd–Vos modified CIE 2-deg CMFs

(e.g., Vos & Walraven, 1971; Smith & Pokorny, 1975), the Stiles and Burch 2-deg CMFs (e.g., Estévez, 1979; Vos et al., 1990; Stockman, MacLeod, & Johnson, 1993), or the CIE 1964 10-deg CMFs (Stockman, MacLeod, & Johnson, 1993). Those tabulated in Table 2.1 (Appendix) are a linear transformation of the Stiles and Burch (1959) 10-deg CMFs adjusted to a 2-deg viewing field, by correcting for the increases in macular pigment density and photopigment optical density. Either the 2-deg or 10-deg Stiles and Burch CMFs are to be preferred because they were directly measured, and are relatively uncontaminated by adjustments introduced by CIE committees. Such changes, although well intentioned, are often unnecessary and lead to unwanted distortions of the underlying color matching data and the derived cone fundamentals. In the remainder of this chapter, therefore, only the Stiles and Burch 2-deg and 10-deg CMFs are considered.

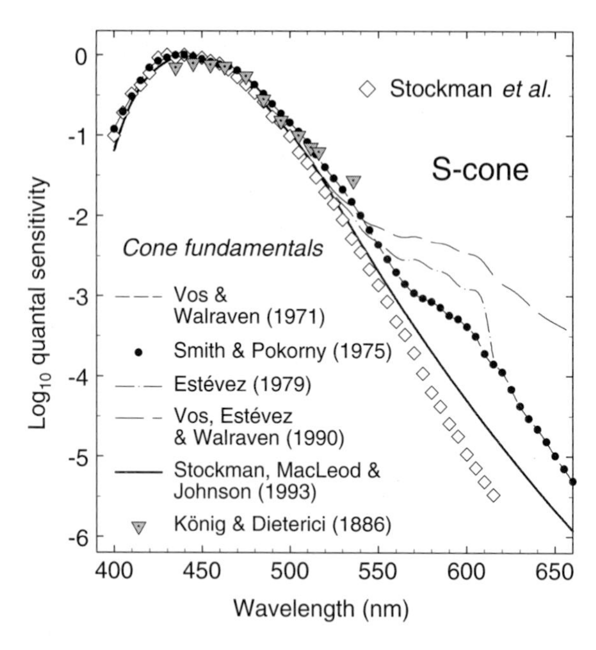

Figure 2.7: Previous estimates of the S-cone spectral sensitivity by König and Dieterici (1886) (dotted inverted triangles); Vos and Walraven (1971) (dashed line); Smith and Pokorny (1975) (filled circles); Estévez (1979) (dot-dashed line); Vos et al. (1990) (long dashed line); and Stockman, MacLeod, and Johnson (1993) (continuous line) compared with the mean S-cone thresholds (open diamonds) of Stockman, Sharpe, and Fach (1999).

Previous S-cone fundamentals. Figure 2.7 shows some of the previous estimates of the S-cone fundamental. Those based on the Judd–Vos modified CIE 2-deg CMFs include the identical proposals of Vos and Walraven (1971, as modified by Walraven, 1974, and Vos, 1978) (dashed line) and Smith and Pokorny (1975) (filled circles). Those based on the Stiles and Burch (1955) 2-deg CMFs include Estévez (1979) (dot-dashed line); Vos et al., 1990) (long dashed line); and two, which are not shown, Smith, Pokorny, and Zaidi (1983) and Stockman, MacLeod, and Johnson (1993). An estimate by Stockman, MacLeod, and Johnson (1993), based on the CIE (1964) 10-deg CMFs adjusted to 2 deg and extrapolated beyond 525 nm, is shown as the continuous line. The estimate by König and Dieterici (1886) (dotted inverted triangles) was discussed previously.

The several estimates of the S-cone fundamental can be compared with the recent S-cone threshold data (diamonds) obtained by Stockman, Sharpe, and Fach (1999). Those data suggest that all of the proposed fundamentals shown in Fig. 2.7 are too sensitive at longer wavelengths.

New S-cone fundamental. The poor agreement between the threshold data and the proposed S-cone fundamentals shown in Fig. 2.7 led Stockman, Sharpe, and Fach (1999) to derive new S-cone fundamentals in the Stiles and Burch (1955) 2-deg and Stiles and Burch (1959) 10-deg spaces. The derivation of the relative S-cone spectral sensitivity in terms of $\bar{r}(\lambda)$, $\bar{g}(\lambda)$, and $\bar{b}(\lambda)$ involves just one unknown, $\bar{s}_G/\bar{s}_B$; thus:

$$\frac{\bar{s}_G}{\bar{s}_B}\bar{g}(\lambda) + \bar{b}(\lambda) = k_s\bar{s}(\lambda). \tag{4}$$

Stockman, Sharpe, and Fach (1999) employed two methods to find $\bar{s}_G/\bar{s}_B$ for the Stiles and Burch (1955) 2-deg CMFs: The first was based on their 2-deg S-cone threshold measurements, and the second was based on the CMFs themselves. The two methods yielded nearly identical results. The second method was also used to find $\bar{s}_G/\bar{s}_B$ for the Stiles and Burch (1959) 10-deg CMFs.

(i) Threshold data: Figure 2.8A shows the mean central S-cone spectral sensitivities (gray circles) measured by Stockman, Sharpe, and Fach (1999). The sensitivities are averaged from normal and blue-cone monochromat data below 540 nm and from blue-cone monochromat data alone from 540 to 615 nm. Superimposed on the threshold data is the linear combination of the Stiles and Burch (1955) 2-deg $\bar{b}(\lambda)$ and $\bar{g}(\lambda)$ CMFs that best fits the data below 565 nm with best-fitting adjustments to the lens and macular pigment densities. The best-fitting function, $\bar{b}(\lambda)$ + $0.0163\,\bar{g}(\lambda)$, produces an excellent fit to the data up to 565 nm; thus $\bar{s}_G/\bar{s}_B = 0.0163$.

(ii) Color matching data: By using the method explained in Stockman, MacLeod, and Johnson (1993), the unknown value $\bar{s}_G/\bar{s}_B$ can be derived directly from the color matching data (see also Bongard & Smirnov, 1954). This derivation depends on the longer wavelength part of the visible spectrum being tritanopic for lights of the radiances that are typically used in color matching experiments. Thus, target wavelengths longer than about 560 nm, as well as the red primary, are invisible to the S-cones (at higher intensity levels than those used in standard color matching experiments; wavelengths longer than 560 nm would be visible to the S-cones). In contrast, the green and blue primaries are both visible to the S-cones. Targets longer than 560 nm can be matched for the L- and M-cones by a mixture of the red and green primaries, but a small color difference typically remains because the S-cones detect the field containing the green primary. To complete the match for the S-cones, a small amount of blue primary must be added to the field opposite the green primary. The sole purpose of the blue primary is to balance the effect of the green primary on the S-cones. Thus, the ratio of green to blue primary should be negative and fixed at $\bar{s}_G/\bar{s}_B$, the ratio of the S-cone spectral sensitivity to the two primaries.

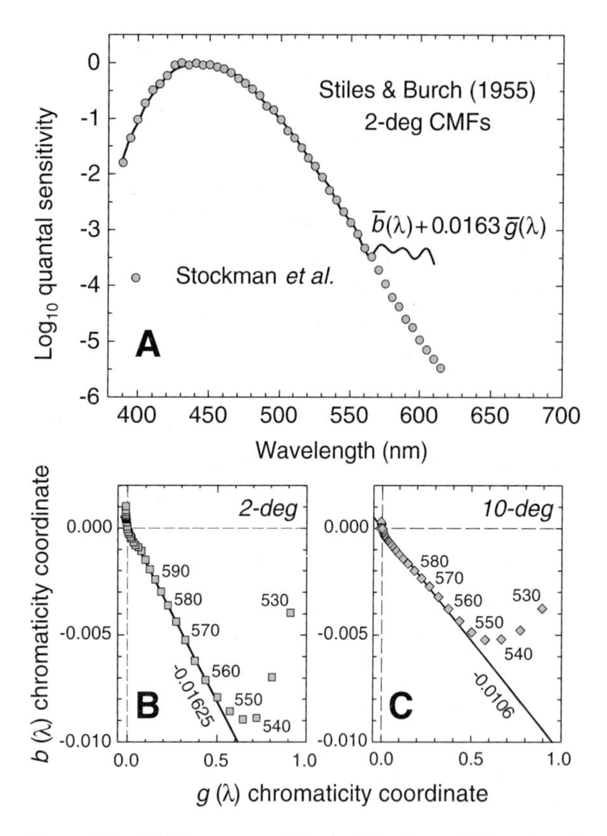

Figure 2.8: (A) Mean central data of Stockman, Sharpe, and Fach (1999) (gray circles) and linear combination of the Stiles and Burch (1955) 2-deg CMFs ($\bar{b}(\lambda)$ + $0.0163\,\bar{g}(\lambda)$, continuous line) that best fits them (≤565 nm), after applying lens and macular pigment density adjustments. (B) Stiles and Burch (1955) $g(\lambda)$ 2-deg chromaticity coordinates plotted in 5-nm steps against the $b(\lambda)$ chromaticity coordinates (gray squares). The best-fitting straight line from 555 nm to long wavelengths (continuous line) has a slope of –0.01625. (C) Stiles and Burch (1959) $g(\lambda)$ 10-deg chromaticity coordinates plotted against the $b(\lambda)$ chromaticity coordinates (gray diamonds). The best-fitting straight line from 555 nm to long wavelengths (continuous line) has a slope of –0.0106.

Figure 2.8B shows the Stiles and Burch (1955) green, $g(\lambda)$, and blue, $b(\lambda)$, 2-deg chromaticity coordinates (gray squares), which are related to the CMFs by $g(\lambda) = \bar{g}(\lambda)/[\bar{r}(\lambda) + \bar{g}(\lambda) + \bar{b}(\lambda)]$ and by $b(\lambda) = \bar{b}(\lambda)/[\bar{r}(\lambda) + \bar{g}(\lambda) + \bar{b}(\lambda)]$. As expected, the function above ~555 nm is a straight line. It has a slope of -0.01625, which implies that $\bar{s}_G/\bar{s}_B$ = 0.01625. This value is very similar to the value obtained from the direct spectral sensitivity measurements, which supports the adoption of $\bar{b}(\lambda)$ + $0.0163\,\bar{g}(\lambda)$ as the S-cone fundamental in the Stiles and Burch (1955) 2-deg space (Stockman, Sharpe, & Fach, 1999)

Stockman, Sharpe, and Fach (1999) also used the second method to determine the ratio of $\bar{s}_G/\bar{s}_B$ directly for the Stiles and Burch (1959) 10-deg CMFs. Figure 2.8C shows the green, $g(\lambda)$, and blue, $b(\lambda)$, 10-deg chromaticity coordinates (gray diamonds) and the line that best fits the data above 555 nm, which has a slope of –0.0106. The color matching data suggest that $\bar{b}(\lambda) + 0.0106\ \bar{g}(\lambda)$ is the S-cone fundamental in the Stiles and Burch (1959) 10-deg space.

To adjust the 10-deg S-cone fundamental to 2 deg, Stockman, Sharpe, and Fach (1999) assumed a peak photopigment density increase of 0.1 (from 0.3 to 0.4; the absolute densities are not critical in this calculation) and a macular density increase from a peak of 0.095 to one of 0.35. These values were based on analyses of the differences between the original Stiles and Burch 2-deg and 10-deg CMFs; the differences between 2- and 10-deg S-, M-, and L-cone fundamentals derived from the two sets of CMFs and our data; and on calculations from the cone fundamentals back to photopigment spectra.

The 2-deg S-cone fundamental based on the Stiles and Burch (1959) 10-deg CMFs is shown in Fig. 2.12 and is tabulated in the Appendix.

Previous M- and L-cone fundamentals. Figure 2.9 shows previous estimates of the M-cone (A) and L-cone (B) cone fundamentals by Vos and Walraven (1971) (dashed lines); Smith and Pokorny (1975) (filled circles); Estévez (1979) (dot-dashed lines); Vos et al. (1990) (long dashed lines); and Stockman, MacLeod, and Johnson (1993) (continuous lines). For comparison, the mean L1M2/L2M3 (white circles, panel A) and L(ser^{180}) (white squares, panel B) data of Sharpe et al. (1998) are also shown. The estimates by König and Dieterici (1886) (dotted inverted triangles) were described previously.

Both the Vos and Walraven (1971) and the Smith and Pokorny (1975) cone fundamentals are based on the Judd–Vos modified CIE 1931 2-deg CMFs. The crucial difference between them is that in deriving the former it was assumed that $V(\lambda) = \bar{l}(\lambda) + \bar{m}(\lambda) + \bar{s}(\lambda)$, whereas in deriving the latter it was assumed that $V(\lambda) = \bar{l}(\lambda) + \bar{m}(\lambda)$ (i.e., that the S-cones do not contribute to luminance). Of the two, the Smith and Pokorny (1975) M- and L-cone fundamentals are much closer to the dichromat data at short wavelengths.

The Vos, Estévez, and Walraven (1990) and the Estévez (1979) M- and L-cone fundamentals are based on the Stiles and Burch (1955) 2-deg CMFs. The Estévez (1979) proposal was an attempt to reconcile dichromat spectral sensitivities with Stiles's π_4 and π_5, but it was a reconciliation for which there was little justification. Vos, Estévez, and Walraven (1990) intended their M-cone fundamental to be consistent with protanopic spectral sensitivities, but clearly it is not. Stockman, MacLeod, and Johnson (1993) also proposed M- and L-cone fundamentals based on the Stiles and Burch (1955) 2-deg CMFs (not shown). Except at short wavelengths, these are similar to the alternative version of the Stockman, MacLeod, and Johnson (1993) M- and L-cone fundamentals that are based on the CIE 1964 10-deg CMFs adjusted to 2 deg, which are shown in Fig. 2.9 (continuous lines).

The comparisons in Fig. 2.9 suggest that the M- and L-cone fundamentals that are most consistent with dichromat data are those of Smith and Pokorny (1975) and Stockman, MacLeod, and Johnson (1993). However, neither estimate agrees perfectly with the new dichromat data provided by Sharpe et al. (1998), even after optimal adjustments in macular and lens pigment densities (Stockman & Sharpe, 1998).

New M- and L-cone fundamentals. The definition of the M- and L-cone spectral sensitivities in terms of $\bar{r}(\lambda)$, $\bar{g}(\lambda)$, and $\bar{b}(\lambda)$ requires knowledge of four unknowns [see Eqn. (3)] $\bar{m}_R/\bar{m}_B$, $\bar{m}_G/\bar{m}_B$, $\bar{l}_R/\bar{l}_B$, and $\bar{l}_G/\bar{l}_B$; thus

$$(5)\qquad \frac{\bar{m}_R}{\bar{m}_B}\bar{r}(\lambda) + \frac{\bar{m}_G}{\bar{m}_B}\bar{g}(\lambda) + \bar{b}(\lambda) = k_m\bar{m}(\lambda)$$

and

$$(6)\qquad \frac{\bar{l}_R}{\bar{l}_B}\bar{r}(\lambda) + \frac{\bar{l}_G}{\bar{l}_B}\bar{g}(\lambda) + \bar{b}(\lambda) = k_l\bar{l}(\lambda).$$

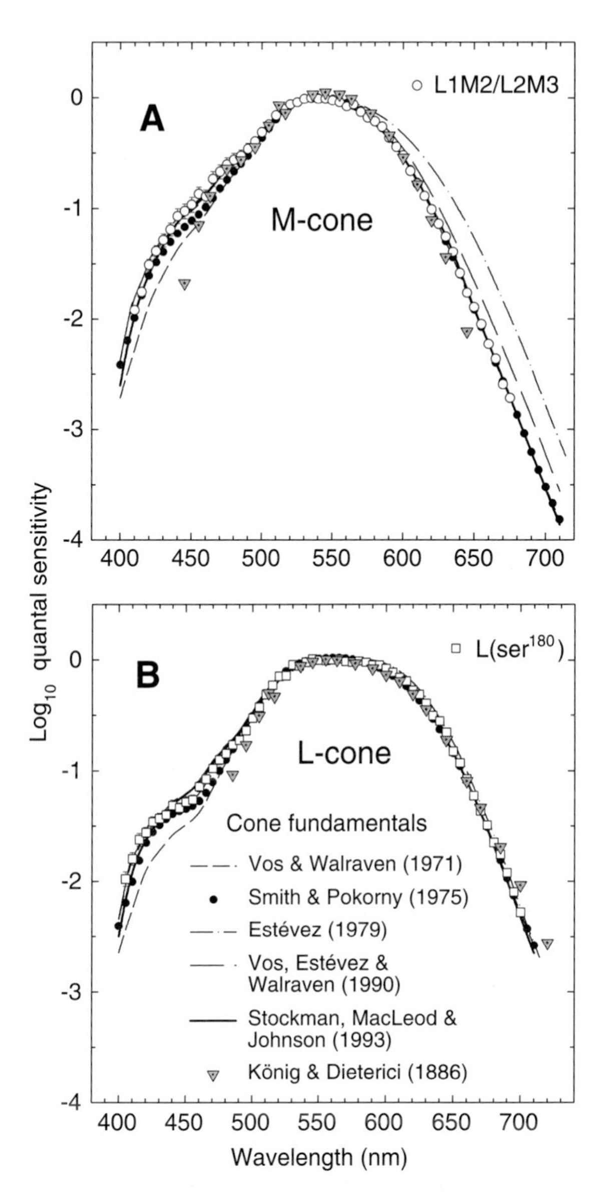

Figure 2.9: Estimates of the M-cone (A) and L-cone (B) fundamentals by König and Dieterici (1886) (dotted inverted triangles); Vos and Walraven (1971) (dashed lines); Smith and Pokorny (1975) (filled circles); Estévez (1979) (dot-dashed lines); Vos et al. (1990) (long dashed lines); and Stockman, MacLeod, and Johnson (1993) (black lines) compared with the mean L1M2/L2M3 (white circles, A) and L(ser^{180}) (white squares, B) data of Sharpe et al. (1998).

Stockman and Sharpe (2000a) used the new red-green dichromat data of Sharpe et al. (1998) to estimate the unknowns in Eqns. (5) and (6), first in terms of the Stiles and Burch (1955) 2-deg CMFs, and then by way of the 2-deg solution in terms of the Stiles and Burch (1959) 10-deg CMFs corrected to 2 deg.

Their strategy, like that of Stockman, MacLeod, and Johnson (1993), was first to find the linear combinations of the Stiles and Burch (1955) 2-deg CMFs that best fit mean spectral sensitivity data. The Stiles and Burch (1955) 2-deg–based cone fundamentals were then used to obtain estimates of the cone spectral sensitivities based on the Stiles and Burch (1959) 10-deg CMFs corrected to 2 deg. In general, color matching data are more precise than threshold data, so it is preferable to define cone spectral sensitivity data in terms of them rather than in terms of the original threshold data.

The linear combinations of the Stiles and Burch (1955) 2-deg CMFs that best fit the mean L(ser^{180}) deuteranope data (open diamonds), L(ala^{180}) deuteranope data (open and filled squared), and L1M2/L2M3 protanope data (open circles) of Sharpe et al. (1998) with macular and lens density adjustments are shown as the continuous lines in Fig. 2.10A. The best-fitting values are given in the figure legend. Figure 2.10B shows the residuals.

The mean Stockman, MacLeod, and Johnson (1993) M-cone data (dotted triangles) and L-cone data (dotted inverted triangles) are also shown in Fig. 2.10A. The M-cone data agree well with the L1M2/L2M3 data, but, at long wavelengths, the L-cone data are slightly steeper than the L(ser^{180}) data and slightly shallower than the L(ala^{180}) data. These differences are expected, because the subjects employed by Stockman, MacLeod, and Johnson have a mixture of L(ser^{180}) and L(ala^{180}) photopigment genes. Consequently, their mean L-cone spectral sensitivity function should be intermediate in spectral position between the mean L(ser^{180}) and L(ala^{180}) functions – as is found.

To derive a normal L-cone spectral sensitivity function from the L(ser^{180}) and L(ala^{180}) data, we needed an estimate of the ratio of L(ser^{180}) to L(ala^{180}) photo-

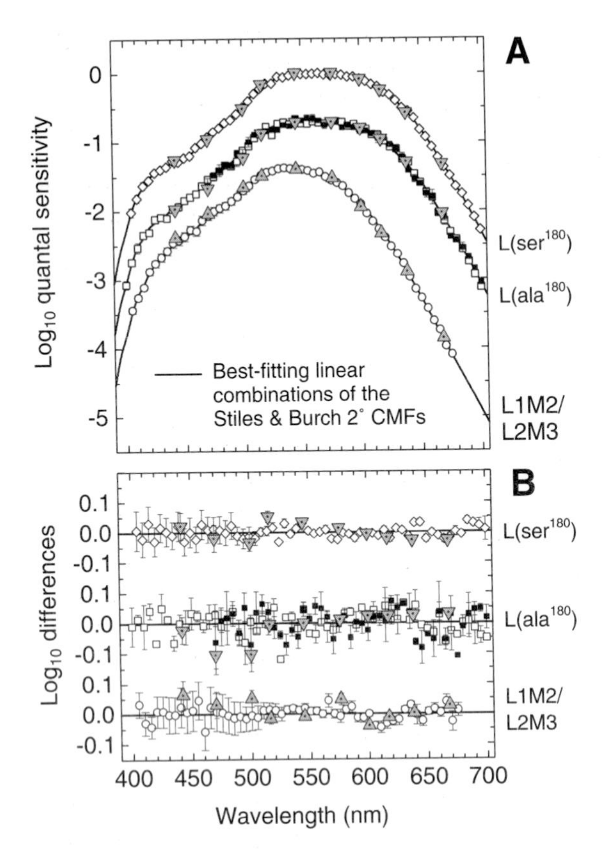

Figure 2.10: Fits of the 2-deg CMFs to dichromat data. (A) Mean L1M2 and L2M3 protanope data (open circles, $n = 9$), and L(ala^{180}) (open squares, $n = 2$; filled squares $n = 3$) and L(ser^{180}) (open diamonds, $n = 15$) deuteranope data from Sharpe et al. (1998), and the linear combinations of the Stiles and Burch (1955) CMFs (continuous lines) that best fit each set of dichromat data. The dichromat data have been adjusted in macular and lens density to best fit the CMFs. The best-fitting values (ignoring the vertical scaling constant) are $\bar{l}_R / \bar{l}_B = 5.28554$ and $\bar{l}_G / \bar{l}_B = 16.80098$ (L(ser^{180}), upper line), $\bar{l}_R / \bar{l}_B = 4.15278$ and $\bar{l}_G / \bar{l}_B = 16.75822$ (L(ala^{180}), middle line), and $\bar{m}_R/\bar{m}_B = 0.29089$ and $\bar{m}_G/\bar{m}_B = 12.24415$ (L1M2/L2M3, lower line). Also shown are the mean M-cone (dotted triangles) and L-cone (dotted inverted triangles) data from Stockman, MacLeod, and Johnson (1993), unadjusted in macular and lens density. (B) Differences between the L1M2/L2M3 data (open circles), L(ala^{180}) data (open squares), and L(ser^{180}) data (open diamonds) and the corresponding linear combination of the CMFs, and between the mean M-cone (dotted triangles) or L-cone (dotted inverted triangles) data and the CMFs. Error bars are ±1 standard error of the mean.

pigments expressed in the normal population. We used the ratio of 0.56 L(ser^{180}) to 0.44 L(ala^{180}) found in 308 male Caucasian subjects (see Table 1.2) to determine the mean L-cone fundamental. That is, we set $\bar{l}_G / \bar{l}_B$ and $\bar{l}_R / \bar{l}_B$ to be 0.56 times the L(ser^{180}) values plus 0.44 times the L(ala^{180}) ones. Thus, $\bar{l}_G / \bar{l}_B = 16.782165$ and $\bar{l}_R / \bar{l}_B = 4.787127$.

Having derived the L- and M-cone fundamentals in terms of the 2-deg CMFs of Stiles and Burch (1955), we next defined them in terms of the Stiles and Burch (1959) 10-deg CMFs corrected to 2 deg. The 10-deg data set of Stiles and Burch (1959) were used to define the cone fundamentals because they represent the most extensive and secure set of color matching data available. The derivation of the 2-deg M- and L-cone fundamentals as an intermediate step produces relatively smooth and noise-free 2-deg functions that can then be fitted with the adjusted 10-deg CMFs.

We derived the 10-deg–based cone fundamentals by a curve-fitting procedure in which we found the linear combinations of the Stiles and Burch 10-deg CMFs that, after adjustment to 2-deg macular, lens, and photopigment densities, best fit the Stockman and Sharpe 2-deg L- and M-cone fundamentals based on the Stiles and Burch 2-deg CMFs. In making these fits, we assumed a macular pigment density change from a peak of 0.095 for the 10-deg CMFs to a peak of 0.35 for the 2-deg CMFs, a change in lens density from the tabulated values in the Appendix for the 10-deg functions to 92.5% of the tabulated values for the 2-deg functions, and a change in peak photopigment optical density from 0.38 to 0.50 from 10 deg to 2 deg. These are all optimized or best-fitting differences. The best-fitting linear combinations are, for M, $\bar{m}_R/\bar{m}_B = 0.168926$ and $\bar{m}_G/\bar{m}_B = 8.265895$ and, for L, $\bar{l}_R / \bar{l}_B = 2.846201$ and $\bar{l}_G / \bar{l}_B = 11.092490$.

Tritanopic color matches and the M- and L-cone fundamentals. Tritanopic matches provide a useful means of distinguishing between candidate M- and L-cone fundamentals. Since tritanopes lack S-cones, their color matches should be predicted, at least approximately, by any plausible M- and L-cone spectral sensitivity estimates. Stockman, MacLeod, and

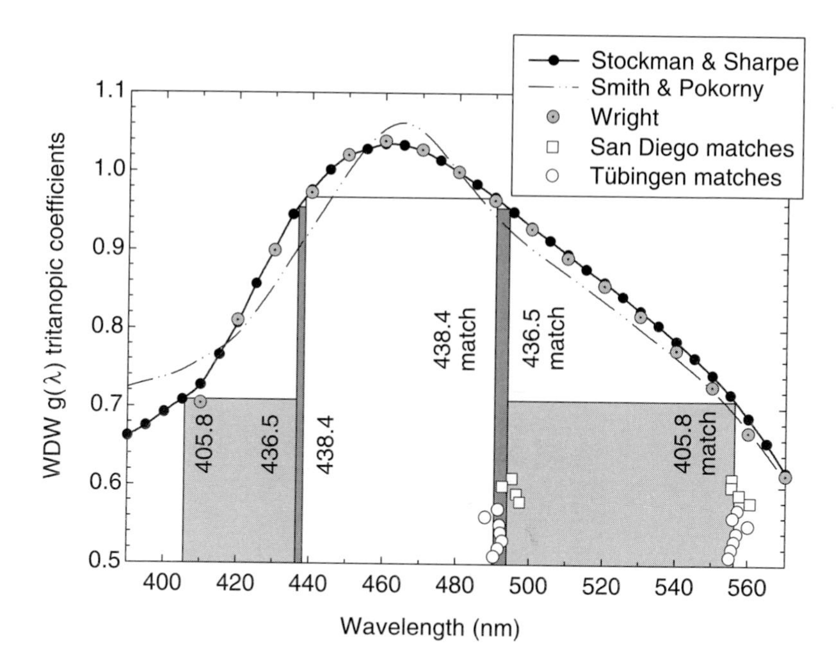

Figure 2.11: Tritanopic color matches and M- and L-cone spectral sensitivities. Tritanopic $g(\lambda)$ predictions of the M- and L-cone fundamentals tabulated in the Appendix (filled diamonds, continuous line); and the wavelengths found by 11 subjects (9 color normals and 2 tritanopes) to match either a 405.8- or a 436.5-nm target light (open squares, San Diego) or a 405.8- or 438.4-nm target light (open circles, Tübingen) under conditions that produce tritanopia in the normals. The matches to the 405.8-, 436.5-, and 438.4-nm lights predicted by the Stockman and Sharpe (2000a) $g(\lambda)$ function are indicated by three large gray or white rectangles. Also shown are Wright's (1952) tritanopic $g(\lambda)$ coefficients (gray dotted circles) and the tritanopic $g(\lambda)$ predictions of the Smith and Pokorny (1975) M- and L-cone fundamentals (dotted-dashed line).

Johnson (1993) actually used the tritanope data of Wright (1952) to substantially adjust their M- and L-cone fundamentals.

Pokorny and Smith (1993) suggested that a simple way to test M- and L-cone fundamentals was to determine the spectral lights that tritanopes confused with the 404.7- and 435.8-nm Hg lines (which, when spectrally isolated, are nearly monochromatic). These two spectral pairs or tritanopic metamers should be predicted by the M- and L-cone spectral sensitivities in question. Following up on their suggestion, Stockman and Sharpe (2000b) carried out matching experiments separately in Tübingen and in San Diego under intense short-wavelength–adapting conditions that produced tritanopia in normals. The matches for five normals and one tritanope measured in Tübingen (open circles) and three normals and one tritanope measured in San Diego (open squares) are shown in Fig. 2.11.

The Hg lines at 404.7 and 435.8 nm are broadened and shifted to longer wavelengths in high-pressure Hg arc lamps (Elenbaas, 1951), so that spectral "lines," after also taking into account the effects of prereceptoral filtering, were 405.8 and 438.4 nm in Tübingen (where a filter nominally of 435.8 nm skewed the spectral line to longer wavelengths) and 405.8 and 436.5 nm in San Diego.

The L- and M-cone spectral sensitivities have been plotted in the form of Wright (WDW) $g(\lambda)$ coordinates by transforming them to Wright's primaries of 480 and 650 nm, normalizing them, and setting them to be equal at 582.5 nm [the WDW $r(\lambda)$ coordinates are simply 1 - $g(\lambda)$]. Two advantages of plotting the estimates in this way are that WDW coordinates are independent of individual differences in macular and lens pigment densities, and that Wright's (1952) tritanope data (dotted, gray circles) are tabulated in the same form, thus allowing straightforward comparisons. Wright's data, however, are for a target that is more than 50% smaller in area than the 2-deg target and so probably reflect slightly higher L- and M-cone photopigment optical densities.

Figure 2.11 shows the $g(\lambda)$ function (filled diamonds and continuous line) calculated from the L- and M-cone fundamentals (Stockman & Sharpe, 2000a) tabulated in the Table 2.1. The tritanopic matches predicted by the $g(\lambda)$ function are any two wavelengths that have the same $g(\lambda)$ value. As can be seen by following the outlines of the three rectangles from 405.8,

436.5, and 438.4 nm, the Stockman and Sharpe $g(\lambda)$ function predicts the tritanopic matches obtained in Tübingen (open circles) and San Diego (open squares) well, with each predicted match lying within the range of measured matches.

Figure 2.11 also shows the $g(\lambda)$ function predicted by the Smith and Pokorny (1975) fundamentals (dot-dashed line) and Wright's (1952) mean data for seven tritanopes (gray dotted circles). The Smith and Pokorny (1975) predictions agree poorly with Wright's data. The problem lies mainly in the Judd–Vos modified CIE CMFs, which, as others have pointed out, are inconsistent with tritanopic color matching data (Alpern, 1976; Estévez, 1979). The Smith and Pokorny (1975) fundamentals predict the 436.5- and 438.4-nm matches obtained by Stockman and Sharpe (2000b) poorly and the 405.8-nm matches very poorly, missing the mean match by about 12 nm. Stockman, MacLeod, and Johnson (1993) optimized their cone fundamentals to be consistent with Wright's (1952) data. Nevertheless, their predictions (not shown) fail to predict the mean 405.8-nm match found by Stockman and Sharpe (2000b) by about 4.5 nm.

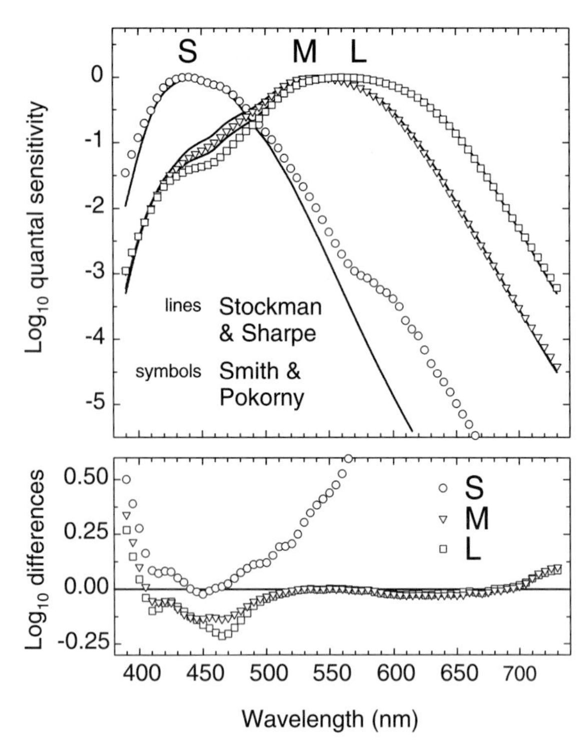

Figure 2.12: (A) S-, M-, and L-cone spectral sensitivity estimates of Stockman and Sharpe (2000a) (continuous lines) compared with the S- (open circles), M- (open inverted triangles), and L- (open squares) cone spectral sensitivity estimates of Smith and Pokorny (1975). (B) Differences between the Smith and Pokorny and Stockman and Sharpe estimates.

New cone fundamentals

The new S-, M-, and L-cone spectral sensitivities (Stockman & Sharpe, 2000a) are shown in Fig. 2.12 as the continuous lines and are tabulated in the Appendix. They are consistent with spectral sensitivities measured in X-chromosome–linked red-green dichromats, in blue-cone monochromats, and in color normals; they are consistent with tritanopic color matches, and they reflect typical macular pigment and lens densities for a 2-deg field.

The S- (open circles), M- (open inverted triangles), and L- (open squares) cone spectral sensitivities proposed by Smith and Pokorny are also shown in Fig. 2.12. Although they agree with our proposed M- and L-cone spectral sensitivities at middle and long wavelengths, they do not agree at short wavelengths. The agreement between the S-cone functions is poor throughout the spectrum. Large decreases in the macular pigment density of the Smith and Pokorny functions can improve the agreement between the two sets of functions, but the implied macular density of the mean Smith and Pokorny (1975) observer before adjustment is implausibly high for a 2-deg field, and largely artificial.

The luminosity function, $V(\lambda)$

Luminance efficiency is a photometric measure that might loosely be described as "apparent intensity" but is actually defined as the effectiveness of lights of different wavelengths in specific photometric matching tasks. Those tasks now most typically include hetero-

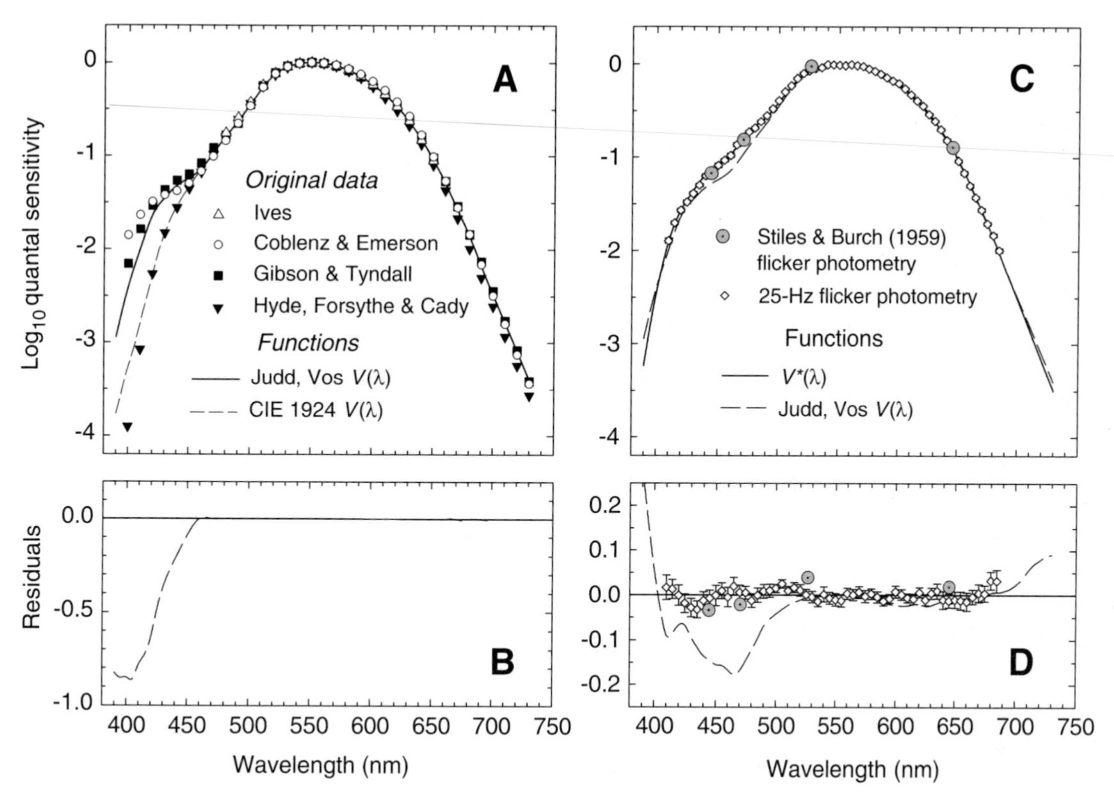

Figure 2.13: (A) The original luminosity measurements by Ives (1912a) (open triangles), Coblentz and Emerson (1918) (open circles), Gibson and Tyndall (1923) (filled squares) and Hyde, Forsythe, and Cady (1918) (filled inverted triangles), which were used to derive the CIE 1924 $V(\lambda)$ function (dashed line), and the Judd–Vos modification to the CIE 1924 $V(\lambda)$ function (continuous line). (B) Differences between the Judd–Vos modified $V(\lambda)$ and the CIE 1924 V(λ) functions. (C) $V^*(\lambda)$ (continuous line), the Judd–Vos modified $V(\lambda)$ (dashed line), recent 25-Hz heterochromatic flicker photometric measurements from 22 color normals (open diamonds) used to guide the choice of $V^*(\lambda)$ (Sharpe et al., unpublished observations), and flicker photometric sensitivity measurements by Stiles and Burch (1959) (dotted circles). (D) Differences between the $V^*(\lambda)$ and the other functions shown in (C).

chromatic flicker photometry (HFP), or a version of side-by-side matching, in which the relative intensities of the two half-fields are set so that the border between them appears "minimally distinct" (MDB). Both tasks minimize contributions from the S-cones and produce nearly additive results (e.g., Ives, 1912a; Wagner & Boynton, 1972). So defined, however, luminance is inseparable from the tasks used to measure it.

Previous luminosity functions. The $V(\lambda)$ function, which was adopted by the CIE in 1924 and is still used to define luminance today, was originally proposed by Gibson and Tyndall (1923). It is shown in Fig. 2.13A (short dashed line). The function was based on data obtained by Ives (1912a) (open triangles); Coblentz and Emerson (1918) (open circles); Hyde, Forsythe, and Cady (1918) (filled inverted triangles); and Gibson and Tyndall (1923) (filled squares). Despite the enormous differences between their own data and the proposed standard, Gibson and Tyndall concluded that: "The [older] Illuminating Engineers Society data in the violet have been accepted by the authors for lack of any good reason for changing them, but the relative as well as the absolute values are very uncertain and must be considered as tentative only." As a result, the 1924 $V(\lambda)$ function deviates from typical luminosity data (e.g., filled squares, open circles) by a factor of nearly ten in the violet, a problem that continues to plague both colorimetry and photometry 75 years later.

In 1951, Judd proposed a substantial revision to the $V(\lambda)$ function in an attempt to overcome the discrepancies at short wavelengths (Judd, 1951). He retained the older photopic sensitivities at 460 nm and longer wavelengths but increased the sensitivity at shorter wavelengths. Unfortunately, while improving the $V(\lambda)$ function in the violet part of the spectrum, this adjustment artificially created an average observer with implausibly high macular pigment density for a 2-deg field. Vos (1978) subsequently made minor adjustments to the Judd modified CIE $V(\lambda)$ function below 410 nm to produce the Judd–Vos modified CIE $V(\lambda)$ or $V_M(\lambda)$ function (continuous line).

Cone spectral sensitivities and the luminosity function. The luminosity function $V(\lambda)$ falls into a quite different category from cone spectral sensitivities, yet it is often treated as if it were a cone spectral sensitivity. Unlike cone spectral sensitivities, the shape of the luminosity function *changes with adaptation* (e.g., DeVries, 1948b; Eisner & MacLeod, 1981). Thus, any luminosity function is only of limited applicability, because it strictly defines luminance only for the conditions under which it was measured. The function is *not* generalizable to other conditions of adaptation – particularly to other conditions of chromatic adaptation – or necessarily to other measurement tasks. In contrast, cone spectral sensitivities (and CMFs, in general) do not change with adaptation, until photopigment bleaching becomes significant (in which case, the changes reflect the reduction in photopigment optical density). Cone spectral sensitivities are receptoral; the luminosity function is postreceptoral.

Both the L- and M-cones contribute to luminance, although their contribution is typically dominated by the L-cones (e.g., Cicerone & Nerger, 1989; Vimal et al., 1989). The contribution of the S-cones to luminance has been somewhat contentious, but it now seems clear that the S-cones do make a small contribution (Eisner & MacLeod, 1980; Stockman & MacLeod, 1987; Verdon & Adams, 1987; Lee & Stromeyer, 1989; Stockman, MacLeod, & DePriest, 1991).

Given that the S-cone contribution is so small, and is also so dependent on temporal frequency and adaptation (Stockman, MacLeod, & DePriest, 1991), it is of practical convenience to assume that the S-cone contribution is zero, which is the assumption that we make in deriving $V^*(\lambda)$, the new luminosity function, and it is the assumption that Smith and Pokorny (1975) made in deriving their cone spectral sensitivities. Although convenient, this assumption restricts the validity of any $V(\lambda)$ function to conditions under which S-cone stimulation is small.

Photometry and physiology. The goal in providing a $V(\lambda)$ luminosity function is to construct a spectral sensitivity function that predicts "effective intensity" over the broadest range of conditions. It should be recognized, however, that $V(\lambda)$ is more of a photometric convention than a physiological reality. The $V(\lambda)$ function predicts behavior on HFP and MDB tasks under neutral adaptation and near-threshold. Under these limited conditions, $V(\lambda)$ may even reflect activity in some postreceptoral pathway (e.g., Lee, Martin, & Valberg, 1988). However, under other conditions (for example, with targets of long duration, with large targets, with strongly suprathreshold targets, with short-wavelength targets, and with chromatic adaptation), $V(\lambda)$ is inappropriate. The incorporation of $V(\lambda)$ into color spaces (see below, Two- and three-dimensional color matching and cone spaces) similarly limits the usefulness of those spaces.

To the extent that there is a luminance mechanism, its spectral sensitivity is not, in general, predicted by $V(\lambda)$, since the spectral sensitivity changes with adaptation. To incorporate changes with chromatic adaptation, the spectral sensitivity of the luminance mechanism could be defined as the combination of weighted cone contrasts (see Chapter 18).

$V^*(\lambda)$ luminosity function. Unlike the CIE 2-deg CMFs, the Stiles and Burch (1959) 10-deg CMFs are purely colorimetric and are not connected to any directly measured luminosity function. In some ways this is fortunate, since it prevents the Stiles and Burch based cone spectral sensitivity functions from being altered to be consistent with the prevailing model of $V(\lambda)$, as has happened with the CIE based functions (e.g., Vos & Walraven, 1971; Smith & Pokorny, 1975). In other ways, however, it is unfortunate, since a knowledge of the appropriate luminosity function for the Stiles and Burch based observer is, in many cases, desirable.

Given the differences between the CIE 2-deg and the Stiles and Burch 2- and 10-deg spaces, the CIE $V(\lambda)$ function cannot be used to define luminosity in the Stiles and Burch spaces without introducing large errors into some calculations (such as in the calculation of the MacLeod–Boynton coordinates; see below, Equal-luminance cone excitation space). The known problems of the CIE $V(\lambda)$ functions aside, the CIE functions were, inevitably, measured in different subjects than those used to measure the Stiles and Burch CMFs, so that individual differences play a role.

Instead of $V(\lambda)$, we could use the CIE 1964 estimate of the luminosity function for 10-deg vision [$\bar{y}_{10}(\lambda)$], which we refer to as $V_{10}(\lambda)$, corrected to 2 deg. However, this function is "synthetic," because it was constructed from luminosity measurements made at only four wavelengths (see Stiles & Burch, 1959). An advantage of the $V_{10}(\lambda)$ function, however, is that it was based on data from some of the same subjects that were used to obtain the Stiles and Burch (1959) 10-deg CMFs.

To define luminance we propose a modified $V(\lambda)$ function, which we refer to as $V^*(\lambda)$, that retains some of the properties of the original CIE $V(\lambda)$ but is consistent with the new cone fundamentals. One property that is retained, which is also a property of the Smith and Pokorny (1975) cone fundamentals and $V(\lambda)$, is that:

$$(7) \qquad V^*(\lambda) = a\,\bar{l}(\lambda) + \bar{m}(\lambda),$$

where a is a scaling constant. The appropriate value of a for the Stockman and Sharpe (2000a) cone fundamentals tabulated in the Appendix could be estimated by finding (i) the linear combination of $\bar{l}(\lambda)$ and $\bar{m}(\lambda)$ that best fits the CIE Judd–Vos $V(\lambda)$ and (ii) the linear combination of $\bar{l}(\lambda)$ and $\bar{m}(\lambda)$ – *before* their adjustment from 10 to 2 deg – that best fits $V_{10}(\lambda)$, both after macular and lens adjustments. The best-fitting values of a [relative to $\bar{l}(\lambda)$ and $\bar{m}(\lambda)$ having the same peak sensitivities] are 1.65 with a standard error of the fitted parameter of 0.15 for the CIE Judd–Vos $V(\lambda)$ and 1.76 with a standard error of the fitted parameter of 0.05 for $V_{10}(\lambda)$. Alternatively, we can find the linear combination of $\bar{l}(\lambda)$ and $\bar{m}(\lambda)$ that best fits experimentally determined FPS data. Such data, recently obtained in one of our labs for 22 male subjects of known genotype [13 L(S180) and 9 L(A180)] using 25-Hz flicker photometry, are shown in Fig. 2.13C. The mean (open diamonds) has been weighted so that, like the cone spectral sensitivities, it represents a ratio of 0.56 L(ser^{180}) to 0.44 L(ala^{180}) (see Table 1.2). After macular and lens adjustments, the best-fitting value of a is 1.50 with a standard error of the fitted parameter of 0.05. (If an S-cone contribution is allowed, the S-cone weight is negative and 0.10% of the L-cone weight.) For consistency with the experimental data, we chose a value of a of 1.50. The definition of $V^*(\lambda)$, therefore, is:

$$(8) \qquad V^*(\lambda) = 1.50\,\bar{l}(\lambda) + \bar{m}(\lambda),$$

again relative to $\bar{l}(\lambda)$ and $\bar{m}(\lambda)$ having the same peak sensitivities. $V^*(\lambda)$ is tabulated in the Appendix and is shown in Fig. 2.13C optimally adjusted in macular and lens densities for agreement with the experimental FPS data. The differences between the macular and lens adjusted $V^*(\lambda)$ and the data are shown in Fig 2.13D.

Figure 2.13C also shows the 2-deg flicker photometric measurements (dotted circles) made at four wavelengths in 26 of the 49 observers of the Stiles and Burch (1959) 10-deg color matching study, which are also more consistent with $V^*(\lambda)$ than with the CIE Judd–Vos $V(\lambda)$.

Photopigment optical density spectra

Calculating photopigment spectra from corneal spectral sensitivities, and vice versa. The calculation of photopigment optical density spectra from corneal spectral sensitivities is, in principle, straightforward, provided that the appropriate values of (i) D_{peak} – the peak optical density of the photopigment, (ii) k_{lens} – the scaling constant by which the lens density spectrum ($d_{lens}(\lambda)$) should be multiplied, and (iii) k_{mac} – the scaling constant by which the macular density spectrum ($d_{mac}(\lambda)$) should be multiplied are known. Starting with the *quantal* spectral sensitivity of, for example, the L-cones ($\bar{l}(\lambda)$), the effects of the lens pigment ($k_{lens}d_{lens}(\lambda)$) and the macular pigment ($k_{mac}d_{mac}(\lambda)$) are first removed, by restoring the sensitivity losses that they cause:

$$(9)\quad \log(\bar{l}_p(\lambda)) = \log(\bar{l}(\lambda)) + k_{lens}d_{lens}(\lambda) + k_{mac}d_{mac}(\lambda).$$

The functions $d_{lens}(\lambda)$ and $d_{mac}(\lambda)$ are the optical density spectra of the lens and macular tabulated in the Appendix. They are scaled to the densities that are appropriate for the Stockman and Sharpe (2000a) 2-deg cone fundamentals that are also tabulated there (0.35 macular density at 460 nm and 1.765 lens density at 400 nm). Thus, the values k_{lens} and k_{mac} are 1.0 for the mean fundamentals, but should be adjusted for individual observers or groups of observers with different lens and macular densities. $\bar{l}_p(\lambda)$ is the spectral sensitivity of the L-cones at the photoreceptor level.

To calculate the photopigment optical density of the L-cones scaled to unity peak ($\bar{l}_{OD}(\lambda)$), from $\bar{l}_p(\lambda)$:

$$(10)\quad \bar{l}_{OD}(\lambda) = \frac{-\log(1 - \bar{l}_p(\lambda))}{D_{peak}}.$$

We assumed D_{peak}, the peak optical density, to be 0.5, 0.5, and 0.4 for the L-, M-, and S-cones, respectively [$\bar{l}_p(\lambda)$ should be scaled before applying Eqn. (10), so that $\bar{l}_{OD}(\lambda)$ peaks at unity]. These calculations from corneal spectral sensitivities to retinal photopigment optical densities ignore changes in spectral sensitivity that may result from the structure of the photoreceptor or other ocular structures and pigments (unless they are incorporated in the lens or macular pigment density spectra).

The calculation of relative quantal corneal spectral sensitivities from photopigment or absorbance spectra is also straightforward, again if the appropriate values (D_{peak}, k_{lens}, and k_{mac}) are known. First, the spectral sensitivity at the photoreceptor level, $\bar{l}_p(\lambda)$, is calculated from the normalized photopigment optical density spectrum, $\bar{l}_{OD}(\lambda)$, by the inversion of Eqn. (10) (see Knowles & Dartnall, 1977):

$$(11)\quad \bar{l}_p(\lambda) = 1 - 10^{-D_{peak}\bar{l}_{OD}(\lambda)}.$$

Then, the filtering effects of the lens and macular pigments are added back:

$$(12)\quad \log(\bar{l}(\lambda)) = \log(\bar{l}_p(\lambda)) - k_{lens}d_{lens}(\lambda) - k_{mac}d_{mac}(\lambda).$$

The lines in Fig. 2.14 are the logarithm of the photopigment optical densities, $\bar{l}_{OD}(\lambda)$, $\bar{m}_{OD}(\lambda)$, and $\bar{s}_{OD}(\lambda)$, calculated using Eqns. (9) and (10) from the cone fundamentals tabulated in the Appendix (Table 2.1). The photopigment optical densities are also tabulated in the Appendix.

Scales. Attempts have been made to simplify cone photopigment spectra by finding an abscissa that produces spectra of a fixed spectral shape, whatever the photopigment λ_{max}. An early proposal was by Dartnall (1953), who proposed a "nomogram" or fixed template shape for photopigment spectra plotted as a function of wavenumber ($1/\lambda$, in units of cm^{-1}). Another proposal was that the spectra are shape-invariant when plotted as a function of $\log_{10}$ frequency or wavenumber [$\log_{10}(1/\lambda)$] (Mansfield, 1985; MacNichol, 1986), which is equivalent to $\log_{10}$ wavelength [$\log_{10}(\lambda)$] or normalized frequency (λ_{max}/λ). For this scale, Lamb (1995) has proposed a template (see Chapter 3). Barlow (1982) has also proposed an abscissa of the fourth root of wavelength ($\sqrt[4]{\lambda}$). The three photopigment spectra (Appendix, Table 2.1) are most similar in shape when plotted against log wavelength.

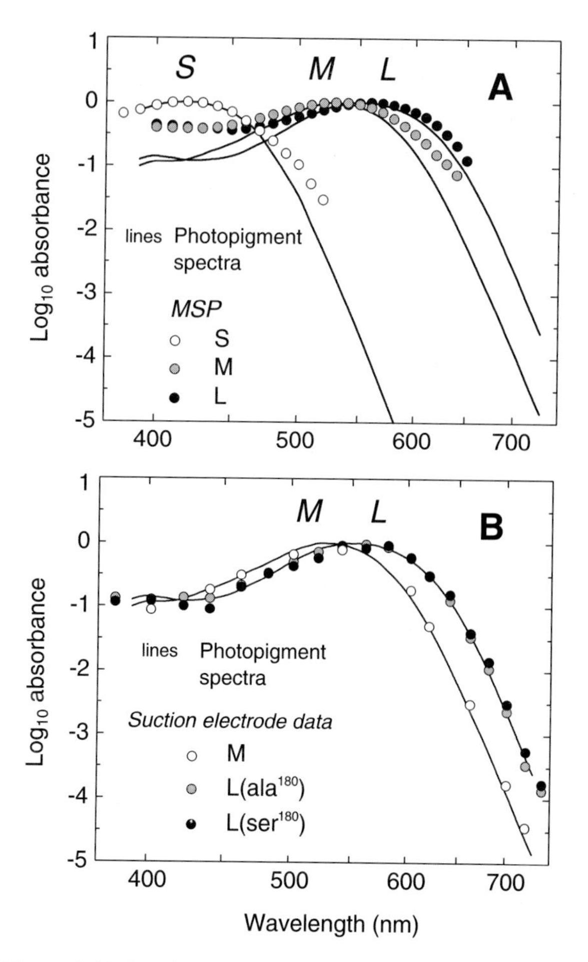

Figure 2.14: Psychophysical estimates of the photopigment optical density spectra compared with direct measurements. Log_{10} S-, M-, and L-cone photopigment spectra calculated from the Stockman and Sharpe cone fundamentals (continuous lines, Appendix) and (A) human S- (white circles), M- (gray circles), and L-cone (black circles) MSP measurements by Dartnall, Bowmaker, and Mollon (1983) or (B) human M-cone (white circles) suction electrode measurements by Kraft (personal communication), and L(ala^{180}) cone (gray circles) and L(ser^{180}) cone (black circles) suction electrode measurements by Kraft, Neitz, and Neitz (1998).

Direct methods of determining photopigment spectra. There are several direct methods of measuring photopigment spectra, only two of which we will consider in any detail. A promising new approach is to produce the cone pigment apoprotein in tissue culture cells transfected with the corresponding complementary DNA clones and then, after reconstitution of the apoprotein with 11-*cis* retinal, to measure the bleaching difference spectrum in solution (Merbs & Nathans, 1992a; Asenjo, Rim, & Oprian, 1994). As yet, the technique is too noisy to be useful in predicting corneal spectral sensitivities. Moreover, the technique adds another level of uncertainty into the reconstruction of corneal spectral sensitivities from photopigment spectra, since, unlike the other techniques described in this section, the photopigment is not embedded in the photoreceptor membrane. The incorporation of the photopigment into the membrane may change its spectral sensitivity.

Two methods have yielded human photopigment spectra that have frequently been compared with corneally measured spectral sensitivity functions: microspectrophotometry and suction electrode recordings.

(i) Microspectrophotometry: In MSP work, the spectral transmission of a small measuring beam passed transversely through the outer segment of a single cone is compared with that of a reference beam passed outside the cone to derive the absorption spectrum of the outer segment (e.g., Bowmaker et al., 1978). Of interest here are the MSP measurements of Dartnall, Bowmaker, and Mollon (1983) of photoreceptors "from the eyes of seven persons." Figure 2.14A, which compares the MSP results (circles) with the photopigment optical density spectra from the Appendix, Table 2.1 (lines), makes it clear that MSP is of little use in estimating cone spectral sensitivities far away from the photopigment λ_{max}, because the MSP functions are much too broad. Nonetheless, the comparisons support the use of MSP for defining photopigment λ_{max}.

(ii) Suction electrode recordings: In suction electrode recordings, a single human or primate cone outer segment is drawn inside a small glass electrode and its current response to light is recorded (e.g., Baylor, Nunn, & Schnapf, 1984; 1987). Spectral sensitivity is obtained by finding, as a function of wavelength, the radiance required to elicit a criterion photocurrent response (see Chapter 4).

Relevant human suction electrode data have so far been obtained only from M- and L-cones (Schnapf, Kraft, & Baylor, 1987; Kraft, Neitz, & Neitz, 1998). Recently, Kraft, Neitz, and Neitz (1998) made measurements in human L-cones known to contain photopigments that are determined either by L(ser^{180}) or by L(ala^{180}) genes, and Kraft (personal communication) has made measurements in human M-cones. Their M (white circles), L(ala^{180}) (gray circles), and L(ser^{180}) (black circles) data are shown in Fig. 2.14B along with the photopigment spectra from the Appendix (lines).

Given the differences between the two methods of obtaining the photopigment spectra and the fact that no attempt has been made to improve the agreement between the data sets, the agreement, particularly at shorter wavelengths, is good. At longer wavelengths, the L(ala^{180}) suction electrode data agree well with the corneally derived L-cone photopigment spectrum, but the M and L(ser^{180}) suction electrode data are slightly shallower than the corneally derived M- and L-cone spectra. Such differences have been encountered before, but they have been minimized by assuming unusually low photopigment densities for the central 2 deg of vision (Baylor, Nunn, & Schnapf, 1987) or by comparing them with corneally measured spectral sensitivities that are unusually shallow at longer wavelengths (Nunn, Schnapf, & Baylor, 1984; Baylor, Nunn, & Schnapf, 1987). The differences at long wavelengths may be due to waveguiding.

Waveguides. In both microspectrophotometry and suction electrode recordings the spectral sensitivity of the photopigment is measured transversely through the isolated cone outer segment, rather than axially along the outer segment, as in normal vision. The disadvantage of both methods is that the results must be adjusted before they can be used to predict corneal spectral sensitivities. Factors such as macular, lens, and photopigment density can be corrected for with some certainty, but other less well-known factors, such as other filters (e.g., Snodderly et al., 1984) or waveguide effects, cannot.

Light is transmitted along the photoreceptor in patterns called waveguide modal patterns (see Fig. 6 of Enoch, 1963). The fraction of the power of each modal pattern that is transmitted inside the photoreceptor to its power outside the photoreceptor decreases with the wavelength of the incident light, so that, in principle, the structure of the photoreceptor can change its spectral sensitivity (see, for example, Enoch, 1961; Enoch & Stiles, 1961; Snyder, 1975; Horowitz, 1981). It is difficult to know precisely how waveguide factors influence the spectral sensitivity for axially incident light in the human fovea because many of the relevant quantities, such as the refractive indices inside and outside the cone outer segment, and the models themselves are uncertain. Assuming values of 1 μm for the diameter of a human foveal cone outer segment (Polyak, 1941) and 1.39 and 1.35, respectively, for the refractive indices inside and outside the cone outer segment (Fig. 6.11 of Horowitz, 1981), standard formulas [Eqn. (7a) and Fig. 9 of Snyder, 1975] suggest a loss of spectral sensitivity for mode η_{11} (the most important mode for axially incident light) of about 0.2 $\log_{10}$ unit for red light relative to violet. Waveguide effects of this magnitude could account for the differences between the suction electrode data and the corneally measured photopigment spectra shown in Fig. 2.14B.

Two- and three-dimensional color matching and cone spaces

The representation of color matching functions and/or cone fundamentals in various two-dimensional (2D) or three-dimensional (3D) spaces can be a vital aid to interpreting and calculating color mixtures, complementary wavelengths, dichromatic confusion colors, chromatic adaptation and discrimination data, and even the behavior of postreceptoral mechanisms (see Chapters 1 and 18).

Color matching data or cone spectral sensitivities are often simplified by plotting them in relative units called chromaticity coordinates. Chromaticity coordinates ($r(\lambda)$, $g(\lambda)$, and $b(\lambda)$) are related to the CMFs ($\bar{r}(\lambda)$, $\bar{g}(\lambda)$, and $\bar{b}(\lambda)$) as follows:

$$r(\lambda) = \frac{\bar{r}(\lambda)}{\bar{r}(\lambda) + \bar{g}(\lambda) + \bar{b}(\lambda)}, \tag{13}$$

$$g(\lambda) = \frac{\bar{g}(\lambda)}{\bar{r}(\lambda) + \bar{g}(\lambda) + \bar{b}(\lambda)}, \text{ and}$$

$$b(\lambda) = \frac{\bar{b}(\lambda)}{\bar{r}(\lambda) + \bar{g}(\lambda) + \bar{b}(\lambda)}.$$

Given that $r(\lambda) + g(\lambda) + b(\lambda) = 1$, only $r(\lambda)$ and $g(\lambda)$ are typically plotted, since $b(\lambda)$ is $1 - (r(\lambda) + g(\lambda))$. [Likewise, $l(\lambda)$, $m(\lambda)$, and $s(\lambda)$ are the cone chromaticity coordinates corresponding to the cone fundamentals $\bar{l}(\lambda)$, $\bar{m}(\lambda)$ and $\bar{s}(\lambda)$.]

RGB (or XYZ) color spaces. Figure 2.15A shows the chromaticity coordinates (continuous line) of the locus of monochromatic spectral lights (or "spectrum locus") in the Stiles and Burch (1955) 2-deg $r(\lambda)$, $g(\lambda)$ chromaticity space. Selected wavelengths are shown as open circles. The Stiles and Burch (1955) based 2-deg cone fundamentals of Stockman and Sharpe (2000a), the derivation of which was discussed above, are plotted in terms of $r(\lambda)$ and $g(\lambda)$ as the dotted diamond (L), circle (M), and square (S).

Although chromaticity coordinates are a convenient way of plotting spectral distributions and predicting color mixtures, they inevitably reduce the available information by projecting the three-dimensional color space onto the two-dimensional plane: $\bar{r}(\lambda) + \bar{g}(\lambda) + \bar{b}(\lambda) = 1$. Figure 2.15B shows the Stiles and Burch (1955) $\bar{r}(\lambda)$, $\bar{g}(\lambda)$, and $\bar{b}(\lambda)$ color matching space and the L- (solid line), M- (long dashed line), and S- (short dashed line) cone vectors. The equal-energy spectrum locus is shown by the solid line, and selected wavelengths are shown by the filled circles. In three dimensions, the relationship between the cone fundamentals and the spectrum locus can be seen more clearly.

LMS color space. Color spaces are much more straightforward and intuitive when they are defined by the cone fundamentals and represent cone stimulation.

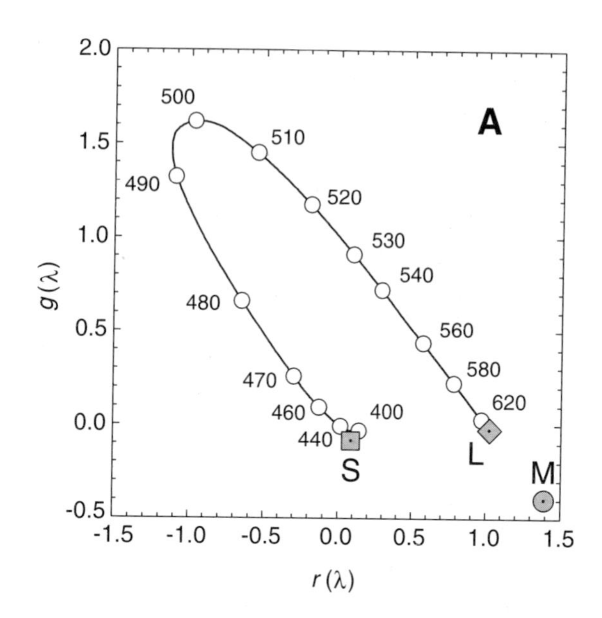

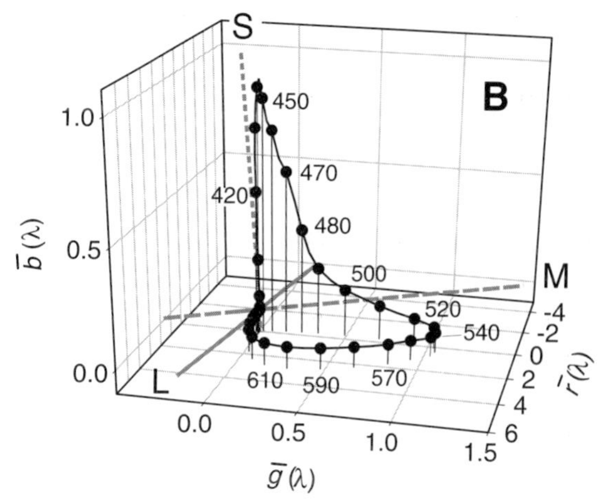

Figure 2.15: Color matching functions and chromaticity coordinates. (A) Spectrum locus (continuous line) and selected wavelengths (open circles) plotted in the Stiles and Burch (1955) 2-deg $r(\lambda)$, $g(\lambda)$ chromaticity space, and the projection of the 2-deg L- (dotted diamond), M- (dotted circle), and S- (dotted square) cone fundamentals of Stockman and Sharpe (2000a). (B) Spectrum locus (continuous line) and selected wavelengths (filled circles) plotted in the Stiles and Burch (1955) 2-deg $\bar{r}(\lambda)$, $\bar{g}(\lambda)$ and $\bar{s}(\lambda)$ space, and the L- (solid gray line), M- (long-dashed gray line), and S- (short-dashed gray line) cone vectors.

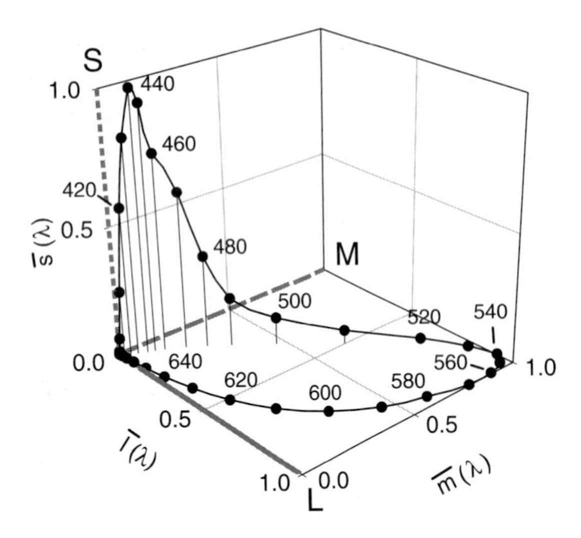

Figure 2.16: Cone fundamentals. Spectrum locus (continuous line) and selected wavelengths (filled circles) plotted in the Stockman and Sharpe (2000a) 2-deg $\bar{l}(\lambda)$, $\bar{m}(\lambda)$, and $\bar{s}(\lambda)$ cone excitation space, and the L- (solid gray line), M- (long-dashed gray line), and S- (short-dashed gray line) cone vectors.

Figure 2.16 shows the spectrum locus plotted in a three-dimensional 2-deg $\bar{l}(\lambda)$, $\bar{m}(\lambda)$, and $\bar{s}(\lambda)$ cone space.

The cone contrast space is a version of the cone fundamental space, in which the cone excitations produced by a stimulus are scaled separately for each cone type, according to Weber's law ($\Delta S/S$, $\Delta M/M$, and $\Delta L/L$, where ΔS, ΔM, and ΔL are the differential cone excitations produced by the stimulus, and S, M, and L are the unchanging cone excitations produced by, for example, a background). This space is useful for understanding postreceptoral mechanisms; it is discussed further in Chapter 18.

Equal-luminance cone excitation space. An oft referred to projection of the LMS cone space is the MacLeod–Boynton equal-luminance plane (Luther, 1927; MacLeod & Boynton, 1979), which is shown in Fig. 18.1B (see Chapter 18). Its popularity rests, in part, on current models about postreceptoral organization and, in particular, on the theory that only L- and M-cones contribute additively to the luminance channel [see above, Luminosity channel, $V(\lambda)$].

The MacLeod–Boynton chromaticity coordinates are defined as:

$$(14) \quad r_{MB}(\lambda) = \frac{\bar{l}(\lambda)}{V(\lambda)},$$

$$g_{MB}(\lambda) = \frac{\bar{m}(\lambda)}{V(\lambda)}, \text{ and}$$

$$b_{MB}(\lambda) = \frac{\bar{s}(\lambda)}{V(\lambda)},$$

where

$$V(\lambda) = \bar{l}(\lambda) + \bar{m}(\lambda).$$

The MacLeod–Boynton coordinates can be calculated from the cone fundamentals and $V^*(\lambda)$, which are tabulated in the Appendix, or they can be obtained from our Web sites (see Appendix).

Other variations of this space have been proposed (Krauskopf, Williams, & Heeley, 1982; Derrington, Krauskopf, & Lennie, 1984). See also the Appendix by Brainard in Kaiser and Boynton (1996).

Concluding remarks

On the basis of extensive measurements in single-gene protanopes and deuteranopes, in blue-cone monochromats and in normals (Sharpe et al., 1998; Stockman & Sharpe, 2000a; Stockman, Sharpe, & Fach, 1999) we present the new L-, M-, and S-cone fundamentals shown in Fig. 2.12 and tabulated in the Appendix. We believe that these represent an improvement to the Smith and Pokorny (1975) fundamentals based on the Judd–Vos modified CIE 2-deg CMFs and a refinement of the Stockman, MacLeod, and Johnson (1993) fundamentals based on the CIE 10-deg CMFs corrected to 2 deg.

In the Appendix we present a consistent set of functions: the three cone fundamentals $\bar{l}(\lambda)$ (L-cone), $\bar{m}(\lambda)$ (M-cone), and $\bar{s}(\lambda)$ (S-cone); the luminosity function, $V^*(\lambda)$; the lens density spectrum, $d_{lens}(\lambda)$; the macular density spectrum, $d_{mac}(\lambda)$; and the three photopigment optical densities, $\bar{l}_{OD}(\lambda)$, $\bar{m}_{OD}(\lambda)$, and $\bar{s}_{OD}(\lambda)$. These together can be used to define normal human color vision.

Acknowledgments

This work was supported by the National Institutes of Health grant EY 10206 awarded to AS, and by the Deutsche Forschungsgemeinschaft (Bonn) under grants SFB 340 Tp A6 and Sh23/5-1 and a Hermann- und Lilly-Schilling-Professor awarded to LTS. We thank Timothy Kraft for providing the suction electrode data, and Rhea Eskew, Jan Kremers, Anne Kurtenbach, Donald MacLeod, and Dan Plummer for comments. Rhea Eskew contributed substantially to the section on luminance. We also thank Sabine Apitz for help and encouragement.

Appendix

Table 2.1 (see next page) contains the proposed 2-deg cone spectral sensitivities: log $L(\lambda)$ or log $\bar{l}(\lambda)$, log $M(\lambda)$ or log $\bar{m}(\lambda)$, and log $S(\lambda)$ or log $\bar{s}(\lambda)$; luminosity function: log $V^*(\lambda)$; photopigment optical densities: log $\bar{l}_{OD}(\lambda)$, log $\bar{m}_{OD}(\lambda)$, and log $\bar{s}_{OD}(\lambda)$; and standard lens and macular densities. All cone spectral sensitivities and the luminosity function are in quantal units and are scaled to (interpolated) unity peak. To convert to energy units, add log(λ) and renormalize. The lens and macular densities are independent of the units used.

The cone fundamentals were calculated using the Stiles and Burch (1959) 10-deg CMFs with $\bar{s}_G/\bar{s}_B = 0.010600$ for S, $\bar{m}_R/\bar{m}_B = 0.168926$ and $\bar{m}_G/\bar{m}_B = 8.265895$ for M, and $\bar{l}_R/\bar{l}_B = 2.846201$ and $\bar{l}_G/\bar{l}_B = 11.092490$ for L. For further details about the long-wavelength S-cone extension after 520 nm, see Stockman, Sharpe, and Fach (1999), who were unable to measure S-cone spectral sensitivity data after 615 nm [after which $S(\lambda)$ is so small that it can reasonably, for most purposes, be set to zero]. The photopigment optical density spectra were calculated using Eqns. (9) and (10) assuming peak photopigment optical densities of 0.40 for S, and 0.50 for L, M, and the tabulated macular and lens densities.

The Stiles and Burch 10-deg CMFs used to calculate the cone fundamentals are from Table I (5.5.4) of Wyszecki and Stiles (1982a), in which they are tabulated in wavelength steps of 5 nm, and from Tables 7 and 8 of Stiles and Burch (1959), in which they are tabulated in wavenumber steps of 250 or 500 cm^{-1}. At shorter wavelengths, we used the CMFs from Table I (5.5.4) (Wyszecki & Stiles, 1982a). Those CMFs, however, are uncorrected for rod intrusion and are tabulated only to four decimal places, which is too imprecise to define cone sensitivities at longer wavelengths. At longer wavelengths, therefore, we have corrected the original CMFs (Table 7 of Stiles & Burch, 1959) for rod intrusion (according to Table 8; Stiles & Burch, 1959) and reinterpolated them at 5-nm intervals. For further details, see Stockman and Sharpe (2000a).

The data contained in Table 2.1, and other information, are available on http://www-cvrl.ucsd.edu (USA) and on http://www.eye.medizin.uni-tuebingen.de/cvrl (Germany).

nm (λ)	log $L(\lambda)$ log $\bar{l}$ (λ)	log $M(\lambda)$ log $\bar{m}$ (λ)	log $S(\lambda)$ log $\bar{s}$ (λ)	log $V^*(\lambda)$	log $\bar{l}_{OD}$ (λ)	log $\bar{m}_{OD}$ (λ)	log $\bar{s}_{OD}$ (λ)	lens $d_{lens}(\lambda)$	macular $d_{mac}(\lambda)$
390	-3.2186	-3.2907	-1.9642	-3.2335	-0.9338	-1.0479	-0.1336	2.5122	0.0453
395	-2.8202	-2.8809	-1.5726	-2.8309	-0.8948	-0.9974	-0.0906	2.1306	0.0649
400	-2.4660	-2.5120	-1.2020	-2.4712	-0.8835	-0.9707	-0.0498	1.7649	0.0868
405	-2.1688	-2.2013	-0.8726	-2.1690	-0.9016	-0.9742	-0.0257	1.4257	0.1120
410	-1.9178	-1.9345	-0.5986	-1.9119	-0.9154	-0.9711	-0.0092	1.1374	0.1365
415	-1.7371	-1.7218	-0.3899	-1.7183	-0.9408	-0.9623	-0.0023	0.9063	0.1631
420	-1.6029	-1.5534	-0.2411	-1.5699	-0.9549	-0.9398	0.0000	0.7240	0.1981
425	-1.5136	-1.4234	-0.1526	-1.4627	-0.9576	-0.8990	-0.0054	0.5957	0.2345
430	-1.4290	-1.3033	-0.0821	-1.3617	-0.9536	-0.8564	-0.0222	0.4876	0.2618
435	-1.3513	-1.1899	-0.0356	-1.2668	-0.9390	-0.8027	-0.0499	0.4081	0.2772
440	-1.2842	-1.0980	-0.0004	-1.1874	-0.9267	-0.7627	-0.0810	0.3413	0.2884
445	-1.2414	-1.0342	-0.0051	-1.1338	-0.9041	-0.7159	-0.1199	0.3000	0.3080
450	-1.2010	-0.9794	-0.0260	-1.0859	-0.8734	-0.6675	-0.1665	0.2629	0.3332
455	-1.1606	-0.9319	-0.0763	-1.0417	-0.8335	-0.6174	-0.2395	0.2438	0.3486
460	-1.0974	-0.8632	-0.1199	-0.9756	-0.7801	-0.5543	-0.3143	0.2279	0.3500
465	-1.0062	-0.7734	-0.1521	-0.8852	-0.7211	-0.4924	-0.4008	0.2131	0.3269
470	-0.9200	-0.6928	-0.2145	-0.8019	-0.6643	-0.4374	-0.5165	0.2046	0.2996
475	-0.8475	-0.6300	-0.3165	-0.7346	-0.6122	-0.3925	-0.6623	0.1929	0.2842
480	-0.7803	-0.5747	-0.4426	-0.6736	-0.5515	-0.3405	-0.8161	0.1834	0.2786
485	-0.7166	-0.5234	-0.5756	-0.6163	-0.4871	-0.2850	-0.9679	0.1749	0.2772
490	-0.6535	-0.4738	-0.7169	-0.5600	-0.4289	-0.2378	-1.1316	0.1675	0.2688
495	-0.5730	-0.4078	-0.8418	-0.4867	-0.3618	-0.1821	-1.2887	0.1601	0.2485
500	-0.4837	-0.3337	-0.9623	-0.4048	-0.3040	-0.1384	-1.4581	0.1537	0.2093
505	-0.3929	-0.2569	-1.1071	-0.3208	-0.2499	-0.0980	-1.6570	0.1463	0.1652
510	-0.3061	-0.1843	-1.2762	-0.2406	-0.2007	-0.0644	-1.8804	0.1378	0.1211
515	-0.2279	-0.1209	-1.4330	-0.1693	-0.1558	-0.0383	-2.0865	0.1293	0.0812
520	-0.1633	-0.0699	-1.6033	-0.1109	-0.1093	-0.0095	-2.2925	0.1230	0.0525
525	-0.1178	-0.0389	-1.7853	-0.0719	-0.0771	0.0000	-2.5009	0.1166	0.0329
530	-0.0830	-0.0191	-1.9766	-0.0438	-0.0550	-0.0037	-2.7142	0.1102	0.0175
535	-0.0571	-0.0080	-2.1729	-0.0243	-0.0332	-0.0082	-2.9241	0.1049	0.0093
540	-0.0330	-0.0004	-2.3785	-0.0071	-0.0095	-0.0146	-3.1409	0.0986	0.0046
545	-0.0187	-0.0035	-2.5882	0.0000	0.0000	-0.0370	-3.3599	0.0922	0.0017
550	-0.0128	-0.0163	-2.8010	-0.0016	-0.0040	-0.0731	-3.5809	0.0859	0.0000
555	-0.0050	-0.0295	-3.0168	-0.0021	-0.0014	-0.1055	-3.8030	0.0795	0.0000

Table 2.1: Proposed 2-deg cone spectral sensitivities (continued on next page).

nm (λ)	log $L(\lambda)$ / log $\bar{l}(\lambda)$	log $M(\lambda)$ / log $\bar{m}(\lambda)$	log $S(\lambda)$ / log $\bar{s}(\lambda)$	log $V^*(\lambda)$	log $\bar{l}_{OD}(\lambda)$	log $\bar{m}_{OD}(\lambda)$	log $\bar{s}_{OD}(\lambda)$	lens $d_{lens}(\lambda)$	macular $d_{mac}(\lambda)$
560	-0.0019	-0.0514	-3.2316	-0.0085	-0.0055	-0.1485	-4.0231	0.0742	0.0000
565	-0.0001	-0.0769	-3.4458	-0.0166	-0.0138	-0.1966	-4.2437	0.0678	0.0000
570	-0.0015	-0.1114	-3.6586	-0.0296	-0.0280	-0.2554	-4.4629	0.0615	0.0000
575	-0.0086	-0.1562	-3.8692	-0.0492	-0.0519	-0.3251	-4.6798	0.0551	0.0000
580	-0.0225	-0.2143	-4.0769	-0.0769	-0.0863	-0.4084	-4.8939	0.0488	0.0000
585	-0.0325	-0.2752	-4.2810	-0.1015	-0.1113	-0.4903	-5.1033	0.0435	0.0000
590	-0.0491	-0.3443	-4.4811	-0.1320	-0.1460	-0.5788	-5.3087	0.0381	0.0000
595	-0.0727	-0.4263	-4.6766	-0.1696	-0.1898	-0.6791	-5.5095	0.0329	0.0000
600	-0.1026	-0.5198	-4.8673	-0.2132	-0.2378	-0.7868	-5.7034	0.0297	0.0000
605	-0.1380	-0.6247	-5.0529	-0.2619	-0.2929	-0.9054	-5.8932	0.0254	0.0000
610	-0.1823	-0.7389	-5.2331	-0.3179	-0.3561	-1.0305	-6.0765	0.0223	0.0000
615	-0.2346	-0.8610	-5.4077	-0.3804	-0.4268	-1.1617	-6.2544	0.0191	0.0000
620	-0.2943	-0.9915		-0.4490	-0.5026	-1.2989		0.0170	0.0000
625	-0.3603	-1.1294		-0.5229	-0.5833	-1.4425		0.0148	0.0000
630	-0.4421	-1.2721		-0.6106	-0.6809	-1.5909		0.0117	0.0000
635	-0.5327	-1.4205		-0.7060	-0.7854	-1.7444		0.0085	0.0000
640	-0.6273	-1.5748		-0.8051	-0.8918	-1.9033		0.0053	0.0000
645	-0.7262	-1.7365		-0.9081	-0.9986	-2.0670		0.0042	0.0000
650	-0.8408	-1.8924		-1.0252	-1.1204	-2.2246		0.0032	0.0000
655	-0.9658	-2.0524		-1.1520	-1.2523	-2.3872		0.0011	0.0000
660	-1.0965	-2.2196		-1.2845	-1.3878	-2.5558		0.0000	0.0000
665	-1.2323	-2.3853		-1.4217	-1.5264	-2.7217		0.0000	0.0000
670	-1.3734	-2.5477		-1.5638	-1.6696	-2.8842		0.0000	0.0000
675	-1.5201	-2.7075		-1.7110	-1.8178	-3.0442		0.0000	0.0000
680	-1.6729	-2.8700		-1.8642	-1.9717	-3.2067		0.0000	0.0000
685	-1.8320	-3.0362		-2.0236	-2.1317	-3.3730		0.0000	0.0000
690	-1.9985	-3.2109		-2.1904	-2.2988	-3.5477		0.0000	0.0000
695	-2.1590	-3.3745		-2.3510	-2.4596	-3.7113		0.0000	0.0000
700	-2.3194	-3.5360		-2.5115	-2.6203	-3.8728		0.0000	0.0000
705	-2.4813	-3.6978		-2.6734	-2.7824	-4.0347		0.0000	0.0000
710	-2.6486	-3.8678		-2.8408	-2.9498	-4.2047		0.0000	0.0000
715	-2.8164	-4.0373		-3.0086	-3.1177	-4.3742		0.0000	0.0000
720	-2.9801	-4.1986		-3.1723	-3.2815	-4.5355		0.0000	0.0000
725	-3.1433	-4.3582		-3.3353	-3.4447	-4.6951		0.0000	0.0000
730	-3.3032	-4.5112		-3.4950	-3.6047	-4.8481		0.0000	0.0000

Table 2.1 (continued).

3

Photopigments and the biophysics of transduction in cone photoreceptors

Trevor Lamb

Photopigments: spectral shape

The properties of color vision are determined to a considerable extent by the absorbance spectra of the retinal cones, and it is not only the *positions*, but also the *shapes*, of the spectral curves that are important. In the first section of this chapter (and also in Chapters 2 and 4), the shape of the absorbance spectra of visual pigments will be analyzed, while Chapter 2 investigated factors that determine the positions of these peaks.

In 1953, Dartnall noticed that the absorbance spectra of different visual pigments had fundamentally the same shape, even when their wavelengths of maximum absorbance (λ_{max}) differed. He further reported that the different spectra appeared most similar when plotted in terms of frequency displacement from the frequency of maximum absorbance, and he introduced a "nomogram" of absorbance plotted against $(1/\lambda - 1/\lambda_{max})$. Subsequent work, however, has shown that the Dartnall nomogram does not provide a particularly good description for pigments with widely different λ_{max} values, and various alternative schemes have been proposed (e.g., Ebrey & Honig, 1977; Barlow, 1982; Mansfield, 1985). Of these, the last is particularly interesting, as Mansfield reported that simply by normalizing the wavelength (or frequency) axis one could bring into coincidence the four different absorbance spectra for monkey photoreceptors (the rods and the three classes of cones).

Frequency scaling. Recently I extended Mansfield's analysis, by examining the spectral curves of photoreceptors from a number of different sources. On normalization of the frequency axis, the spectral sensitivity curves for a range of vertebrate photoreceptors were found to be indistinguishable from one another. Furthermore, it was possible to obtain a template curve that provided a good description of the common shape (Lamb, 1995).

The spectral curves that were analyzed (and that will be presented here) were taken from eight studies, published over a period of more than 60 years. The studies used a variety of methods, including psychophysical sensitivity measurements, single-cell electrophysiology, and pigment absorbance measurement. Where necessary, sensitivities were converted from energy units to quantal units, and corrections were made for preretinal filtering and for pigment self-screening; for details of these corrections, see Lamb (1995, p. 3085).

The classical study is that of Goodeve (1936), who measured the visibility of red stimuli presented to the fovea of a dark-adapted human observer (i.e., photopic sensitivity). Similar measurements by Griffin, Hubbard, and Wald (1947) are also included, as are exhaustive measurements of scotopic sensitivity made by Crawford (1949). Single-cell electrophysiological measurements on mammalian photoreceptors were not made until much more recently, and those included here are for monkey rods and cones (Baylor, Nunn, & Schnapf, 1984; 1987), squirrel cones (Kraft,

1988), and human rods (Kraft, Schneeweis, & Schnapf, 1993); these experiments are described in detail in Chapter 4. Over the years, numerous measurements of absorbance spectra have been made from pigments (both in vivo and extracted). However, a limitation with absorbance measurements is that they typically suffer from a substantial *absolute* level of noise, and this prevents the spectrum from being followed down more than about 1.5 log units below the maximum. The one spectrophotometric study selected for analysis here is that of Partridge and De Grip (1991) using extracted bovine rhodopsin.

The spectral sensitivities, *S*, from these studies are plotted in Fig. 3.1 against normalized frequency, $\nu/\nu_{max} = \lambda_{max}/\lambda$ on a logarithmic vertical scale. Note that in this graph the left side of the plot (low frequency) corresponds to the red end of the spectrum. Details are given in Table 3.1 of the studies from which these results have been taken, together with identification of the plotting symbols and the scaling factors (λ_{max}) used in the frequency normalization.

A striking feature in Fig. 3.1 is the close similarity in the shape of the spectral curves obtained from the different sources. In this plot, the points from different experiments are indistinguishable from each other over a range of at least 7 $\log_{10}$ units of sensitivity. Hence this figure extends Mansfield's finding of a common spectral shape down a further 5 $\log_{10}$ units in sensitivity than could be examined with his absorbance measurements. It also indicates that the common shape appears to be applicable to results obtained from single-cell electrophysiology, from spectrophotometry on extracted rhodopsin, and from psychophysical sensitivity.

Another clear result of Fig. 3.1 is that in the long-wavelength region the decline of log sensitivity for all the points is well fitted by a straight line. Such a form of decline was first reported for an individual spectrum by Goodeve (1936), and it has subsequently also been found to hold for the individual raw spectra of primate photoreceptors (Baylor, Nunn, & Schnapf, 1984, 1987). The results of Fig. 3.1 extend those observations by showing that different pigments all exhibit the *same* slope when plotted against normalized frequency. The magnitude of the limiting slope, shown by the broken line, is 70 $\log_e$ units (or 30.4 $\log_{10}$ units) per unit of normalized frequency.

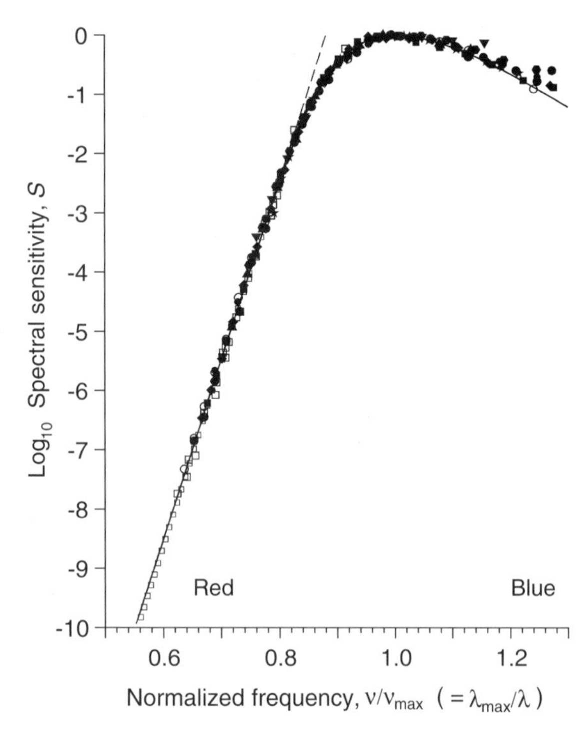

Figure 3.1: Spectral sensitivities on a logarithmic ordinate scale, plotted as a function of normalized wavelength. The symbols, identified in Table 3.1, have been taken from raw measurements of photoreceptor electrophysiology (monkey, squirrel, and human) and spectrophotometry (bovine rhodopsin), and by conversion from energy-based psychophysical sensitivity results (photopic and scotopic); for details, see Lamb (1995). The solid curve plots Eqn. (1) with the standard parameters given in the text. The broken straight line plots just the first term in Eqn. (1), i.e., a slope of 70 $\log_e$ units (or 30.4 $\log_{10}$ units) per unit of normalized frequency. Modified from Lamb (1995).

The equality of the limiting slopes in a plot against normalized frequency units implies that, when the results are plotted on a *raw* scale of absolute frequency, the limiting slope for each pigment will be directly proportional to λ_{max}. For example, in monkey red- or long-wave–sensitive cones (■) fitted with λ_{max} = 561 nm the limiting raw slope would be 30.4 × 561 nm = 17.1 $\log_{10}$ units μm, whereas in monkey

Symbol	λ_{max} (nm)	Measurement	Species	Method	Source of data
□	561	Photopic sensitivity	Human	Ψ	Goodeve (1936)
▫	561	Photopic sensitivity	Human	Ψ	Griffin, Hubbard, & Wald (1947)
○	496	Scotopic sensitivity	Human	Ψ	Crawford (1949)
■	561	Red-cone sensitivity	Monkey	Φ	Baylor, Nunn, & Schnapf (1987)
◆	533	Green-cone sensitivity	Monkey	Φ	Baylor, Nunn, & Schnapf (1987)
▲	431	Blue-cone sensitivity	Monkey	Φ	Baylor, Nunn, & Schnapf (1987)
⬡	523	Green-cone sensitivity	Squirrel	Φ	Kraft (1988)
▼	440	Blue-cone sensitivity	Squirrel	Φ	Kraft (1988)
●	496	Rod sensitivity	Monkey	Φ	Baylor, Nunn, & Schnapf (1984)
•	496	Rod sensitivity	Human	Φ	Kraft, Schneeweis, & Schnapf (1993)
✶	498	Rod absorbance	Bovine	Abs	Partridge & De Grip (1991)

Table 3.1: The open symbols were obtained from psychophysical (Ψ) experiments on dark-adapted human observers; see Lamb (1995) for details of the conversions applied to these psychophysical values. The filled symbols were obtained from electrophysiological (Φ) experiments on single photoreceptors from monkey, squirrel, or human retina; further details of these experiments are given in Chapter 4. The stars (✶) were obtained from absorbance measurements (Abs) on extracted bovine rhodopsin; see Fig. 3.2. The second column (λ_{max}) gives the wavelength scaling used in Figs. 3.1–3.3 to obtain a common shape.

green- or middle-wave–sensitive cones (◆) fitted with a λ_{max} = 533 nm the limiting slope would be 30.4 × 533 nm = 16.2 $\log_{10}$ units µm. As shown subsequently in Fig. 3.4, this finding provides a good description for the perceived "yellowing" of red light that Brindley (1955) reported to occur at very long wavelengths.

Template curve. It would be extremely useful to obtain a template curve describing the common spectral shape shown in Fig. 3.1. Ideally, such a curve should be based on firm theoretical principles, but as yet a comprehensive theory of light absorption by visual pigments does not appear to have been developed. In 1948, Stiles put forward a theory of absorption at long wavelengths that took into account the thermal energy of vibration of the pigment molecule and that accounted for a straight-line dependence of log sensitivity on frequency at long wavelengths. Subsequently, Lewis (1955) incorporated a description of the effects of multiple vibrational modes of the molecule that provided a reasonable fit from long wavelengths up to the peak. However, such models have not been able to account for the decline in sensitivity at short wavelengths. In the absence of a complete physical theory, it seemed that the logical starting point in finding a template was to describe the constant-slope region at long wavelengths and then try to obtain a simple equation describing the remainder of the spectral form.

To examine the behavior near the peak, it is convenient to plot the results on a linear ordinate scale. Of the various spectra analyzed here, the results that display the smallest fractional error (near the peak) are the absorbance measurements for bovine rhodopsin (✶, Partridge & De Grip, 1991). Those results are replotted on a linear ordinate scale in Fig. 3.2, and in these coordinates it is clearer that the spectrum exhibits the classical shape, with a relatively narrow half-width (full width at half-height ≈ 0.2 normalized frequency units).

A curve that provides the correct limiting behavior at long wavelengths (in the plot of Fig. 3.1), together with a reasonable fit near the peak (in the plot of Fig. 3.2), is

$$(1) \quad S(x) = \{\exp a(A - x) + \exp b(B - x) + \exp c(C - x) + D\}^{-1},$$

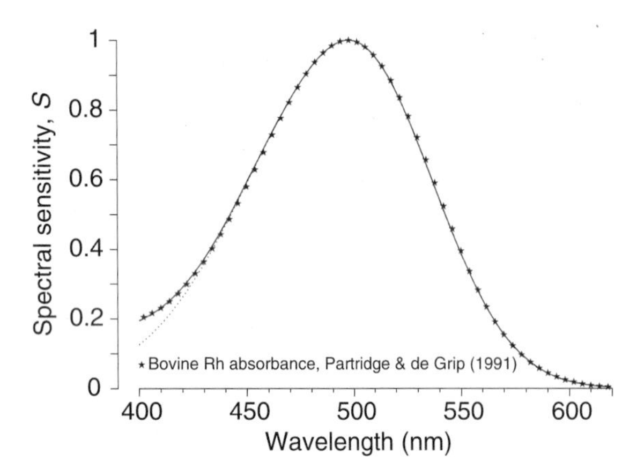

Figure 3.2: The measurements of absorbance for extracted bovine rhodopsin obtained by Partridge and De Grip (1991), shown by ✶ in Fig. 3.1, are replotted on a linear ordinate scale against wavelength. The dotted curve plots Eqn. (1) with $\lambda_{max} = 498$ nm. The solid curve additionally includes a representation of absorption by the β-band, by adding in a second term with the same form as Eqn. (1), but with $\lambda_{max} = 375$ nm and with an amplitude of 10% of the primary peak.

where $x = \nu/\nu_{max} = \lambda_{max}/\lambda$ is the normalized frequency. This equation is plotted as the solid curve in Fig. 3.1 and as the dotted curve in Fig. 3.2. Although it may appear that this equation has seven free parameters, two of these are redundant because of the normalization (i.e., the peak occurs at $x = 1$, with unit amplitude $S = 1$ and zero slope $dS/dx = 0$). Furthermore, the limiting slope as $x \rightarrow 0$ is firmly established as $a = 70$. So, in practice, four parameters remain to describe the *shape* of the curve. In all of the results described here, the parameters have been fixed at $a = 70$, $b = 28.5$, $c = -14.1$, and $A = 0.880$, $B = 0.924$, $C = 1.104$, $D = 0.655$.

It should be emphasized that Eqn. (1) represents an exercise in curve-fitting, since the equation has no known physical significance apart from the asymptotic slope at low frequencies. But it proved very difficult to find any other curve that provided a good fit both in the logarithmic plot of Fig. 3.1 as well as in the linear plot of Fig. 3.2. Put another way, the combination of sensitivity measurements (applicable out to the far red) with absorbance measurements (having high accuracy in the vicinity of the peak) provides a tough test for the fit of any candidate expression for the spectral shape.

Equation (1) is intended only as a description of the primary absorption band (the α-band) of the photopigment, and for this reason the fitted curve has deliberately been chosen to pass below the points at higher frequencies ($x > 1.15$) in Figs. 3.1 and 3.2; this corresponds to fitting only at wavelengths above 440 nm for a pigment with $\lambda_{max} = 500$ nm. Absorption by the subsidiary β-band (or *cis*-band) of retinal has been reported to exhibit a broadly similar shape to that of the primary band (Stavenga, Smits, & Hoenders, 1993), and so it would seem reasonable to introduce a second similar term to describe the absorbance at short wavelengths. Indeed, P. Walraven (personal communication) has recently done this. By adding in a second term of the same form as Eqn. (1), with an amplitude of about 10% of the primary peak but with $\lambda_{max} < 400$ nm, he has obtained a good fit to the human red- and green-cone spectra over the entire visible spectrum. The sum of two such terms is plotted as the solid curve in Fig. 3.2, and it provides a good fit.

Fit to the raw spectra. The collected spectra from the literature that were plotted on a normalized frequency axis in Fig. 3.1 are now replotted in Fig. 3.3 on a scale of raw wavelength. The seven sets of filled symbols are from electrophysiological experiments on photoreceptors, whereas the three sets of open symbols are from psychophysical experiments on human observers. The curves near the symbols represent the template curve of Eqn. (1), scaled to the appropriate wavelength of peak absorbance given in Table 3.1, and with an added β-band component, at 395 nm, of 10% of the height of the main band. In each case the curve provides a reasonable description of the data.

The three sets of symbols near the right-most curve apply to the primate red-cone system and are closely similar to each other. The corrected psychophysical measurements of human photopic sensitivity (□, ▫) from Goodeve (1936) and Griffin, Hubbard, and Wald (1947), respectively, are quite close to the electrophysiological spectra obtained for monkey cones (■) by Baylor, Nunn, and Schnapf (1987). The similarity is consistent with the idea that, in the range 650–900 nm,

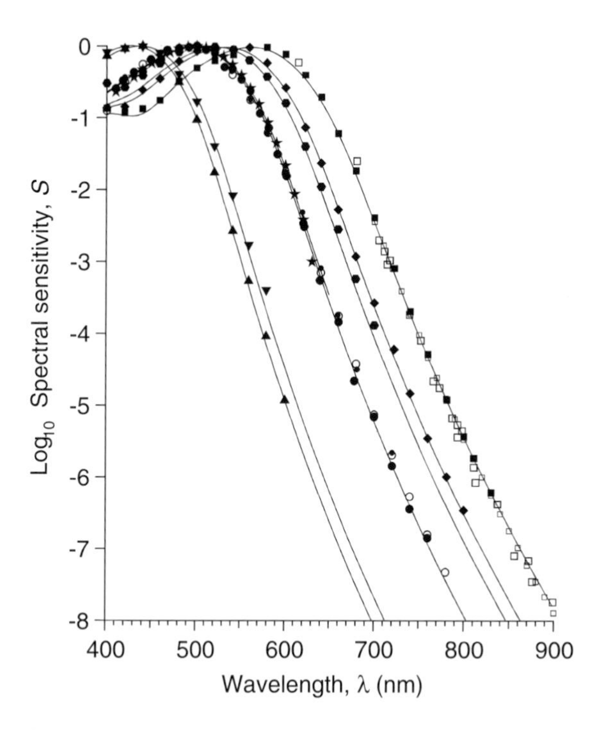

Figure 3.3: Individual sets of measurements that were presented in the normalized frequency coordinates of Fig. 3.1 are replotted on a raw wavelength scale. Solid curves plot the sum of two terms of the form of Eqn. (1), representing absorption by the α- and β-bands. The main term has unit amplitude and λ_{max} given by the values in Table 3.1; the secondary term has an amplitude of 0.1 and a λ_{max} of 395 nm.

human photopic sensitivity is determined exclusively by the responses of the red-cones.

A result that may be inconsistent with frequency scaling has been reported by Stockman, MacLeod, and Johnson (1993). In deriving the cone fundamentals from psychophysical color matching functions, they found that the red-cone spectrum was very slightly broader than the green-cone spectrum. Although this finding may represent a small departure from Mansfield scaling, it is conceivable that it might alternatively result from a slight difference in the waveguide properties between the inner segments of red- and green-cones, or perhaps from the expression in (some) red-cones of a second pigment with a slightly displaced λ_{max}. But in terms of the plots in Figs. 3.1 and 3.3, this difference is fairly small.

In the past it has been difficult to extract reliable measures of the spectral sensitivity of the blue- or short-wave–sensitive cones, but the very recent work of Stockman, Sharpe, and Fach (1999; see Chapter 2, Figs 2.1, 2.2, and 2.8), as well as McMahon and MacLeod (1998), indicates that the human blue-cone sensitivity is quite accurately described by Eqn. (1).

Perceived "yellowing" at long wavelengths. In addition to providing a visually acceptable fit to the spectra, the template in Eqn. (1) is able to account accurately for a rather obscure psychophysical effect that occurs at very long wavelengths. As the wavelength of monochromatic light is lengthened beyond 640 nm, the perceived color becomes more deeply red. However, for wavelengths beyond 700 nm (i.e., in the near-infrared), Brindley (1955) reported that the perceived color reverted from deep red towards yellow-orange – a phenomenon that may readily be checked by an observer with access to an infrared light source.

Beyond 600 nm the sensitivity of the blue-cones is so low (see Fig. 3.3) that our color vision is essentially dichromatic, and the perceived color will therefore be determined by the ratio of red-cone to green-cone sensitivity. This ratio, denoted *R/G*, is plotted in Fig. 3.4A as a function of wavelength. The symbols are the experimental recordings of Baylor, Nunn, and Schnapf (1987) from monkey cones, while the curve is the prediction of Eqn. (1), using the λ_{max} values of 561 and 533 nm given in Table 3.1 for the red- and green-cones. In both cases the ratio *R/G* is unity near 550 nm, and it progressively increases with wavelength until reaching a maximum near 700 nm, after which it declines.

On the basis that, for $\lambda > 600$ nm, the perceived color is determined by the ratio of red-cone to green-cone stimulation, then any two wavelengths that elicit the same ratio should appear the same color (although it will of course be necessary to make the light at longer wavelength more intense for it to appear equally bright). For example, Fig. 3.4A predicts that a wavelength of 800 nm (which elicits an *R/G* ratio of about 12) should appear the same color as a wavelength of 663 nm. The form of the theoretical curve in Fig. 3.4A has been replotted parametrically in Fig.

3.4B, to show more clearly the expected equivalence of different wavelengths. These predicted equalities, including the reversal just beyond 700 nm, are in close agreement with the experimental findings of Brindley (1955), giving further credence to the accuracy of Eqn. (1).

Biophysics of transduction in cone photoreceptors

The transduction of light into a neural signal in photoreceptors is now well understood at a molecular level. We begin this section by describing the molecular nature of the G-protein cascade of reactions underlying transduction. Then the reactions will be modeled, by using both stochastic and analytical techniques, the onset phase of the cone's electrical response will be predicted, and the predictions compared with experimental measurements. In Chapter 4 the later phase of the cone's electrical response will additionally be described, and the properties of light adaptation will be investigated. For other reviews of phototransduction, the reader is referred to McNaughton (1990), Pugh and Lamb (1993), Yau (1994), Baylor (1996), Helmreich and Hofmann (1996), and Koutalos and Yau (1996).

The G-protein cascade of phototransduction. The G-protein cascade of reactions underlying transduction in rod and cone photoreceptors is illustrated schematically in Fig. 3.5. The membrane of the outer segment contains the three principal proteins of the cascade: the photopigment receptor molecule, R (rhodopsin, or porphyropsin), a G-protein, G (transducin), and an effector protein, E. Absorption of a photon of light converts the photopigment to an active form, R*, that has the ability to activate the G-protein from G to G*. In turn, G* is able to activate the third protein from E to E*. This kind of G-protein cascade is common to numerous cellular signalling systems, with the classical example being the β-adrenergic receptor system where the effector protein is adenylyl cyclase. In the photoreceptor, in contrast, the effector protein is the cGMP-phosphodiesterase (PDE). When activated to E* (=PDE*), this protein hydrolyzes cGMP in the cytoplasm, causing closure of cGMP-gated ion channels in the plasma membrane and thereby eliciting the reduction in circulating electrical current that comprises the cell's electrical response to light.

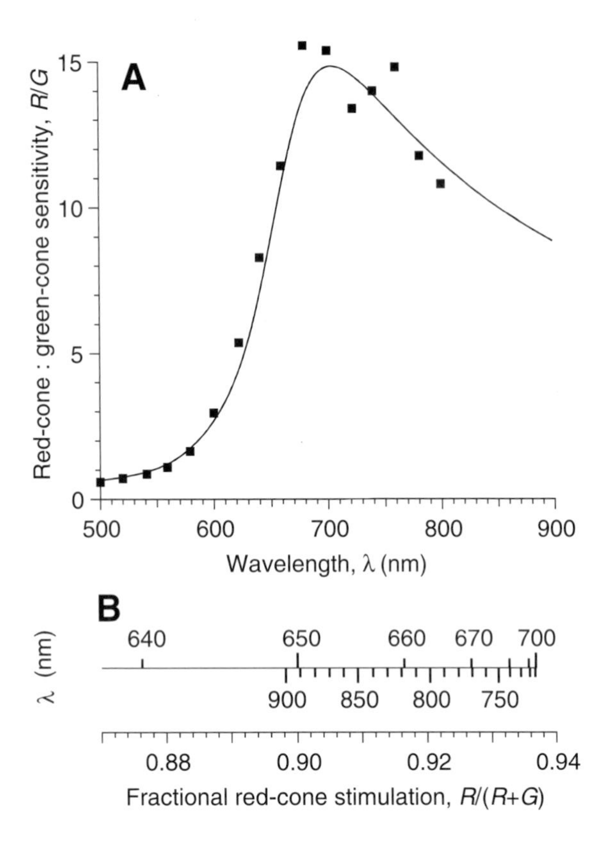

Figure 3.4: (A) Ratio of red-cone to green-cone sensitivity, R/G, as a function of wavelength. The symbols are from the electrophysiological recordings of Baylor, Nunn, and Schnapf (1987), Table 3.1, and the curve is predicted by Eqn. (1) with λ_{max} equal to 561 and 533 nm for the red- and green-cones, respectively. (B) An alternative parametric plot of the predictions of Eqn. (1). The fractional red-cone stimulation, $R/(R+G)$, predicted by Eqn. (1) with the above values of λ_{max} is shown as a function of the wavelength of stimulation. As may be seen in panel A, reversal occurs at a wavelength of about 704 nm. Pairs of wavelengths that elicit an equal R/G ratio in panel A lie opposite each other in panel B. Reproduced with permission from Lamb (1995).

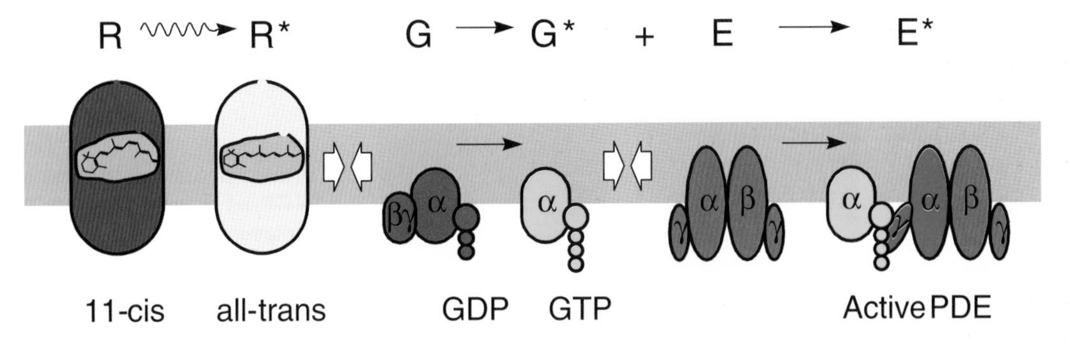

Figure 3.5: Schematic diagram of activation in the G-protein cascade of phototransduction. The receptor protein, rhodopsin or porphyropsin (R; colored gray), is an integral membrane protein with characteristic 7-*transmembrane* segment structure; it is activated to R* (metarhodopsin II; white) within a few milliseconds of photon absorption. The G-protein transducin (G; green) is a heterotrimer, which in the quiescent state has a molecule of GDP bound to its α-subunit. Upon diffusional contact between R* and G (open arrows), the G-protein is activated to G* (yellow) in a process involving the exchange of the bound GDP for a GTP; this process is catalytic since the R* is released unaltered. The PDE effector protein E (blue) is tetrameric, comprising two closely similar hydrolytic α- and β-subunits, each with an inhibitory γ-subunit. Upon contact between G* and E, the G* binds to a γ-subunit and relieves its inhibitory influence, thereby activating E to E* (red). Because of its double-unit structure, the PDE can bind two G*s to form E**, but in the analysis presented here an E** is considered to be identical to two separate E*s. The activated PDE, E*, hydrolyzes cGMP in the cytoplasm, reducing the concentration of cGMP, and thereby causing closure of cGMP-gated channels in the plasma membrane. Reproduced with permission from Lamb (1996).

Activation of the cascade. How does this G-protein cascade work? What is the gain, and what are the kinetics, of activation of the proteins G* and E* in response to a single photon hit?

In different G-protein systems the mechanism of activation appears to be closely similar. The proteins of the cascade are membrane-associated, and biological membranes are quite fluid. In fact, the photoreceptor membrane is extremely fluid, with a viscosity similar to that of olive oil. As a result, the protein molecules exhibit Brownian motion, jostling around randomly in the plane of the membrane, and thereby undergo lateral diffusion (indicated by the open arrows in Fig. 3.5). On contact between the R* and a G, the G-protein can be activated to G* through a ubiquitous mechanism involving the release of the bound GDP and the uptake of a molecule of GTP from the cytoplasm. Having activated one molecule of G to G*, the R* is released unaltered and is free to continue diffusing laterally in the membrane. In this way a single R* is able sequentially (or catalytically) to activate numerous molecules of G to G*, thereby providing the first stage of amplification in the cascade. Each activated G* similarly diffuses laterally, until it encounters an effector E, whereupon the two molecules bind together to form E*. However, no amplification accompanies this second step, as it involves a 1:1 binding of G* to E.

Hence the first step (R* $\rightarrow$ G*) is catalytic and provides high gain, whereas the second step (G*$\rightarrow$ E*) simply involves binding and provides no further gain. From this perspective, G-proteins can be regarded as universal "facilitators" that permit an activated receptor protein both to achieve high amplification and to couple to the appropriate effector protein (which will be a molecule specialized for the particular signalling system).

A second stage of amplification is provided by the activated effector E*. In the photoreceptor, as in most G-protein cascades, the effector is an enzyme: In this case, E* is the phosphodiesterase PDE*, which catalyzes the hydrolysis of cGMP in the cytoplasm. Phototransduction therefore comprises two cascaded

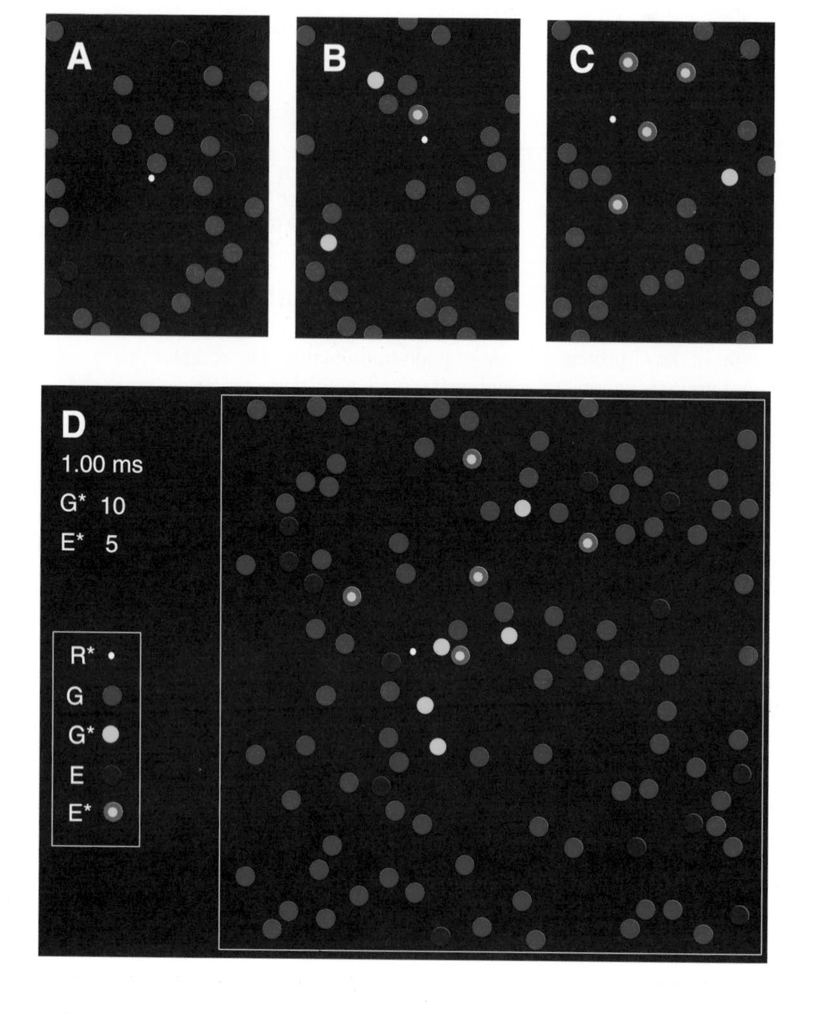

Figure 3.6: Simulation of the activation of G* and E* in the G-protein cascade of phototransduction. Each panel depicts an area of membrane at successive times after the activation of a single molecule of R*: (A) time zero; (B) 0.2 ms; (C) 0.4 ms; (D) 1.0 ms. The molecular species are identified in the inset to panel (D). The single white R* was initially placed at the center of the region, and the green Gs and blue Es were randomly distributed, with mean concentrations 2500 and 500 μm^{-2}. At time zero, the molecules began undergoing lateral diffusion, simulated by Monte-Carlo methods, using the estimated lateral diffusion coefficients for R*, G, G*, and E of 0.7, 1.2, 1.5, and 0.8 $\mu m^2\ s^{-1}$, respectively; for details see Lamb (1994). Scale: the region of simulation was approximately 200 nm square, with nonabsorbing boundaries, and the diameters of the R*, G, and E molecules were 3, 6, and 7 nm. Reproduced with permission from Lamb (1996). The computer program "WALK" that performed these simulations is available on the Internet; see text.

stages of enzymatic gain, by R* and by E*, permitting the attainment of very high overall amplification.

The subsequent steps involved in *inactivation* of the response are less well understood than the activation steps outlined above, and they will not be addressed here; for reviews, see for example Koutalos and Yau (1996). The following analysis will instead be restricted to times sufficiently early in the response that the effects of inactivation and light adaptation have not yet set in. It therefore provides a description of only the *rising* phase of the electrical response.

Stochastic simulation of the reactions underlying activation. The diffusional interactions of the proteins have been simulated by using Monte-Carlo techniques (Lamb, 1994), an illustrative example of which is shown in Fig. 3.6. The molecular species are identified in the inset at the lower left (using the same color code as in Fig. 3.5), and the four panels depict areas of membrane at successive times after the activation of a single molecule of R*. Initially (panel A) the molecules were distributed randomly, with the G-protein (green) and effector protein (blue) present at their biochemically determined concentrations and with a single R* (white) at the center of the region.

For $t > 0$, the molecules underwent two-dimensional diffusion, according to the estimated lateral diffusion coefficients of the respective molecular species (see legend of Fig. 3.6). After 0.2 ms (panel B), the R* had contacted and activated three molecules of G-protein to the excited form, G* (yellow); of these three G*s, one had contacted and bound to an effector molecule to produce an E* (red). Subsequently, at 0.4 ms (panel C) five G*s had been activated, of which four had contacted effectors to produce E*s. Finally, after 1 ms, a total of ten G*s and five E*s had been generated (panel D, illustrating the whole area of the simulation).

Although one is able to see the stochastic nature of the activation reactions by inspection of the snapshots captured in the panels of Fig. 3.6, a much more intuitive appreciation can be gained by viewing a *dynamic* simulation, such as that provided by the computer program "WALK" that was used to generate this figure.[1]

The simulation in Fig. 3.6 suffers a significant limitation due to the small area of membrane modeled and the consequent small number of molecules present. At 1 ms, the activated molecules have already traversed a significant fraction of the area under consideration. To extend the simulation time to 100 ms, it is necessary to expand the area by more than 100-fold, to include in excess of 10,000 molecules. Because of the need to increase the area in line with the maximum time to be analyzed, the computation time is in practice found to increase in proportion to the square of the time simulated.

The simulated time course of activation of G* and E* is shown in Fig. 3.7 for a standard set of parameters for an amphibian rod outer segment (Lamb, 1994). The upper panel superimposes the traces obtained for 20 simulations and demonstrates the degree of variability obtained in successive trials due to the stochastic nature of the reactions. The lower panel plots the average behavior determined from 100 such simulations. For G*, the average simulated time course (solid trace) is very similar to the prediction (dashed trace) obtained from the macroscopic model that was developed by Lamb and Pugh (1992), based on analogy with the diffusion of heat in two dimensions.

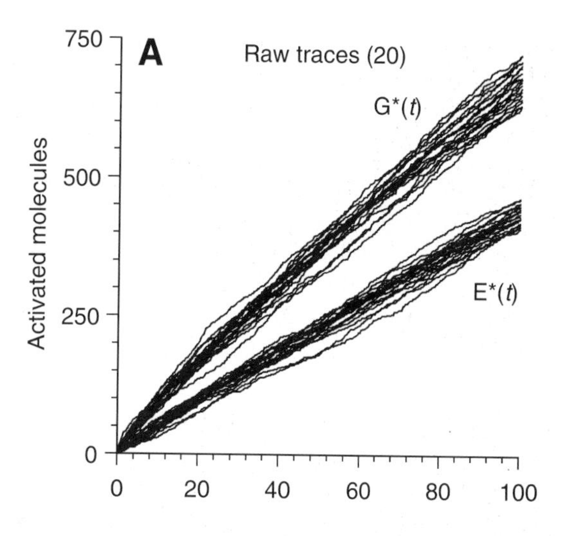

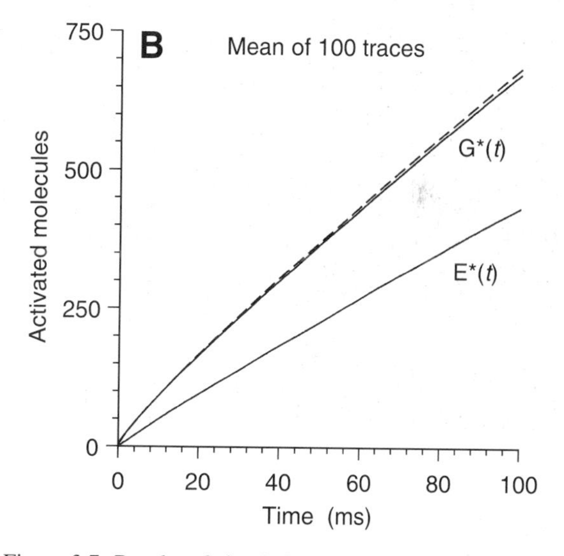

Figure 3.7: Results of simulations obtained using the same physical parameters as in Fig. 3.6 but with a larger area of disc membrane. (A) Raw traces. The individual responses for G* and E* from the first 20 simulations are plotted. (B) Mean responses. The averages from 100 simulations are plotted as the solid traces. The broken trace plots the theoretical prediction for G*, from the analytical solution to a corresponding problem for the diffusion of heat in two dimensions; no analytical solution for E* is yet available. Reproduced with permission from Lamb (1994), where further details are given.

[1]A Windows ('95 or NT) version of this program is available on the Internet at: http://www.physiol.cam.ac.uk/staff/Lamb/Walk.html.

Rising phase of the electrical response. The mathematics of coupling from G* to E* is complicated, and no analytical model is yet available for describing the time course of activation of E*. However, an important characteristic of the stochastic simulations (Fig. 3.7) is that the number of activated molecules of E* is found to rise approximately linearly with time after photon arrival. This finding permits the derivation of a simple expression for the initial phase of the electrical response to flashes (Lamb & Pugh, 1992).

Figure 3.8 outlines the approach. In response to the arrival of a single photon at time zero, a single molecule of R* is produced with little delay. Since the shut-off reactions are being ignored, this R* remains present, and while it is present the quantities of G* and E* approximately "ramp" with time. The action of E* (= PDE*) is to hydrolyze cGMP in the cell's cytoplasm, and so, as the quantity of E* rises, the concentration of cGMP will drop. If the time course of E* activation is known (which it is), then it is straightforward to calculate the time course of the decline in cGMP concentration, and from this the form of the electrical response can readily be obtained. For the simple case that E* rises linearly with time, it may be shown that (to a good approximation) the fractional electrical response $F(t)$ will be a delayed Gaussian function of time,

(2) $$F(t) = \exp\{-\Phi\, 0.5\, A\, (t - t_d)^2\},$$

for $t > t_d$.

In this equation, Φ is the number of photoisomerizations, A is a constant, which we call the *amplification constant* of transduction, t is time after the flash, and t_d is a small delay time representing the net effect of several reaction steps within the system. The form of this equation is sketched as the solid trace in Fig. 3.8C; the fractional current $F(t)$ begins at unity in darkness, and it declines with time after the flash.

An important feature of this analysis is that all of the parameters in the molecular model (the diffusion coefficients, protein concentrations, enzyme activities, etc.) are found to condense down into a single constant

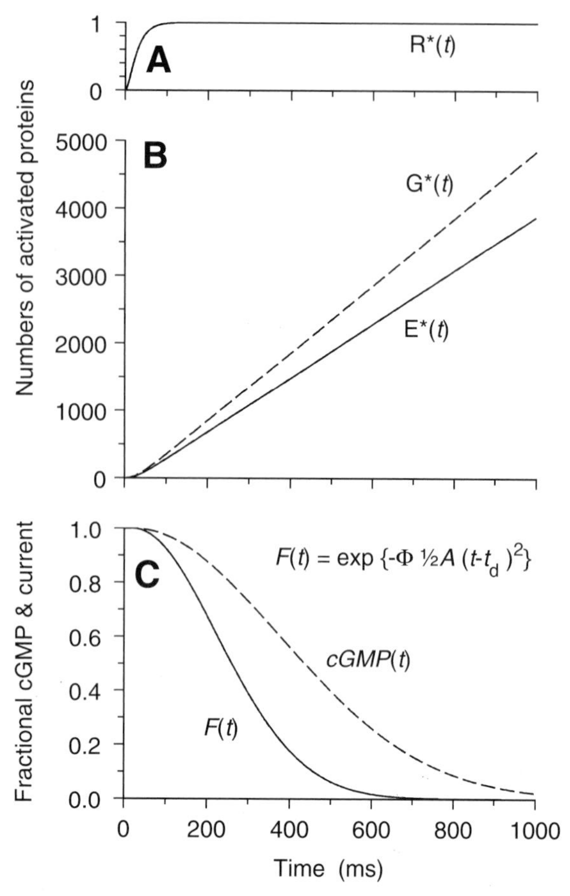

Figure 3.8: Predicted kinetics of protein activation, cGMP concentration, and electrical response when inactivation reactions are ignored. (A) Following photoisomerization, a single R* is activated with a delay of a few milliseconds. (B) Levels of G* and E* rise approximately linearly with time after R* activation, and the ratio of their slopes can be defined as the coupling efficiency of the reaction step G* + E → E*. (C) The cGMP concentration declines as E* is activated, and consequently cGMP-gated channels close, so that the fractional circulating current $F(t)$ declines according to Eqn. (2). Responses in (C) are shown for a flash delivering Φ = 300 photoisomerizations, with A = 0.08 s^{-2}. Reproduced with permission from Lamb (1996).

A, the "amplification constant" in Eqn. (2). This means that *Eqn. (2) provides a single-parameter description of the onset phase of the fractional electrical response to light* (provided that the short delay t_d can be ignored). Accordingly, after normalization of responses to the maximum, it should be possible to fit

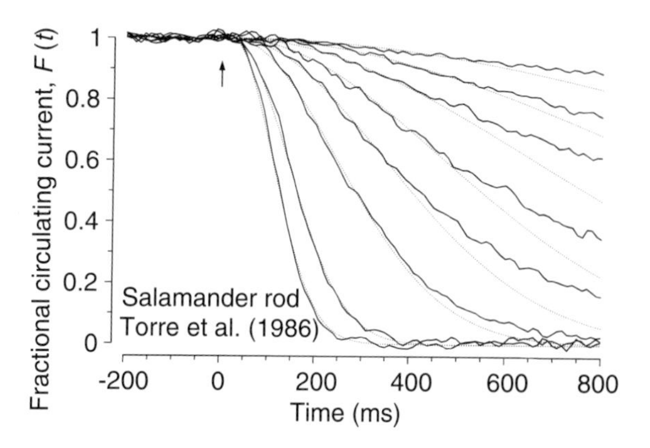

Figure 3.9: Comparison of experiment and theory for the onset of the electrical response to light. Suction pipette recordings of the circulating current of a salamander rod from the experiments of Torre, Matthews, and Lamb (1986). The solid traces show the experimentally measured signal, normalized to its level in darkness, and the dotted traces plot $F(t)$ predicted by Eqn. (2). To minimize light-induced changes in calcium concentration, the calcium buffer BAPTA had been incorporated into the cytoplasm via a whole-cell patch pipette. Flashes, presented at time zero, delivered from 10 to 2,000 photoisomerizations, Φ. Dark current, 29 pA. Amplification constant, $A = 0.077\ s^{-2}$; delay time, $t_d = 20$ ms. Reproduced with permission from Lamb (1996).

the rising phase of a whole family of electrical responses to flashes of increasing intensity by using Eqn. (2), with just a single value of A, together with the measured flash intensities Φ.

Comparison with experiment. The predictions of Eqn. (2) are compared with experiment in Fig. 3.9, in suction pipette recordings from a salamander rod. In this cell the cytoplasmic Ca^{2+} concentration had been buffered by the incorporation of a chelator, to minimize the influence of calcium-dependent inactivation reactions (Torre, Matthews, & Lamb, 1986). Under these conditions the electrical response was well described using the single parameter $A = 0.077\ s^{-2}$ in conjunction with the measured flash intensities. Equation (2) has also been shown to provide a good description of the rising phase of the flash response of rod and cone photoreceptors from a variety of vertebrate species under control conditions, that is, in the absence of a calcium chelator (e.g., Lamb & Pugh, 1992; Kraft, Schneeweis, & Schnapf, 1993; Pugh & Lamb, 1993).

In Fig. 3.10, experimental recordings of the *a*-wave of the human electroretinogram (ERG), obtained under rod-isolated (panel A) and cone-isolated (panel B) conditions are compared with the predictions of a slightly modified form of Eqn. (2) (Smith & Lamb, 1997). The modification was simply to incorporate the cell's capacitive time constant explicitly, rather than to lump it into the total delay time t_d. In both panels of Fig. 3.10 the responses are plotted in μV (i.e., without normalization), and the theoretical curves for $F(t)$ have therefore been scaled by the estimated maximal response amplitude, a_{max}, of $-350\ \mu$V for the rods and $-110\ \mu$V for the cones. The modified equation provides a good description of the rising phase of the whole family of responses, for both the rod signals and the cone signals.

For the rod responses, the fit has been obtained with an amplification constant of $A_{rod} = 4.5\ s^{-2}$ and a time constant of $\tau_{rod} = 1.1$ ms. In a previous analysis of rod *a*-waves (Breton et al., 1994) the original equation failed to fit well at the very earliest times, with extremely intense flashes, because the cell's capacitive time constant was not taken into account. But when allowance for the capacitive time constant is included the rising phase in Fig. 3.10A is well described, even for flashes delivering up to 250,000 photoisomerizations per rod. Similar results have also been obtained by Cideciyan and Jacobson (1996).

Cone parameters and rod/cone comparisons. To fit the cone *a*-wave responses, account must be taken of the existence of two classes of longer wavelength sensitive cones, the red- and green-cones, since the blue-cones make no significant contribution with red stimuli delivered in the presence of a bright blue background. The curves in Fig. 3.10B were obtained on the assumption that the two sets of cones exhibited identical parameters and that they contributed equally to the maximal response, with the only difference being caused by the different numbers of photon absorptions resulting from the respective spectral sensitivities (see also Hood & Birch, 1995). Thus the maximal *a*-wave

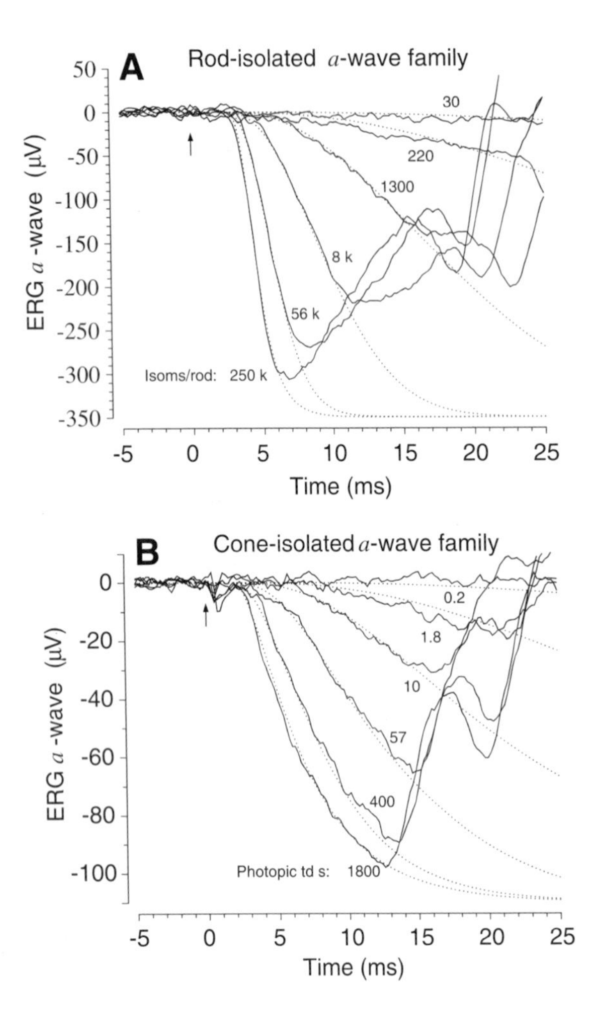

Figure 3.10: Comparison of experiment and theory for the a-wave of the ERG recorded from a human subject. Panel (A) shows rod-isolated responses, and panel (B) shows cone-isolated responses. The solid traces show the experimentally measured signals for ganzfeld stimulation with very brief flashes, and the dotted traces plot the predictions of the modified form of Eqn. (2) that takes account of the membrane capacitive time constant. (A) Rod-isolated responses obtained under fully dark-adapted conditions by presenting blue flashes and then subtracting the responses obtained with photopically matched red flashes; the numbers indicate the estimated photoisomerizations per rod (Φ) for each flash intensity. Parameters used in the fit were: maximal amplitude, $a_{rod} = -350$ μV; amplification constant, $A_{rod} = 4.5\ s^{-2}$; capacitive time constant, $\tau_{rod} = 1.1$ ms; and pure delay, $t_d = 2.3$ ms. The factor used to convert troland values to photoisomerizations was $K_{rod} = 8.6$ isomerizations per scotopic td s. (B) Cone-isolated responses obtained by delivering red test flashes on a rod-saturating blue background of 48 scotopic cd/m^2; measured pupil diameter, 3.3 mm. The numbers indicate the retinal illuminance of each flash in photopic td s. (The transient seen about 1 ms after the flash represents an artefact caused by electromagnetic coupling from the xenon flash unit; it was not seen in panel A because of the subtraction method used there.) Fitted parameters were: maximal red- and green-cone response amplitudes, $a_{red} = a_{green} = -55$ μV; capacitive time constant, $\tau_{cone} = 4.4$ ms; pure delay, $t_d = 1.5$ ms; and sensitivity, $K_{red} A_{cone} = 1340\ s^{-2}$ / td s. If the intensity conversion factor is taken as $K_{red} = 200$ isomerizations per photopic td s, then this sensitivity yields a cone amplification constant of $A_{cone} = 6.7\ s^{-2}$. Reproduced with permission from Smith and Lamb (1997).

response of −110 μV was attributed equally to the red- and green-cones, with $a_{red} = a_{green} = -55$ μV, and in addition a common cone membrane time constant of $\tau_{cone} = 4.4$ ms was used.

The membrane time constant needed to fit the cone responses (ca. 4–5 ms in collected results) is considerably greater than that needed for the rods (ca. 1 ms). The difference is presumably caused by the much greater capacitance of the cone outer segment membrane that results from the greater surface area of the patent sac structure of the cone outer segments, compared with the sealed-off disc structure of the rods. Although this longer capacitive time constant is expected from anatomical considerations, it comes as something of a surprise in the cells that mediate our rapid daylight vision.

To fit the family of experimental responses, a value for the sensitivity of the red-cones of $K_{red} A_{cone} = 1340$ s^{-2}/td s was required (in units of incident light intensity). To convert this value to an amplification constant for the cones it was necessary to obtain an accurate estimate of light collection by cones in the peripheral retina, because it is from the periphery that most of the ERG signal originates. Although it is clear that the diameter of the cone inner segments in the periphery is very much greater than in the fovea (Polyak, 1941), there remains considerable uncertainty about the degree of funnelling of the incident light. However, when reasonable estimates of cell dimensions are employed, the extent of light collection by peripheral cones is calculated to be about ten times greater than for foveal cones, and the estimated intensity conversion factor is obtained as $K_{red} \approx 200$ isomerizations per red-sensitive cone per photopic td s (Smith & Lamb, 1997). Substitution of this value into the above measurement of sensitivity then gives the cone amplification constant as $A_{cone} = 6.7 \text{ s}^{-2}$. From collected results, the cone amplification constant was estimated as $A_{cone} \approx 3\text{–}7 \text{ s}^{-2}$.

It may seem surprising that this amplification constant obtained for the cones is so similar to that obtained above for the rods (4–5 s^{-2}). What this finding indicates is that, at early times after a flash, the rising phase of the response to each photon is similar in rods and cones. Thus it seems that the transduction process has been optimized to operate at just as high a gain in the cones as in the rods (Pugh & Lamb, 1993). However, a significant difference between rods and cones concerns the speed with which the response turns off; the cones recover very rapidly from a flash of light, whereas the rods recover much more slowly. This difference may alternatively be viewed as an inverse measure of the time over which the photoreceptor is able to integrate up the effects of light.

From these results and other information, one can summarise the similarities and differences in the electrical response of rods and cones as follows. (1) The early rising phase of the response per photon is similar in rods and cones. (2) The time-to-peak of the flash response is shorter in cones than in rods, permitting the cones to recover faster but allowing the rods to integrate the effects of light for longer. (3) The rods exhibit a "dark light" that is orders of magnitude smaller than that in the cones (Barlow, 1972; Lamb & Simon, 1977), and this enables the rod system to function down to extremely low intensities. (4) The cones are able to adapt over a very wide range of background intensities (Burkhardt, 1994), whereas the rods can only adapt over a few log units of intensity. The cone system is thus able to function adequately even in the presence of backgrounds so intense that they "bleach" most of the photopigment, whereas the rods are saturated if just a small fraction of the pigment is bleached. (5) Following extinction of intense bleaching illumination, the cones recover much more rapidly than the rods; that is, they "dark adapt" much more rapidly.

Summary

1. The applicability of the frequency normalization procedure proposed by Mansfield (1985) has been tested for spectral sensitivity measurements from photoreceptors from a number of mammalian sources (rods and cones; human, monkey, squirrel, and cow). Within the accuracy of the available measurements, this frequency normalization procedure appears to generate spectra with a common shape.

2. A simple analytical equation [Eqn. (1)] has been presented that provides a reasonably accurate description of this common spectral shape. In particular, the equation shows the correct asymptotic slope for log sensitivity at long wavelengths, and in addition it provides an adequate description of the region near the peak in a linear plot. Furthermore, it accounts for the "yellowing" of lights at very long wavelengths, which was reported by Brindley (1955).

3. We now have a comprehensive understanding of the molecular steps involved in the activation stages of phototransduction. This understanding allows us to model the activation reactions kinetically and to generate a single-parameter prediction for the onset phase of the electrical response to light [Eqn. (2)].

4. The predictions of the model accurately describe the rising phase of the electrical response of photoreceptors in a variety of vertebrate species, recorded either by single-cell techniques or noninvasively by the *a*-wave of the ERG.

5. Somewhat surprisingly, the gain of transduction is quite similar in rods and cones, despite the considerable anatomical and physiological differences between them. The cones are specialized for rapid response, for the ability to avoid saturation during the presence of intense light, and for rapid recovery following intense exposures.

Acknowledgement

This work was supported by grants from the Wellcome Trust (034792) and the Human Frontiers Science Program (RG-62/94).

4

Electrophysiology of cone photoreceptors in the primate retina

Julie L. Schnapf and David M. Schneeweis

The retinal cones transform light energy into electrical signals and play a key role in the extraction of information about color in the visible world. The three-dimensionality or trichromacy of our color world is thought to be a reflection of the number of retinal cone types with distinct absorption spectra, while the shapes of those spectra underlie the basic nature of our color experience (see Chapters 1 and 2). Other aspects of color perception are undoubtedly determined in part by other characteristics of the cones, such as their temporal response properties, adaptational properties, and patterns of synaptic interactions. In this chapter we describe some of these properties and where possible relate them to visual experience.

The absorption spectra of photoreceptors have been estimated several ways. By using microspectrophotometry, absorption spectra have been measured directly from single isolated photoreceptors of the primate retina (Bowmaker, Dartnall, & Mollon, 1980; Dartnall, Bowmaker, & Mollon, 1983). These spectra are limited to wavelengths for which absorption is strong and consequently cannot be used to determine the shape of the absorption function at the extreme ends of the visible range. Indirect estimates of absorption spectra have been obtained from an analysis of psychophysical color matching data of normal and color blind observers (Smith & Pokorny, 1975; see Chapters 1 and 2).

An alternative approach to measuring absorption spectra is to generate photoreceptor action spectra from the light-evoked electrical signals of rods and cones. The action spectrum is obtained by determining, as a function of wavelength, the intensity of light required to evoke responses of some criterion size. Assuming that the processes that convert light absorption to electrical response are wavelength-independent, then action spectra and absorption spectra will be proportional (Naka & Rushton, 1966; Cornwall, MacNichol, & Fein, 1984). The electroretinogram (ERG), a massed electrical response recorded at the corneal surface of an intact eye, can be used to measure cone action spectra if the ERG is dominated by a single cone type, as is the case for light-adapted retinas of dichromats. This method has been used to characterize the spectral properties of cones in human and nonhuman primates (Neitz & Jacobs, 1984; Neitz, Neitz, & Jacobs, 1991; see Chapter 6).

Action spectra have also been derived from single-cell recordings in primates. Two types of techniques have been employed, photocurrent recording (Baylor, Nunn, & Schnapf, 1984,1987; Schnapf, Kraft, & Baylor, 1987; Kraft, Schneeweis, & Schnapf, 1993) and photovoltage recording (Schneeweis & Schnapf, 1995). The photocurrent (light-evoked changes in the membrane current) can be recorded from a single rod or cone outer segment drawn into a suction electrode (Baylor, Lamb, & Yau, 1979; Schnapf & McBurney, 1980). An example of this kind of recording is shown in Fig. 4.1A.

Figure 4.1A is a photomicrograph of a small piece of retina from a human eye. The suction electrode at

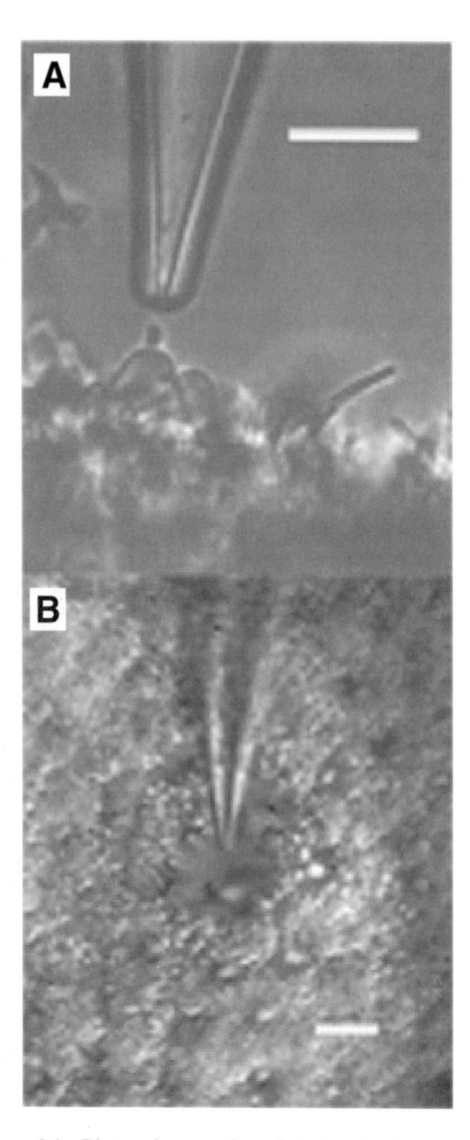

Figure 4.1: Photomicrographs of isolated primate retina preparations. (A) Photocurrent recording from a small piece of retina from a human eye. A suction electrode lies above a single cone. Normally the cone would be invisible, buried within a sea of rod outer segments, but in this portion of the preparation the retinal piece was denuded of all rod outer segments save one to the right of the cone. (B) Photovoltage recording from an isolated "flat-mount" retina of a macaque monkey. The patch electrode is blowing away the rod outer segments on the surface, allowing access to the cell body of a cone for recording. Calibration bars corresponds to 20 μm in both panels.

the top of the figure is pointing toward a cone. The large diameter of the cone inner segment with its stubby outer segment is characteristic of cones from the retinal periphery of primates. The photocurrent primarily reflects membrane conductance changes that are driven by a photochemical cascade within the outer segment (see Chapter 3). Due to the biophysical properties of the light-sensitive channels, and their specific localization to the outer segment, the photocurrent is largely insensitive to other neuronal mechanisms within the photoreceptor such as voltage-dependent conductances and synaptic input from neighboring cells. As a result, this method provides an electrical signal that reflects absorption and phototransduction within single outer segments.

In contrast, the photovoltage (the light-evoked change in the membrane potential) is regulated by several factors, including the light-sensitive conductance of the outer segment, voltage-dependent conductances in the inner segment, and synaptic input (Baylor, Fuortes, & O'Bryan, 1971; Bader, MacLeish, & Schwartz, 1979). To study the effects of these influences, intracellular electrodes have been used to record photovoltages from the photoreceptors of cold-blooded vertebrates (e.g., Baylor, Fuortes, & O'Bryan, 1971). More recently, patch electrodes have been used to make measurements from primate photoreceptors (Schneeweis & Schnapf, 1995). The patch electrode, which forms a tight seal against the cell's plasma membrane (Fig. 4.1B), is filled with a pseudointracellular solution as well as an antibiotic. The antibiotic incorporates into the plasma membrane underlying the electrode, forming perforations (Horn & Marty, 1988). These perforations establish electrical continuity between the cell interior and the electrode, but they prevent the loss of large intracellular molecules that are important for phototransduction and other cellular processes.

Photocurrent recordings

Action spectra. An example of a cone photocurrent recording is shown in Fig. 4.2A from the retina of

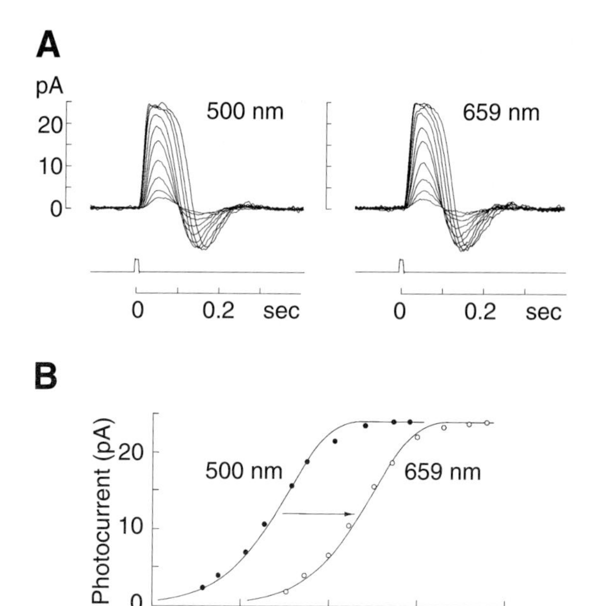

Figure 4.2: Intensity and wavelength dependence of photocurrent in a red-cone in the macaque retina. (A) Superimposed responses to flashes of increasing strength at wavelengths of 500 (left) or 659 nm (right). Note the similarities in the waveforms at the two wavelengths. A flash monitor is shown below the current traces. (B) Dependence of peak response amplitude on flash strength for the responses illustrated in panel A. The points plot the measured amplitudes at 500 (●) and 659 nm (○). The smooth curves are exponential saturation functions, $r = 24$ pA $[1 - \exp(-ki)]$, where r is the peak amplitude, i is the flash photon density, and the constant k is of 2.77×10^{-4} (500 nm) or 3.01×10^{-5} photons^{-1} μm^2 (659 nm). Adapted from Baylor, Nunn, and Schnapf (1987).

an Old-World monkey, the *Macaca fascicularis*. Psychophysical studies in this species (DeValois et al., 1974) suggest that macaque photoreceptors are similar to those in human retinas.

The traces plot photocurrents elicited in a red-cone to flashes of light of increasing intensity at wavelengths of 500 nm (left) or 659 nm (right). The families of responses at the two wavelengths are virtually identical to each other, supporting the idea that the cone photocurrent obeys the *Principle of Spectral Univariance* (Naka & Rushton, 1966): The response depends on the number of photons absorbed and not the wavelength of the absorbed photons. Wavelength does however affect the probability that an incident photon will be absorbed. The relationship between the probability of absorption and the wavelength is assessed by generating an action spectrum, as illustrated in Fig. 4.2B. The points plot the peak amplitude of the light responses in Fig. 4.2A as a function of flash photon density; the smooth curves plot the best-fitting exponential saturation functions. The curves have the same shape and are separated on the abscissa by 0.96 log unit, indicating that this cone is about ten times more sensitive to light of 500 nm than that of 659 nm. The difference in sensitivity is due to an approximately tenfold difference in absorption probability. When the intensities of flashes at the two wavelengths are adjusted so as to produce equal photon capture, the responses are identical in form, as evidenced by the responses in Fig. 4.2A.

Action spectra obtained in this way from photoreceptors of primate retina are illustrated in Fig. 4.3. Figure 4.3A plots average spectral sensitivities from the macaque retina (Baylor, Nunn, & Schnapf, 1984; 1987), and Fig. 4.2B from human (Schnapf, Kraft, & Baylor, 1987; Kraft, Schneeweis, & Schnapf, 1993). The peaks of the spectra lie at about 430 nm (Δ), 530 nm (□), and 560 nm (○) for the three cone types and at about 493 nm (●) for rods. Spectra can be characterized reliably over more than seven orders of magnitude of absorption probability because the signal-to-noise ratio of the measurement remains constant over the entire range. The limit to further exploration of the spectrum is the availability of enough light to stimulate at wavelengths of lower absorption probability.

The spectra obtained from humans and *Macaca fascicularis* are nearly identical, as seen in Fig. 4.4. The open symbols plot the spectra obtained from humans and the closed symbols from the corresponding cell types in the macaque eye. The close correspondence is consistent with the similarity of color vision psychophysics measured in humans and macaques (DeValois et al., 1974).

Recordings from blue- (or short-wave–sensitive) cones have not yet been obtained in the human eye,

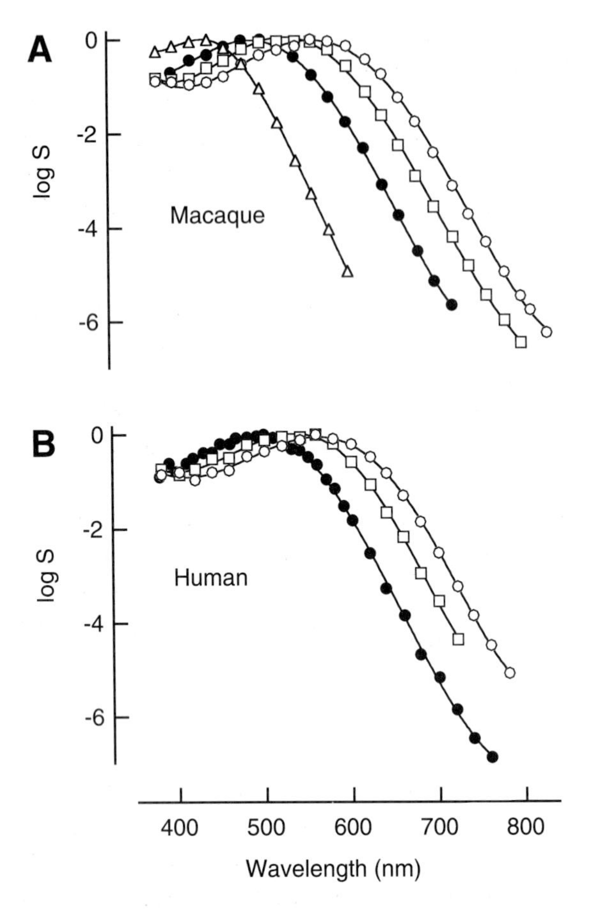

Figure 4.3: Action spectra obtained from photocurrent recordings in macaque (A) and human (B) retinas. The points plot log relative spectral sensitivity as a function of wavelength for rods (●), blue- (or short-wave–sensitive) cones (Δ), green- (or middle-wave–sensitive) cones (□) and red- (or long-wave–sensitive) cones (○). The smooth curves are sixth-order polynomials, with the polynomial constants chosen to minimize the differences between the points and the curves.

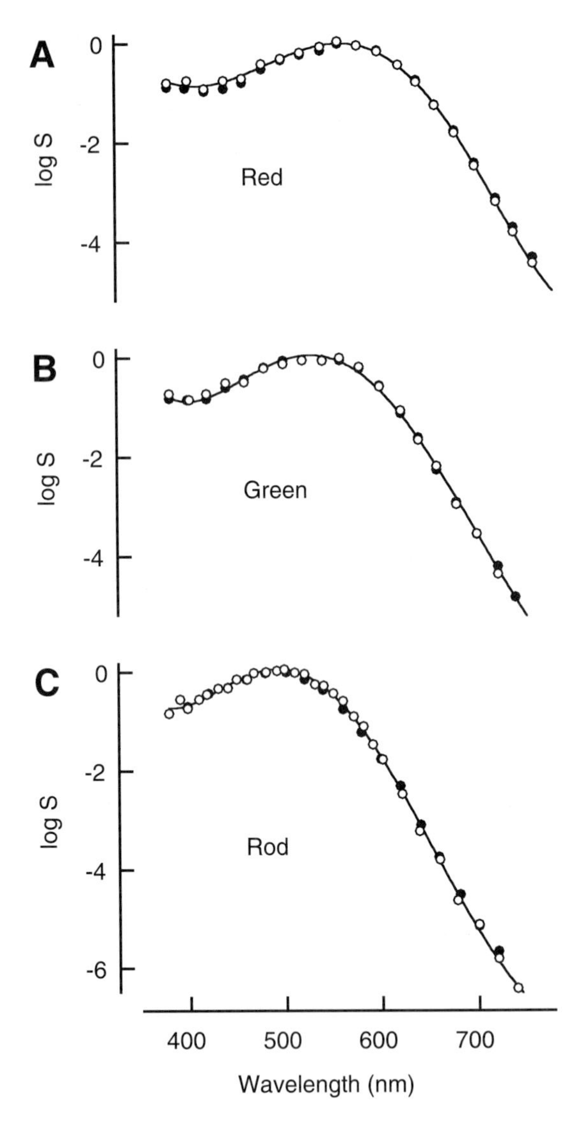

Figure 4.4: Comparison of human and macaque action spectra of red-cones (A), green-cones (B), and rods (C). Human spectra are plotted with open symbols and macaque spectra with filled symbols. The smooth curves are sixth-order polynomials, with the polynomial constants chosen to minimize the differences between the points and the curves.

owing to the relatively low number of blue-cones (only a few percent) and the difficulty in obtaining viable human retinas. It will be interesting to see if there is also a close correspondence of the human blue-cone spectrum with that of the macaque. The success with which the macaque spectra are able to predict human color matching data (Stiles & Burch, 1955;1959; Baylor, Nunn, & Schnapf, 1987) suggests that the agreement might be good. On the other hand, one estimate of the human blue-cone spectrum, calculated from psychophysical adaptation experiments, is shifted to shorter wavelengths as compared with that of the macaque (Stockman, MacLeod, & Johnson, 1993).

Among observers with normal color vision, a common polymorphism has been identified at amino acid position 180 of the red- (or long-wave–sensitive) cone photopigment. Substantial molecular and psychophysical evidence (see Chapters 1 and 2) indicates that this serine/alanine polymorphism is associated with a 3–5-nm difference in spectral tuning of the red-cone photopigment. This polymorphism implies that two individuals with "normal" color vision do not necessarily share the same trichromatic color experience, as evidenced by recent action spectra obtained from human red-cones (Kraft, Neitz, & Neitz, 1998). Furthermore, other experiments suggest that the retinas of some color normal males contain multiple variants of red- or green- (middle-wave–sensitive) cone photopigments (Neitz & Jacobs, 1986; Neitz, Neitz, & Jacobs, 1993; Neitz & Neitz, 1995). In a single retina, a cone may contain one of *more than* three photopigments, challenging the simplest assumptions for the basis of trichromacy (but see Chapter 1).

Comparison to cone fundamentals and photopic luminosity function. Cone fundamentals give an estimate of the spectral sensitivity of cones to light incident on the cornea of the intact human eye. One set of estimates derived from human psychophysical color matching experiments (Smith & Pokorny, 1975) is illustrated in Fig. 4.5.

The Smith and Pokorny L (○), M (□), and S (Δ) functions, plotted by the points in both the upper and lower panels, are estimates of the sensitivities of red-, green-, and blue-cones, respectively, to corneal illumination. To compare these estimates with the cone action spectra described above, the action spectra must first be adjusted to account for absorption in the intact eye by the lens and macular pigment, and for pigment self-screening within the cone outer segment. The smooth curves plot the "corrected" action spectra of the macaque cones (A) and the human cones (B) obtained from adjustments of the fitted curves in Fig. 4.3. The positions of the curves have been shifted on the log ordinate to best coincide with the points. The match is quite good between the two sets of data, particularly for the red- and green-cones.

The efficacy with which light stimulates the visual system varies with the wavelength of light. Under photopic conditions, this wavelength dependence (the "photopic luminosity function") is a function of the action spectra of the cones, their relative numerosity in the retina, and their relative synaptic strengths and gains. The blue-cones are thought to add relatively little to luminosity. The curve in Fig. 4.6 plots the log photopic luminosity function of Vos (1978), and the symbols plot the log of the weighted sum of the human red- and green-cone spectra, corrected as in Fig. 4.5 for corneal illumination (Schnapf, Kraft, & Baylor, 1987).

A weighting factor of 1.7:1 for the red/green spectra was chosen to minimize the squared difference between the psychophysical and physiological sensitivities (on a linear scale). The fitting was done for wavelengths greater than 500 nm to minimize possible errors associated with the uncertainties of preretinal screening. The weighting suggests that stimulation of red-cones contributes roughly twice as much to the luminosity of a stimulus as does stimulation of green-cones.

Light adaptation of cone photocurrent. In cold-blooded vertebrates, the presence of steady background illumination decreases both the amplitude and the duration of photon signals in cones (e.g., Baylor & Hodgkin, 1974). What is the effect of background illumination on primate cones? To assess this, responses of macaque cones to dim flashes of light were measured both in the presence and absence of steady background lights; sensitivity was defined as the reciprocal of the flash intensity needed to evoke a criterion size response (Schnapf et al., 1990).

The symbols in Fig. 4.7 plot the dependence of sensitivity on background intensity. As background intensity increases, flash sensitivity decreases. The functional dependency is fitted by the smooth curve, a plot of Weber's Law: $S_F/S_F^D = I_0/(I_0 + I)$, where S_F is flash sensitivity, I is background intensity, and S_F^D and I_0 are normalizing constants (see legend to Fig. 4.7). This function has the same form as the intensity dependence of photopic desensitization in human psychophysical experiments (Hood & Finkelstein, 1986). The reduction in flash sensitivity of single cones cannot, however, account for the observed psychophysical

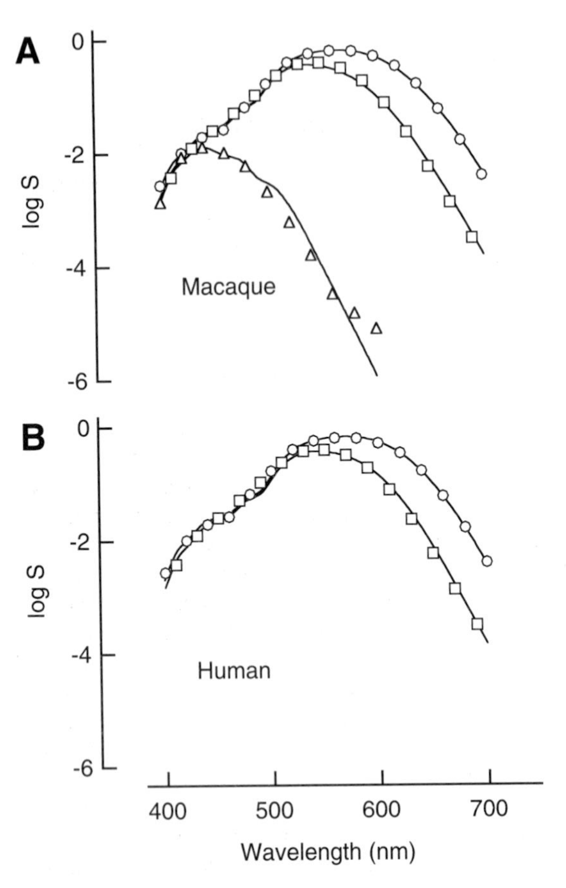

Figure 4.5: Comparison of psychophysical cone fundamentals and photocurrent cone action spectra. The points plot the Smith and Pokorny (1975) cone fundamentals, L (○), M (□), and S (Δ), as tabulated in Wyszecki and Stiles (1982a, p. 614) and are shown in 20-nm intervals. The smooth curves plot the photocurrent action spectra from macaque (top) and human (below) recordings, corrected for preretinal screening and pigment self-screening. The corrected spectra were obtained as follows: The uncorrected action spectra were taken as the polynomial approximations in Fig. 4.3. Self-screening was accounted for by Beer's Law, assuming a peak axial outer segment density of 0.27 (Baylor, Nunn, & Schnapf, 1987). Preretinal absorption was calculated by scaling the absorption spectra for the lens and macular pigment tabulated in Wyszecki and Stiles (1982a, p. 719) to give the best fit between the corrected action spectra and the cone fundamentals. The best scaling was obtained with density values for the lens at 400 nm of 1.5 (macaque) and 1.7 (human); the density values for the macular pigment at 460 nm were 0.55 (macaque) and 0.61 (human). The positions of the curves were shifted on the log ordinate to align with the peaks of the cone fundamentals.

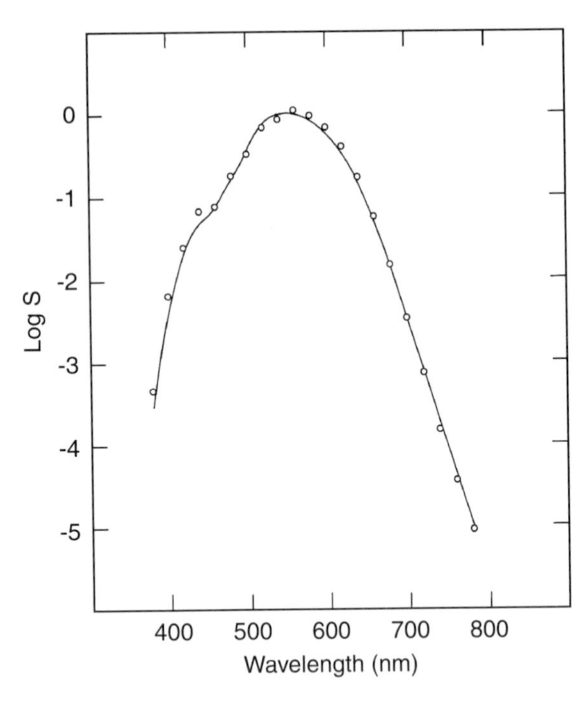

Figure 4.6: Comparison of cone action spectra to photopic luminosity. The smooth curve is the log photopic luminosity function as calculated by Vos (1978). The points plot the log of the weighted sum of the human red- and green-cone spectral sensitivities (Fig. 4.3), corrected for the effects of preretinal screening and pigment self-screening (as described in Fig. 4.5). The relative weighting of the underlying cone spectra was 1.7 red and 1.0 green. Reprinted from Schnapf, Kraft, and Baylor (1987).

desensitization. Psychophysical sensitivity is halved by a background light intensity of 1 to 1.5 log trolands (Hood & Finkelstein, 1986). Illumination of this intensity would have virtually no effect on the sensitivity of single cones. As indicated in Fig. 4.7, the half-desensitizing intensity for a single cone is about 100 times greater, 3.3 log trolands (corresponding to about 26,000 photoisomerizations s^{-1}).

Light adaptation is also known to alter the kinetics of photopic vision, preferentially reducing psychophysical sensitivity to low temporal frequency stimulation. No corresponding change in the kinetics of the cone photocurrent has been observed (Schnapf et al., 1990). The traces in Fig. 4.7 show sample flash responses from two cones in both dark-adapted (heavy

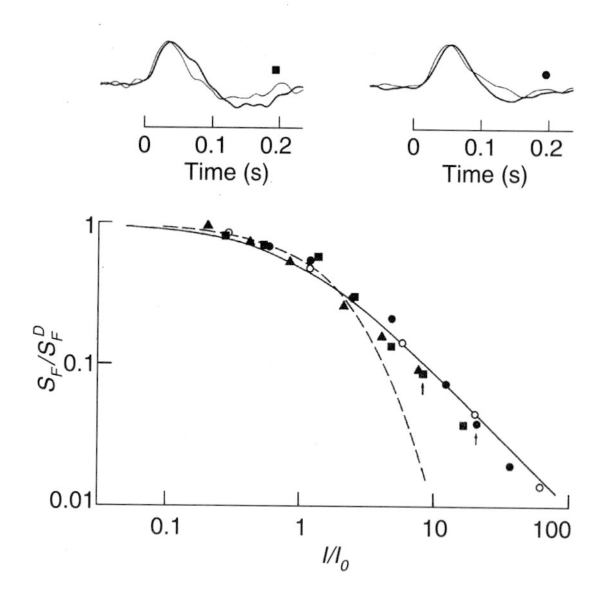

Figure 4.7: The dependence of the flash sensitivity of macaque cones on background light intensity. Normalized flash sensitivity is plotted as a function of normalized background intensity. Each kind of symbol represents data from a different cone. The smooth curve is Weber's Law, as described in the text. S_F is flash sensitivity on the background of intensity I, S_F^D is the dark-adapted flash sensitivity, and I_0 is the intensity that reduces the dark sensitivity twofold. The average value for I_0 in the four cells illustrated was 2.6×10^4 photoisomerizations s^{-1}. *Top*: Photocurrent traces from two cones illustrate the invariance of response waveform with adaptation. The heavy lines are dark-adapted responses and the thin lines light-adapted responses, all scaled to a common peak amplitude. Reprinted from Schnapf et al. (1990).

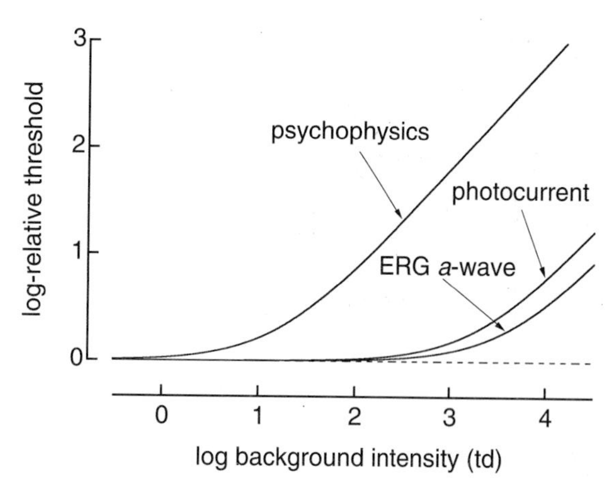

Figure 4.8: Comparison of psychophysical increment threshold and cone thresholds. Log-relative threshold is plotted as a function of background light intensity, in log photopic trolands. The line labeled "photocurrent" was obtained from sensitivity measurements from single cones in the macaque retina. This curve is derived from the continuous curve in Fig. 4.7, where threshold is the reciprocal of sensitivity and units of background intensity are converted to trolands. The curve "ERG *a*-wave" plots the threshold for the leading edge of the *a*-wave measured in the intact human eye (Hood & Birch, 1993). The "psychophysics" curve shows the intensity dependence of human incremental threshold as measured in psychophysical tests obtained with long-duration, large-diameter test flashes (Hood & Finkelstein, 1986). Adapted from Hood and Birch (1993).

lines) and light-adapted (thin lines) states, normalized to the same peak amplitude. Although the background lights were sufficient to lower light-adapted flash sensitivity 20- to 30-fold, the waveform of the flash responses remained unaltered.

The characteristics of light adaptation in single macaque cones described above are contrary to the commonly held view that alterations in sensitivity and kinetics of photopic vision with light adaptation are due to related alterations in the cones themselves. One possible explanation is that the photo*voltage* as opposed to the photo*current* of cones displays adaptational characteristics that are more similar to those of the psychophysics. But, as will be described below, the adaptational properties of the cones are not substantially altered with the transformation from current to voltage. A second possibility is that the adaptational properties of cones in the isolated macaque retina are different from those in an intact human eye. Evidence to the contrary comes from measurements of cone responses derived from in vivo ERG recordings in humans (Hood & Birch, 1993). Cone responses were estimated from the leading edge of the *a*-wave of the ERG, after subtraction of the rod contribution.

Plotted in Fig. 4.8 is the variation in threshold (reciprocal of sensitivity) with background intensity. The increase in *a*-wave threshold with increasing background intensity ("ERG *a*-wave") is similar to the intensity dependence estimated in isolated macaque cones (curve labeled "photocurrent"), and both differ

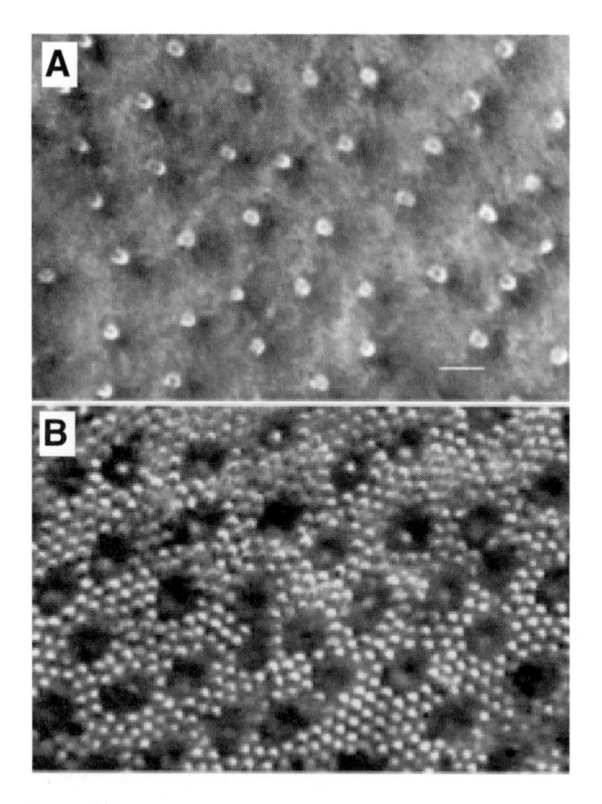

Figure 4.9: Photomicrograph of isolated macaque retina used for patch recordings from photoreceptors. In (A) the tips of the cone outer segments are in focus, while in (B) the more distally positioned tips of the rod outer segments are in focus. Same magnification in (A) and (B); scale bar in (A) is 10 μm.

dramatically from a sample psychophysical increment threshold function (labeled "psychophysics"). This further supports the idea that photopic adaptation largely reflects processes other than alterations in the gain of phototransduction. It should be pointed out as a caveat that, while the ERG data are similar to those of isolated cones, the ERG could estimate only the rising phase of the underlying cone response but not its duration or peak amplitude.

Photovoltage recordings

Photovoltage recordings were made from single rods and cones in isolated retinas of the macaque monkey (Schneeweis & Schnapf, 1995). Figure 4.9 is a photomicrograph of the isolated retina preparation used, with focus at the tips of the outer segments of cones (A) or rods (B).

Rod/cone interactions. Unlike the cone photocurrent described above, the cone photovoltage does not obey the Principle of Spectral Univariance. This implies that more that one photoreceptor type contributes to the voltage signal in a single cone. Evidence for a lack of univariance is demonstrated in Fig. 4.10.

Photovoltages of a macaque green-cone are plotted for 500-nm stimulation (solid lines) and 660-nm stimulation (dashed lines). Both wavelengths hyperpolarize the cone, but the waveform of the photovoltage is wavelength-dependent (in contrast to the photocurrent records in Fig. 4.2). When intensities at the two wavelengths were adjusted to evoke responses of comparable peak amplitude (Fig 4.10A), the 500-nm response displayed a larger and longer lasting afterhyperpolarization. An analysis of the spectral sensitivity of the photovoltage reveals that this late component originated in rods. When light intensities were readjusted to equate absorption in rods (Baylor, Nunn, & Schnapf, 1984), the responses at late times were well matched (Fig. 4.10B). Rodlike spectral sensitivities have been observed in the late components of both red- and green-cone responses (Schneeweis & Schnapf, 1995).

The intrusion of rod signals in the cone photovoltage is not just restricted to the late phase of the response, but is apparent at the peak of the cone response as well. Figure 4.10 shows that the peak amplitudes at the two wavelengths were well matched with a 300-fold difference in light intensities, as compared with the 160-fold difference expected from outer segment recordings of single green-cones (Baylor, Nunn, & Schnapf, 1987). This "blue-shift" in the peak photovoltage action spectrum was observed for both red- and green-cones. While it might be expected that rod signals are too slow to contribute to the cone peak, the rod photovoltage speeds up dramatically with increasing intensity. At the intensities used to measure peak cone spectral sensitivity, the time-to-peak of the

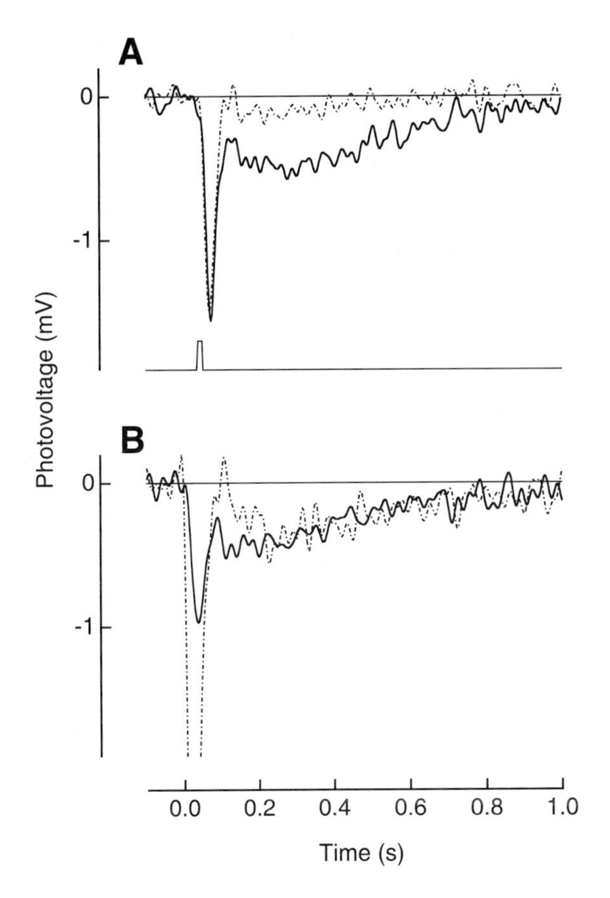

Figure 4.10: Cone photovoltage is not spectrally univariant. Photovoltage recordings from a green-cone in the macaque retina elicited by 500- (solid curves) or 660-nm light (dashed curve). (A) Flash strengths were adjusted to evoke responses of equal peak amplitude: 178 photons μm^{-2} (500 nm) and 5.4×10^4 photons μm^{-2} (660 nm). (B) Flash strengths were equalized for absorption in rods: 94 photons μm^{-2} (500 nm) and 4.3×10^5 photons μm^{-2} (660 nm). Reprinted from Schneeweis and Schnapf (1995).

rod photovoltage is only 30–40 ms (Schneeweis & Schnapf, 1995).

Gap junctions have been observed between rods and cones and between cones and cones in the primate retina (Raviola & Gilula, 1973; Tsukamoto et al., 1992). In the location of the retinal periphery where the photovoltage recordings were made, rods outnumber cones 15 or 20 to 1 (Fig. 4.9). This may explain why rod signals contribute to the cone response but

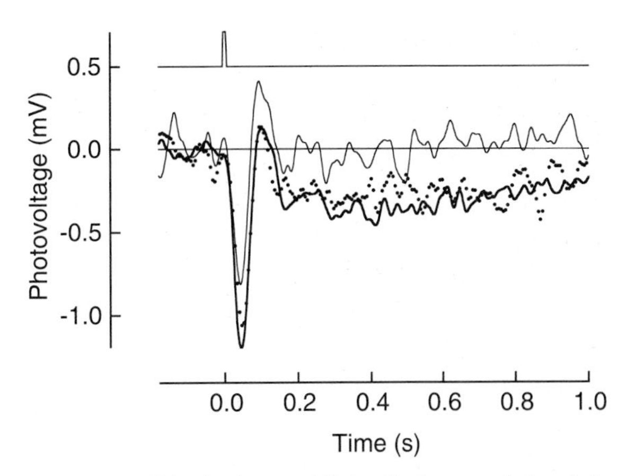

Figure 4.11: Dim background light eliminates rod signals in cones. Photovoltage recording of a green-cone. Responses to a test flash in the presence or absence of dim background lights matched for photon absorption in the cone outer segment. Flash monitor shown above voltage traces. Test flash: 500 nm, 262 photon μm^{-2}. Backgrounds: heavy line, no background; light line, 500 nm, 290 photon $\mu m^{-2}\, s^{-1}$; dots, 660 nm, 43,000 photon $\mu m^{-2}\, s^{-1}$.

cone signals are not observable in the rod photovoltage (Schneeweis & Schnapf, 1995). Rod signals have also been seen in photovoltage recordings from cones in the cat retina (Nelson, 1977). Whether or not cones, particularly cones of different spectral types, are functionally connected to one another is still an open question. Any effects of mixed cone coupling on the dark-adapted cone photovoltage are apparently small, however, since red-cone input to a green-cone would produce a red-shift of the spectrum of the green-cone photovoltage, as opposed to the blue-shift observed (Fig. 4.10).

Light adaptation of cone photovoltage

Rod and cone photocurrents desensitize over different ranges of background light intensity. The half-desensitizing intensity is about 100 photoisomerizations s^{-1} for rods (Baylor, Nunn, & Schnapf, 1984; Kraft, Schneeweis, & Schnapf, 1993) and 26,000 photoisomerizations s^{-1} for cones (see above). Dim illumination would be expected to selectively reduce the rod component of the cone photovoltage, as illustrated in Fig. 4.11.

The photovoltage of a green-cone evoked by a dim flash is shown under three states of adaptation. The heavy curve is the dark-adapted response, with a prominent rod component visible at late times. The other curves plot the responses to the same test flash but in the presence of a dim background light of either 500 nm (light line) or 660 nm (dots). Background intensities were chosen to evoke roughly equal rates of photoisomerization for green-cones (about 160–170 photoisomerizations s^{-1}) but unequal rates in rods (about 290 photoisomerizations s^{-1} at 500 nm and 11 photoisomerizations s^{-1} at 660 nm). Based on photocurrent recordings, neither background would be expected to desensitize cones, but the 500-nm background would be expected to significantly reduce rod signals. As predicted, the 660-nm background had no effect, but the 500-nm background eliminated the slow putative rod component of the flash response. The remaining "pure" cone response is quite similar to the biphasic waveform of the photocurrent measured from cone outer segments (Schnapf et al., 1990). Note that its peak amplitude is smaller than that of the dark-adapted response, which is further evidence that the rod contribution can arise at early times in the cone photovoltage.

Background desensitization of the *late* component has both the spectral and intensity dependence expected from a rod-generated mechanism, while the intensity dependence of desensitization of the *peak* cone photovoltage appears to have combined rod- and conelike components (data not shown). Preliminary results suggest that the rod and cone signals that drive the cone photovoltage adapt by independent mechanisms with intensity dependencies similar to those characteristic of photocurrent measurements.

Summary

The colored appearance and brightness of mixtures of wavelengths agree well with the action spectra derived from measurements from cone outer segments. Cone spectra are altered in later processing at the cone synapse by the addition of rod signals. Rod signals adapt independently of cones, so that under photopic conditions the rod contribution to cones is expected to be minor. The magnitude of photopic (psychophysical) desensitization with background illumination cannot be accounted for by a reduction of either cone photocurrent or photovoltage. It is not known if the gain of synaptic transfer at the cone synapse is altered by adaptation. If significant synaptic adaptation occurs, this might allow for cone-specific desensitization of comparable magnitude to that reflected in psychophysical experiments.

5

The trichromatic cone mosaic in the human eye

David R. Williams and Austin Roorda

In 1802, Young stated definitively that human color vision depends on three fundamental channels. For many years psychophysical observations, which had been Young's guide, led the way toward characterizing the spectral properties of these channels (see Chapter 1). The color matches of dichromats (e.g., Smith & Pokorny, 1975; Stockman & Sharpe, 1998) and psychophysical methods to isolate cone classes in trichromats (e.g. Stiles, 1949; Stockman, MacLeod, & Johnson, 1993) have established and continue to refine estimates of the cone spectra (see Chapter 2). Microspectrophotometry (e.g., Dartnall, Bowmaker, & Mollon, 1983) and suction electrode recordings (Schnapf, Kraft, & Baylor, 1987) of single S-, M-, and L-cones provided satisfying objective confirmation of the psychophysical estimates (see Chapters 2 and 4).

Less is known about the topography of the three cone classes in the normal human retina. We know the most about the S-cone mosaic. Thought to have diverged from the L/M-cones perhaps 500 million years ago (Nathans et al., 1986; see Chapter 1), S-cones differ in a number of ways from their longer wavelength cousins. There are now a host of methods to map S-cone topography based on morphology, in situ hybridization, immunocytochemistry, histochemical markers, selective uptake of dyes, and psychophysics (Marc & Sperling, 1977; Williams, MacLeod, & Hayhoe, 1981b; de Monasterio et al., 1985; Ahnelt et al., 1987, 1990; Wikler & Rakic, 1990; Curcio et al., 1991; Diaz-Araya et al., 1993; Bumsted & Hendrickson, 1999). The S-cone mosaic accounts for about 7% of the cone population (Curcio et al., 1991) and is absent in many humans from an irregular zone at the foveal center about 20 min of arc in diameter (Williams, MacLeod, & Hayhoe, 1981a; Curcio et al., 1991).

The human L- and M-cone submosaics have proven much more difficult to distinguish. No morphological differences distinguish them, and the similarity in their photopigments has made it impossible so far to selectively label them. The techniques that have been successful in providing topographic information have usually relied on differences in the absorption spectrum of the M- and L-cone photopigments. Mollon and Bowmaker (1992) measured the spectra of cones one by one in intact patches of retina of the talapoin monkey. Packer, Williams, and Bensinger (1996) developed an imaging method in which photopigment absorptance in excised primate retinas could be measured simultaneously in many receptors. However, neither of these approaches has been successfully applied to humans. Gowdy and Cicerone (1998) estimated the arrangement of human M- and L-cones psychophysically by using vernier stimuli seen by either the M-cones, the L-cones, or both. Hagstrom, Neitz, and Neitz (1998) have demonstrated changes in the ratio of L- and M-cones across the human retina with an mRNA analysis. In this chapter, we describe a method in which high-resolution retinal imaging (Liang, Williams, & Miller, 1997) is combined with retinal densitometry (Campbell & Rushton, 1955) to obtain the first images of the arrangement of S-, M-, and L-cones in the living human eye. This work was first reported by Roorda and Williams (1999).

High-resolution imaging of the living retina

Liang, Williams, and Miller (1997) showed that a high-magnification fundus camera, equipped with adaptive optics, can provide sharper images of the fundus than had been possible before. Adaptive optics was originally proposed as a method to overcome blur caused by atmospheric turbulence in ground-based telescopes (Babcock, 1953). Atmospheric turbulence skews the otherwise parallel rays of light from a star, so that they cannot be focused to a compact point in the image plane of the telescope. The aberrations in the cornea and lens of the human eye blur images of the retina just as atmospheric turbulence blurs telescopic views of the sky. The most dominant monochromatic aberrations of the eye are defocus and astigmatism, which can be corrected by conventional spectacles, but a host of additional aberrations also contribute to the blur and these require adaptive optics to correct.

Figure 5.1 shows the essential method for correcting them. If the eye were perfect, the light that originated at a single point on the retina would form parallel rays as it emerged from the pupil. In physical optics terms, the wavefront emerging from the eye would be planar. However, in real eyes, aberrations warp this plane wave into a distorted shape as shown in the figure. If this warping were to go uncorrected, image blur would result. Under these conditions, the fine grain of the cone mosaic generally cannot be resolved, although Miller, Williams, Morris, and Liang (1996) showed that it is occasionally possible in eyes with excellent image quality. The shape of this aberrated wavefront, which is different for each eye, can be measured with a Shack-Hartmann wavefront sensor (see Liang, Grimm, Goelz, & Bille, 1994, and Liang & Williams, 1997, for details). A small spot was imaged on the retina (not shown) and the returning light was directed into the wavefront sensor. A computer converted the measured wave aberration into a control signal for a deformable mirror. The deformable mirror is

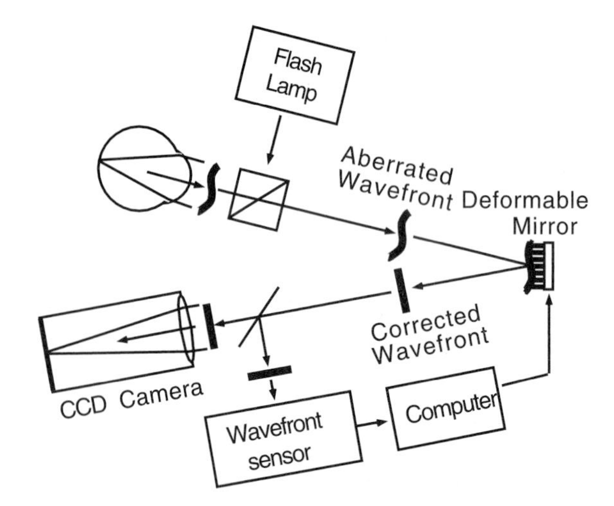

Figure 5.1: Simplified diagram of apparatus for high-resolution imaging of the retina with adaptive optics. The wave aberration of the eye, shown emerging from the pupil, is recorded with a Shack-Hartmann wavefront sensor. A computer converts the wave aberration into a control signal for each of the actuators of the deformable mirror. The mirror takes on the appropriate shape to flatten the wave aberration into an approximation of a plane wave, which is what is desired for diffraction-limited imaging. After adaptive correction, a flash lamp delivers light to the retina and a high-resolution image of the cone mosaic is recorded by a scientific-grade CCD array.

a plate of aluminized glass similar to a conventional first-surface mirror except that it has 37 ceramic actuators or pistons in a square grid mounted on its back surface. The actuators can independently push or pull on the mirror, shaping it into the correct shape to compensate for the distortion imposed by the optics of the eye.

Once the process of flattening the wavefront was complete, we then collected images of the retina. A krypton flash lamp delivered a flash of light to the retina, and the light returning from the retina reflected off the deformable mirror and was directed into a CCD camera (instead of the wavefront sensor), which recorded an image. The mosaic was illuminated with a 4-ms flash (1-deg diameter, ~0.3 μJoules) through a 2-mm entrance pupil. The flash duration was chosen to be short enough to prevent image blur from eye movements. Images were obtained with a 6-mm exit pupil at an eccentricity of 1-deg in the nasal retina. This retinal location was chosen because we have consistently obtained the clearest images there with our high-resolution camera. The cone spacing at the foveal center is almost two times smaller, taxing the resolution of our system. Interestingly, cones at larger eccentricities, although larger in diameter, are less easily imaged, perhaps because the mechanism that reflects light back out of the receptors is less efficient or perhaps because the thicker, highly vascularized inner retina reduces image quality (Weinhaus, Burke, Delori, & Snodderly, 1995). Figure 5.2 shows sample images of the cone mosaic obtained in this way for three subjects, JW, AN, and MM. JW and AN are normal trichromats and MM is a protanope, whose color deficiency was verified psychophysically and by genetic screening. There are no obvious differences between the color normal and color defective retinas in these monochromatic images, with cone spacing about the same.

We have been unable so far to find evidence for rod photoreceptors in images such as these. This is probably because they are sparse, small, and relatively ineffective at sending light back through the pupil. At 1-deg eccentricity, we expect about 4.5 times as many cones as rods (Curcio et al., 1990). Inspection of micrographs from Curcio et al. (1990) suggests that rods near the fovea have inner segments that are approximately one-third of the diameter of cones, which presses the resolution of our present camera. The fact that rods have broad directional sensitivity relative to cones reduces their visibility further because a smaller fraction of the light that they radiate passes through the pupil and into the camera.

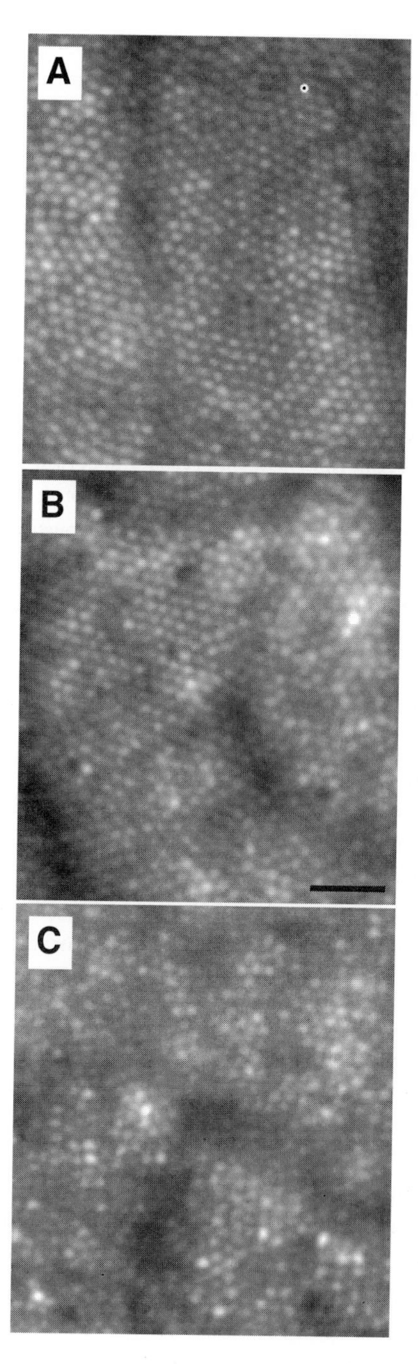

Figure 5.2: Images of the right eyes of subjects (A) JW (temporal retina), (B) AN (nasal retina), and (C) MM (nasal retina). Panel (A) is a registered sum of 51 images taken after a full bleach, (B) the registered sum of 57 images, and (C) the registered sum of 20 images. The scale bar represents 5 min of arc.

Retinal densitometry of single cones

All images were taken with 550-nm light to maximize absorptance by L- and M-cone photopigments. Individual cones were classified by comparing images when the photopigment was fully bleached with those taken when it was either dark-adapted or exposed to a light that selectively bleached one photopigment. Images of fully bleached retinas were obtained following exposure to 550-nm light (70-nm bandwidth, 37×10^6 td-s). Images of dark-adapted retinas were taken following five minutes of dark adaptation. From these images, we created absorptance images defined as 1 minus the ratio of a dark-adapted or selectively bleached image and the corresponding fully bleached image. The virtue of absorptance images is that they reveal only the distribution of the photolabile pigments that distinguish the cone classes and do not contain static features common to the bleached and dark-adapted images. The minimum light was used for each image to minimize the amount of pigment bleached by the imaging light. For this reason, absorptance images had a large pigment signal but high photon noise. This photon noise was reduced by adding multiple images together. For this experiment, averages of about 50 images were required to obtain a sufficient signal-to-noise ratio. Fixational eye movements translated the image from flash to flash, requiring registration with cross-correlation before averaging.

To distinguish S- from M- and L-cones, we obtained absorptance images by combining fully bleached images (Fig. 5.3A) and dark-adapted images (Fig. 5.3B). Because the S-cones absorb negligibly, whereas the M- and L-cones absorb strongly, at the imaging wavelength of 550 nm the S-cones appear as a sparse array of dark cones in the absorptance image of Fig. 5.3C while the M- and L-cones appear bright. Variations in absolute pigment absorptance due to, for example, systematic changes in outer segment length, prevented us from identifying S-cones using a single absorptance criterion across the entire patch of retina. Instead, S-cones were identified as cones whose absorptance was substantially lower than cones in the same neighborhood. Once the sparse population was identified, they were removed from analysis so that the M- and L-cones could be distinguished.

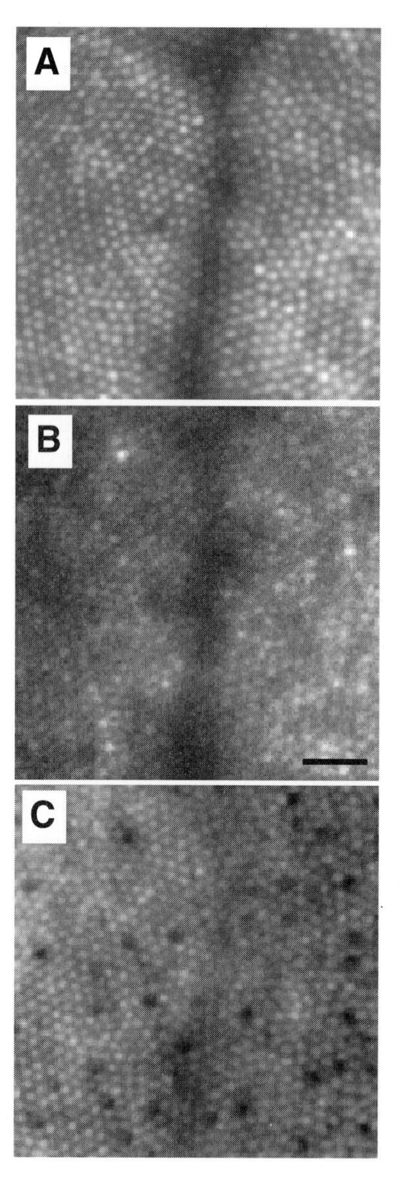

Figure 5.3: A fully bleached image (A) and a dark-adapted image (B) are used to generate an absorptance image (C). The absorptance is calculated as 1 minus the ratio of the dark-adapted image over the fully bleached image. Panel (C) reveals a sparse array of S-cones, which appear dark, due to their low absorptance at 550 nm. The scale bar represents 5 min of arc.

To distinguish L- from M-cones, we took images immediately following either of two bleaching conditions, illustrated in Fig. 5.4. In the first bleaching condition, a dark-adapted retina was exposed to a 650-nm light that selectively bleached the L-cone pigment. In the second, a 470-nm light selectively bleached the M-cone pigment. The absorptance image for the 650-nm bleach at two locations for JW and one location for AN, shown in Figs 5.5A, 5.5C, and 5.5E, reveal dark, low-absorptance L-cones that have been heavily bleached and bright, highly absorbing M-cones spared

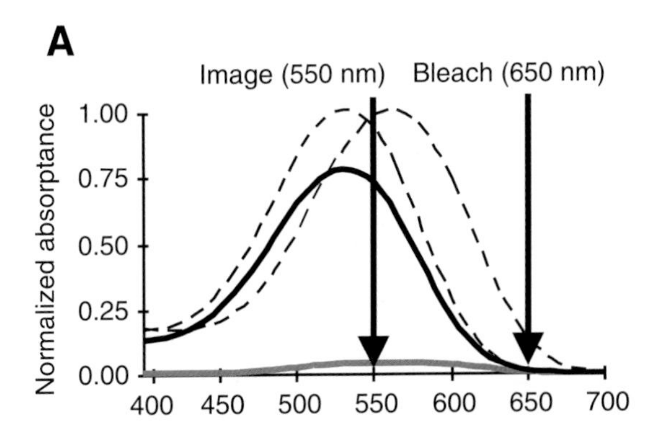

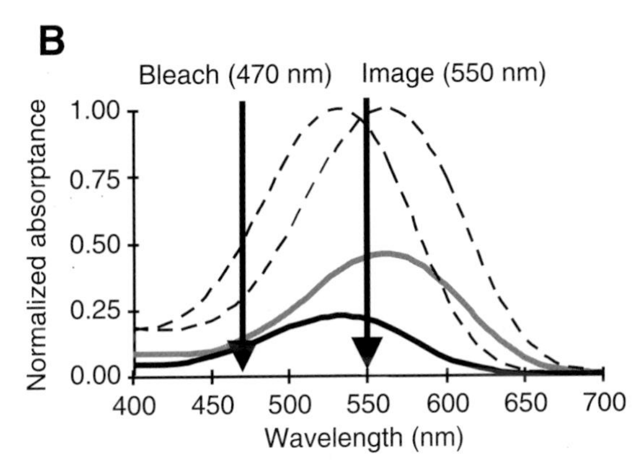

Figure 5.4: Method for selective bleaching. The imaging wavelength of 550 nm was used because it is near the peak of both M- and L-cone spectra, shown as dashed lines. (A) A 650-nm bleaching light that is used to selectively bleach away the L-cone pigment, the spectrum for which is shown in gray, sparing the M-cone pigment, the spectrum for which is shown in black. (B) A 470-nm bleaching light that is used to selectively bleach the M-cone pigment prior to taking an image. As illustrated, the 470-nm bleach is less selective due to the smaller difference in pigment at the blue end of the spectrum.

from bleaching. The absorptance images for the 470-nm bleach, shown in Figs 5.5B, 5.5D, and 5.5F, show selective bleaching of M-cones but the effect is smaller because of the similarity of M- and L-cone action spectra at this wavelength. The absorptance images for both of the bleaching conditions have higher pigment absorptance toward the fovea caused by the increase in outer segment length. Moreover, in the 470-nm bleaching condition, the increase in macular pigment toward the foveal center reduces bleaching at smaller retinal eccentricities, contributing to a higher pigment density there.

Bleaching levels had to be carefully set to maximize the difference in concentration between the L- and the M-cone classes because too much bleaching at any wavelength would leave no concentration of pigment in either cone. Using the wavelength of the bleaching light, and our best knowledge of the spectral absorptance of the L- and M-cones, we calculated their respective bleaching rates. Then we calculated at what point the maximum difference in concentration occurs as well as the total concentration of the pigment at that point. Given this concentration, we calculated the expected reflectance of the retina relative to the dark-adapted and fully bleached reflectance. We determined the optimal bleaching levels empirically, by regulating the bleach energy until the optimal retinal reflectance relative to the fully bleached and dark-adapted retinal reflectance was obtained. One caveat is that the optimal reflectance is dependent on the relative proportions of the L- and M-cones, which were unknown. In our experiment, we initially assumed an L/M-cone ratio of 2:1; as the experiment progressed we altered the bleaching energy depending on the properties of each subject's eye.

Relative numbers of the three cone classes

Figure 5.6 shows scatter plots for each of the three subjects. Each point in each scatterplot indicates the absorptances of a single cone following 470- and 650-nm bleaches. For the two trichromats, the cloud of points produces a bimodal distribution. The bimodality can be seen more easily in the corresponding histograms below each scatterplot. Each histogram shows the numbers of cones as a function of angle in the scatterplots. If these modes represent L- and M-cones, then only a single mode should be observed in a similar experiment on a protanope, who lacks the L-cone pigment. This prediction is confirmed by the data shown in Figs 5.6E and 5.6F. As with the trichromats, bleach-

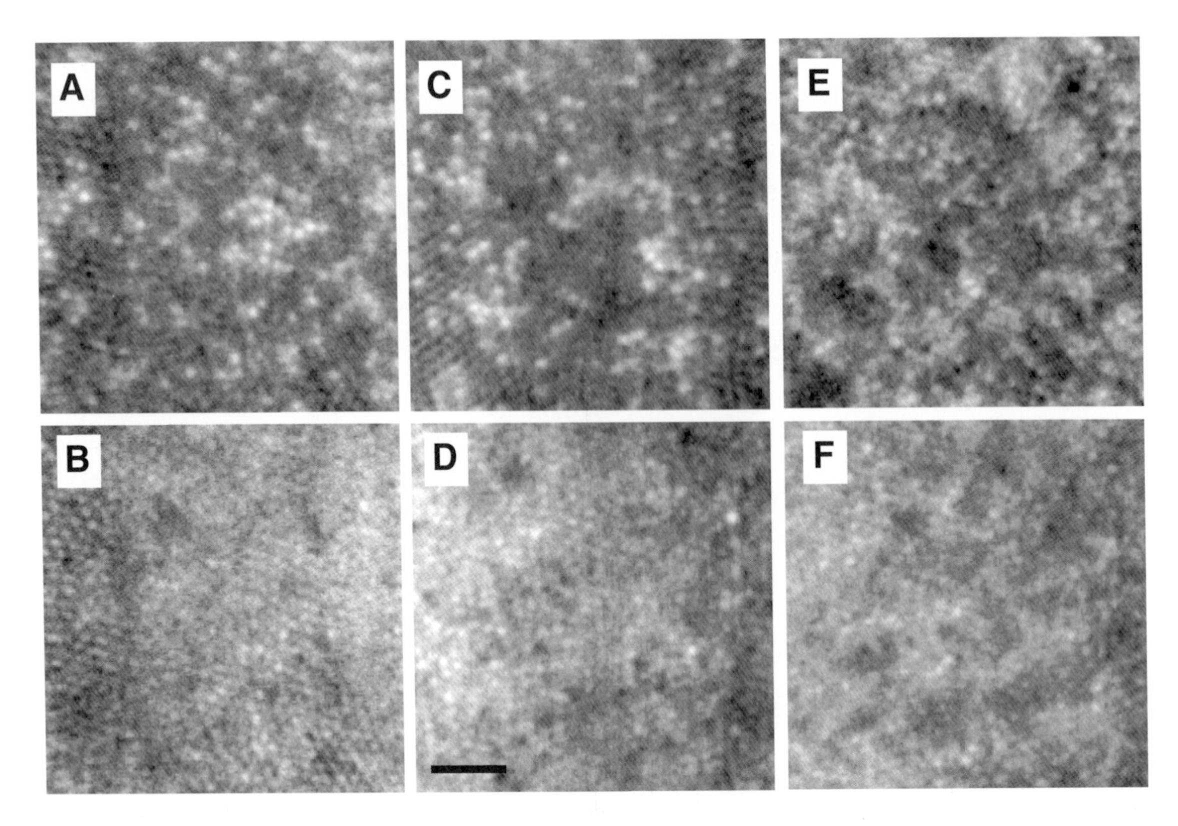

Figure 5.5: Absorptance images for JW temporal (A, B), JW nasal (C, D), and AN nasal (E, F). The upper row shows the absorptance image following a 650-nm bleach. The lower row shows the absorptance images following a 470-nm bleach. The scale bar represents 5 min of arc.

ing levels were chosen to optimize the chances of distinguishing two pigments.

For the trichromats, we fitted the sum of two Gaussians to the histograms. The angle corresponding to the intersection of the Gaussians was used to categorize L- and M-cones. The overlap of these Gaussians provided an estimate of the fraction of cones that were misidentified, which was 2.1% for JW and 5.6% for AN. Actually, these are only approximate estimates of the error since the distribution shapes are subject to such deterministic factors as optical blur and are not truly Gaussian. Table 5.1 summarizes the numbers of S-, M-, and L-cones for the two trichromats. The relative number of L- and M-cones differs greatly between them. In two patches of retina, one from the nasal and one from the temporal fovea, subject JW had a mean ratio of L- to M-cones of 3.79 whereas AN had a ratio of 1.15. Subject JW was not selected for this experiment based on prior knowledge of his color vision. However, AN was selected because measurements of his spectral ERG (Jacobs & Brainard, personal communication) had suggested that he was unusually green-sensitive, which was confirmed by our imaging observations. A comparison of the L- to M-cone ratios found by imaging with psychophysical measures and spectral ERG will be published elsewhere.

Arrangement of the three cone classes

The arrangement of the S-, M-, and L-cones for subjects JW and AN is shown in Fig. 5.7. Cones not labeled corresponded either to cones that lay beneath a capillary or were difficult to see in every image. The

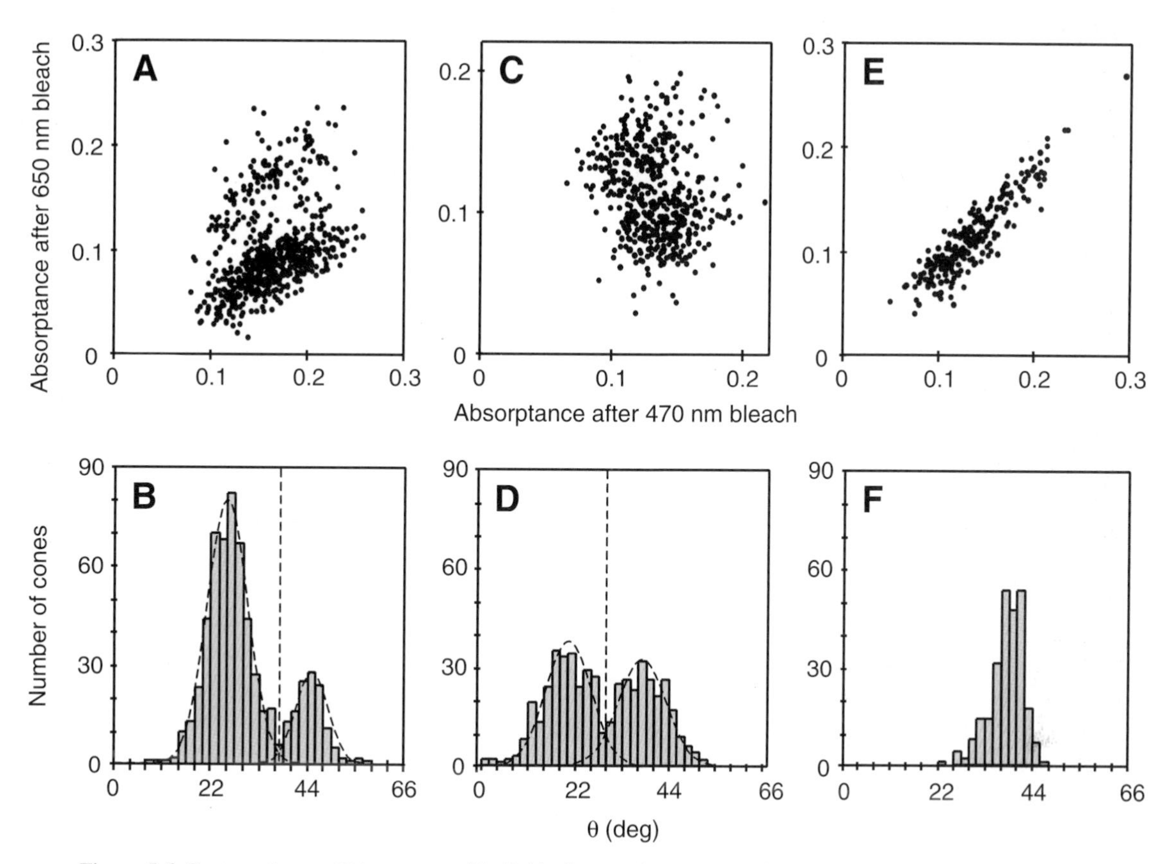

Figure 5.6: Scatter plots and histograms of individual cone absorptances. Scatter plots show the absorptance of each cone in a contiguous patch after the 470- and 650-nm selective bleaches. Cone absorptance was taken as the average value computed within a 0.4 arc min square region centered on the cone. Histograms show the distribution of cones as a function of angle in the scatter plots. For JW and MM, the origin from which we measured the angle was the point 0,0 on the scatterplot; but for AN, the best separation in the histogram was obtained when we chose an origin of 0.0 horizontal and 0.04 vertical. We justified shifting the origin because the presence of the macular pigment, which affected only the 470-nm bleach (see text), tended to elongate the distributions horizontally in the scatterplot. Panel (A) is derived from the absorptance images of Figs. 5.5A and 5.5B with the S-cones removed. Panel (C) is similarly derived from Figs. 5.5E and 5.5F.

distribution of the sparse S-cones is not significantly different from random in either trichromat. This agrees with Curcio et al. (1991), who found evidence of a developmental mechanism in humans that spaces S-cones in a regular, nonrandom manner in the peripheral retina. However, this mechanism is not evident near the fovea. Perhaps the extensive cone migration during the formation of the fovea disrupts the mechanism that otherwise spaces the S-cones more evenly.

To assess whether the packing arrangement of the cones was clumped, random, or regular, we performed a statistical test of the mosaic as described by Diggle (1983). For each patch of cones that we measured, we generated a histogram of all of the intercone distances between the M-cones. We generated a similar histogram from 100 simulations in which the same number of M-cones were randomly assigned within the same cone mosaic. An average histogram was generated from the 100 random simulations. The total number of intercone distances was the same between the experimental data set and the 100 random simulations, so the area under the histograms was the same. The advan-

Subject	JW	AN
Number of cones	1462	522
L-cone (%)	75.8	50.6
M-cone (%)	20.0	44.2
S-cone (%)	4.2	5.2
Error (%)	2.1	5.6
L:M-cone ratio	3.79	1.15

Table 5.1: Cone numbers for two subjects. The error in the assignments of L- and M-cones is taken as the fractional area of the intersection of the sum of two Gaussians used to fit the histograms in Fig. 5.6.

tage of this analysis over a nearest-neighbor analysis, for example, is that it allows us to examine departures from randomness at any distance scale. If the measured value in the histogram lied within 95% of the range spanned by the random simulations for each distance in the histogram, we concluded that the array was not significantly different from random. If there were clumping in the measured cone array, there would be a preponderance of short distances that would cause an increase in the frequency of lower distance values compared to the random histogram. A regular arrangement, on the other hand, would have an absence of short distances that would cause an increase in the frequency of intermediate and longer distances.

In neither subject were the M-cones uniformly dispersed among the L-cones. JW's array was no different from random at either location. AN's array showed significant clumping of the data ($p < 0.01$), but the possibility of a random assignment cannot be ruled out. The reason for this is that a small fraction of the cones are misidentified, perhaps 6% or so in AN. Because of residual optical blur in the images that we obtained, the cones that are most likely to be misidentified are those that are surrounded by cones of the opposite type. For example, the image of an M-cone in a field of L-cones will contain a substantial amount of light from its neighbors, which will make it look more like an L-cone. Similarly, an M-cone in a field of M-cones is least likely to be misidentified. One can see from this argument that optical blur will encourage clumps of cones of like class. Simulations in which we have identified 6% of cones at the centers of clumps in AN's array and deliberately changed their identity make the distributions of cones that we observed indistinguishable from a random assignment rule. It is conceivable that the clumping could arise if, for example, progenitor cells biased the decisions of their progeny to express either L- or M-cone opsin, but this will have to await measurements on AN's eye with further improvements in image quality. What is quite clear is that neither subject showed a packing geometry that is very different from a random one, and there is no evidence of a regular packing scheme in which M-cones are uniformly dispersed among L-cones. This result is consistent with the available data on primate M- and L-cone topography. Microspectrophotometry (Mollon & Bowmaker, 1992) on small foveal patches of excised talapoin retina indicated a random assignment of L- and M-cones, and photopigment transmittance imaging in a macaque peripheral retina (Packer et al., 1996) showed if anything a tendency toward aggregation. Calkins et al. (1994) found a bimodal distribution of excitatory synapses from bipolar cells onto midget ganglion cells in primate retinas. Under the assumption that the two modes correspond to L- and M-cone input, they inferred that the L- and M-cone mosaic was random. These data taken together suggest that the lack of a regular packing scheme for these cone classes may be ubiquitous among Old-World primates.

Implications

The three cone classes are interleaved in a single mosaic, so that at each point in the retina only a single class of cone samples the retinal image. As a consequence, observers with normal trichromatic color vision are necessarily color blind on a local spatial scale (Williams et al., 1991). The limits that this places on vision depend on the relative numbers and arrangement of cones. What are the implications of this arrangement for vision? The coarse grain of the cone

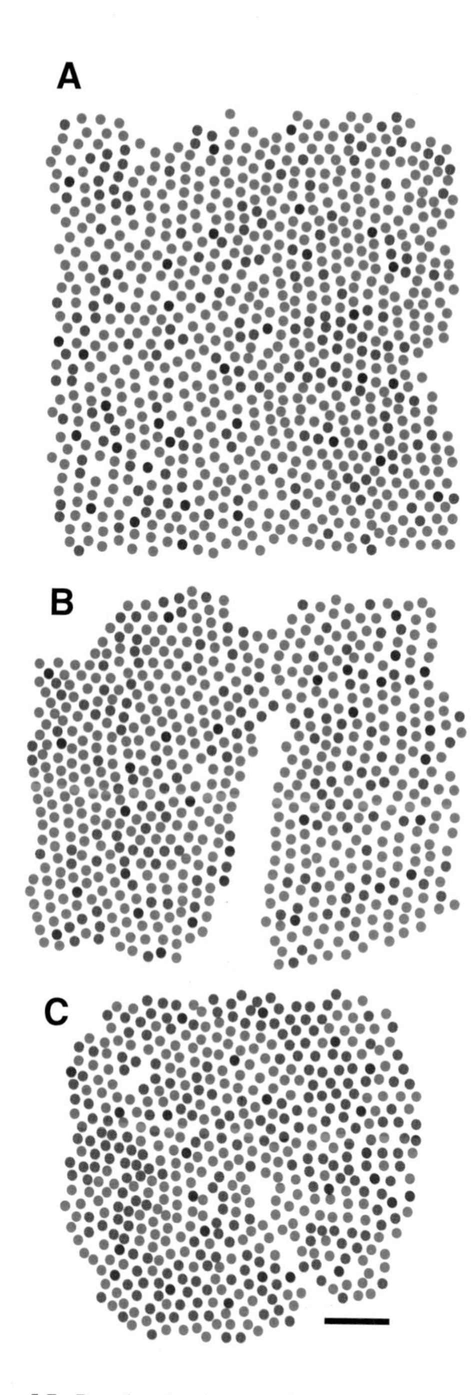

Figure 5.7: Pseudocolor image of the trichromatic cone mosaic. Blue, green, and red colors represent the S-, M-, and L-cones, respectively. (A) and (B) Subject JW, temporal and nasal retina, 1-deg eccentricity, respectively. (C) Subject AN, nasal retina, 1-deg eccentricity. The scale bar represents 5 min of arc.

submosaics causes fluctuations in the color appearance of tiny, monochromatic light flashes (Holmgren, 1884; Krauskopf, 1964; Cicerone & Nerger, 1989) because the relative excitation of different cone classes depends on the location of the flash. A related illusion is Brewster's colors, the perception of irregular splotches of pastel colors, while viewing periodic, black and white patterns of high spatial frequencies (Brewster, 1832; Williams et al., 1991). Similarly, red-green isoluminant gratings with spatial frequencies above the resolution limit look like chromatic and luminance spatial noise (Sekiguchi, Williams, & Brainard, 1993). All of these perceptual errors are examples of aliasing that is produced when the three cone submosaics inadequately sample the retinal image. They are akin to the errors in images taken with digital cameras that have interleaved pixels of different spectral sensitivities (Williams, Sekiguchi, & Brainard, 1993).

The S-cone mosaic can afford to sample the visual scene more coarsely than the M- and L-cones because axial chromatic aberration usually causes that mosaic to see a smoother retinal image that contains less energy at the high spatial frequencies that are prone to aliasing. Nonetheless, it is remarkable that the central 20 min of the fovea can typically lack S-cones without leaving a trace in ordinary visual experience, suggesting the existence of sophisticated interpolation machinery downstream from the receptors (Brainard & Williams, 1993).

The fact that M- and L-cones are more or less randomly assigned exacerbates sampling errors for some stimuli. Both subjects have retinal patches of 5 arc min or more across that contain only one of the two longer wavelength-sensitive cone classes. Whereas these patches imply that the trichromat will sometimes misjudge the color appearance of tiny objects, they may provide a benefit for recovering high-frequency luminance patterns because cortical neurons tuned to high spatial frequencies are more likely to be fed by contiguous cones of the same class (Gowdy & Cicerone, 1998; Roorda & Williams, 1999). This may explain the observation that, in some normal trichromats, there is little or no difference in resolution for gratings seen only with M-cones, L-cones, or when both cone classes operate together (Williams, 1990). Only when one cone class is greatly underrepresented, however, as in some heterozygous carriers for congenital X-linked protanopia, is resolution clearly mediated by the more dense submosaic (Miyahara et al., 1998). Even at high spatial frequencies, where aliasing might become a threat to clear vision, perhaps a number of factors conspire to make the L- to M-cone ratio relatively unimportant. The statistics of natural scenes make high-contrast, high spatial frequency signals rare events, optical blurring in the eye reduces the potential for aliasing, and clever postreceptoral processing based on prior information about natural visual scenes may also tend to hide the apparently haphazard organization of the trichromatic mosaic.

The large individual difference in L/M-cone ratio between the two trichromats is consistent with the variability found using psychophysical methods (Rushton & Baker, 1964; Vimal et al., 1989), spectral ERG (Jacobs & Neitz, 1993; Jacobs & Deegan, 1997), microspectrophotometry (Bowmaker & Dartnall, 1980; Dartnall et al., 1983), and mRNA analysis (Hagstrom et al., 1998; Yamaguchi et al., 1998). Evolution has not driven all human eyes toward some optimum proportion of M- and L-cones. We do not know what genetic and developmental factors are responsible for these individual differences, but it is striking how little difference the ratio of L- to M-cones makes in color vision, except when the ratio becomes very extreme. Rushton and Baker (1964) provided evidence that individual differences in the relative amounts of M- and L-cone pigments assessed with retinal densitometry were correlated with variations in the shape of the photopic luminosity function measured with flicker photometry. It seems likely that an individual's photopic luminosity function directly reflects L- and M-cone numerosity (see Chapters 2 and 6). All the same, these changes are not substantially larger than variations in other factors, such as preretinal absorption and the cone spectra themselves, so that it is rather difficult to accurately estimate cone ratios from flicker measurements. The photopic luminosity function changes relatively little with the L/M-cone ratio. Jacobs and Deegan (1997) showed that the L- and M-cone contributions to ERG flicker photometry average about 1-to-1 in macaque monkeys and 2-to-1 in humans. This twofold difference in the L/M-cone ratio produces only slightly more than a 0.1 log unit difference in sensitivity from 550 to 660 nm between the two species.

Pokorny, Smith, and Wesner (1991) argue that the perception of redness and greenness at low spatial frequencies where aliasing is not an issue, depends rather little if at all on the L- to M-cone ratio. They suggest that the variation in unique yellow observed in the population would predict variability in the L/M-cone ratio about three times smaller than the variability in L/M-cone ratios expected from heterochromatic flicker photometry. This would suggest that the zero crossing of the red-green opponent mechanism is set by some other factor than cone numerosity. Pokorny and Smith (1987) proposed that the normalization of the chromatic mechanism may depend on the average white of the individual's environment. Setting the red-green zero point to correspond to the ambient chromaticity of natural scenes would be a more efficient choice for the visual system, because it would minimize the metabolic cost, in terms of action potentials, of transmitting variations in redness and greenness. However, this view is not universally shared. Cicerone (1990) found a correlation between measurements of unique yellow and the L/M-cone ratio estimated from detection experiments with small flashes. The method described here allows us to address issues such as this because it is now possible to assess visual performance in eyes for which the topography of the trichromatic mosaic is known.

Acknowledgments

The authors wish to thank David Brainard, Dennis Dacey, Jerry Jacobs, Junzhong Liang, Donald Miller, and Orin Packer for their assistance. We acknowledge financial support from the Fight for Sight research division of Prevent Blindness America to AR and from the National Eye Institute and Research to Prevent Blindness to DRW.

6

The ecology and evolution of primate color vision

Jan Kremers, Luiz Carlos L. Silveira, Elizabeth S. Yamada, and Barry B. Lee

Color vision is the ability to distinguish between lights exclusively on the basis of differences in spectral distributions. Two requirements have to be met for color vision to occur: Two or more different photoreceptor types with different absorption spectra must be present in the eye; and postreceptoral mechanisms (in the eye or more centrally) must be present that can compare the outputs of the photoreceptor types (Jacobs, 1993). For human vision, Young (1802) and later Helmholtz (1867) postulated the existence of three different photoreceptor types in the eye to account for color mixture experiments, in which almost all colors can be achieved by mixing three primaries. The idea of distinct postreceptoral mechanisms is implicit in Hering's proposal (1878) of red-green and blue-yellow opponent processes, which can be viewed in terms of comparisons of receptor outputs. Although it has become apparent that Hering's opponent processes do not precisely correspond to the comparisons of receptor outputs that occur at the retinal level, the Young–Helmholtz and Hering proposals can be seen to refer to the first and second requirements for color vision, respectively.

In mammals, trichromatic color vision is only present in primates. Recent behavioral and genetic analyses of receptor types in different primate species have much advanced our knowledge of the evolution of color vision within this group. Excellent overviews of these developments may be found elsewhere (Mollon, 1989; 1991; 1996; Jacobs, 1993; Tovée, 1994). In this chapter, we summarize the behavioral and genetic results and also discuss comparative aspects of postreceptoral color processing. Although the genetic basis of the neural systems involved is unknown, the evolution of neural mechanisms of color vision must have occurred in parallel to the evolution of different receptor types and is of equivalent interest.

Within the vertebrates, the requirements for color vision have been met in different ways, suggesting that color vision has developed independently several times during evolution (Bowmaker, 1991b). Major differences can be found concerning the chemistry of photopigments and photoreceptor number and structure. For instance, the presence of appropriate metabolic pathways for vitamin A allows fishes, amphibia, and reptiles, but not birds and mammals, to possess two classes of photopigments, vitamin A_1–derived rhodopsins and vitamin A_2–derived porphyropsins (Bowmaker, 1991b, 1991c). Tri-, tetra-, and even pentachromatic vision has developed in fish, amphibians, reptiles, and birds. In addition, in animals belonging to these taxons, color vision is influenced by the presence of oil droplets or ellipsosomes, which are absent in mammals (Bowmaker, 1991b, 1991c). Further information as to links between color vision in different species has been derived from comparing amino-acid sequences of their pigments (e.g. Dulai et al., 1994). For postreceptoral mechanisms, less information is available, but major differences are certainly present. For example, in fishes and reptiles, horizontal cells of the retina are the first stage in color processing (Kolb & Lipetz, 1991). However, this does not seem to be the case in primates (Dacey et al., 1996; see also Chapter 9).

Besides rod rhodopsins with maximal absorption between 490 and 500 nm, almost all mammals possess one cone photopigment with peak absorption in the short-wavelength (S) range, between 420 and 450 nm, and at least one other cone photopigment with maximal absorption between 500 and 570 nm (Jacobs, 1993). This consistency suggests a common ancestral pattern. Rod signals pass through rod bipolar cells and the AII amacrine cells to ganglion cells via the cone bipolars. This convergence on the cone system implies that the rod pathway may be phylogenetically younger than the cone system (see Chapter 7). Evidence as to postreceptoral mechanisms comparing cone outputs is sparse in mammals other than primates, but some ganglion cells in the cat show evidence of cone opponency with excitation from the S-cone and inhibition from the longer wavelength cone (Cleland & Levick, 1974).

Jacobs (1993) pointed out that the correlation between the number of cone types and behavioral evidence of color vision, that is, between genotype and phenotype, is very strong, so that it is commonly assumed that color vision occurs when more than one cone type is present and that the dimension of color vision (di-, tri-, tetra-, or pentachromacy) coincides with the number of cone types (two, three, four, or five, respectively). However, it should be stressed that this remains an assumption. Some female carriers of color vision deficiencies in humans might be an exception (Jordan & Mollon, 1993; see Chapter 1).

Physiological and anatomical nomenclature

In the trichromatic Old-World monkeys, the short-, middle-, and long-wavelength–sensitive cones are often termed the S-, M-, and L-cones, respectively. The term "S-cone" is appropriate for short-wavelength cones of all primates (and indeed for other mammals), but in view of the polymorphism in New-World monkeys we will identify their pigments by their peak absorption wavelengths and refer to them generically as "long-wavelength" pigments.

In the Old-World monkey, two parallel pathways begin in the midget and parasol ganglion cells and pass through the parvocellular and magnocellular layers of the lateral geniculate nucleus (LGN), respectively. These cells and pathways are often designated P and M, respectively, a classification based on a combination of anatomical and physiological observations (Shapley & Perry, 1986). One aim of this chapter is to establish that this close correspondence between anatomy and physiology also extends to New-World monkey species, justifying the use of the same terminology. In addition, ganglion cells with S-cone input form a separate system; they have been identified as the small-field bistratified cells of the retina. A more extensive account of these pathways is given elsewhere in this volume (Chapters 9 and 10).

Experimental techniques

The resurgence of interest in the evolution of color vision in primates stems from molecular biological and behavioral studies on New-World monkeys (Jacobs, Neitz, & Neitz, 1993; Mollon, Bowmaker, & Jacobs, 1984; Williams et al., 1992; Hunt et al., 1993). Sequences of visual pigment genes can be determined by standard molecular biological techniques. The genomic structure itself is complex (see Chapter 1), with frequent multiple copies of the M-gene in humans. Multiple copies of the M-gene are present in other Old-World monkey species (Ibbotson et al., 1992; Dulai et al., 1994). Comparisons of base sequences further indicate that the amino-acid sequences and the inferred absorption spectra of the L- and M-photopigments do not differ greatly between different Old-World monkey species (Dulai et al., 1994).

Behavioral studies involve training animals to distinguish or detect different wavelengths of light (Jacobs, 1981). Such behavioral experiments can be laborious, and an electroretinographic technique that requires brief anesthesia of the subject, has proved to be a very effective way of screening large numbers of animals (Neitz & Jacobs, 1984; Jacobs, Neitz, &

Krogh, 1996). For anatomical and physiological experiments, standard methods developed on Old-World monkeys have been employed in New-World monkeys as well (Lee, Martin, & Valberg, 1988; Lee et al., 1990).

Primate evolution

The first prosimian-like animals occurred about 60 million years ago (MYA) in North America and later in Europe (Napier & Napier, 1985). Recently, Martin (1993) argued that a sparsity of fossils might lead to an underestimation of the time the first primates occurred. He suggested that primates occurred about 80 MYA. The origins of prosimians and anthropoids (or simians) is still unresolved due to sparsity in the fossil record. Prosimians probably originated on the African continent, and it is now believed that the common ancestor of the New-World monkeys (platyrrhines) and the Old-World monkeys, apes, and humans (catarrhines) also originated in Africa about 40 MYA. The invasion of the South American continent by the platyrrhine ancestor is thought to have taken place by rafting and island hopping, because South America and Africa had already separated. But, again, the sparsity of fossils might have led to an underestimation of the time at which catarrhines and platyrrhines separated. More recent estimates suggest that the ancestral anthropoids already evolved about 55 MYA (Martin, 1993). According to Martin (1990), the common ancestor of the modern anthropoids had an arboreal life-style and nocturnal habits. A more recent viewpoint (Kay, Ross, & Williams, 1997) is that the stem anthropoids were relatively small animals with a diurnal life-style and a combined diet of insects and fruit. An active arboreal life-style is thought to have brought about the forward rotation of the orbits and neural developments for stereoscopic vision. The ancestral New-World monkeys have radiated extensively but have remained largely arboreal. Traditionally, two groups have been commonly recognized: the callitrichids, a group of small-bodied species including marmosets and tamarins; and the cebids, which include squirrel, capuchin, and spider monkeys. But according to Fleagle (1988) the taxon of the cebids only includes the squirrel and capuchin monkeys (and possibly the owl and titi monkeys); a third group, the atelids, would include all others. The cebids are possibly more closely related to the callitrichids than to the atelids. The catarrhines have not remained exclusively arboreal and have invaded savanna and other habitats. They are divided into the cercopithecids (including macaques and baboons) and the hominoids (the apes and humans). Figure 6.1 gives the classification of living primates according to Napier and Napier (1985), Martin (1990; 1993), and Fleagle (1988). Although the term "Old-World monkey" would in the proper sense not include humans and apes, we use it in this chapter as a synonym for catarrhines.

Pigments in living primates

The living primates can be subdivided into three major groups: the prosimians or strepsirhini, the tarsiers, and the anthropoids. In the following we discuss each group briefly. Nowadays, tupaias are no longer considered to be primates; instead, they are classified as a separate order: the scandentia. They will therefore not be considered in this chapter.

Although different authors define prosimians in different ways, we use here the definition of Napier and Napier (1985), in which prosimians include only the infraorders of lorisiformes and lemuriformes. Tarsiers are not included in this group. Although the term "prosimian" might give the impression of a primitive group, many members of this suborder are highly specialized.

All prosimians studied so far are not trichromatic (Jacobs, 1993). The nocturnal *Galago* has cones, but their density is very low and S-cones are lacking (Wikler & Rakic, 1990). Thus, *Galago* is probably a monochromat after a secondary loss of S-cones; the same appears to be true of *Aotus* (the owl monkey), the only nocturnal anthropoid (Wikler & Rakic, 1990). The diurnal lemurs are probably dichromats (Jacobs, 1993).

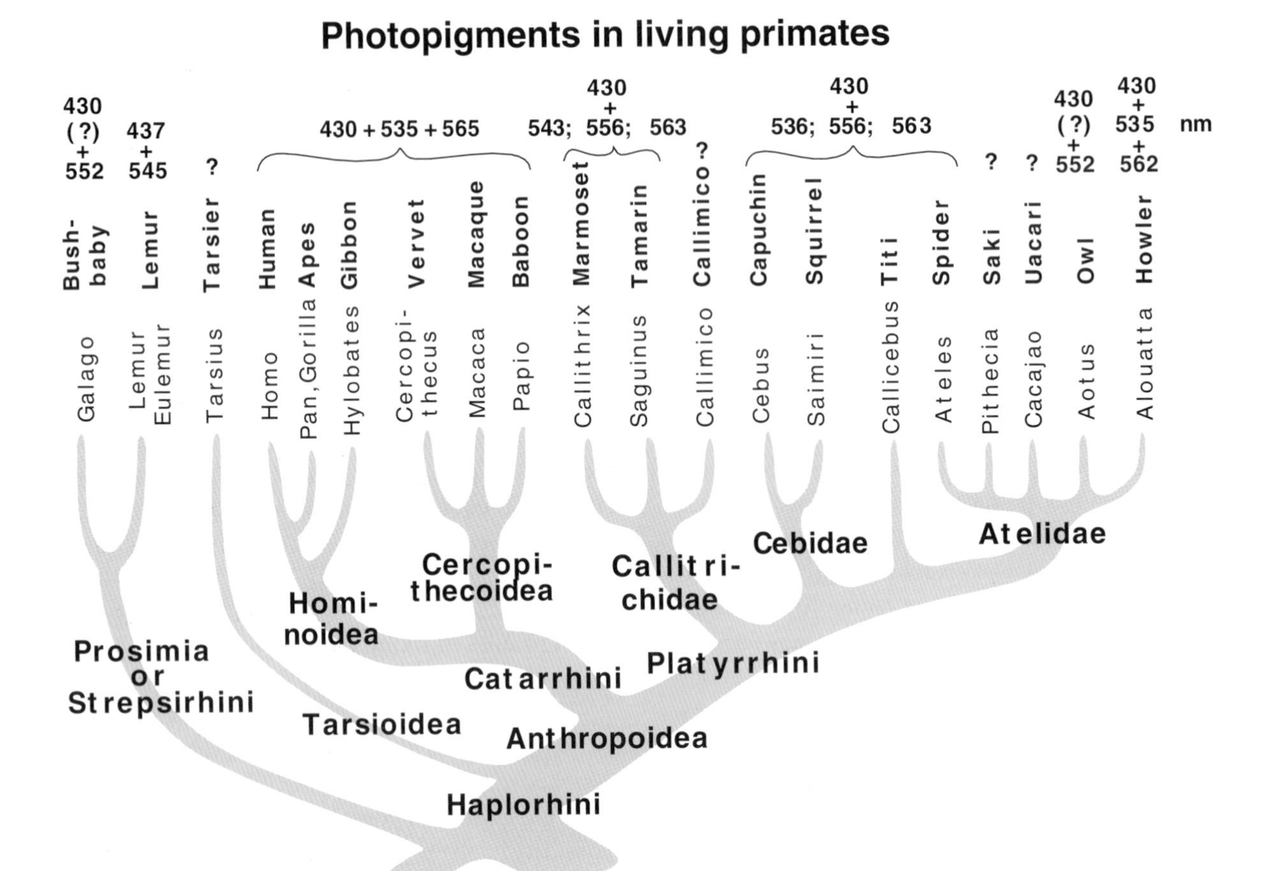

Figure 6.1: The classification of living primates based on Napier and Napier (1985), Martin (1990; 1993), and Fleagle (1988). A summary of photopigments found in the various groups is included. For each taxon, some representative genera are presented, with emphasis on the New-World primates because of the different pigment combinations within this group. The photopigment absorption spectra were determined using different techniques, including microspectrophotometry, electroretinography, and behavioral techniques. The pigment data were gathered from several sources (Dartnall et al., 1965; Dartnall, Bowmaker, & Mollon, 1983; MacNichol et al., 1983; Baylor, Nunn, & Schnapf, 1987; Harosi, 1987; Jacobs & Neitz, 1987a, 1987b; Jacobs, Neitz, & Crognale, 1987; Petry & Harosi, 1990; Bowmaker, 1991a, 1991b; Tovée, Bowmaker, & Mollon, 1992; Jacobs et al., 1993, 1996a; Jacobs & Deegan, 1993b; Jacobs & Deegan, 1993a; Jacobs et al., 1996b).

The Tarsoidea form a separate suborder and may have developed independently of prosimians and anthropoids (Napier & Napier, 1985). Another view is that the tarsiers and the anthropoids belong to a monophylogenetic group, the haplorhines, in contrast to the strepsirhines, in which the prosimians are represented (Fleagle, 1988; Martin, 1993). Tarsiers are exclusively nocturnal and are the only primates that feed exclusively on prey. Nothing is known yet about color vision in these animals. The lateral geniculate body has an organization that is not found in any other group of primates (Rosa et al., 1996): The eye input to the parvocellular layers is reversed. In other primates, the superficial parvocellular layer receives input from the contralateral eye and the second layer from the ipsilateral eye. In contrast, the superficial parvocellular layer of the tarsiers receives input from the ipsilateral eye and the second layer from the contralateral eye. In view of the strong similarities in retinal organization between Old- and New-World monkeys, detailed in later sections, it would be of great interest to extend the morphological and physiological comparisons to tarsiers and prosimians.

Color vision in anthropoids (simians) is unique among mammals through the abundant occurrence of trichromacy. All Old-World monkeys are trichromats; there are no records of dichromatic catarrhines, with the exception of color defective humans (Jacobs, 1993). The visual pigments of catarrhines studied so far are relatively uniform (Tovée, 1994). The pigments of the platyrrhines are more divergent, and dichromacy and trichromacy can be present within the same species. For a summary of possible pigments, see Fig. 6.1. A more extensive overview of known absorption maxima of photopigments in primates is given by Tovée (1994).

Rods absorb maximally close to 500 nm; S-cones absorb maximally between about 420 and 435 nm. The gene for the S-cone opsin is located on chromosome 7. It is only ~40% homologous to genes for pigments in the long-wavelength range (see Chapter 1), and Mollon and Jordan (1988) argue convincingly that the chromatic system associated with the S-cone is phylogenetically more ancient than the one involving exclusively the L- and M-cones. This is now supported by evidence about the postreceptoral mechanism for this system, as will be discussed in a later section (see also Chapters 9 and 10).

The absorption maxima of the cones in the long-wavelength spectral region span a range between ~532 and ~565 nm. In Old-World monkeys, the genes for the L- and M-cone opsins both lie on the q-arm of the X-chromosome and are 97% homologous, indicating more recent divergence as compared with the S-cone gene (Dulai et al., 1994; see also Chapter 1). Their absorption maxima lie near 535 nm and 565 nm. Sequences only diverge at a few sites in those species of Old-World monkeys so far studied (Ibbotson et al., 1992; Dulai et al., 1994). The effect of different amino-acid substitutions within the chromophore on spectral tuning of pigments in human is discussed elsewhere in this volume (Chapter 1).

Platyrrhines display a polymorphism in their pigment complement. In most species, males and some females are dichromatic, whereas most females are trichromatic. The X-chromosome contains only one gene coding for a photopigment. This gene has three different alleles, and as a result there are three dichromatic and three trichromatic phenotypes (Mollon, Bowmaker, & Jacobs, 1984; Jacobs & Neitz, 1985). All males and females with the same gene on their two X-chromosomes fall into one of the dichromatic types. Females with different genes on their X-chromosome enjoy trichromacy (Jacobs, 1983, 1984; Mollon, Bowmaker, & Jacobs, 1984; Jacobs & Neitz, 1985). The three pigment alleles in the callitrichids have different absorption maxima than those of the cebids.

This pattern has until recently been thought to be standard for all platyrrhines apart from the owl monkey (*Aotus*). This nocturnal species appears to have only one allele and no functional S-cones due to a defect in the S-cone opsin gene (Wikler & Rakic, 1990; Jacobs et al., 1993). However, howler monkeys (genus: *Alouatta*; family: Cebidae) have recently been shown to be fully trichromatic (Jacobs et al., 1996a). This unexpected finding indicates that a search for other exceptions among the platyrrhines would be worthwhile.

Comparative anatomy and physiology of retino-geniculo-striate pathways of primates

Most reviews on the chromatic systems of primates have concentrated on photopigment spectral tuning and related phenotypes. Recently, some aspects of retinal anatomy and physiology that are relevant for color vision have been subjected to comparative studies. The retinal anatomy of macaques and humans resemble each other closely, which is probably why both species perform similarly in many basic visual tasks. These tasks include spatial and temporal visual acuity and contrast sensitivity as well as chromatic vision (DeValois, Morgan, & Snodderly, 1974; Harwerth & Smith, 1985), although some differences were observed in vernier acuity (Kiorpes, 1992). Some psychophysical findings may have their physiological and anatomical bases on the properties of different cell classes that are found in both macaques and humans. The physiological properties of macaque M-pathways can explain human performance in tasks such as heterochromatic flicker photometry (Lee, Martin, & Valberg, 1988), the minimally distinct border (Kaiser et al., 1990), and vernier acuity (Lee et al., 1993b, 1995). This indicates that the visual system of catarrhines is uniform across species and human visual functions can be understood in terms of the results obtained in macaques and other Old-World monkeys. For a more extensive review on the retinal anatomy and physiology concerning macaque color vision, see Chapters 9 and 10 in this book.

The question remains whether there are major differences in the retinal anatomy and physiology between Old- and New-World monkeys. This is important for several reasons. First, it addresses the evolution of the primate visual system, particularly how the different forms of color vision in platyrrhines and catarrhines may have changed retinal structure and function. Moreover, since dichromacy is common in many New-World monkeys but absent or at least very rare in Old-World monkeys, dichromatic platyrrhines might be good animal models for dichromatic human vision, whereas some trichromatic platyrrhines might be comparable with human anomalous trichromats (see Chapter 1).

To obtain a better understanding of the similarities and differences in the organization of the visual system within primates, it is necessary to compare the available anatomical and physiological data of these animals. In the next sections we restrict ourselves to the retina and the lateral geniculate nucleus. For comparative studies on the organization of the visual cortex we refer the readers to recent reviews (Sereno & Allman, 1991; Kaas & Krubitzer, 1991).

Retinal ganglion cell morphology

Anatomical studies of the retinal ganglion cell morphology in Old-World monkeys, apes, and humans have identified two major ganglion cell classes of the retino-geniculo-cortical pathway: the parasol, magnocellular (M-) cells (also called A- or P_{α}-cells) and the midget, parvocellular (P-) cells (B- or P_{β}-cells) (Dogiel, 1891; Polyak, 1941; Boycott & Dowling, 1969; Leventhal, Rodieck, & Dreher, 1981; Perry & Cowey, 1981; Rodieck, Binmoeller, & Dineen, 1985; Watanabe & Rodieck, 1989; Dacey & Petersen, 1992; Kolb, Linberg, & Fisher, 1992). The cell bodies of these cells are intermingled in the retinal ganglion cell layer, but their dendrites end in distinct levels of the inner plexiform layer where they are contacted by distinct types of bipolar cells (Watanabe & Rodieck, 1989), and their axon terminals segregate in different LGN layers, the magno- and parvocellular layers, respectively (Leventhal, Rodieck, & Dreher, 1981; Perry, Oehler, & Cowey, 1984). Furthermore, LGN magno- and parvocellular cells project to separate layers of the striate cortex (V1) (Lund, 1988; Henry, 1991). It has thus become common to distinguish two visual pathways (the M- and P-pathways) connecting the retinal ganglion cell layer through the LGN to the V1 input layers. These cell pathways are not only anatomically distinct; they also possess very distinct physiological properties, and the nomenclature P and M was suggested (Shapley & Perry, 1986) based on a combination of anatomical and physiological features. Although possession of magno- and parvocellular laminae is a characteristic primate feature, it is of considerable interest whether ganglion cells of New-World monkeys meet all of the diverse classification criteria for M- and P-cells.

Parasol retinal ganglion cells of Old-World monkeys have large cell bodies, thick axons, and large dendritic trees with radiated branching pattern, whilst midget cells have small cell bodies, thin axons, and small dendritic trees with a more bushy and dense branching pattern. Both parasol and midget cells occur in two subclasses, each having dendritic branching in the inner or outer half of the inner plexiform layer, cor-

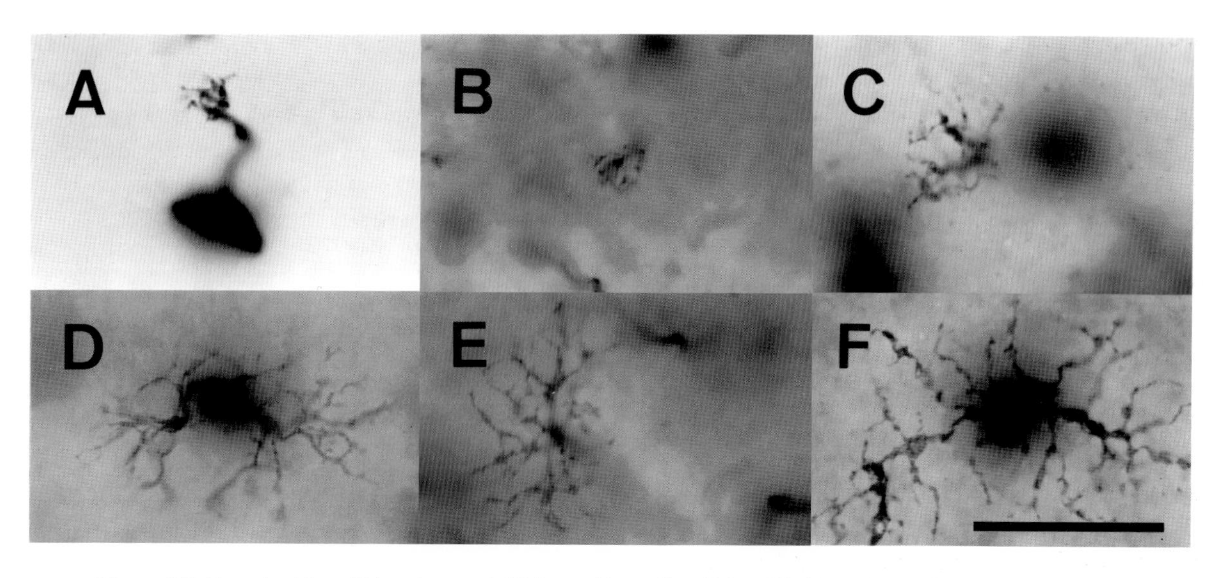

Figure 6.2: Foveal midget (P-) and parasol (M-) ganglion cells of New-World monkey retinas. All cells shown in this and the next two figures were retrogradely labeled with biocytin deposited into the optic nerve, using methods described in Piçanco-Diniz et al. (1992). (A) *Cebus* midget P-ON cell, 1-mm nasal to the fovea. (B) *Callithrix* midget P-ON cell, 0.4-mm temporal to the fovea. (C) *Aotus* midget P-OFF cell, 0.22-mm dorsal-temporal to the fovea. (D) *Cebus* parasol M-ON cell, 0.9-mm temporal to the fovea. (E) *Callithrix* parasol M-ON cell, 0.4-mm temporal to the fovea. (F) *Aotus* parasol M-OFF cell, 0.13-mm ventral-temporal to the fovea. Scale bar = 25 μm. Reproduced from Yamada et al. (1996) with permission from the Academia Brasileira de Ciências.

responding to the "ON" and "OFF" physiological subclasses (Dacey & Lee, 1994a).

In New-World monkeys, cells with a close anatomical resemblance to parasol and midget cells have been found, thus meeting the criteria for M- and P-cells. They have been identified in all species so far investigated: the diurnal squirrel monkeys, *Saimiri* (Leventhal et al., 1989; Leventhal, Thompson, & Liu, 1993); capuchin monkeys, *Cebus* (Lima, Silveira, & Perry, 1993; 1996; Silveira et al., 1994; Yamada, Silveira, & Perry, 1996); marmosets, *Callithrix* (Ghosh et al., 1996; Goodchild, Ghosh, & Martin, 1996; Yamada et al., 1996); and the nocturnal owl monkeys, *Aotus* (Lima, Silveira & Perry, 1993; Silveira et al., 1994; Yamada et al., 1996). In Fig. 6.2, we show examples of cells with midget and parasol morphologies from central retinal regions of the *Cebus*, *Callithrix,* and *Aotus*. Midget, P-like cells are shown in the upper row and parasol, M-like cells in the lower row. In these and subsequent illustrations retinal ganglion cells were retrogradelly labeled after optic nerve implantation of biocytin, a neurotracer that reveals their entire dendritic tree (Piçanco-Diniz et al., 1992). Cells from dichromatic males of two diurnal genera, *Cebus* (panels A and D) and *Callithrix* (panels B and E), are smaller in size than their homologues from the nocturnal *Aotus* (panels C and F).

With increasing retinal eccentricity, both cell classes decrease in density and increase in size. Examples of peripheral P- and M-cells in both diurnal and nocturnal New-World monkeys are shown in Figs 6.3 and 6.4, respectively.

Figure 6.5 displays the dendritic field size of parasol, M-like and midget, P-like cells as a function of retinal eccentricity for different species. The curves shown are the best-fitting second- or third-order polynomials. Dendritic field size is expressed as the diameter (in μm) of the circle with equivalent area, and eccentricity is given in degrees of visual angles to normalize for differences in eye size. Morphological data for *Cebus* (Yamada, Silveira, & Perry, 1996) and *Callithrix* (Yamada et al., 1996) were obtained from male

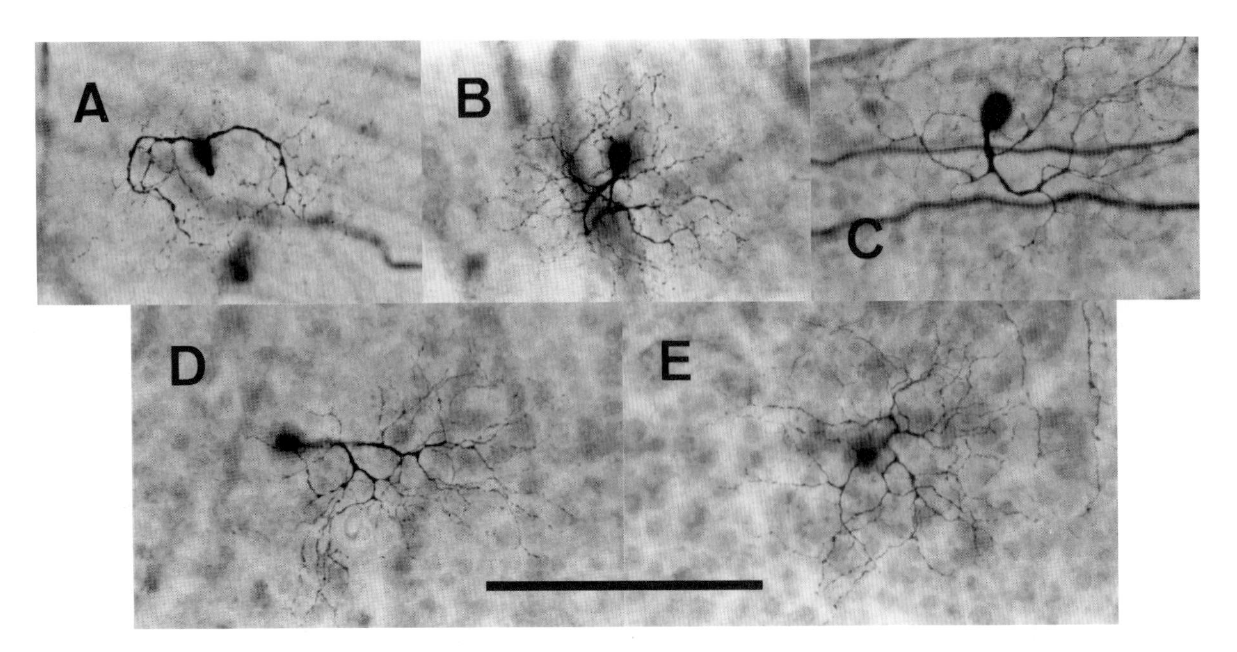

Figure 6.3: Peripheral P-cells of New-World monkey retinas. (A) *Cebus* midget P-OFF cell, 12.3-mm dorsal to the fovea. (B) *Cebus* midget P-ON cell 11.4-mm dorsal to the fovea. (C) *Callithrix* midget P-ON cell, 7.8-mm dorsal to the fovea. (D) *Aotus* midget P-OFF cell, 9.9-mm dorsal to the fovea. (E) *Aotus* midget P-OFF cell, 9.6-mm dorsal to the fovea. Scale bar = 100 μm. Reproduced from Yamada et al. (1996) with permission from the Academia Brasileira de Ciências.

dichromats. We also plot available data for *Macaca* (Perry, Oehler, & Cowey, 1984) and humans (Dacey & Petersen, 1992) to compare platyrrhine parasol and midget cells with their homologues in catarrhines. At all (temporal and nasal) retinal eccentricities, diurnal New-World anthropoids (*Cebus* and *Callithrix*) as well as diurnal Old-World anthropoids (*Macaca*) have comparable parasol and midget dendritic field sizes. However, ganglion cells of the nocturnal platyrrhine *Aotus* have larger dendritic trees. Interestingly, with the exception of central midget cells, all dendritic field sizes of human ganglion cells are larger than those of the other diurnal anthropoids.

Thus, both dichromatic platyrrhines and trichromatic catarrhines seem to possess parasol and midget ganglion cells with very similar morphologies and retinal distribution profiles. Consistent with this similar pattern, it has recently been shown that there are no differences in ganglion cell morphology and photoreceptor convergence between dichromatic and trichromatic marmosets (Ghosh et al., 1996; Goodchild, Ghosh, & Martin, 1996).

Foveal midget cells of diurnal New-World monkeys, such as *Cebus* and *Callithrix*, have very small dendritic trees, ranging from 5.5 to 11 μm (Ghosh et al., 1996; Goodchild, Ghosh, & Martin, 1996; Yamada, Silveira, & Perry, 1996; Yamada et al., 1996). This is about the same size as in the Old-World monkey, where midget cells are of the right size to connect with the axon terminal of a single midget bipolar cell (Polyak, 1941; Boycott & Dowling, 1969; Kolb & Dekorver, 1991; Calkins et al., 1994). In the central retina of Old-World monkeys, there is one ON- and one OFF-midget bipolar cell per cone. Midget bipolar cells are not only present in the central retina but also in the far retinal periphery (Polyak, 1941; Boycott & Dowling, 1969; Boycott & Hopkins, 1991; Martin & Grünert, 1992), where several bipolar cells converge onto one ganglion cell. The midget, P-like ganglion cell morphology in New-World monkeys suggests the

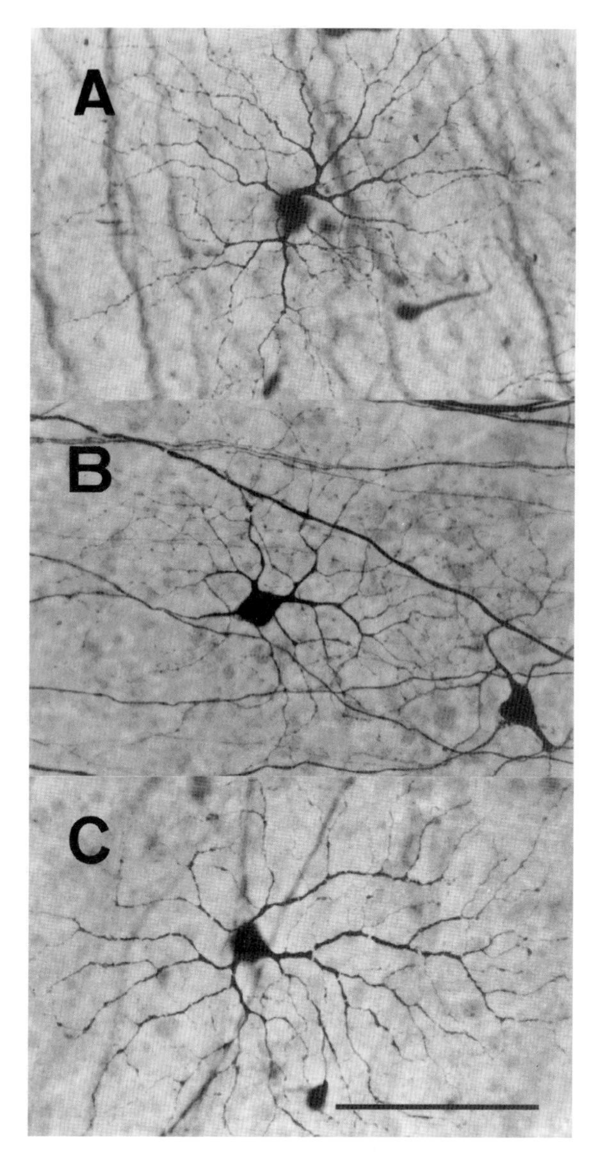

Figure 6.4: Peripheral M-cells of New-World monkey retinas. (A) *Cebus* parasol M-OFF cell, 11-mm dorsal to the fovea. (B) *Callithrix* parasol M-ON cell, 7.8-mm dorsal to the fovea. (C) *Aotus* parasol M-ON cell, 10-mm nasal to the fovea. Scale bar = 100 μm. Reproduced from Yamada et al. (1996) with kind permission from the Academia Brasileira de Ciências.

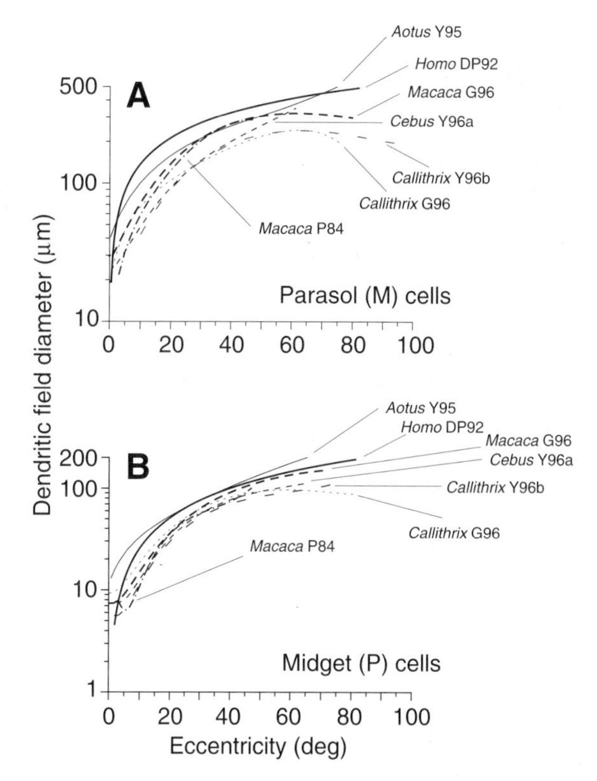

Figure 6.5: (A) Dendritic field diameter of parasol ganglion cells for different primate species. Best-fitting polynomials are shown. Eccentricity was transformed to angular distance (degrees of visual angle) after applying the following equations. *Cebus*: $y = 0.04\,x^2 + 4.8\,x + 0.2$ (Yamada, Silveira, & Perry, 1996). *Aotus*: $y = 0.046\,x^2 + 5.09\,x + 0.25$. *Callithrix*: $y = 0.0538\,x^3 - 0.542\,x^2 + 9.396\,x - 0.75$ (Troilo, Howland, & Judge, 1993). *Macaca*: $y = 0.038\,x^2 + 4.21\,x + 0.1$ (Dacey & Petersen, 1992). In these equations, x represents linear distance in mm, and y angular distance in degrees. (B) Dendritic field diameter of midget ganglion cells for the same species.

presence of midget bipolar cells in these species; this has recently been confirmed (Silveira et al., 1998). Thus it is likely that in both dichromatic and trichromatic diurnal New-World monkeys, as in trichromatic diurnal Old-World monkeys, foveal midget cells receive input through "private lines" from cones to midget bipolar cells, and the cone-specific signal thus provided for the central retina persists far into the periphery.

The platyrrhine nocturnal owl monkey (*Aotus*) seems to be a partial exception to this general picture. Owl monkeys differ from the other New-World monkeys in that they are monochromats: They lack S-cones and do not exhibit the polymorphism of the long-wavelength–sensitive visual pigments (Wikler &

Rakic, 1990; Jacobs et al., 1993). It is likely that the lower cone and ganglion cell density in this genus, particularly in the central retina, represents a secondary adaptation related to a nocturnal life-style. *Aotus* foveal midget cells have larger dendritic trees than their homologues in *Cebus* and *Callithrix* retinae (Silveira et al., 1994; Yamada et al., 1996). Nevertheless, the dendritic fields of *Aotus* foveal midget retinal ganglion cells have a similar size to the axon terminals of putative midget bipolar cells (Ogden, 1974). Also, the ratio of midget ganglion cell to cone density in the central retina is close to 2:1 (Wikler & Rakic, 1990; Silveira, Perry, & Yamada, 1993). Although Ogden (1974) described midget bipolar cells connected to two cones, a single cone–midget bipolar–ganglion cell pattern might also be present in nocturnal *Aotus*, but this requires further verification. Unfortunately, it is not known yet whether generally one or two cones are connected to midget bipolar cells in platyrrhines. This might provide some additional information about the role of color vision on the retinal wiring.

A third retinal ganglion cell class of catarrhines has recently aroused interest: the small-field bistratified cells, which have dendrites branching in both the inner and outer halves of the inner plexiform layer (Rodieck, 1991; Dacey, 1993a; Rodieck, Brening, & Watanabe, 1993) and which receive bipolar cell synaptic input conveying information from S-cones (Dacey & Lee, 1994b). They correspond to the blue-ON cells recognized electrophysiologically. It is argued elsewhere in this chapter (see also Chapters 9 and 10) that these cells are not part of the parvocellular system; for example, there is evidence that blue-ON cells in the LGN are located in the intercalated layers (Martin et al., 1997), which in turn project to V1 layers 1–3 (Hendry & Yoshioka, 1994). This indicates that small-field bistratified cells form part of a third information processing pathway going from the retina to specific LGN layers and V1 compartments (Casagrande, 1994; Hendry & Yoshioka, 1994; Yoshioka & Hendry, 1994; Martin et al., 1997).

Diurnal platyrrhines have small-field bistratified cells similar to those described in catarrhines (Ghosh et al., 1996; Yamada et al., 1996). Electrophysiological recordings of *Cebus* retina (Lee et al., 1996a; Silveira et al., 1997) and marmoset LGN cells (Kremers et al., 1997) have shown the presence of tonically responding blue-ON ganglion cells with properties similar to those recorded in macaques. In addition, it seems that diurnal platyrrhines also have a distinct retino-geniculo-striate pathway that is specialized for relaying S-cone information to the higher visual centers (Martin et al., 1997). Other than in *Cebus* and *Callithrix*, biocytin retrograde labeling of *Aotus* retinal ganglion cells failed to reveal a ganglion cell class morphologically similar to the small-field bistratified cell (Yamada, 1995). This seems to be consistent with the other data showing that *Aotus* lacks functional S-cones (Wikler & Rakic, 1990; Jacobs et al., 1993) and implies that in the absence of S-cones the associated retinal pathways fail to develop.

Thus it appears that the morphological criteria for M- and P-cells are well met by the ganglion cells of the New-World monkeys. We now turn to the physiology of these cell systems.

Physiology of retinal ganglion cells and cells in the LGN of catarrhines and platyrrhines

To further test for homology between M- and P-pathways in the retina and LGN in catarrhines and platyrrhines, we compare results from macaque retinal ganglion cells with those from retinal ganglion cells in the platyrrhine capuchin monkey (*Cebus apella*) or with data from LGN cells in the common marmoset (*Callithrix jacchus*). Generally, responses of retinal ganglion cells and LGN cells are similar, but in certain cases there appears to be some temporal filtering of transmission in the LGN (Kaplan, Purpura, & Shapley, 1987; Kaplan, Mukherjee, & Shapley, 1993). Although this book is mainly concerned with color vision of primates, we have included other physiological properties that are relevant to the role of color vision in determining retinal wiring and in the evolution of the primate retina. Recordings from platyrrhine species require the study of both dichromatic and trichromatic phenotypes, which is a rather hit or miss affair without prior genetic investigation. Data from trichromats remain restricted to the marmoset.

Chromatic properties in trichromats. To study the chromatic properties of cells, the paradigm of Smith et al. (1992a) is useful. They measured the responses of macaque retinal ganglion cells as a function of relative phase of temporally modulated red and green light sources (light-emitting diodes, LEDs). This results in pure luminance modulation when the LEDs modulate in phase and in isoluminant chromatic modulation when the diodes modulate in counterphase. At all of the other relative phases, both luminance and chromaticity are modulated. Figure 6.6 (left panels) shows the amplitude and phase of the cell responses of a macaque retinal P-cell as a function of relative diode phase at four different temporal frequencies. The solid lines are the best fits with a model based on a vector addition of cone inputs (Smith et al., 1992a). Figure 6.6 (right panels) shows the responses of a P-cell in the LGN of a trichromatic marmoset to the same stimulus. The vector addition model can also describe these data (Yeh et al., 1995).

Several features of the data indicate that the processing of chromatic stimuli in platyrrhines and catarrhines is very similar. First, at lower frequencies the cell gives a maximal response to chromatic modulation and a minimal response to luminance modulation. This suggests well-balanced cone inputs (this is, however, not always the case in the marmoset cell sample). Second, as temporal frequency increases, the location of the minimum moves away from a phase of zero (i.e., luminance modulation). This is consistent with a small center/surround latency difference of a few milliseconds, and this behavior is similar in the two species. A number of other chromatic response features are also similar (Yeh et al., 1995); for example, P-cells display a high degree of linearity (e.g., there is little harmonic distortion of responses and they lack contrast gain control mechanisms; see below) and M-cells show a nonlinear chromatic response component. It therefore seems that the processing of chromatic information is similar in the retinae of macaques and of trichromatic marmosets.

Rod inputs. Using the same stimulus type, the interaction of rod and cone signals has been studied in the retina of trichromatic macaques (Lee et al., 1996b) and dichromatic capuchin monkeys (unpublished data) and in the LGN of dichromatic marmosets (Yeh et al., 1995). In all of these measurements, the data can be described satisfactorily with the same model by assuming vector addition of rod and cone signals. Rod signals are difficult to detect in P-cells of macaques and capuchin monkeys, but are clearly present in M-cells below 20 td (Lee et al., 1997a). However, LGN cells in the dichromatic marmoset receive stronger rod input, although large differences are found between individual cells (Weiss, Kremers, & Zrenner, 1995; Kremers et al., 1997; 1998; Weiss, Kremers, & Maurer, 1998). Rod input is present in most P- and M-cells and is detectable in some cells at mean retinal illuminances up to 800 td. This might reflect a difference between marmosets and other species, although retinal eccentricity is not strictly matched in the retinal and LGN recordings.

Spatial properties. The receptive field dimensions of macaque retinal ganglion cells and LGN cells increase with increasing retinal eccentricity (Lee, 1996; see Chapter 10). Generally, receptive field center size and dendritic tree size of ganglion cells are well correlated (Wässle & Boycott, 1991), and this is also the case for M-cells. However, measurements of M- and P-cells yield similar center sizes despite midget and parasol ganglion cells having different dendritic tree sizes (see previous section). The reason for this discrepancy remains unresolved; either optical or physiological factors may be involved (see Chapter 10).

Receptive field centers of marmoset LGN cells have been measured using bipartite field stimuli and moving gratings (Kremers & Weiss, 1997). The bipartite field stimulus consists of two counterphase modulating fields either side of an edge. The response of the cells strongly depends on the position of the edge in the receptive field, and this can be used to derive the sizes of the receptive field centers and surrounds. Center sizes were used to predict the spatial resolution, based

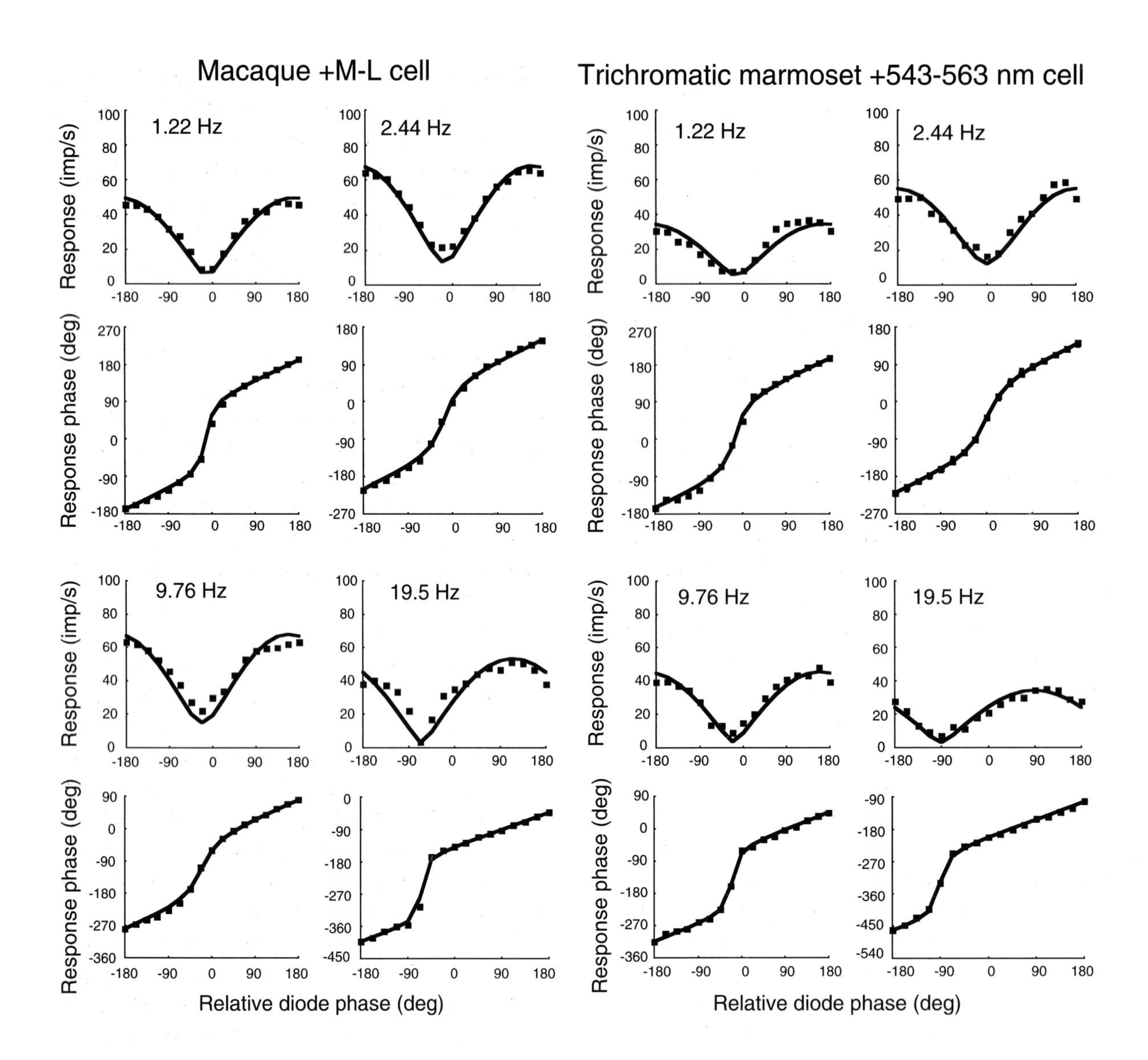

Figure 6.6: *Left panels*: Responses of P-cell in the macaque retina to a red and a green modulating light emitting diode (LED). The abscissa gives the relative phase of the two LEDs. A relative phase of 0 deg is an in-phase modulation of the two diodes and results in pure luminance modulation. A relative phase of 180 deg is a counterphase modulation of the two diodes and results in isoluminant chromatic modulation. All other conditions are a combined luminance and chromatic modulation. At four different temporal frequencies the response amplitudes (upper graphs) and response phases (lower graphs) are shown (filled squares). *Right panels*: The responses of an LGN P-cell in a trichromatic female marmoset having the 543- and 563-nm photopigments to the same stimuli. The response behaviors of both cells can be described well with a model in which the signals of the different cone vectors add in the response of the cells. Best fits with this model are given by the drawn curves.

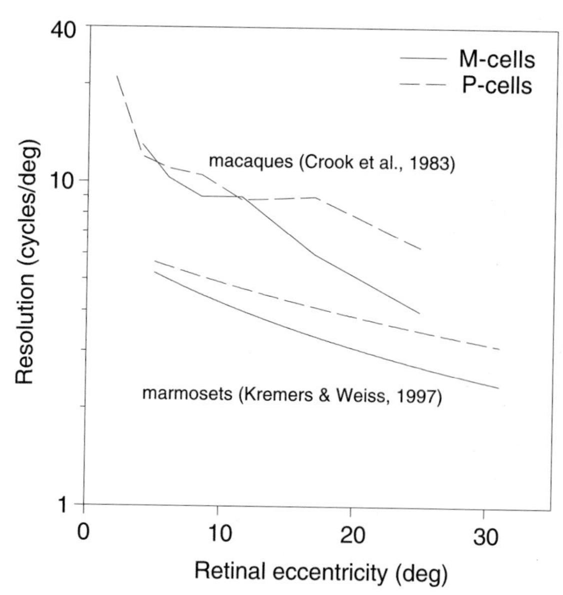

Figure 6.7: The spatial resolution of P- and M-cells as function of retinal eccentricity measured directly in macaques (Crook et al., 1988) and calculated from center sizes of marmoset LGN cells (Kremers & Weiss, 1997). The resolution of the marmoset cells is about a factor 1.5 smaller than that of macaque retinal ganglion cells. This can be fully explained by the ca. 1.6 times smaller marmoset eye. After Kremers and Weiss (1997). (Reproduced with permission from Elsevier Science.)

on fitting 1.8 cycles fitting into the centers at the resolution limit (Peichl & Wässle, 1979). Direct measurements of the spatial resolution with gratings validated this approach. Figure 6.7 shows a comparison of computed spatial resolutions of marmoset LGN cells (Kremers & Weiss, 1997) and macaque retinal ganglion cells (Crook et al., 1988). The resolution of marmoset LGN cells is a constant factor of about 1.5 lower than the resolution of macaque retinal ganglion cells. Apart from this, the dependency of the resolution on retinal eccentricity is very similar for P- and M-cells in both species. The lower resolution of the marmoset cells can be fully explained by the size of the marmoset eye, which is about 1.6 times smaller than that of the macaque (Troilo, Howland, & Judge, 1993). Measurements of surround properties also give similar results. M-cells in the marmoset LGN are spatially more nonlinear than P-cells, although there is a considerable amount of overlap. This has also been found in macaque retinal ganglion cells and LGN cells. In conclusion, the electrophysiological and anatomical data (Ghosh et al., 1996; Goodchild, Ghosh, & Martin, 1997; Wilder et al., 1996) indicate that spatial organization is very similar in macaque and marmoset retinas, implying a similarity of neuronal wiring for spatial processing.

Temporal properties. It is convenient to measure cell temporal response properties employing sinusoidal modulation at different temporal frequencies. At each temporal frequency the responses are measured for several modulation contrasts, which is a measure for stimulus strength. Contrast is described as the Michelson contrast (= $100\% \times (L_{\max} - L_{\min})/(L_{\max} + L_{\min})$, where L_{max} and L_{min} are the maximal and minimal luminance in the stimulus). The contrast gain is the initial slope of the response as a function of contrast and is a measure for the responsivity at the given temporal frequency. Figure 6.8A shows the contrast gains as a function of temporal frequency of luminance modulation for P- and M-cells in the macaque retina (Lee et al., 1990); similar results have been obtained by others (Kremers, Lee, & Kaiser, 1992; Purpura et al., 1990). Both P- and M-cells display bandpass characteristics to luminance modulation, but P-cells are much less responsive.

Figure 6.8B displays the mean temporal responsivities of several ganglion cells in the retina of dichromatic capuchin monkeys (*Cebus*). Although it is not straightforward to classify P- and M-cells in these animals, because P-cells do not show cone opponent responses, it is possible to distinguish between two cell classes on the basis of temporal properties. Moreover, those cells classified as M-cells show response saturation and phase advancement with increasing stimulus contrast that is indicative of a contrast gain control mechanism (Shapley & Victor, 1978, 1981). This is also found in macaque M-cells. Macaque P-cells do not show saturation and their responses show no signs of a contrast gain control mechanism.

Yeh et al. (1995) reported similar data for P- and M-pathways in dichromatic marmosets. Figure 6.8C

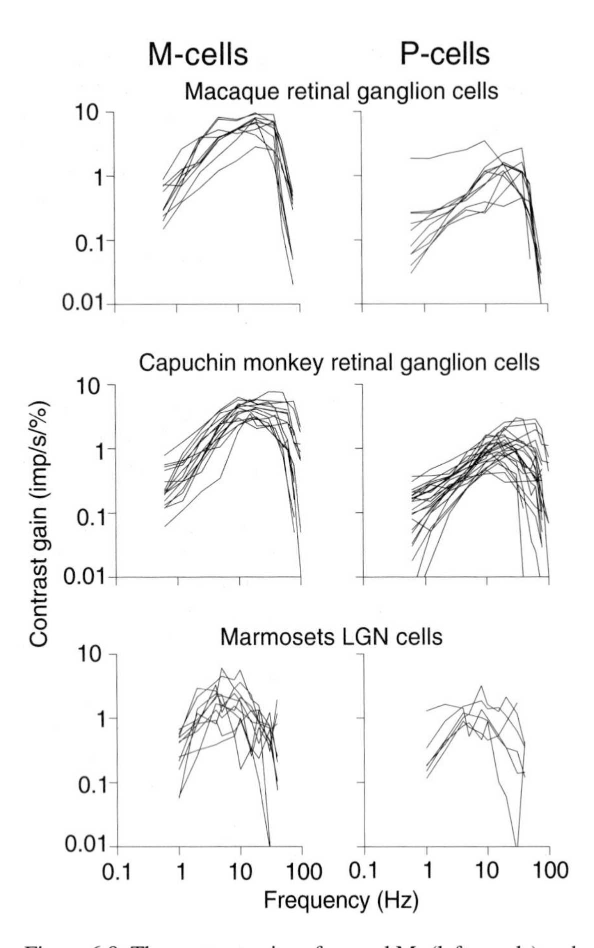

Figure 6.8: The contrast gains of several M- (left panels) and P-cells (right panels) as a function of temporal frequency in the retina of trichromatic macaques (*Macaca fascicularis*), dichromatic capuchin monkeys (*Cebus apella*), and in the LGN of dichromatic marmosets (*Callithrix jacchus*). Basically the cells of the same types have similar contrast gains in all three species. The LGN cells have a stronger high-frequency roll-off than the retinal ganglion cells, probably due to processes at the synaptic transmission in the LGN (Kremers, Weiss, & Zrenner, 1997). The cells of the capuchin monkeys can be classified less easily as P- or M-cells, since the color opponency in P-cells, which is their main characteristic in trichromats, is absent in dichromats. The macaque and capuchin monkey data are from Lee et al. (1996a).

shows the contrast gain as a function of temporal frequency measured in the LGN of dichromatic marmosets (Kremers, Weiss, & Zrenner, 1997). Qualitatively, the same characteristics are obvious in these cells as in the macaque retinal ganglion cells. For instance, M-cells have a stronger attenuation for low temporal frequencies than P-cells as described for macaque retinal ganglion cells (Purpura et al., 1990).

Temporal responses to chromatic modulation have been recorded in trichromatic marmosets. In macaques, P-cells are very responsive to red-green chromatic modulation and demonstrate a low-pass temporal response (Lee et al., 1990; Kremers, Lee, & Kaiser, 1992). This is also the case in trichromatic marmoset P-cells (Yeh et al., 1995).

From this comparison we conclude that M- and P-cells have very similar temporal response properties across the different species. However, a closer comparison between the data of Figs 6.8A and 6.8C reveals that there are some quantitative differences. The retinal ganglion cells respond to higher temporal frequencies than the LGN cells. Possibly, this difference is due to temporal filtering in the LGN (Sherman & Koch, 1986; Kaplan, Purpura, & Shapley, 1987; Kaplan, Mukherjee, & Shapley, 1993) rather than species differences, because marmoset retinal ganglion cells seem to be responsive to high temporal frequencies (Yeh et al., 1995).

A novel way to show the difference in temporal characteristics of P- and M-cells in dichromatic marmosets is stimulation with rotating hemicircles (Kremers, 1996). The hemicircles are located on different locations within the receptive field. Figure 6.9A shows the response of OFF-center M-cells to the rotation of a hemicircle at location L on the receptive field. The orientation of the hemicircle during the response is sketched on the right side. In Fig. 6.9B responses of a P- (left) and an M-cell (right) are shown. The responses are displayed as line segments, the midpoints of which coincide with the locations of the hemicircles on the receptive field. The length of the line segments encodes the response amplitude, and the angle of the line segments with the abscissa gives the response phase. Owing to this convention and because of the low temporal frequency (1 s for a complete rotation), the orientation of the line segments is very similar to the orientation of the straight edge of the hemicircle when the cell responds maximally. Clearly the P- and the M-cell show a very different picture,

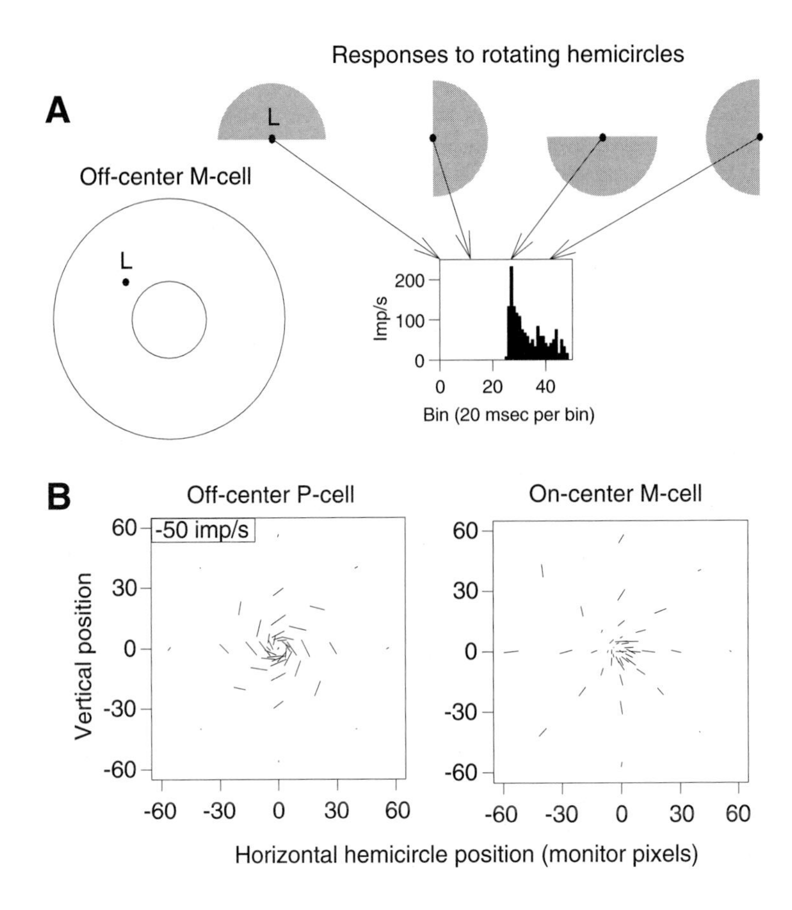

Figure 6.9: Responses of LGN cells in a dichromatic marmoset to rotating hemicircles. A rotation was completed after 1 s. (A) The response of an OFF–center M-cell to a rotating hemicircle with its midpoint at location L in the receptive field. The cell response is given as a PSTH. The orientation of the hemicircle at four instances within the response of the cell are drawn. (B) Responses of an OFF-center P-cell (left) and an ON-center M-cell (right) to rotating hemicircles for different locations of the hemicircles in the receptive field. The midpoints of the line segments indicate the position of the hemicircles' midpoints. The length of the line segment indicates the response amplitude at each hemicircle position. The orientation of the line segment indicates the response phase and is by definition similar to the orientation of the straight edge of the hemicircles, when the cell responds maximally. The orientations of the line segments indicate that the P-cell responds more to illuminance levels, whereas the M-cell responds when the edge moves through the receptive field center.

which is consistent for all measured cells. The line segments for the M-cell responses all more or less point toward the receptive field center. This means that the cell responds maximally when the straight edge of the hemicircle crosses the receptive field center. Thus the cell mainly responds to the illuminance change. The P-cell responds when the center is maximally (for ON-center cells) or minimally (for OFF–center cells) illuminated. This experiment shows that even in dichromats P- and M-cells have clearly different temporal response properties. It seems as if the M-cells encode the changes in the visual scene whereas P-cells respond proportionally to the levels of illumination.

All of these physiological data suggest that there are only minor differences in retinal processing between New- and Old-World anthropoids, which is consistent with the conclusions from the anatomical comparisons. It seems likely that this conclusion also holds for other species than those studied. Thus, magno- and parvocellular pathways indeed appear to be strongly homologous across all anthropoids, although data from the exceptional platyrrhines, the nocturnal *Aotus* and the fully trichromatic *Alouatta*, will be of considerable interest.

Color vision and the evolution of the primate visual system

The S-cone pathway. On the basis of anatomical and physiological data, it is argued elsewhere in this book that a separate anatomical and physiological pathway is associated with the S-cones (see Chapters 9 and 10; also see the section on retinal ganglion cell morphology). This is consistent with the argument that S-cones form part of a phylogenetically older chromatic system (Mollon & Jordan, 1988).

The system carrying S-cone signals has specific anatomical substrates in primates: Bipolar cells connecting exclusively to S-cones have been found in macaques (Mariani, 1984; Wässle et al., 1994; Kouyama & Marshak, 1992; Marshak et al., 1990) and recently in marmosets (Ghosh, Martin, & Grünert, 1997), and ganglion cells carrying S-cone signals have a distinct morphology as small-field bistratified cells in both platyrrhines and catarrhines (Dacey, 1993a; Dacey & Lee, 1994b; Ghosh, Martin, & Grünert, 1997; Silveira et al., 1997). LGN cells with S-cone input seem to be mainly located in the interlaminar regions (Martin et al., 1997). All of these data strongly suggest that the pathway carrying S-cone signals is a chromatic system that is distinct from the parvo- and magnocellular pathways. The S-cone pathway may be part of a third (koniocellular) pathway, which has also been identified in prosimians (Casagrande, 1994). Cat ganglion cells with physiological properties resembling the primate blue-ON cells have been described (Cleland & Levick, 1974). This suggests that the system carrying the S-cone signals indeed is present in most mammals, but it needs to be resolved yet whether there is true homology between the systems of the different taxa.

Evolution of trichromacy. It is currently held likely that full trichromacy evolved in two stages. One stage is a point mutation of the long-wavelength pigment gene causing a shift in the absorption spectrum. The other stage is a gene duplication, so that two pigment genes come to be present on the same X-chromosome (see Chapter 1). These two stages could have occurred in either order (Mollon, 1991). A point mutation prior to gene duplication would provide a situation in which 50% of females are potential trichromats. A further point mutation could then give rise to a situation corresponding to the standard platyrrhine pattern, with two-thirds of females possessing trichromacy. The platyrrhine polymorphism could thus be an earlier stage in the evolution of full trichromacy. The second alternative, gene duplication before point mutation, would be a modus that is often assumed in the evolution of multigene families. However, it leaves the status of the platyrrhines unresolved.

As discussed by Mollon (1991), there are three possible alternative scenarios for the evolution of color vision from ancestral anthropoids. The first possibility is that they were dichromats (no point mutation and gene duplication had occurred yet) and trichromacy in platyrrhines and catarrhines developed independently. Shyue et al. (1995) suggested that the polymorphism in callitrichids and cebids also evolved independently. The second possibility is that, like platyrrhines, the stem anthropoids were polymorphic (so that at least one point mutation already had occurred) and catarrhines became fully trichromatic through a gene duplication. Kay, Ross, and Williams (1997) propose that the stem anthropoids were, similar to the platyrrhines, small arboreal animals with a diurnal activity pattern who lived on a combined diet of fruit and insects. If diet and activity pattern influence the diversity of cone types, then stem anthropoids were also possibly polymorphic. It seems very likely that a gene duplication was also responsible for the development of full trichromacy in *Alouatta* (Jacobs et al., 1996a). As a last possibility, it could be that the ancestral anthropoid was a full trichromat and platyrrhines have lost full trichromacy.

Postreceptoral mechanisms to compare cone signals must have appeared in parallel with these genetic changes, either by using existing mechanisms or by secondary evolutionary changes in the visual system. As was pointed out previously, the presence of parvocellular and magnocellular laminae in the LGN is a characteristic primate feature, even in prosimia. The nature of these ancestral systems remains obscure. They may have been involved in the processing of spatial and temporal information, so that magnocellular and parvocellular pathways might have covered different areas in spatio-temporal stimulus space, as with X- and Y-cells of the cat. Electrophysiological data on LGN cells of the prosimian bush baby (*Galago*) are in accordance with this view (Norton & Casagrande, 1982). The remarkable similarity, both anatomically and electrophysiologically, between P-cells of catarrhines and platyrrhines strongly suggests that this cell system acquired its current characteristics before the divergence of these two lines.

It is usually held (Derrington, Krauskopf, & Lennie, 1984; Shapley & Perry, 1986; see also Chapter 10, but see Chapter 8 for a dissenting viewpoint) that the P-pathway forms the substrate for an M/L-cone–opponent, red-green chromatic system. Cone-specific receptive field centers are provided by the one-to-one cone–midget bipolar–midget (P-) ganglion cell synaptic connectivity. This automatically provides a cone-opponent signal; whether receptive field surrounds are cone-specific (Lennie, Haake, & Williams, 1991) is not relevant at this point.

The evolution of a postreceptoral cone-opponent mechanism must have been triggered by the presence of two (or more) visual pigments in the middle-to-long wavelength range. Any scenario must be speculative, but certain constraints may be envisaged. For example, if a receptive field center summed input from many cones of mixed provenance, then it would be difficult to see how cone specificity could arise; a low cone-to-ganglion cell ratio would seem necessary. This could occur as follows. If the ancestral anthropoid had an arboreal habit, moving in the trees and jumping from branch to branch will have imposed strong adaptational pressure for an improvement in the acuity of spatial vision. This would lead to a decrease in retinal ganglion cell center sizes, and Wässle and Boycott (1991) have argued convincingly that a ganglion cell array for high-resolution spatial vision, with high density relative to the cone matrix, would provide an opportunity for development of something approaching a midget system.

However, a one-cone–one-ganglion-cell connection is not necessary for optimal spatial vision; optical blur stops center sizes below 3–4 cone diameters from improving the spatial resolution of single cells, and for deconfounding stimulus intensity and position some overlap of neighboring receptive fields is even desirable (see Chapter 10, but see Chapter 7 for an alternative interpretation). This would imply that the midget, P-cell system underwent further miniaturization associated with the processing of cone-opponent signals. If cone convergence were already low in an ancestral species, a degree of cone specificity could be improved by patchiness in the cone mosaic. This would be especially marked in a polymorphic animal where mosaicism may be present in the cone array, as possibly occurs in human female carriers of defective color vision (Cohn, Emmerich, & Carlson, 1989; see Chapter 1). Once a degree of chromatic specificity was present in the retinal output, then the selective advantage so endowed would favor the development of more cone-specific connections. In New-World monkeys it is still ambiguous whether a foveal midget bipolar cell is connected to one cone (see the anatomy section). If in these species the midget bipolar cells are connected to more than one cone, a one-cone–one-ganglion-cell connection in Old-World monkeys might indeed be considered as an adaptation to full trichromacy.

If the ancestral anthropoid was a dichromat, the one-cone–one-ganglion-cell connectivity, and the similarities in physiology, must have evolved in parallel in platyrrhines. If the ancestral anthropoid were a full trichromat, then no parallel evolution would be assumed, but it is unclear why full trichromacy should have been lost in the polymorphic platyrrhines. Moreover, this would have involved an extra genetic step: a gene deletion, which is not a very likely event, although events of this sort have occurred to produce human red-green color deficiencies (see Chapter 1). With the hypothesis of a polymorphic common ancestor, there is neither a necessity to assume a parallel evolution nor an improbable genetic change; a P-cell system with cone opponency would already be present in the ancestral anthropoid. However, if one assumes that the polymorphism in the stem anthropoid was similar to that in New-World monkeys (thus a single gene with two or three alleles for the middle- and long-wavelength–sensitive pigment on the X-chromosome), it would imply that this further specialization of the P-cell system evolved only to make cone-opponent processing available for a minority of individuals, since only the heterozygotic females would have been trichromats (25% of the animals with two alleles and 33% with three different pigment alleles).

In sum, the physiological and anatomical data suggest that some kind of trichromacy had evolved before the divergence of the catarrhine and plathyrrhine lines. However, these data raise further questions about the evolution of postreceptoral mechanisms; further studies, for instance, on the prosimian P- and M-systems, might resolve some of them.

Evolution and ecology. Mollon (1989) has argued for coevolution between primate trichromacy and trees with yellow and orange fruit. Several tree species are almost exclusively disseminated by primates (Mollon, 1996), and fruits change color as they ripen so that they can most effectively be detected by a system with two pigments in the long-wavelength range (Osorio & Vorobyev, 1996; Mollon, 1996). A common complaint of color deficient human observers is difficulty in finding small fruit such as cherries and raspberries amidst foliage, which provides a dappled background that precludes the use of luminance cues.

The argument for coevolution is convincing, but it is uncertain whether frugivory provided the sole evolutionary pressure toward trichromatic vision or whether other factors may have contributed. Primates are of course not exclusively frugivorous. Apart from fruit, their diet often contains leaves, seeds, nuts, tree gum, insects, and other prey (Jolly, 1985). Further, there seems to be a strong correlation in primates between body size and diet (Martin, 1990; Fleagle, 1988; Terborgh, 1983). Smaller primates (such as marmosets and tamarins) require relatively more energy than larger species and tend to rely more on insects for their diet. Insects are an excellent source of nutrients, and they meet the requirement of small animals of a rapid energy turnover (Martin, 1990; Kay, 1984). Fruits are not as ideal because of their low protein content; even larger primates spend a large proportion (15–50%) of their time foraging for insects (Terborgh, 1983). Terborgh (1983) proposes that monkey species cannot rely on insects when their size exceeds the size of a capuchin monkey (*Cebus*), and these species have to supplement their diet with other sources of protein such as young leaves, buds, shoots, or nuts. Eventually, in the largest primates, leaves make up a substantial fraction of the diet. Figure 6.10 summarizes the change in primate diet with increase in body weight, passing through insectivory, insectivory-frugivory, frugivory, frugivory-folivory, and folivory (Martin, 1990). Interestingly, platyrrhines are generally smaller than catarrhines (Fleagle, 1988), and they therefore probably rely more on insects than on leaves for their protein requirements.

Species that are to some extent folivorous as well as fruit-eating (the catarrhines and the platyrrhine howler monkeys) are full trichromats. It is possible that folivory provided additional evolutionary pressure toward trichromacy. For these animals it would not only be advantageous to detect ripe fruit from some distance but also to distinguish tree leaf coloration (see, e.g., Hendley & Hecht, 1949). Spectral measurements of the color changes that leaves undergo when they age could determine if such changes are better recognized by a red-green opponent system or by a blue-yellow system; human dichromats are notoriously impervious to the diversity of fall colors.

Apart from the advantages of trichromacy for frugivory or folivory, the reasons for such a variety of patterns of color vision in New-World monkeys remain obscure. For example, the callitrichids show a smaller difference in the peak absorption of their pigments than the cebids. It is difficult to see why this should be desirable, except that bringing pigments closer together in peak absorption might bring about optical benefits for the smaller eye of the smaller species. Furthermore, the owl monkey is thought to be secondarily nocturnal. It is also active under twilight conditions (Tyler, 1991), so it is uncertain why it apparently has abandoned the usual platyrrhine pattern for having a single long-wavelength opsin gene. However, the most intriguing question is why full trichromacy does not occur more often amongst these species. Jacobs et al. (1996a) propose that full trichromacy evolved in the howler monkey through gene duplication within the usual platyrrhine pattern. It might be that this gene duplication is a very rare event. However, there are other possible scenarios, such as kin selection: The social structure of a group with genetically related animals might provide a system in which all animals have

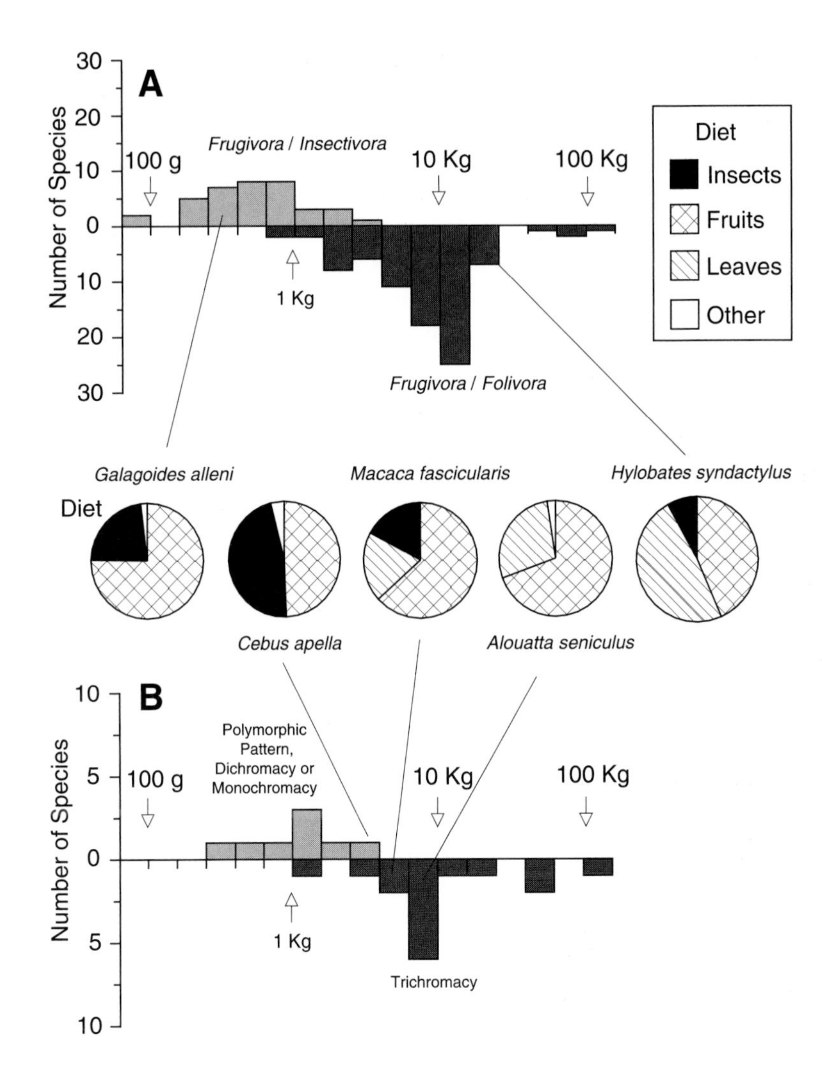

Figure 6.10: (A) The number of species with insectivorous/frugivorous and with a folivorous/frugivorous diet as a function of their weight. Adapted from Fleagle (1988). Prosimians and platyrrhines are generally smaller and supplement their diet with insects. The larger species (mainly catarrhines) rely more on leaves as a protein source. Middle panels show pie diagrams of diets in five different primate species (Fleagle, 1988): one prosimian (*Galagoides alleni*), two New-World monkeys (*Cebus apella* and *Alouatta seniculus*), and two Old-World monkeys (*Macaca fascicularis* and *Hylobates syndactylus*). (B) The numbers of species with a monochromatic, dichromatic, or polymorphic color vision and with full trichromacy as a function of their weight. Comparison of the two distributions suggests that only the larger species (Old-World monkeys and platyrrhine howler monkeys), often with a folivorous/frugivorous diet, have acquired full trichromacy.

the advantages of both the di- and the trichromatic systems (Tovée, Bowmaker, & Mollon, 1992). Dichromats might detect a perceptual organization based on texture more easily than trichromats when the scene is masked by hue differences (Tovée, Bowmaker, & Mollon, 1992; Shyue et al., 1995). Platyrrhines might take advantage of a dichromatic system when searching for insects that may use camouflage. In that case however, it is difficult to see why all species with the standard pattern have three pigment alleles. The only obvious advantage is an increase of trichromatic females from 50 to 66%. It is interesting that the howler monkeys so far studied have only possessed the 535- and 563-nm pigment genes; apparently when full trichromacy had been acquired, the middle allele became redundant and was lost.

Finally, in many catarrhines, colors play an important social role. Colored faces (in *Cercopithecus* and the mandrill, *Mandrillus sphinx*) and genitals (in *Cercopithecus* but also chimpanzees, *Pan troglodytes*) strongly suggest a social function of these colors. Among platyrrhines, a spectacular example of such coloration is the uacari (*Cacajao*), of which males have a bright red faces and heads, which are thought to play a role in sexual selection. As Mollon (1989) points out, colors might also give indications of emotional state or health. It is possible that the social function of color developed secondarily after trichromacy. But once color gained a social function, it might have added to the pressure for trichromacy to be retained or extended.

Concluding remarks

In this chapter we have stressed the comparison of the anatomy and the physiology of the peripheral visual system in anthropoids. There is compelling evidence that the parvo- and magnocellular systems in the retina and LGN of anthropoids are anatomically and physiologically very similar. Whatever the original role of these subsystems, they are likely to have evolved prior to the divergence of the platyrrhine and catarrhine lines. More data on homologous cell types in prosimians might be helpful to come to a definitive conclusion as to their evolutionary history.

Old-World monkeys are all trichromats, whereas the chromatic system of New-World monkeys is highly polymorphic. Prosimians are, as far as is known, dichromats or monochromats. There seems to be common agreement that the combination of daily activity pattern and diet influenced the dimension of color vision in primates. More information about habitat and diet of the ancestral anthropoid primate might give information as to what features played the key role in the evolution of color vision. However, monkeys are highly visually oriented animals, and their color vision might also have been influenced by other important factors in their normal environment, such as social interactions or the avoidance of predators. These ecological factors should also be taken into consideration before coming to definite conclusions on the evolutionary forces on color vision.

Acknowledgments

Jan Kremers was supported by DFG Grant Zr 1/9-3 and a DFG Heisenberg-Fellowship. Luiz Silveira was supported by FINEP/FADESP #4.3.90.0082.00, CNPq #52.1749/94-8. Luiz Silveira, Elizabeth Yamada, and Barry Lee were supported by CNPq/Max Planck Gesellschaft #91.0234/94-9. We would like to thank B. Boycott, G. Jacobs, P.R. Martin, W. Kaumanns, and L.T. Sharpe for comments on the manuscript.

Part II: Retinal Circuitry

7

Parallel pathways from the outer to the inner retina in primates

Heinz Wässle

It is well established that the information processing in the primate visual system occurs in parallel. In the cortex as many as 30 highly interconnected areas are involved in visual processing (Van Essen, Anderson, & Felleman, 1992; Merigan & Maunsell, 1993; Van Essen & Gallant, 1994; Salin & Bullier, 1995). The projection from the retina to the cortex is organized in parallel routes. The parvocellular (P) and magnocellular (M) pathways through the lateral geniculate nucleus are well established (Kaplan, Lee, & Shapley, 1990; Lee, 1996) and recently an additional projection through the interlaminar regions has been described (Hendry & Yoshioka, 1994). In the primate retina, like other mammalian retinas, there may well exist as many as 10–20 different ganglion cell types that cover the retina homogeneously with their dendritic fields. They represent 10–20 specific filters that encode in parallel different aspects of the image projected onto the retina. Not all of them send their axons to the thalamus, but some – in addition to P- and M-cells – almost certainly provide input to the visual cortex (Rodieck, Brening, & Watanabe, 1993). However, they form sparse mosaics and so are unlikely to be involved with spatial vision.

Ganglion cells receive specific inputs from bipolar and amacrine cells in the inner plexiform (IPL) layer. The IPL is precisely stratified and the different ganglion cell types have their dendrites at specific levels within the IPL. The overall subdivision is into the ON and OFF layers. Dendrites of OFF-ganglion cells stratify in the outer half of the IPL and those of ON-ganglion cells in the inner half of the IPL (Nelson, Famiglietti, & Kolb, 1978; Peichl & Wässle, 1981). Within this ON/OFF dichotomy further subdivisions occur. Dendrites of ON- and OFF-parasol cells keep a very narrow level of stratification close to the center of the IPL. In contrast, dendrites of ON- and OFF-midget ganglion cells stratify more diffusely and are found more towards the outer and inner IPL, respectively. The dendritic tiers of the small bistratified blue ganglion cells stratify even further towards the outer and inner edges of the IPL (Dacey, 1994). This suggests that the neurally encoded retinal image is different at different levels of the IPL, depending on the stratification of the various bipolar, amacrine, and ganglion cells. For instance, directional selective processing almost certainly would costratify with the level that is occupied by the processes of cholinergic amacrine cells (Masland, 1988). The rod signal is relayed in the inner part of the ON-sublamina, where rod bipolar cell axons and AII-amacrine cell dendrites meet (Wässle et al., 1991, 1995). Application of molecular markers such as antibodies against different subunits of transmitter receptors has recently provided evidence for a very precise subdivision of the IPL (Greferath et al., 1995; Brandstätter et al., 1995).

This raises the question of how specific aspects of the light signal are transferred from the outer to the inner retina. This might be the role of the nine to ten different types of bipolar cells (Sterling et al., 1995). For color vision one has to ask, which bipolar cells transfer the cone-specific signals into the IPL and at what level do their axons terminate? For achromatic phasic and tonic signals or directional-selective responses, different sets of bipolar cells are probably involved with axons terminating at different levels of the IPL. Multiple signal pathways from the outer towards the inner retina imply that parallel processing starts immediately after the cone pedicle, the first synapse of the retina. One has to ask how an individual cone provides enough output synapses to feed all of these different bipolar cells.

These and related questions will be dealt with in this chapter; we try to answer them by comparing the circuitry of the primate retina with that of the rat retina. Since rats, like all other mammals, are dichromats (Jacobs, 1993; see also Chapter 6), we hope that this comparison will reveal some details of the circuitry that subserves trichromacy in primates. We will speculate as to how this circuitry might have evolved from a "standard" mammalian retina.

Methods

Golgi staining of macaque monkey retinas. The retinas were those used in Boycott, Hopkins, and Sperling (1987); Wässle, Boycott, and Röhrenbeck (1989); and Boycott and Wässle (1991). All quantitative measurements were made from whole retinas, mounted photoreceptor side up. The descriptions are exclusively from Golgi-Colonnier–stained material. Two of the whole mounts contained patches from between ~1 and 7 mm eccentricity, where bipolar cells were stained well. By changing the plane of focus it was possible to observe the dendrites, cell bodies, and axon terminals of the bipolar cells as well as the cone mosaic. Thus is was possible to analyze the cone contacts of the bipolar cells shown in Figs. 7.3 and 7.4. Altogether, ten different bipolar cell types could be distinguished. They differ in their dendritic branching pattern, in the number of cones contacted, and in the shape and branching level of their axons. The diagram in Fig. 7.1B was constructed from the observation of the cells in retinal whole mounts and, therefore, represents a schematic and idealized vertical view.

Immunocytochemical labeling of S-cone and midget bipolar cells. Details of the immunocytochemical procedures are given in Wässle et al. (1994). S-cone bipolar cells were immunostained in horizontal sections of macaque monkey retinas with antibodies directed against cholecystokinin (CCK) (Kouyama & Marshak, 1992). Their dendrites and cell bodies were drawn at a magnification of ×1000 by using a drawing apparatus attached to the microscope (Fig. 7.2). Flat midget bipolar cells (FMB) were immunostained in vertical (Fig. 7.5) and in horizontal sections (Fig. 7.6) with antibodies directed against recoverin (Milam, Dacey, & Dizhoor, 1993). Cones and FMB cells were counted from fields at different eccentricities drawn at a final magnification of ×1000 (Fig. 7.7). Cones and axon terminals were drawn at a magnification of ×2000 (Fig. 7.6).

Electron microscopy of cone pedicles. As described in detail by Chun et al. (1996), macaque monkey retinas were optimally fixed for electron microscopy. Serial horizontal sections were cut through the cone pedicle layer in a piece close to the fovea and in a peripheral piece. The sections were photographed in the electron microscope and electron micrograph montages at a final magnification of ×14000; 17600 were made. Reconstructions of the cone pedicle synaptic complex from the serial sections were made by hand on tracing paper (Fig. 7.8).

Intracellular staining of rat bipolar cells. As described in detail by Euler and Wässle (1995), bipolar cells of the rat retina were injected in fixed vertical sections under visual control with Lucifer Yellow and Neurobiotin. The injected cells were drawn from the vertical sections at a final magnification of ×1000 and were classified according to their dendritic branching patterns and the shapes and branching levels of their axons. In further experiments patch clamp recordings were performed from bipolar cells in a slice preparation of the rat retina, and their response to glutamate was measured (Euler, Schneider, & Wässle, 1996).

Results

Bipolar cells of the mammalian retina. Bipolar cells of the mammalian retina can be subdivided according to their morphology into many different types. Cajal (1893) recognized rod bipolar cells as a separate type (Fig. 7.1: RB). Their dendrites make invaginating contacts with rod spherules and their axons terminate in the innermost part of the IPL (Kolb, 1970; Boycott & Kolb, 1973; Dacheux & Raviola, 1986; Greferath, Grünert, & Wässle, 1990). Many types of cone bipolar cells have been recognized in different mammalian species. In the rabbit retina, nine types have been described from Golgi-staining (Famiglietti, 1981). Recently, two of these types were confirmed by intracellular dye injection and an additional type was described (Mills & Massey, 1992; Merighi, Raviola, & Dacheux, 1996). In the cat retina, eight to ten different types of cone bipolar cells have been recognized (Famiglietti, 1981; Kolb, Nelson, & Mariani, 1981; McGuire, Stevens, & Sterling, 1984; Cohen & Sterling, 1990a,1990b).

The diagram in Fig. 7.1 compares the bipolar cells of the rat retina (Fig. 7.1A) with those of the peripheral macaque monkey retina (Fig. 7.1B). The nine putative cone bipolar cell types (labeled 1–9) and the rod bipolar cell (RB) of the rat retina are arranged according to the stratification level of their axon terminals in the IPL. The cells were drawn from vertical sections following intracellular injections (Euler & Wässle, 1995). The cone contacts of the nine cells have not yet been analyzed in detail, but they contact several neighboring cone pedicles with one exception: Bipolar cell 9 has a wide dendritic tree that is cone-selective and therefore a putative S-cone (BB) bipolar cell.

The rat retina is considered rod-dominated because only 1% of its photoreceptors are cones (Szél, Röhlich, & van Veen, 1993). However, the perspective changes if one examines the absolute number of cones. The cone density is between 4000 and 5000 cones/mm^2, which is similar to peripheral cat, rabbit, and macaque monkey retinas. Thus the cone bipolars of peripheral macaque monkey retina (Fig. 7.1B) are from a region of cone density that is comparable to that of the rat retina (Fig. 7.1A).

The bipolar cells of the monkey retina, which are shown schematically in Fig. 7.1B, were determined initially from Golgi-stained whole mounts (Boycott & Wässle, 1991). There is a striking similarity between the rat and the monkey bipolar cells with respect to the shapes and stratification levels of their axons; however, there is also a clear difference: Midget bipolar cells (FMB, IMB) are only found in the monkey retina. FMB and IMB cells have dendritic trees that are restricted to a single cone pedicle and their axons terminate at different levels within the IPL (Polyak, 1941). Following the nomenclature of Polyak (1941), we named bipolar cells with dendritic trees contacting several neighboring cone pedicles "diffuse" bipolar cells (DB1–DB6) (Boycott & Wässle, 1991).

In summary, these studies suggest that there are about ten types of cone bipolar cells in the mammalian retina and their major defining features are the shape and stratification of their axons in the IPL and in some instances their cone contacts in the OPL (Hopkins & Boycott, 1996, 1997).

Cone contacts of midget and diffuse bipolar cells. The synaptic terminal of cone photoreceptors, the cone pedicle, contains three different kinds of synaptic specializations. First, the pedicle has gap junctions for electrical contacts to other cone pedicles and to rod spherules (Cohen, 1965; Baylor, Fuortes, & O'Bryan, 1971; Raviola & Gilula, 1973; Tsukamoto et al., 1992). Second, flat (basal) contacts with putative OFF-bipolar cells are found at the cone pedicle base (Missotten, 1965; Dowling & Boycott, 1966; Kolb, Boycott, & Dowling, 1969; Kolb, 1970). Third, the pedicle contains invaginating contacts (Missotten, 1965), which usually contain a presynaptic ribbon and three invaginating processes: two lateral elements that are horizontal cell dendrites, and a central element, which in mammals is an ON-bipolar cell dendrite (Dowling & Boycott, 1966). This synaptic arrangement has been named a triad. An individual cone pedicle has been shown to contain many triads (Missotten, 1965).

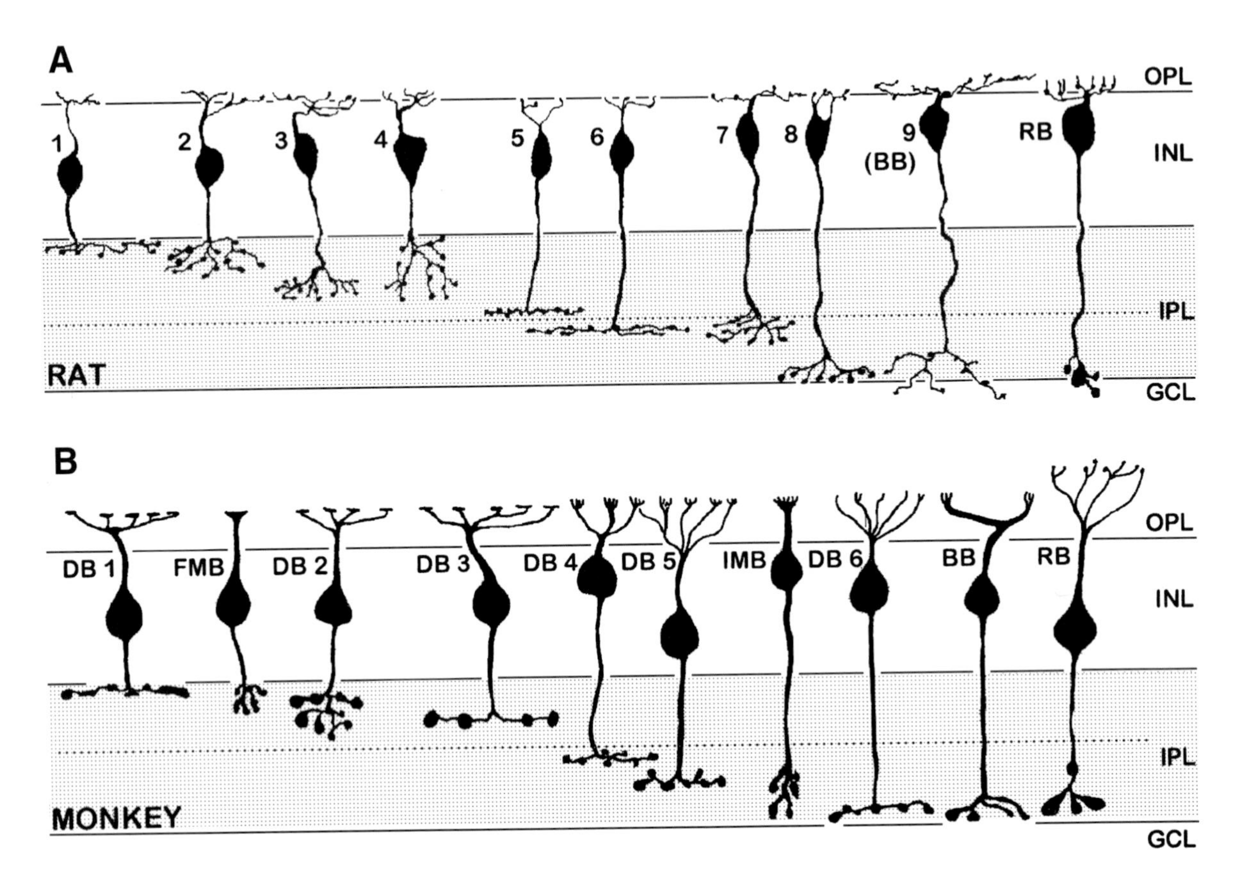

Figure 7.1: Summary diagrams of the bipolar cells described in the rat retina (A) by Euler and Wässle (1995) and in the macaque monkey retina (B) by Boycott and Wässle (1991). (A) The cells were injected intracellularly with Lucifer Yellow or Neurobiotin in vertical slices of the retina and were drawn with the aid of camera lucida. The nine putative cone bipolar cell types (labelled 1–9) are arranged according to the stratification level of their axon in the IPL. Cell 9 (BB) is a putative S-cone bipolar cell and cell RB is a rod bipolar cell. The horizontal line subdividing the IPL represents the border between the OFF- (upper) and ON- (lower) sublamina. Cells 1–5 are, therefore, putative OFF-bipolar cells, cells 6–9 and RB cells are putative ON-bipolar cells (Euler, Schneider, & Wässle, 1996). (B) Schematic diagram of the macaque monkey bipolar cells, which were observed and classified in a Golgi-stained whole mount of a macaque monkey retina (eccentricity: 6–7 mm). DB1–DB6 cells are diffuse bipolar cells that contact several neighboring cone pedicles. FMB and IMB cells are flat and invaginating midget bipolar cells that contact a single cone pedicle. BB cells are S-cone bipolar cells and RB cells are rod bipolar cells.

Reconstructions of Golgi-impregnated midget bipolar cells of the primate retina by serial electron microscopy showed a clear dichotomy: IMB cells made exclusively invaginating contacts while FMB cells made only flat contacts (Kolb, Boycott, & Dowling, 1969; Kolb, 1970). An individual IMB cell can make up to 25 contacts with its cone pedicle, an FMB cell as many as 50 (Boycott & Hopkins, 1991). Since the axons of IMB and FMB cells terminate in the ON- and OFF-stratum, respectively, it was thought that they represent ON- and OFF-bipolar cells. Subsequently this was formulated as a general rule: ON-bipolar cells make invaginating contacts, while OFF-bipolars make flat contacts (Raviola & Gilula, 1975; Stell, Ishida, &

Lightfoot, 1977). However, later studies and results from different species have challenged this correlation as a general rule (see Kolb & Nelson, 1995).

Reconstructions of the cone contacts of Golgi-impregnated diffuse bipolar cells by electron microscopy have revealed more complex cone-to-bipolar cell connections in midperipheral primate retina (Boycott & Hopkins, 1991, 1993; Hopkins & Boycott, 1992, 1995, 1996, 1997). Bipolar cells DB1, DB2, and DB3, which have their axon terminals in the outer IPL (Fig.7.1B), make nearly exclusively flat contacts with the cone pedicles, an average of 20 contacts per pedicle, and thus follow the FMB cell scheme. Bipolar cells DB4 and DB5, which have their axon terminals in the inner IPL (Fig. 7.1B) make, as expected, an average of seven invaginating synapses per pedicle; contrary to expectation, both, especially DB4, have 40% flat contacts with cone pedicles. DB6 also has both flat and invaginating synapses. Reconstructions of the cone contacts of foveal bipolar cells are reviewed by Calkins (Chapter 8).

We have recently measured the response of rat bipolar cells to glutamate, the putative cone transmitter (Euler, Schneider, & Wässle, 1996). Bipolar cells in Fig. 7.1A with axon terminals in the outer part of the IPL expressed a conventional ionotropic kainate/AMPA type of receptor and are therefore OFF-bipolar cells. Bipolar cells with axon terminals in the inner part of the IPL (Fig. 7.1A) expressed a metabotropic glutamate receptor (mGluR6, L-AP4-receptor) and are therefore ON-bipolar cells. If this result is also valid for the diffuse bipolar cells of the primate retina, it would suggest that DB1, DB2, and DB3 are OFF-bipolar cells while DB4, DB5, and DB6 are ON-bipolar cells. Thus the termination of the axon in the IPL seems to be a good predictor for the physiological type while the type of synapse, flat or invaginating, may be more ambiguous.

S-cone bipolar cells of the primate retina. Mariani (1983, 1984) described bipolar cells selective for S-cones in the macaque monkey retina. They have long, smoothly curved dendrites and contact between one and three cone pedicles; they are clearly cone-selective. Their axons terminate in rather large varicosities in the innermost part of the IPL, close to the ganglion cell layer (see BB cells in Fig. 7.1). Recently, S-cone bipolar cells have been quantified by selective labeling with an antibody against cholecystokinin (CCK) (Kouyama & Marshak, 1992; Wässle et al., 1994). Figure 7.2 shows a horizontal view of the S-cone bipolar cells in a small field (170 × 350 μm) of midperipheral retina. They were immunostained with antibodies against CCK; their cell bodies are irregularly shaped and stippled in Fig. 7.2. The S-cone pedicles were located at points where the dendrites of neighboring S-cone bipolar cells converged and made their invaginating contacts (Kouyama & Marshak, 1992). These are indicated in Fig. 7.2 by the regularly rastered circles. From the analysis of such fields as shown in Fig. 7.2 we were able to estimate the numerical connectivity. Individual S-cone bipolars contacted 1.6 (range 1–3) cone pedicles on average (convergence). The average number of S-cone bipolar cells postsynaptic at individual pedicles, the so-called divergence, was 2.4 (range 1–5). The S-cone bipolar cell density exceeded the S-cone density by a factor of approximately 1.5.

As expected for ON-type bipolar cells, S-cone bipolar cells make invaginating contacts at the cone pedicle base (Mariani, 1983, 1984; Kouyama & Marshak, 1992) and their axons terminate close to the ganglion cell layer. It has been shown that they provide synaptic input to the inner tier of the dendritic tree of the small bistratified ganglion cells (see Chapter 8). Small bistratified ganglion cells give blue-ON, yellow-OFF responses (Dacey & Lee, 1994).

It has to be emphasized that S-cone pedicles are not exclusively contacted by S-cone bipolars. There is evidence from Golgi-staining that diffuse cone bipolar cells also contact S-cone pedicles (Boycott & Wässle, 1991). Electron microscopy shows that blue cones also contact DB1, DB2, DB3, and OFF-midget bipolar cells (see Chapter 8); hence there are parallel routes from the OPL to the IPL for the S-cone signal.

Among primates, the blue-yellow pathway has been electrophysiologically demonstrated in diurnal simians of both Old-World monkeys (Wiesel & Hubel,

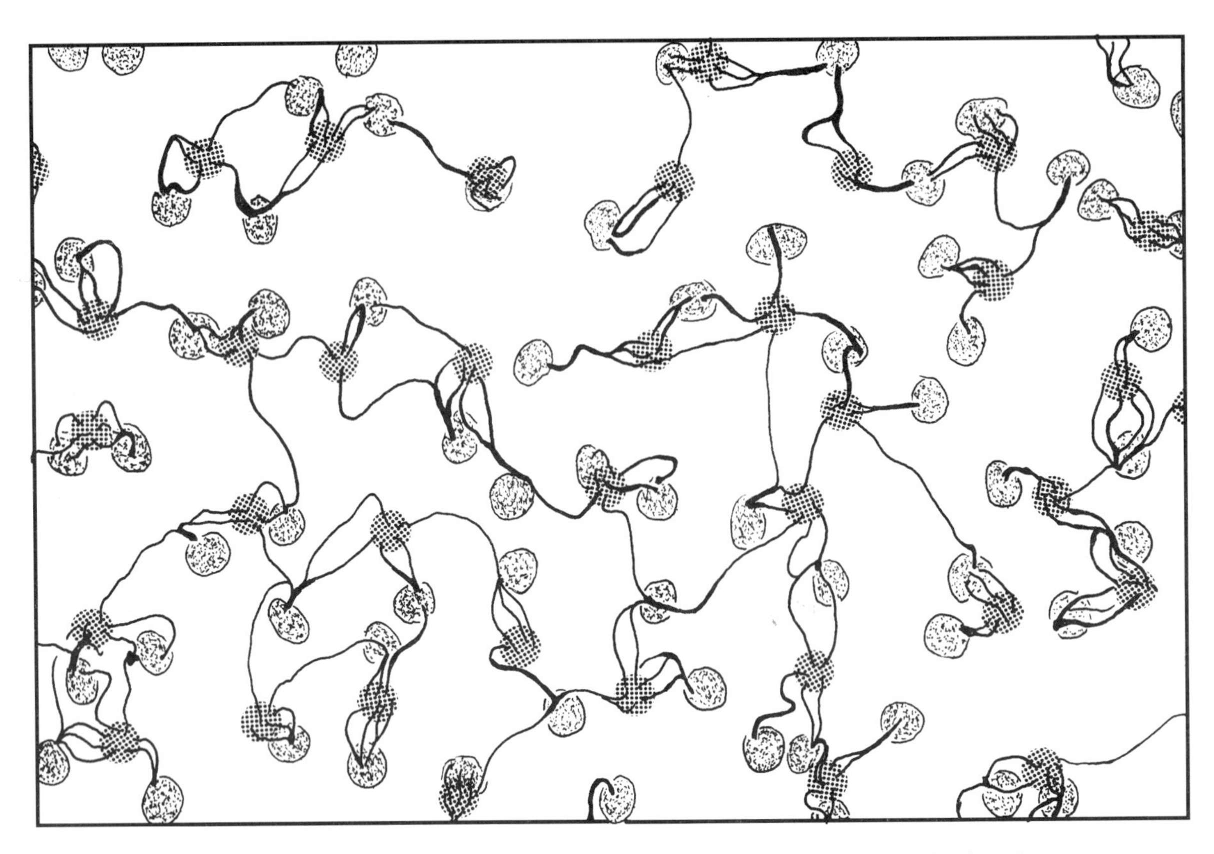

Figure 7.2: The mosaic and the cone contacts of S-cone bipolar cells (BB) in the peripheral retina of a macaque monkey. Horizontal sections were immunostained with antibodies against CCK and the BB cells were drawn from these sections. The cell bodies of BB cells are the stippled, more irregular profiles. The putative S-cone pedicles are the regularly rastered circles. The dendrites of individual BB cells can contact several neighboring S-cone pedicles ("convergence"). Individual S-cone pedicles can also contact several neighboring BB cells ("divergence").

1966; de Monasterio, 1978; Dacey & Lee, 1994) and New-World monkeys (*Saimiri*: Jacobs & De Valois, 1965; *Cebus:* Lee et al., 1996a). Consistent with this is that S-cone bipolar cells and small-field bistratified ganglion cells are found in these species. However, these cells are not found in the only living nocturnal simian, the *Aotus* (Yamada et al., 1996a), which lacks S-cones (Wikler & Rakic, 1990; Jacobs et al., 1993).

Recent immunostaining with antisera specific for the S-cone opsin has shown that S-cones constitute approximately 10% of the cones in most mammalian retinas (Szél, Diamantstein, & Röhlich, 1988; Szél, Röhlich, & van Veen, 1993). Putative S-cone selective bipolar cells have also been described in the rabbit and in the rat retina (Famiglietti, 1990; Euler & Wässle, 1995). S-cone input has also been found in color-coded ganglion cells of the cat retina (Daw & Pearlman, 1970; Cleland & Levick, 1974). They had an ON-response to blue light and an OFF-response to red and green light, which is very similar to the responses of blue-ON ganglion cells of the primate retina (Dacey & Lee, 1994). In the retina of the ground squirrel, blue-ON and green-OFF as well as blue-OFF and green-ON ganglion cells have been described (Michael, 1969).

Diffuse cone bipolar cells of the primate retina. Six different types of diffuse bipolar cells are illustrated in Fig 7.1B. Since Golgi-staining is capricious,

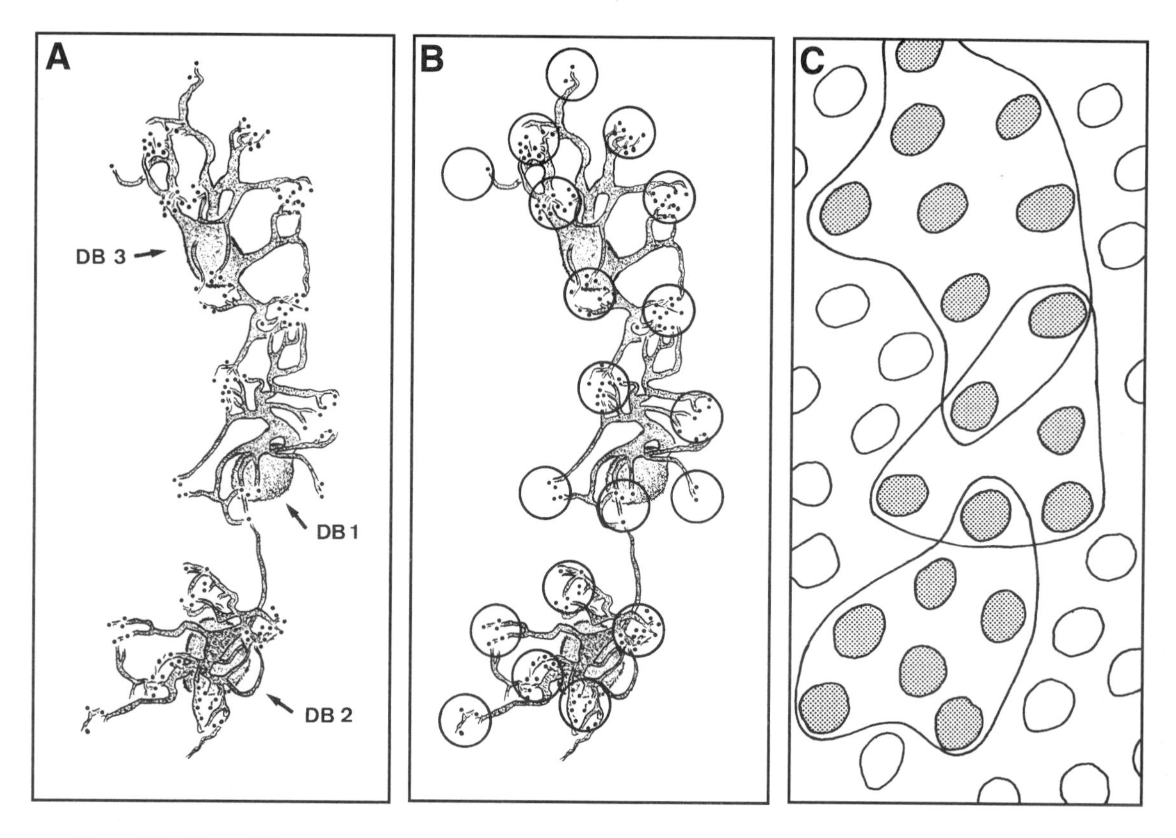

Figure 7.3: Three diffuse bipolar cells drawn from a Golgi-stained whole-mount of a macaque monkey retina. The cells are viewed from the photoreceptor side, and their dendrites and cell bodies are shown. (A) The dendritic trees of the three bipolar cells are classified as DB1, DB2, and DB3 according to their axon terminals (not shown). The dendrites terminate in small punctate varicosities representing their contacts at the cone pedicle base. (B) The terminals grouped by circles representing the cone pedicles. (C) The cone mosaic was visible above the dendrites in this whole mount and was also corrected for the Henle-fibre shift. Cones contacted by the stained DB cells are hatched and outlined for each cell type (frame width: 50 μm).

it is possible that even more than those six types exist in the primate retina. However, the six are not only established by Golgi-staining; at least three of them (DB2, DB3, and DB4) have also been stained by immunocytochemical markers and their populations have been studied (Grünert, Martin, & Wässle, 1994).

The dendritic trees and cell bodies of three different types of diffuse bipolar cells, which were stained by the Golgi method, are shown in Fig. 7.3A. The cells were drawn from a retinal whole-mount as viewed from the photoreceptor side. The dendrites terminate in small punctate varicosities representing the contacts at the cone pedicle bases. In Fig. 7.3B the contacts are grouped by circles that depict the cone pedicles. Although light microscopy indicates that different numbers of contacts are made with individual cone pedicles, only electron microscopy can provide an accurate estimate of these numbers (Hopkins & Boycott, 1995). The cone array was also visible in this retinal whole-mount and is shown in Fig. 7.3C. The hatched cones are those that are likely to be contacted by the three bipolar cells. Figure 7.3 makes several points. First, it is difficult to separate the different diffuse bipolar cell types according to their dendritic morphology. This was possible in Fig. 7.3 only by comparing the shapes and stratifications of the axon terminals (not shown). Second, the number of cones contacted varies between six and nine (DB1:6; DB2:7; DB3:9). Third, the bipolar cells are apparently nonselective since they contact all cones within the reach of their dendrites. Fourth, neighboring cells can contact the same cone pedicles.

With respect to the cone selectivity, one proviso has to be added. We observed many diffuse bipolar cells and their cone contacts and found that they usually contacted between five and ten cones, which were found within their dendritic tree. Since L- and M-cones of the monkey retina are randomly distributed and present in equal numbers (Mollon & Bowmaker, 1992), the chances that a diffuse bipolar cell that contacts seven cones has a pure L- or M-cone input are less than 1%. Hence, an individual diffuse bipolar cell almost certainly contacts both L-and M-cones. The situation is different in the case of S-cone input, however. Given the low number of S-cones, many diffuse bipolar cells could miss S-cones. To find out whether certain diffuse bipolar cell types actually avoid S-cones, double labeling experiments will have to be performed in which S-cone pedicles and diffuse bipolar cells are specifically labeled. Recently such experiments have been performed to estimate the S-cone input to horizontal cells (Goodchild, Chan, & Grünert, 1996).

It was possible to selectively stain three of the six diffuse bipolar cell types by using immunocytochemical markers (Grünert, Martin, & Wässle, 1994). Antibodies to calbindin (CaBP D28K) labeled the diffuse bipolar cell type DB3, and their mosaic was analyzed in more detail. The number of DB3 cells was compared with the number of cones; the ratio was 0.15. DB3 cells in Golgi-stained preparations contacted between eight and ten cone pedicles (Boycott & Wässle, 1991). Multiplication of the cone contacts (8–10) with the relative proportion of DB3 cells (0.15) gives a cone divergence of 1.2–1.5, which means that an individual cone pedicle makes contacts with 1.2–1.5 DB3 cells. Therefore, despite their relatively low density, the DB3 dendrites completely cover the cone mosaic.

An antibody against the α isoenzyme of PKC labels the diffuse bipolar cell type DB4 as well as rod bipolar cells (Grünert, Martin, & Wässle, 1994). The ratio of DB4 cells to cones was 0.37, and this ratio was constant throughout the retina. Multiplication of the cone contacts of DB4 cells (7–8) with the relative proportion of DB4 cells (0.37) gives a cone divergence of 2.6–3, which means that an individual cone pedicle makes contacts with 2–3 DB4 cells.

	Central	Mid-peripheral	Far peripheral
Total number	102	45	28
One cone contact	98	38	–
Two cone contacts	4	6	6
Three cone contacts	–	1	17
Four cone contacts	–	–	5
Eccentricity (mm)	0.7–2.1	6–8	14–15.7

Table 7.1: Analysis of Golgi-stained midget bipolar cells at three different eccentricities. The top row gives the number of cells observed. The following four rows show how many midget bipolar cells contact how many cones. The bottom row gives the eccentricity.

For the remaining types of diffuse bipolar cells a first approximation of their relative numbers was derived from Golgi-staining (Boycott & Wässle, 1991) and suggested a minimum divergence of 1.5 and a maximum divergence of 3–4. This predicts that an individual cone pedicle makes contacts with 10 to 15 diffuse bipolar cells, including all 6 different types. The axons of the diffuse bipolar cells terminate within specific layers of the IPL (Fig. 7.1). Hence signals from every individual cone are transmitted in parallel from the OPL to the IPL along at least six different DB-cell channels.

Midget bipolar cells of the primate retina. Polyak (1941) described bipolar cells in the primate

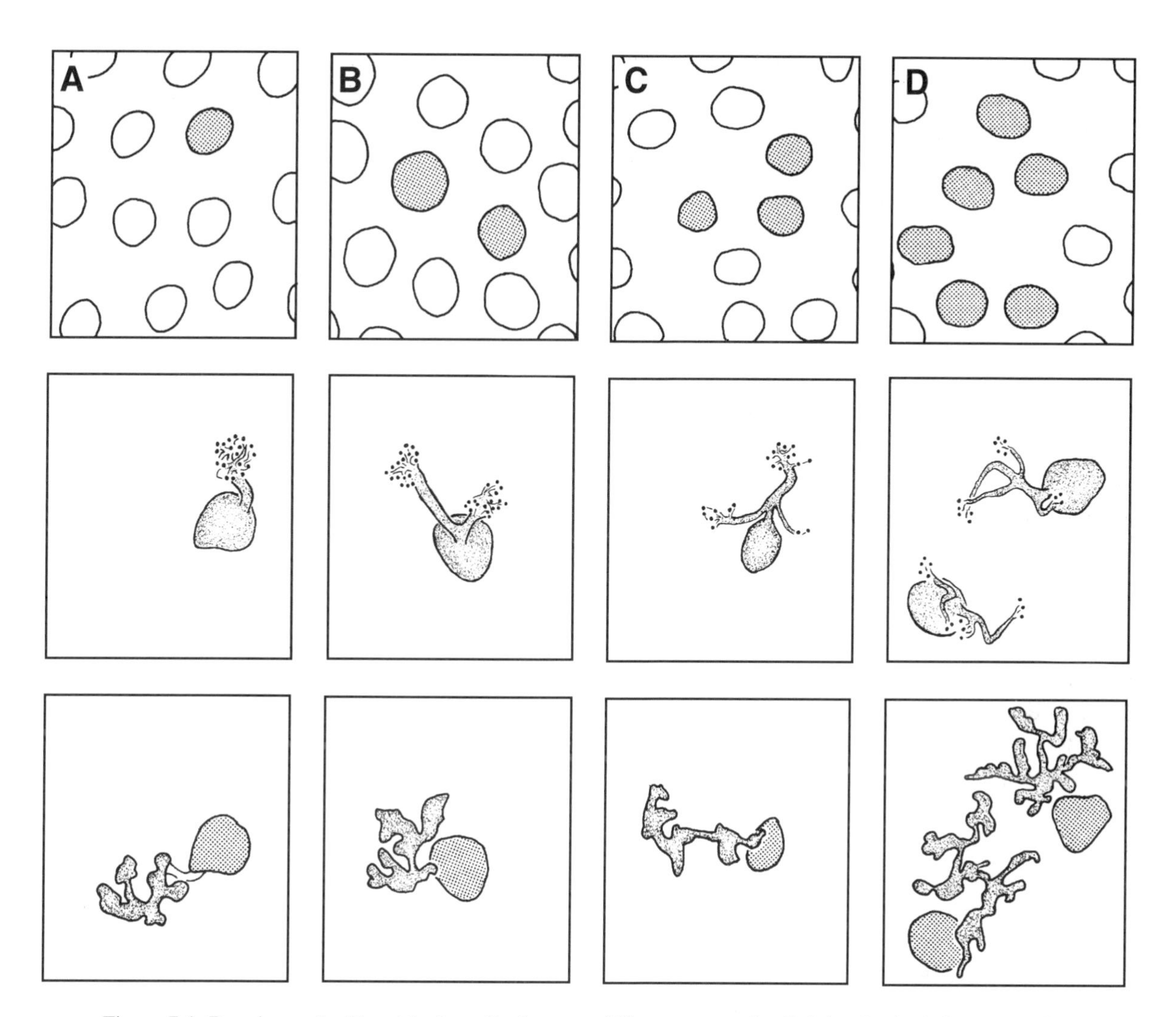

Figure 7.4: Drawings of midget bipolar cells fom two different areas of a Golgi-stained whole mount of a macaque monkey retina. The top row shows the cone mosaic. The cones contacted by the cells in the center row are indicated by hatching. The center row shows the cell bodies, dendrites, and dendritic terminals. The bottom row shows the cell bodies and axon terminals (frame size 40 μm × 35 μm). (A) IMB cell, which contacts a single cone pedicle. (B) "Double headed" IMB cell, which contacts two neighboring cone pedicles. (C) "Triple headed" IMB cell, which contacts three neighboring cone pedicles. (D) Two neighboring IMB cells, which contact three cone pedicles each.

retina, which synapse with only one cone pedicle, and called them midget bipolar cells. Midget bipolar cells have been observed from the fovea all the way to the periphery in Golgi-stained primate retinas (Polyak, 1941; Rodieck, 1988; Boycott & Hopkins, 1991, Boycott & Wässle, 1991). In one retinal whole-mount (M130L) midget bipolar cells were stained exceptionally well and could be studied at different eccentricities (Wässle et al., 1994). The vast majority of midget bipolar cells in the central and midperipheral retina contact a single cone pedicle (Fig. 7.4A) and only occasionally are two neighboring cones contacted (Fig. 7.4B). However, in peripheral retina there was an increased incidence of three and more cone contacts

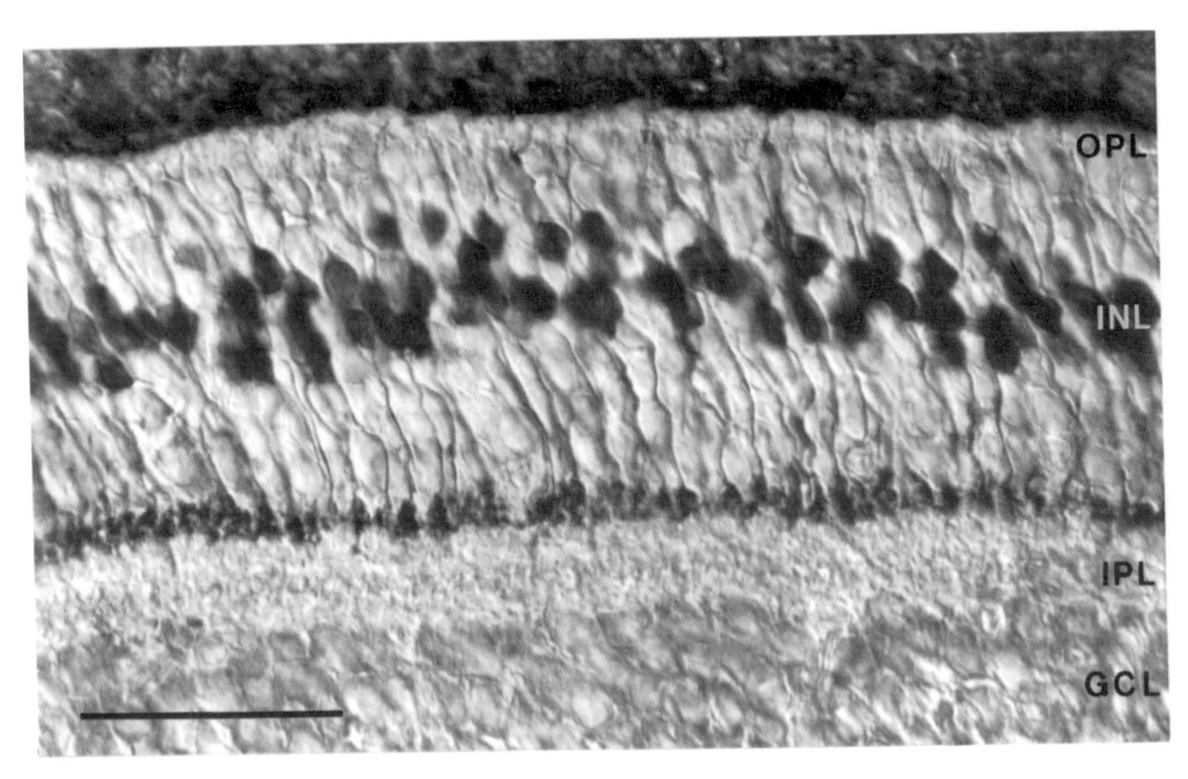

Figure 7.5: Light micrograph of a vertical frozen section through a macaque monkey retina stained immunocytochemically with antibodies against recoverin. FMB cells are labeled. Their axons terminate in small varicose swellings in the outer part of the IPL. Recoverin is a calcium-binding protein that was first isolated from photoreceptors (Lambrecht & Koch 1991), and this is the reason for the strong labeling on top of this micrograph. (OPL: outer plexiform layer; INL: inner nuclear layer; IPL: inner plexiform layer; GCL: ganglion cell layer; scale bar: 50 μm).

without any evidence of cone selectivity (Figs. 7.4C and 7.4D). Table 7.1 shows this more quantitatively.

Recently, FMB and IMB cells were stained by using immunocytochemical markers (Milam, Dacey, & Dizhoor, 1993; Wässle et al., 1994). Antibodies against recoverin were used to stain and study the retinal distribution of FMB cells. IMB cells were stained with antibodies against CCK. A vertical section that was immunostained with antibodies against recoverin is shown in Fig. 7.5. The cell bodies of FMB cells occupy the center of the inner nuclear layer (INL), and their axons terminate in small varicosities in the IPL. The FMB cell mosaic was studied in recoverin-labeled 70-μm-thick horizontal sections. Cell bodies and axon terminals were well stained in these sections, and the cone mosaic could also be detected.

Figure 7.6 shows an analysis of three fields from different eccentricities in the nasal retina. The respective cone densities are shown on top and are 12,000/mm^2 (left), 6,500/mm^2 (middle), and 3,800/mm^2 (right). The corresponding axon terminals are shown in the center of the figure, and their individual areas are outlined at the bottom. The axon terminal density follows the cone density, and the sizes and shapes of individual axon terminals – due to packing constraints – adjust to the respective densities. The axon terminals are small and look relatively more varicose at high densities. They adapt their shapes to the available space without any overlap and thus show "territorial" behavior. The densities of FMB cells within the three fields of Fig. 7.6 can be directly compared with the cone densities. Although in the left column cone and FMB cell densities are very similar, FMB cells occurr at 80% of the cone density in the middle column and at only 60% of the cone density in the right column. We also measured cone and FMB cell densities along an intersect from the optic nerve head toward the upper retina; the results are shown in Fig. 7.7. The cone density in Fig. 7.7A decreases from $8000/mm^2$ at 5 mm eccentricity to $3000/mm^2$ at 14 mm eccentricity. The FMB cell density decreases from $8000/mm^2$ at 5 mm to less than $2000/mm^2$ at 14 mm. In the central retina there is one FMB cell for every cone; in the far peripheral retina FMB cells occur at only 60% of the cone density. The 1:1 correspondence of cones to FMB cells seems to hold up to an eccentricity of approximately 9 mm, where the cone density is $4000/mm^2$. The Golgi data (Fig. 7.4) show that midget bipolar cells in the far peripheral retina contact more than one cone pedicle, which could compensate for their lower proportion. Immunocytochemical- and Golgi-staining are therefore mutually supportive and suggest that every L- and M-cone throughout the primate retina synapses with at least one FMB cell. The same result has also been found for the IMB cells by combining Golgi-staining with immunocytochemistry using antibodies against CCK (Wässle et al., 1994). Hence, L- and M-cone pedicles synapse with at least one FMB and one IMB cell.

Synaptic organization of the cone pedicles of the primate retina. The question arises as to whether there are enough synapses on a cone pedicle to accommodate the dendritic tips of the flat (FMB) and invaginating (IMB) midget bipolar cells and all six types of diffuse bipolar cells (DB1–DB6). To address this my colleagues and I have serially sectioned cone pedicles for electron microscopy and measured the number of ribbons and invaginating bipolar cell processes in two patches of a macaque monkey retina. One patch was from the midperipheral retina (6–7 mm eccentricity) and the other was from the foveal retina. The midperipheral patch corresponds to the eccentricity at which the bipolar cells in Fig. 7.1B were classified.

A reconstruction of a peripheral cone pedicle from horizontal sections is presented in Fig. 7.8. The diagram in Fig. 7.8A shows the outline of the pedicle and the 46 synaptic ribbons. The reconstruction of the lateral elements, formed by two horizontal cell processes flanking each ribbon, is shown in Fig. 7.8B. The very long ribbons marked by the arrowheads in Fig. 7.8A actually have four lateral elements and therefore represent two triads in Fig. 7.8B. The number of triads was therefore 48. The diagram in Fig. 7.8C shows the invaginating processes and their location with respect to the ribbons. A total of 104 invaginating processes were found in this particular cone pedicle; of these, 24 were thicker and filled with organelles in the electron micrographs. They are possibly the dendrites of an invaginating midget bipolar cell. On average, each triad of the pedicle reconstructed in Fig. 7.8C contained two invaginating processes. Unfortunately, it was not possible to also reconstruct the flat contacts from the horizontal sections through the cone pedicle. However, published micrographs from vertical sections through cone pedicles suggest their number to be substantially higher than those of invaginating contacts. Missotten (1965) estimated that there might be as many as 500 flat contacts on a cone pedicle base. The drawing in Fig. 7.8D shows a reconstruction of two sections slightly inner to the triads and is an attempt to estimate all of the processes connected to that particular cone pedicle: Some 450 processes were found. The ribbons and triads of 14 peripheral cone pedicles were reconstructed. The minimum number of triads was 38, the maximum 48, and the average 41.8 (±3). Chun et al. (1996) recently performed a detailed quantitative comparison between the number of synaptic sites available at peripheral cone pedicles and the demands for synaptic contacts of all of the different bipolar cells; there are enough synaptic sites available to accommodate all of the different bipolar cells.

The situation is different for cone pedicles in the fovea. The ribbons and triads of a total of 25 central cone pedicles were counted. Each pedicle contained between 18 and 24 triads, with an average of 21.4 (±1.6 s.d., $N = 25$). This result is in good agreement with the estimate of 16–22 ribbons in foveal cone pedicles reported by Esfahani et al. (1993) and by Calkins, Tsukamoto, and Sterling (1996). Since foveal IMB

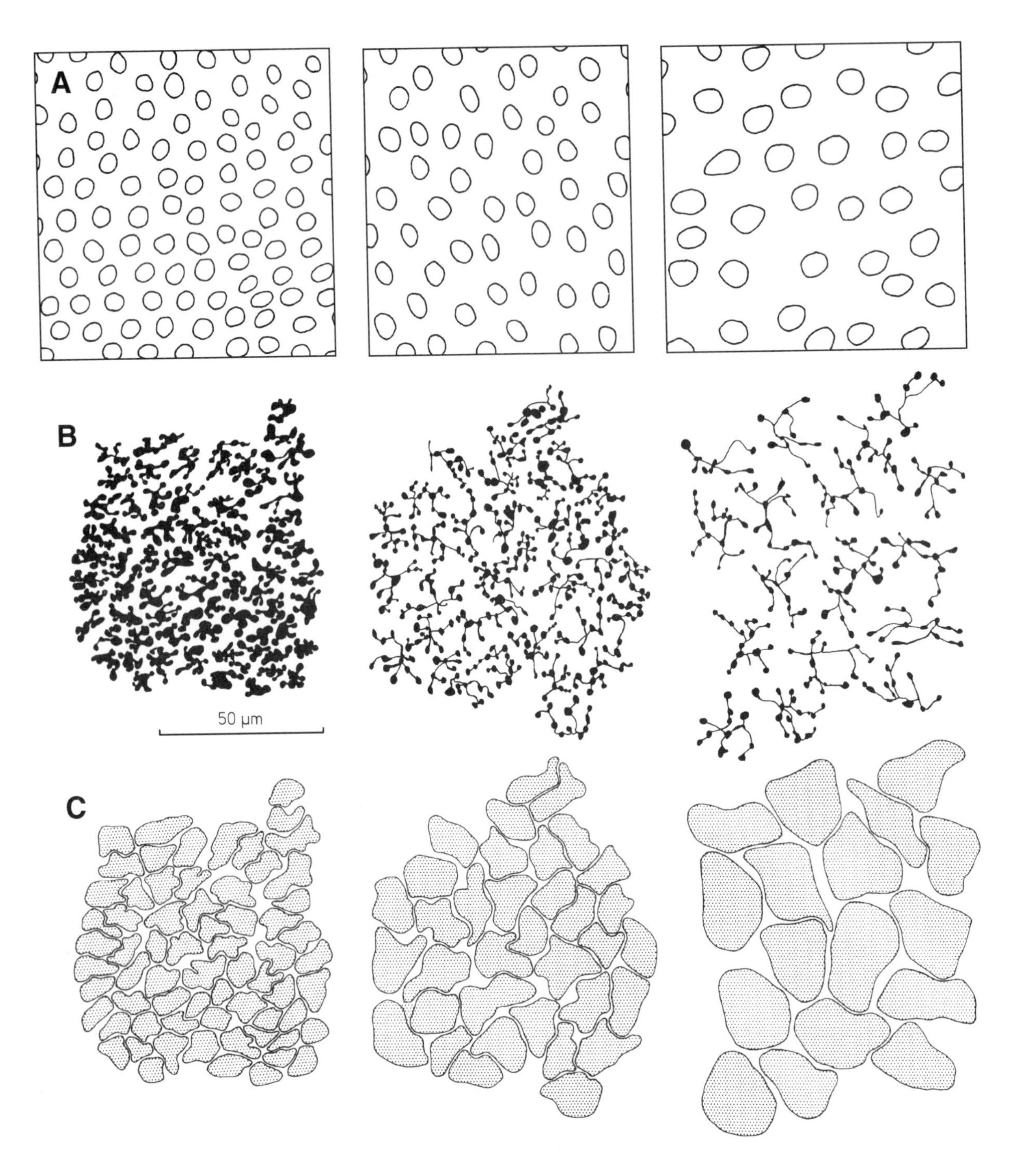

Figure 7.6: Spatial densities of cones and recoverin immunoreactive FMB cells. (A) The cone pattern at three different eccentricities. (B) The FMB cell axon terminals at the same retinal locations. (C) The encircled axon terminals. The axon terminals show a "territorial behavior" and fill the available space without any overlap.

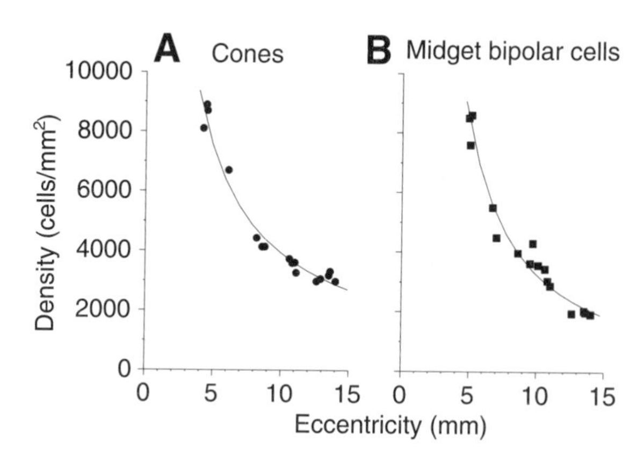

Figure 7.7: Comparison of the cone density (A) and the FMB-density (B) measured from horizontal sections of a macaque monkey retina that was immunostained for recoverin. The cells were sampled along an intersect extending from the optic nerve head towards the upper peripheral retina.

cells have been shown to occupy approximately 20 triads (Kolb, 1970; Calkins, Tsukamoto, & Sterling, 1995), this leaves very few slots for all of the different types of invaginating diffuse bipolar cells.

There are two possible solutions to this dilemma: Either there are no DB4, DB5, and DB6 cells close to the fovea, or their dendrites make – contrary to the current dogma – flat rather than invaginating contacts. We think that the first possiblity is unlikely because there is converging evidence from Golgi staining, from immunocytochemistry, and from electron microscopy for the presence of diffuse bipolar cells in the fovea. Furthermore, Calkins, Tsukamoto, and Sterling (1996) have shown that DB4–DB6 cells of the fovea preferentially make flat or triad-associated contacts but not fully invaginating contacts. This is not completely unexpected because, as was mentioned before, Hopkins and Boycott (1995, 1996) have shown that DB4, DB5, and DB6 cells in the midperipheral retina make a small number of flat contacts. It is possible that, for foveal cones, because of the limitations of space for triads in their smaller cone pedicles, DB4, DB5, and DB6 cells make more flat and only a few invaginating contacts.

Which bipolar cells transfer chromatic signals from the outer to the inner retina? A definitive answer to this question can only be obtained from electrophysiological recordings. Light responses have to be measured for all of the different bipolar cell types and the chromatic modulation of the responses when stimulating the different cone types has to be worked out. These experiments have not yet been done, and therefore firm answers cannot be given; however, there are some predictions and constraints from the anatomy with respect to bipolar cells and their chromatic selectivity (Kolb, 1991, 1994).

There is not much doubt that S-cone bipolar (BB) cells must transfer a blue-ON signal into the IPL. Their selectivity for S-cone pedicles, their invaginating contacts, and the termination of their axons in the inner IPL are all predictors for this functional role (Dacey & Lee, 1994). They are not restricted to the primate retina but seem to be present in all mammals. Axon terminals of BB cells contact the inner dendritic tier of the small-field, bistratified blue ON-ganglion cell (Dacey, 1993a, 1994; Dacey & Lee, 1994). These ganglion cells give OFF-responses when L- and M-cones are stimulated, and this input is probably derived from an OFF-diffuse bipolar cell synapsing onto the outer tier of the bistratified dendritic tree (Dacey, 1994, 1996). DB1, DB2, and FMB axon terminals overlap with the outer tier of the dendrites of the small, bistratified ganglion cells, and the question arises as to which of the three bipolar cells provides the input. In the case if it is derived from DB1 or DB2 cells, cone selectivity becomes an issue (see Chapter 8). Are the cone contacts of DB1 or DB2 cells restricted to L- and M-cones, or do they, nonselectively, contact all three cone types?

There is also no doubt that midget bipolar cells, which contact a single cone, transfer the chromatic signature of this cone to the inner retina. As we have shown, there is a 1:1 correspondence between both IMB and FMB cells and the cones in both the central and midperipheral retina. Thus the cone mosaic is transferred in complete detail into both the ON- and the OFF-sublamina of the IPL. The situation is different in far peripheral retina, where midget bipolar cells

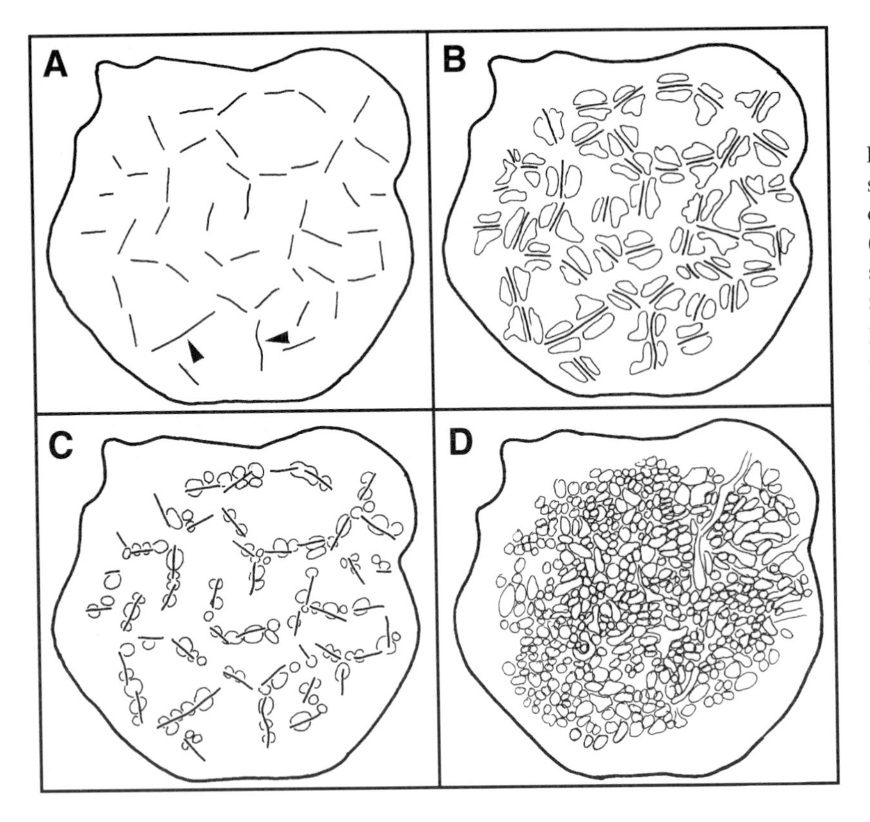

Figure 7.8: Reconstruction of the synaptic complex of a cone pedicle of the macaque monkey retina (eccentricity: 5–6 mm). The reconstruction was performed from horizontal serial sections, observed, and photographed by electron microscopy. The frame width represents 10 μm. (A) The solid circumference shows the outline of the pedicle. The short lines inside the pedicle represent the synaptic ribbons. Two rather long ribbons are marked by arrowheads. (B) Ribbons and lateral horizontal cell processes forming the triads are shown. The long ribbons marked in panel A are engaged with two triads each. (C) Central invaginating processes and their location with respect to the ribbon are shown. (D) Profiles of the dendritic processes underneath the cone pedicle.

contact several neighboring cone pedicles (Figs. 7.4C and 7.4D). This does not necessarily exclude cone selectivity since M- and L-cones might have a clumped distribution (Mollon & Bowmaker, 1992), but at the border of such patches a mixed cone input may occur and the chromatic signal might be degraded.

In the *fovea* it has been shown by electron microscopy that ON- and OFF-midget ganglion cells (P-cells) receive their major excitatory input from a single IMB or FMB cell (Kolb, 1991, 1994; Kolb & Dekorver, 1991; Calkins et al., 1994). Here the center response of the receptive field would be dominated by whichever (L- or M-) cone the MB cell happened to contact. Two models of the surround response can be considered. For the surround to be strictly color-selective, it should receive input from only one cone type (Reid & Shapley, 1992; Smith et al., 1992a; Kremers, Lee, & Yeh, 1995). Horizontal cells, because of their nonselective cone connections, cannot provide such a surround inhibition (Boycott, Hopkins, & Sperling, 1987; Wässle, Boycott, & Röhrenbeck, 1989; Dacheux & Raviola, 1990; Dacey et al., 1996). Small-field amacrine cells have recently been shown to receive input from both L- and M-cones (Calkins & Sterling, 1996), and hence, like horizontal cells, they cannot provide a pure surround (see Chapter 9). This favors the second model of a mixed cone input to the surround of midget ganglion cells (P-cells) (Paulus & Kröger-Paulus, 1983; Lennie, Haake, & Williams, 1992).

The situation becomes more complicated in peripheral retina, where there are more cones than there are ganglion cells (Wässle et al., 1990). Midget ganglion cells in midperipheral retina have extended dendritic trees (Watanabe & Rodieck, 1989; Dacey & Petersen, 1992; Dacey, 1993b); thus, several midget bipolar cells must convergence to create the receptive field center of a peripheral midget ganglion cell (Croner & Kaplan, 1995). If peripheral midget ganglion cells have "pure cone" centers, their dendrites must make selective contacts with MB-axon terminals; however, if peripheral midget ganglion cells have "mixed cone" centers, they can contact all MB-axon terminals within their reach. Recent physiological recordings (Dacey, 1996a) suggest that peripheral midget ganglion cells have mixed L- and M-cone inputs to their receptive field centers, which is consistent with nonselective wiring. L- and M-cones are clustered (Mollon & Bowmaker, 1992), however, and there might be an occasional peripheral midget ganglion cell with a "pure cone center." These few cells and their chromatic signal might be sufficient to account for the residual color sensitivity of the peripheral visual field. Mullen and Kingdom (1996) performed psychophysical experiments measuring the peripheral color sensitivity in human observers and compared their results with the model of nonselective wiring by using a "hit and miss" analysis. They showed that the losses of red-green color sensitivity across the human visual field can be accounted for by a nonselective wiring of the peripheral midget ganglion cells.

The functional role of diffuse bipolar cells. There are two possible mechanisms by which diffuse bipolar cells gain color-specific light responses. The first model assumes cone-selective contacts. One type of diffuse bipolar cell might avoid L-cones and contact only M-cones, while another type might avoid M-cones and contact solely L-cones. The second model is more sophisticated and proposes that diffuse bipolar cells contact all cones within their dendritic field but express specific glutamate receptors: At synapses with L-cones an ionotropic (kainate/AMPA) receptor and at synapses with M-cones a metabotropic (mGluR6, L-AP4) receptor, or vice versa.

There is a precedent in the fish retina for the first model, that is, for cone-selective contacts of bipolar cells (Scholes, 1975; Ishida, Stell, & Lightfoot, 1980). There is also a precedent in the fish retina for the second model, that is, for specific glutamate receptors on single bipolar cells, which subserve separate, functionally defined synaptic inputs (Nawy & Copenhagen, 1987, 1990). For horizontal cells of the fish retina it was found that the transmitter released from S-cones decreases the membrane conductances of H1 horizontal cells, whereas the synapses made by L- and M-cones are of a classical excitatory type. The question arises as to whether similar mechanisms are also present in the mammalian and, more specifically, in the primate retina.

Boycott and Wässle (1991) did not find any indication of cone-selective connections of DB1–DB6 cells, although, as mentioned before, the possibility cannot be excluded that they avoid S-cones. Furthermore, given the clumped distribution of M- and L-cones, there might be an occasional DB-cell with a pure cone input; however, because of the convergence of diffuse bipolar cells onto ganglion cell dendrites in the IPL, it is to be expected that this chromatic signal would be lost.

The second model, cone-specific expression of glutamate receptors, cannot be excluded at present, but it is unlikely for the following reason. Electrophysiological recordings from dissociated cone bipolar cells of the cat retina (de la Villa, Kurahashi, & Kaneko, 1995) and from cone bipolar cells in rat retinal slices (Euler, Schneider, & Wässle, 1996; Hartveit, 1996) did not reveal two opponent glutamate responses within the same bipolar cell. When kainate or glutamate were applied to putative OFF-bipolar cells, nonselective cation channels opened and the cells depolarized. Such cells did not respond to L-AP4, the specific glutamate agonist at ON-bipolar cells. When L-AP4 or glutamate were applied to putative ON-bipolar cells, nonselective cation channels closed and the cells hyperpolarized. Such cells did not respond directly to kainate. It would be surprising if this ON/OFF dichotomy did not hold for primate diffuse bipolar cells.

Comparable to other mammalian retinas, primate diffuse bipolar cells are most likely involved with the transfer of a luminosity signal to the IPL. Those terminating in the outer half of the IPL are most likely OFF-bipolar cells, while those terminating in the inner half of the IPL are ON-bipolar cells. But why are there several types? Their precise functional role still has to be elucidated; however, they almost certainly provide input to different ganglion cell classes. Those cells terminating close to the center of the IPL, such as DB2, DB3, DB4, and DB5, probably provide input to the dendrites of parasol cells (Calkins, Schein, & Sterling, 1995; Jacoby & Marshak, 1995, 1996). Parasol cell light responses are phasic and parasol cells have a high contrast sensitivity (Lee, Martin, & Valberg, 1989a; Kaplan, Lee, & Shapley, 1990; Lee et al., 1994, 1995, Lee, 1996). DB2 and DB5 cells are most likely candidates for providing input to parasol cells since these two bipolar cells have a high coverage factor (Boycott & Wässle, 1991) and make many contacts with cone pedicles (Hopkins & Boycott, 1995), which is consistent with a high contrast sensitivity. Whether they give phasic light responses still has to be shown. Of course, it is also possible that other DB cells contribute inputs to parasol cells.

In the rat retina it has been shown that bipolar cells 5 and 6 costratify with the dendrites of the displaced cholinergic amacrine cells (Fig. 7.1A) and thus may provide direct and indirect input to directional selective ganglion cells (Brandstätter et al., 1995). DB4 and DB5 cells could be the homologous types of the primate retina and could also provide input to directional selective ganglion cells.

Dacey and Lee (1994) have shown that the "blue-ON" ganglion cells have a bistratified dendritic tree, with the outer tier of dendrites close to the amacrine cell layer. There they should receive an OFF input from a diffuse bipolar cell (Dacey, 1996a). DB1 cells, which have their axon terminals in that stratum, are the most likely candidates to provide that input.

More generally, there seems to be a tendency in different mammalian retina for phasic ganglion cells to stratify toward the center of the IPL, whereas tonic ganglion cells have their dendrites further away from the center. This holds for cat α (phasic) and β (tonic) ganglion cells (Wässle & Boycott, 1991), and it has also been found in the case of monkey parasol (phasic) and midget (tonic) ganglion cells (Dacey, 1994). One might predict that the diffuse bipolar cells terminating in the respective strata also have phasic or tonic properties. The more general idea is that the different bipolar cells are subdivided according to their temporal bandwith (Sterling et al., 1995). Perhaps the different bipolar cells express glutamate receptors with different temporal properties, that is, with different desensitization characteristics (see, for example, Hollmann & Heinemann, 1994). Alternatively, the change in the temporal properties of the bipolar cells might occur in the IPL through different feedback circuits in the IPL (reciprocal synapses) or through intrinsic differences in the time course of transmitter release. Whatever the final answer, the existence of so many diffuse bipolar cells suggests that each of them transmits only a small part of the information available in the cone array. Thus parallel processing of the visual image occurs after the first synapse, the earliest possible point in the visual system.

Evolution of primate color vision. Phylogenetic trees of mammalian photoreceptor opsins show that a primordial visual pigment evolved first into ancestral L- and S-cone pigments (Okano et al., 1992; Chiu et al., 1994; Hisatomi et al., 1994). Therefore, the dichromatic cone system, which is common to the majority of mammals, is phylogenetically the older color system (Mollon & Jordan, 1988; Jacobs, 1993; Mollon, 1996). Rhodopsin evolved from the ancestral S-cone opsin only later (Okano et al., 1992). This suggests that scotopic vision developed after photopic vision, and this may explain the complex way in which the rod circuitry is superimposed on the cone circuitry in the mammalian retina (Kolb & Famiglietti, 1974). The high homology and close vicinity of primate (Old-World monkey, human) L- and M-opsin genes suggest that they are a recent event in evolution and are the result of a duplication of an ancestral L-opsin gene (Nathans, Thomas, & Hogness, 1986; Neitz & Neitz, 1994, 1995). This probably happened 30–40 million years ago, some time after the divergence of Old-World and New-World monkeys. As two recent studies have shown, trichromacy has a great adaptive advantage for frugivorous primates (Osorio & Vorobyev, 1996; Regan et al., 1996).

The question arises then as to why, amongst mammals, have only primates acquired trichromacy?

Clearly for other mammals trichromacy would also be advantageous, and mutations of the cone pigments must have happened given the perfect adaptation of mammalian cone pigments to their habitats (Jacobs, 1993). For a mutation in the cone pigment to be advantageous to an animal the brain must have direct access to the information carried by the mutant. For instance, if the pigment of all L-cones shifts to a wavelength that is more appropriate to the habitat, the diffuse bipolar cells might give stronger signals and this information could be used by the brain. The situation is different, however, if a split of cone pigments into L- and M-pigments comparable to the mutation in primates occurs. The split might result in equal numbers of L- and M-cones randomly distributed across the retina. In mammals other than primates, bipolar cells pool the signals of several neighboring cones and, in turn, ganglion cells pool the signals from several converging bipolar cells. In such a highly convergent system the chromatic information introduced into the cone mosaic by the L/M mutation would be lost within the retina and thus would never reach the brain.

We would argue that the situation 30 million years ago was different in Old-World monkeys. The primate eye and retina had been "optimized" during evolution for highest spatial resolution, which requires a high cone density and a low cone-to-ganglion cell convergence in the "acuity pathway." The anatomical limits for this optimization are reached when each cone is connected through a midget bipolar cell to a midget ganglion cell, thus establishing a "private line" to the brain. We suggest that only after this private line of cones in the central retina had evolved did a subsequent mutation in the L-cone pigment create approximately equal numbers of L- and M-cones at random spatial locations (Mollon & Jordan, 1988; Wässle & Boycott, 1991; Wässle et al., 1994). The midget system of the central retina was able to transmit this chromatic information to the brain where it could be used, for example, to detect red fruit amongst green leaves (Osorio & Vorobyev, 1996; Regan et al., 1996). Later, the selective advantage of trichromatic vision must have led to a proliferation of color pathways in cortical and perhaps even subcortical centers. The well-known plasticity of the brain and the subcortical visual pathways (Shatz, 1996) might have been the mechanism for such changes. This "midget theory" of the evolution of trichromacy in Old-World monkeys has the advantage that it is based on the general mammalian retinal wiring diagram. It is not necessary to postulate additional specific mutations that change the cone selectivity of bipolar cells, or the expression of glutamate receptors, or the selectivity of ganglion cells. The idea that trichromacy "piggybacks" on the high acuity system also postulates that midget ganglion cells perform a double duty in visual signaling, an idea that has been promoted for some years (Ingling & Martinez-Uriegas, 1983a, 1983b; Lennie, 1984; Mollon & Jordan, 1988; Merigan, 1989).

The "midget theory of trichromacy" predicts that primates, without a fovea or a midget system are dichromats. This seems to be true in the case of prosimians (Ringtail Lemurs, Brown Lemurs: Jacobs & Deegan, 1993a). An interesting case is the New-World monkeys. They probably separated from Old-World monkeys before the acquisition of trichromacy but after the evolution of the fovea and the midget system. Capuchin monkeys (Silveira et al., 1989; Silveira et al., 1994; Lima, Silveira, & Perry, 1993; 1996; Yamada et al., 1996; Yamada, Silveira, & Perry, 1996), squirrel monkeys (Stone & Johnston, 1981; Leventhal et al., 1989; Leventhal, Thompson, & Liu, 1993) and marmoset monkeys (Ghosh, Martin, & Grünert, 1997; Goodchild, Ghosh, & Martin, 1996; Wilder et al., 1996) have well-developed foveae and ganglion cells that are similar to those of the macaque monkey retina. New-World monkeys show an extreme polymorphism of cone pigments (Jacobs, 1993, 1996) with some species showing a sex-linked dimorphism, the females being trichromats and males being dichromats. Within this genetic diversity a true trichromat recently has been described (Jacobs et al., 1996a). There is so much variation in New-World monkeys that they can be seen as a currently evolving population whose study could be a test system for some of the above-formulated ideas.

Summary

Parallel processing in the mammalian retina begins at the cone pedicle, the first synaptic relay. Approximately ten types of cone bipolar cells contact every individual cone pedicle and transfer different aspects of the light signal to the inner plexiform layer. There the bipolar cell axons terminate in different sublayers and contact the dendrites of specific ganglion cells. The primate retina follows this general scheme; however, the midget bipolar cell channel is optimized for highest spatial resolution: A midget bipolar cell contacts a single cone pedicle and also makes a one-to-one connection with a single midget ganglion cell. Because of this private line, midget bipolar and midget ganglion cells are capable of transferring chromatic information to the brain. This is probably the only channel in primates that carries chromatic signals from L- and M-cones. There is growing evidence from molecular genetics that the system that compares the absorptions in the ancestral L- and S-cones, which is common to most mammals, is the phylogenetically older system. Here it is argued that the evolution of the midget system in primates, that is, of a high acuity system, allowed the brain to take advantage of the additional chromatic information offered by a subsequent split of the ancestral L-cone pigment gene into L- and M-cone pigment genes. Thus trichromacy essentially "piggybacks" on the phylogenetically older, high-acuity system.

Acknowledgments

The results presented here are the collaborative effort of many co-workers during the years. I am most grateful to Dr. B. B. Boycott, Dr. M.-H. Chun, Dr. T. Euler, Dr. U. Grünert, and Dr. P. Martin. I would like to thank Dr. B. B. Lee, Dr. D. Calkins, and Dr. D. Dacey for many interesting discussions. I am also grateful to I. Odenthal for typing the manuscript, to F. Boij for help with the illustrations, and to Dr. W. R. Taylor for improving the English text.

8

Synaptic organization of cone pathways in the primate retina

David J. Calkins

The great diversity of cell types contained within the broader classes of retinal neuron reflect the critical role of local circuits in the early encoding of visual information. The hierarchical structure of the retina lends itself to the study of these circuits, and for the most part this has been pursued by an intense focus on the cell types that comprise their individual components. Investigations that focus on either the morphology or physiology of these cell types have led directly to specific inferences or hypotheses regarding their connectivity, in particular, their presynaptic circuitry. A major avenue of our research has been to exploit the union of electron microscopy with digital computing to test rigorously and quantitatively these specific hypotheses about synaptic relationships by reconstructing retinal circuits in their entirety, that is, not in single sections through the retina, but rather in small *volumes*. The emphasis in these investigations is (1) to establish synaptic contact between cell types, (2) to quantify the extent of this contact by reconstructing complete neurons and scoring all of their synapses, and (3) to measure the sampling of these circuits by reconstructing their common, neighboring components. Conversely, our reconstructions of complete circuits have also revealed a tremendous capacity to *generate* hypotheses not only about the physiology of specific cell types, but also about their role in encoding the information that ultimately results in visual perception.

The study of color vision for us has become a study of the differences in how the short- (S-), middle- (M-), and long- (L-) wavelength–sensitive cone types parcel their information along distinct postsynaptic pathways. This study has necessitated first and foremost the identification of S-, M-, and L- cones and has proceeded generally into an investigation of how the cells that collect from different cone types differ themselves. Although these architectural differences may have broader implications for visual channels that are not directly involved in coding color, our investigation has evolved into a critical examination of the sites of convergence of signals from S-, M-, and L-cones, in particular those sites where these signals converge with opposite signs. This constraint arises naturally from how we perceive color. While the range of hues we can experience is quite diverse, we cannot perceive certain pairs of hues simultaneously: They are mutually exclusive or *opponent* percepts. These opponent pairs are blue or yellow (B/Y) and red or green (R/G); while we can perceive color combinations between these pairs, for example, blue and red yielding a violet percept that is in part both, we cannot perceive combinations within a pair (Hurvich & Jameson, 1957; Krantz, 1975; Calkins, Thornton, & Pugh, 1992; for a review see Lennie & D'Zmura, 1988).

A large and diverse body of psychophysical data implies that the critical neural event that underlies color opponency is antagonism between different cone types. Thus, for B/Y opponency signals from S-cones are combined antagonistically with those from M- and L-cones (abbreviated as S/(M+L), where "/" indicates antagonism), while for R/G opponency signals from L-cones are combined antagonistically with those from M-cones (abbreviated as L/M). These combinations represent the minimal conditions required of putative neural pathways consistent with the psychophysical properties of the color channels. That the critical antagonism between cone types within the S/(M+L) and L/M pathways is established in the retina and not at sites of convergence further along in the visual streams is an inference supported by the vast physiological literature demonstrating cone antagonism with similar spectral properties within the receptive fields of many ganglion cells providing input to parvocellular pathways from the lateral geniculate nucleus to the visual cortex (e.g., de Monasterio & Gouras, 1975; also see Chapters 9 and 10).

At issue in this chapter is the nature of those pathways, in particular the *critical locus* of color opponency: where in the visual pathways cone signals converge antagonistically to code opponent color percepts (Teller & Pugh, 1983; Teller, 1990). Color opponency and its neural basis have been studied most extensively for the central visual field, and a large portion of the visual cortex represents this same region (Wässle et al., 1990). Thus, this chapter focuses on the pathways collecting from cones in the central most retina or *fovea*. The vast majority of retinal ganglion cells serving this region of the visual field have receptive fields that are nominally *opponent*: Their net spectral sensitivity to full-field stimulation is cone-antagonistic, most often either S/(M+L) or L/M, although some other combinations also have been found (de Monasterio & Gouras, 1975). Thus, the axons of these ganglion cells are expected to carry information that results either in the perception of blue or yellow, or of red or green. The neural locus of the cone antagonism that is critical to the establishment of the color opponent channels is therefore presumed to be embedded in their presynaptic circuitry.

The cone antagonism inherent to color opponency is both spatially coextensive and temporally coincident: For example, excitation of L- and M-cones together causes less activity within the L/M pathway than excitation of either cone type alone (Calkins, Thornton, & Pugh, 1992). In this chapter we explore two general possibilities for wiring such cone anatgonism in a ganglion cell receptive field. The first is through the convergence of inhibitory and excitatory inputs with different spectral sensitivities. For example, the excitatory center of a foveal midget (or P) ganglion cell is derived from a single M- or L-cone, while its inhibitory surround is thought to be derived from cones of the other type, either L or M, perhaps via GABA-ergic or glycinergic lateral connections. The ganglion cell's full-field spectral sensitivity reflects the difference between these different cone inputs. The second general possibility is that cones of different types converge on a ganglion cell strictly through excitatory cells that respond to light with opposite polarity, that is, OFF versus ON. This sort of wiring is thought to underlie the spectral sensitivity of the S-ON/(M+L)-OFF bistratified ganglion cell (see below; Rodieck, 1991; Dacey & Lee, 1994).

Thus, this chapter has two goals: to describe (1) our investigations of how postsynaptic pathways from different cone types differ fundamentally themselves and (2) our more pointed investigations of the wiring schemes of cone antagonism. To understand the "black box" between the input element of the circuit (the cones) and the output element (the ganglion cell) requires determining the types of cells in a circuit, enumerating and classifying the synapses between pre- and postsynaptic elements, and establishing how these specific pathways sample the cone mosaic. To this end we constructed a library of electron micrographs from serial sections through the primate fovea and traced through this library entire circuits from cone to ganglion cell; much of this work has been published in detail (Tsukamoto et al., 1992; Calkins, 1994; Calkins & Sterling, 1996; Calkins et al., 1994, 1996). This chapter will describe some of this effort and, in doing so, provide a detailed look at different neural pathways leading from S-, M-, and L- cones and how this circuitry might provide a framework for antagonistic convergence of cone signals in a ganglion cell's dendritic tree.

Methods

The arrangement of the mosaics of different classes of retinal neurons compels certain inferences about the connectivity between cell types (see Chapters 7 and 9). However, to prove that a synaptic relation exists between cell types involves: (1) positive identification of the cells and (2) resolution of individual synapses and their pre- and postsynaptic elements; our goal was to completely quantify entire circuits from cone to ganglion cell.

We isolated a portion of the fovea from the retina of an adult male *Macaca fascicularis* and prepared it for electron microscopy using standard methods (Tsukamoto et al., 1992; Calkins, Tsukamoto, & Sterling, 1996). Consecutive sections (319) were cut vertically at a thickness of 90 nm along the horizontal meridian just nasal of the center fovea, photographed at 2000–12,000× and printed with an additional magnification of at least 2.8×. At these magnifications, the pre- and postsynaptic elements of a circuit and the electron-dense material that marks either conventional or "ribbon" synapses between them are readily distinguished (Figs. 8.1 and 8.8).

The cone terminals and bipolar, ganglion and amacrine cells studied from this series were contained in a region spanning 640–500 μm eccentricity just nasal of the center fovea (as shown in Tsukamoto et al., 1992). Within this narrow band, a bipolar cell dendritic tree and the cone terminals contacting it were generally displaced 40–50 μm toward the center fovea from the axon terminals of that particular bipolar cell. The lengths of the axons connecting each cone terminal to its inner segment ranged from 350 μm for terminals at the peripheral edge to 295 μm for terminals at the foveal edge, and the magnification factor for this retina was 216 μm/deg. Thus, the inner segments providing input to the ganglion cell circuits studied here were located 1.2–0.8 deg from the center fovea.

Reconstruction of a retinal circuit from electron micrographs of serial sections involves identifying the neuronal components of the circuit in their entirety and the synaptic contacts between them. The reconstruction of an individual neuron is a matter of locating its more obtrusive elements, that is, its soma, primary dendritic stalk or axon, and then tracing these elements through the series, tracking new pieces as they are identified by continuity of the cell membrane. This is "forward" reconstruction beginning with and testing the null hypothesis that neuronal processes neighboring the cell in question are not part of the cell. The evidence for complete cells is also strengthened by "reverse" reconstruction. For ganglion cells, once a presynaptic bipolar cell was identified, the other processes postsynaptic to that bipolar cell were traced to determine whether they too joined the ganglion cell in question, the null hypothesis being that they did not. For reconstructing a bipolar cell axon terminal, once a postsynaptic ganglion cell was identified, other bipolar cell terminals presynaptic to that ganglion cell were traced to determine whether they joined the bipolar cell in question, the null hypothesis once again being that they did not. Thus, reconstruction of a circuit that is sufficient within the capacity to resolve cell membranes involves a continuous process of forward and reverse tracing and of adopting new working null hypotheses. The key at each turn is to weigh the visual evidence in favor of the null hypothesis and make decisions based on the result.

Once we identified all of the pieces of a circuit and the synaptic contacts between them, we traced the pieces onto acetate sheets aligned on a cartoonist's jig. We maintained alignment between serial sections on the acetate by also tracing around the pieces of interest numerous micro- and macromarkers: vesicles, mitochondria, microtubules, other neuronal processes, nuclei, and soma. We then digitized the tracings from the acetate sheets to computer using a bit pad and a Unix-based software package (*Montage*: Smith, 1987).

Results

ON-bipolar cell pathways of S-cones differ from those of M- and L-cones. Synaptic active zones in cone terminals are marked by an electron-dense "ribbon" that is thought to serve as a docking site for

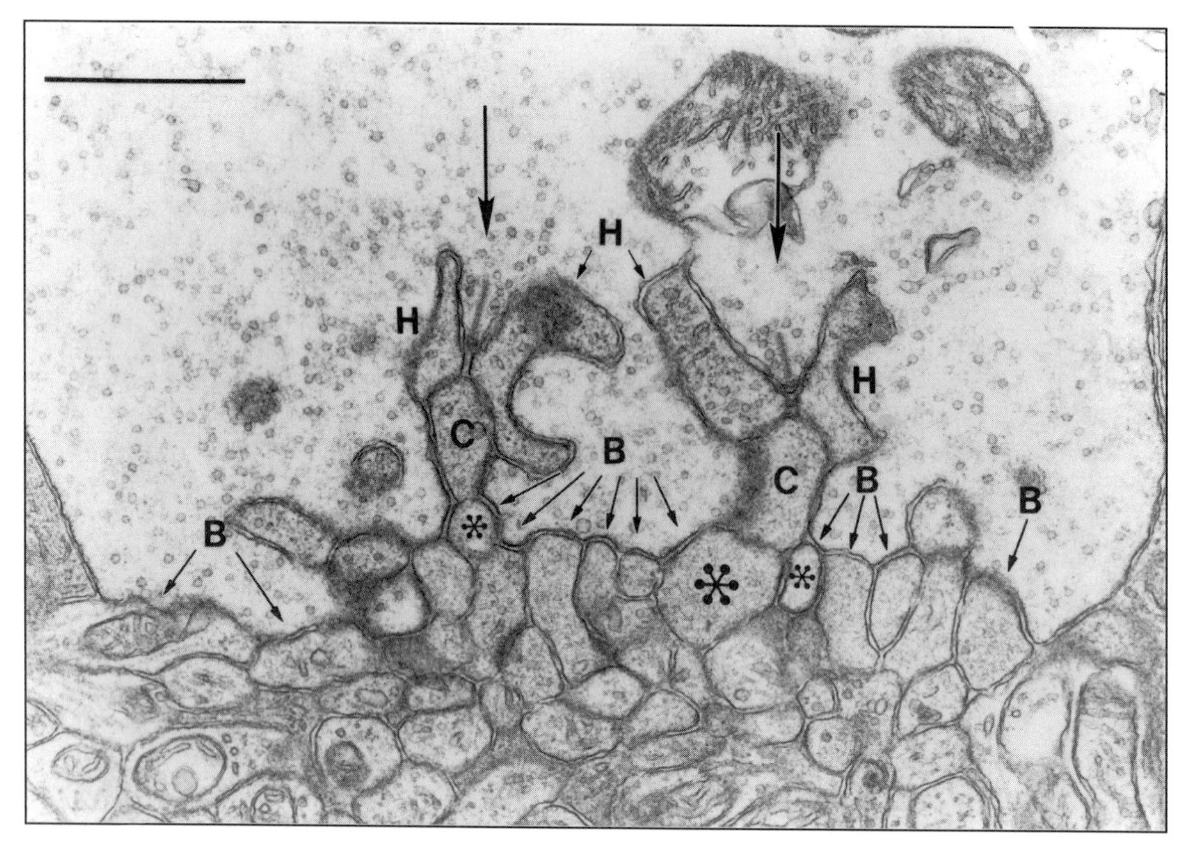

Figure 8.1: Electron micrograph of the base of a macaque cone terminal in vertical view. Two active zones, each marked by a synaptic ribbon, are shown (large arrows). Each ribbon points between a pair of horizontal cell processes (H) to an invagination of the terminal membrane that houses a bipolar cell dendrite as a central element (C); this arrangement is called a "triad" (see text). For foveal L- and M-cones, a midget ON-bipolar cell contributes nearly every central element (triad on the right), while diffuse ON-bipolar cells do so only rarely (triad on the left). Sites of "basal" contact (B) with bipolar cell dendrites occur adjacent to the central elements of triads ("triad-associated", asterisks) or outside of the invagination ("nontriad-associated"), depending on the type of bipolar cell (see text). Scale bar represents 1 μm.

glutamate-containing synaptic vesicles (Missotten, 1965; Dowling & Boycott, 1966; Rao-Mirotznik et al., 1995; Calkins, Tsukamoto, & Sterling, 1996). In vertical sections, these ribbons are most commonly observed oriented in a highly specific geometry, pointing between a pair of horizontal cell processes to an invagination of the terminal membrane that houses a central, bipolar cell dendrite (Fig. 8.1).

This arrangement has been called a "triad" (Missotten, 1965; Dowling & Boycott, 1966); often, however, an additional bipolar cell dendrite forms a second central element at the invagination (Chun et al., 1996; Calkins, Tsukamoto, & Sterling, 1996). In primate retina, a bipolar cell with axon terminals stratifying in the *b* or ON-sublamina of the IPL contributes the central element(s) of each triad. These bipolar cells provide the direct feed-forward connections that presumably form the excitatory centers of ON-ganglion cells. Thus, they likely express one or more isoforms of metabotropic glutamate receptor (mGluR) and therefore depolarize with light onset. It is presumed moreover that they are the sole conveyer of an ON-response

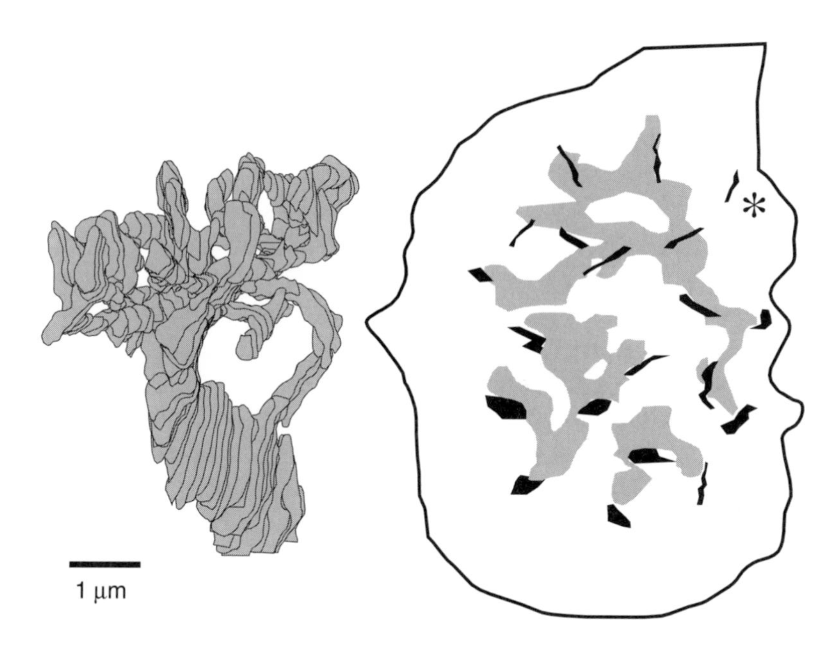

Figure 8.2: Reconstruction of the dendritic tree of a midget ON-bipolar cell in radial view (left) and of the cone terminal contacting it in horizontal view (right). The dendrites of the midget bipolar cell (gray) form a central element at all but one (asterisk) of the 20 ribbon synapses (black); only 3 other central elements are contributed by diffuse ON-bipolar cells. Modified from Calkins, Tsukamoto, and Sterling (1996).

to the IPL (for a review, see Massey, 1990; Hopkins & Boycott, 1995; Kolb & Nelson, 1995; Calkins, Tsukamoto, & Sterling, 1996).

Nearly every central element of a foveal M- or L-cone invagination arises from the dendritic tree of a single midget ON-bipolar cell, with only a few central positions occupied by the dendrites of other ON-bipolar cell types (Table 8.1; Fig. 8.2; Kolb, 1970; Calkins, Tsukamoto, & Sterling, 1996; Chun et al., 1996). Moreover, in our reconstructions of the cone complex and the midget bipolar cell dendrites forming the central elements of their triads, we did not detect a systematic pattern in which the number of central elements contributed by the midget ON-bipolar cell of M- and L-cones differs, although we have distinguished these cone types by other means (see below). Thus, with regard to their connections with midget and other ON-bipolar cells, M- and L-cones can be considered as one group.

In contrast, in an early survey of the cone terminals in our material and their connections with midget bipolar cells, Klug et al. (1992; 1993) found a small group of terminals (about 5% of the sample) that each lacked contact with a midget ON-bipolar cell altogether. The central elements of the triads in these terminals instead arose from a type of cell whose dendrites streamed beneath the layer of cone terminals to collect exclusively from on average two of these cones. The morphology and connectivity of the dendritic tree of this bipolar cell, and both the morphology and stratification of its axon terminal in the ON-sublamina of the IPL, correspond to that of the so-called blue-cone bipolar cell that receives selective input from S-cones (Figs. 8.3A and 8.9B; Mariani, 1984; Kouyama & Marshak, 1992). Since these are the only

Cone ribbon synapses		Postsynaptic invaginating process		
Cone	Ribbons	Midget ON	Diffuse ON	Total
1	21	20	2	22
2	18	17	4	21
3	20	16	5	22
4	20	19	3	21
Mean	19.8	18.0	3.5	21.5
s.d.	1.3	1.8	1.3	0.6

Table 8.1: Organization of four M- and L-cone terminals in macaque fovea.

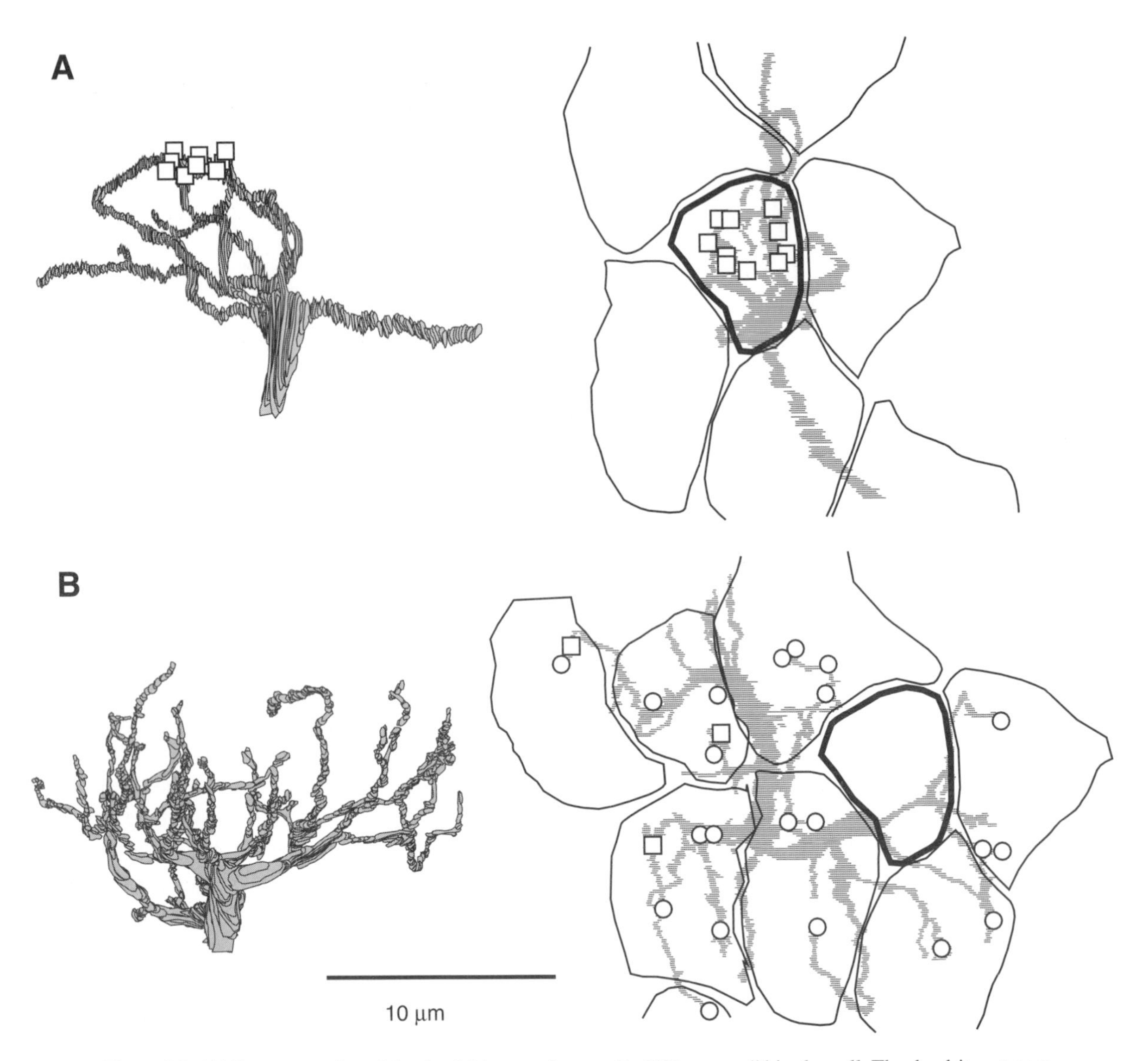

Figure 8.3: (A) Reconstruction of the dendritic tree of a so-called "blue-cone" bipolar cell. The dendrites stream under the terminals of L- and M-cones to receive exclusive contact from an S-cone (thick outline) strictly as the central elements of invaginations (squares). Two dendrites ran out of the series. (B) Vertical (left) and horizontal (right) views of the dendritic tree of a diffuse ON-bipolar cell, modified from Calkins, Tsukamoto, and Sterling (1996). The dendrites of these cells receive contact nonselectively from L- and M-cones most often at TA basal sites (circles) and only rarely as the central elements of triads (squares). The dendrites skip underneath an S-cone (thick outline). Thus, "blue-cone" bipolar cells may be the only means through which S-cone signals can penetrate the ON-sublamina of the IPL (see text).

bipolar cells in the mammalian retina whose dendrites "skip" underneath cone terminals to receive select contact from a few widely spaced cones, they form a conspicuous marker by which we were able to identify positively S-cones (Klug et al., 1992, 1993; Calkins et al., 1994; Herr et al., 1996).

There is some evidence that every invagination of the S-cone terminal houses as a central element the dendrite of a blue-cone bipolar cell and, conversely, that every central dendrite is contributed by this same type of bipolar cell (Herr et al., 1996). We found that "diffuse" bipolar cells with axon terminals in the ON-sublamina collect synapses from every M- and L-cone within reach of their dendritic trees, but not from the cones identified as S-cones (Fig. 8.3B; Calkins, Tsukamoto, & Sterling, 1996). Thus, the blue-cone bipolar cell may be the only feed-forward means through which S-cone signals in the fovea can penetrate the ON-region of the IPL and, in this sense, may be called without ambiguity the S-ON-bipolar cell. In contrast, M- and L-cones appear to contact not only a midget ON, but also the dendritic trees of multiple types of diffuse ON-bipolar cell (Fig. 8.3B; see also Chapter 7).

OFF-bipolar cell pathways of S-cones are similar to those of L- and M-cones. There are perhaps more than 100 separate bipolar cell dendritic "twigs" that abut the membrane of a foveal cone terminal at one or more sites of "basal" contact (Fig. 8.1; Calkins, Tsukamoto, & Sterling, 1996; Chun et al., 1996). These are the only sites of contact from cones to bipolar cells with axon terminals that stratify in the *a* or OFF-sublamina of the IPL. Thus, in primates, each dendrite postsynaptic to a cone at a basal contact is presumed to express one or more isoforms of ionotropic glutamate receptor (GluR) and therefore to hyperpolarize with light onset. It is often presumed that they are the sole conveyer of the OFF-response to the IPL (for review, see Hopkins & Boycott, 1995; Kolb & Nelson, 1995). However, for each M- and L- (but not S-) cone, a few basal sites are occupied by the dendritic twigs of diffuse bipolar cells with axon terminals in the *b* sublamina of the IPL; in fact, in the fovea, this is the primary site of contact to these putative ON-bipolar cells (Figs. 8.1 and 8.3B; Calkins, Tsukamoto, & Sterling, 1996). Thus, "basal" is not necessarily synonymous with a postsynaptic hyperpolarizing response to light. Nevertheless, the large number of basal contacts at each cone terminal indicates a potentially vast divergence to OFF-pathways.

Two types of basal contact are distinguished based solely on their proximity to an invagination of the cone terminal and not on any morphological or ultrastructural differences (Boycott & Hopkins, 1991; Hopkins & Boycott, 1995). Triad-associated (TA) basal contacts occur on the inside wall of each invagination and are directed to dendritic twigs that run adjacent to the central dendrite of the invagination. This contact also has been called "semi-invaginating," and it is a few of these for each M- and L-cone that are occupied by diffuse ON-bipolar cells (Figs. 8.1 and 8.3B; Calkins, Tsukamoto, & Sterling, 1996). Nontriad-associated (NTA) basal contacts occur outside of the invaginations, along the lower base of each cone terminal. These are occupied primarily by the dendrites of diffuse OFF-bipolar cells, which collect input from every cone within reach of their dendrites.

Each foveal M- and L-cone not only contacts the invaginating dendrites of a midget ON-bipolar cell, but also the dendrites of a midget bipolar cell whose axon terminals stratify diffusely throughout OFF-sublamina of the IPL. Cone contact to the dendrites of this cell in the fovea occur primarily at TA basal sites (Kolb, 1970; Boycott & Hopkins, 1991), and it is this bipolar cell that presumably conveys an OFF response to the corresponding midget ganglion cell (Calkins et al., 1994).

Some of our preliminary data show that the cones identified as S based on their connections with S-ON-bipolar cells also appear to contact a midget OFF-bipolar cell that in turn contacts a midget OFF-ganglion cell (Fig. 8.4; Klug et al., 1992, 1993). Thus, in this regard, S-, M-, and L-cones are not qualitatively different, and this midget pathway may correspond to the few ganglion cells identified physiologically as having relatively narrow S-OFF receptive field centers (de Monasterio & Gouras, 1975).

On the other hand, the scarcity of physiological recordings from ganglion cells demonstrating an OFF response to S-cone stimulation (see Dacey & Lee, 1994) could be interpreted as indicating a scarcity of S-cone contacts to other OFF-bipolar cells. However, electron micrographs indicate a plethora of basal processes receiving contact from S-cones identified histochemically (Kouyama & Marshak, 1992). We tested this hypothesis by partially reconstructing the den-

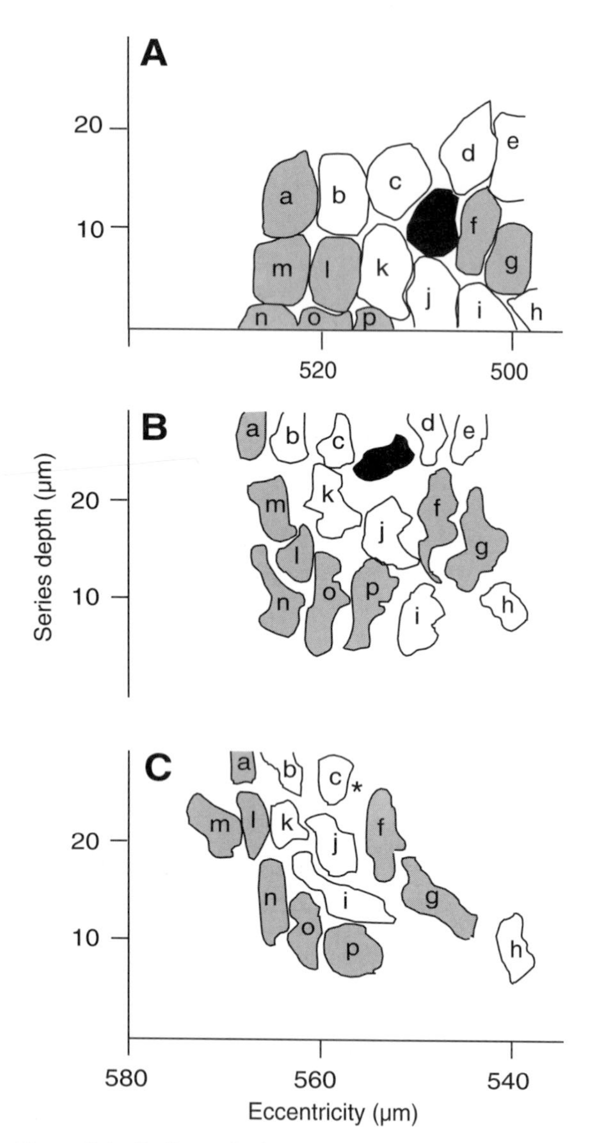

Figure 8.4: Outlines of a few neighboring cone terminals in our series. Some cones contacted midget pathways with about 30 bipolar → ganglion cell synapses (white), while others contacted midget pathways with about 50 bipolar → ganglion cell synapses (gray). (A) Single S-cone is in this patch (black). (B) Outlines of the axonal terminals of the midget OFF-bipolar cells from the cone terminals in panel A, including the S-cone (black). (C) Outlines of the axonal terminal of the midget ON-bipolar cell from all but the S-cone terminal in panel A. The asterisk marks the location where the missing midget ON-bipolar cell terminal would be expected. Modified from Calkins (1994).

dritic trees of diffuse OFF-bipolar cells branching near an identified S-cone. Dendritic trees from two types of OFF-cell are shown in Fig. 8.5. Both types receive contact from every cone within reach of the dendritic tree, including the identified S-cone. The dendrites of these bipolar cells, unlike those of the S-ON-bipolar cell, do not course through the OPL for definite contact with S-cones, but rather receive contact from any S-cone that fortuitously is within the dendritic field. However, the pattern of connectivity differs for the two types of cells. The first type has a fairly flat dendritic tree that is contacted strictly at NTA basal sites at each cone; its axon terminal stratifies narrowly along the border of the OFF- and ON-sublamina in the IPL (Fig. 8.5A). This cell resembles the DB3 cell of Boycott and Wässle (1991; see Chapter 7). The second type has a dendritic tree that is contacted roughly equally at both TA and NTA basal sites at nearly every cone; its axon terminal is more diffusely stratified throughout the proximal portion of the OFF-sublamina (Fig. 8.5B). The cell resembles the DB2 cell of Boycott and Wässle (1991; see also Hopkins & Boycott, 1997).

To summarize our findings thus far, S-cones appear to have only a single type of ON-bipolar cell pathway to the inner retina via the S-ON-bipolar cell, while M- and L-cones have multiple ON-bipolar cell pathways via the midget ON- and the various types of diffuse ON-bipolar cells. In contrast, the three cone types appear to have common types of OFF-bipolar cell pathways via the midget OFF-cells and at least two types of diffuse OFF-bipolar cells. Thus, the range of feed-forward pathways from foveal S-cones to the OFF-region of the IPL is quite diverse, indicating that multiple types of ganglion cells may hyperpolarize in response to S-cone stimulation. Indeed our reconstruction of a parasol OFF-ganglion cell indicates a mixture of synaptic input from not only amacrine cells, but also more numerously from DB2 and DB3 cells (Fig. 8.6; compare with Jacoby et al., 1996). So some parasol-OFF cells should respond to S-cone modulation. Next, as outlined in the introductory paragraphs, we turn to other ganglion cell types and how their presynaptic circuitry relates to the cone antagonism underlying color opponency.

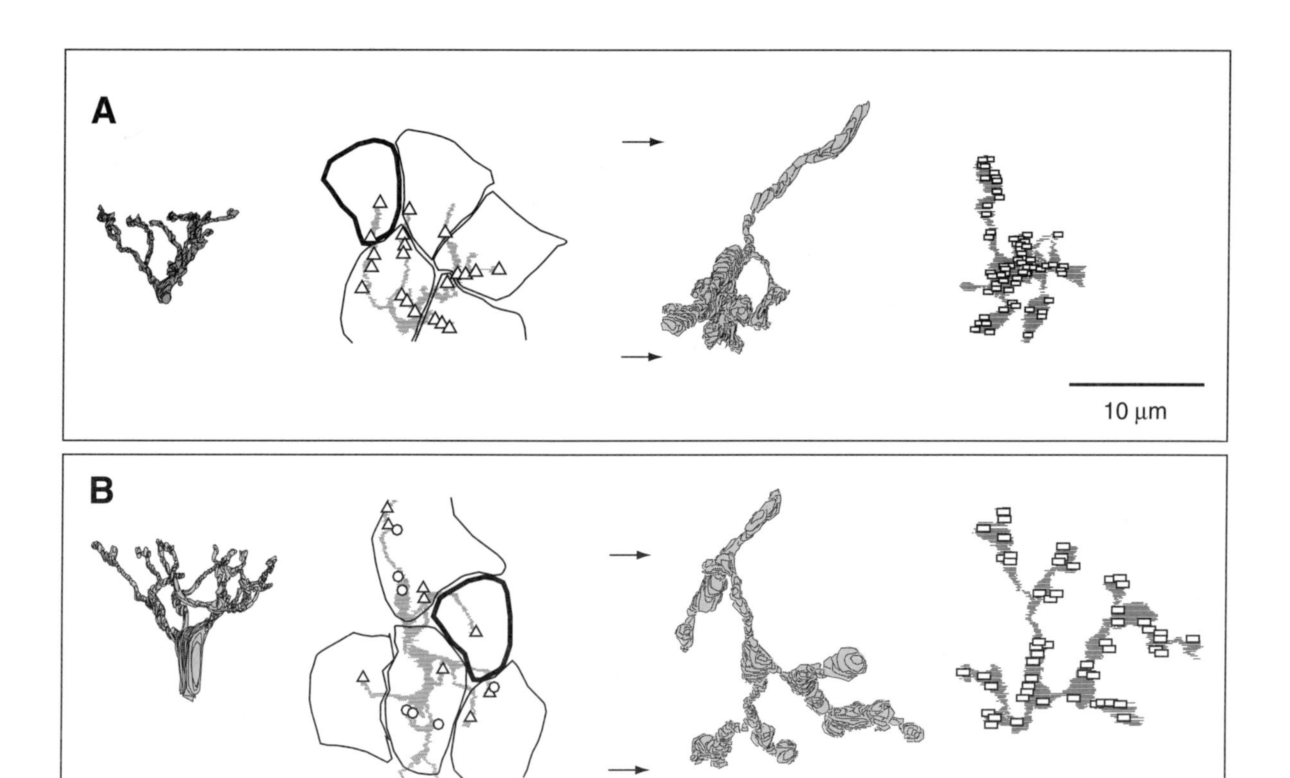

Figure 8.5: (A) Reconstruction of the dendritic tree of a "DB3" diffuse OFF-bipolar cell; an S-cone (thick outline) was identified by its contact with S-ON-bipolar cells. This cell receives nonselective contact strictly at NTA basal sites (triangles). A vertical view of an axon terminal from the same cell type is also shown; arrows mark the complete depth of the OFF-sublamina of the IPL. The horizontal view of this axon terminal marks the locations of its 70 ribbon synapses (rectangles). (B) The dendritic tree of a "DB2" diffuse OFF-bipolar cell with the same S-cone indicated. This cell is postsynaptic at a mixture of both NTA and TA sites (circles) for nearly every cone. The axon terminal of a neighboring DB2 cell contains 54 ribbon synapses.

M- and L-cone midget ganglion cells differ. Each foveal M- and L-cone contacts a midget ON- and a midget OFF-bipolar cell that each collect from only a single cone; each bipolar cell in turn contacts a single midget ganglion cell (Fig. 8.7A; see also Polyak, 1941; Chapters 7 and 9). Our reconstructions of the OFF- and ON-midget pathways leading from the patch of cones in our series divide the group of M- and L-cones into two distinct populations based on differences in how their midget bipolar cells contact their midget ganglion cells (Fig. 8.7B; Calkins et al., 1994). Each cone in one population has OFF- and ON-midget pathways with about 30 synapses between the bipolar and ganglion cell, all marked by ribbon synapses pointing to a "dyad" of two postsynaptic processes. In contrast, each cone in the second population has OFF- and ON-midget pathways with about 50 synapses between the bipolar and ganglion cells.

Since there is neither convergence nor divergence between the cone and a midget ON- or OFF-bipolar cell, and since the synaptic junctions between the bipolar and ganglion cells are identical for ON- and OFF-midget pathways, this difference in the number of ribbon synapses essentially partitions the cone mosaic into two types of cones (Fig. 8.7C). Naturally, we have hypothesized that these types are in fact M- and L-cones, although we cannot say yet which is M- and which is L-cone. This hypothesis is supported by other observations. The two types of cones in our material are about equally numerous and randomly distributed into small clusters of like types (Fig. 8.7; Calkins et al., 1994). A similar result was found for L- and M-cones in the fovea from another Old-World species, *Cercopithecus talapoin*, using direct measurements of the spectral sensitivity of cones in intact patches of living retina (Mollon & Bowmaker, 1992, but compare with Packer, Williams, & Bensinger, 1996). This hypothesis has not yet been tested, but we have adopted it for the moment to investigate other aspects of how M- and L-cones contribute to the presynaptic circuitry of ganglion cells.

Lateral connections to midget ganglion cells are nonselective. Nearly all (88%) of the dyads postsynaptic to a midget bipolar cell ribbon synapse consist of a midget ganglion cell dendrite paired with an amacrine cell process (Calkins et al., 1994). Resolving the circuitry of these amacrine cells and how they may contribute to the receptive field of the midget ganglion cell is critical to our understanding of color vision. This is because the L/M spectral antagonism within the receptive field of a foveal midget ganglion cell is linked with spatial antagonism between its concentric center and surround. Since each center is derived from a single M- or L-cone, the cone antagonism within the receptive field is thought to be derived from an inhibitory surround that is either completely (Reid & Shapley, 1992; see also Chapter 10) or predominantly (Lennie, Haake, & Williams, 1991) driven by the other cone type. The "midget hypothesis" of R/G opponency places the critical locus of the underlying L/M antagonism within the circuitry producing the inhibitory interaction between the center and surround of the midget ganglion cell (see also Chapter 7) .

There are two possible sources of lateral inhibitory connections that might contribute to the formation of spectrally pure surrounds for foveal midget cells. In the OPL, horizontal cells (which are GABA-ergic) could collect widely from cones and feed an averaged signal as inhibition back to the center cone and/or for-

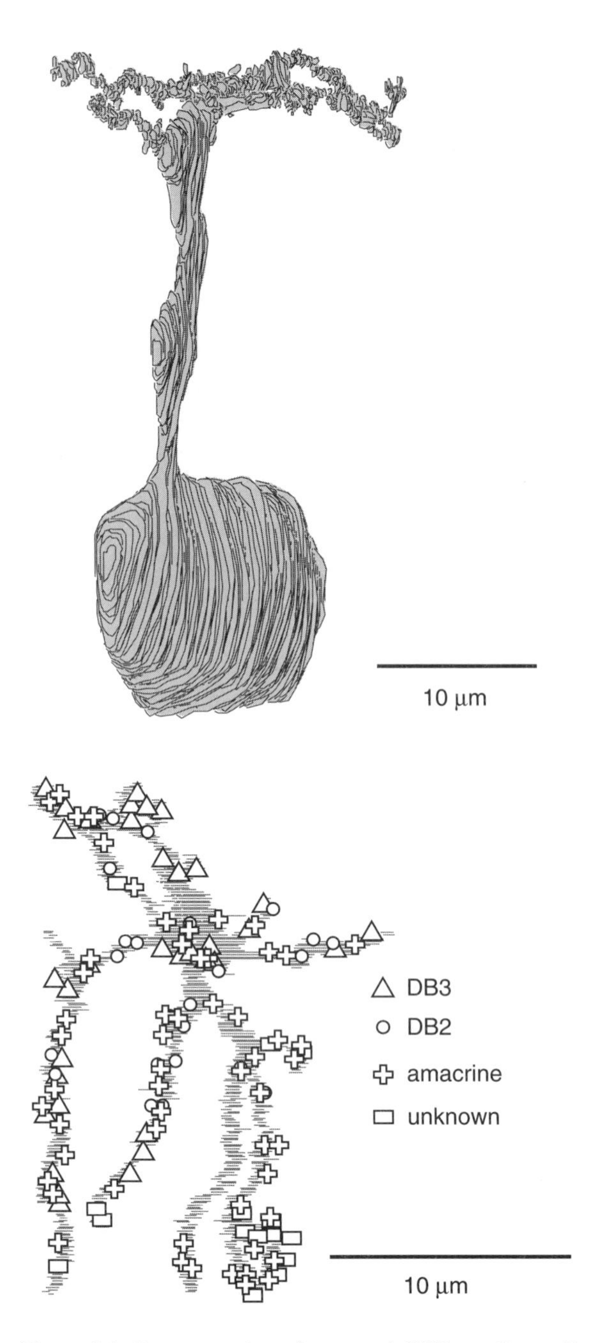

Figure 8.6: Reconstruction of a parasol OFF-ganglion cell. The dendritic tree, shown below in horizontal view, receives a mixture of synaptic input from DB3 (triangles) and DB2 cells (circles), as well as from amacrine cell processes (crosses). Some synapses were from an unknown number of DB3 or DB2 cells (rectangles).

ward to its midget bipolar cells (Grünert & Wässle, 1990; Vardi et al., 1994; Vardi & Sterling, 1994). In the IPL, the amacrine cells postsynaptic to each midget bipolar cell dyad could collect widely from other bipolar cells and feed their averaged signal as inhibition back to this "center" midget bipolar axon terminal and/or forward to its ganglion cell (e.g., Kolb & Dekorver, 1991).

Horizontal cells are known to be spectrally nonselective for M- and L-cones, both from direct observation of their contacts with cone terminals and from intracellular recordings of their light responses (Boycott, Hopkins, & Sperling, 1987; Dacheux & Raviola, 1990; Dacey et al., 1996; see Chapter 9). Amacrine cells form dense conventional synapses on the midget bipolar axon terminal and midget ganglion cell dendrite, providing contacts to both postsynaptic structures that are equal in number to the number of bipolar cell to ganglion cell synapses (Kolb & Dekorver, 1991; Calkins et al., 1994). Most of these contacts are probably inhibitory because: (1) the vast majority of amacrine cells in the primate retina are either GABA-ergic or glycinergic (Hendrickson et al., 1988; Grünert & Wässle, 1990; Davanger, Ottersen, & Storm-Mathisen, 1991; Crooks & Kolb, 1992; Koontz et al., 1993) and (2) there is direct evidence for GABA-ergic input to monkey ganglion cells (Grünert & Wässle, 1990; Koontz & Hendrickson, 1990; Zhou, Marshak, & Fain, 1994).

Our reconstructions of amacrine cells that connect to midget bipolar and ganglion cells show two important results (Calkins & Sterling, 1996). First, nearly every amacrine cell process that forms a conventional synapse to a midget bipolar cell and its midget ganglion cell is also postsynaptic to the very same bipolar cell at one or more dyads (Fig. 8.8). Thus, a large portion of the amacrine cell input to the midget circuit must be provided by the very same cone that would contribute to the receptive field center. Second, we found that these amacrine cell processes collect from *all* of the midget bipolar cells within their field, both the putative M- and L-cone cells discriminated by the presence of either 30 or 50 ribbon synapses in the axon terminal (see above). Moreover, these same amacrine cells also collect from various diffuse bipolar cells whose dendrites collect indiscriminately from L- and M-cones (Calkins & Sterling, 1996; Boycott & Wässle, 1991). Thus, our reconstructions indicate no trace of spectral selectivity in the amacrine cell connections of midget ganglion cells.

A circuit for S/(M+L) cone antagonism. Emerging models suggest that the spatially coextensive S-cone and M+L-cone perceptive fields of B/Y color opponency might have their basis in the retina, where the receptive field of the small bistratified ganglion cell is composed of spatially coextensive regions of excitation to onset of S-cone stimulation and to offset of M+L-cone stimulation (Rodieck, 1991; Dacey & Lee, 1994; see Chapter 9). This S-ON/(M+L)-OFF ganglion cell's bistratified morphology suggests a basic circuitry underlying its light response: Dendrites in the ON-sublamina of the IPL costratify with the axon terminals of S-ON-bipolar cells, while dendrites in the OFF-sublamina could feasibly intermingle with the axon terminals of OFF-bipolar cells collecting from M- and L-cones. In this circuit, the critical locus is not at the point of convergence of excitatory and inhibitory inputs (as might be the case for L/M antagonism in the midget ganglion cell receptive field), but rather is embedded at the point of convergence of both ON and OFF *excitatory* pathways in the dendritic tree of this ganglion cell.

We have cataloged the 50 or so nonmidget ganglion cells whose dendritic trees breached a roughly 2500-mm^2 plane above the GCL in our material; 18 of these cells sent dendrites to both the ON- and OFF-sublamina of the IPL. There is precedence for describing all such cells as "bistratified," implying dendrites confined to two narrow strata of the IPL. However, in the fovea, where our sample was taken, each of a group of six nominally small bistratified cells did indeed send dendrites to a narrow portion of the ON-sublamina, but they also branched more diffusely throughout the proximal half of the OFF-sublamina (Fig. 8.9A). Interestingly, the stratification of these same dendrites for cells in the peripheral retina appears nearer to the INL (Dacey, 1993a). Thus, these six putative S-ON/(M+L)-

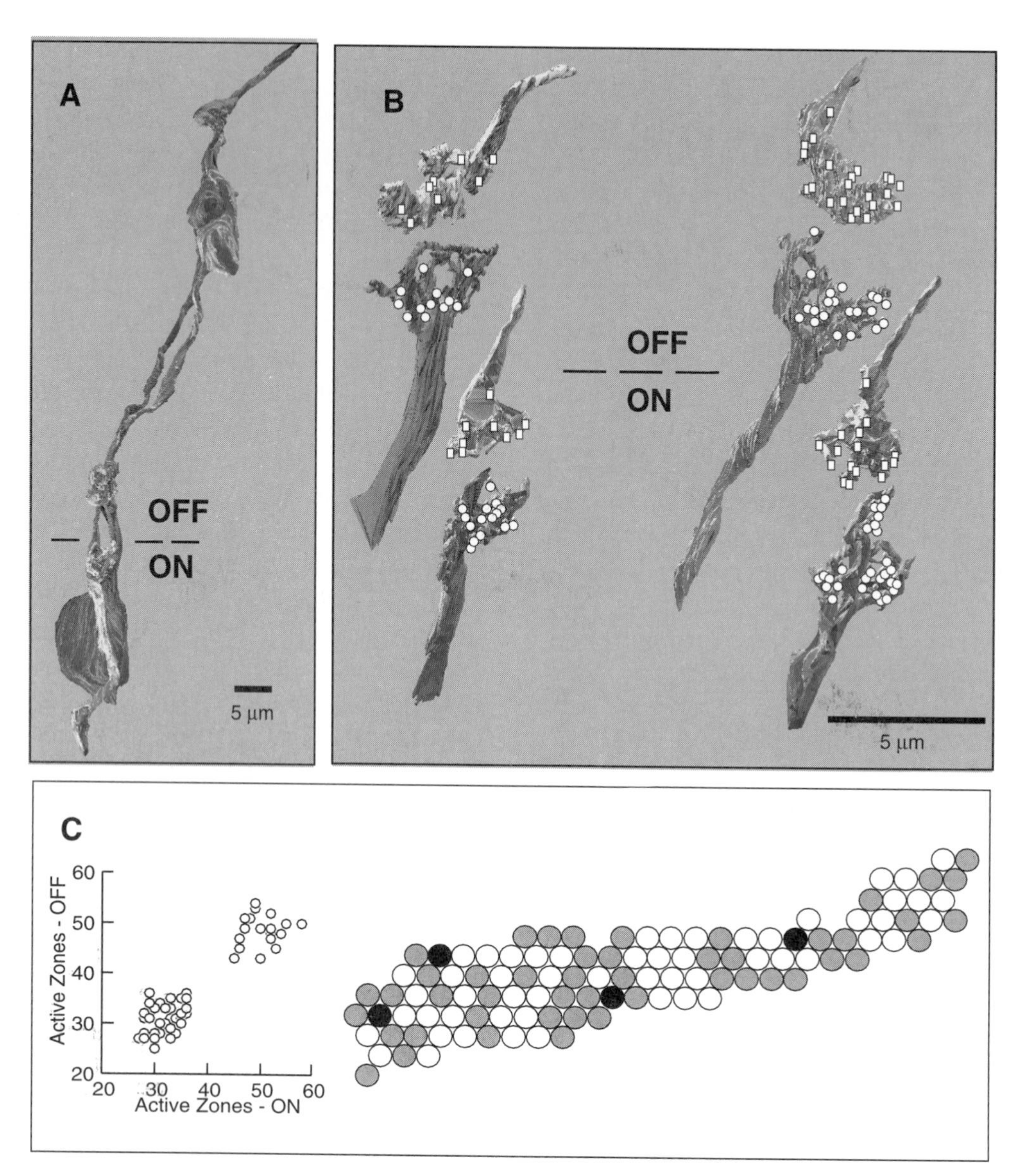

Figure 8.7: (A) Reconstructions of the midget OFF and ON pathways from a single cone terminal (top). (B) Paired midget OFF and ON pathways from one cone (left). The bipolar cell terminals contained about 30 ribbon synapses (squares), most of which were directed to the corresponding midget ganglion cells (circles). Paired midget OFF and ON pathways from a neighboring cone (right). For this cone, the number of synapses between the bipolar and ganglion cells was about 50. Some synapses are hidden; dotted line separates OFF from ON pathways. (C) Mosaic of foveal cones whose midget pathways were examined from our EM series. S-cones (black) were identified based on their connections to S-ON-bipolar cells (Klug et al., 1992, 1993). Cones whose midget pathways contained about 30 synapses between the bipolar and ganglion cells (white) and cones whose midget pathways contained about 50 synapses between the bipolar and ganglion cells (gray) were about equally numerous and distributed randomly into small clusters of like type (modified from Calkins et al., 1994; see also Herr et al., 1995). This classification was possible since the OFF and ON midget pathways from the same cone had similar numbers of ribbon synapses (inset).

OFF ganglion cells resembled the "shrub" cell described by Polyak (1941).

Each of these six ganglion cells was postsynaptic in the ON-sublamina along the IPL/GCL border to two to three sparsely spaced terminals of S-ON-bipolar cells. The axon terminals of three complete S-ON-bipolar cells in our material formed 42 ± 1.5 ribbon synapses (mean ± SEM; Fig. 8.9B). The total number of synapses from S-ON-bipolar cells to the two ganglion cells in Fig. 8.9 is modest: 34 and 32. These synapses are not evenly distributed among the bipolar cell terminals; rather, for the two ganglion cells shown in Fig. 8.9, about 65% of the S-cone input is conveyed by one of the two to three presynaptic S-ON-bipolar cells.

We were able to link about 30% of the processes postsynaptic to the identified S-ON-bipolar cells with the same six S-ON/(L+M)-OFF ganglion cells; these were the only ganglion cells that we could confirm were postsynaptic to the S-ON-bipolar cells. Whether the S-ON/(L+M)-OFF ganglion cell accounts for all of those receiving input from S-ON-bipolar cells is not yet known. However, since S-cones lack a midget ON-bipolar cell and diffuse ON-bipolar cells probably do not receive contact from S-cones (Fig. 8.3; Calkins, Tsukamoto, & Sterling, 1996; Herr et al., 1996), ganglion cells that would presumably depolarize with S-cone stimulation are conceivably restricted to this cell type.

The dendritic tree in the OFF-sublamina of the IPL is smaller than that in the ON region (Dacey, 1993a; Dacey & Lee, 1994) and, for foveal cells, receives only about 15 synapses from 3–4 diffuse bipolar cells that comprise two types (Fig. 8.5). The first type, providing about 70% of the synapses, has an axon terminal with 65–70 ribbon synapses narrowly stratified along the border of the OFF- and ON-sublamina of the IPL; this cell corresponds to the DB3 cell described in Fig. 8.5. The second type, providing the remaining input, has an axon terminal with 45–50 ribbon synapses diffusely stratified throughout the lower half of the OFF region; this cell corresponds to the DB2 cell. Thus, the presynaptic circuitry of these ganglion cells is consistent with the coextensive S-ON and (M+L)-OFF receptive field regions identified physiologically.

Interestingly, not every bistratified ganglion cell that we identified was postsynaptic to an S-ON-bipolar cell. These ganglion cells were not a uniform group, but rather they demonstrated a variety of sizes and morphologies and in all likelihood represent different cell types. Of these cells, only about one in three was confirmed to receive contact from S-ON-bipolar cells – about one ganglion cell for every S-cone that we identified (Klug et al., 1992, 1993). Thus, "bistratified" is not necessarily synonymous with S-cone input.

Discussion

Pathways from S-cones. Our results support the hypothesis that the sole conveyer of an ON response to S-cone excitation is the "blue-cone" bipolar cell. S-cones, unlike M- and L-cones, lack contact with a midget ON-bipolar cell and with at least one type of diffuse ON-bipolar cell (Figs. 8.3 and 8.5; see also Klug et al., 1992, 1993; Herr et al., 1996). Thus, the blue-cone bipolar cell may in fact be the only S-ON-bipolar cell. On the other hand, our reconstructions indicate that S-cones do contact both a midget OFF-bipolar cell and multiple types of diffuse OFF-bipolar cells at basal sites (Figs. 8.4 and 8.5; Klug et al., 1992, 1993). In this regard, S-cones are not qualitatively different from M- and L-cones.

So far the only ganglion cell we have found to be postsynaptic to the S-ON-bipolar cell is the small bistratified cell whose S-ON/(M+L)-OFF response to light is described elsewhere (see Chapter 9). Thus, this may be the only ganglion cell that depolarizes in response to S-cone modulation. In contrast, our small sample of reconstructions of diffuse OFF-bipolar cells indicate that S-cones form basal contacts with both DB2 and DB3 cells. Thus, some parasol OFF-ganglion cells, which are postsynaptic to DB2 and DB3 cells (Fig. 8.6; see also Jacoby et al., 1996), ought to show some response to S-cone stimulation. Presumably, the numbers of such parasol OFF-cells would be small, since not every diffuse OFF-bipolar cell is in contact with an S-cone. This, coupled with a complete lack of S-cone input to other ON-pathways, might explain

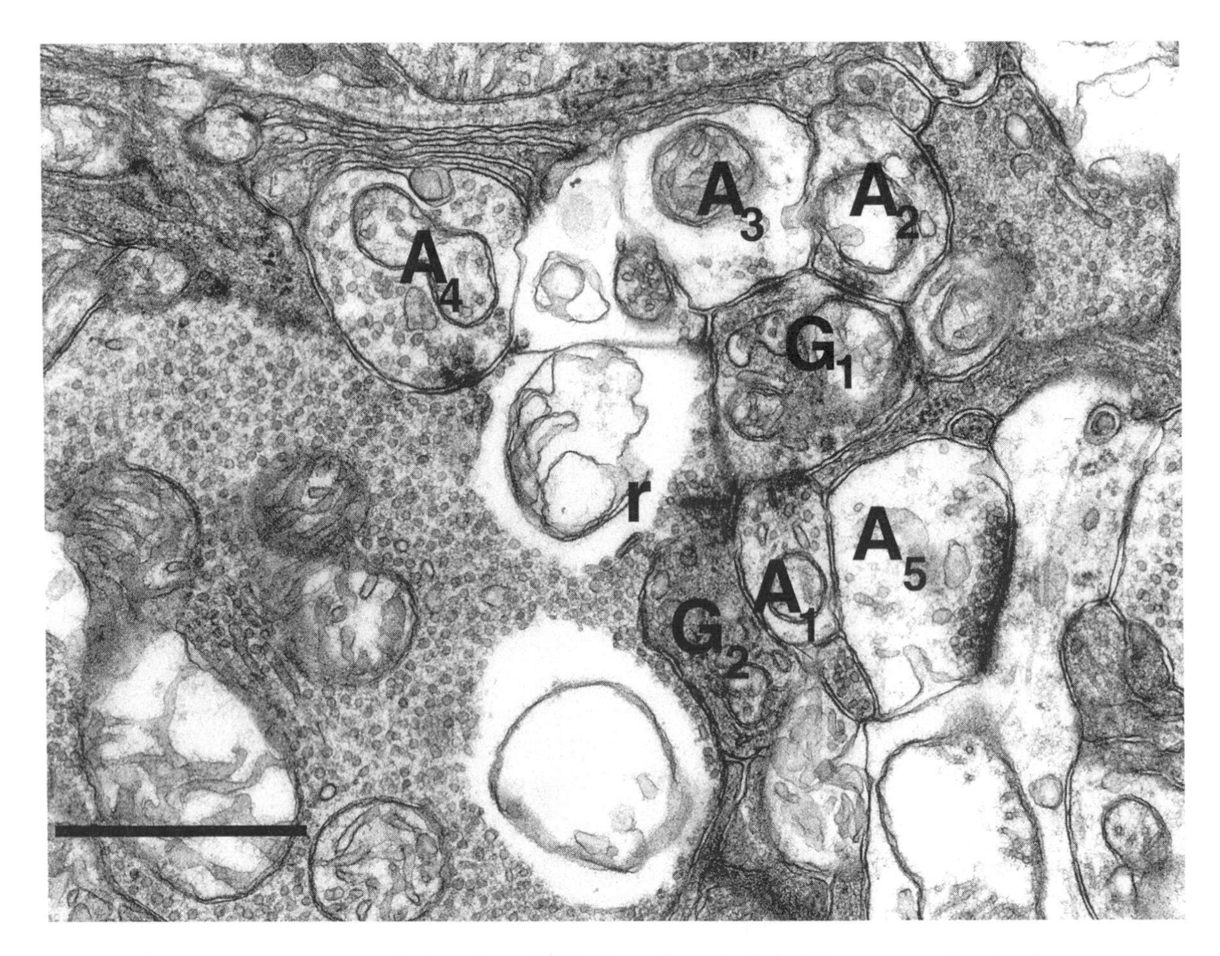

Figure 8.8: Electron micrograph of a midget ON-bipolar cell terminal. A ribbon synapse (r) points between a dyad consisting of a midget ganglion cell dendrite (G_1) and an amacrine cell process that is making a "feed-forward" conventional synapse to the ganglion cell (A_1). We found that most amacrine cell processes that contact the ganglion cell (A_{1-3}) or the bipolar cell terminal (A_4) are also postsynaptic to the same terminal (Calkins & Sterling, 1996). The midget ganglion cell dendrite G_2 is postsynaptic to the terminal in a subsequent section. Another amacrine cell process forming a conventional synapse is shown (A_5). Scale bar represents 1 μm.

why in vitro recordings indicate a lack of S-cone input to any but the S-ON/(M+L)-OFF ganglion cell (Dacey & Lee, 1994).

From quite another perspective, the S-cone contacts to diffuse OFF-bipolar cells – albeit sparse –challenge our understanding of the S-ON/(M+L)-OFF ganglion cell. This ganglion cell type in the fovea is postsynaptic to both DB3 and DB2 cells (Fig. 8.9). However, the number of contacts from an S-cone to any single diffuse OFF-bipolar cell is very small compared with that to an S-ON-bipolar cell (compare Figs. 8.5 and 8.3). Moreover, the synaptic input from the S-cone bipolar cells to the ganglion cell (about 30 synapses from 2–3 cells) is far greater than the input from the diffuse OFF-bipolar cells (12 synapses from 2–3 cells). Thus, the overwhelming S-cone signal to the ganglion cell is depolarizing, and any small hyperpolarization from S-cone stimulation is presumably masked.

About 33% of the nominally bistratified ganglion cells in our series were confirmed to be postsynaptic to S-ON-bipolar cells. Each of these demonstrated a similar morphology and circuitry consistent with the phys-

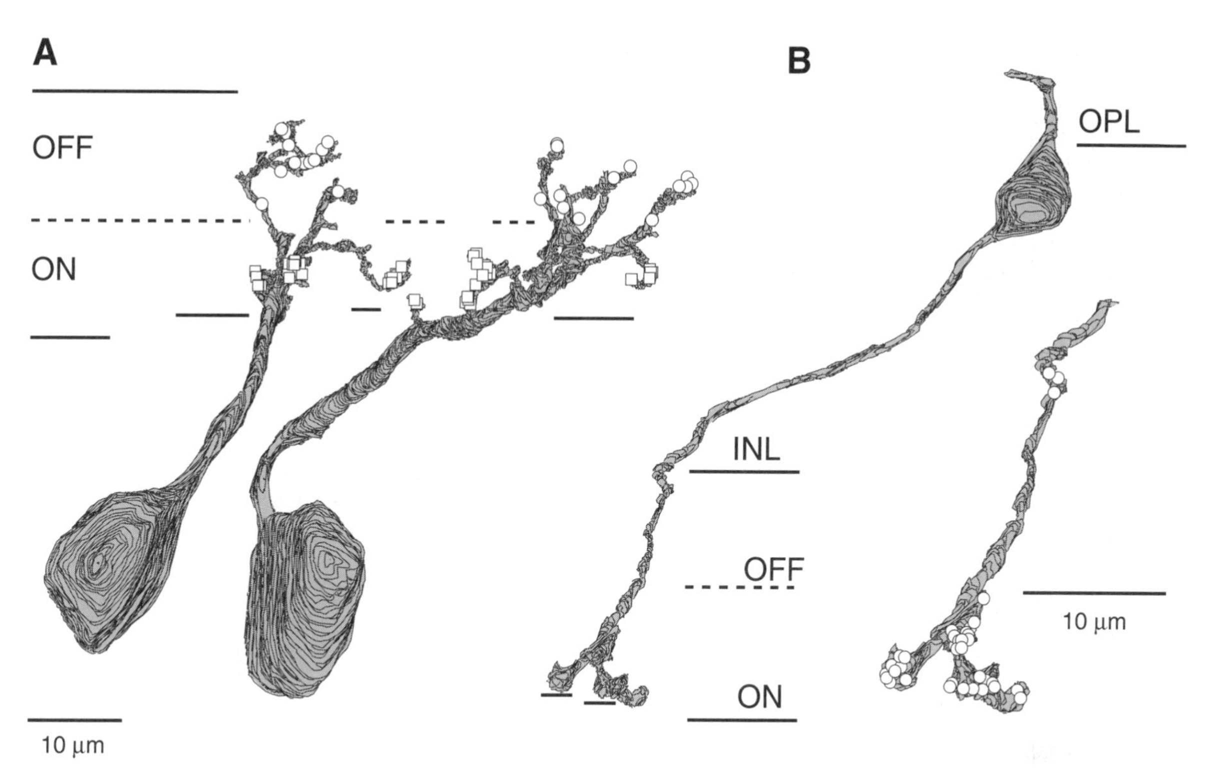

Figure 8.9: (A) Reconstructions of two neighboring ganglion cells whose dendrites ramify not only in the ON-sublamina, but also more diffusely through the OFF-sublamina of the IPL. In the ON-sublamina, the cells receive bipolar cell synapses exclusively from the terminals of the S-ON-bipolar cells (squares), while in the OFF-sublamina the cell receives contact from two types of diffuse OFF-bipolar cell (circles). (B) Reconstructed S-ON-bipolar cell that contacts both ganglion cells in panel A; the OFF- and ON-sublamina of the IPL, INL, and OPL are indicated. Its dendritic tree, which was traced to an S-cone, is only partially shown. The inset shows the axon terminal of the cell with its 39 ribbon synapses (circles). Three complete terminals contained 42 ± 1.5 ribbons (mean ± SEM). These numbers do not include the 2–4 synaptic ribbons found in small boutons of the axon as it penetrates the OFF region of the IPL (top circles).

iological properties of the S-ON/(M+L)-OFF cell (see Chapter 9). The number of these ganglion cells – about one for every S-cone in our material – is sufficient to account for the spatial resolution of S-cone–isolating patterns, which is about the sampling limit of the S-cone mosaic (Curcio et al., 1991; Sekiguchi, Williams, & Brainard, 1993). If we assume that these spatial discrimination tasks are indeed tapping the B/Y opponent channel, then the sampling of the S-ON/(M+L)-OFF ganglion cell in our material is sufficient to account for its spatial acuity.

On the other hand, spatial discrimination tuned to a specific spectral range is best served by sampling a given point with both an OFF- and an ON-ganglion cell. In this sense, one might expect to find an S-OFF/(M+L)-ON cell contributing with equal sampling density to the B/Y channel. Each S-cone does access a midget OFF-ganglion cell via the midget OFF-bipolar cell (Klug et al., 1992, 1993); this cell may correspond to an S-OFF/(M+L)-ON ganglion cell described in other physiological recordings (e.g., de Monasterio & Gouras, 1975; Derrington, Krauskopf, & Lennie, 1984). Thus, the signal from each foveal S-cone may be represented in both an ON- and OFF-ganglion cell mosaic with equal spatial density.

The postsynaptic connections of M- and L-cones. Each foveal M- and L-cone contacts both a single midget OFF- and a single midget ON-bipolar cell, each of which in turn contacts a midget ganglion cell (Fig. 8.7A). Our reconstructions indicate that while the OFF- and ON-midget circuits from the same cone are highly similar in the number of bipolar → ganglion cell synapses, the number of synapses in circuits from different cones varies highly (Fig. 8.7C). Some cones have OFF- and ON-midget circuits with about 30 synapses between the bipolar and ganglion cells, while other cones have circuits with about 50 synapses (Fig. 8.7B). Since there is neither convergence nor divergence at either the cone → bipolar cell synaptic junction or the bipolar → ganglion cell junction (Calkins et al., 1994), this difference in the number of bipolar → ganglion cell synapses effectively partitions the non–S-cone mosaic into two cone types. These two types are about equal in number and randomly distributed (Fig. 8.7C). Thus, we have concluded that these are M- and L-cones distinguished not by histochemical means, but by differences in their postsynaptic connections; indeed, we can think of no tenable alternative hypothesis.

The finding that M- and L-cone midget pathways differ structurally raises the issue of whether the expression of a particular opsin accompanies modifications in the cone's postsynaptic pathways (see Chapter 1). This would imply that trichromacy itself – the expression of multiple pigment types in the middle- to long-wavelength range – entails not only the addition of a pigment type, but also differences in how the additional cone type sends its signals to the brain. In this sense, the hypothesis that the surround of a midget ganglion cell in the fovea is wired selectively for M- or L-cones is no more or no less tenable than the hypothesis that the wiring of either an L- or M-cone *center* entails either 30 or 50 synapses between the midget bipolar and ganglion cells. Both hypotheses, although addressing fundamentally different aspects of receptive field organization, imply a strict relationship between the type of cone contacting the midget ganglion cell via the midget bipolar cell and the structure of the ganglion cell's presynaptic circuitry.

Nevertheless, the accumulating evidence suggests that neither horizontal nor amacrine cells provide a straightforward anatomical basis for a surround derived solely from M- or L-cones (Fig. 8.8). Thus, the physiological evidence that indicates complete separation of L- and M-cone inputs to the midget ganglion cell receptive field remains to be explained (Reid & Shapley, 1992; see also Chapter 10; Kaplan, Lee, & Shapley, 1990; Lee, 1996). It is possible that whatever degree of L/M antagonism is present between the center and surround of a foveal midget ganglion cell is due to the random distribution of those cones surrounding the center cone (Lennie, Haake, & Williams, 1991). Such a nonselective wiring scheme would explain the large degree of variation in spectral sensitivity across nominally R/G opponent ganglion cells and in the spectral neutral points of LGN type I cells (Zrenner & Gouras, 1983; Derrington, Krauskopf, & Lennie, 1984; Lennie, Haake, & Williams, 1991).

But do M- and L-cone postsynaptic pathways differ in any other ways? An important inference from the extant Golgi studies of primate bipolar cells is that diffuse bipolar cells collect from every cone within reach of their dendritic fields (Boycott & Wässle, 1991). We have found from our reconstructions that this is indeed so, with the notable exception that S-cones seem to lack contact with diffuse ON-bipolar cells (Fig. 8.3). Thus, it is generally presumed that the spectral information conveyed by diffuse bipolar cells to the IPL is "broadband" and, therefore, that these collectively provide the sole input to the magnocellular system and are not involved in color vision at all.

This seems to be the case for the DB3 and DB2 cells, which provide the bipolar cell input to the parasol OFF-cell (Fig. 8.6). In the fovea, both M- and L-cones contact the DB2 cell at a mixture of TA and NTA basal sites and the DB3 cell strictly at NTA basal sites (Fig. 8.5). Similarly, both cone types contact the small sample of diffuse ON-bipolar cells that we have examined primarily at TA sites (Table 8.1; Fig. 8.3); these cells presumably provide the bipolar cell input to the parasol ON-ganglion cell. However, we have begun to examine the detailed connections from cones to a third type of diffuse bipolar cell (Calkins, Schein, & Sterling, 1995). This cell has a broad axonal terminal stratifying narrowly in the OFF-sublamina along the INL/IPL border and therefore probably corresponds to the DB1 cell of Boycott and Wässle (1991; see also Hopkins & Boycott, 1996). In the fovea its terminals are beyond the reach of both the parasol and the midget OFF-ganglion cells, as well the S-ON/(M+L)-OFF ganglion cell (Fig. 8.9). Thus, whatever ganglion cells in the fovea are postsynaptic to the DB1 are yet to be reported from in vitro injections or recordings.

Because of the location of its axon terminal, DB1 is nominally an OFF-bipolar cell. However, the types of contacts made by M- and L-cones to two examples of this cell in our material are not the same. Cones whose midget pathways form 50 synapses between the midget bipolar and ganglion cells – that we think are one type of M- or L-cone – contact this diffuse cell at TA basal sites. In contrast, cones whose midget pathways form 30 synapses – that would correspond to the other M- or L-cone type – contact this diffuse bipolar cell strictly at NTA basal sites.

The significance of this subtle difference in connectivity is unknown, nor do we have a measure of how reliable this difference is across a greater sample of foveal cells. In the periphery, the type of basal contact is apparently independent of the type of presynaptic cone (Hopkins & Boycott, 1996). Perhaps the difference in the type of basal contact made (TA vs. NTA) by the two cone types (M- versus L-cone, or vice versa) reflects different responses to light, that is, depolarizing versus hyperpolarizing. This would require the dendrites postsynaptic at TA sites to express an mGluR and those postsynaptic at NTA sites to express a GluR.

Summary

Cones in the primate fovea access a variety of postsynaptic pathways that are distinguished by the synaptic organization of their constituent cell types. M- and L-cones contact a variety of ON-bipolar cells, while S-cones appear to contact only a single type that is selective for S-cones. In contrast, S-, M- and L-cones alike contact diverse types of OFF-bipolar cells. These bipolar cells in turn contact different types of ganglion cells. Thus, M- and L-cones access a midget ON-ganglion cell, while all three cone types may access a midget OFF-ganglion cell. Similarly, an ON response to S-cone stimulation appears to be signalled via only the S-ON/(M+L)-OFF, small bistratified ganglion cell, while all three cone types contact the parasol OFF-ganglion cell via two types of diffuse OFF-bipolar cells.

There are two general schemes for wiring cone antagonism in ganglion cell receptive fields. Excitatory input from the midget bipolar cell converges with inhibitory inputs via horizontal and amacrine cells in the dendritic tree of the midget ganglion cell; we found no trace of spectral selectivity in the connections from amacrine cells. Thus, whatever degree of L/M–cone anatgonism may be present in the midget cell's spectral response is probably due to the random distribution of the cones forming its inhibitory surround; the possibility that another cell type (DB1) is L/M cone antagonistic is discussed. The S/(M+L) spectral sensitivity of the small bistratified cell is formed through the convergence of excitatory inputs of opposite response polarity: S-ON via the S-ON-bipolar cell versus (M+L)-OFF via the DB2 and DB3 OFF-bipolar cells. Thus, R/G and B/Y opponency arise in the retina via fundamentally different circuits.

Acknowledgments

The author would like to thank his colleagues for their valuable contributions to this work and for their continued collaboration: Y. Tsukamoto, S. J. Schein, K. Klug, S. Shrom, A. Shrom, and P. Masarachia. Much of this work was completed while the author was supported by grant EY 08124 to Professor Peter Sterling at The University of Pennsylvania Medical School, for whom the author reserves special thanks for his kind support as teacher, friend, and mentor.

9

Functional architecture of cone signal pathways in the primate retina

Dennis M. Dacey and Barry B. Lee

Ten or more signal pathways, comprising about 80 neuronal cell types, emerge from the complex circuitry of the cone axon terminal. In this chapter we summarize current knowledge of the correspondence between morphology and physiology for some of the cell types of these cone signal pathways, with special regard to the neural basis for spectral opponency. We describe an approach that uses an in vitro preparation of the macaque monkey retina in which morphologically identified cell types are selectively targeted for intracellular recording and staining under microscopic control. The goal is to trace the physiological signals from the long- (L-), middle- (M-), and short- (S-) wavelength–sensitive cones to identified cell types that participate in opponent and nonopponent signal pathways. The methods of heterochromatic modulation photometry and silent substitution are used to characterize L-, M-, or S-cone input to the receptive fields of distinct horizontal cell, bipolar cell, ganglion cell, and amacrine cell types. The majority of the retinal cell types awaits detailed analysis, and knowledge of the mechanisms of opponency remains largely incomplete. However, results thus far have established the following: (1) Horizontal cell interneurons make preferential connections with the three cone types but cannot provide a basis for spectral opponency based on cone type–selective connections. (2) A morphologically distinctive bistratified ganglion cell type transmits a blue-ON yellow-OFF spectral opponent signal to the parvocellular division of lateral geniculate nucleus. The morphology of this ganglion cell type suggests a simple synaptic mechanism for blue-yellow opponency via converging input from an ON-cone bipolar cell connected to S-cones and an OFF-cone bipolar cell connected to L- and M-cones. (3) Midget ganglion cells of the parafovea are known to correspond to red-green opponent cells. However, midget ganglion cells of the retinal periphery show a nonopponent, achromatic physiology; the underlying circuitry for red-green opponency thus remains speculative. (4) Successful recordings from identified bipolar and amacrine cells in macaque retina suggest that a more complete accounting of opponent circuitry is a realistic goal.

Cell types, circuits, and spectral opponency

The vertebrate retina, often considered a model for a simple and well-understood neuronal system, displays a diversity of cell types and functionally distinct synaptic pathways that rivals that of the cerebral cortex (Masland, 1996). Much of this anatomical diversity has been reviewed in earlier chapters of this book and elsewhere (Rodieck, 1988; Sterling, 1990; Vaney, 1990; Wässle & Boycott, 1991): At least two horizontal and ten bipolar cell populations transmit signals in parallel from photoreceptors to ganglion cells. In turn, the ganglion cells further subdivide into an estimated 20–25 anatomically distinct populations that project in parallel to about a dozen target structures in the midbrain and thalamus. Still more complex, the link between ganglion cells and bipolar cells is modulated by the amacrine cell types. To a first approximation there appears to be a number of distinct amacrine cell types dedicated to a given bipolar–ganglion cell pathway, with the total number of amacrine cell populations currently estimated to be at least 40 (Vaney, 1990; Wässle & Boycott, 1991). Thus the "retinal circuit" is not a single circuit but many "microcircuits," comprising on the order of 80 distinct cell types. One purpose of these multiple microcircuits is to create the characteristic physiological properties of the parallel visual pathways from the retina to a diverse array of target structures in the brainstem. The challenge, then, is to systematically characterize each of the retinal cell types with the goal of clarifying the structure and function of the circuitry dedicated to each of the central visual pathways.

The neural code for color begins in the retina. The probability that a photon is absorbed by a cone cell is a function of both wavelength and the density of photons incident on the photoreceptor. The cone light response is therefore independent of wavelength (see Chapters 1, 2, and 4). The wavelength-independent responses of the L-, M-, and S-cones are transformed into wavelength-sensitive responses of certain retinal ganglion cells (Kaplan, Lee, & Shapley, 1990; see Chapter 10). Our goal in this chapter is to briefly review what is currently known of the physiology of postreceptoral cell types in the macaque monkey with special reference to the generation of color-related signals. The diversity of retinal cell types raises a question as to which cell types display spectral opponency or other physiological properties (such as input from only a single cone type) that would suggest a critical role in wavelength coding. One recent approach to untangling the circuitry related to color uses anatomical reconstruction of retinal cells and synaptic pathways and is reviewed in this book (see Chapter 8). The functional hypotheses that are generated by this method must be tested by directly linking morphology and physiology. The classical approach to this problem has been to combine an intracellular recording of the light response of a retinal cell with intracellular staining to reveal the morphology of the recorded cell (e.g., Dacheux & Raviola, 1990). Recently we applied this basic technique to an in vitro preparation of the macaque retina, modified so that identified cell types could be targeted for intracellular recording and staining under direct microscopic control (Yang & Masland, 1994). Before reviewing the results, critical aspects of the method are outlined in the next section.

Linking morphology and physiology on a microscope stage

In the primate retina intracellular recording and staining for the purpose of directly relating morphology to physiology was first attempted with limited success in the intact eye of the anesthetized macaque monkey (de Monasterio, 1979; Zrenner, Nelson, & Mariani, 1983). A superfused eyecup preparation was later used to obtain the first recordings of a non-spiking interneuron, the H1 horizontal cell (Dacheux & Raviola, 1990). This approach was also limited as the retinas were only physiologically viable for a few hours. More successfully, an isolated retina, maintained in vitro, has now been used to make whole cell patch electrode recordings of rod and cone photoreceptor light responses (Schneeweis & Schnapf, 1995; see Chapter 4). In a variant of this last approach we have developed an in vitro preparation of the macaque retina in which the retina, retinal pigment epithelium (rpe) and choroid layers are dissected intact from the eyecup and mounted flat in a superfusion chamber (Dacey & Lee, 1994) (Fig. 9.1A). A significant advantage of this dissection is that the anatomical relationship of the photoreceptors to the rpe is undisturbed; since the reactions of the visual cycle that return all-*trans* retinol to the photosensitive 11-*cis* configuration take place in the rpe, the retina-rpe-choroid can be maintained even at high photopic levels and retains the ability to regenerate photopigment in vitro.

A second advantage of the retina-rpe-choroid preparation is that retinal cell types can be directly observed and targeted for intracellular recording under microscopic control. The recording chamber is mounted on the stage of a light microscope; retinal

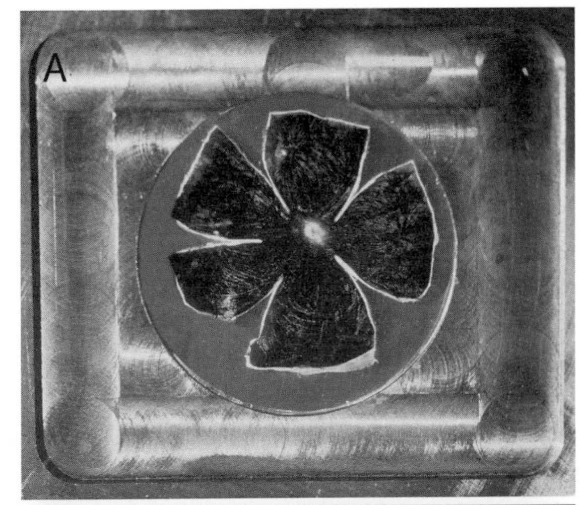

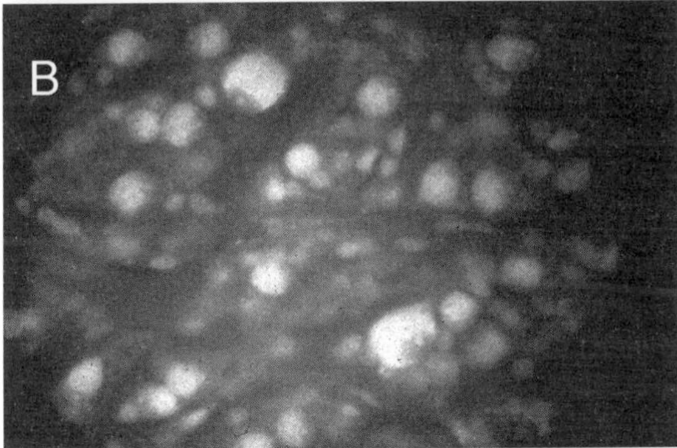

Figure 9.1: In vitro retina-rpe-choroid preparation of the macaque monkey retina. (A) Photograph of a Macaca nemestrina retina in vitro just prior to the start of an experiment. The retina was dissected with retinal pigment epithelium and choroid intact and placed in a stainless steel superfusion chamber. Several radial cuts were made in the retina so that it could lie flat on the glass bottom of the chamber. A "sticky" molluscan protein coats the glass and serves to glue the retina in place. The dark, striated choroid is evident through the transparent retina except at the optic disc. (B) After placing the chamber on the stage of a light microscope the retinal cells can be observed at high magnification in vitro using a water immersion objective lens. In this micrograph the retina has been stained with the dye acridine orange and the plane of focus is on the ganglion cell layer. The two largest cell bodies in this field, about 20-μm diameter, correspond to the parasol-type ganglion cell.

cells are selectively stained with a variety of vital fluorescent markers and observed with high-resolution water immersion optics and episcopic illumination. Because photopigment is regenerated, the retinal cells retain normal light responses after visual targeting. Figure 9.1B shows cell bodies in the ganglion cell layer of the macaque retina in vitro stained with the dye acridine orange (observed as they would be through the microscope) prior to selecting a particular cell for intracellular recording.

To deliver visual stimuli to the in vitro retina, a small optic bench is mounted above the microscope. Light passes down through the camera port and the optics of the microscope are used to bring the light into focus as a small spot on the retinal surface. Red, green, and blue light-emitting diodes (LEDs) with dominant wavelengths of 638, 554, and 445 nm, respectively, serve as light sources. The output of each LED is an independently modulated waveform (square or sine) that can be adjusted for radiance, temporal frequency, phase, or modulation depth (Swanson et al., 1987). The size and shape of a given stimulus can be adjusted by passing the light through slits, annuli, or apertures of different sizes; the position of this stimulus relative to the receptive field center of a cell can be further controlled with a motorized x–y-positioner mounted on the microscope camera port.

Heterochromatic modulation photometry. Stimulus protocols were generated to address two related questions: (1) What are the cone inputs to the recorded cell, and (2) which cells show red-green or blue-yellow spectral opponency? To identify red-green spectral opponency in retinal cells we used the paradigm of heterochromatic modulation photometry (HMP) (Pokorny, Smith, & Lutze, 1989). In this variant of heterochromatic flicker photometry (HFP) a pair of chromatic lights is presented in temporal alternation. In HFP, the relative luminances of the two lights are varied until the perception of flicker is minimized or eliminated; at this point the two lights are considered equal in luminance. The spectral sensitivity function ($V(\lambda)$) that derives from such a method is consistent with a

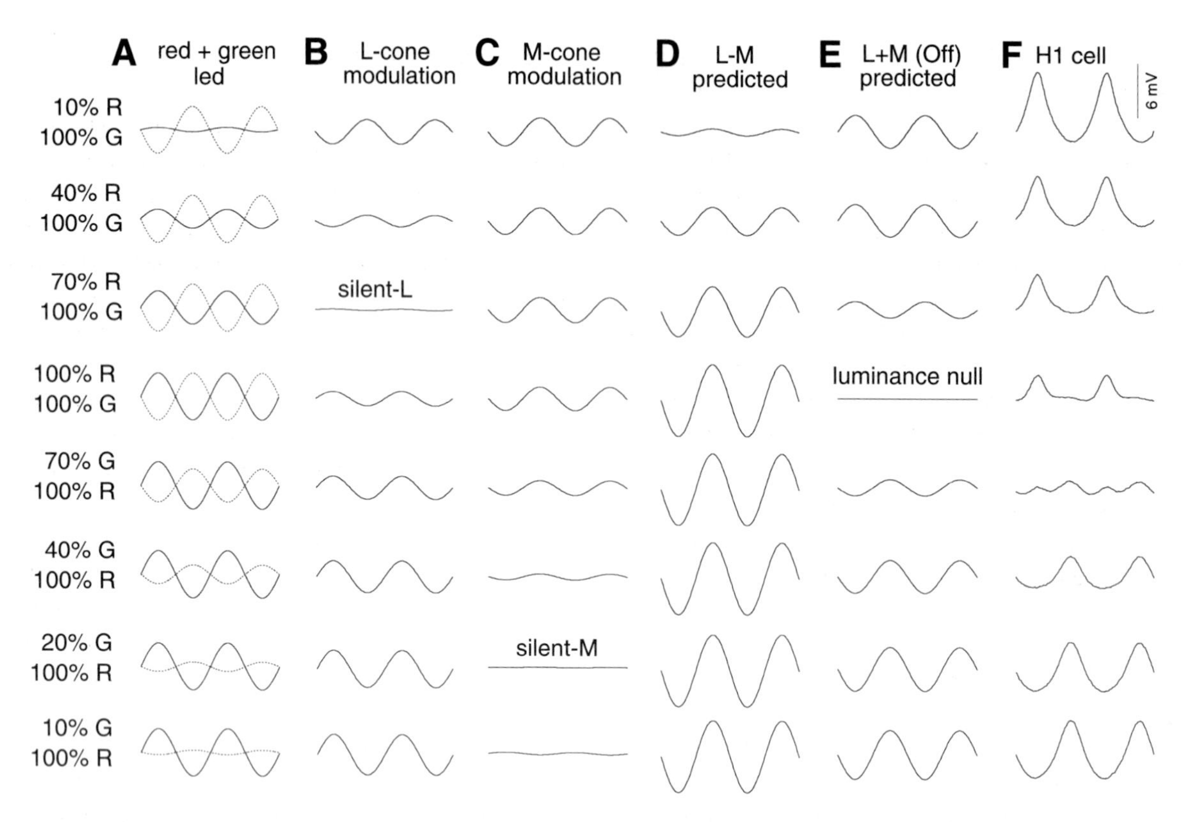

Figure 9.2: Characterization of spectral sensitivity using the paradigm of heterochromatic temporal modulation photometry. (A) Red (dominant wavelength: 638 nm) and green (dominant wavelength: 553 nm) are modulated in counterphase; relative modulation depths of LEDs are varied. 100% R, 100% G point corresponds to equiluminance for the nominal human observer with the sensitivity of the spectral luminosity function $V(\lambda)$. (B) and (C) Relative L- and M-cone modulations in response to LEDs; L-cone is modulated in phase with the red LED, M-cone modulates in phase with the green LED; silent-L and silent-M conditions are indicated. (D) Predicted response of an L – M opponent neuron with equally weighted L- and M-cone inputs. Note that silent-L and silent-M conditions elicit responses of the same phase (ie., ON-response to M-cone modulation and OFF-response to L-cone modulation). (E) Predicted response of an L + M nonopponent neuron. Response null is expected at equiluminance; silent-L and silent-M conditions elicit responses in counterphase (i.e., ON-response to M-cone modulation and ON-response to L-cone modulation). (F) Voltage response of an H1 horizontal cell to this set of stimuli; cell responds in phase to both L- and M-cone modulation with a response minimum near the equiluminant point (see Fig. 9.3 for more detail).

postreceptoral process that sums input from L- and M-cones and receives little or no S-cone contribution. In HMP the mean luminance of the two lights is held constant, but the modulation amplitude around the mean is varied to acquire a threshold for flicker detection. Thus an advantage of HMP over conventional flicker photometry is that mean luminance and chromaticity (and presumably the relative state of adaptation of the L- and M-cones) remain constant. To study the spectral sensitivity of retinal neurons using HMP, we first determine the equiluminance for the red and green LEDs by obtaining a perceived flicker minimum for human subjects observing the LEDs as they are projected onto the retinal surface. The dominant wavelengths of the red and green LEDs are such that excitation of the S-cone is negligible. An observer adjusts the luminances of the LEDs until the perception of flicker is minimized; this setting is taken as the human equiluminant point. The equiluminant settings used in these experiments were also checked radiometrically and corresponded closely to that of the standard observer (Lennie, Pokorny, & Smith, 1993). The relative modulation depth of the LEDs is then adjusted to produce a series of stimuli that varies in the degree to which the L- and M-cones are modulated (Fig. 9.2). The relative stimulation of the L- and M-cones is calculated from the Smith–Pokorny cone fundamentals (Smith & Pokorny, 1975) given a mean retinal illuminance estimated at 2000 trolands and the spectral luminous efficiency function of the Judd standard observer. To illustrate the HMP protocol, Fig. 9.2 shows the voltage responses of a recorded H1 horizontal cell to one stimulus series (more detail on the light response and cone input to macaque horizontal cells is given in the next section). Note that the neuron gives a hyperpolarizing light response to stimuli that modulate either the L- or M-cones in isolation, and that there is a response null between these two points. These cells thus receive additive input from both L- and M-cones. The response minimum is interpreted as the point at which the strength of the L-cone + M-cone signals are equal for both lights, so as to cancel when modulated in counterphase. By contrast, a neuron that receives antagonistic input from L- and M-cones will show a large response to chromatic modulation since L- and M-cones will be modulated in phase and combine additively.

To determine whether a given retinal neuron receives S-cone input, the method of silent substitution was also used to generate an S-cone–isolating stimulus. To silence both the L- and M-cones, the blue LED is first adjusted to be equal in luminance to the red and green LEDs by flicker photometry as described above. The blue LED is then run in phase with the red diode and counterphase to the green diode. Relative modulation depths for the three diodes are then adjusted to silence L- and M-cones, leaving a remaining S-cone contrast of ~85% (Lee & Yeh, 1995).

Horizontal cells

Horizontal cells are the lateral interneurons of the outer retina; their dendritic processes innervate the axon terminals of the photoreceptors and form the lateral elements of the synaptic triad (see Chapters 7 and 8). Horizontal cells of a given type are electrically coupled to one another by gap junctions and form a widespreading electrical syncytium; they provide a negative feedback signal to photoreceptors and play an important role in the generation of receptive field surrounds in bipolar cells and ganglion cells either via the feedback pathway and/or by a direct feedforward connection to bipolar cell dendrites.

Intracellular recordings from horizontal cell types in teleost retina (Svaetichin & MacNichol, 1958) provided initial and very exciting evidence that there could be a simple physiological basis for an opponent process stage in human color vision (Hurvich & Jameson, 1957). The underlying mechanisms for the complex spectral opponent responses found in fish horizontal cells, and their functions in the vision of fish, remain controversial (Burkhardt, 1993b; Kamermans & Spekreijse, 1995), although results in fish pointed to the horizontal networks in the trichromatic primate retina as a possible locus for an opponent transformation.

In primates, including man, the anatomy of horizontal cells and speculation about their function has been controversial. Wässle and Boycott (1991) concluded that, in the macaque monkey, the dendritic terminals of two distinctive horizontal cell types, the H1 and H2 cells, made nonselective contact with all cone types (Wässle, Boycott, & Röhrenbeck, 1989). It was suggested that in primates the horizontal cells might sum inputs from all cones and play no role in cone-type–specific spectral opponency. The first intracellular recordings of the H1 cell light response supported this interpretation of the anatomy (Dacheux & Raviola, 1990). By contrast, other anatomical data, principally from human retina, indicated that the primate horizontal cells could make preferential connections with cones. In this scheme H2 cells made increased contact with S-cones, H1 cells connected to all three

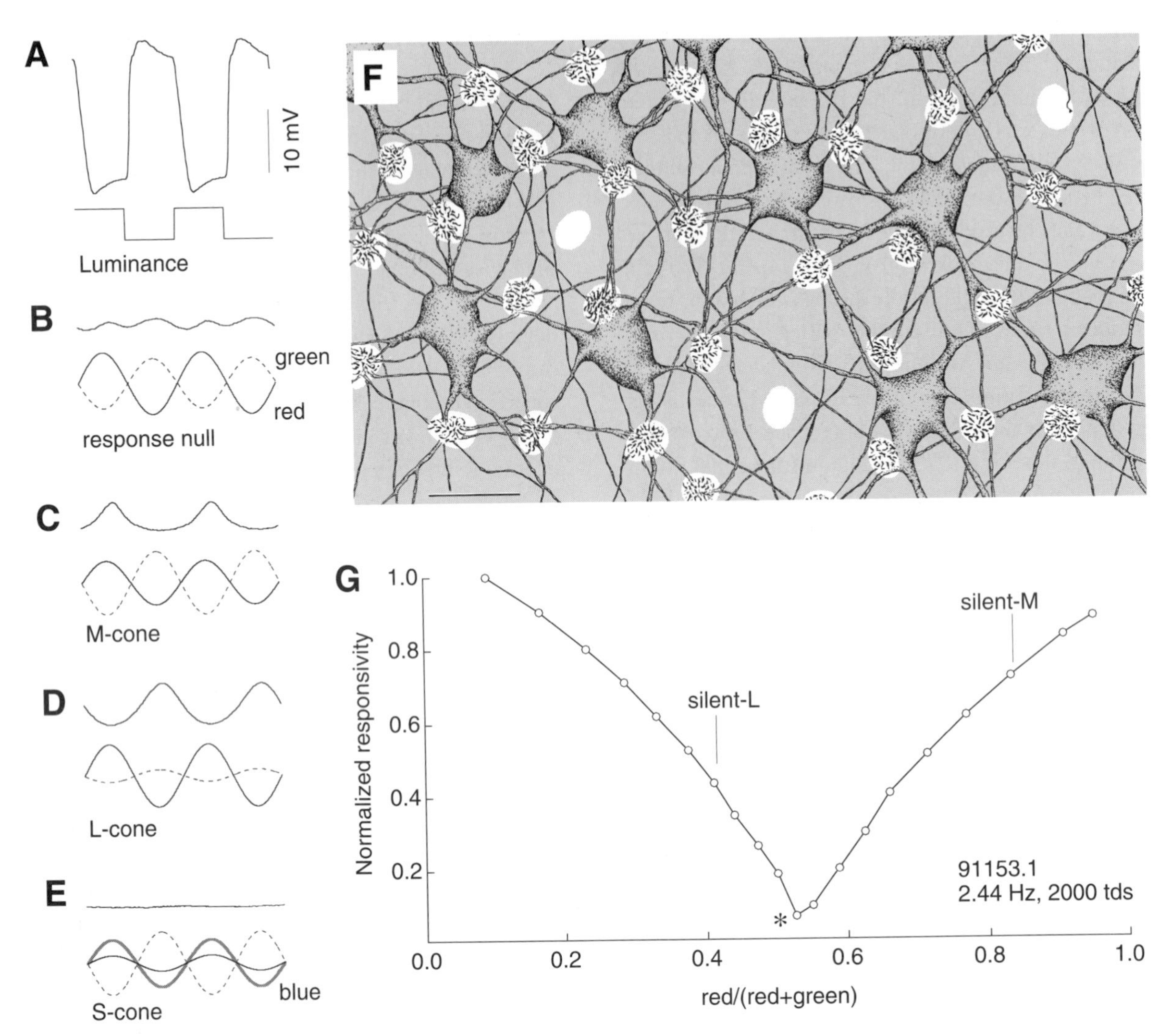

Figure 9.3: Physiology and cone connectivity of the H1 horizontal cells. (A) Light response of H1 cell network to square wave luminance modulation (red and green LEDs in phase). (B) Response minimum to heterochromatic temporal modulation (red and green LEDs in counterphase modulation; green: 0.7, red: 1.0). (C) Response to silent–L-cone condition (M-cone modulation in phase with green LED). (D) Response to silent M-cone condition (L-cone modulated in phase with red LED). (E) Lack of response to selective modulation of S-cone. Blue diode (thicker gray sinusoid) is added in phase with red LED. (F) Camera lucida tracing of the H1 cell network revealed by intracellular injection of Neurobiotin. H1 cell bodies and dendritic processes are shown by stippling; white "holes" in the gray background indicate approximate position of cone axon terminals (pedicles). Majority of cones are densely innervated by dendritic terminals. Three cones in this field (~10%) lack innervation and correspond to S-cones. (G) Plot of responsivity as a function of modulation depth of red and green LEDs. Silent–M- and silent–L-cone conditions are indicated. Equiluminance for the nominal observer (indicated by the asterisk) corresponds to a red/(red+green) ratio of 0.5. Temporal modulation 2.44 Hz; luminance estimated at 2000 trolands.

cone types indiscriminately, and a third horizontal cell type was identified that was nearly identical in morphology to the H1 cell but avoided contacting S-cones (Ahnelt & Kolb, 1994b). From these data the possibility was considered that spectral opponency, like that present in fish horizontal cells, might be found in primates if one looked carefully enough.

The in vitro preparation allowed us to reexamine the cone connections and light responses of macaque horizontal cells (Dacey et al., 1996). To selectively target these cells for recording in vitro we labeled them with a fluorescent marker. The nuclear stain, diamidinophenylindole (DAPI), that was used previously to stain various retinal cell types in other mammals (e.g., Tauchi & Masland, 1984; Mills & Massey, 1992) labeled the macaque horizontal cell nuclei clearly. Two populations could be distinguished based on differences in nuclear size and staining intensity. An orderly mosaic of large, brightly stained cells corresponded to the H1 population and a less regular, lower density array of smaller, less brightly labeled nuclei belonged to the H2 cell population. When the low molecular weight tracer Neurobiotin was intracellularly injected into the horizontal cells, it passed through the gap junctions that couple cells of the same type so that a local patch of either the H1 or H2 mosaic was revealed (Figs. 9.3F and 9.4F).

H1 cells hyperpolarize to light across the spectrum and show a spectral sensitivity like that of the photopic luminosity function, strongly dominated by L- and M-cone input (Dacheux & Raviola, 1990). Two major questions remained unanswered. First, does a second population of H1-like cells exist that shows selectivity in the degree to which they contacted the three cone types, and, if so, do these cells show spectral opponency? Second, what is the nature of the light response of H2 cells? Do these cells receive a major physiological input from S-cones as some anatomical data have suggested? In recordings from now over 200 cells with the characteristic morphology of H1 cells, a single characteristic response to cone-isolating stimuli was found: H1 cells hyperpolarize to light that modulates either the L- or M-cone in isolation but do not respond to selective S-cone modulation (Fig. 9.3). The spectral sensitivity of the H1 cells as measured with the HMP protocol (Figs. 9.2 and 9.3G) reflected additive input from L- and M-cones and showed a response minimum near the equiluminance. The spectral sensitivity as shown by the HMP data reinforces the anatomical observation that H1 cells draw input indiscriminately from L- and M-cones and show a spectral sensitivity like the photopic luminosity function $V(\lambda)$. Since $V(\lambda)$ is well characterized by a 1.6 to 1 ratio in the relative number of L- and M-cones (or the relative synaptic gain of the two cone types), the HMP data provide indirect evidence that, at least in the retinal periphery, the L- and M-cone mosaic is similarly organized in macaque and human retinas. This is consistent with the same conclusion derived from analysis of the spectral sensitivity of M-ganglion cells (Lee, Martin, & Valberg, 1988) and is discussed further below (see also Chapters 1, 2, 4, and 10).

An anatomical basis for the lack of response to S-cone stimulation was revealed in the anatomy of the connections of the H1 cell dendrites with cone axon terminals. The majority of cone pedicles was densely innervated by the H1 cells as previously observed, but a small percentage of cones were "skipped" over. The density and spacing of these skipped cones indicated that they corresponded to the S-cones. This conclusion has been recently confirmed by directly labeling both the H1 cells and the S-cones with an S-cone opsin specific antibody (Goodchild, Chan, & Grünert, 1996).

What of the anatomical distinction made between H1 cells, which were suggested to contact all cone types indiscriminately, and the "H3 cells," which were suggested to avoid contact with S-cones (Ahnelt & Kolb, 1994a)? The large-bodied cells that we have termed H1 show properties of both of the previously described H1 and H3 cells in that they have a nonopponent light response and a broad spectral sensitivity derived from additive L- and M-cone input (H1 cells), but they make little or no contact with S-cones (presumed "H3" cells). There is now direct anatomical evidence that some H1 cells do make a reduced contact with a small percentage of S-cones (Goodchild, Chan, & Grünert, 1996); thus we conclude that the previous distinction made between H1 and H3 cells does not

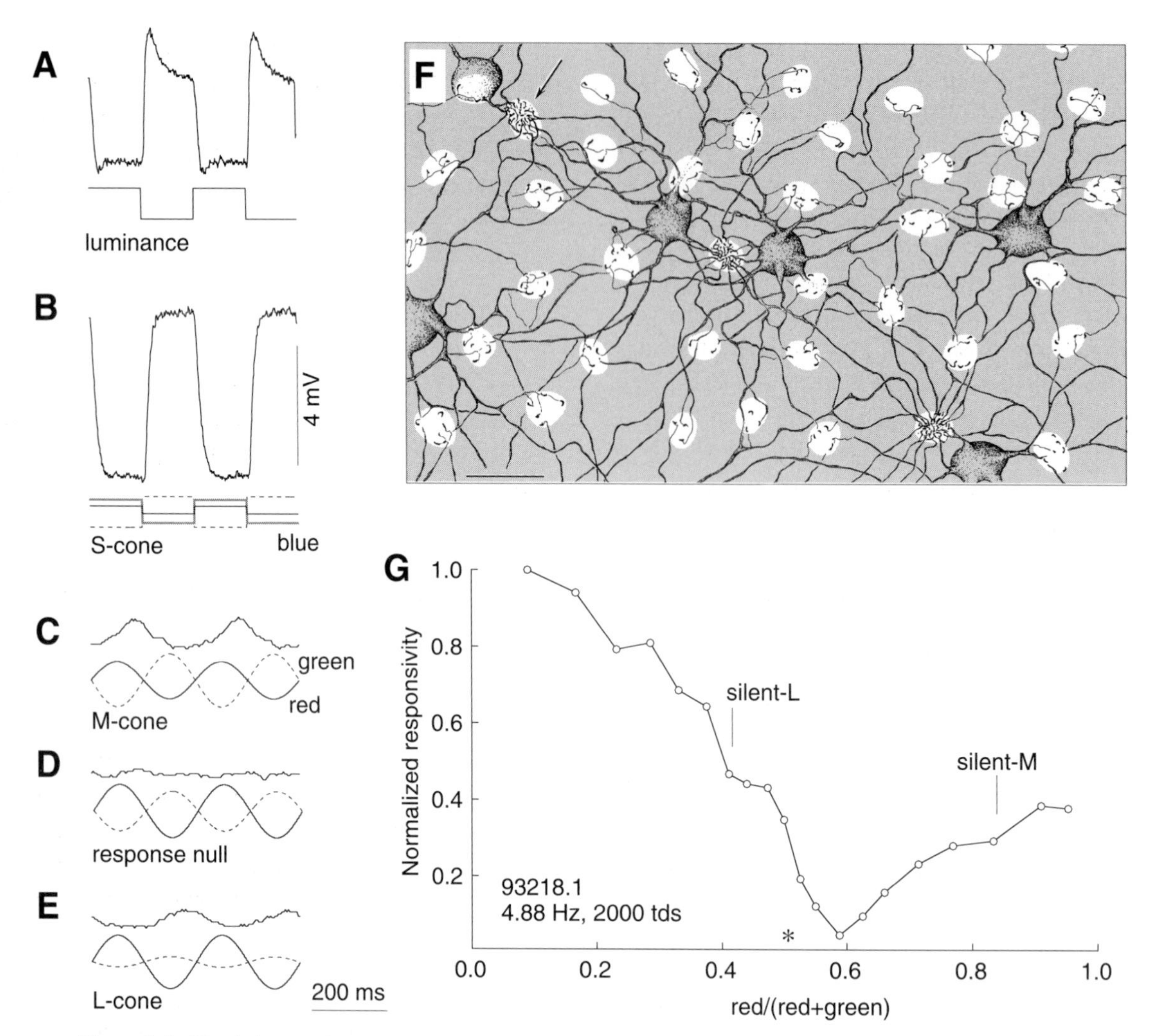

Figure 9.4: Physiology and cone connectivity of the H2 horizontal cells. (A) Hyperpolarizing light response of an H2 cell to luminance modulation (red and green LEDs in phase). (B) Large hyperpolarizing response to a stimulus that modulates the S-cones in isolation. S-cone is modulated in phase with the blue LED (thicker, gray square wave). (C) Silent-L condition; cell hyperpolarizes in phase with M-cone modulation. (D) Response null at 0.7 green, 1.0 red. (E) Silent-M condition; cell hyperpolarizes in phase with L-cone modulation. (F) Camera lucida tracing of the H2 cell network shown by intracellular injection of Neurobiotin during physiological recording. Format as in Fig. 9.3; the H2 cell dendrites converge on and densely innervate three cone pedicles in the field (~10% of total, arrow in upper left indicates one such pedicle); these pedicles have the spacing and density of S-cones. The other cone pedicles are only sparsely innervated. (G) Plot of responsivity as a function of modulation depth of red and green LEDs for the cell whose light response is shown in panels A–E. Conventions as described for Fig. 9.3. Temporal modulation 4.88 Hz; luminance estimated at 2000 trolands.

reflect the existence of two anatomically and functionally distinct horizontal cell mosaics; instead there is a single, electrically coupled mosaic of large-bodied horizontal cells – H1 – that shows a greatly reduced connection to the S-cone mosaic, and this is reflected in the response of the overall network to our cone-isolating stimuli.

Like the H1 cells, the H2 horizontal cells also receive hyperpolarizing input from both L- and M-cones; the null point for these cells in the HMP protocol was similar to that of the H1 cells (Fig. 9.4G), suggesting that the H2 cell network does not show any preference for selecting L- or M-cone input. Unlike the H1 cell, however, the H2 cells also give a large response to the S-cone stimulus (Figs. 9.4A–E). The response to S-cone modulation however is also hyperpolarizing, so that no spectral opponency is conferred by this additional cone input.

A particularly striking anatomical basis for the light response of the H2 network was discovered in H2 connections with the cones. The majority of the L- and M-cone population was innervated only sparsely – with a few dendritic terminals entering each pedicle. However, the H2 cell dendrites also converged on and densely innervated a sparse array of cones, with the expected spacing and density (~7% of the total) of the S-cones, the result being a near reverse image of the H1 network pattern of cone contacts (Fig. 9.4F). The identity of these densely innervated pedicles as S-cones has also been directly confirmed (Goodchild, Chan, & Grünert, 1996).

Even though the S-cones make up less than 10% of the cone population, they appear to have a stronger anatomical, and correspondingly physiological, input to the H2 cells than either the L- or M-cone population. The significance of this S-cone–dominated horizontal cell pathway is not clear, but it raises a problem for understanding the nature of the presumed feedback signal from horizontal cells to cones. Given that the H2 cell contacts L- and M-cones, it would be expected that strong S-cone stimulation should have a depolarizing effect on the H1 cell via a feedback pathway, but this is not observed. One alternative is that the two horizontal cell types function in parallel and their output is primarily feedforward to bipolar cell dendrites.

In sum, evidence is now strong that the primate, like other mammals, has two horizontal cell types; these types are capable of selectively avoiding or seeking out S-cone axon terminals. However, no selectivity is found for L- or M-cones, and all cone inputs to both cell types are hyperpolarizing. Thus, unlike the horizontal cells in certain nonmammalian retinas, primate horizontal cells do not show spectral opponency and cannot provide a simple mechanism for cone-type–specific signals to the receptive field surround of inner retinal neurons – the bipolar, amacrine, or ganglion cells.

Bipolar cells

The retinal bipolar cells convey photoreceptor signals to the amacrine and ganglion cells, yet despite this central position in the retinal circuitry very little is known about their responses to light. In nonmammalian retina, bipolar cells show a distinct center-surround receptive field organization (eg., Kaneko, 1973). Some nonmammalian bipolar cells also show spectral opponency (Kaneko & Tachibana, 1983). In mammalian retina, bipolar cell physiology has been studied mainly at the biophysical level via recordings made from dissociated cells or slice preparations (e.g., Yamashita & Wässle, 1991; Euler et al., 1996); light responses from the intact retina have been rarely recorded, and then only briefly, probably due to the small size of these interneurons. There is little evidence that mammalian cone bipolar cells show a center-surround receptive field organization (Nelson & Kolb, 1983; Dacheux & Raviola, 1986).

In primates we now have a detailed picture of the morphology and synaptic connections of the diverse bipolar cell types, reviewed in this book in Chapters 7 and 8. Briefly, primate cone bipolar cells can be divided into two main classes: diffuse bipolar cells, which are distinguished by connections to multiple cones, and midget bipolar cells, which are distinguished, over much of the retinal area, by a "private line" connection to a single cone axon terminal. The diffuse bipolar cells have been classified at the light microscopic level into six distinct types, according to the depth of stratification of the axon terminal in the inner plexiform layer (Boycott & Wässle, 1991); DB1–3 stratify in the outer portion of the IPL (presumed OFF or hyperpolarizing cells) cells, and DB4–6 stratify in the inner portion of the IPL (presumed ON

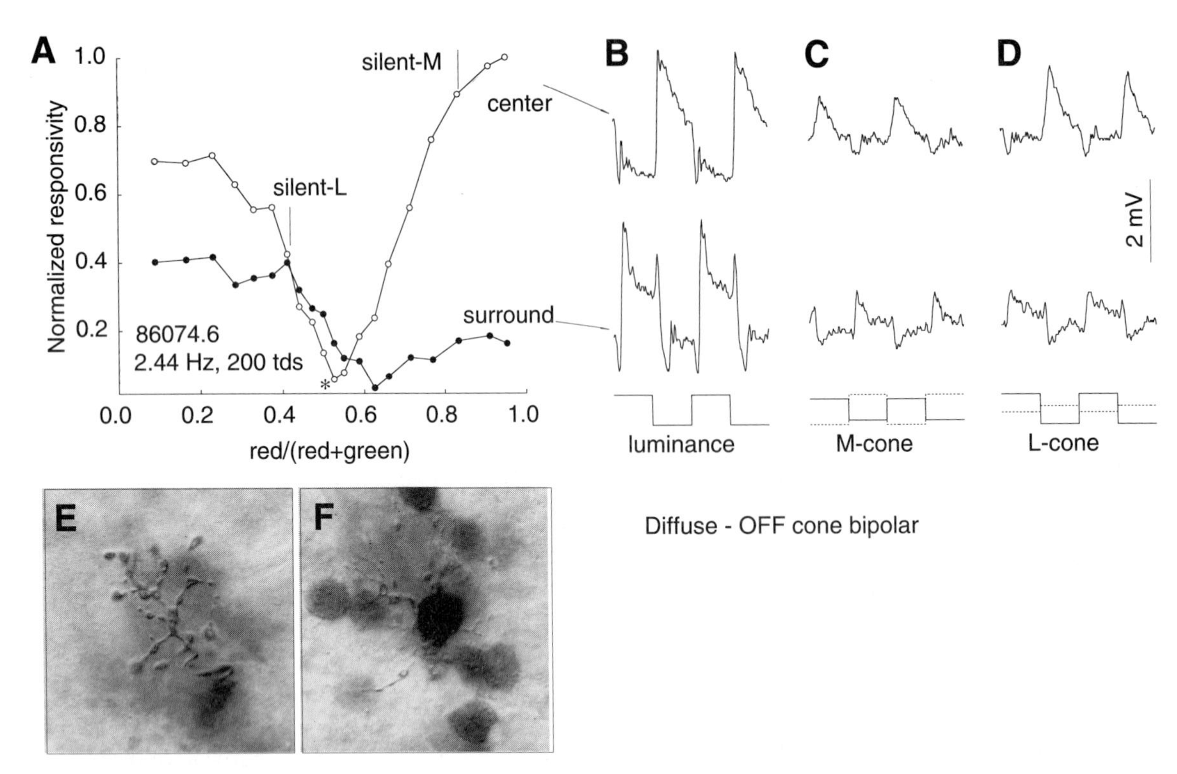

Figure 9.5: Light response of an OFF-center diffuse cone bipolar cell. This cell was tentatively identified as a DB3 cell based on the depth of stratification of the axon terminals in the inner plexiform layer. (A) Spectral sensitivity determined from HMP paradigm shows additive input from L- and M-cones for both center- (1-deg stimulus field) and surround-dominated (10-deg stimulus field) light responses. Response minimum was located near equiluminance (*) for both center and surround responses. (B) Response to square wave luminance modulation shows a depolarizing OFF-center response to a 1-deg stimulus field (top trace) and surround-dominated response to a 10-deg stimulus field (bottom trace). (C) Center- and surround-dominated responses to an M-cone isolating stimulus. (D) Center and surround responses to an L-cone isolating stimulus. (E–F) Micrographs of the recovered Neurobiotin intracellular filling, demonstrated by HRP histochemistry, seen in flatmount. (E) Focus at the level of the axonal tree in the IPL and (F) at the level of the dendritic arbor in the OPL.

or depolarizing cells). The midget bipolar cells also comprise inner and outer stratifying types. In addition, a bipolar cell that selectively connects to S-cones, called the blue-cone bipolar, has recently been characterized (Kouyama & Marshak, 1992).

The richness of detail now available for the morphology of macaque bipolar cells is a necessary starting point for a detailed analysis of the bipolar cell light response. A first question relevant to this review is, Do primate cone bipolar cells show strong center-surround receptive field organization, and, if so, does this provide for a red-green or blue-yellow opponency in one or more types? In the in vitro preparation cone bipolar cells can be seen in DAPI-stained retinas. Study of the physiology of these macaque bipolar cells is just beginning, however, results thus far – 17 diffuse bipolar cells and a single, presumed midget bipolar cell have been recorded – show clearly that both diffuse and midget bipolar cells have strong center-surround organization.

Diffuse bipolar cells: center-surround organization. A center-surround structure could be demonstrated by changing the size of the stimulus field. Fig-

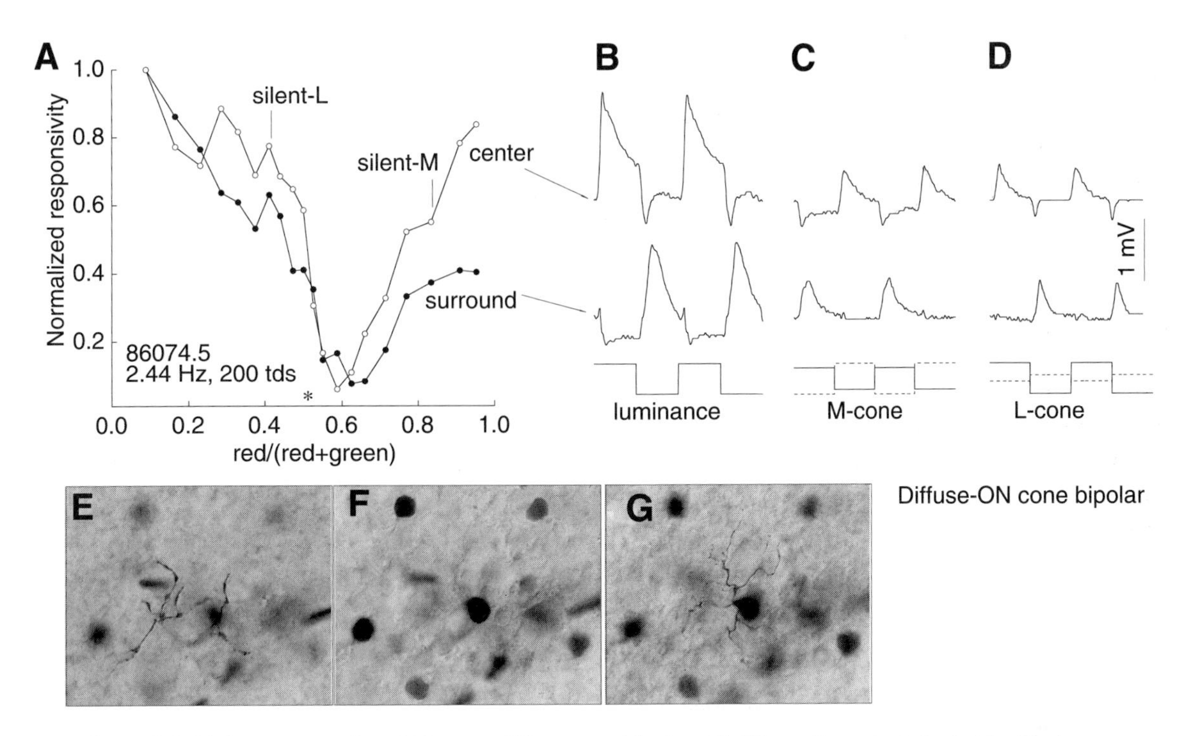

Figure 9.6: Light response of an ON-center diffuse cone bipolar cell. This cell was tentatively identified as a DB5 cell based on the depth of stratification of the dendrites in the inner plexiform layer. (A) Spectral sensitivity determined from HMP paradigm shows additive input from L- and M-cones for both center- (1-deg stimulus field) and surround-dominated (10-deg stimulus field) light responses. Response minimum was located near equiluminance (*) for both center and surround responses. (B) Response to square wave luminance modulation shows a depolarizing ON-center response to a 1-deg stimulus field (top trace) and a hyperpolarizing surround-dominated response to a 10-deg stimulus field (bottom trace). (C) Center- (top trace) and surround-dominated (bottom trace) responses to an M-cone isolating stimulus; M-cone input to center and surround, respectively, depolarize and hyperpolarize the cell. (D) Center- and surround-dominated responses to an L-cone isolating stimulus; L-cone input to center and surround also, respectively, depolarize and hyperpolarize the cell. (E–G) Morphology of recorded cell after intracellular injection of Neurobiotin and HRP histochemistry, wholemount preparation: (E) Plane of focus on the axonal tree in the inner portion of the IPL. (F) Focus shifted to cell body; note tracer coupling to mosaic of neighboring cells. (G) Focus shifted to level of dendritic tree in OPL.

ures 9.5 and 9.6 illustrate center-surround interaction for an OFF and ON diffuse bipolar cell, respectively. Small fields that were 1 deg in diameter or less evoked a sustained, center-dominated response. By contrast, larger stimulus fields (10 deg in diameter) almost completely inverted the polarity of the light response, although a transient component of the center response remained. Diffuse bipolar cells received additive input from both L- and M-cones to both the center and surround of the receptive field. This was shown by obtaining the HMP null point for both center- and surround-dominated responses (Figs. 9.5A and 9.6A). The location of the response minimum was similar to that found for the H1 cells and indicates that diffuse bipolar cells draw input indiscriminately from the L- and M-cones. This is in agreement with the pattern of cone connections for diffuse bipolar cells determined anatomically (Boycott & Wässle, 1991). Thus far no significant S-cone input has been found for diffuse bipolar cells, although an anatomical connection with S-cones has been suggested (Boycott & Wässle, 1991). Physiological recording of diffuse bipolar cell light responses is just beginning, and no attempt has yet been made to reach conclusions about the specific type (DB1–6) of diffuse bipolar cell recorded. No recordings have yet been made from an identified blue-cone bipolar cell.

Midget bipolar cell: red-green spectral opponency? The physiology of the midget bipolar cell may hold the key to understanding the mechanisms of red/green opponency. Individual midget bipolar cells receive synaptic connections from a single cone over much of the retinal area (from fovea to ~50 deg eccentricity) (Milam et al., 1993; Wässle et al., 1994). This private line connection has the potential of delivering a cone-type–specific input to the receptive field center of a midget ganglion cell. Does the midget bipolar cell show spectral opponency? If so, what is the nature of the cone input to the receptive field that generates the opponent response? Thus far only one midget bipolar cell has been recorded, but the response showed clear red-green spectral opponency (Fig. 9.7). This cell was a hyperpolarizing, OFF-center cell. The light response was dominated by a hyperpolarizing M-cone input when a 1-deg-diameter stimulus field was used (Figs. 9.7B–C). When a 5-deg stimulus field was used a depolarizing L-cone input was added to the response (Fig. 9.7E). These two inputs act synergistically in response to red-green chromatic modulation, since the L- and M-cones are modulated in counterphase (see Fig. 9.2), resulting in a large red-ON, green-OFF response (Fig. 9.7D). The response to a luminance-modulated square wave was also characteristic of a spatially as well as chromatically opponent cell. The cell showed a strong OFF response to stimulation of the receptive field center; when the surround was added the L- and M-cone inputs (now modulated in phase) acted antagonistically to greatly diminish the light response (Fig. 9.7B). It was not possible in this cell to determine whether the cone input to the surround was derived entirely from L-cones or from both L- and M-cones.

In sum, the first recordings from primate cone bipolar cells show evidence for strong center-surround receptive field organization. Diffuse bipolar cells studied thus far sum L- and M-cone inputs to both center and surround and show, like the H1 cells, a spectral sensitivity that derives from L- and M-cone input with a weighting like that of the photopic luminosity function. Much more will need to be done on a large number of morphologically identified diffuse types before any conclusions can be reached about the physiological properties and functional role of each type. A single, midget bipolar cell has been recorded. It showed a spectral opponent M-cone–ON, L-cone–OFF center-surround light response.

Ganglion cells

To understand the circuitry that gives rise to spectral opponency, it is necessary to clearly identify the ganglion cell types that transmit wavelength-selective signals. In an early attempt to directly link morphology to physiology using intracellular recording and staining methods in the intact primate eye, de Monasterio addressed the question of which ganglion cell types transmitted spectral opponent signals (de Monasterio, 1979). He suggested that two common types, the parasol and midget ganglion cells, first described and named by Polyak (1941), might transmit blue-yellow and red-green opponent signals, respectively, to the lateral geniculate nucleus. It is now well understood that the parasol ganglion cells project to the magnocellular layers of the LGN, where cells with nonopponent light responses are recorded. The parasol cells have for this reason been considered the anatomical counterpart of the M-cells that have been well characterized in extracellular recordings of macaque ganglion cells (see Chapters 10 and 11) and show a broad spectral sensitivity that matches the photopic luminosity function (Lee, Martin, & Valberg, 1988). Midget ganglion cells whose axons project to the parvocellular layers are considered equivalent to the spectrally opponent P-cells that also have been studied intensively by extracellular recording in both retina and LGN (review: Kaplan, Lee, & Shapley, 1990). However, ganglion cell types other than midget and parasol cells project to the LGN (Rodieck & Watanabe, 1993), calling into

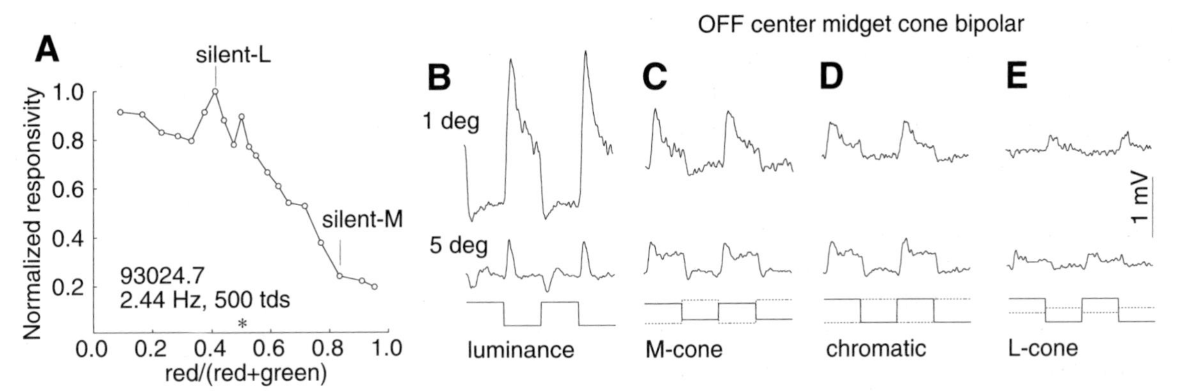

Figure 9.7: Light response of a presumed OFF-center midget bipolar cell shows red-green spectral opponency. This cell was tentatively identified as a midget cell based on the morphology of the cell as observed in vitro after dye staining; however, the cell was not recovered by HRP histochemistry for later analysis. (A) Spectral sensitivity determined from HMP paradigm for a 5-deg field shows strong input from M-cones and a strong response to equiluminant chromatic modulation; response minimum was located near silent-M condition. (B) Response to square wave luminance modulation shows a hyperpolarizing OFF-center response to a 1-deg stimulus field (top trace) but only small transient responses at stimulus ON and OFF for a 5-deg stimulus field (bottom trace). (C) Center- (top trace) and surround-dominated (bottom trace) responses to an M-cone isolating stimulus; cell response is M-cone OFF to both stimulus fields. (D) Center- and surround-dominated responses to equiluminant red/green chromatic modulation; cell hyperpolarizes in phase with green LED. (E) Center- and surround-dominated responses to L-cone isolating stimulus. A 1-deg field elicits a very small hyperpolarizing response in phase with red LED; a 5-deg field elicits a larger depolarizing response: The cell is L-cone–OFF in response to a small field and L-cone–ON in response to a large field. Thus, this cell is an M-cone–dominated OFF-center and L-cone–dominated ON-surround opponent cell.

question the simple one-to-one correspondence between these two ganglion cell types and the physiologically defined magnocellular and parvocellular pathways. To determine which of the many ganglion cell types (Rodieck, 1988; Kolb, Linberg, & Fisher, 1992) project to the LGN (and contribute to color vision), it is necessary to directly link ganglion cell dendritic morphology, physiology, and central connection. As a first step toward this goal, we have recorded the light responses of parasol cells, midget cells, and the small-bistratified cell, a recently characterized ganglion cell type (Dacey, 1993a) that is retrogradely labeled after tracer injections are made into the parvocellular LGN (Rodieck, 1991).

Parasol cells: nonopponent cells of the magnocellular pathway. Parasol ganglion cells have large cell bodies that are easily identified in the in vitro retina after staining with acridine orange (Fig. 9.1B). Intracellular staining of recorded cells shows the distinctive dendritic morphology of parasol cells observed previously (Fig. 9.8A) (Perry, Oehler, & Cowey, 1984; Watanabe & Rodieck, 1989; Dacey & Petersen, 1992). Consistent with their projection to the magnocellular LGN parasol cells show nonopponent physiology with a response minimum near the equiluminant condition, as expected (Figs. 9.8B, D–F). Using a small spot or an annulus it is possible to isolate a relatively pure center and pure surround response (Fig. 9.8C); both components show a similar spectral sensitivity (Fig. 9.8B). The recordings from identified parasol cells confirm their identity as a type of M-cell, but they do not determine whether all magnocellular projecting ganglion cells are parasol cells. To address this question it will be necessary to use the intracellular approach in combination with retrograde labeling (e.g. Pu, Berson, & Pan, 1994; Yang & Masland, 1994).

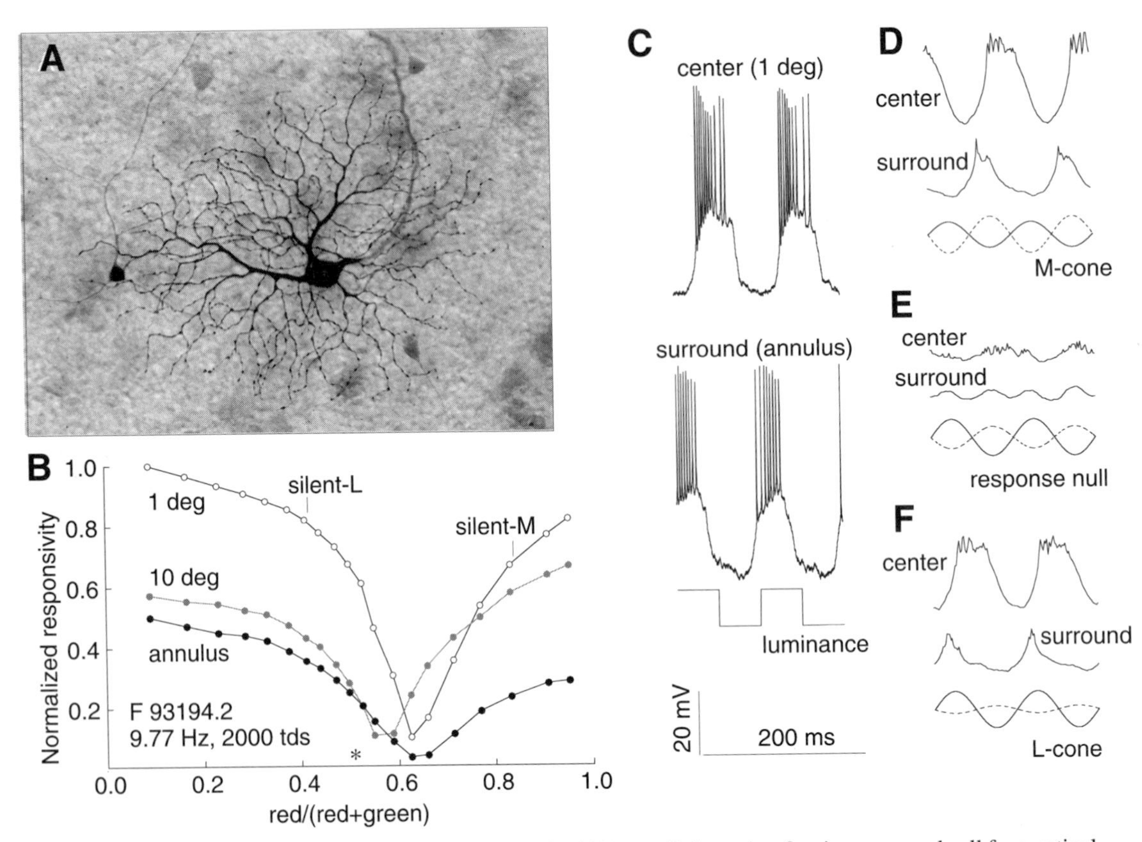

Figure 9.8: Light response of a parasol ganglion cell. (A) Intracellular stain of an inner parasol cell from retinal periphery. (B) Plot of responsivity (1st harmonic amplitude of averaged synaptic potential) of the cell shown in panel A as a function of red and green LED modulation depth. Separate plots are shown for 1-deg stimulus field (open symbols), 10-deg stimulus field (gray symbols), and an annulus with a central occluding disc ~1.5-deg diameter that elicted a pure surround response (solid symbols). Synaptic potentials for the silent-L, response null, and silent-M points in this plot are shown in panels D–E. Other conventions are as described in Fig. 9.3. (C) Response to 9.77-Hz square wave luminance modulation. The top trace shows a sustained-ON response (peak discharge rate (280 spikes/s) to a 1-deg stimulus field; the lower trace shows a sustained-OFF response to an annulus. (D) Synaptic potential, averaged over ~ 5-s stimulus modulation, in response to M-cone isolating stimulus (silent-L condition indicated in B) for both ON-center and OFF-surround responses. (E) Averaged synaptic potential at response minimum (red:1.0, green:0.6) for both ON-center and OFF-surround responses. (F) Averaged synaptic potential, in response to L-cone isolating stimulus (silent-M condition indicated in panel B) for both ON-center and OFF-surround responses.

Midget ganglion cells and the circuitry for red-green opponency. It is generally accepted that the midget ganglion cells form the substrate for the red-green opponent pathway (see Chapters 7, 8, and 10); the issue that remains controversial is the nature of the circuitry that accomplishes the L/M-cone opponent transformation. The recording from a single midget bipolar cell illustrated in Fig. 9.7 suggests that opponency is already present at the bipolar cell level. If a midget ganglion cell receives input from a single midget bipolar cell, as occurs in the central 7–10-deg eccentricity, then the opponent response of the midget bipolar cell can be simply transmitted to the midget ganglion cell. This is consistent with the large number of red-green opponent cells that are recorded in the parafoveal retina, where the private-line midget system dominates. However, with increasing distance from the fovea midget ganglion cells increase greatly in dendritic field diameter relative to the midget bipolar cells. Do these peripheral midget ganglion cells have the ability to make selective connections with either the L- or M-cone–connecting bipolar cells? Or is there a lack of cone-type selectivity with the consequent loss of opponency in the periphery? In the in vitro preparation it has been possible to answer this question by recording from midget cells in the periphery of the retina and mapping the cone inputs to both the receptive field center and surround (Dacey, 1996a; Dacey & Lee, 1997).

In the retinal periphery midget ganglion cells can be as large as 150–200 μm in diameter (Fig. 9.9A) and must receive convergent input from a large number of midget cone-bipolar cells. We have found that these peripheral midget ganglion cells receive input from both L- and M-cones to the receptive field center as well as the receptive field surround (Figs. 9.9C–F). This results in a nonopponent light response similar to that found for all other cells, including the parasol ganglion cell, that receive additive L- and M-cone input and lack significant S-cone input (Fig. 9.9B). Spatial maps of the cone inputs to the receptive field also very clearly reveal L- and M-cone input to both the center and the surround components (Fig. 9.10). These results suggest that the surround of the midget ganglion cell, whether it derives exclusively from the midget bipolar or via amacrine cell input or both, is not cone-type–selective and that opponency is manifest when the number of cone inputs to the center is reduced to one, as occurs in the parafovea. This is consistent with the class of nonselective, "mixed-surround" and "hit or miss" models (Buchsbaum & Gottschalk, 1983; Lennie, Haake, & Williams, 1991; DeValois & DeValois, 1993) and with psychophysical evidence for a gradual decline in the sensitivity of red-green color vision with increasing distance from the fovea (e.g., Mullen & Kingdom, 1996).

Small-bistratified cells: blue-ON/yellow-OFF cells of the parvocellular pathway. A number of early recording studies pointed out large differences in the physiology of S-cone opponent cells (eg., Zrenner & Gouras, 1981), so with hindsight and current knowledge of the great diversity of ganglion cell types it is not surprising that cells with S-cone input correspond to a distinctive anatomical type that projects to the LGN in parallel with the parasol and midget ganglion cells. The small-bistratified cells were recently recognized by intracellular staining in both macaque and human retinas (Dacey, 1993a). These cells were also shown to project to the parvocellular layers of the LGN (Rodieck, 1991; Rodieck & Watanabe, 1993), suggesting a role in color coding. The distinctive bistratification of the dendritic tree (Figs 9.11C–E) suggested that these cells might convey S-cone signals: The inner dendritic tree appeared to costratify with the axon terminals of the blue-cone bipolar type, whose dendrites connect exclusively to S-cones (Kouyama & Marshak, 1992). The relatively large and distinctive cell body of the small-bistratified cell was identified in the in vitro macaque retina stained with acridine orange, and recordings clearly showed a blue-ON/yellow-OFF opponent light response (Dacey & Lee, 1994) (Fig. 9.11A). The S-cone isolating stimulus elicits a strong ON-depolarization and spike discharge in phase with the S-cone modulation, in striking contrast to either the parasol or midget ganglion cells that show no response to the same stimulus.

The receptive field structure of the blue-ON cell taken together with its bistratified morphology suggests a simple underlying mechanism for the cone opponent response. First, the S-cone–ON and L+M-cone–OFF responses do not represent a classical center-surround organization, but instead appear as two fields that are spatially similar in size and nearly coextensive (Fig. 9.12) (Dacey, 1996a). Second, the inner dendritic tier (Fig. 9.11D), as mentioned above, is well suited to receive direct depolarizing input from the blue-cone bipolar cell, and, in addition, the more sparsely branching outer dendritic tree (Fig. 9.11E) is stratified close to the amacrine cell layer, in the outer, presumed OFF portion of the inner plexiform layer. Recent electron microscopic reconstruction (see Chapter 8) indicates that these outer dendrites are the sites of synaptic input from a diffuse cone bipolar that is

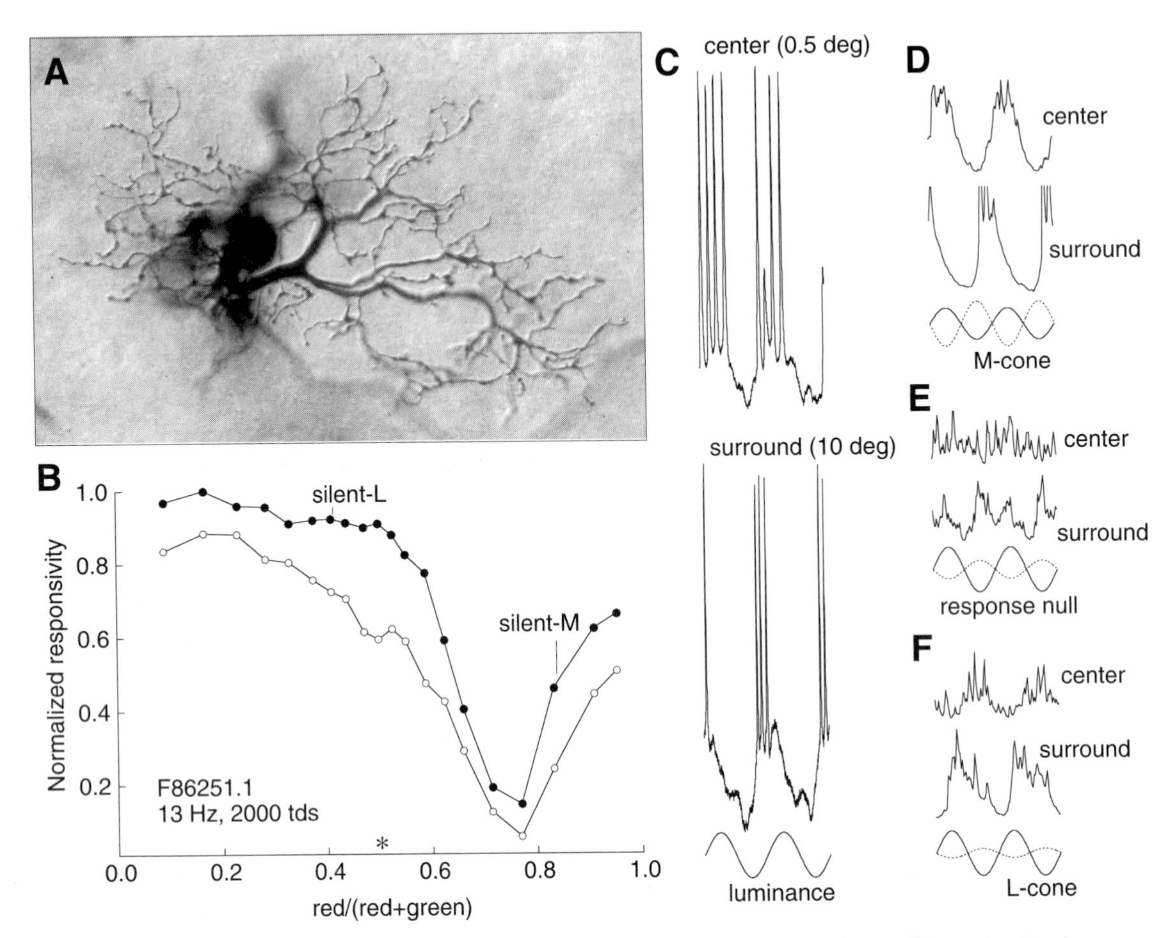

Figure 9.9: Light response of a midget ganglion cell in the retinal periphery. (A) Intracellular stain of an inner midget ganglion cell at 11-mm eccentricity in the temporal retina. (B) Plot of responsivity (1st harmonic amplitude of averaged synaptic potential) of the cell shown in panel A as a function of red and green LED modulation depth. Separate plots are shown for 0.5-deg stimulus field (open symbols) and 10-deg stimulus field (filled symbols). Cell receives additive input from both L- and M-cones, but the response to M-cone modulation is stronger, shifting the response null toward the silent-M condition. Synaptic potentials for the silent-L, response null, and silent-M points in this plot are shown in panels D–F. Other conventions are as described in Fig. 9.8. (C) Response to 13-Hz square wave luminance modulation. The top trace shows a sustained-ON response (peak discharge rate, 100 spikes/s) to a 1-deg stimulus field; the lower trace shows an OFF surround-dominated response to a 10-deg stimulus field. (D) Synaptic potential, averaged over ~ 5-s stimulus modulation, in response to M-cone isolating stimulus (silent-L condition indicated in panel B) for both center- and surround-dominated responses. (E) Averaged synaptic potential at response minimum (red:1.0, green:0.3) for center- and surround-dominated responses. (F) Averaged synaptic potential, in response to L-cone isolating stimulus (silent-M condition indicated in panel B) for center- and surround-dominated responses.

connected nonselectively to L- and M-cones. Thus the spatially coextensive blue-ON and yellow-OFF fields appear to derive from separate ON and OFF excitatory cone-bipolar pathways.

Amacrine cells

The lack of spectral opponency in the H1 and H2 horizontal cells focuses attention on the amacrine

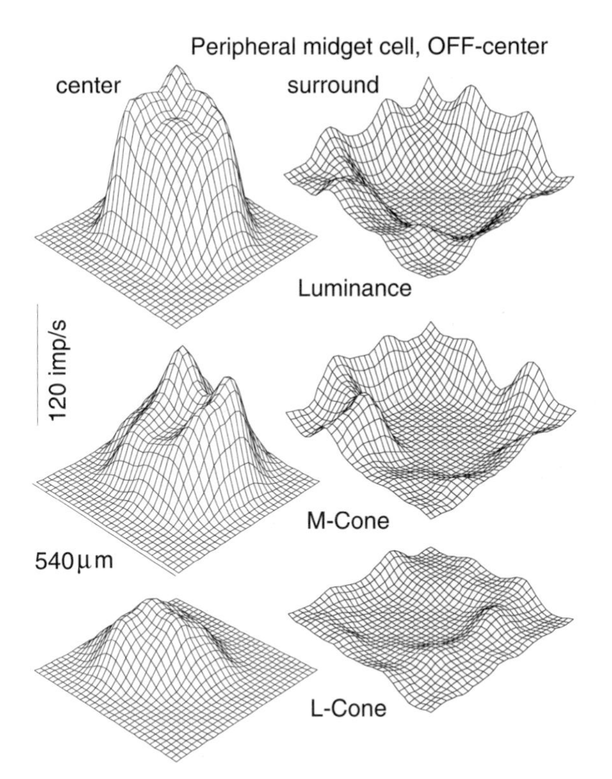

Figure 9.10: Map of cone inputs to receptive field center and surround of an OFF-center midget ganglion cell in the retinal periphery. Cone inputs were mapped using luminance modulated, M- and L-cone isolating spot stimuli, 40 µm in diameter and modulated at 2.44 Hz temporal frequency. The spot was moved to successive locations in a 13 × 13 grid covering a 540-µm square field. Center and surround responses were clearly identified by a ~180-deg shift in response phase. The left-hand column shows two-dimensional mesh plots of the location and amplitude of center-OFF responses to each of the three stimulus conditions (surround response locations were given zero values). The receptive field center received additive input from M- and L-cones. The right-hand column shows mesh plots of the surround-mediated ON response to the same stimuli (center response locations were given zero values); these responses were strongest around the edges of the center. As for the center, the surround received additive input from M- and L-cones, with the M-cone input the stronger of the two.

cells, the laterally connecting interneurons of the inner retina, as a basis for red-green cone-opponent circuitry. One possibility, for example, is that a "midget" amacrine cell exists, that is, an amacrine cell type that receives input from a single midget bipolar connected to, say, an L-cone, and directs inhibitory output exclusively to a midget ganglion cell that receives excitatory M-cone input from another midget bipolar cell; such a circuit would require a great deal of connectional specificity in the midget system. This question has recently been addressed by reconstructing amacrine cell processes that synapse onto the private line connection of a midget bipolar cell and midget ganglion cell (Calkins & Sterling, 1996) (see Chapter 8). No anatomical evidence was found to suggest that an amacrine cell could receive input from only one cone type so as to provide a simple basis for a cone-type–specific inhibitory pathway into the midget system. However, the physiology of mammalian amacrine cells is complex (Masland, 1988) and little understood. Given the presence of up to 40 distinct amacrine cell populations, most of which have yet to be characterized physiologically, the specific role of identified amacrine cell types in an opponent transformation remains an unsettled and difficult question.

In macaque retina use of the in vitro retina-choroid preparation has permitted the first physiological studies of identified amacrine cell types (Dacey, 1996b; Stafford & Dacey, 1997). Thus far, it has been possible to determine the cone inputs and light responses of two distinctive amacrine types in macaque; these results are briefly reviewed below.

The A1 cell, a spiking, axon-bearing amacrine type. It is possible to divide the amacrine cells physiologically into two broad groups: those cell types that spike and those, like the other retinal interneurons, that show only graded changes in potential in response to a synaptic input. For the spiking amacrines the nature of the spiking behavior varies greatly and the exact origins of the spike trigger zones have not been defined for any type. Several mammalian amacrine types show both dendritic trees and distinctive axonlike arbors that project for long distances within the inner retina (reviewed in Vaney, 1990; Wässle & Boycott, 1991). In general, these axon-bearing types are good candidates for long-range, spiking inhibitory connections that could contribute to the properties of the receptive

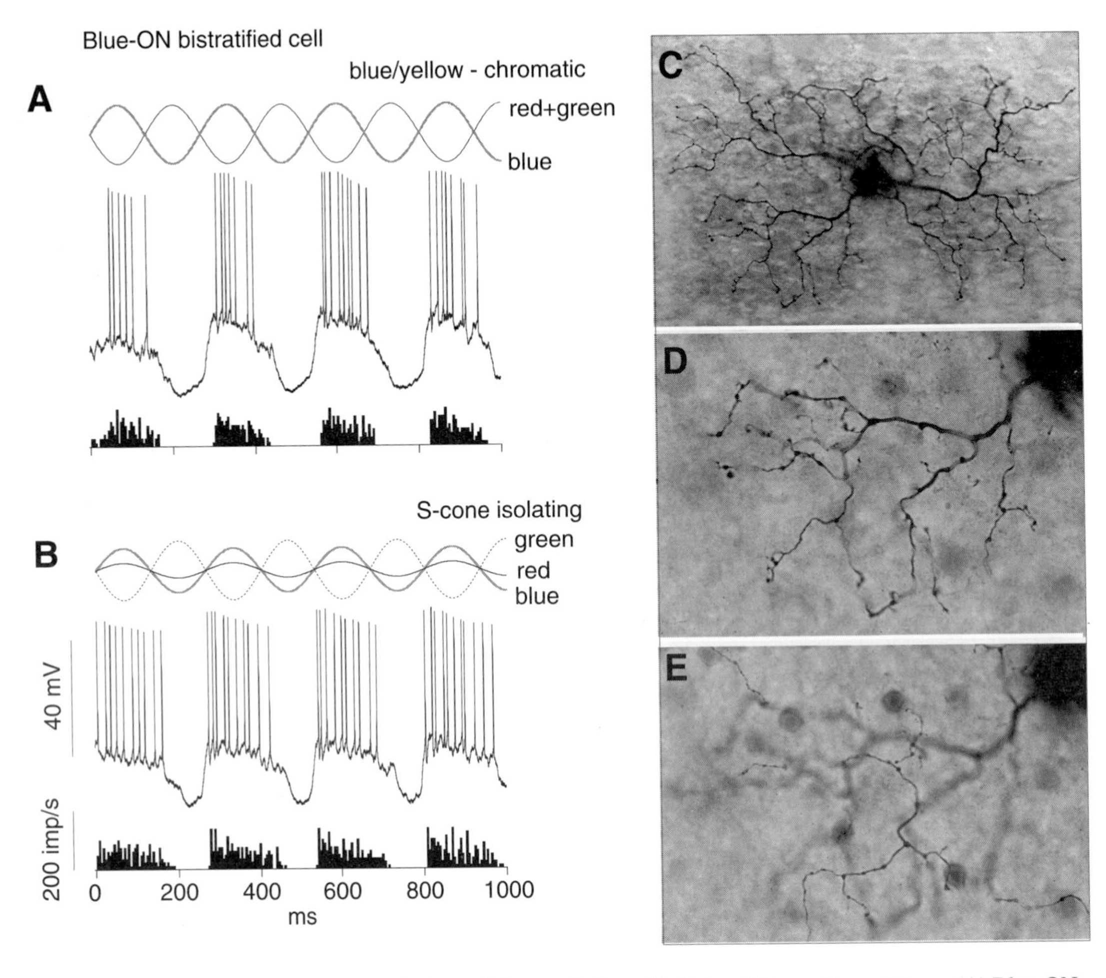

Figure 9.11: Light response and morphology of the small-bistratified blue-ON ganglion cell type. (A) Blue-ON response to equiluminant blue-yellow modulation. *Top*: Stimulus waveform, blue LED output is run in counterphase to red and green LED output. *Middle trace*: Intracellular voltage response shows strong blue-ON depolarization and spike discharge. Poststimulus time histogram of spike discharge, shown under voltage record, is averaged over 6-s stimulus modulation. (B) S-cone–mediated ON-response. *Top trace*: Stimulus waveform, blue LED is modulated in phase with red LED and counterphase to green LED. Modulation depths were set to silence L- and M-cones; S-cones are modulated in phase with the blue LED. *Middle trace*: Voltage response shows strong depolarization and spike discharge in phase with S-cone modulation. Poststimulus time histogram as in panel A. (C) Dendritic morphology of cell whose light response is shown in panel A. The dendritic tree is more sparsely branching than the parasol cell shown in Fig. 9.8. Morphology demonstrated by intracellular injection of Neurobiotin and subsequent HRP histochemistry. (D) Higher magnification of a small portion of the dendritic tree indicated by the box in panel C; plane of focus on the inner tier of dendrites. (E) Same field as in panel D, but plane of focus shifted to the outer tier of dendrites.

field surrounds of bipolar and ganglion cells. One of these types in the primate retina, the A1 cell, is a particularly striking example of an axon-bearing type (Dacey, 1989) (Fig. 9.13). In addition to a distinctive, long-range, axonlike arbor, the dendritic tree of this cell type stratified broadly across the ON–OFF subdivision of the IPL, suggesting that this type might show a spiking, ON–OFF light response.

Because the A1 cell body diameter is one of the largest in the amacrine cell layer, it was easily targeted for intracellular recording in the in vitro retina stained with the nonselective vital dye acridine orange. We found the A1 to truly be a spiking cell, with a phasic ON–OFF discharge. The diameter of the A1 receptive field showed a close correspondence with the size of the dendritic tree (shown in white in Fig. 9.13A), suggesting that the spikes originate at the axonal origins near the soma and propagate distally for long distances. The A1 cells received additive input from both L- and M-cones to both ON- and OFF-components of the receptive field and lacked any significant response to an S-cone stimulus (Figs. 9.13C–F). The HMP protocol showed a response null near the equiluminant point (Fig. 9.13B), as for the other cell types described thus far that sum L- and M-cone inputs. Thus this cell type has a spectral sensitivity like the photopic luminosity function and is excluded from a role in transmitting a cone-type–specific inhibitory signal to bipolar or ganglion cells.

The AII cell: cone inputs to a rod-amacrine type. The AII amacrine is one of the best characterized amacrine types of the mammalian retina. It reaches the highest spatial density of any amacrine population and is considered the pivotal cell type for the transmission of rod photoreceptor signals to ganglion cells; the circuitry that accomplishes this has been well studied and reviewed in detail (e.g.,Wässle, Boycott, & Röhrenbeck, 1991). In brief, depolarizing (ON) rod bipolar cells synapse not on ganglion cell dendrites but on the inner dendrites of the AII amacrine. The AII is a bistratified cell (Figs. 9.14A and B) that makes sign-inverting inhibitory synapses with cone bipolar cells in the OFF part of the IPL and sign-conserving gap junctions with cone bipolar cells in the ON portion of the IPL; the AII cell thus "piggybacks" on the cone bipolar cells to transmit both ON- and OFF-rod signals to ganglion cells. Why, then, consider the possibility that the AII rod amacrine plays a role in photopically driven cone signal pathways and color coding? Surprisingly, the AII cell also receives significant synaptic input from OFF-cone bipolar cells (Dacheux & Raviola, 1986). In primate retina it has recently been shown that presumed OFF-midget bipolar cells synapse on the AII cell (Grünert & Wässle, 1996). In addition, it might be expected that under photopic conditions bidirectional

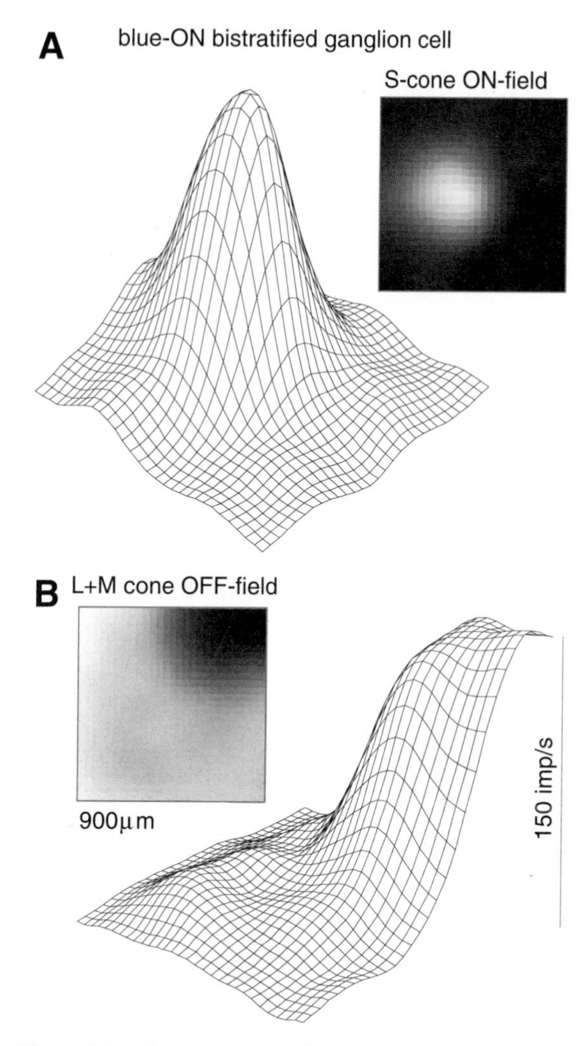

Figure 9.12: Spatial map of S-cone–ON and L+M-cone–OFF field for a blue-ON, small-bistratified ganglion cell. Map was made as described in Fig. 9.10 using an S-cone selective stimulus and a luminance-modulated stimulus (red and green LEDs run in phase; S-cone modulation is negligible). (A) S-cone–ON response is shown as both a surface plot and an image plot. (B) L+M-cone–OFF field is shown as both a surface plot and an image plot. The two fields are similar in size but do not overlap completely; as a result one edge of the OFF-field was not mapped with the grid shown here.

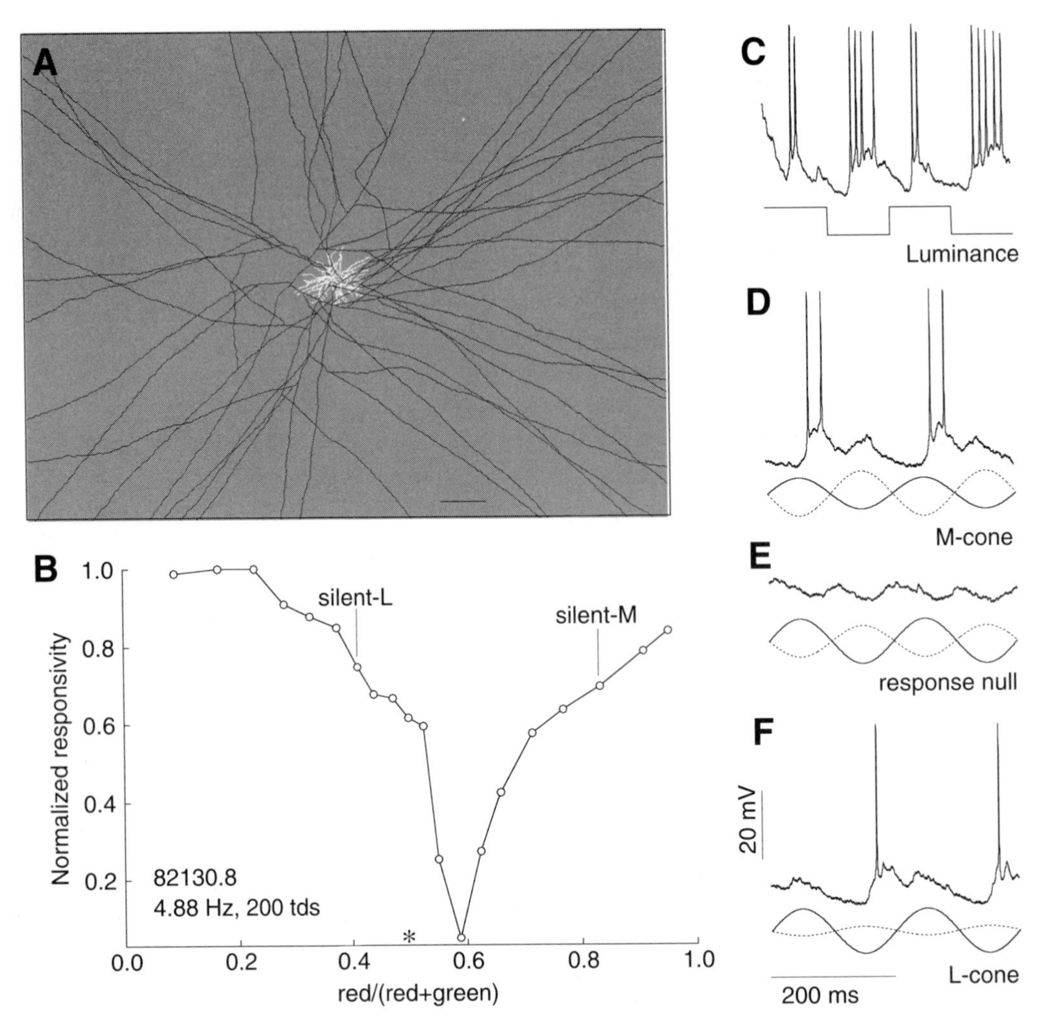

Figure 9.13: Light response and morphology of the A1 amacrine cell. (A) Morphology of the A1 cell in the retinal periphery. The dendritic tree (shown in white) is ~400 μm in diameter and moderately branched; a distinctive axonlike arbor arises near the cell body and gives rise to a second, more sparsely branched arbor (shown in black) that extends for long distances (4–5 mm) in the inner plexiform layer; scale bar = 300 μm. (B) Plot of responsivity (1st harmonic amplitude of averaged synaptic potential) of the cell shown in panel A as a function of red and green LED modulation depth. This amacrine cell type receives additive input from both L- and M-cones with a response minimum in near equiluminance. (C) ON–OFF spiking light response to square wave luminance modulation. Responses for the silent-L, response null, and silent-M points taken from the plot in panel B are shown in panels D–F.

gap junctions with ON-cone bipolar cells might function to transmit photopic signals to the AII. It is possible then that the AII could introduce cone-type–selective inhibition into the midget pathway.

To address these questions the AII cell was identified in vitro by selective staining with DAPI applied under somewhat different conditions than those used to label the horizontal cell nuclei (Dacey, 1996b). In the in vitro preparation high-intensity light is used to observe the fluorescent dyes that label retinal cells, and this produces a strong bleach of the rods; we found that a threshold rod response in an AII cell required ~30

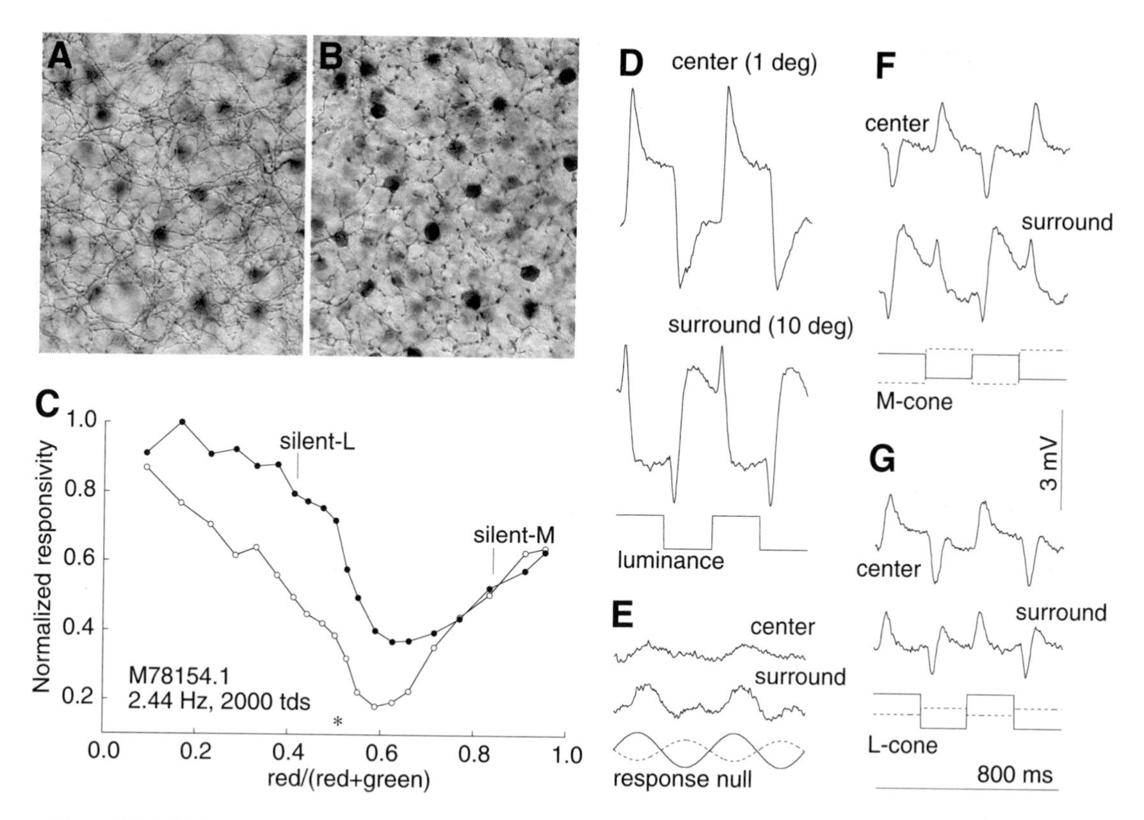

Figure 9.14: Light response and morphology of the AII amacrine cell. (A) Intracellular Neurobiotin injection reveals characteristic network of AII cells; plane of focus on inner, "arboreal" dendrites. (B) Plane of focus shifted to level of AII cell bodies and outer, "lobular" dendrites. (C) Plot of responsivity as a function of red and green LED modulation depth. Center (1-deg field) and surround-dominated (10-deg field) responses are shown by the open and filled symbols, respectively. This cell type receives additive input from L- and M-cones and shows a response minimum near equiluminance (asterisk). (D) Light response to luminance-modulated square wave for both 1-deg (top trace) and 10-deg (lower trace) diameter fields. Strong hyperpolarizing surround components are elicited by the 10-deg stimulus. (E–G) Center- and surround-dominated responses taken from the plot in panel C, for response null (red = 1.0, green, 0.7), and for M- and L-cone–isolating conditions.

min of dark adaptation. However, after the rod saturation, AII cells show a strong cone-driven light response at light levels 3–4 log units above cone threshold. This response is similar in amplitude to the light response of cone bipolar cells and, like the cone bipolar cells, shows a strong center-surround receptive field organization that is easily observed by varying the diameter of the stimulus field (Fig. 9.14D). The waveform of the light response to a large field stimulus that engages both center and surround is complex and multiphasic, showing a transient depolarization at light-ON, followed by a more sustained surround-mediated hyperpolarization; a second, more phasic hyperpolarization appears at light-OFF, followed by a weaker, sustained depolarization. Responses to cone-isolating stimuli show that, like the A1 amacrine, the AII cell receives additive input from L- and M-cones but lacks a significant S-cone input to both the center and the surround of the receptive field (Figs. 9.14E–G). Characterizing the spectral sensitivity with the HMP paradigm gave response minimums for both center and surround near the equiluminant condition, similar to all other cell types recorded thus far that combine L- and M-cone input.

Summary and conclusions

One major problem for fully characterizing the retinal origins of spectral opponency is the great diversity of cell types in the primate retina whose physiology and morphology remain largely uncharacterized. A fresh approach to this problem is the use of an in vitro preparation of the macaque monkey retina that permits a systematic study of the cone inputs to morphologically identified ganglion cells and retinal interneurons. Horizontal cell types, which have long been suggested as key components in opponent circuitry, are shown not to play such a role in macaque retinas. The first recordings from cone bipolar cell types demonstrate a strong center-surround receptive field organization, but the specific roles of the distinct types of cone bipolar cells in the transmission of spectral opponent signals remains to be established. Cone inputs to three morphologically identified ganglion cell types whose axons project via the retino-geniculate pathway – the magnocellular-projecting parasol cells, and the parvocellular-projecting small-bistratified and midget ganglion cells – have been determined. Parasol ganglion cells, as anticipated from previous extracellular recordings, receive additive input from L- and M-cones, lack significant S-cone input to both center and surround of the receptive field, and thus show nonopponency characteristics of cells in the magnocellular visual pathway. The small-bistratified cells are ON–OFF cells that receive a major excitatory input from S-cones and transmit a blue-ON/yellow-OFF opponent signal. The mechanism for the cone opponency appears to be the segregation of S-cone and L+M-cone signals, respectively, into ON- and OFF-bipolar cell channels. Midget ganglion cells of the retinal periphery, expected to show red-green opponency, surprisingly receive additive input from L- and M-cones to both center and surround of the receptive field and give nonopponent light responses; the precise circuitry for red-green opponency thus remains elusive. Finally, the first recordings from two identified amacrine cell types have been made in the macaque: the spiking, axon-bearing A1 cell and the AII amacrine cell, a key amacrine type in the rod-signal pathway. Both amacrine types receive additive input from L- and M-cones to all parts of the receptive field and can be excluded from any role in a cone-opponent transformation in the inner retina. The great majority of identified amacrine cell types still remains to be studied, so no conclusions can yet be reached about their specific roles in the generation of opponent circuitry. It is likely, though, that the list of physiologically and anatomically identified retinal cell types will continue to grow and to provide new insights into the basic structure and function of the primate retina and its role in color vision.

Acknowledgments

We thank Keith Boro and Toni Haun for technical assistance. We are especially grateful to Joel Pokorny, Vivianne Smith, Donna Stafford, and Steve Buck for their collaborative contributions, and to Kate Mulligan for her continued support. The research reported here was funded by NIH grants EY06678 (D.M.D), EYO1730 (Vision Research Core), and RR00166 to the Regional Primate Center at the University of Washington.

10

Receptor inputs to primate ganglion cells

Barry B. Lee

The receptive field structure of retinal ganglion cells is a major determinant of the structure of signals used in visual perception. This chapter is concerned with physiological evidence as to receptor inputs to centers and surrounds of ganglion cell receptive fields. Anatomy and modeling can suggest what to look for in cells' receptive fields, but physiological measurement provides the critical test of such predictions. The organization of primate ganglion cell receptive fields, and their receptor inputs, has not yet been fully resolved (see Chapters 7, 8, and 9), but it is critical when considering the origin of chromatic signals emanating from the retina. The purpose of this chapter is to summarize physiological evidence as to receptor input and the spatial extent of receptive fields. The size of chromatic receptive fields is small enough that the optics of the eye must be taken into account, as well as methods of field measurement. Some functional implications of receptive field structure are also considered. More extensive reviews of structure and function may be found elsewhere (Kaplan, Lee, & Shapley, 1990; Merigan & Maunsell, 1993; Lee, 1996).

Ganglion cell types

Cell systems projecting through the lateral geniculate nucleus (LGN) to the visual cortex are usually divided into the magnocellular (M) and parvocellular (P) pathways. Parasol retinal ganglion cells project to the magnocellular layers of the LGN. These M-cells may be of ON- or OFF-center types. ON- and OFF-center cells have dendritic trees laminating in the inner and outer sublayers of the inner plexiform lamina (IPL), respectively, as in other species (Famiglietti & Kolb, 1976). Thus two separate mosaics (ON and OFF) of M-cells are present. Centers and surrounds receive input only from middle- (M-) and long- (L-) wavelength–sensitive cones. There is good evidence that this system forms the physiological substrate of a psychophysical luminance channel in flicker photometry and similar tasks (Lee, Martin, & Valberg, 1988; Kaiser et al., 1990).

Midget ganglion cells project to the parvocellular layers of the LGN, and almost all show cone opponency, with excitation from the M- and inhibition from the L-cone, or vice versa (Derrington et al., 1984; Lee et al., 1987). Within the central region of the retina, there are one ON- and one OFF-center ganglion cell per cone, forming two mosaics (Wässle & Boycott, 1991). Consequently, red and green ON-center cells (together with their bipolar cells) share the inner mosaic, and their OFF-center counterparts share the outer mosaic. This mixture of functional properties (e.g., red ON-center and green ON-center) within the same ganglion cell mosaic is unusual in mammalian retinas. The M/L-cone opponent cells are likely to form the physiological basis for the detection of chromatic modulation in a red-green direction, an issue that will be discussed in more detail in a later section.

Pathways carrying information from S-cones have usually been subsumed into the P-pathway but should probably be considered a separate retino-geniculo-striate system. Two cell types are present, receiving either excitatory or inhibitory input from the short-wavelength–sensitive (S) cone. The S-cone input is opposed by some combination of the other two cone types. Those cells receiving excitatory S-cone input have been identified as small-bistratified cells (Dacey & Lee, 1995). Those cones receiving inhibitory S-cone input form a very distinctive cell class physiologically (Valberg, Lee, & Tigwell, 1986) but have not yet been anatomically identified. The small-bistratified, S-cone excitatory cells and the S-cone inhibitory cells are likely to provide the physiological basis for detecting chromatic modulation involving the S-cone.

It has been argued that color vision involving this cone is phylogenetically more ancient than M/L-cone opponency (Mollon, 1991). Consistent with this view, small-bistratified ganglion cell axons probably terminate in the intercalated layers rather than in the main parvocellular laminae of the lateral geniculate nucleus (Hendry & Yoshioka, 1994). The term "P-cells" will be used here only to refer to the M,L-cone opponent cells.

These eight cell mosaics give rise to the great majority of axons projecting to the LGN, and they are the subject of this chapter. There are other, sparser mosaics present, probably projecting to other LGN relay cells (Hendry & Yoshioka, 1994) as well as to the superior colliculus and midbrain structures. Detailed information about these other cell types is not yet available.

The data cited here were obtained either through recordings in the LGN or directly from the retina. The former approach has the advantage of unambiguous cell classification based on the lamina of recording, at least as far as P- and M-cells are concerned. Retinal recording offers the advantages of greater physiological and mechanical stability and a very flexible choice of recording sites on the retina. However, cell classification has to be based on physiological response properties.

P-cells: cone inputs and center size

Anatomically, a single cone appears to provide input to a single midget ON-bipolar and thence to a single midget ON-center ganglion cell (Wässle & Boycott, 1991). Similarly in the OFF-pathway, the same cone connects to a midget OFF-bipolar and a midget OFF-center ganglion cell. If this anatomical inference is correct, the center size of M,L-cone–opponent P-cells should be determined by the sampling aperture of a cone, rather than by the ganglion cell dendritic tree diameter. Dendritic tree size is usually closely correlated with center size (e.g., Peichl &Wässle, 1979), since the dendritic tree is the main site of spatial summation for most ganglion cells, but this is not the case in the midget system.

Single-cone centers have consequences for coverage of the retina by the receptive field mosaic, and for processing of spatial information. This is illustrated in Fig. 10.1 for parafoveal retina.In the upper panel is sketched a one-dimensional parafoveal cone array, based on anatomical measurements of cone density and inner segment diameter from the literature (Packer, Hendrickson, & Curcio, 1989). Sampling apertures of cones at 3.8-deg eccentricity can be described by a Gaussian profile with a radius of ca. 0.32 min of arc (MacLeod, Williams, & Makous, 1992), giving the profile array shown in the middle panel. With such nonoverlapping receptive fields, the location and intensity of stimuli A–C cannot be deconfounded. To be able to localize stimuli with a precision of a fraction of the cone diameter, as in the hyperacuities, integration and comparison over small groups of cones are necessary.

It is plausible that the necessary signals for such comparisons are normally provided at the retinal level by ganglion cell receptive field overlap. This overlap can be expressed as the coverage factor (the number of receptive fields looking at a given point in the visual field), which is usually 3–6 (Wässle & Boycott, 1991). The lower sketch shows how such an array, with the same density as the cone array and a coverage factor of 4, can distinguish the stimuli A–C by comparison of the activity of the neighboring cells A'–C'. The den-

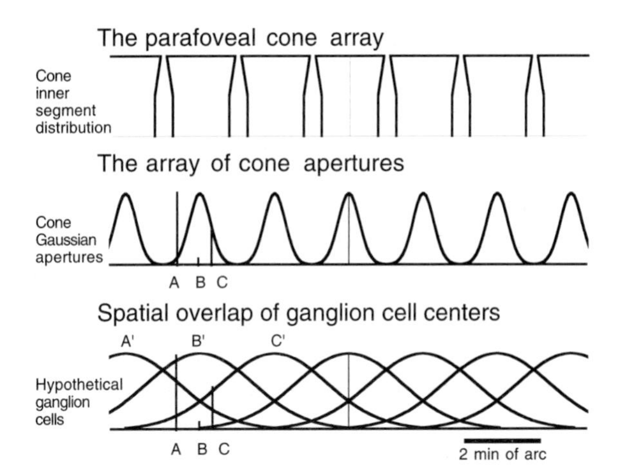

Figure 10.1: Single-cone centers and spatial vision. For a parafoveal cone array, the upper panel sketches the anatomical distribution of inner segments (Packer, Hendrickson, & Curcio, 1989). Diameter is ca. 1.6–1.8 min of arc. Psychophysical measurements of cone sampling aperture suggest a Gaussian profile with a standard deviation of 0.3–0.4 min of arc in the parafovea (MacLeod, Williams, & Makous, 1992), and the corresponding array of profiles is drawn in the center panel. An array of such cells or cones would be unable to distinguish between the three stimuli A, B,C since only a single cone is activated. An array of ganglion cells of the same density that have larger receptive fields to give a coverage factor of 3–4 (Wässle & Boycott, 1991) can distinguish the stimuli A,B,C by comparing activities of ganglion cells A',B',C'.

dritic trees of ganglion cells would seem a good location for generating such overlapped signals, before any noise is added to the visual signal by the spike generation mechanism.

Why has the midget system apparently abandoned typical ganglion cell coverage factors of 3–4 for a coverage factor of 1 (or less, since rods are found between cones outside the fovea)? One possibility is that the purpose of the midget connectivity pattern is not to make very small receptive fields but to preserve a cone-specific path into the inner retina. However, an additional complication is generated by the optics of the eye, which will spread the image of any point source onto several cones, and so effectively enlarge the center size. The extent of this enlargement is discussed below.

Similar arguments to those sketched in Fig. 10.1 apply through the central retina, including the fovea. At greater eccentricities, neighboring midget ganglion cells begin to receive input from multiple midget bipolars and thus multiple cones. However, they still show no overlap of their dendritic fields (Dacey, 1993b). The reasons for this interesting morphology are uncertain, as is the extent of M,L-cone opponency in peripheral retinas (see Chapter 9).

Single-cone centers are often cited in the literature as if they are physiologically demonstrable, but measurements of P-cell center size have yielded much larger diameters. Figure 10.2A shows estimates of center size gathered from the literature. They are plotted as center Gaussian radius as a function of retinal eccentricity. There is reasonable agreement among the various estimates. In most studies, gratings have been used to map receptive field dimensions. Gaussian center radius was derived by fitting a difference-of-Gaussians model to spatial frequency tuning curves (Derrington & Lennie, 1984; Croner & Kaplan, 1995). In other studies, only cell visual resolution was provided (Blakemore & Vital-Durand, 1986; Crook et al., 1988), but it is possible to estimate Gaussian center radius from this measure (Peichl & Wässle, 1979; see below). Alternatively, center radius may be mapped directly by using discrete stimuli such as spots or edges (de Monasterio & Gouras, 1975; Lee, Kremers, & Yeh, 1998).

Also included in Fig. 10.2A are estimates of Gaussian radii of single cones (cone prediction curve), based on anatomical measures of inner segment diameter (Packer, Hendrickson, & Curio, 1989) and psychophysical results (Coletta & Williams, 1987; MacLeod, Williams, & Makous, 1992; MacLeod & He, 1993). The cone curves have been terminated at 10 deg, since beyond this eccentricity the strict midget connectivity pattern breaks down. Physiological Gaussian center radius estimates are much larger than expected on the basis of a single-cone input. A major contributory factor will be optical blur, and it is thus desirable to estimate how far single-cone centers might be enlarged by optical factors.

Before considering the effect of the eye's optics, it is useful to consider the relation between center size, cell resolution, and measurement techniques. Figure

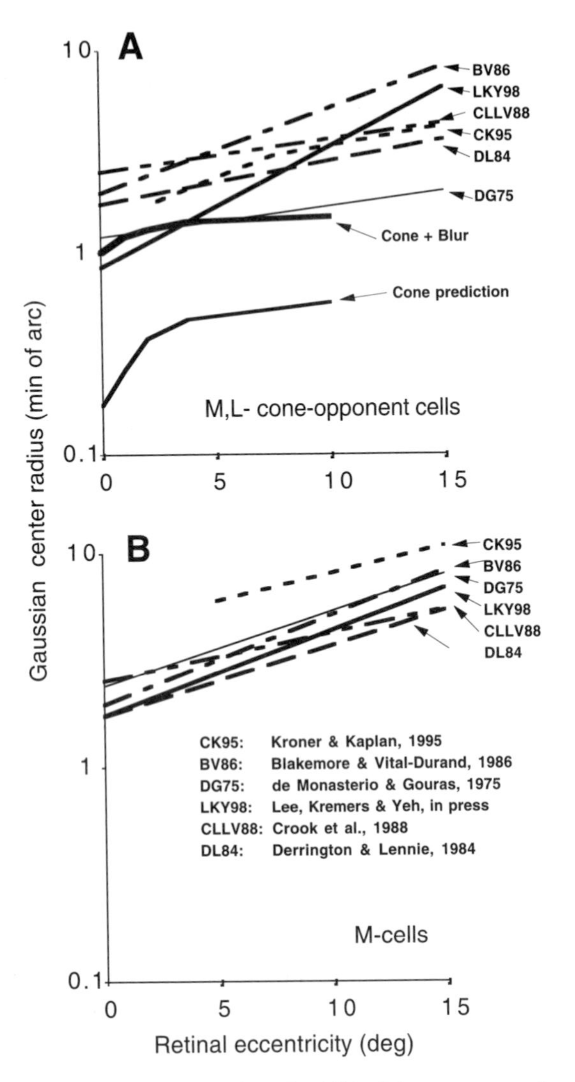

Figure 10.2: (A) A collation of published data as to receptive field center Gaussian radii as a function of retinal eccentricity for P-cells. For data in which the function $\sigma_{Ctr} = \exp(mx + c)$ (where σ_{Ctr} is center radius and x eccentricity) was fitted, this function has been used. In other cases, the function was fitted to published data, except for Croner and Kaplan (1995), where their median values were taken. For the single-cone curve, Gaussian radii have been estimated from anatomical measures of inner segment diameter (Packer, Hendrickson, & Curcio, 1989) in relation to psychophysical estimates of cone Gaussian radius from MacLeod, Williams, and Makous (1992). The cone + blur curve was derived as described in Fig. 10.4. (B) Published data as to receptive field Gaussian radii as a function of retinal eccentricity for M-cells. Estimates are quite consistent, except for those of Croner and Kaplan (1995).

10.3A shows a spatial frequency tuning curve derived from a difference-of-Gaussians profile. The spatial frequency axis has been expressed in cycles per Gaussian center radius. It can be seen that the spatial frequency cutoff of the cell (defined in this example when responsivity falls one log unit from the peak) occurs when about one grating period corresponds to the Gaussian radius. In fitting data with a difference-of-Gaussians model, the center radius parameter is largely determined by the falling limb of the tuning curve, that is, the visual resolution of the cell. Gaussian center radius could thus be estimated from cell resolution in those studies when only these data are provided (Blakemore & Vital-Durand, 1986; Crook et al., 1988); at the cell's resolution limit about one period length corresponds to the radius. At the resolution limit about three grating cycles are expected to fit into the field center (Fig. 10.3B).

Methods of determining center size using discrete stimuli include area summation curves, small spots, or edges. For example, if an area summation curve is constructed using spots of different sizes, cell threshold decreases as a function of stimulus area, reaches a minimum, and then may increase again due to surround antagonism, as sketched in Fig. 10.3C. The so-called "equivalent center size" (Peichl & Wässle, 1979) is calculated from the spot area defined by the intersection of the straight lines derived from the slope of the linear part of the curve and the minimum threshold. The equivalent center corresponds to a spot of a diameter of about two Gaussian radii. The spot diameter that fills the center to yield minimum threshold and maximum responsivity (full center) corresponds to about three Gaussian radii, as sketched in Fig. 10.3D.

The optical transfer function of the eye as a function of retinal eccentricity is currently only available for the human eye (Navarro, Artal, & Williams, 1993), and one must assume that the macaque eye has similar optics. Figure 10.4A shows the optical modulation transfer function (MTF) of the eye for three selected eccentricities. Figure 10.4B shows the expected effect of optical blur on spatial frequency tuning for a hypothetical midget ganglion cell from the fovea. The single-cone center curve shows the spatial frequency

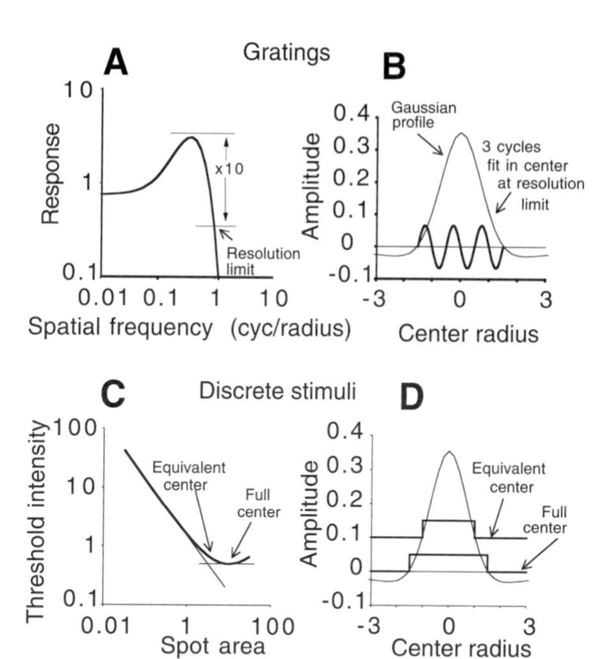

Figure 10.3: Methods of determining center size. Center size may be derived from spatial frequency tuning curves to sinusoidal gratings, by fitting a difference-of-Gaussians model. (A) A cell spatial frequency tuning curve with spatial frequency expressed in cycles per Gaussian radius. The cell's spatial frequency cutoff (one log unit below peak responsivity) occurs when one cycle fits the Gaussian radius. Transferring this onto a sketch of the receptive field profile (B) shows that the cell responds until about three cycles fit into the center. Center size may also be determined with discrete stimuli. (C) For example, area summation curves may be constructed by plotting cell threshold against stimulus area. Space has again been expressed in terms of Gaussian radius. The equivalent center is defined from the intersection of the straight-line segments drawn in on the figure. The full extent of the center can be estimated from the lowest threshold point. (D) The sketch shows these dimensions mapped onto a receptive field.

tuning curve expected for a one-cone center. A 5-min-of-arc radius surround has been assumed, although this is not critical. The cell responds up to several hundred cycles per degree. This is consistent with aliased fringes in the foveal cone mosaic, which are visible up to these frequencies (Williams, 1985). The tuning curve is multiplied by the optical transfer function, and the result is shown as the "one cone + blur curve," which has a resolution limit of ca. 40–50 cyc/deg at one log unit down from the peak. Refitting with a Gaussian model gives an estimate of the center radius after optical blur has been taken into account.

This analysis was applied at different retinal eccentricities and the resulting curve is drawn into Fig. 10.2A (cone + blur). The predicted center radii now form a lower bound to the physiological data. If the macaque eye, with its smaller size, is optically inferior to the human eye, then this curve would be closer to the grating data. It should be stressed that these optical limitations apply to the intact eye, not just to those prepared for physiological purposes. A similar analysis can be applied to the use of discrete stimuli like spots and edges, with similar results.

The Gaussian fit in Fig. 10.4B is not very satisfactory. In particular, the descending limb of the cone + blur curve is shallower than the very steep curve expected of the difference-of-Gaussians model, because it is largely determined by the optical MTF. In published data Gaussian fits usually appear satisfactory. However, in earlier measurements we (Crook et al., 1987) did find that the descending limbs of some P-cell tuning curves were shallow, while others had the steeper slope characteristic of the difference-of-Gaussians model.

This analysis suggests that single-cone centers do not confer a great advantage to spatial vision due to their effective enlargement through optical blur. This supports the hypothesis that single-cone centers may represent a development designed to provide a cone-specific signal to the inner retina, rather than a development for spatial vision.

In addition to optical factors, with receptive fields only a few cones across, field structure is likely to interact with the distribution of cones in the mosaic. Figure 10.5 shows a patch of simulated cone mosaic with cones randomly assigned as L or M in a proportion of 2:1. By chance, some cones are surrounded by cones of the other type. It would seem likely that the corresponding midget ganglion cell must have a center derived from just that cone. Other cones are surrounded by cones of the same type. Cone-specific coupling at any stage through the retina might result in a degree of "neural blur," leading to an enlargement of

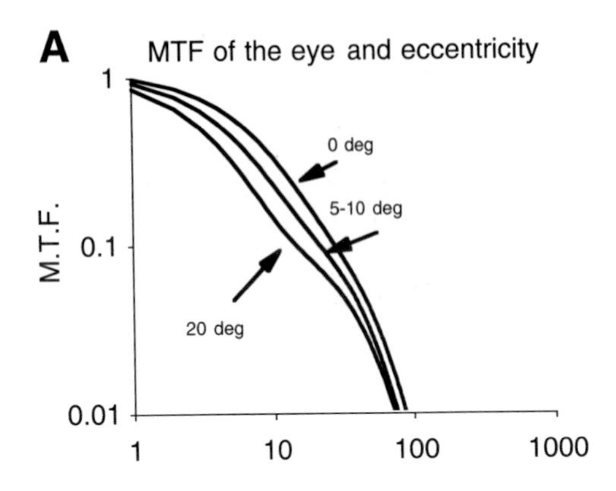

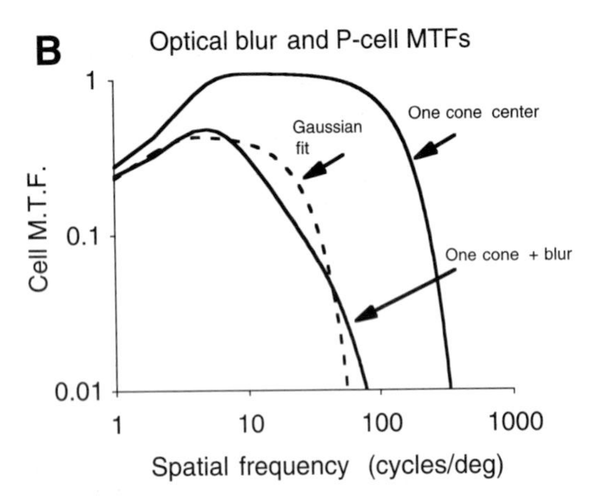

Figure 10.4: (A) The optical transfer function (MTF) of the human eye as a function of spatial frequency, for three different eccentricities (Navarro, Artal, & Williams, 1993). (B) Hypothetical cell tuning curves. Three curves are shown: that expected of a single-cone center with a surround of Gaussian radius 5 arc min; a curve indicating the effect of the optical MTF on such a field; and an attempt at a Gaussian fit to the cone + blur curve.

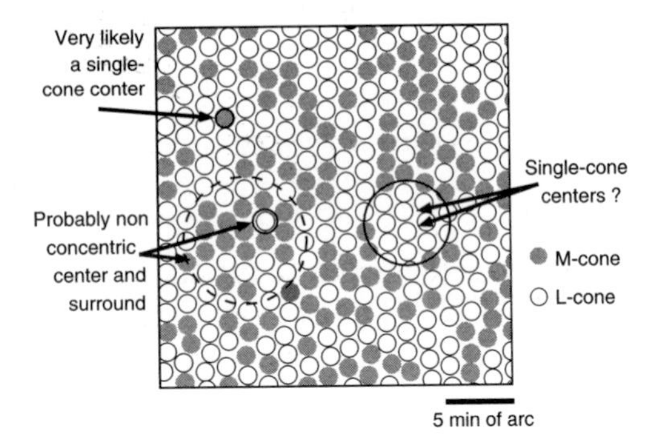

Figure 10.5: Possible interactions of field structure with the cone mosaic. A sketch of P-cell receptive field structures superimposed on the cone mosaic. A patch of parafoveal mosaic (ca. 4 deg) was simulated and cones randomly assigned as L or M in a ratio of 2:1. A single cone of one type surrounded by a patch of the other type seems very likely to lead to a single-cone center. On the other hand, other cones sit among cones of the same type or a mixed population. Any cone-specific interactions through the retina will complicate the receptive field structure and perhaps enlarge centers beyond a single-cone diameter. The grain of the cone mosaic may also cause the center and surround mechanisms to be nonconcentric.

the physiological field size beyond single-cone dimensions. Measurements of P-cell center radii show considerable intercell variability (Derrington & Lennie, 1984; Lee, Martin, & Valberg, 1989a), perhaps due to effects of this kind, although anatomical evidence for cone-specific connectivity in the IPL is so far lacking (see Chapter 8).

Thus, what appears from the anatomy as a straightforward result becomes a complex issue. Although a promising way to resolve center structure is to use interference fringes as stimuli, and so bypass the eye's optics, these experiments have yielded more complex data than anticipated (McMahon et al., 1995).

M-cells: cone inputs and center size

Published data as to the Gaussian center radius of M-cells of the macaque is plotted as a function of eccentricity in Fig. 10.2B. Data from all sources are in good agreement, except for the larger estimates of Croner and Kaplan (1995), whose cell sample was small. It should be noted that M-cell center radii are small enough in the central retina to permit them to respond to high spatial frequencies and are scarcely larger than measurements for P-cells. Regression lines for the different studies intersect the ordinate at a Gaussian radius of 1.5–2 min of arc. As the Gaussian radius corresponds to one period at the resolution limit, in the fovea M-cells are expected to respond to 30–40 cycles/deg, which is comparable to the behavioral visual resolution of the macaque (e.g., Cavonius & Robbins, 1973).

In relation to cone density (Packer, Hendrickson, & Curcio, 1989), a center diameter three times the Gaussian radius would be about 6–8 cones across, so that a center would cover about 30–40 cones in the fovea and parafovea. This is comparable to anatomical estimates (Goodchild, Ghosh, & Martin, 1996). Direct comparison of center size with the dendritic tree of the same cell is consistent with this interpretation (Dacey & Lee, unpublished observations).

This is illustrated in Fig. 10.6A, which shows a patch of cones hexagonally packed to mimic the cone distribution in the fovea. A center three times the M-cell center Gaussian radius has been drawn in, and the number of cones under the center can be counted. However, with a center of Gaussian profile the cones in the middle of the center will dominate its spectral sensitivity. There is some variability in M-cell spectral sensitivity (Kaiser et al., 1990), and it is worthwhile to consider whether this variability is consistent with a random sampling of M- and L-cones through a Gaussian center.

Microspectrophotometric mapping of patches of the cone mosaic has indicated that cone distributions are random, rather than, for example, a regular alternation (Mollon & Bowmaker, 1992). Each cone in the matrix was therefore randomly assigned as an L- or M-cone. Each cone was given a weight appropriate to its distance from the center of the Gaussian, and the integrated L:M weighting was calculated over the whole center. This was repeated for a large sample of fields set on the cone matrix. Figure 10.6B shows measured cone weightings from a sample of M-cell centers (22) compared with the distributions of cone weightings expected on the basis of L:M ratios of 2:1 (solid curve) and 1:1 (dashed curve).

The variability in weightings in the experimental data is clearly consistent with the M-cell center being drawn from a random 2:1 L:M-cone matrix, but it is inconsistent with that expected of a 1:1 L:M-cone ratio. This is a reflection of the fact that the spectral sensitivity of the M-pathway in the macaque is similar to the human $V(\lambda)$ function (Lee, Martin, & Valberg, 1988), which can be described with a 2:1 L/M-cone weighting ratio. The usual explanation for this weighting of the $V(\lambda)$ function is that it reflects cone numerosity (e.g., Pokorny, Smith, & Wesner, 1991). However, cumulative microspectrophotometric samples in the macaque are more in favor of an L:M ratio for the macaque of 1:1 rather than 2:1 (Bowmaker, 1991). This raises the interesting possibility that the 2:1 L:M cone weighting of the $V(\lambda)$ function reflects not only cone numerosity but is also influenced by the relative synaptic weights of M- and L-cones to M-cells (see also Chapters 1, 2, 4, and 5).

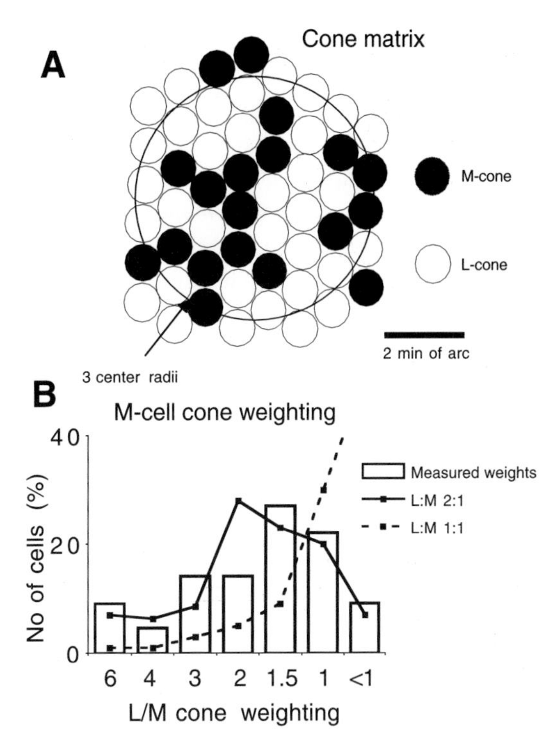

Figure 10.6: (A) M-cell cone weighting and the cone matrix. A patch of parafoveal cone mosaic with the receptive field size of an M-cell is superimposed on it. If the center size is taken as three times the Gaussian radius, then about 40 cones are touched by the field. (B) Predicted variability in M-cell spectral sensitivity obtained from superimposition of centers onto cone mosaics as in panel A, for 2:1 and 1:1 L:M-cone ratios, compared with data taken from Kaiser et al. (1990). The cell sample was restricted (N=22), and agreement is satisfactory for the 2:1 ratio case, but is clearly inconsistent with a 1:1 ratio.

S-cone cells

Data as to field size of cells with S-cone input are sparse. In vitro measurements (see Chapter 9) indicate that the S- and M/L-opponent cone inputs are coextensive and match the dendritic tree. In the peripheral retina, trees of parasol and small-bistratified cells are similar in size. This would mean that each small-bistratified cell receives input from just a few S-cones (2–5), assuming an S-cone fraction of 10–20% (see Chapter 8).

Receptor inputs to receptive field surrounds

Although in early work the M/L-cone opponency of P-cells was assumed to be generated by excitation from one cone and inhibition by the other (de Monasterio & Gouras, 1975), there has recently been much interest in an alternative model in which receptive field surrounds receive mixed-cone input (Paulus & Kröger-Paulus, 1983; Lennie, Haake, & Williams, 1991; see Chapter 8). This is very attractive from a developmental viewpoint, since with cone-specific input to the center conferred by the midget system no further wiring specificity would be required to provide an opponent signal.

With full-field stimuli, it is not possible to distinguish between cone-specific and cone-mixed surrounds from response amplitude data alone (Lennie, Haake, & Williams, 1991). By using both amplitude and phase data, it is possible to distinguish between the two alternatives because of center-surround latency differences, and such an analysis supported the cone-specific model (Smith et al., 1992a). The direct method of distinguishing the two hypotheses is to map the receptive field structure with cone-isolating stimuli. Reid and Shapley (1992) made such maps, and their data favored cone-specific surrounds. However, their maps only had a resolution of 7 arc min, which is low in comparison to the grain of the cone mosaic.

The physiological evidence against mixed surrounds stands against the lack of an anatomical substrate (see Chapters 7 and 8). However, further evidence in favor of the cone-specific surround model (Lee et al., 1997a) is shown in Fig. 10.7. We reinvestigated field structure on a finer spatial scale by using a bipartite field stimulus in which fields on either side of an edge were modulated in counterphase (Lee, Kremers, & Yeh, 1998). For a spatially linear cell, a response null will occur when such an edge bisects the receptive field. From the response amplitude and phase, as the edge is stepped away from the null, receptive field properties may be deduced. Receptive fields were mapped using luminance, chromatic, and M- and L-cone–isolating stimuli. Figure 10.7A sketches the maps expected from a red ON-center cell for cone-specific and cone-mixed models. For the cone-specific model, the response amplitude increases monotonically on either side of the null under both M- and L-cone–isolating conditions, which is consistent with an excitatory L-cone center of smaller diameter than an inhibitory M-cone surround. There is no center-surround structure manifest in the L-cone response, but such a structure immediately becomes apparent when luminance-modulated stimuli are used; response peaks occur on either side of the null because if the edge is just to one side of the center, the surround reinforces the center response.

With the cone-mixed model with a 2:1 L:M-cone ratio, a center-surround structure appears for the L-cone, and the luminance modulation condition indicates strong center-surround interaction. Parameters were constrained to yield similar full-field responses as for the cone-specific surround.

Figure 10.7B shows actual data obtained from a red-ON-center cell. The pattern of responses clearly resembles the cone-specific model. The solid curves represent the fit of the model to the data, the same parameters being used for all four conditions. The model also predicts a response phase. Data at higher temporal frequencies (when center-surround delays become significant) also support a cone-specific model.

This analysis provides further physiological support for the hypothesis of cone-specific surrounds, accentuating the conflict with anatomical data. No substrate for cone-specific surrounds can be found in

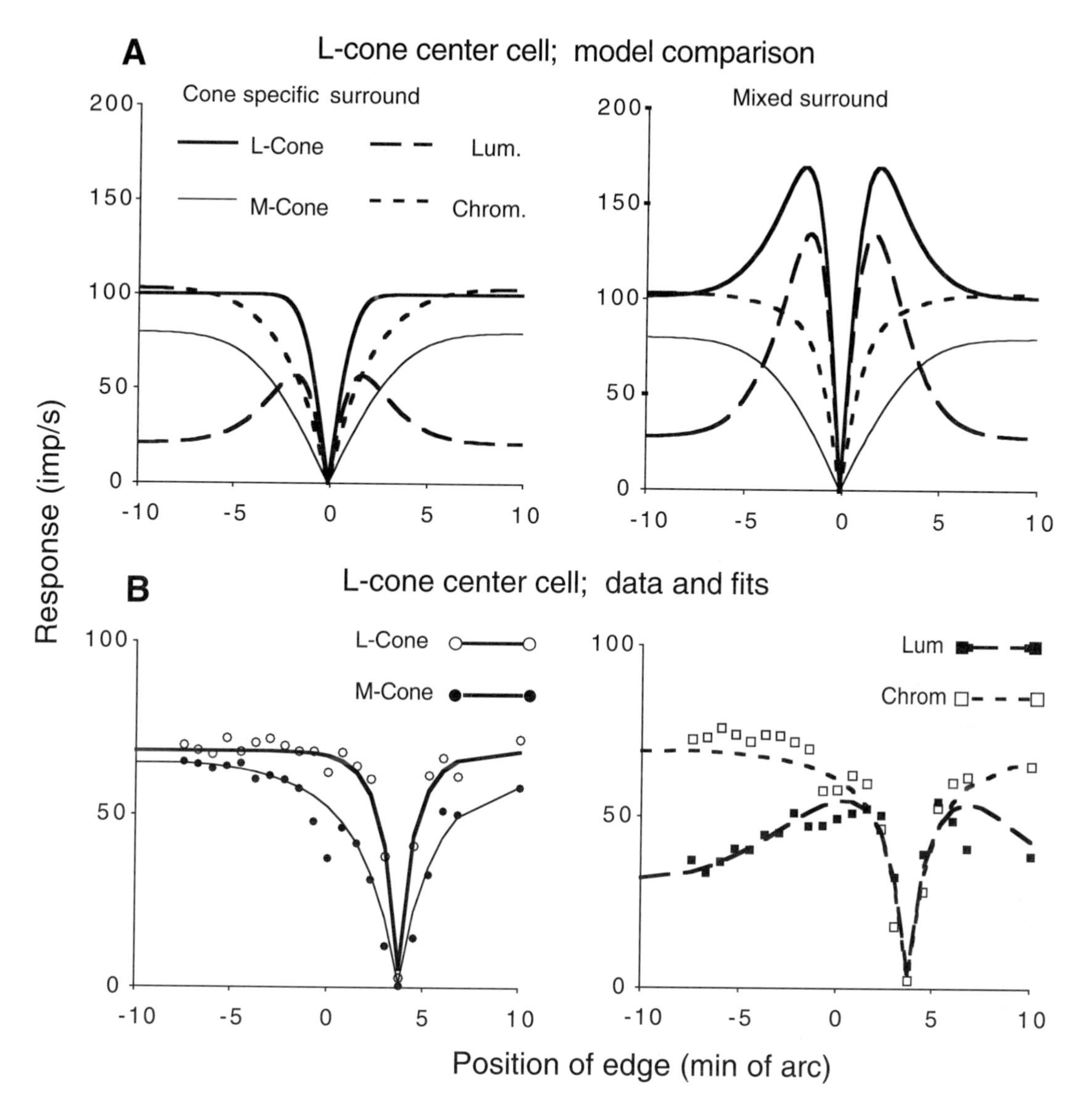

Figure 10.7: (A) Predictions of two different models for the surround of P-cells. The stimulus is a bipartite field, counterphase-modulated at low frequency. Four different stimulus conditions are considered: L-cone–isolating and M-cone–isolating, luminance, and red-green chromatic modulation. A response null is expected when the edge bisects the field. For the cone-specific model, no center-surround organization is visible except for luminance modulation. Center-surround organization is visible as a response peak when the edge is situated just to one side of the center, since then the surround response to the other side of the edge reinforces the center response. For the mixed surround, model parameters have been constrained to give the same full-field stimulus amplitudes. The L-cone response shows indications of center-surround organization, and a strong center-surround structure is seen in the luminance response. (B) Actual data and model fits for a L-cone center cell. They correspond to the cone-specific pattern. Data for the four conditions were obtained simultaneously, and the same model parameters were used for all data sets. Cell eccentricity was 3.8 deg; modulation frequency was 2.44 Hz. Six seconds of activity were averaged for each data point. First-harmonic amplitudes were extracted by Fourier analysis of the response histograms. For further details, see Lee et al. (1998).

the outer plexiform layer. Horizontal cells of the primate retina consist of only two types (Dacey et al., 1996). The HI cell collects input indiscriminately from M- and L-cones, while avoiding S-cones, and is likely to be associated with the triadic contacts associated with midget bipolars (see Chapter 9). Although this could provide a substrate for mixed surrounds, receptive fields of HI cells are larger than the surround mechanism in Fig. 10.7B (Dacey, personal communication). Cone-specific surround could be located in the IPL, but attempts to find an anatomical substrate have so far been unsuccessful (Calkins & Sterling, 1996; see Chapter 8).

Although the physiological data are against the strong form of the cone-mixed model, in which surrounds draw their input indiscriminately from all cones in the neighborhood, a smaller degree of cone mixture would be difficult to detect physiologically, and so a weaker form of the hypothesis, with some mixing, would be physiologically tenable and presumably compatible with the anatomy. This alternative, however, is less attractive from a developmental viewpoint. Lastly, it is worth noting that cone-specific surrounds represent the most efficient means of generating a cone-opponent signal. When expressed in terms of cone contrast, the contrast gain of P-cells to chromatic modulation is similar to that of M-cells. This implies that both cell types make close-to-optimal use of cone signals. With cone-mixed surrounds, a large part of the center cone response would be canceled by its surround contribution, which seems an inefficient use of cone signals. Although subtraction of the surround from center signals occurs in many types of ganglion cells, it usually serves to enhance and sharpen the response to contrast borders. These local contrast enhancement effects are difficult to demonstrate in chromatic vision.

Such measurements as those in Fig. 10.7 give an estimate of the surround radius of P-cells. On average, surrounds were only a few times larger than the center. This is small enough to cause some interaction with the grain of the cone mosaic. This is also illustrated in Fig. 10.5; a center of the midget ganglion cell associated with the cone indicated would draw its surround from a nonconcentric patch of cones. In the measurements of Fig. 10.7, it was common for the fields of M- and L-cones to be displaced. Just in the one dimension tested, 50% of cells showed displacements greater than 1 min of arc and 14% greater than 3 min of arc. The original distinction between Type I and Type II P-cells is now thought to reflect the extremes of a continuum (Derrington & Lennie, 1984). The variability in field structure implied by the Type I–Type II distinction may reflect receptive field patterns imposed on cells by the local distribution of cone types.

Receptor inputs to M-cell surrounds are less easy to define, since quite complex characteristics appear to be present. As well as an achromatic surround component (Derrington & Lennie, 1984; Croner & Kaplan, 1995), chromatic stimuli reveal additional, chromatic inputs. One appears to derive from an M/L-cone opponent signal (Smith et al., 1992a) and is presumably related to earlier reports in which large red fields suppress responses of M-cells (Wiesel & Hubel, 1966). A second harmonic response is also present to chromatic modulation (Lee, Martin, & Valberg, 1989a). It appears more complex than a simple nonlinearity of cone summation, since it is present when cone-isolating stimuli are used (Lee et al., 1993a). This might suggest that it derives from the nonlinear summation of inputs that are themselves cone-opponent. The relation (if any) between these two chromatic inputs and their anatomical substrates remains obscure.

For the S-cone pathway, the dendritic tree of the small-bistratified cell ramifying in the outer sublamina of the IPL would seem a likely site for inhibitory input from M- and L-cones through one of the diffuse bipolar systems (Dacey & Lee, 1994). However, as noted above, S-cone and M/L-cone fields seem coextensive (Type II, Dacey & Lee, 1994), so that a center-surround structure is not present.

Rod input to primate ganglion cells

As well as cone inputs, ganglion cells receive signals from rods. Reports in the literature about rod input to primate ganglion cells have been mixed. Wiesel and

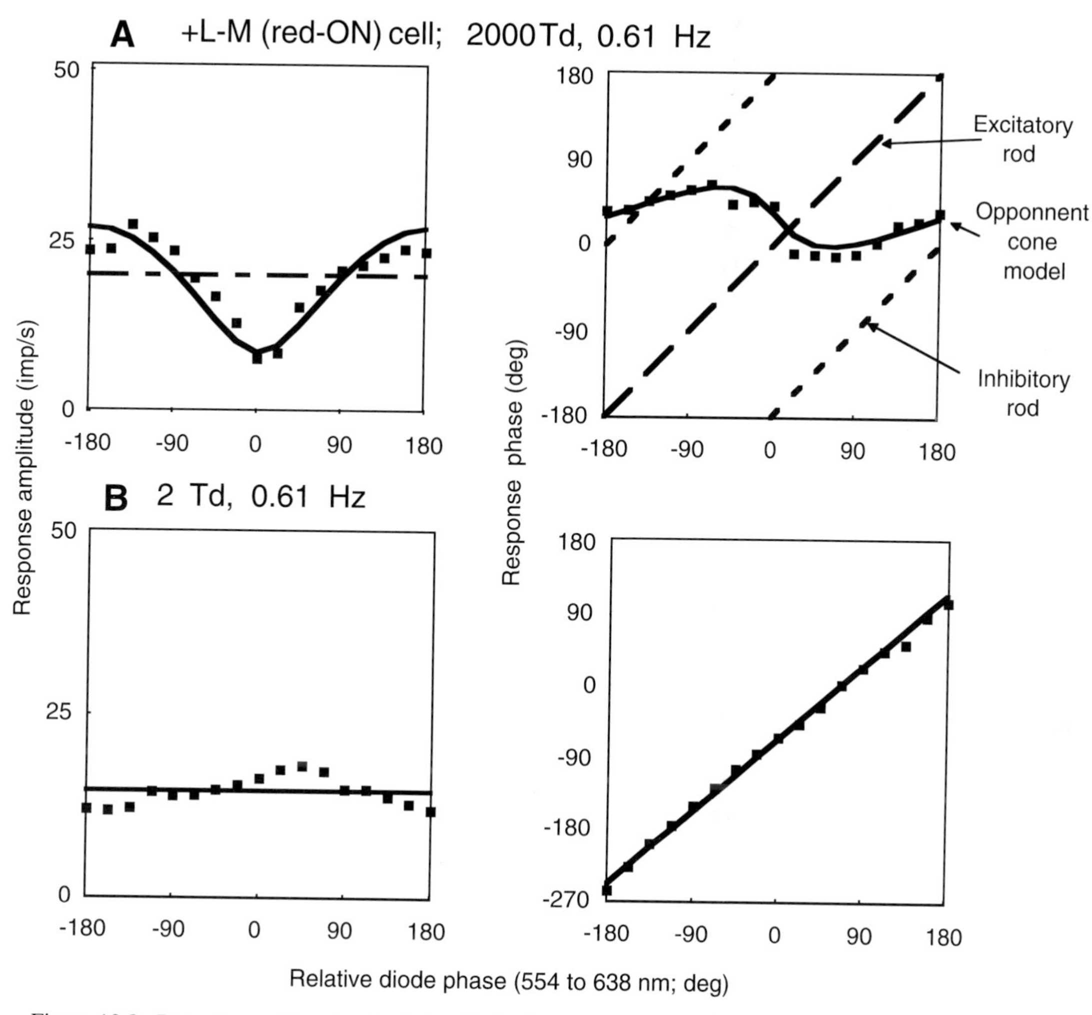

Figure 10.8: Detecting rod input and relative diode phase. Response amplitude and phase on altering the relative phase of heterochromatically modulated lights provides a powerful means of detecting rod input. (A) Amplitude and phase of response of a +L–M red-ON cell fitted with an opponent cone model (Smith et al., 1992a). (B) If the cell were to become rod-driven, then only the green light evokes a response. The amplitude template becomes flat and the response phase falls along the diagonal, as sketched in the upper panels and shown to occur in the lower panels.

Hubel (1966) reported rod input to a fraction of P-cells and little rod input to M-cells, whereas Gouras and Link (1966) reported strong rod input to M-cells. Virsu and Lee (1983) reported strong rod input to M-cells and variable rod input to P-cells. The data of Purpura, Kaplan, and Shapley (1988) imply strong rod input to M-cells and little or no rod input to P-cells, although the authors did not measure spectral sensitivities (and thus could not ascertain if a Purkinje shift had occurred).

We have recently reexamined this issue by using a method capable of detecting the Purkinje shift even if responses are very weak (Lee et al., 1996b). Figure 10.8 illustrates results obtained by this method. Red (638 nm) and green (554 nm) lights were sinusoidally modulated, and the response amplitude and phase

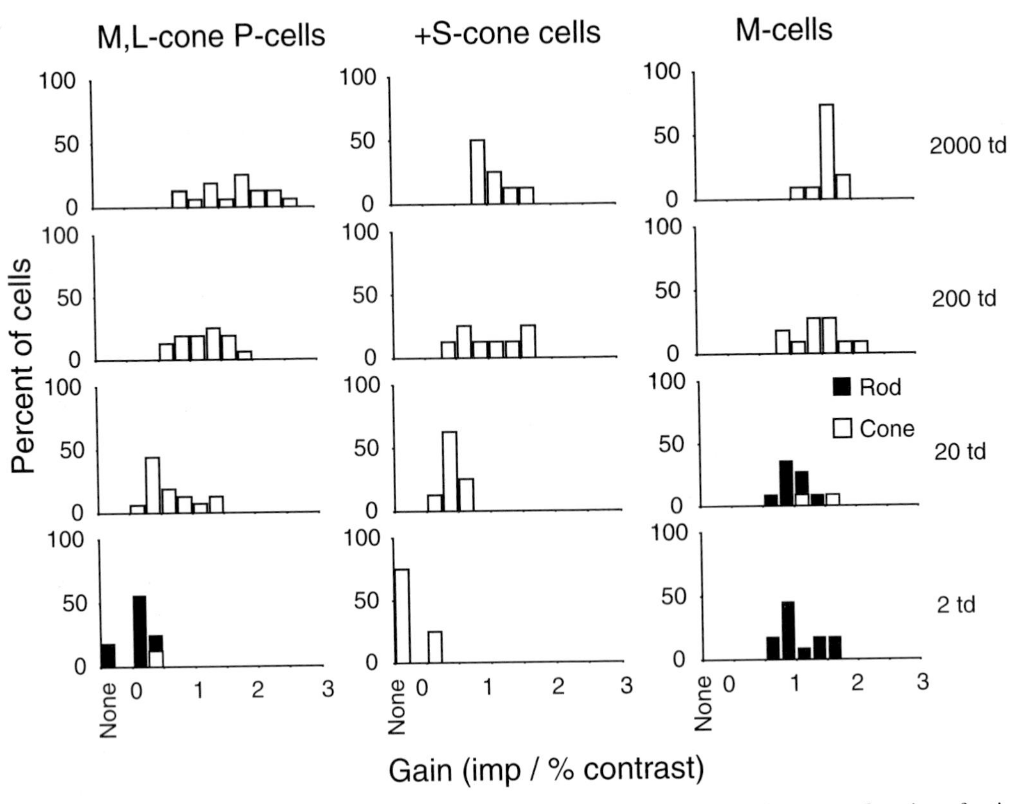

Figure 10.9: Receptor gain in terms of impulses per percent receptor contrast is shown as a function of retinal illuminance for different cell types. Receptor gain was derived from fits to data of curves as in the previous figure. Rod and cone inputs were identified as in the previous figure, and also checked using the phase paradigm and heterochromatic modulation of blue and green lights.

were measured as a function of the relative phase of the lights. If the responses are cone-driven, the response amplitude and phase can be modeled on the basis of addition or subtraction of M- and L-cone signals (Smith et al., 1992a). This is shown in Fig. 10.8A, which contains data and model fits from a red-ON cell at 2000 td. Rods are only activated by the 554-nm light, so that if the response is rod-driven, amplitude becomes independent of diode phase and the response phase as a function of diode phase acquires a slope of 45 deg, as is drawn in on the 2000-td data. At 2 td in Fig. 10.8B the red-ON cell has become rod-driven, although the response is weak. The advantage of this method is that response amplitudes can be very weak and noisy, but the response phase will still deliver a characteristic signature if rod input is present.

The strength of rod input to P-cells, M-cells, and blue-ON cells, expressed in terms of receptor contrast gain, is summarized in Fig. 10.9 as a function of retinal illuminance. The gain of M,L-cone–opponent P-cells decreases with retinal illuminance and a shift to rod spectral sensitivity occurs at 2 td, although responses are very weak, with only a few impulses per second response to 100% contrast. Rod input to blue-ON cells was very difficult to detect, but rod input to M-cells was always strong, cutting in at about 20 td, and receptor gain remained similar throughout the illuminance range tested. These data go some way to reconcile conflicting reports in the literature; there is rod input to P-cells as Wiesel and Hubel reported, but it is very weak, as suggested by other data.

The vigorous response of M-cells at scotopic levels presumably passes through the AII amacrine system, as in other mammals (Wässle & Boycott, 1991). The weak response of P-cells could derive from other sources, such as rod–cone coupling (Schneeweis & Schnapf, 1995), although AII amacrine cells do make contact with midget ganglion cells (Grünert & Wässle, 1996). It is possible that the weak rod input to P-cells arises simply from the restricted size of their dendritic trees, resulting in few AII contacts. Alternatively, substantial interference of rods with chromatic mechanisms might be undesirable, and so rod intrusion into cone-opponent cells has been suppressed.

Rods do interact with color perception. They can interfere with color matching (Wyszecki & Stiles, 1982a) and affect perceived colors (Knight & Buck, 1993). It is still uncertain how rod intrusion into color vision can be reconciled with their weak and indiscriminate input to all P-cell types (e.g., excitatory rod input to both red-ON–center and green-ON–center cells) and their lack of input to blue-ON cells. In any event, it is difficult to see how the very few impulses that rods evoke in P-cells could be a convincing substrate for scotopic spatial vision, as argued on a sampling theory basis (Lennie & Fairchild, 1994). Summation over many P-cells might provide a signal to influence color perception, but such summation weakens sampling theory arguments in the case of spatial vision.

Receptor inputs and psychophysics

There is now compelling evidence that M-cells provide the physiological substrate for performance on a variety of photometric tasks such as flicker photometry (Lee, Martin, & Valberg, 1988; Smith et al., 1992a) and the minimally distinct border (Kaiser et al., 1990), and thus form the physiological basis of a luminance channel. Although it has been argued that a combination of P-cell activities could yield the $V(\lambda)$ function (Lennie, Pokorny, & Smith, 1993), this may be possible in only a few paradigms, such as the detection of small stimuli on an achromatic background (King-Smith & Carden, 1976; Crook et al., 1987). When a sensation is minimized or abolished at equal luminance, as, for example, in flicker photometry, it is much more difficult to construct a plausible model to synthesize P-cell activities to build a satisfactory mechanism (Lee, 1991).

As discussed in connection with Fig. 10.5, the averaged spectral sensitivity of macaque M-cells conforms closely to the human $V(\lambda)$ function, implying a 2:1 L-to-M–cone ratio. Microspectrophotometric evidence that points to a 1:1 ratio (Bowmaker, 1991) is thus puzzling; one solution would be if the 2:1 L:M ratio in the luminosity function represented a measure of synaptic efficacy rather than cone numerosity.

There is no evidence of any contribution of the S-cone to M-cell centers; a contribution based on relative S-cone numerosity, as might occur if M-cells sampled every cone within their dendritic fields, should be physiologically apparent. This is consistent with the lack of S-cone contribution to the $V(\lambda)$ function (Smith & Pokorny, 1975; see Chapter 2). It has recently been shown that HI horizontal cell dendrites avoid making contact with S-cones (Dacey et al., 1996). This precedent raises the possibility that the diffuse bipolar cells that provide input to M-cells might also be able to avoid S-cone contacts.

There is also good evidence that P-cells form the basis for a red-green chromatic detection mechanism (Lee et al., 1990, 1993a). In studies in which the weight of M,L-cone inputs were estimated by modulation in some kind of cone space, the balance of M,L-cone inputs is found to be closely matched, as illustrated in Fig. 10.10A (Lee et al., 1987). +M–L-cone and +L–M-cone cells have been plotted separately, and each has been further subdivided into ON- and OFF-center types. For ON-center cells the excitatory cone tends to be dominant and for OFF-center cells the inhibitory cone, but the distribution is continuous. A very similar distribution was reported by Derrington, Krauskopf, and Lennie (1984). The consequence of such balanced cone inputs is a maximal response to chromatic modulation and a minimum response to luminance modulation.

It is convenient to transform this result into an L,M-cone contrast space, for then it becomes possible to directly compare the properties of ganglion cells with psychophysical data. In these coordinates, luminance and chromatic modulation are represented by vectors in the directions indicated in Fig. 10.10B. Under certain conditions, psychophysical thresholds (squares) plotted in such a space can be described by combinations of straight-line segments, as shown in Fig. 10.10B, which is redrawn from Stromeyer, Cole, and Kronauer (1985). When cell thresholds are plotted in these coordinates, data such as those in Fig. 10.10C result (replotted from Lee et al., 1993a). Thresholds of M-cells (filled squares) are lowest in the luminance quadrant and can be described by a straight-line segment in this region. Thresholds of M,L-cone–opponent P-cells (open circles) are lowest in the chromatic quadrant, and their thresholds again fall on a straight-line segment of orthogonal slope to the line segments for M-cells. Combining the cells' threshold segments accounts well for the psychophysical data. Thus only midget ganglion cells of the P-pathway are required to support the M,L-cone–opponent chromatic detection mechanism of human vision; there is no need to postulate another cell type (Rodieck, 1991). In addition, thresholds of M,L-cone–opponent P-cells can account for the detection of large chromatic stimuli on a white background (Crook et al., 1987), stimulus conditions that are thought to reveal the activity of chromatic mechanisms (Sperling & Harwerth, 1971; King-Smith & Carden, 1976). Also, the adaptation behavior of M,L-cone–opponent cells shows many features attributed to second-site effects in the psychophysical adaptation of chromatically opponent channels (Yeh, Lee, & Kremers, 1996).

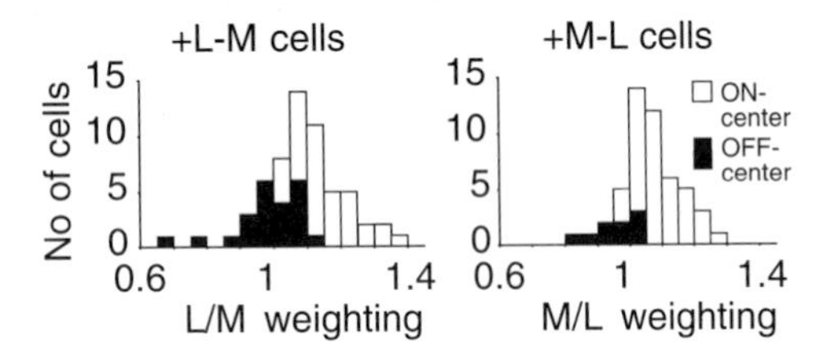

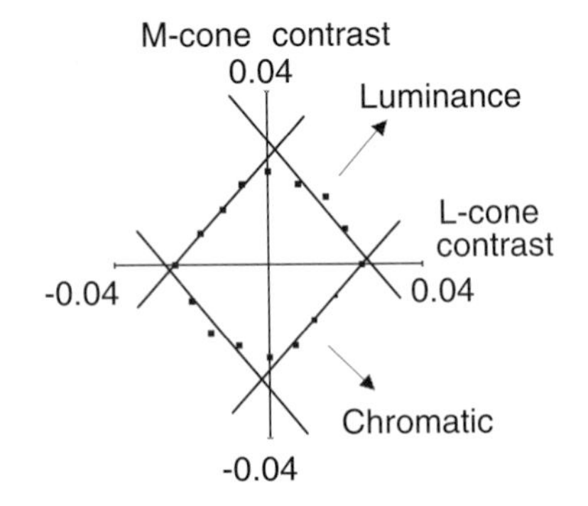

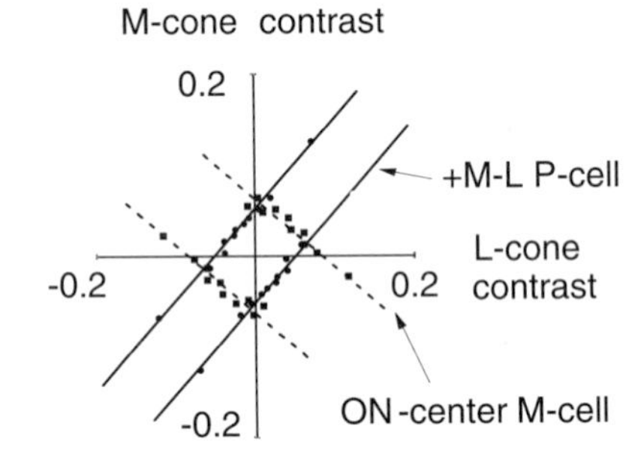

Figure 10.10: (A) Cone weightings of P-cells derived from Lee et al. (1987). For both ON- and OFF-center cells the center cone tends to be dominant but distributions show considerable overlap. These distributions result in a cell in which there is a maximal response to chromatic modulation and a minimal response to luminance. (B) Psychophysical thresholds plotted in M/L-cone coordinates resemble combinations of straight-line segments. (C) Plots of some kind of cell threshold measure in this cone space indicated that M-cells can be responsible for detection in the luminance quadrant and P-cells in the chromatic quadrant.

Cone and ganglion cell mosaics and signals: functional considerations

The anatomy and physiology of primate ganglion cells are closely linked to trichromatic vision, and it is interesting to consider the pattern of receptor inputs in an evolutionary context. The evolution of color vision is discussed elsewhere in this volume (see Chapters 1, 6, and 7). I have argued here that single-cone centers are unnecessarily small when compared to the eye's optics and are even disadvantageous for the localization of objects to a precision greater than the grain of the cone mosaic. This implies that the midget system evolved as a means of providing a cone-specific signal to the inner retina.

This is not necessarily inconsistent with the hypothesis that a high-resolution system for spatial vision developed into the M,L-cone–opponent system on the appearance of a second, long-wavelength pigment (see Chapter 7). Such a precursor of the midget system might have had only a few cones per ganglion cell dendritic field, especially in the rod-dominated retina of a nocturnal species. A degree of cone selectivity may then have occurred by sampling from a random cone mosaic. Any selective advantage conferred by such a rudimentary color signal could then promote the evolution of the midget system as a more reliable way of generating chromatic information.

If the midget system has become specialized for M,L-cone opponency, it becomes more likely that M-cells form the physiological substrate not only for a luminance channel in temporal tasks, but also for spatial vision. As was pointed out in relation to Fig. 10.5, these cells respond to gratings up to close to the psychophysical resolution limit, and combined physiological and psychophysical evidence suggests that they provide the physiological signal responsible for the hyperacuities (Lee et al., 1993b, 1995). Although M-cell sampling density is low in relation to the Nyquist limit for spatial vision, this may not be such a serious drawback as is usually assumed (Lennie, 1991). Each orientation column in area 17 handles a particular orientation for an elongated line or edge segment falling on a patch of retina, and so for that orientation the two-dimensional array of ganglion cells in the patch will collapse into one dimension. This may provide an adequate sampling density for gratings of the appropriate orientation. Other acuity tests, such as Landolt's C, only require the localization of a gap within one quadrant, which would not require such a high sampling density.

Conclusions

Measurements of the center size of M,L-cone–opponent P-cells have not confirmed input from a single cone, as is so strongly suggested by anatomy. A major reason for this discrepancy is blur due to the eye's natural optics. It remains to be seen, however, whether physiological factors might also enlarge centers beyond what might be expected from the interaction of optical blur with a single cone's sampling aperture.

M-cell centers are small enough to respond to several tens of cycles per degree in the fovea, about the same as P-cells. Random sampling of M- and L-cones by their dendritic trees should and does lead to a variability in their spectral sensitivity.

Maps of receptive field surrounds of P-cells suggest that they are cone-specific, or at least that they do not indiscriminately sample from all cones. However, anatomical data, which have failed to find a substrate for cone specificity, are in conflict with this hypothesis.

Rod inputs to M-cells are very pronounced, to P-cells they are very weak, and to S-cones they are excitatory; blue-ON cells are absent. This implies a major role for the M-pathway in scotopic spatial vision.

It seems likely that M-cells provide the physiological substrate for most if not all photometric tasks, implying that they form the foundation for the luminance channel in human psychophysics. Physiological properties of midget ganglion cells of the P-pathway strongly suggest that they form the physiological substrate of an M,L-cone–opponent detection mechanism of human psychophysics.

Acknowledgments

The author would like to thank Dennis Dacey, Daniel Osorio, Joel Pokorny, and Vivianne Smith for permission to draw on unpublished data. Paul Martin, Lukas Rüttiger, and Luiz Silveira offered useful comments upon the manuscript.

Part III: Cortical Processing

11
Parallel retino-cortical channels and luminance

Robert M. Shapley and Michael J. Hawken

There are two visual channels that connect the primate lateral geniculate nucleus (LGN) to the primary visual cortex, V1. These neural pathways are usually called P and M, after the parvocellular layers and the magnocellular layers of the LGN, through which they travel. Neurons in the different pathways have very different visual properties. Neurons in the P-channel are more sensitive for color than they are for black–white modulation; M-cells are just the opposite, with a higher sensitivity for achromatic than for chromatic patterns. There are several other functional differences: receptor inputs, sampling density on the retina, dynamics of response, and functional connections with the circuitry of the visual cortex (Shapley & Perry, 1986; Lennie et al., 1990b; Lennie, 1993).

One major outstanding issue is whether the psychophysically defined *luminance* channel is identical with the M-channel or whether both P- and M-signals contribute to what psychophysically is called luminance. In this chapter we address this issue by discussing the color-processing properties of P- and M-neurons in the retina and LGN and then considering similar processing in complex cells in layer 4B of the striate cortex. Layer 4B in the striate cortex is a crucial stage of the M-channel since signals from 4B cells course to the extrastriate cortex and in particular to the middle temporal area MT that is involved in motion perception. The similarity of some complex cells located in cortical layer 4B to M-cells in the retina and LGN, in their responses to color-exchange stimuli, is evidence in support of the hypothesis that the M-channel is the neural basis for luminance and that it responds weakly or not at all to chromatic equiluminant stimuli. A separate question is how much P- and M-signals contribute to visual perception of motion, depth, and shape. This question is related to the nature of the luminance channel because there are specific perceptual deficits associated with equiluminant stimuli – those stimuli that silence the luminance mechanism. However, we focus here exclusively on the issue of the neural basis of the luminance mechanism.

The luminance mechanism

Human sensitivity to light as a function of wavelength across the visible spectrum, under bright daylight conditions, is called the photopic luminosity function and is denoted $V(\lambda)$. It might be thought that the easiest, and certainly the most straightforward, way to determine $V(\lambda)$ would be to measure psychophysically the sensitivity for increments of light of different wavelengths on a bright background. However, the photopic luminosity function is not measured in this way, mainly because such measurements are variable between and within observers (Sperling & Harwerth, 1971; King-Smith & Carden, 1976). Rather, the procedure known as heterochromatic flicker photometry has been employed. Monochromatic light of a given wavelength is alternated with a white light at a frequency of 20 Hz or above, and the radiance of the monochromatic light is adjusted until the perception of flicker disappears or is minimized (Coblentz & Emerson, 1917). This technique exploits the fact that neural mechanisms that can respond to the color of monochromatic light are not able to follow a fast flicker. The photopic luminosity function has been measured more recently using contour distinctness (Wagner & Boynton, 1972) and minimal motion (Cavanagh, MacLeod, & Anstis, 1987) as response criteria. These measurements agree with the luminosity function determined by flicker in the same subjects.

The luminance of a light source is its effectiveness in stimulating the visual neural mechanism that has as its spectral sensitivity $V(\lambda)$, the photopic luminosity function. Thus, the luminance of any light may be computed by multiplying its spectral radiance distribution, wavelength by wavelength, by $V(\lambda)$, and then summing all of the products. The neural mechanism that is the basis for luminance therefore must be close to linear, broadband in its spectral sensitivity, and very rapid in its response to flickering lights (see Wyszecki & Stiles, 1982a, Ch. 4). There is an ongoing controversy about the nature of this neural mechanism and even whether there is a single luminance mechanism (Lennie, Pokorny, & Smith, 1993).

As we will demonstrate, magnocellular neurons in the LGN and their cortical targets have many of the properties of the luminance mechanism so that, on the face of it, it seems plausible that the neural basis for luminance is the population of M-cells. However, the opposing view, that the P-cell channel could support luminance, has also been advanced (Gouras & Zrenner, 1979; Lennie & D'Zmura, 1988). Gouras and Zrenner found that at high flicker rates P-ganglion cells had a broadband spectral sensitivity that they believed approximated the spectral sensitivity of $V(\lambda)$. However, the flicker rates that they needed to demonstrate the change in spectral sensitivity function of P-cells was very high, and at lower flicker rates near to the temporal frequencies used to measure $V(\lambda)$ the spectral sensitivity of P-cells is narrow and color-opponent (Derrington et al., 1984). Another hypothesis for a P-cell luminance channel was based on the spatial summation of P-cell signals. Lennie and D'Zmura (1988) observed that signals from parvocellular color-opponent neurons could be summed to produce a broadband neuron. Thus, a neuron that sums signals from many parvocellular neurons might have a $V(\lambda)$ spectral sensitivity. In this way, the P-channel might be the neural substrate for luminance without the requirement for M-pathway input. This argument was subsequently withdrawn (Lennie, Pokorny, & Smith, 1993) in the face of mounting physiological evidence of the kind described later in this chapter. One problem for the P-cell–summation hypothesis is that it would require that there be a broad distribution of spectral sensitivities among the population of summing neurons, or else extremely precise adjustment of the functional weights of the summed inputs. However, as was described already, luminance is characterized by a nulling technique and the behavioral null can be very sharp around the equiluminant point (Pokorny, Smith, & Lutze, 1989). Therefore, the neurons that make up the luminance pathway must be very uniform in their spectral sensitivity functions. This very stringent criterion is met by cells in the M-pathway.

One of the reasons that Derrington, Krauskopf, and Lennie (1984), Lennie and D'Zmura (1988), and others sought to find a role for P-cells in luminance is that they believed that the luminance pathway had to possess high spatial frequency resolution. It is well known that achromatic patterns can be detected at much higher spatial frequencies than colored patterns, and it was thought that this sensory capacity depended on the luminance mechanism. Since M-cells in the retina are relatively sparse, they cannot sample the visual stimulus at a fine enough scale to support the high spatial frequency resolution of achromatic vision (Lennie et al., 1990b; Lennie, 1993). However, there is no direct evidence about the spatial frequency resolution of the neural mechanisms that support the psychophysical tasks that define luminance, like heterochromatic flicker photometry, for instance. It might well be that the luminance mechanism does not possess high spatial frequency resolution. Indeed, its high temporal frequency sensitivity suggests that it might well be a low- or midresolution system in spatial frequency, since (psychophysically) high temporal sensitivity is usually associated with lower spatial frequency sensitivity

(Graham, 1989). It is highly likely that there are several spatio-temporal mechanisms involved in visual sensitivity for achromatic patterns and the luminance mechanism is only one of them. The stringent criterion that the luminance mechanism must possess high spatial frequency resolution seems unsupported and unlikely. Therefore, we repeat that the neural substrate of luminance should be close to linear, broadband in its spectral sensitivity, and very rapid in its response to flickering lights; furthermore, the neural population making up the luminance channel must be very uniform in its spectral properties. It will become clear that we believe that there is strong evidence that the M-channel fits this description.

P- and M-pathways in the retina and LGN

The concept of parallel neural pathways in the primate visual system emerged from the study of the layering of nerve cells in the LGN in monkeys and humans. In an Old-World primate's LGN there are six clearly segregated layers of cells. The four more dorsal layers are composed of small cells and are named the parvocellular layers. The two more ventral layers, composed of larger neurons, are called magnocellular layers. Research on the visual properties of LGN neurons has revealed that cells in the different layers of the LGN have markedly different visual capacities (Wiesel & Hubel, 1966; Shapley, Kaplan, & Soodak, 1981; Kaplan & Shapley, 1982, 1986; Hicks, Lee, & Vidyasagar, 1983; Derrington, Krauskopf, & Lennie, 1984; Blakemore & Vital-Durand, 1986). For example, the color-opponent cells of the macaque's LGN reside exclusively in parvocellular layers (DeValois, 1965; Wiesel & Hubel, 1966).

The spatial and chromatic properties of the macaque monkey's LGN neurons are inherited from their ganglion cell inputs. The visual properties of each parvocellular neuron are determined by its retinal ganglion cell input from ganglion cells of the P type (also called midget ganglion cells; see Rodieck, Brening, & Watanabe, 1993). Magnocellular neurons receive input from M-ganglion cells (also called parasol cells; see Rodieck, Brening, & Watanabe, 1993). Anatomical and physiological evidence supports the idea that there is almost complete segregation of the M-ganglion cell input to the magnocellular layers and P-ganglion cell input to the parvocellular layers (Leventhal, Rodieck, & Dreher, 1981; Perry, Oehler, & Cowey, 1984; Kaplan & Shapley, 1986). The two pathways already begin to diverge at the level of bipolar cells in the retina (Rodieck, Brening, & Watanabe, 1993). The visual functional differences between P- and M-pathways can be characterized by the pattern of photoreceptor connections to the different classes of retinal ganglion cells and, through highly specific retino-geniculate transmission, to cells in the different LGN layers. Next we discuss how the cone inputs can be characterized.

Color-exchange experiments and cone inputs to P- and M-cells

Color exchange, or silent substitution (Estévez & Spekreijse, 1974, 1982) is a technique for identifying contributions from particular photoreceptors or spectral response mechanisms. Consider any spectral sensitivity function, and any two lights with different spectral distributions within the spectral band of the sensitivity function. One can create a color-exchange stimulus by alternately stimulating with one of the colored lights and then the other. By adjusting the relative intensity of the lights, one can create a "silent" or null response – no response at all to the color exchange. For example, if one chooses two monochromatic lights such that they are equally effective for the (long-wavelength) L-cone (Smith & Pokorny, 1975; Bowmaker & Dartnall, 1980; Bowmaker et al., 1980; Schnapf et al., 1988), then temporal alternation between these two lights should cause no variation in the response of the L-cone. The same argument works for the (middle-wavelength) M-cone. Silent substitution will work for any visual mechanism that is an additive combination of cone signals. This is a statement that we will prove below, but before the proof we will try to make it clear intuitively here. A visual mechanism that adds cone signals will have a spectral sensitivity function that is some additive combination of the cone spectral sensitivities. One can think of it as if it were another kind of cone, with a spectral peak intermediate between the two peaks of spectral sensitivity of the cones. Two lights of different colors that are equivalent in exciting this hypothetical additive mechanism will produce no response when exchanged. An important example of such an additive mechanism is the luminance mechanism, which is a putative neural mechanism that receives additive inputs from L- and M-cones. It can also be nulled by heterochromatic color exchange. Two lights that, when exchanged, produce no response from the luminance mechanism are called equiluminant.

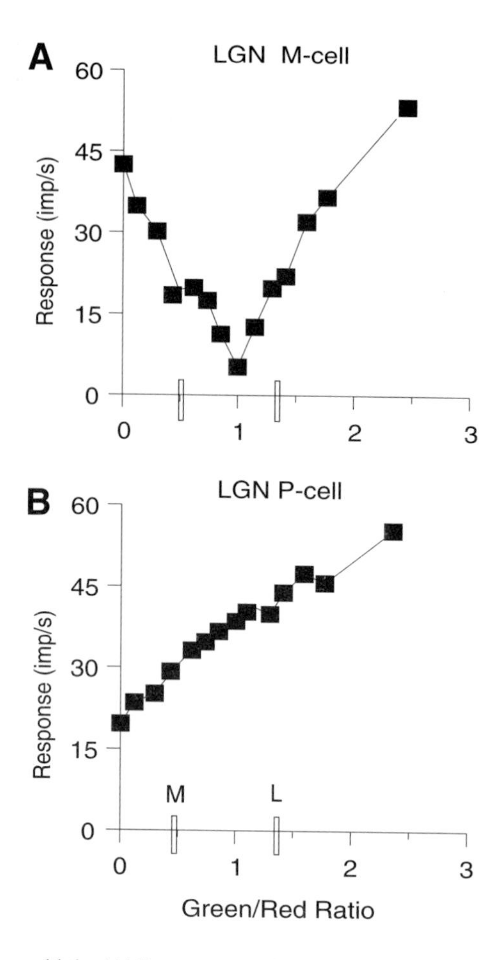

Figure 11.1: (A) Response vs. G/R contrast ratio in a color-exchange experiment. Red gun contrast was held fixed at 1.0, and green gun contrast varied from 0 to −1.0. The green and red modulation were 180 deg out of phase, so that the pattern appeared as a reddish-greenish color exchange. In fact the stimuli were drifting heterochromatic gratings. The data are from a typical magnocellular neuron driven by drifting red/green gratings at 4 Hz. The spatial frequency was 1 c/deg in this experiment. (B) Color exchange on a parvocellular M+ L− neuron. Here the stimulus was an optimal 1 c/deg grating of the red-green heterochromatic type, so that color exchange could be used to test spectral properties. The grating was drifting uniformly at 4 Hz. The data are well fit by a prediction based on pure subtractive cone interaction with equal weights for the L- and M-cones. The cone nulls are indicated as vertical bars along the green/red ratio axis: the M null at 0.5 and the L null at 1.3.

A luminance neuron. To add to an intuition about color-exchange experiments, we present the results of color-exchange experiments on a magnocellular neuron and on a parvocellular neuron in Fig. 11.1. The spectral distributions of the light sources were those of the red and green phosphors on standard color television sets, P22 phosphors. The red phosphor is narrowband and centered around 630 nm. The green phosphor is more broadband and centered around 530 nm. Such colored lights have been used in many experiments on color vision (e.g., Derrington et al., 1984; Livingstone & Hubel, 1987, among many others). The experiment is color exchange between the red (denoted capital R) and green (denoted capital G) phosphors. The mean intensities of the R gun, the G gun, and the (blue) B gun were adjusted to approximate the chromaticity coordinates of an equal-energy white, so that the screen appeared white when there was no modulation (technically speaking, the screen white was metameric with a white light source that has a flat energy spectrum). When a chromatic stimulus was produced to modulate the response of the neurons, the R gun modulation was fixed at a high modulation depth, 1.0, and then the G modulation depth was varied from 0 to −1.0. The G gun's modulation depth was called negative because it was modulated in antiphase with R. For example, if the pattern on the R phosphor was a sine grating, the pattern on the green phosphor was the same sine grating shifted by exactly 180 deg to be in antiphase with the red. Such a stimulus appeared to human observers as a red-green heterochromatic sine grating. In the graph of Fig. 11.1, the x-axis is the ratio of the G modulation depth to the R modulation depth in units of luminance modulation, and is called the G/R ratio. Thus, when the G/R ratio was 1.0, the G phosphor modulation was the same modulation in luminance as the R phosphor's, but in antiphase, so the net luminance modulation, when G/R = 1.0, was zero. As a consequence of the way in which this red/green color-exchange experiment was designed, the null points of specific cone mechanisms are located at widely spaced points along the x-axis. When the luminance modulation of the green phosphor was approximately 0.5 that of the red (G/R ratio 0.5), the response of the M-cones was nulled. When the G/R ratio was about 1.3, the L-cone response was nulled.

Now we have reached the point in this chapter where we will prove that a spectral mechanism that sums the responses of M- and L-cones, like the luminance mechanism, for example, will have a null in a color-exchange experiment at a G/R ratio between the nulls of the two cones.

Color exchange in cones

To begin with, we must define terms. The spectral sensitivity of the L-cone is $L(\lambda)$; the spectral sensitivity of the M-cone is $M(\lambda)$. The effectiveness for the L-cone of a broadband light E with spectral energy distribution $E(\lambda)$ will be the sum across wavelength of the spectral sensitivity multiplied by the energy at that point in the spectrum:

$$R_{E,L} = \int d\lambda\ E(\lambda)L(\lambda) = \langle E, L \rangle ,$$

and, for the M-cone,

$$R_{E,M} = \int d\lambda\ E(\lambda)M(\lambda) = \langle E\ M \rangle ,$$

but in our experiments all lights are sums of the phosphor illuminants R, the red gun with spectral energy distribution $R(\lambda)$, and G, the green gun, with $G(\lambda)$ and the blue gun with spectral energy distribution $B(\lambda)$. Thus, the average illumination on the eye, E_0, had spectral energy distribution:

$$E_0(\lambda) = [aR(\lambda)+bG(\lambda)+dB(\lambda)].$$

In fact, in our experiments we used equal-energy white, meaning that $a = b = d$. Therefore, $E_0(\lambda) = R(\lambda) + G(\lambda) + B(\lambda)$. The modulated colored pattern in the red-green color-exchange experiments was produced by modulation of the red and green guns only and can be represented as:

$$(1) \qquad E = E_0 + E_1 = [R(\lambda) + G(\lambda)+B(\lambda)] + R(\lambda)\ c_r\ S(x,t) + G(\lambda)\ c_g\ S(x,t),$$

where c_r is the contrast of the red gun modulation, equal to $\Delta R/R$, and c_g is the contrast of the green gun. $S(x,t)$ is the spatiotemporal pattern, for example, a drifting sine wave grating pattern as in the experiments of Fig. 11.1. In the color-exchange experiment, the object is to keep c_r fixed and vary c_g until the response is nulled out. When a null is reached, the ratio of c_g/c_r is the G/R ratio of the null point. For simplicity's sake, we can choose $c_r = 1$; then the G/R ratio is simply $-c_g$.

To calculate the response of the cones in a color-exchange experiment we need to make the initial assumption that the cone responses are linear functionals of the stimulus modulation around the constant level set by E_0. Evidence for this assumption will be considered below. Under the assumption of linearity, the modulation of the L-cone's response will have an amplitude proportional to

$$(2) \qquad \langle E_1, L \rangle = c_r \langle R, L \rangle + c_g \langle G, L \rangle,$$

and similarly for the M-cone:

$$(3) \qquad \langle E_1, M \rangle = c_r \langle R, M \rangle + c_g \langle G, M \rangle.$$

The L-cone null will happen when $\langle E_1, L \rangle = 0$, and this implies from Eqn. (2) that $c_r \langle R, L \rangle = -c_g \langle G, L \rangle$ or, equivalently,

(4) $c_{g,L} = -<R, L> / <G, L>$,

given that $c_r = 1$.

The M-cone null will occur when $<E_1, M> = 0$, implying from Eqn. (3) that

(5) $c_{g,M} = -<R, M> / <G, M>$.

Because $<G, L>$ and $<G, M>$ are approximately equal (see below), the relative magnitudes of $c_{g,L}$ and $c_{g,M}$ are determined by the relative values of $< R\ L >$ and $<R, M>$, respectively. The magnitude of $<R, L>$ is always much bigger than $<R, M>$, and so $|c_{g,L}|$ must be correspondingly larger than $|c_{g,M}|$. Therefore, $|c_{g,L}| \gg |c_{g,M}|$.

Color exchange applied to mixed spectral mechanisms

Additive color mechanisms. Now we consider the case of a mixed spectral mechanism that sums the two cone types, L and M, for example, a mechanism like luminance. Then its spectral sensitivity will be

(6) $V(\lambda) = pL(\lambda) + qM(\lambda)$,

where $p + q = 1$. The modulation produced by E_1 as in Eqns. (2) and (3) will be

(7) $<E_1, V> = p\,[< R,\ L> + c_g <G,\ L>]$

$$+ q\,[<R, M> + c_g <G, M>],$$

and the null point of the color-exchange experiment will take place when $<E_1, V> = 0$, implying that

(8) $c_g = [p <R, L> + q <R, M>] / [p <G, L>$

$$+ q <G, M>].$$

Let us note that, for the usual green primary light, $<G, L> = <G, M>$ approximately. This is an interesting approximation. For most lights categorized as "green," including the green phosphor on a CRT as here, the photon catch for the L- and M-cones is approximately the same.

Therefore, Eqn. (8) simplifies to

(9) $c_{g,V} \sim -[p<R, L>/<G, L> + q<R, M> / <G, M>]$

$$\sim [p\, c_{g,L} + q\, c_{g,M}],$$

and

(9a) $|c_{g,M}| < |c_{g,V}| < |c_{g,L}|$.

This is the goal of this section, since it follows from Eqns. (9) and (9a) that contrast of the green gun at the null point for V lies between the null points for L and M and is a weighted average of their values.

It follows from Eqn. (9) that if the spectral sensitivity of the summing mechanism is $S(\lambda) = pL(\lambda) + qM(\lambda)$, then when p approaches zero the magnitude of the green contrast at the color-exchange null is greater than the magnitude at the M-cone null $|c_{g,M}|$ and approaches $|c_{g,M}|$ from above, along the G/R-axis. When p approaches 1, the magnitude of the contrast at the color-exchange null approaches $|c_{g,L}|$, the magnitude of contrast at the L-cone null, from below, along the G/R-axis. The green contrast at the null of the luminosity curve in Fig. 11.1 is such an example. For that curve the ratio of L-cone strength to M-cone strength is approximately 2, and so the p and q coefficients that sum to 1 are 2/3 and 1/3. Thus,

$$V(\lambda) = 0.66\, L(\lambda) + 0.33\, M(\lambda),$$

and therefore

(10) $c_{g,LUM} = [p c_{g,L} + q c_{g,M}] = 0.66\, c_{g,L} + 0.33\, c_{g,M}$.

The above theory has to include the condition that the cone signals being summed have the same time course. The existence of sharp "V's" in the response magnitude versus G/R ratio curves near the equiluminant point in color-exchange experiments on M-ganglion cells and magnocellular cells is reasonably good evidence that L- and M-cone responses have similar time courses under the conditions of those experiments (Lee, Martin, & Valberg, 1988; Shapley & Kaplan, 1989). Otherwise, L- and M-cone signals could not cancel each other out at the nullpoint.

An example of magnocellular LGN cell data is shown in Fig. 11.1A. The points plotted in that figure are from the magnitudes of response of a magnocellular cell in response to drifting red-green, heterochromatic gratings. The response magnitude plummeted near a G/R ratio of unity, the equiluminant point. This is the usual result: Cells in the M-pathway respond weakly or not at all to equiluminant color exchange under these stimulus conditions.

Color-opponent mechanisms. Next we consider what happens in a color-exchange experiment on a color-opponent neuron. In such a cell, L- and M-cone signals are not added but instead are subtracted. The results of Fig. 11.1B for the parvocellular neuron would occur. This is the response of a cell in which the strength of L- and M-cone signals is equal but the sign is opposite – the response is proportional to (M–L). By a derivation similar to Eqns. (2)–(9), we can infer that such an opponent neuron will have no null response between the cone nulls along the G/R axis, but rather the nullpoints must lie outside the region of the axis between the two cone nullpoints. This result is general for any red-green color-opponent neural mechanism that has a spectral sensitivity of the following form:

$$S(\lambda) = pL(\lambda) - qM(\lambda);$$

where $p + q = 1$.

When p goes to zero and q approaches 1, the magnitude of the green gun's contrast at the null of the mechanism will approach $|c_{g,M}|$, the value at the M-cone null, from below; as p approaches 1 and q zero, the magnitude of contrast of the green gun at the null of the mechanism will approach $|c_{g,L}|$, that of the L-cone null, from above. There will be no local minimum of response for such an opponent mechanism along the green/red axis between the two cone null points. As before, all of these statements hinge on linearity and the identity of temporal response properties for M- and L-cones. Similarity of response time courses in neural signals driven by L- and M-cones was found in parvocellular color-opponent neurons by Gielen, van Gisbergen, and Vendrik (1982), who used color exchange to isolate responses of the different cones. There have been several demonstrations of small-signal linearity in P- and parvocellular neurons (Shapley, Kaplan, & Soodak, 1981; Kaplan & Shapley, 1982; Derrington & Lennie, 1984).

As an example of data from P-neurons, data from an M+L– parvocellular LGN neuron are plotted in Fig. 11.1B. The stimulus in this experiment was a red-green heterochromatic grating at the optimal spatial frequency for the cell, drifting at 4 Hz. Notice that the response amplitudes exhibited no local minimum at any G/R ratio, and that the cell responded to all color exchanges. Such a neuron must have approximately equal input from M- and L-cones, but the cone inputs must be subtracted from each other.

Responses to equiluminant stimuli

One particular color-exchange experiment has become crucial, namely, measuring the responses of P- and M-neurons to equiluminant color exchange. The interesting result is that for M-ganglion cells and LGN magnocellular neurons studied with stimuli that produce responses from the receptive field center mechanism, the position of the null on the color-exchange axis is close to that predicted from the human photopic luminosity function (Lee, Martin, & Valberg, 1988; Shapley & Kaplan, 1989; Kaplan, Lee, & Shapley, 1990), as in Fig. 11.1A for the magnocellular neuron. We and Lee et al. (1988) have found little variance in the G/R ratio at the magnocellular equiluminant point – no more variability in the position of the color-exchange null in the neurophysiological data than there is in psychophysical experiments on the luminosity function in humans (Crone, 1959) or in behavioral experiments on macaques (DeValois et al., 1974). While Cavanagh (1991) and Cavanagh and Anstis (1991) have hypothesized the existence of variability of the equiluminant null points in populations of M-cells, to explain their psychophysical data, we have been impressed by how tightly clustered the cells' equiluminant points are. Now we wish to turn to new experiments in the macaque monkey primary visual cortex (V1) utilizing color exchange that imply that the M-channel has little variability in its equiluminant point, from neuron to neuron. This new evidence for the uniformity of M-neurons supports the notion that the M-channel is the neural basis for luminance.

Neurons in layer 4B of monkey striate cortex

To find out whether cells in the M-channel are uniform or heterogeneous in their spectral sensitivities, we measured the responses of neurons in layer 4B of the striate cortex of macaque monkeys. While this may seem at first glance a rather indirect approach, we believe that it answers the question about M-cell color heterogeneity definitively; the answer is that magnocellular cells form a very uniform population with little variation in the equiluminant point from one M-cell to another.

The reason we measured in layer 4B is that we encountered highly responsive complex cells there, some of which had very high contrast gain and very sharp nulls in a color-exchange experiment, like M-cells in the retina and LGN. However, the size and nature of their receptive fields made it certain that these neurons summed M-cell inputs from many magnocellular cells, and this summation was only after an intervening synapse within layer 4C. As was shown by Lund and her colleagues, layer 4B neurons receive a predominant synaptic input from neurons in 4Cα, the afferent recipient zone for magnocellular inputs from LGN (Fitzpatrick, Lund, & Blasdel, 1985; Lund, 1988). We have found that many 4B neurons are complex cells with high contrast gain (Hawken & Parker, 1984). It is the combination of nonlinear spatial summation (in these complex cells) together with relatively pure M-cell inputs that allows us to answer the question about the M-cell uniformity of spectral sensitivity.

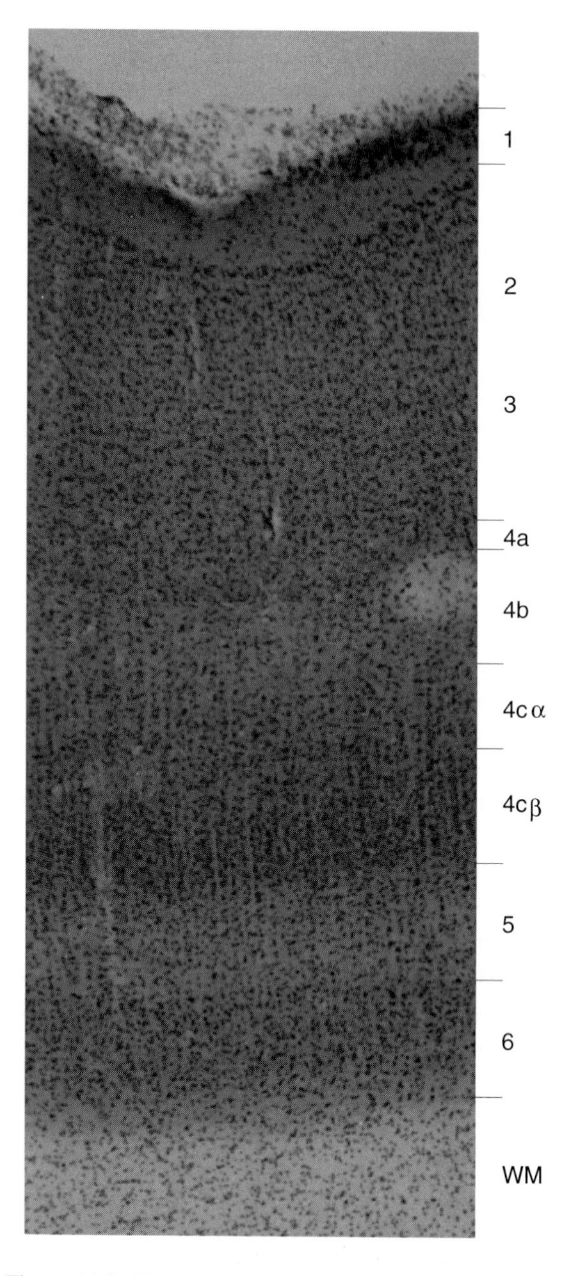

Figure 11.2: Electrode track through the macaque's striate cortex stained with cytochrome oxidase to illustrate the layering pattern. There is a pale round lesion in layer 4B just at the spot where we recorded the neuron's activity that is shown in Figs. 11.3–11.6. Layer 4B gets most of its input from layer 4Cα, which in turn receives its input from magnocellular layers in the LGN.

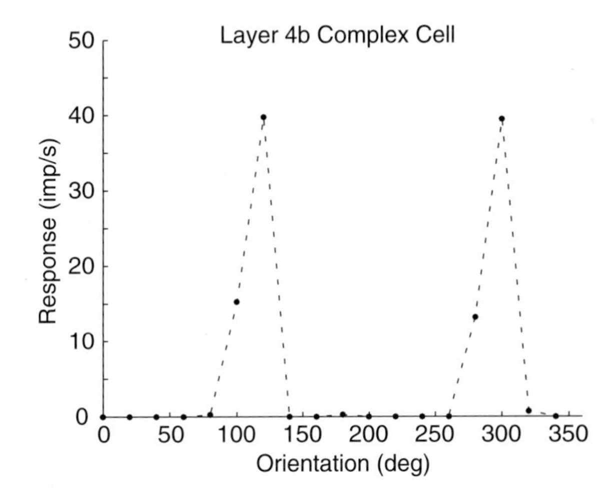

Figure 11.3: Orientation tuning. This is the response to a drifting sine grating of optimal spatial frequency at high contrast (64%), as a function of orientation of the stimulus. Responses were averaged over 4 s of stimulus presentation. The response of this complex cell to a drifting grating was mainly an elevation of the mean spike rate. The average spike rate during the 4-s stimulus epoch is plotted as a function of orientation. Like many layer 4B cells in the striate cortex, this neuron was sharply tuned for orientation.

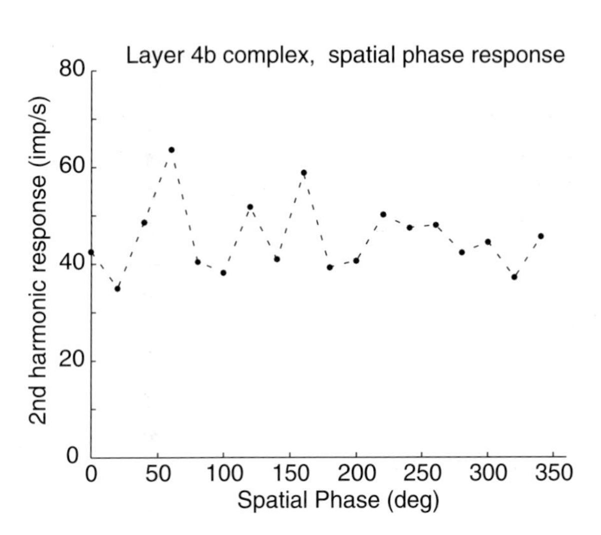

Figure 11.4: Spatial phase dependence. This is the response to contrast reversal of a high-contrast sine grating at 4-Hz modulation frequency. The response of the cortical complex cell was a frequency-doubled modulation of the spike rate, with the largest response component at 8 Hz. The contrast reversal was repeated at many spatial phases of the sine grating, and the response amplitude of the 8-Hz component is plotted as a function of spatial phase. The response amplitude was approximately constant at all spatial phases of the stimulus, and this implies that the neuron summed over many subcortical inputs (see text).

First let us consider a special case of one of these M-like layer 4B cells. It was located in layer 4B, as indicated by the histological section in Fig. 11.2, which shows a lesion at the site of recording of this neuron, right in the middle of layer 4B of the cortex. The cell was an orientation-tuned unit, as indicated by Fig. 11.3. It is a complex cell, as indicated by the spatial phase plot in Fig. 11.4. The dependence on spatial phase is a crucial experiment, so let us consider the implications of Fig. 11.4. Here we used a stimulus that was a contrast-reversing black–white sine grating of optimal temporal frequency, orientation, and spatial frequency. The spatial phase was varied in steps of 22.5 deg and covered a complete 360 deg of spatial phase. In linear neurons, for example, simple cells in the striate cortex, there is a marked dependence of response amplitude and phase on stimulus spatial phase – resembling the behavior of X-cells in the retina (Enroth-Cugell & Robson, 1966; Hochstein & Shapley, 1976; Movshon, Thompson, & Tolhurst, 1978; DeValois, Albrecht, & Thorell, 1982). However, complex cells often exhibit spatial phase independence, along with frequency-doubled responses to contrast reversal (Movshon, Thompson, & Tolhurst, 1978; DeValois, Albrecht, & Thorell, 1982). The frequency doubling is an indication of nonlinear summation (Spitzer & Hochstein, 1985). While it is not always the case that the response versus spatial phase curve is as flat as in Fig. 11.4, the weak dependence of the complex cell response amplitude on spatial phase is usually interpreted to mean that there is significant pooling across many receptive field subregions, and that there is a nonlinearity like a threshold before the pooling stage (Spitzer & Hochstein, 1985). Other models have been considered, but they also require pooling across subregions of different types. Therefore, when one measures a spatial phase curve like the one seen in Fig. 11.4, a reasonable interpretation is that

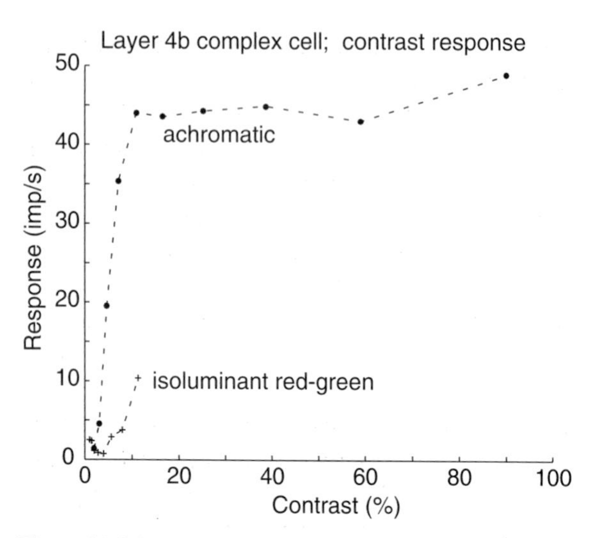

Figure 11.5: Response vs. contrast. This shows the response to drifting gratings as a function of achromatic and chromatic contrasts for the same cell as in Figs. 11.3 and 11.4. The stimulus was a sine grating at the optimal orientation, temporal frequency, and spatial frequency drifting for 4-s epochs. Contrast was increased from low to high, and the mean spike rate during the stimulus epoch was counted and is graphed here. Chromatic equiluminant gratings were also used and are graphed as well.

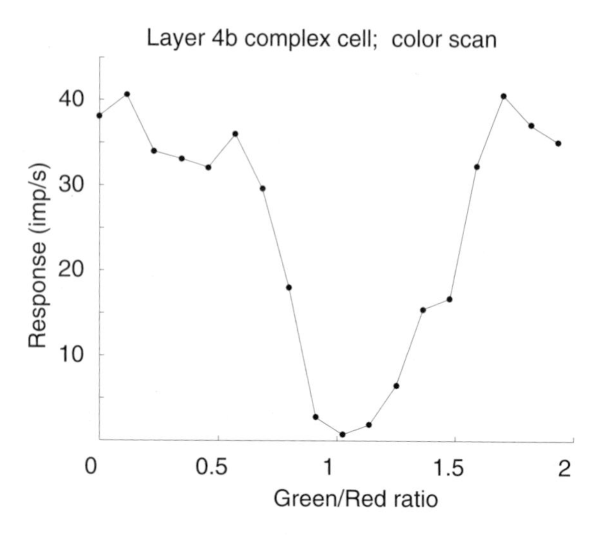

Figure 11.6: Color-exchange experiment. This is a similar experiment to the experiment that generated the results in Fig. 11.1, but performed here on the layer 4B cell from Figs. 11.3–11.5. Modulation was around equal-energy white. Red gun contrast was held fixed at 1.0, and green gun contrast varied from 0 to 1.0. The green and red modulation were 180-deg out of phase, so that the pattern appeared as a reddish-greenish, drifting, heterochromatic grating. Spatial frequency, orientation, and temporal frequency were all optimal for the neuron. Response was measured as the mean spike rate during the 4-s stimulus epoch. When the G/R ratio is 1, this is the equiluminant point, since modulation is expressed here in luminance units. Note that the axis is the same as in Fig. 10.1.

the neuron is pooling many visual inputs in a nonlinear way to produce the spatial phase insensitivity. A precise numerical estimate is not available. Fitzpatrick, Lund, and Blasdel (1985) estimated a convergence factor of about 50 in layer 4Cα/4B based on receptive field sizes and numbers of neurons, and this seems a reasonable conjecture.

The 4B neuron in question also has a very steep contrast response function for achromatic drifting gratings, as seen in Fig. 11.5. This is further evidence for predominance of M-channel input to this neuron since it has been shown that M-ganglion cells and magnocellular neurons have high contrast gain for achromatic patterns and that P-cells (and their parvocellular LGN targets) do not (Shapley, Kaplan, & Soodak, 1981; Kaplan & Shapley, 1982; Blasdel & Fitzpatrick, 1984; Hawken & Parker, 1984).

Finally we consider the response of this 4B cortical cell to a color-exchange experiment like those discussed earlier (Fig. 11.6). The stimulus was an optimal (temporal frequency, orientation, and spatial frequency) drifting sine grating. The grating was heterochromatic, red-green on an equal-energy white background. Response amplitude (spike rate elevation, or F0, the zeroth Fourier component) is plotted in Fig. 11.6 versus G/R ratio in units of luminance as in Fig. 11.1. It can be seen immediately that the 4B complex cell has a very steep drop in amplitude of response around equiluminance, and furthermore it is most important that the response amplitude is almost zero near equiluminance. This pattern of response, taken together with the high contrast gain of the neuron and its nonlinearity of spatial summation, force the conclusion that all of the M inputs to this neuron are nulling at very nearly the same equiluminant point.

If a neuron is summing multiple inputs only after a threshold nonlinearity, as seems to be the case for the 4B neuron we are considering, then pooling across a population with a diverse distribution of equiluminant null points would cause the null to be smeared out and the "V" shape to broaden into a "U" shape. Even more important, the response would not drop to zero near equiluminance. One can see this intuitively because, if the null points are spread out along a continuum, then wherever one input is nulled another one is not and the summed response will surely be greater than zero. It is possible to suppose that a high neural threshold might bring the responses down to zero, but the measured contrast-response function of Fig. 11.5 does not support the supposition of a high threshold – significantly large responses were produced for very low contrasts. One can calculate how sharp the distribution of M-cell null points must be to be consistent with the data in Fig. 11.6. Consider the ratio of L/M-cone weights for the different M-cells (the ratio p/q with p and q defined as relative weights as in Eqn. (6)) that are ultimately providing the input to the layer 4B neuron. The sharpness of the color-exchange null means that the variation in the ratio of cone weights must be less than 20%.

Note that our argument depends on the fact that this neuron is a complex cell and that it sums nonlinearly over many inputs. If it were a simple cell, one could believe that the linear summation of many P-cell inputs might generate a luminance-like spectral sensitivity, as was postulated by Lennie and D'Zmura (1988). This is not a possible explanation of the complex cell behavior seen in the layer 4B cell. The known predominance of 4Cα inputs to layer 4B, and the known predominance of M-channel inputs to layer 4Cα (Lund, 1988), also tend to support our interpretation of the cell's response as driven by M-cell inputs exclusively.

Our findings of complex cells in V1 that are silenced at equiluminance are not new, although our interpretation of this phenomenenon is original. Lennie et al. (1990a) showed an example of the same kind of neuron that we exhibit here as Fig. 11.7. Again, it is a complex cell in V1 that is nulled in the equiluminant plane of color space. In the Lennie et al. paper, the laminar location is given as layer 2/3, but the neural data resemble very closely the results of our layer 4B complex cell. This also was a cell sharply tuned in orientation and did not respond to low spatial frequency, as shown in the figure. There are similar neurons in the extrastriate cortical area MT that show a very similar pattern of response in color-exchange experiments (Gegenfurtner et al., 1994), as seen in Fig. 11.8. MT neurons are also complex cells, meaning that they sum inputs from receptive field subregions in a nonlinear manner and so, by the same argument we used for the V1 complex cell in layer 4B, their sharp-nulling behavior in color-exchange experiments further reinforces the conclusion that the M-channel is very uniform in its spectral properties and behaves like a luminance channel.

Functional implications

We believe these new results support the notion that, under stimulus conditions in which the M-cells' receptive field centers are isolated by the stimulus, the entire M-cell channel can be silenced by using equiluminant stimuli. This does not mean that all of layer 4B will go silent at equiluminance – not all of the cells in 4B are as M-cell–driven as the one shown in Figs. 11.3–11.7. And not all of MT goes silent at equiluminance (Gegenfurtner et al., 1994). Nevertheless, to the extent that M-cell signals are kept segregated (DeYoe & Van Essen, 1988) in the cortex, one could investigate a functional lesion of the M-channel by using equiluminant stimuli.

There is a discrepancy about this issue between the work, on the one hand, of Logothetis et al. (1990) and, on the other hand, our own work and that of others on the responses of M- and P-cells at equiluminance. First, Logothetis et al. reported that a large fraction of magnocellular neurons in their LGN sample were not silenced at equiluminance, to a variety of stimuli. They infer from this result that magnocellular neurons as a group are not silenced at equiluminance. This may simply be a difference of opinion in interpretation of results. Logothetis et al. state that they used a variety

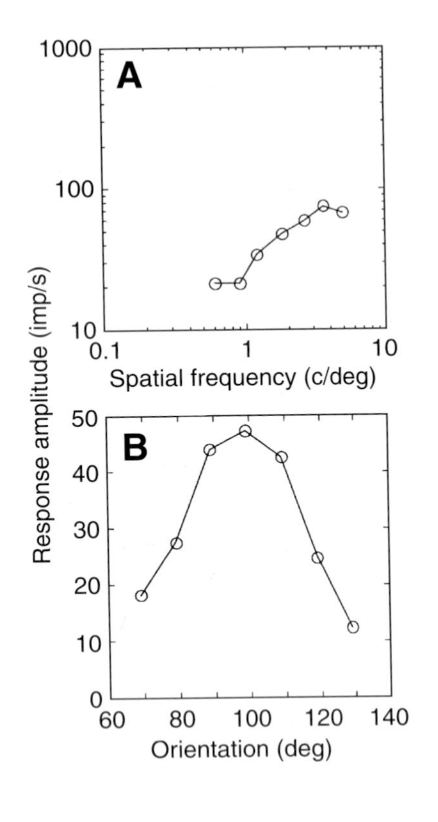

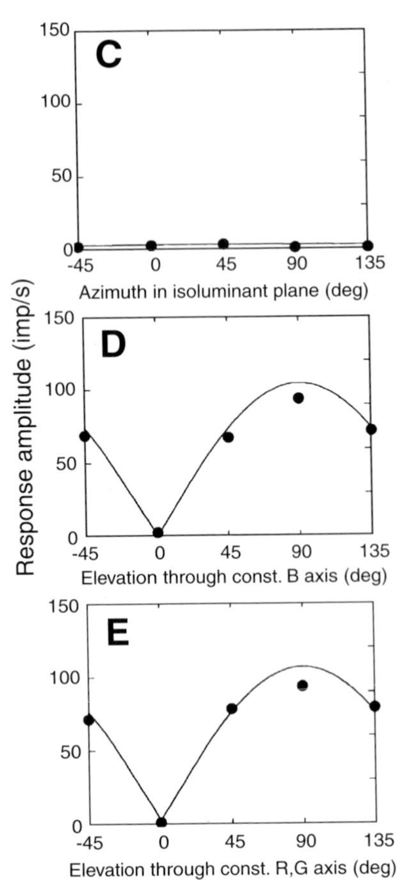

Figure 11.7: Another cortical luminance neuron from macaque striate cortex. This neuron was reported by Lennie et al. (1990a). The cell was located in layer 2/3. (A) Spatial frequency tuning. (B) Orientation tuning. (C) Color tuning in the isoluminant plane. (D and E) Responses to color modulation around the equiluminant point reveal sharp nulls at equiluminance – where the elevation in the Derrington et al. (1984) color space is zero. Thus these data are consistent with the results for the cell in Figs. 11.3–11.6. Figure reproduced from Lennie et al. (1990a) with permission.

of spatiotemporal stimuli to evaluate equiluminant responses, but the spatial pattern of the stimulus can have a crucial effect on whether magnocellular neurons are silenced at equiluminance. Our conclusion from a large body of work is that the M-neuron population can be silenced with equiluminant stimuli that stimulate only the M-cell receptive field centers, for example, midspatial frequency grating patterns. Therefore, using a variety of spatiotemporal stimuli, some of which isolate receptive field center responses and some of which do not, as was done by Logothetis et al. (1990), is irrelevant to the issue of whether a specific class of spatial stimuli would enable M-signals to be nulled out by equiluminance.

Another result in the Logothetis et al. (1990) paper is more problematic because there seems to be a conflict with the data obtained by three other laboratories. This concerns the behavior of P-channel neurons at equiluminance. Logothetis et al. state that a significant fraction of parvocellular neurons were silenced at equiluminance, contrary to other work (Derrington et al., 1984; Lee, Martin, & Valberg, 1988; Shapley, Reid, & Kaplan, 1991; Reid & Shapley, 1992). Our work on the cone connections to parvocellular neurons shows that P-cells are designed specifically to respond at equiluminance, because of spatially overlapping inputs from different cone types of opposite sign (Reid & Shapley, 1992). From our theoretical argument earlier in this chapter, it follows that in a cone-opponent neuron one predicts no minimum in the response magnitude in a color-exchange experiment for green/red ratios that lie between the cone null points. This is basically consistent with the results obtained by several experiments on P-ganglion cell and parvocellular

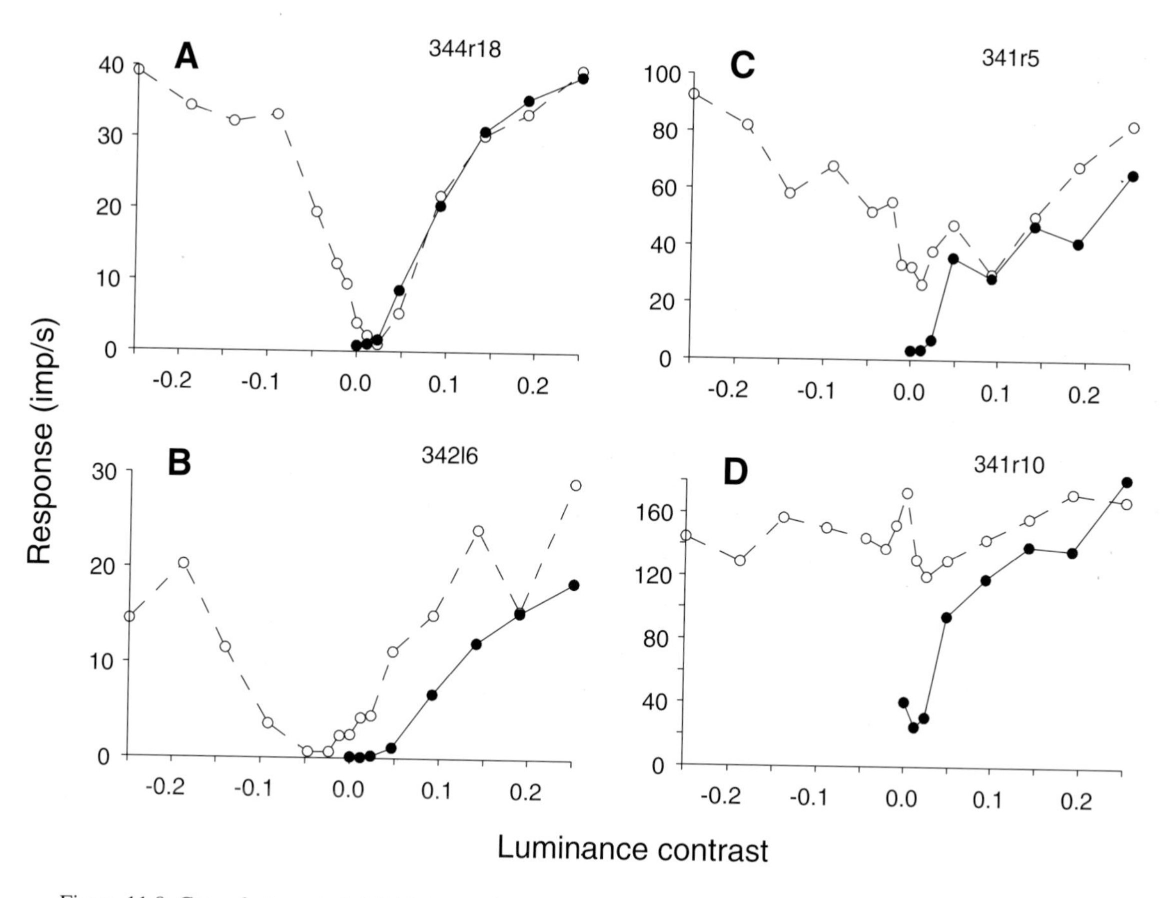

Figure 11.8: Gegenfurtner et al. (1994) found that MT neurons also have sharp nulls in their responses to color exchange around equiluminance. Particularly the results from the cell 344r18 in panel A, which was "representative of the most common type," showed a sharp null in its response at equiluminance. For discussion, see text. Redrawn from Gegenfurtner et al. (1994).

LGN cells (Derrington et al., 1984; Lee, Martin, & Valberg, 1988; Shapley, Reid, & Kaplan, 1991; Reid & Shapley, 1992). Therefore, it is difficult to understand how Logothetis et al. obtained contrary results. This is not a trivial point – from this result Logothetis et al. concluded that the parvocellular population response went through a dip at equiluminance, too, and that therefore a failure of a central response at equiluminance means nothing about P- and M-inputs to that response. From the preponderance of the evidence, we would conclude that Logothetis et al. were wrong in this conclusion, and that a dip in response at equiluminance is likely to be an indication of magnocellular neural input.

It is worth being careful about what experiments with equiluminant stimuli mean, as with any experiment of the "lesion" type. Lack of care in design and interpretation has led to false conclusions about the meaning of these experiments. If purely equiluminant stimuli are not useful for some function, say apparent motion, one would be wrong to conclude that "color does not support apparent motion." This is because the chromatic system may have quite a different "window of visibility" in the dimensions of size or time duration, or spatial or temporal frequency. Therefore, a change of performance at equiluminance, keeping all other spatial and temporal variables fixed, may be misleading. One might incorrectly assume that neural signals about color have no influence on perception of motion, for instance, from a single experiment in which motion perception was abolished at equiluminance. But with different stimuli one can elicit motion percepts using equiluminant stimuli, so parvocellular signals about color can be used somewhere in the brain for neural computations about motion. This is one field where careful parametric studies of spatial and temporal dependencies under luminance and equiluminant conditions are required before one can conclude anything reliable about the importance of chromatic or achromatic inputs to perception (see Chapter 15). Our own experience has taught us that many neurons in the striate cortex receive convergent input from the P- and M-channels, and that multiple mechanisms with possibly different connectivities to P- or M-cells may be involved in the neural computations that underlie perceptions about surfaces and objects.

Acknowledgments

This work was supported by the U.S. National Eye Institute through grants EY 01472 and EY 08300.

12
Color coding in the cortex

Peter Lennie

One of the striking things about color vision is the contrast between what we understand about peripheral mechanisms and what we understand about central mechanisms. The physiologist who studies peripheral mechanisms has these advantages: The expected properties of neurons have been sharply specified by psychophysicists; cells generally respond reliably to chromatic stimuli; most neurons of a particular class behave in very much the same way; and most neurons behave linearly (or have well-understood nonlinearities), so that their properties can be well characterized with relatively few measurements. In contrast, the physiologist who studies the cortex has much less firmly stated expectations of what to find, sees relatively few cells that look as though they are interested in color, finds great heterogeneity in the properties of cells, and encounters marked nonlinearities that often make it difficult to characterize cells.

All of this makes it hard to say anything precise about color coding in the cortex. The major problem is not that we know too little about the behavior of cortical neurons – I think that we know quite a lot – but that we have relatively ill-defined notions about what these neurons ought to be doing for color vision. To get to grips with this problem, I want initially to step back from experimental observations and ask rather generally what we might reasonably expect cortical mechanisms to do. Since color vision is about making objects more distinguishable and more readily identifiable, our questions need to focus on how cortical chromatic mechanisms are involved in object vision. This raises the following questions, which I want to explore in turn.

How important are chromatic attributes in identifying object structure (segmenting surfaces)? Most studies that have looked at this have focused on the performance of tasks that measure performance with objects or scenes defined by isoluminant variations in color. Much of this kind of work concludes that color provides a poor cue to structure.

What kind of code do we use to represent chromatic attributes so as to make surfaces distinguishable, and does this code vary with the perceptual task? A great deal has been said about this, mostly in studies that have examined chromatic habituation and induction. We have a convincing body of evidence that at some stage there are more than three postreceptoral channels, but we do not have solid agreement on how many there are or at what level in the visual pathway they arise.

Are color cues to object structure analyzed together with, or independently of, other cues? Much of the work on this question is from studies that have looked at the stimulus specificity of aftereffects or have examined the anatomical and physiological organizations of parallel pathways in the visual system. The preponderant view is that different attributes of the image are analyzed relatively independently.

How do we make the code robust to accidental variations in color signal? A good deal has been said about this, in work that has examined mechanisms of color constancy. We know a lot about the requirements of color constancy, and how it might be achieved, but rather less about how well it works in real visual systems, and where the mechanisms might reside. In what follows, I touch on each of these questions.

Color as a cue to structure

It is not absolutely necessary that, in providing a cue to the identity of an object, color gives us information about *shape*. One could imagine circumstances in which the chromaticities of points on a surface might be distributed distinctively in color space, in a way that could signify an object, regardless of their spatial arrangement on the surface; indeed computational models have been developed that identify objects by the ways in which surface samples cluster in color space (Swain & Ballard, 1991). Nevertheless, because we never dissociate color and shape perceptually – we do not see disembodied colors – and because chromaticity provides a potentially robust cue to surface continuity in the face of accidental variations in luminance (Rubin & Richards, 1982), it would be odd if the role of chromaticity in identifying objects were not through providing information about shape.[1]

The most widely used approach to understanding color as a cue to shape has been to confine variations in stimulus contrast to the isoluminant plane, so that mechanisms that are only sensitive to luminance variations are left unable to contribute to the task. One can then measure basic spatiotemporal sensitivities or performance on more complex tasks.

It is widely thought that the visual system lacks, or has relatively impoverished, machinery for analyzing object properties defined by chromatic differences (Livingstone & Hubel, 1987). An alternative account is that the visual system actually has less information available to it when objects are defined by only color variations, because color variations produce smaller modulations of cone signals than do brightness variations. Ideal observers allow one to see what could be done with the information available at the photoreceptors; and when one takes account of this, performance on basic measures depends little on whether structure is defined by chromatic or achromatic contrast. Sekiguchi, Williams, and Brainard (1993) found that, at worst, the spatial contrast sensitivity for chromatic gratings was 1.8 times poorer than for achromatic, and mostly was on par. Indeed, at low spatial frequencies the system probably detects chromatic contrast more efficiently than it does achromatic contrast. Scharff and Geisler (1992) showed that stereopsis is equally efficient whether structure is defined by chromatic or achromatic contrast.

Relatively few studies have used ideal observers to characterize the performance of observers on other kinds of tasks, but if one uses threshold contrast as a rough means of normalizing for peripheral factors, one finds that on a range of measures – bandwidths of orientation-selective mechanisms revealed by habituation (Bradley et al., 1988) or masking (Switkes et al., 1988), orientation and spatial frequency discrimination (Webster et al., 1990), vernier acuity (Krauskopf & Farell, 1991), colinearity judgments (Kingdom et al., 1992), movement thresholds (Metha et al., 1994; Hawken, Gegenfurtner, & Tang, 1994), texture segmentation (McIlhagga et al., 1990), and depth from motion parallax (Cavanagh, Saida, & Rivest, 1995) – the machinery works with chromatically defined objects about as well as it does with achromatically defined ones.

We almost never have to depend on purely chromatic contrasts to identify objects. Most objects are articulated redundantly, with structure defined by contrast in any or all of the dimensions of movement, color, brightness, disparity, texture, and so on. The reliability of these cues will vary with circumstances, but the visual system will rarely be required to rely on only one.[2] Perhaps the questions that we need to focus

[1]Chromaticity is evidently such a powerful cue that it is exploited in camouflage.

[2]Where color and brightness covary in defining visual stimuli, although detectability appears to be little improved, suprathreshold performance is often better than when stimuli vary along a single dimension (Gur & Akri, 1992; Mullen, Cropper, & Losada, 1997; Simmons & Kingdom, 1997).

on are: What is the *special* value of color differences in distinguishing and recognizing objects? How well does the visual system deal with objects defined by color differences in the face of noisy or weak cues on other dimensions? The likeliest circumstance under which color will help articulate surfaces is when there are accidental variations in brightness of the kind caused by shadows or irregularities in pigment density. We might therefore want to examine performance on tasks that require observers to use color to segment surfaces in the presence of noise.

Andrea Li and I (Li & Lennie, 1999) have looked at this by constructing a task in which an observer views a two-dimensional array of texture elements that simulates a surface; elements can differ in chromaticity, luminance, or both. In the simplest case the array contains only two kinds of elements that differ in color (red or green), brightness (light or dark), or both (light red and dark green, or dark red and light green). In a 32 $\times$ 32 element array that contains equal numbers of each kind of element, we vary the *proportions* of the two kinds of elements displayed in the left and right halves, or the *contrasts* of the elements in the left and right halves, to find the least asymmetry the observer can detect. Within each half, elements are randomly distributed. To simulate accidental variations in color and brightness within and across surfaces, variations of the kind that might be caused by shadows and interreflections, we can add brightness and/or color noise by perturbing the value in color space of each texture element. Figure 12.1 shows two examples of arrays in which element values are perturbed by noise: In Fig. 12.1A, the left and right halves differ in the proportions of elements of the two kinds; in Fig. 12.1B, the left and right halves differ in the contrasts of their elements.

Figure 12.1: A 32 $\times$ 32 array contains equal numbers of two kinds of elements (typically light/dark or red/green) that can be distributed differentially in the left and right halves. The observer's task is to detect an asymmetry in the distribution. Chromatic noise varying along one, two, or three dimensions can be added to the elements. In this illustration chromatic noise varying along three dimensions is added to (A) an array in which light and dark elements are completely segregated in the left and right halves, and (B) an array in which high- and low-contrast elements are segregated in the left and right halves.

Under a range of experimental conditions, but particularly when stimuli are of low contrast, we found that observers detect asymmetries defined by variations in color more efficiently than asymmetries defined by variations in brightness. Moreover, an observer's capacity to segment an array by color is essentially unaffected by the addition of noisy luminance variations, although the reverse is not the case.

This kind of result adds weight to the view that color provides a robust and efficiently used cue to the surface properties of objects.

The fact that performance on a variety of tasks is similar when information is made available through color variations or brightness variations leaves open the question of where the limitations to performance arise. It might mean that the same machinery articulates objects defined by color differences and those defined by brightness differences, or it might mean that separate mechanisms have common limitations. In the next two sections I explore some of the evidence that bears on this question, considering first the code for the chromatic attributes of objects and then how the encoding of color is connected with the encoding of other image attributes.

Code for color attributes

The problem here is reasonably well defined: Modern psychophysical evidence points to the existence of more than three postreceptoral channels, yet physiological observations on mechanisms in retina and LGN identify, at best, three.

The psychophysical evidence comes from studies of the aftereffects habituation (Krauskopf, Williams, & Heeley, 1982; Krauskopf et al., 1986; Webster & Mollon, 1994), spatiotemporal noise-masking (Gegenfurtner & Kiper, 1992), texture segmentation (Li & Lennie, 1997), induction (Krauskopf & Zaidi, 1986), and detection of coherent movement (Krauskopf, Wu, & Farell, 1996), all of which demonstrate chromatically selective mechanisms tiling color space more densely than could be achieved by the three classical color-opponent mechanisms.

How many mechanisms are there? Webster and Mollon (1994) suggested an arbitrarily large number of densely overlapping mechanisms. There might indeed be a huge number, but it is useful to ask what is the minimum number of mechanisms that can explain the results, and what would be their properties?

Li and Lennie (1997) found that, for stimuli confined to the chromatic plane, four broadly tuned mechanisms, two tuned along the L-M- and S-axes and two tuned along intermediate directions, did a good job of predicting the discriminability of textures defined by purely chromatic differences. Jim Müller and I (Müller & Lennie, 1995) have shown that only a small number of mechanisms are required to account for a range of results on the change in threshold and color appearance brought about by habituation and chromatic induction. Our model postulates broadly tuned mechanisms (with Gaussian or half-squared cosine sensitivity distribution). A mechanism's sensitivity is reduced by habituation or induction in proportion to its excitation by a habituating or inducing stimulus. The number of mechanisms, their sensitivities, and their preferred directions in the chromatic plane can vary or can be constrained. As few as four mechanisms (two more sensitive ones tuned along the cardinal axes and two less sensitive ones tuned along intermediate directions) can account well for the change in *appearance* brought about by habituation (Webster & Mollon, 1994), the change in *threshold* brought about by habituation (Krauskopf, Williams, & Heeley, 1982), and the change in appearance brought about by induction (Müller & Lennie, 1995). Figure 12.2 shows the model fits to measurements of Webster and Mollon (1994) and Fig. 12.3 the fits to the measurements of Müller and Lennie.

Permitting up to eight mechanisms accounts for essentially all of the variance in the experimental results. I want to emphasize that a simple model like this puts a *minimum* bound on the number of chromatically selective mechanisms; there might well be more. The model does for mechanisms of color vision what others (e.g., Watson, 1990) do for spatially and orientationally selective mechanisms. One attraction is the simple way in which it handles the distinctiveness of the cardinal axes (see Chapter 16): They are special because mechanisms tuned to them are more sensitive (or there are more of them).

We need to look for such mechanisms in the cortex, because the LGN has only two kinds of chromatically selective cells (Derrington et al., 1984) that are tuned to the cardinal chromatic axes. Moreover, LGN neurons do not habituate (Lennie, Lankheet, & Krauskopf,

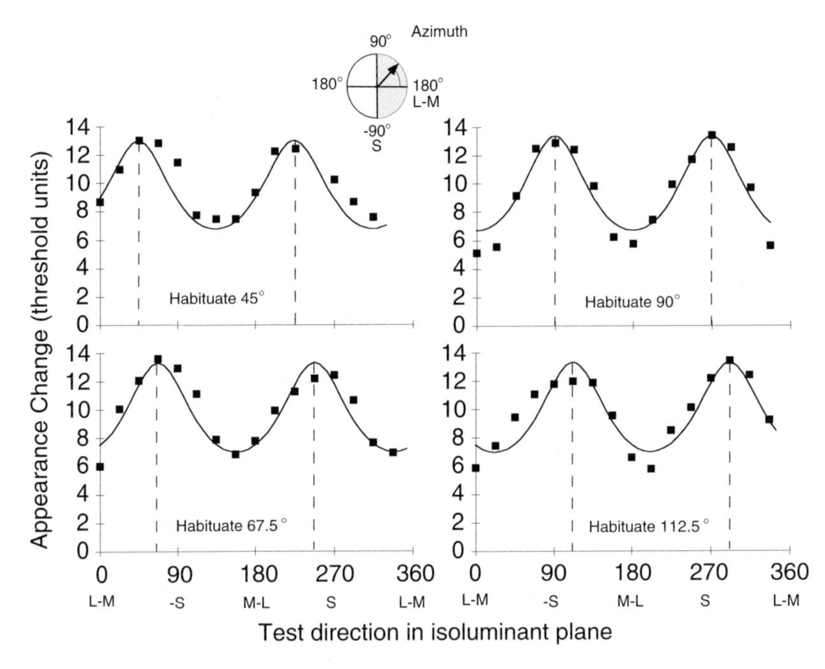

Figure 12.2: The change in chromaticity of a comparison light that matched test lights lying in different directions, following habituation to chromatic modulation along four directions through the white point in the isoluminant plane. The inset shows the convention used to identify directions of modulation. Results are from observer MW of Webster and Mollon (1994).

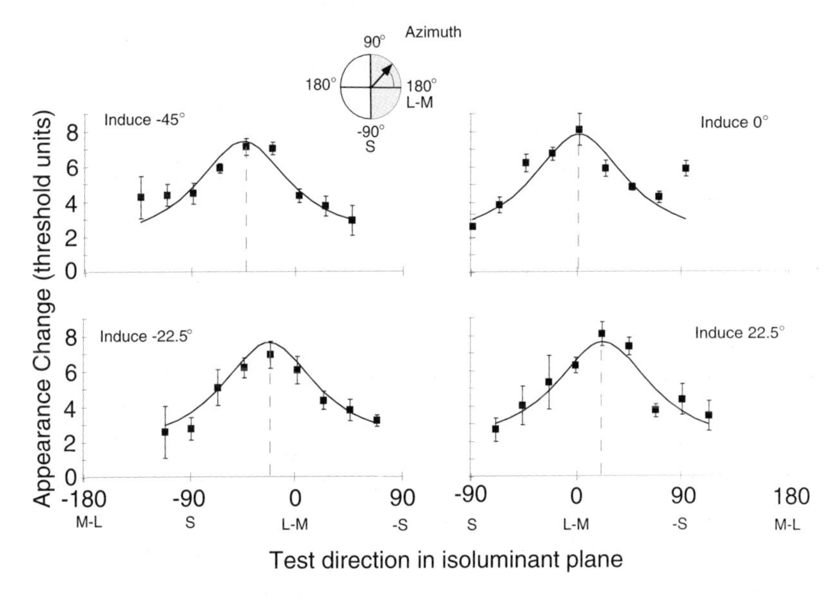

Figure 12.3: The change in chromaticity of a comparison light that matched test lights in different directions, while changes in the appearance of the test were induced by modulation along four directions through the white point. Results are from observer JM of Müller and Lennie (1995).

1994), and they lack spatial properties that are probably important for induction. In the following paragraphs I examine how well to the properties of cortical neurons fit our expectations from psychophysics.

Chromatic tuning. The two clearly distinguishable classes of neurons in parvocellular LGN, which have different but well-defined chromatic tuning, have no similarly distinctive counterparts in the cortex (Vautin & Dow, 1985; Lennie, Krauskopf, & Sclar,

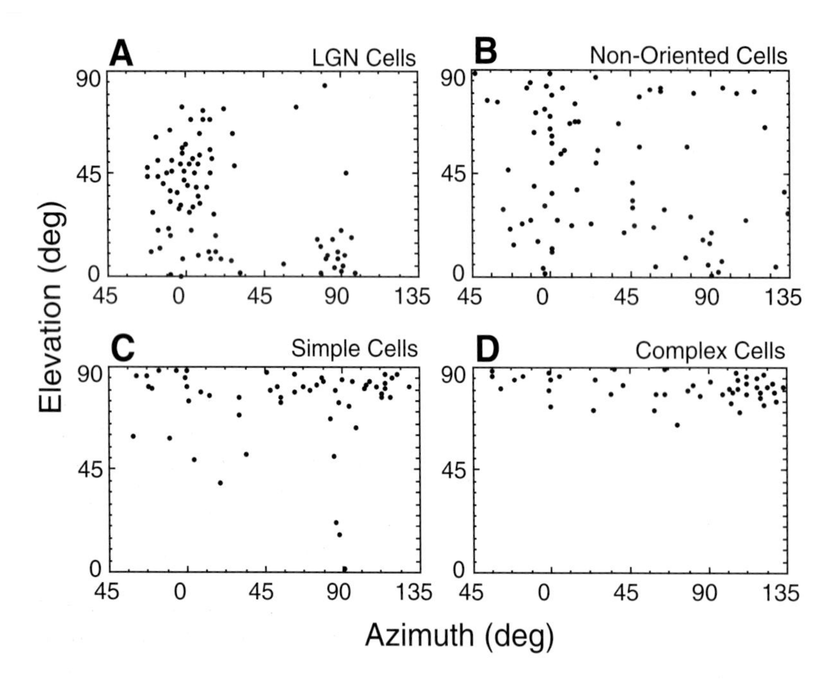

Figure 12.4: Distributions of chromatic preferences among parvocellular LGN neurons (A) and three classes of neurons in area V1 of macaque: nonoriented cells (B); simple cells (C); complex cells (D). In this diagram the azimuthal directions are referred to the L-M- (0–180 deg axis). Elevations are referred to as the isoluminant plane and are plotted without regard to sign. Neurons in parvocellular LGN fall into two distinct classes, while cortical neurons do not. Redrawn from Lennie, Krauskopf, and Sclar (1990).

1990; Leventhal et al., 1995). A simple model that combines cone signals linearly explains the chromatic tuning of LGN cells and also does a good job for V1 cells (with rectification for complex cells) (Lennie, Krauskopf, & Sclar, 1990). If one uses best-fitting values of model parameters to characterize cells, one finds no clear evidence, even among populations of cells drawn from particular layers or falling into groups with other distinguishing characteristics (e.g., simple or complex), that distinct *chromatic* classes exist in the striate cortex. Figure 12.4 shows that, although there is a modest tendency for the chromatic preferences of cortical neurons to lie along the chromatic directions that preferentially excite the LGN neurons (pure L-M-cone modulation, or pure S-cone modulation), the distribution of chromatic preferences is broad (Lennie, Krauskopf, & Sclar, 1990; Leventhal et al., 1995).

Even among neurons that provide the most plausible substrate for color vision – those showing the strongest chromatic opponency and responding best to low spatial frequencies (where psychophysical sensitivity to chromatic modulation is greatest) – there is a broad distribution of chromatic properties. The chromatic preferences among V1 neurons plainly do not fall into four neat classes of the sorts postulated to account for psychophysical observations.

There is little sign that the chromatic properties of neurons become more sharply tuned or more tightly clustered in higher occipital areas. Kiper, Levitt, and Gegenfurtner (Chapter 13) have found that the tuning of most neurons in V2 and nearly all in V3 can be reasonably well described by the model that works well for LGN and V1 (some neurons in V2 had narrower tuning), and that the distribution of chromatic preferences is broad, as in V1. In V4, Lennie, Krauskopf, and Müller (unpublished) have confirmed the earlier observations of de Monasterio and Schein (1982) that the chromatic selectivities of neurons in V4 are generally no sharper than those of ganglion cells and hence can be characterized by the model used in LGN and V1. The polar plots in Fig. 12.5 show the chromatic tuning of sample neurons in different regions to stimuli of different chromaticities. The chromatic tuning of neurons in the temporal cortex does seem to be narrower than elsewhere (Komatsu et al., 1992).

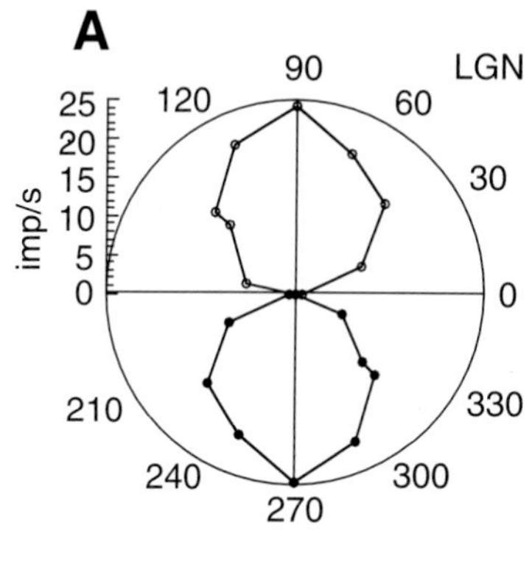

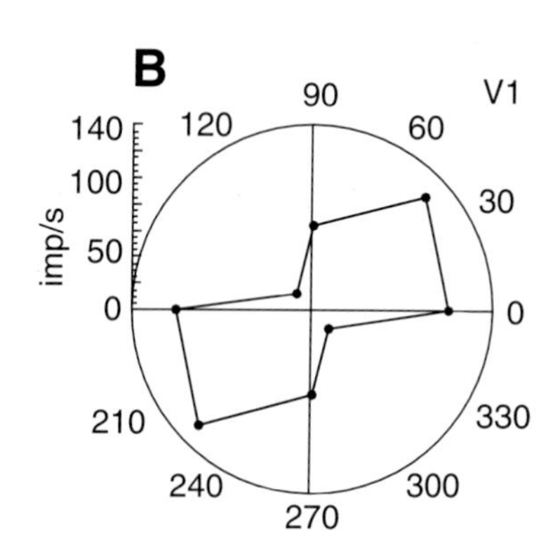

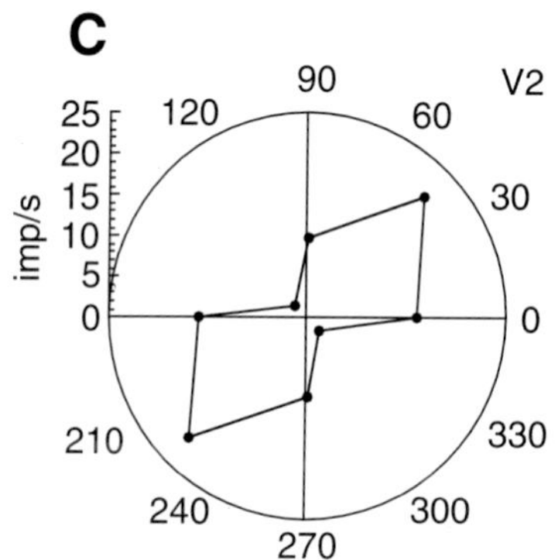

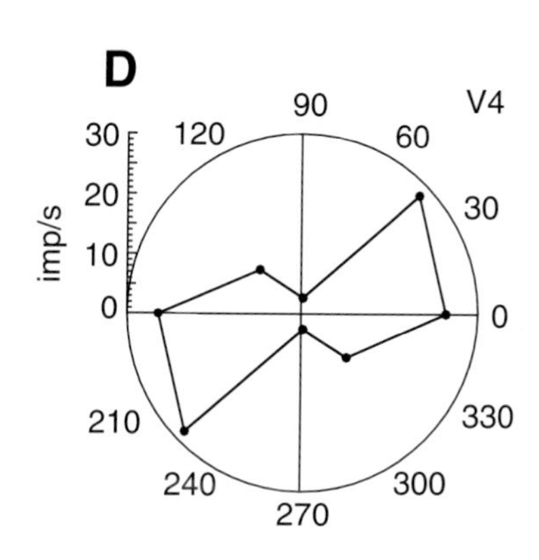

Figure 12.5: Polar plots showing chromatic tuning of sample neurons in (A) LGN (Derrington, Krauskopf, & Lennie, 1984), (B) V1 (Lennie, Krauskopf, & Sclar, 1990), (C) V2 (Levitt, Kiper, & Movshon, 1994), and (D) V4 (Müller et al., unpublished results). Each plot shows how the amplitude of the response varies with the azimuthal direction (referred to as the L-M-axis) of stimulus modulation through the white point in the isoluminant plane. There are no clear differences in the sharpness of tuning among neurons in the different areas.

Within occipital cortical areas there is a modest tendency for preferred directions to be clustered around the cardinal axes – especially the L-M one. This provides a hint, consistent with Müller's (Müller & Lennie, 1995) model, about why the cardinal axes might be special in psychophysical tasks: More neurons are tuned to them than to other directions in color space. This might be an issue that can be resolved by functional imaging.

Habituation. Lankheet, Krauskopf, and I (Lennie, Lankheet, & Krauskopf, 1994) examined chromatic habituation in single neurons and confirmed earlier observations (Derrington & Lennie, 1984) that those in the LGN do not habituate. Many neurons in the striate cortex are affected by prolonged exposure to habituating stimuli. Some behave as one might expect from psychophysical experiments (Fig. 12.6A), becoming less sensitive (usually to all stimuli that can excite them), but in other cases the responsivity can be *increased* with habituation, especially for stimuli of intermediate effectiveness. Figure 12.6B shows some examples of this behavior. Similar behavior occurs in some neurons after prolonged exposure to achromatic gratings (Sclar, Lennie, & DePriest, 1989). Susceptibility to chromatic habituation does not identify any special population of neurons in V1.

Induction. To account for induction one wants a mechanism that is capable of producing such edge or gradient effects as Mach bands. One such mechanism could be a neuron with a so-called double-opponent receptive field, of the kind shown in the caricature in Fig. 12.7. This receptive field has a center and surround, within each of which there is a chromatically opponent mechanism.

Concentrically organized receptive fields such as these are never seen in the LGN, although they have been encountered in the striate cortex (Hubel & Wiesel, 1968; Livingstone & Hubel, 1984; Ts'o & Gilbert,

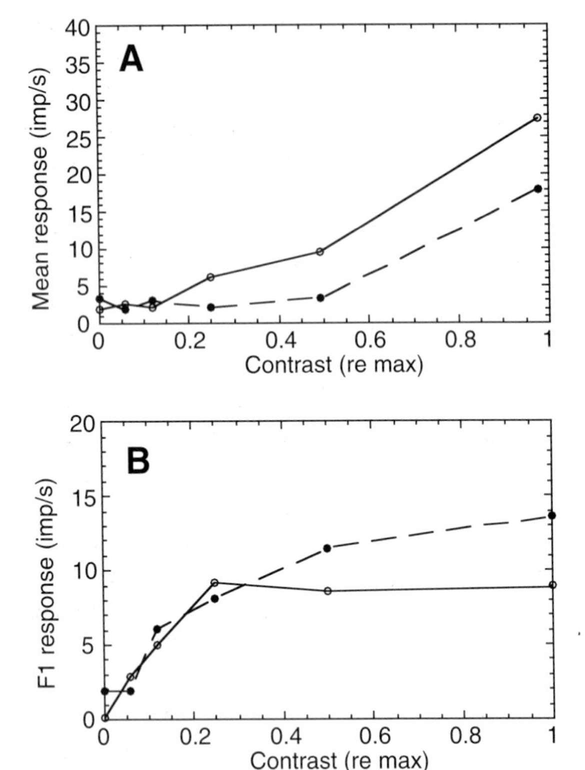

Figure 12.6: Contrast-responses curves to a chromatically modulated (L-M) grating of optimal orientation and spatial frequency, measured before and after habituation to a high-contrast grating of the same color, orientation, and spatial frequency. Open circles: prehabituation responses; filled circles: posthabituation responses. (A) Complex cell whose responses were reduced. (B) Simple cell whose responses to high-contrast gratings were increased. Among neurons whose responses were reduced by habituation, the loss was often greatest for weak responses (which tends to improve selectivity). From Lennie, Lankheet, and Krauskopf (1994).

1988). Investigators disagree about the incidence of double-opponent cells – at least of those with the simple receptive-field structure illustrated in Fig. 12.7. Recent quantitative analyses show them to be very rare. Nevertheless, the general property required to explain chromatic induction – the capacity for spatial differentiation in the chromatic domain – is possessed by many simple and complex neurons in the striate cortex (Thorell, DeValois, & Albrecht, 1984; Lennie, Krauskopf, & Sclar, 1990).

Implications. A variety of studies in several cortical areas indicates that the chromatic code does not change discernibly as one moves from the striate into the extrastriate cortex. Physiological observations implicate neurons in the striate cortex as the machinery of the multiple channels required to account for psychophysical observations. We do not find clusters of neurons tuned to a small number of distinct directions in color space; the broad distributions that we do find everywhere must therefore be equivalent to a simpler organization in which chromatic preferences are clustered (Geisler & Albrecht, 1997).

Independence of analysis

The foregoing discussion of the physiology of color vision, in common with most others, implicitly treats color as an abstract property analyzed in isolation from an object's other attributes – shape, position, movement, and so on – but ultimately reconciled with them. We need to consider this position explicitly, along with the alternative possibility that the analyses of the image along different dimensions are close-coupled and depend on common machinery.

Let me spell out more precisely the contrast that I have in mind. On the one hand, there might be a set of visual mechanisms, one devoted to the analysis of color, another to depth, and so on, but each of which is indifferent (or very broadly tuned) to the stimulus attributes on any other dimension. For example, the color mechanism might be indifferent to the spatial frequency content of the image. Since the analyses along different dimensions are independent, the different classes of analyzers could exist in special concentrations or in separate areas (Fig. 12.8A). On the other hand, if the analyses along all dimensions are close-coupled, one can think of the image as being described in a multidimensional space, all regions of which are tiled by visual analyzers. This kind of scheme encourages us to think of each cortical neuron as having some defined selectivity on all dimensions along which the image is analyzed (Fig. 12.8B). The space need not be tiled uniformly on all dimensions – for example, one

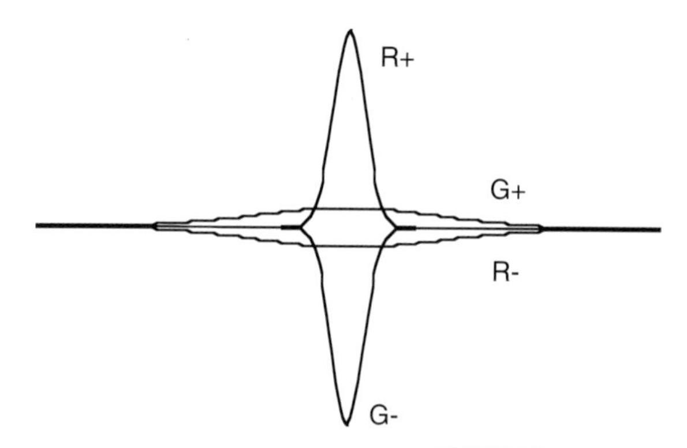

Figure 12.7: Schematic drawing of a 'double-opponent' receptive field. Center and surround each contain chromatically opponent mechanisms that generate opposite polarities of response: The center is excited by long-wavelength light and inhibited by middle-wavelength light; the surround is excited by middle-wavelength light and inhibited by long-wavelength light. The most effective visual stimulus will be one that produces spatial modulation of a chromaticity matched to the dimensions of center and surround. Neurons with this simple receptive field structure are exceedingly rare in V1.

would not expect the system to have machinery for representing chromatic contrasts at high spatial frequencies where it can still represent achromatic contrasts. The more popular view is that the analyses along different image dimensions proceed independently, so I start by examining some arguments for it.

Independent analysis. One of the principal psychophysical arguments for the independent analysis of attributes is that color contrast supports impoverished object vision: The visual system does not use information about color differences as effectively as it does information about contrasts along other image visual dimensions (Livingstone & Hubel, 1987); hence, distinct machinery is involved. For reasons discussed in an earlier section, I believe that that argument is wrong, because it fails to acknowledge the nature of the signals that are available at the cones. When one takes account of these signals, performance with stimuli defined chromatically is seen to be on a par with that found for stimuli defined by other means.

Evidence from single-unit recording has come mainly from studies that have found color-opponent neurons to occur in clusters, either locally within a visual area or as an overall preponderance in an area. A concentration of color-opponent cells has been found in V1 blobs (Livingstone & Hubel, 1984; Ts'o & Gilbert, 1988) and in the thin stripes in V2 to which they project (Livingstone & Hubel, 1983). The thin stripes in turn project to V4. These observations, taken in conjunction with Zeki's (1973, 1977) discovery of a substantial concentration of color-opponent cells in V4, have suggested a special pathway. The argument for a special color pathway that is indifferent to the other attributes of objects is bolstered by the observation that receptive fields of cells in V1 blobs are generally orientationally isotropic and often spatially uninteresting (Livingstone & Hubel, 1984). There is, however, a good deal of contrary evidence from quantitative studies of cortical neurons. First, measurements in V1 show that color-opponent cells are found no more often in blobs than outside of them (Lennie, Krauskopf, & Sclar, 1990; Leventhal et al., 1995); second, there is only the slightest tendency for color-opponent cells to be concentrated in thin stripes in V2 (Levitt, Kiper, & Movshon, 1994; see Chapter 13); third, there is no special preponderance of color-opponent cells in V4 (Schein, Marrocco, & de Monasterio, 1982; Schein & Desimone, 1990). This last point is corroborated by studies of the effects of experimental lesions in macaque V4. The general finding is that all aspects of object vision are seriously impaired, and certainly not color vision more than others (Heywood & Cowey, 1987; Heywood, Gadott, & Cowey, 1992).

Other evidence comes from cases of achromatopsia following localized injury to the occipital cortex. The key issue here is the selectivity of the deficit. Most cases of acquired achromatopsia are accompanied by visual deficits of other kinds in the same region of the visual field; the most specific losses have resulted from damage to the lingual and fusiform gyri (Zeki, 1990). Even if we assume that the loss is confined to color vision, we need to consider how it is expressed: as a failure to discern structure defined by purely chromatic differences, or as a failure to identify colors. Failures of the former type are seldom found among cases studied thoroughly (e.g., Mollon et al., 1980; Victor et al.,

1989; Barbur, Harlow, & Plant, 1994). Achromatopsic subjects can often use chromaticity to segment surfaces and sometimes have near-normal hue discrimination, but they are much impaired in naming colors and grouping items of similar color. Such failures implicate an area that is specialized for chromatic analysis, but when they occur without a corresponding loss of capacity to discern object structure defined by color differences they point to damage at a high level in the system – perhaps at a site where the chromatic dimension is *abstracted* from some object description formed by lower levels. Findings like these provide little reason to suppose that signals about color are conveyed in a private pathway originating in V1 and passing upward through the extrastriate cortex.

Some functional imaging studies, using PET (Zeki et al., 1991) and fMRI (Sakai et al., 1995) to identify areas in the human cortex that are especially well activated by chromatic changes, have implicated the cortex near the fusiform gyrus, a region thought to be homologous to monkey V4 (Clarke & Miklossy, 1990). On the whole, these do not make a persuasive case for a color area. First, the assessment of significant activity in a particular area depends substantially on identifying that area initially as a candidate and confining attention to it. When one looks beyond the fusiform gyrus, recent fMRI work (Engel, Zhang, & Wandell, 1997) makes it clear that cortical areas V1 and V2 are substantially activated by chromatically modulated stimuli. Second, the controls used in most functional imaging studies have involved comparing activity evoked by the presentation of an array of colored surfaces with the activity evoked by the presentation of the same elements at the same luminance, but now set to be gray. Visibility and salience can be quite different.

Overall, the evidence for a specialized color pathway is not strong. Let me continue the argument by developing the alternative view that in fact the analyses of the image along different dimensions are tightly coupled.

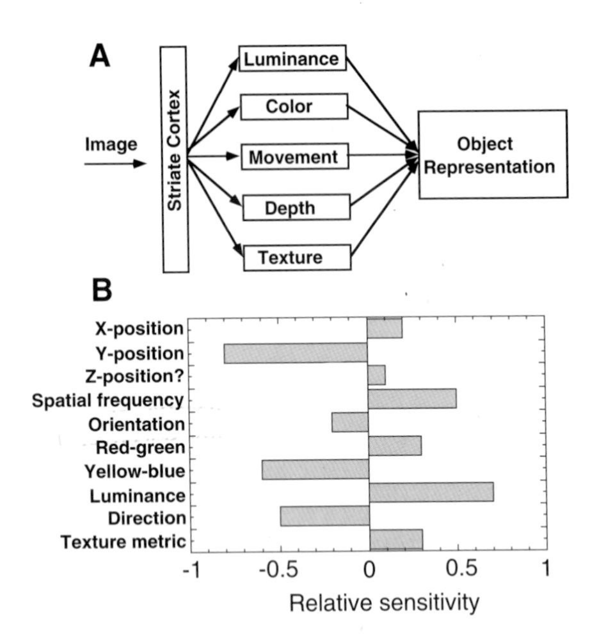

Figure 12.8: Hypothetical schemes for the analysis of different image dimensions by cortical analyzers. (A) Separated analysis, whereby a different set of analyzers exists to extract and represent variations along each dimension of variation. Each set contains analyzers that are selective on the dimension of interest, but are indifferent to (or broadly tuned along) other dimensions of variation. (B) Close-coupled analysis, in which each analyzer has some specified selectivity on every dimension along which the image is analyzed. The dimensions represented here are bipolar, so the sensitivity of a neuron can be represented as a positive or negative value.

Close-coupled analysis. The first line of evidence is from psychophysical studies of contingent aftereffects. Here we have clear evidence for mechanisms that are tuned on multiple dimensions. The McCollough effect (McCollough, 1965) is perhaps the most celebrated example, but there are many others, including different tilt aftereffects that are contingent on viewing different colored patterns (Flanagan, Cavanagh, & Favreau, 1990), and orientation-selective and spatial-frequency–selective habituations that are also color-selective (Bradley, Switkes, & DeValois, 1988). Studies of contingent aftereffects have generally been divorced from mainstream work on the mechanisms of color vision, which has seldom been concerned with the spatial attributes of the stimulus. But the connection is close: In experiments now being completed in my laboratory, Eric Lichtenstein has shown that chromatic habituation of the kind used by Krauskopf, Williams, and Heeley (1982) reveals chromatically selective mechanisms that are *also* orientation-selective.

The second line of evidence is from single-unit studies. By and large these show that every neuron in the cortex is tuned on multiple dimensions; one does not find in V1 (or anywhere else in the occipital cortex) neurons that have well-defined selectivity along one stimulus dimension and lack it along others. Most neurons in V1 are orientation-selective, spatial-frequency–selective, and chromatically selective; many also are directionally selective. A cell that is selective on all of these dimensions can no more be called a color cell than it can be called a spatial-frequency cell or an orientation cell. Polydimensional selectivity is at least as pronounced an attribute of neurons in higher occipital areas (see Chapter 13).

I do not want to argue from this that all neurons in V1 are important for color vision or contribute equally to it. We should not expect the visual system to possess machinery for analyzing color variations at high spatial frequencies, since the dominant components of the chromatic signal are carried in low spatial frequencies and, in any case, chromatic aberration in the eye ensures that high-frequency variations in chromaticity will not be represented faithfully on the retina.

A neuron tuned along several dimensions can be thought of as occupying a particular position in a multidimensional stimulus space; each neuron occupies a different characteristic position. The population of neurons that covers the space is arranged in a two-dimensional cortical map, and there is good reason to suppose that the mapping of these dimensions will be orderly (Swindale, 1991). We know a lot about the mapping of eye input and orientation, but a good deal less about the mapping of other dimensions along which the image might be analyzed; it is perfectly reasonable that, within the larger organization that accommodates ocular dominance and preferred orientation, the map is tessellated to accommodate the orderly mapping of other dimensions. Neurons with similar tuning on several dimensions will therefore naturally occur in small (perhaps very small) clusters. We should not be too eager to see this clustering as the origin of distinct streams that convey different kinds of information to different parts of the extrastriate cortex.

If the multiple extrastriate areas are not to be viewed as modules undertaking analyses of different specialized attributes of the image, what are they doing? I have argued elsewhere (Lennie, 1998) that, at least within the pathway connecting V1 to V4, one should view them as undertaking analyses of the image at successively more complex levels of aggregation (the units of analysis are more complex object elements) rather than along different stimulus dimensions.

Stability of code (color constancy)

If the main function of color vision is to discover object properties (that are expressed through surface reflectance functions), then the information to be recovered is potentially unreliable, because it varies with the chromaticity of the illuminant. Formal analyses of this problem of color constancy show that a trichromatic visual system cannot recover an accurate three-variable description of a surface when the spectral composition of the illuminant is unknown (Maloney, 1986; Maloney & Wandell, 1986) – too many variables are required to characterize adequately both the surface reflectance function and the spectral power distribution of the illuminant. Models that solve this problem start with the fact that the reflectance functions of natural surfaces are constrained and can be described by a small number of underlying basis functions (Maloney, 1986), as can the spectral power distributions of illuminants (Judd, MacAdam, & Wyszecki, 1964). Some of these models (Buchsbaum, 1980; D'Zmura & Lennie, 1986; Lee, 1986) rely on the observer being able, in a single view, to estimate the spectral composition of the illuminant from light reflected by the scene; others (D'Zmura & Iverson, 1993) rely on the observer being able to compare the same scene under two different illuminants.

Formal accounts of color constancy characterize mechanisms that perform better than human observers: Human color constancy is imperfect (Evans, 1948; Brainard, Brunt, & Speigle, 1997). Moreover, to the extent that color constancy exists for natural scenes, it can be equally well explained by assuming that signals from the three classes of cones pass through independent gain controls, or by assuming that the visual system computes an imperfect estimate of the illuminant (Brainard, Brunt, & Speigle, 1997; see Chapter 20).

Gain controls in the separate cone pathways are a well-known feature of retinal organization and are expressed both psychophysically (Stiles, 1939; see Chapter 19) and physiologically in retinal ganglion cells (Gouras, 1968). For our present purposes the important question is therefore whether there is physiological evidence for postretinal mechanisms that might promote color constancy.

Cortical mechanisms of color constancy. Most of the computational schemes for color constancy depend on capabilities that could exist only in the cortex. An analysis that extracts illuminant information from specular highlights requires these highlights to be identified in the image. An analysis that compares a set of objects under two illuminants requires that the objects be matched correctly in the views. An analysis that extracts the average chromaticity in the scene as a way to estimate the illuminant (which can be done by assuming that surface reflectance, averaged across a scene, is flat) requires large-scale spatial integration of a sort unknown in the retina or LGN. In V1, long-range spatial influences on neurons are evident in inhibitory and excitatory interactions that arise when patterned stimulation of regions well outside the classical receptive field is combined with stimulation confined to the receptive field, but such interactions have not been shown to be color-sensitive.

One V1 mechanism postulated to contribute to color constancy is a neuron with a "double-opponent" receptive field (Fig. 12.7). A change in the chromaticity of a stimulus covering the whole receptive field will not evoke a response, because center and surround signals are roughly equal and opposite; the upshot is that the cell will tend to discount a consistent bias in the color of illumination. Double-opponent cells in the striate cortex have a precarious standing. Where they have been the subject of special attention (Livingstone & Hubel, 1984; Ts'o & Gilbert, 1988), they have seldom been found to possess the properties of the canonical form shown in Fig. 12.7. Instead, receptive field surrounds seem to be generally suppressive, but not chromatically opponent. This kind of organization is not obviously useful for color constancy.

Zeki (1983a, 1983b) has suggested that area V4 has an important role in color constancy. He found cells in V4 (but not in V1) whose responses to an element of a Mondrian pattern depend on the color composition of elements surrounding it, in ways that paralleled the appearance of the element to a human observer; if the experimenter changed the chromaticity of the element, and changed the surround so that the element looked the same to a human observer, the response of the cell stayed constant. This behavior evidently reflects some sort of center–surround interaction. Schein and Desimone (1990) have explored this, and have shown that the receptive field is enclosed by a large surrounding region in which stimulation alone has no effect, but that it can antagonize or boost the response to a stimulus falling within the receptive field, depending on its chromaticity. Schein and Desimone suggest that this might provide a mechanism for color constancy, working in much the same way as would a double-opponent receptive field. One of the striking things about V4 cells is that the suppressive surround interaction occurs for multiple dimensions of the stimulus, including spatial frequency and, notably, orientation (Müller, Krauskopf, & Lennie, unpublished observations). One therefore needs to entertain the possibility that the V4 receptive field is designed more generally to seek contrast in the local structure of images, perhaps as part of a mechanism for finding boundaries. By tending to filter out image attributes that extend uniformly over large regions, the receptive field organization necessarily helps with the color constancy problem, but that is probably not its principal function.

Conclusions

In moving from one cortical area to another, we should not be surprised that the chromatic properties of cells change little. If one takes a detached view of how the image is analyzed at different levels, a great deal of the work that needs to be done to analyze color – most of what we know is ever done with color, including the normalization of signals that results from chromatic adaptation – is done before the cortex, and probably much of the rest is done in V1. In contrast, remarkably little of the analysis along other dimensions of the image – leading to the production of orientation selectivity, directional selectivity, binocular combination – and the ensuing analysis of second-order attributes such as texture, grouping, and connecting like regions is done before the cortex. The complexities of these tasks presumably account for the bulk of the cortical machinery.

The kind of view that I have developed here, of color being only one among several dimensions along which the image is analyzed locally, discourages us from looking for color areas and does not permit us to conclude that an area that contains a lot of chromatically driven cells is important for color vision. More importantly, it discourages us from looking for color cells – that is, members of some special class designed expressly for carrying information about the chromatic attributes of the image. One should not really think about mechanisms of color vision at all (any more than one should think about mechanisms of orientation vision), but rather one should think about the chromatic (or movement, or depth) attributes of mechanisms that are solving problems in image analysis. In fact, I think that we tend to promote the wrong emphasis by talking about the cortical mechanisms of color vision – it encourages us to study color as an abstraction. In contemplating the analysis undertaken in the cortex, we ought first to ask about the chromatic attributes of mechanisms that underlie object vision. We can then go on to ask how the brain, when faced with the task of characterizing color appearance, abstracts that information from the responses of mechanisms used for object analysis.

Acknowledgments

The work described in this chapter was supported by the National Institutes of Health through grants EY 04440 and EY 01319. The author is grateful to Andrew Metha for critical comments.

13

Chromatic signals in extrastriate areas V2 and V3

Daniel C. Kiper, Jonathan B. Levitt, and Karl R. Gegenfurtner

In primates, color opponency is already observed in the retinal ganglion cells. They combine the inputs from the short-, middle-, and long-wavelength–sensitive cones linearly in different fashions, resulting in three classes of color-opponent cells. These early "cardinal direction" mechanisms (commonly called red-green, blue-yellow, and luminance) are preferentially represented in the lateral geniculate nucleus (LGN) as well (DeValois, Abramov, & Jacobs, 1965; Derrington, Krauskopf, & Lennie, 1984). There, neurons in the magnocellular layers have high sensitivity for luminance contrast and low sensitivity to chromatic contrast. Neurons in the parvocellular layers, on the other hand, have lower sensitivity to luminance-defined stimuli but respond well to stimuli defined solely by color (for reviews, see Livingstone & Hubel, 1987, and Shapley, 1990). It is thus frequently assumed that the cells in the magnocellular (M-cells) layers represent the physiological basis for the luminance channel, while those in the parvocellular layers (P-cells) are the substrate for the color-opponent channels. The spatial and spectral characteristics of these mechanisms alone can explain a surprisingly large number of psychophysical results, and they correspond remarkably closely to the second-stage opponent mechanisms revealed psychophysically (Lee, Martin, & Valberg, 1988; Krauskopf, Williams, & Heeley, 1982; see Chapter 16).

While a great deal of knowledge has been accumulated about the processing of color information at these early stages of the visual pathways, little is known about the chromatic properties of cells in most visual areas of the cerebral cortex. A number of experimenters have specifically investigated the treatment of color information in the striate visual cortex (V1) of primates (Hubel & Wiesel, 1968; Dow & Gouras, 1973; Gouras, 1974; Yates, 1974; Michael, 1978a, 1978b, 1978c, 1979; Livingstone & Hubel, 1984; Thorell, DeValois, & Albrecht, 1984; Ts'o & Gilbert, 1988; Lennie, Krauskopf, & Sclar, 1990; see Chapter 12) and in extrastriate area V4, which has been suggested to play an important role in the cortical analysis of color information (Zeki, 1980; Schein, Marrocco, & de Monasterio, 1982; Schein & Desimone, 1990). Early studies of the primate primary visual cortex reported only a very small proportion of chromatically responsive cells (Hubel & Wiesel, 1968). Since then several studies showed that many cells that respond to luminance variations also respond to color variations, bringing the overall proportion of color-selective cells to about 50% in the striate cortex of macaque monkeys (Dow & Gouras, 1973; Gouras, 1974; Yates, 1974; Thorell, Albrecht, & DeValois, 1984). Yates (1974) and Gouras (1974) reported the presence of an unspecified number of cells sensitive to a narrow range of wavelengths in the striate cortex, a specificity that is absent at earlier stages.

In addition, several studies reported the existence of a large population of double-opponent cells in the primary visual cortex of primates (Michael, 1978a, 1978b, 1978c, 1979; Livingstone & Hubel, 1984).

These cells have a spatially and chromatically antagonistic center-surround organization, and were hypothesized to play an important role in achieving color constancy (Zeki, 1980). In a more recent and quantitative study Lennie, Krauskopf, and Sclar (1990) reported that less than 2% of the cells in V1 are narrowly tuned to wavelength (but see Cottaris & DeValois, 1998). Furthermore, they and others (Ts'o & Gilbert, 1988) failed to demonstrate the existence of double-opponent cells in this area. Their experiments show that, as in the LGN, cells in the primary visual cortex also tend to combine cone inputs linearly. They also observed color preferences of individual neurons in V1 that do not exist in LGN neurons. Whereas in the LGN the cardinal directions, red-green and blue-yellow, are preferentially represented, all possible color combinations are found in V1 (Lennie, Krauskopf, & Sclar, 1990).

Narrow-wavelength selectivity and color constancy are also the properties that some studies have found in V4 of macaques (Zeki, 1980) and that have led to the characterization of area V4 as the most important area for color processing. Although more recently the interpretation of these results has been questioned on methodological grounds (de Monasterio & Schein, 1982; Schein, Marrocco, & de Monasterio, 1982), it is generally accepted that a major proportion of the neurons in area V4 selectively respond to color (Schein & Desimone, 1990). Thus, it appears that there are a number of controversial issues concerning the processing of color information in cortical visual areas, and that our understanding of these areas is relatively poor. In particular, very little is known about the treatment of chromatic information in areas V2 and V3, which are located between V1 and V4 in the cortical hierarchy.

In addition, little is known about the way in which chromatic signals are analyzed and integrated with the processing of other visual attributes such as form or motion. One possible scheme is that the treatment of various visual attributes is performed in parallel, along functionally segregated pathways, and that these different types of information are ultimately recombined at an unspecified later stage of the visual system. Another possibility is that information about each visual attribute is available at all stages of visual processing, each performing a particular type of analysis that involves several visual attributes. The segregation hypothesis has been favored in recent years based on a variety of anatomical, physiological, and psychophysical findings. It has been proposed that the anatomical and functional segregation found in the LGN continues in cortical areas V1, V2, and beyond, with information processed in separate streams for the analysis of form, color, and motion (Livingstone & Hubel, 1984; DeYoe & Van Essen, 1985; Hubel & Livingstone, 1987; Zeki & Shipp, 1988). According to the most restrictive version of that scheme, the information carried by the P- and M-cells, whose terminal arbors span different layers of V1, take completely separate routes within the visual cortical areas. They eventually give rise to two functionally distinct pathways, a "what" (or "color and form") pathway in the temporal visual areas, whose activity is mostly dominated by P-cells, and a "where" (or "motion") pathway (Ungerleider & Mishkin, 1982; Livingstone & Hubel, 1988), dominated by M-cell inputs, in the parietal areas. It has also been proposed that these pathways through areas V1 and V2 are correlated with the staining pattern of the metabolic enzyme cytochrome oxidase (CO). Figure 13.1 summarizes the major feed-forward connections among the early visual areas and the relationship of these pathways to the pattern of CO-defined subregions in areas V1 and V2. Areas or compartments that receive primarily magnocellular inputs are shaded dark, and those with primarily parvocellular inputs, light. There seems to be a consensus now that area MT is dominated by magnocellular inputs but also receives a weak parvocellular input (Maunsell, Nealey, & DePriest, 1990). Area V4, in contrast, seems to receive equally strong inputs from magno- and parvocellular streams (Ferrera, Nealey, & Maunsell, 1992, 1994). A strict segregation of geniculate projections seems to occur only in the input layers of V1, 4Cα, and 4Cβ. In the upper layers of V1 the information originating in magno- and parvocellular LGN compartments can no longer be disentangled.

The question is whether this anatomical segregation corresponds to a segregation of visual functions, as

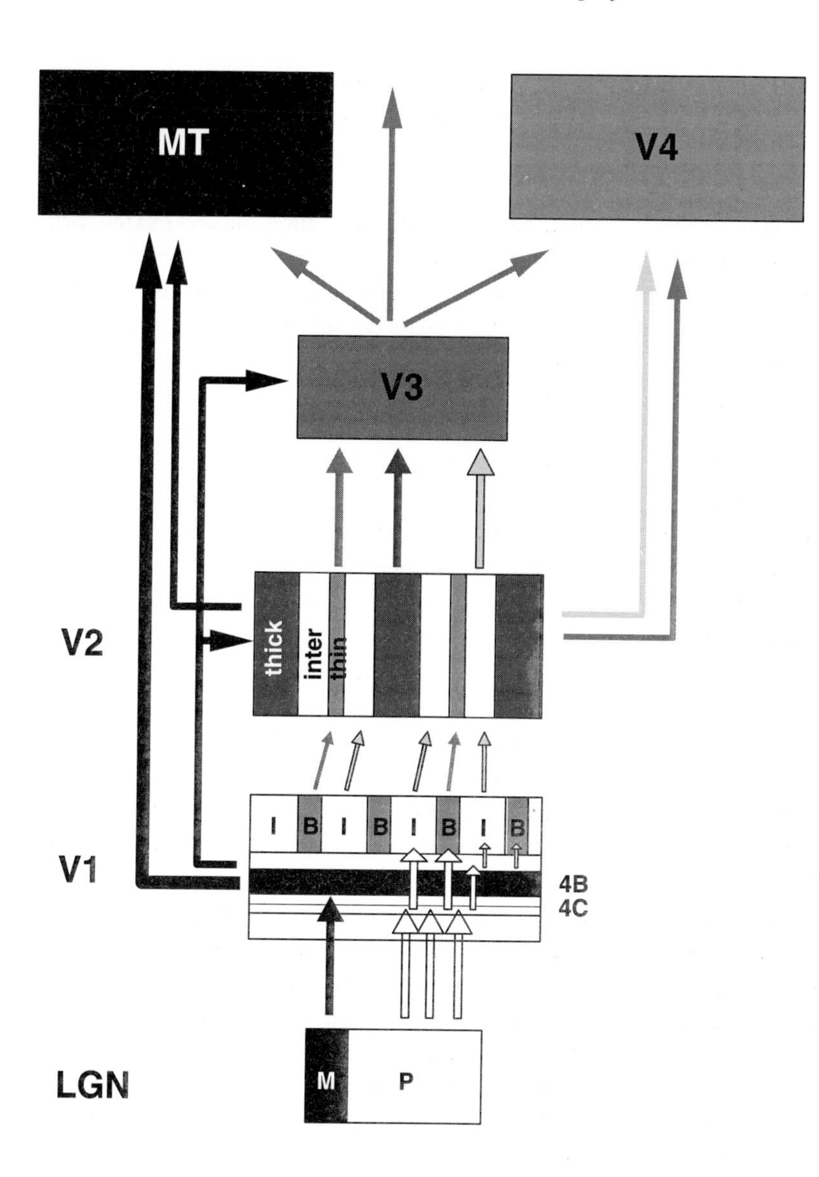

Figure 13.1: Schematic diagram of the major feedforward connections between early visual areas. Areas/projections dominated by magnocellular (M) signals are shaded dark and those by parvocellular (P) signals are shaded bright. Intermediate shades of grey indicate mixed P- and M-signals. M and P LGN inputs target different divisions of layer 4C in V1. From there, they are relayed to layer 4B or the cytochrome oxidase (CO) defined blob (B) and interblob (I) compartments of layers 2/3. These different V1 compartments then relay to different CO-defined compartments in area V2: the CO-rich thick and thin stripes, and the CO-poor interstripes. Areas V1 and V2 also relay to extrastriate areas V3, V4, and MT (V5), although the direct projection from V1 to V4 is much less prominent and essentially restricted to the foveal representation.

was originally suggested (see Merigan & Maunsell, 1993, for a review). It is clear that the representations of different stimulus attributes must ultimately be combined somewhere to form a unified percept of the world. This might reasonably be accomplished by the convergence of signals from specialized areas onto neurons in higher cortical areas, or by the interaction between these specialized pathways earlier in the visual system. Indeed, the existence of anatomical substrates for such interaction between different functional pathways is already well known, both within areas V1 and V2 (Lachica et al., 1992; Levitt, Yoshioka, & Lund, 1994; Yoshioka et al., 1994) as well as between certain specialized areas themselves (reviewed in Felleman & Van Essen, 1991).

We focused our research onto two central questions concerning the treatment of chromatic information within the cortical visual areas. First, we examined the relationship between the chromatic information and information about other visual stimulus attributes such as form or motion. Second, we examined the nature of the processing of color per se that goes on within these areas. We studied areas V2 and V3, which are located between V1 and V4 in the cortical hierarchy. Therefore, it is important to understand the nature of the functional processing of color information that takes place in these areas. These areas are well suited to studying the relationships between the processing of color, form, and motion. In area V2, the streams for the processing of motion, color, and form are assumed to correspond to the thick, thin, and interstripe regions as defined by cytochrome oxidase (CO) staining (DeYoe & Van Essen, 1985; Shipp & Zeki, 1985; Hubel & Livingstone, 1987). This organization of adjacent parallel strips of tissue would seem to provide a good substrate for interaction, making it an ideal candidate to study the relationships between color and other stimulus attributes. The role of cortical area V3 is also particularly interesting because this area, like V2, is known to have prominent connections with both cortical functional streams, in particular with areas V4 and MT. Unlike V2, no anatomically or functionally distinct compartments have been found in V3. Area V3 may therefore represent an important site for the coordinated activity of both pathways, as well as for the integration of different stimulus attributes for visual perception.

We used standard electrophysiological techniques to analyze the chromatic properties of cells in the various CO compartments of V2 and in V3, and compared their selectivity for color to that for other attributes, namely, orientation, motion, and size.

Methods

These experiments were performed on young adult macaque monkeys (*Macaca fascicularis* or *Macaca mulatta*, whose cones' spectral sensitivities and early visual systems resemble those in the human (DeValois et al., 1974). The activity of single cells was recorded extracellularly by means of a microelectrode. The detailed methodology for these experiments has been published elsewhere (Levitt, Kiper, & Movshon, 1994; Gegenfurtner, Kiper, & Fenstemaker, 1996; Gegenfurtner, Kiper, & Levitt, 1997; Kiper, Fenstemaker, & Gegenfurtner, 1997). In brief, each animal was anesthetized to prevent pain and paralyzed to prevent eye movements. A craniotomy allowed access to the cortical region of interest (V2 or V3), and a tungsten-in-glass microelectrode was lowered into the brain. Electrode penetrations were directed toward the posterior bank of the lunate sulcus for V2 recordings, or toward the fundus of the lunate sulcus and annectant gyrus buried within the lunate sulcus for V3 recordings. We used the stereotypical sequence of gray matter–white matter transitions to help us determine when we had entered V2 or V3. Once the activity of a single cell was isolated, visual stimuli were displayed on a computer screen placed in front of the animal. A series of experiments to characterize the cell's receptive field properties was then performed, under the control of a microcomputer. Each penetration was marked by small electrolytic lesions to allow subsequent histological reconstruction. At the end of the experiments, the animal was perfused and the brain fixed and sectioned to allow identification of each penetration.

Anatomical methods. Tissue sections were stained for cytochrome oxidase (CO), the Cat-301 antibody, or myelin. Earlier studies relied on visual inspection of the CO stripe width to determine the boundaries of particular CO compartments. While this may suffice in some primate species (such as squirrel monkeys), the pattern of CO staining in macaque V2 can be quite irregular; stripe width alone is insufficient to reliably differentiate between "thick" and "thin" stripes. We therefore thought it important to use an independent marker, Cat-301, which preferentially labels the "thick" CO stripes that receive input from V1 layer 4B and project to area MT (DeYoe et al., 1990). This allowed us to distinguish the two different CO-rich compartments in V2 reliably and independently of our physiological assessment. It was then possible to determine the laminar position of the recorded cells as well as their location within a given cortical area or CO compartment. These standard electrophysiological and histological techniques allowed us to obtain precise, quantitative measures of receptive field properties as well as to gain insight into the functional architecture of the visual areas under study.

Visual stimulation. Stimuli were displayed on a BARCO color television monitor driven by an AT TrueVision Vista Graphics board. Color modulations were defined along the three cardinal directions of color space introduced by Krauskopf, Williams, and Heeley (1982). At the origin is a neutral white with C.I.E. (x,y)-coordinates (0.31, 0.32) and a mean luminance of 37.5 cd/m^2. Along one axis (L–M), the excitation of the L- and M-cones covaries so as to keep their sum constant. Along another axis (S–(L+M)), only the excitation of the S-cones varies. Along the "luminance" axis, the excitation of all three cone types varies in proportion to their excitation at the white point. A stimulus in this space can also be defined by its azimuth, which is the angle made by its projection on the isoluminant plane with the L–M-axis and by its angular elevation above the isoluminant plane (see Chapters 12 and 16 for further description of this color space).

Stimuli were drifting sine-wave gratings whose color was modulated around the neutral white point. To measure selectivity for spatial frequency, temporal frequency, orientation, direction, and size, we used black and white gratings modulated in luminance. To assess the chromatic properties, we used two different types of stimuli. In the first set of experiments, where we determined the relative selectivity for color, we used square-wave gratings with alternating colored and black bars. The different colors used in this experiment are illustrated in the C.I.E. diagram shown in Fig. 13.1A. All stimuli had an equal luminance of 37.5 cd/m^2. As a control, a stimulus consisting of black and gray bars was used, with the luminance of the gray bars equalling that of the colored bars (37.5 cd/m^2).

For the experiments concerning the intrinsic chromatic properties of the cells, we used sine-wave grating stimuli whose color was modulated around a neutral white point. In the color space introduced above and shown in Fig. 13.2A, the set of stimuli was varying in azimuth and had a fixed elevation of 10 deg.

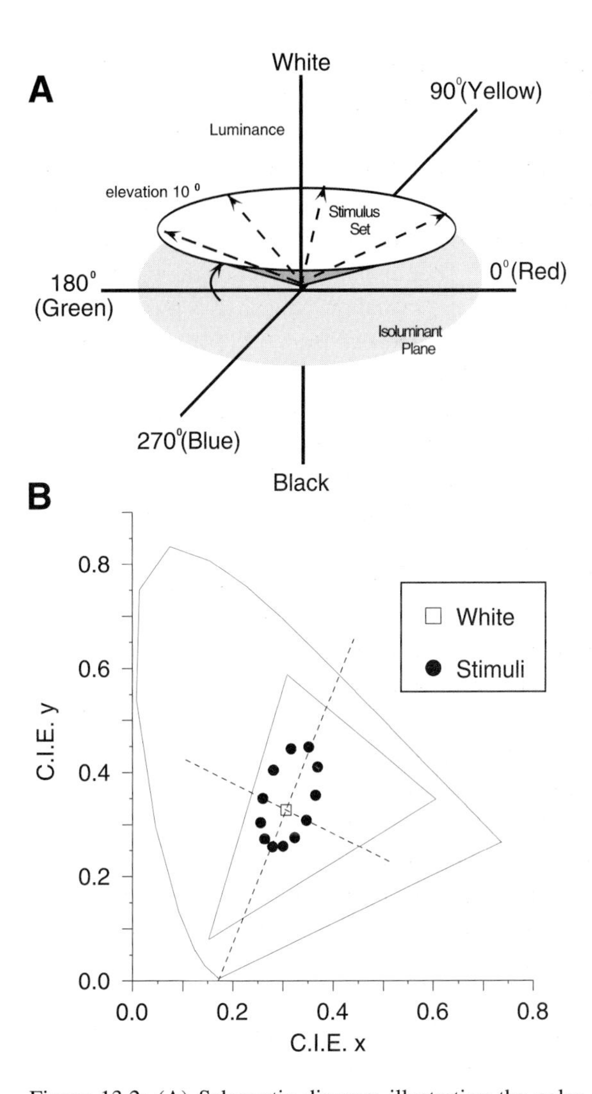

Figure 13.2: (A) Schematic diagram illustrating the color stimuli used in the color experiments. At the origin is a neutral white. Along the L–M-axis, the excitation of the L- and M-cones covaries so as to keep their sum constant. Along the S–(L+M)-axis, only the excitation of the S-cones varies. Along the "luminance" axis, the excitation of all three cones varies in proportion to their excitation at the white point. The inverted cone represents the set of stimuli with an elevation of 10 deg and an azimuth varying between 0 and 360 deg. (B) C.I.E. xy chromaticity diagram illustrating the same color stimuli. The dashed lines indicate the placement of the axis of the color space described above. The triangle indicates the C.I.E. coordinates of the phosphors of our monitor.

Figure 13.2B shows the colors used in the more traditional C.I.E. diagram. The inverted cone in Fig. 13.2A represents the set of stimuli with an elevation of 10 deg and an azimuth varying between 0 and 360 deg. All of these stimuli were sine-wave gratings modulated in this space around the origin, an equal-energy white of 37.5 cd/m^2. They all had an elevation of 10 deg, which gave all of the stimuli a luminance contrast of 12%. The azimuth was varied from 0 to 360 deg, in 30-deg steps. We chose an elevation of 10 deg to excite purely luminance as well as chromatically opponent cells. The scaling of the three axes in this space is arbitrary; in our case, the maximum modulation along the luminance axis was 100% luminance contrast, 85% S-cone contrast along the S–(L+M)-axis, and 13% M-cone modulation along the L–M-axis. All of the stimuli had equal length according to that particular scale. All of the cells gave good responses to at least some of these stimuli. A black and white luminance stimulus with the same contrast of 12% was included in the stimulus set as a control to differentiate between luminance and chromatic contributions to the response. Finally, we used sine-wave gratings made of equiluminant colors in experiments where we wanted to characterize the responses of the cells in the absence of a luminance component.

Parallel pathways

We performed quantitative experiments on 140 individual neurons in area V2 and 171 neurons in area V3 (of *M. mulatta* and *M. fascicularis*). Over 90% of the units sampled in both areas had receptive fields centered within 5 deg of the fovea. We first determined each neuron's spatial and temporal frequency tuning. Then we measured the tuning and responses to otherwise optimal stimuli varying in orientation, direction of motion, size, and color. Responses were compiled into average histograms synchronized to each temporal cycle of the stimulus. These histograms were then Fourier analyzed to calculate the average response (DC) and the response at the fundamental stimulus frequency (F1). For complex cells, which respond with an unmodulated elevation in discharge rate, we used the DC response (mean firing rate minus spontaneous activity) as the response measure; for simple cells we used the F1 response. From the tuning curves we computed indices that served as measures of the neuron's selectivity to a given stimulus attribute.

Basic experiments. Figure 13.3A shows responses of a cell (3781012) from a V2 thick CO stripe to black and white sinusoidal gratings moving at different orientations. The gratings were of optimal spatial frequency, temporal frequency, and size. This cell responded very briskly to gratings oriented at an angle of approximately 60 deg (upward and to the right) and did not give any response at the orthogonal or opposite directions. The dashed horizontal line shows the spontaneous firing of the cell in response to a blank screen. Standard errors of the response are shown, but for this cell they were smaller than the symbols in most cases. We fitted a smooth function through the data points (shown by the solid curve) and defined a direction index:

(1) $DI = 1 - (R_{opp} - b) / (R_{pref} - b)$,

where R_{pref} is the response in the preferred direction, R_{opp} is the response in the direction opposite to the preferred, and b is the baseline response. By using the same data, we also defined an orientation index:

(2) $ORI = 1 - (R_{ortho} - b) / (R_{pref} - b)$,

where R_{ortho} is the average response in the two directions orthogonal to the preferred direction.

Cells with a $DI > 0.7$ were classified as directionally selective and cells with an $ORI > 0.7$ were classified as oriented. These criteria require at least a threefold increase in firing at the optimum (relative to the opposite direction or orthogonal orientation) for a cell to be classified as directionally selective or orientation-selective. For the cell in Fig. 13.3A, the values of DI and ORI are both larger than 1, indicating that the response was slightly inhibited in the opposite direction and at the orthogonal orientation.

Figure 13.3B shows the responses of a different V2 cell (372r006) from an interstripe region to black and white sinusoidal gratings in different window sizes. In our experiments we restricted ourselves to square stimuli, and size refers to the width or height of the stimuli. All of the other parameters were set to optimize the cell's response. Endstopping was defined as:

(3) $ES = 1 - (R_{large} - b) / (R_{pref} - b)$,

where R_{large} is the response to the largest stimulus used and R_{pref} is the response to the stimulus with the preferred size.

Cells with an $ES > 0.5$ were classified as being end-stopped. This particular cell was almost completely inhibited by large stimuli ($ES = 0.96$).

For yet another V2 cell from an interstripe region (3681007), responses to bars of different colors are shown in Fig. 13.3C. The stimuli in this experiment were square-wave gratings made of alternating black and colored bars. Orientation, spatial frequency, temporal frequency, and size were set to the cell's preferred values. The solid horizontal line shows the response to a white bar of the same luminance (37.5 cd/m^2) as the colored bars. We defined color responsivity as

(4) $CR = (R_{col} - b) / (R_{white} - b)$,

where R_{col} is the best response to any of the colored bars, and R_{white} is the response to the white bar.

Cells with a $CR > 1.4$ were classified as color-selective. For this cell we found that the response to a green bar was six times as great as the response to the white bar (after subtracting the baseline); it gave hardly any response to the white bar.

Selectivity to form, color, and motion. If there is a functional segregation of different visual functions, we would expect the tuning properties of neurons in the different CO compartments in V2 to reflect that fact. Subregions in V2 can be distinguished on the basis of staining for the metabolic enzyme cytochrome oxidase (CO); these take the form of parallel stripes of

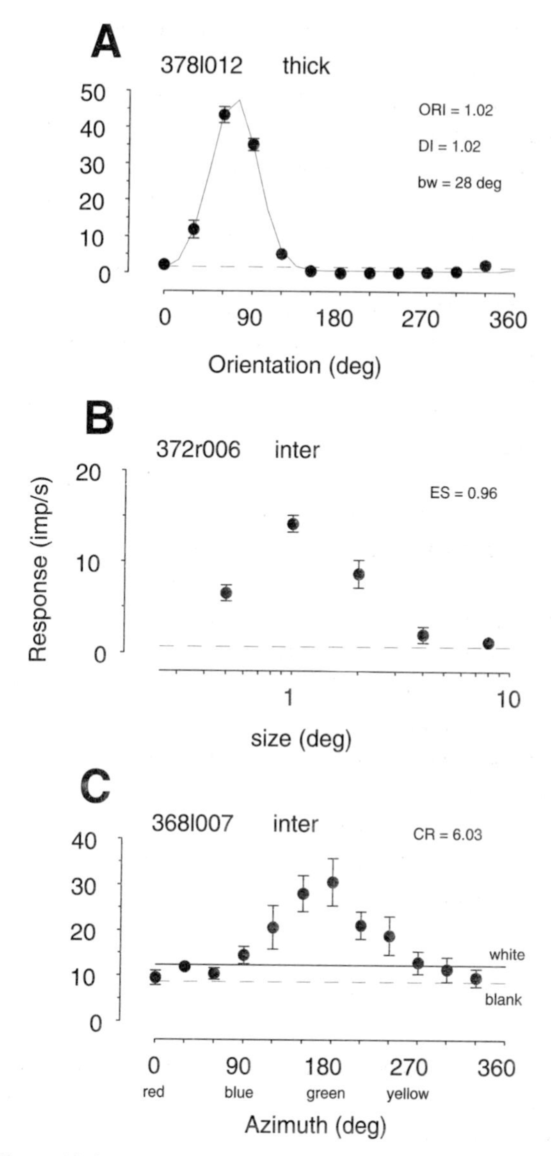

Figure 13.3: Basic measurements of receptive field properties. Response properties of three representative V2 cells. The dashed horizontal line in each panel indicates the response to a uniform gray field with the same mean luminance as the stimuli. (A) Responses of a cell from a thick stripe to black/white sinusoidal gratings of different orientations. The thin line represents a smooth function fit to the data. (B) Responses of a cell from an interstripe region to sinusoidal grating patches of different sizes. (C) Responses to bars of different colors for a cell from an interstripe region. Stimuli were alternating black and colored bars. The solid horizontal line shows the response to a white bar of the same luminance (37.5 cd/m^2) as the colored bars.

tissue, the thick and thin stripes of intense CO activity separated by less reactive pale interstripes. It is also known that anatomical pathways from V1 and on to higher parietal or inferotemporal cortical visual areas are segregated into different V2 CO stripes (Livingstone & Hubel, 1984, 1987; DeYoe & Van Essen, 1985; Shipp & Zeki, 1985, 1989a; Zeki & Shipp, 1989). If visual pathways through V2 were indeed strictly segregated, then one might expect (as has been suggested by others; see Hubel & Livingstone, 1987) that the functional properties of neurons such as sensitivity for color, direction, or form (i.e., orientation or size) might also reflect this segregation. We therefore examined whether the neurons selective for particular stimulus attributes were segregated or clustered into different CO stripes in V2. Although previous studies have concluded that functional segregation in V2 is at best incomplete (DeYoe & Van Essen, 1985; Peterhans & von der Heydt, 1993; Levitt, Kiper, & Movshon, 1994; Tamura et al., 1996), these studies did not investigate the chromatic properties of neurons in any great detail.

Area V3 also projects to both parietal and temporal visual areas, but it is not known whether the anatomical pathways through area V3 remain separate as they do in area V2. Furthermore, very little is known of the physiological properties of V3 neurons (Felleman & Van Essen, 1987). Since there is no anatomical evidence for the segregation of specialized pathways through area V3, it is of similar interest to determine physiologically whether single V3 neurons are selective for particular stimulus attributes in a way that might mirror anatomical specificity. Alternatively, there might be no functional segregation in the area; single V3 neurons might be selective along several stimulus dimensions. As in V2, there is little description of the chromatic properties of V3 neurons and how selectivity for stimulus color relates to that for other stimulus attributes.

Figure 13.4 shows the proportion of neurons selective for different stimulus attributes in each V2 CO compartment and in V3 . There were some tendencies toward functional segregation, for example, that color selectivity was most common in the thin stripes, size selectivity in the interstripes, or that orientation selectivity was somewhat less common in the thin stripes. However, despite the evidence for some degree of segregation, there was clearly no absolute segregation of selective sensitivity to color, form, or motion information into different pathways. Neurons showing selectivity to any attribute could be found in each of the CO compartments. Furthermore, these differences did not depend on the particular criteria used to classify cells, and they were essentially the same in all cortical layers (Gegenfurtner, Kiper, & Fenstemaker, 1996). Figure 13.4 also compares the overall proportion of neurons in areas V2 and V3 that are selective for each of these stimulus attributes, using identical classification criteria in each area. It is clear that the only notable difference between V2 and V3 is the greater incidence of direction selectivity in V3 (roughly 40% versus 20% in V2). In both areas, approximately 85% of the population was orientation-selective, 25% size-selective (endstopped), and 50% color-selective.

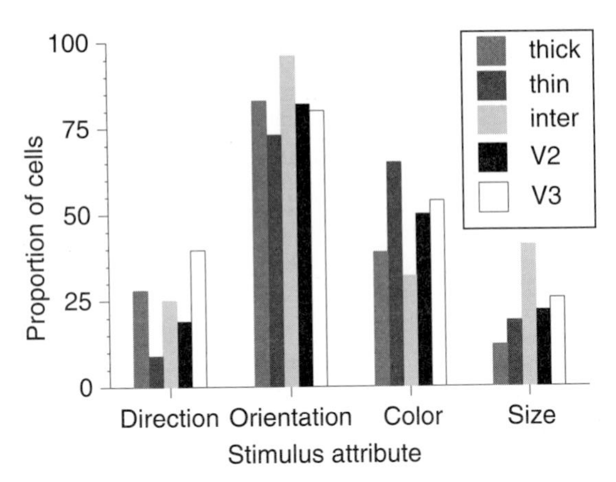

Figure 13.4: Proportion of neurons selective to direction, orientation, color, or size. Different shadings indicate different CO compartments in area V2 (thin: dark gray; thick: medium gray; inter: light gray), area V2 as a whole (filled bars), and area V3 (open bars). The criteria used to classify cells as selective for each attribute were $DI > 0.7$, $ORI > 0.7$, $ES > 0.5$, and $CR > 1.4$.

Association between attributes. So far we have described selectivities to particular stimulus attributes across the total sample and in relationship to the CO architecture. An issue of equal interest is the extent of integration and correlation among the response properties themselves; we therefore investigated the relationships among the tuning characteristics of V2 and V3 neurons for different stimulus attributes. If different stimulus attributes were processed independently, then one might expect neurons to show selectivity primarily to one stimulus attribute, but not to several attributes simultaneously. Alternatively, it could also be that each attribute has the whole range of other visual attributes associated with it. Figure 13.5 shows scatterplots of orientation, direction, and endstopping indices versus color responsivity for all V2 and V3 cells for which we were able to measure these pairs of characteristics. The solid horizontal and vertical lines indicate our criterion values for classifying a cell as selective to that particular attribute. In both V2 and V3, there was no significant correlation between any of these selectivities. For example, in both areas there were cells that were highly selective for both stimulus color and orientation, or for stimulus color and size. These are the cells that fall above and to the right of the criterion lines in Fig. 13.5. Where V2 and V3 did seem to differ was in the association between color and direction selectivity. In V3 we observed a population that was highly selective for both of these stimulus attributes; this group seemed to be absent from V2.

The scatter plots in Fig. 13.5 also indicate that all four indices are distributed continuously through the V2 and V3 samples. Thus, any criterion used to classify cells as selective for a certain attribute is arbitrary. Nonetheless, we used the classification criteria defined above to compute a statistical measure for the degree of association between each pair of stimulus attributes. Fisher's exact test for probabilities (see Hays, 1981, pp. 552–555) confirmed that the probability with which a given neuron was color-selective did not depend on whether the cell was also selective for stimulus orientation, direction of motion, or size. In other words, these data do not support the hypothesis that the different stimulus attributes are processed in parallel in V2 and V3; rather, it seems that there are neurons tuned to any possible combination of attributes. The stimulus space spanned by color, orientation, direction, and size seems to be covered densely by the population of neurons.

Responses to isoluminant stimuli. Another way of illustrating that color information is not represented independently from other attributes is to compare the tuning for different stimulus parameters using both luminance and isoluminant stimuli. For 20 V2 cells and 9 V3 cells that gave robust responses to isoluminant chromatic stimuli, we measured the cells' tuning for orientation, spatial and temporal frequency, as well as its contrast response function. For orientation and spatial and temporal frequency, the contrast of the isoluminant gratings was always the maximum that we could achieve on our display monitor in the particular color direction (azimuth) that the cell preferred. We then compared these measurements to the earlier measurements using achromatic gratings of maximum contrast (near 100%).

Figure 13.6A shows a scatterplot of the optimal orientations determined using luminance and isoluminant chromatic stimuli. The correlation coefficient between these two measures was 0.98, and the mean difference in optimal orientation was 3.5 deg. Similarly, selectivity for orientation as defined by the orientation index was not significantly different for the two types of stimuli ($t_{27} = 1.18, p > 0.1$) and was highly correlated ($\rho = 0.68$). For some V2 cells we looked at other aspects of spatiotemporal tuning as well. For 19 V2 cells the median optimal spatial frequency was 1.16 c/deg at isoluminance versus 1.03 c/deg for luminance stimuli; this small difference was not statistically significant ($t_{18} = 1.47, p > 0.05$), and the correlation between spatial frequency optima was high ($\rho = 0.87$). There was only a small difference in the optimal temporal frequency between the two stimulus types (2.25 Hz at isoluminance versus 2.69 Hz to luminance). There was no variation in the spatial or temporal bandwidths with stimulus type. Finally, we were able to hold long enough only nine V2 cells to measure their contrast responses under both conditions; for these we observed no systematic differences in the steepness or shape of the contrast responses.

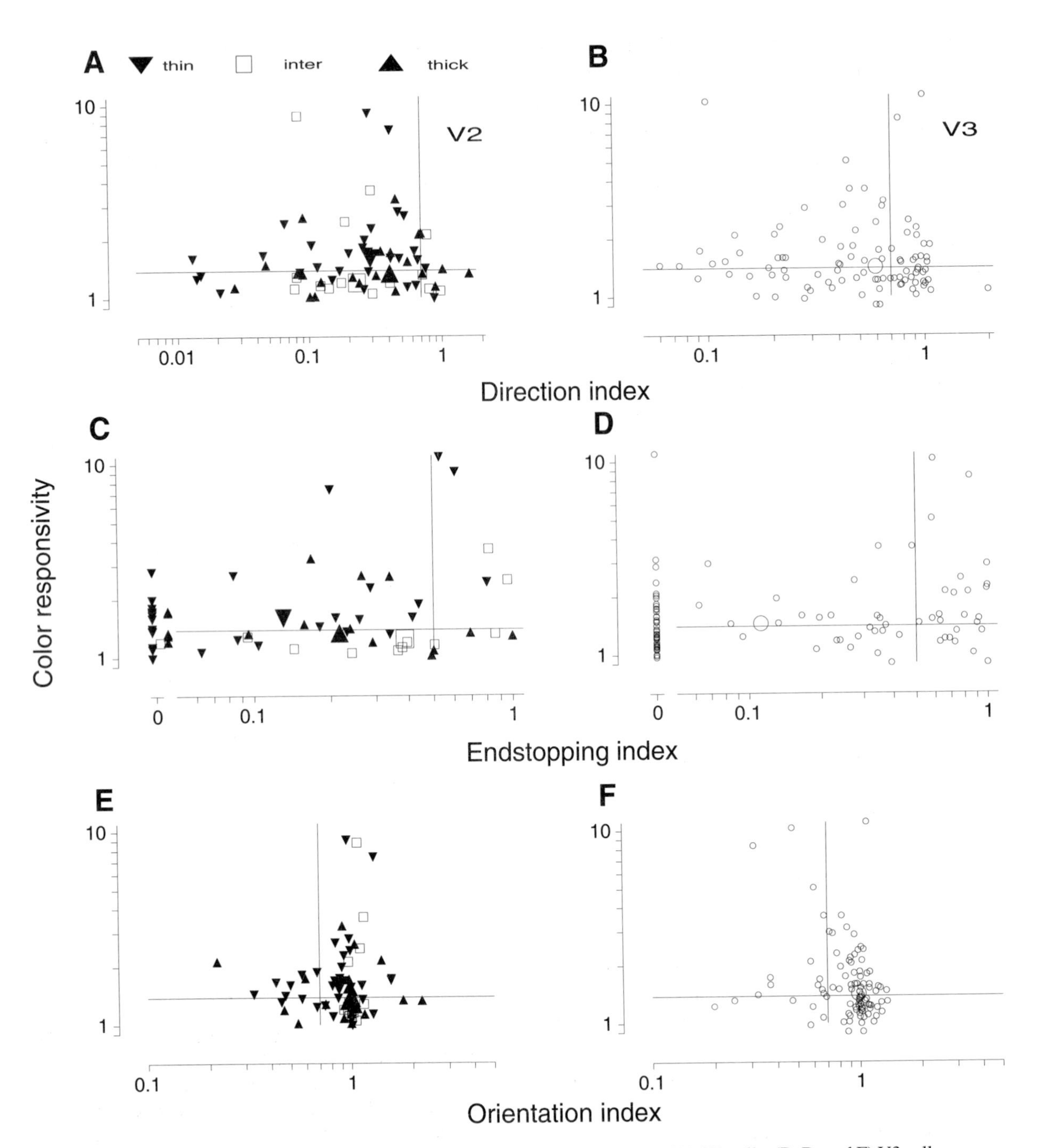

Figure 13.5: Association between different response selectivities. (A, C, and E) V2 cells. (B, D, and F) V3 cells. Scatterplot of color responsivity values vs. orientation, direction, and endstopping indices. Solid lines indicate criterion values above which cells were classified as selective for a particular attribute. In panels A, C, and E different symbols and shadings indicate V2 cells in different CO compartments, and large symbols indicate the median values for the three groups. In panels B, D, and F the large open symbol indicates the population mean for each parameter in V3.

These results show a strong similarity of the tuning for luminance and isoluminant stimuli, but responses at isoluminance do not necessarily have to be due to color-opponent inputs. Previous research (Gegenfurtner et al., 1994; Dobkins & Albright, 1995) has shown that the isoluminant point can vary slightly from cell to cell, and this can lead to a luminance-based response to nominally, that is, photometrically, isoluminant stimuli. To establish the relative magnitude of the response at or near isoluminance, we tested cells with a range of stimuli at different elevations around zero (isoluminance). In this way we could detect response minima or response nulls even in cells that do not strictly adhere to photometric isoluminance.

Two sets of stimuli were used: black and white achromatic gratings of increasing contrasts and heterochromatic gratings, which consisted of an isoluminant colored grating to which we added a black and white achromatic grating. The azimuths of the heterochromatic gratings were chosen to be the ones where the cells gave the best response in the previous experiment. Because adding luminance contrast amounts to changing the elevation of the stimulus, we refer to this set of stimuli as the "elevation" experiment. If a cell simply responds to the luminance component of the stimulus, its response will be the same to the achromatic and the heterochromatic grating of the same luminance contrast. Thus, when the heterochromatic grating is isoluminant, the response will be zero. This zero, or point of isoluminance for each cell, frequently did not correspond to that predicted by the human photopic luminance sensitivity curve $V(\lambda)$. For these cells a small amount of luminance contrast needs to be added or subtracted to obtain a zero response. The response curves for chromatic and achromatic stimuli will be parallel but shifted horizontally. On the other hand, cells that receive color-opponent inputs should behave in a different way. They should respond well to all color stimuli regardless of their luminance contrast. These cells should not have a null response for any of the chromatic stimuli in this experiment.

We ran this experiment on cells for which we had previously established the preferred azimuth in our color space. Since psychophysical experiments using

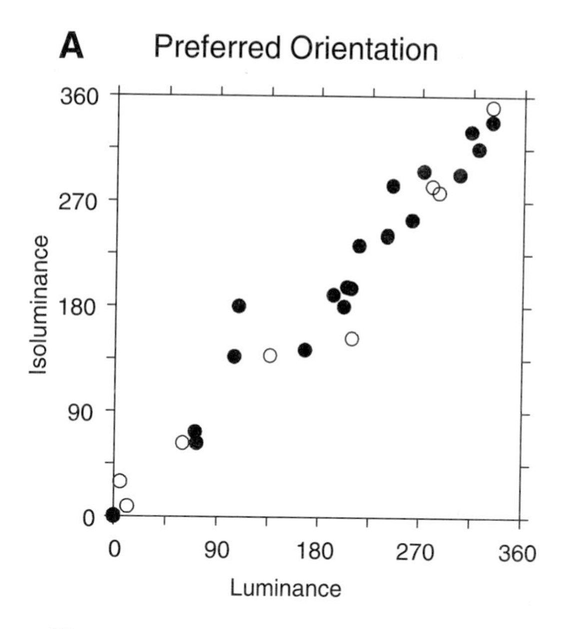

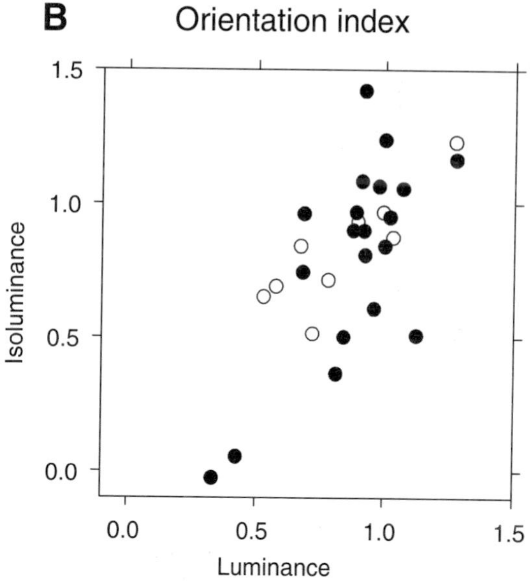

Figure 13.6: (A) Scatterplot of the preferred orientation of 20 V2 cells (filled symbols) and 9 V3 cells (open symbols) for chromatic gratings versus achromatic gratings. Preferred orientations were derived from fits of a smooth function to the data. (B) Scatterplot of the orientation indices for the same cells.

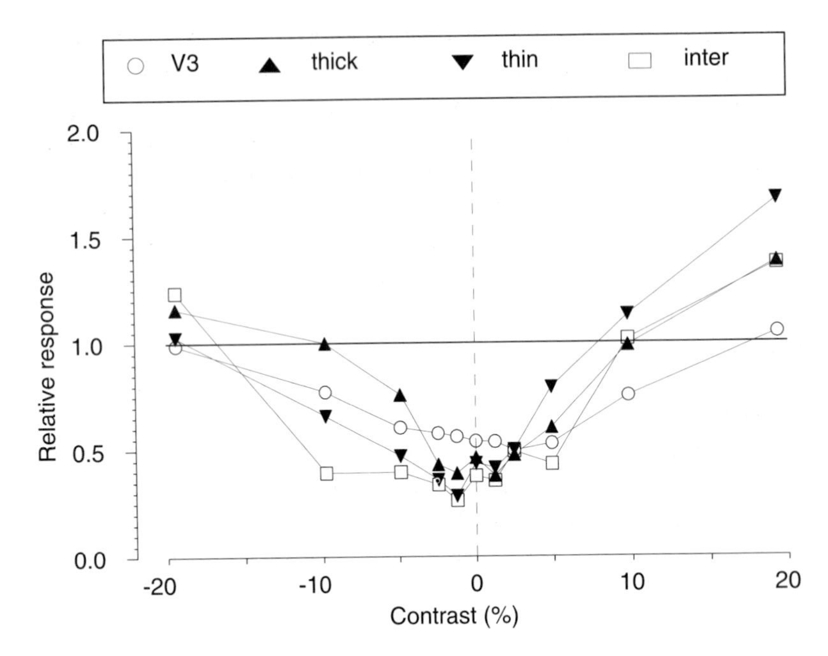

Figure 13.7: Averaged responses to stimuli with a fixed chromatic contrast and differing amounts of luminance contrast (x-axis). Each neuron's baseline response was subtracted before averaging. The response is plotted relative to the response to a luminance grating of 18% contrast, indicated by the horizontal line. Averaged population responses of neurons from V3 (open circles) and the three CO compartments of V2 (thick: upward pointing triangles; thin: downward pointing triangles; inter: squares) are indicated by the different symbols.

isoluminant stimuli typically use photometrically isoluminant stimuli, it was of great interest to determine the overall response of cells for that particular stimulus. For 14 of 33 cells (42%) in V2 and 19 of 71 cells in V3 (26.8%) the response at isoluminance was less than 10% of the maximum response to any of the achromatic gratings. The median response at isoluminance was at 19.1% of the achromatic maximum in V2 and 40.5% in V3. These characteristics are visible in the population's average response, shown in Fig. 13.7.

In both V2 and V3, the response to the chromatic stimuli exceeded that to the achromatic stimuli at all of the contrasts that we used. Moreover, the response curves to the chromatic stimuli did not have a null, although they had a minimum near photometric isoluminance (dashed vertical line). Even though individual null points were shifted leftward or rightward, the average curve was centered at photometric isoluminance. These averaged responses illustrate the fact that the chromatic content of visual stimuli does contribute significantly to the responses of cells in V2 and V3, and that there exists no single color-luminance combination that can totally silence the responses of these areas as a whole. At photometric isoluminance the population response in area V2 was at 44.04% of the maximum response to any of the achromatic stimuli, and it was at 54% in V3. The same held for individual CO compartments of V2. Figure 13.7 also shows the response relative to the highest luminance response (at 18% contrast) of populations of neurons from each of the three different CO compartments. The average response at isoluminance was at 46.3% of the maximum achromatic response in the thick stripes ($N = 8$), at 43.7% in the thin stripes ($N = 19$), and at 37.8% in the interstripes ($N = 6$). In all of the compartments the chromatic component had a facilitative effect on the population response. For comparison, in area MT, an area with primarily magnocellular input, the average response at isoluminance was at 27% of the luminance response (Gegenfurtner & Hawken, 1996b).

Our results suggest that there are no strictly separate pathways in areas V2 and V3. While area V2 does indeed show some degree of segregation along the different CO compartments, no segregation at all is evident in area V3. This is compatible with the notion that area V3 plays an important role in combining information from different pathways. Since color-selective cells were quite common in both V2 and V3, we will now describe in more detail the chromatic properties of neurons in these areas and show how their chromatic selectivities differ from neurons earlier in the visual pathway.

Chromatic properties

Chromatic selectivity. To measure chromatic selectivity we used a set of sinusoidal stimuli with varying azimuths and a fixed elevation of 10 deg, as illustrated in Fig. 13.2A. The inverted cone represents the set of stimuli with an elevation of 10 deg and an azimuth varying between 0 and 360 deg. In contrast to the square-wave stimuli used earlier, all of these stimuli were sine-wave gratings modulated in this space around the origin, an equal-energy white of 37.5 cd/m^2. They all had an elevation of 10 deg, which gave all of the stimuli a luminance contrast of 12%. A black and white luminance stimulus with the same luminance contrast of 12% was included in the stimulus set to distinguish between the luminance and chromatic contributions to the response. The azimuth was varied from 0 to 360 deg, in 30-deg steps. We chose an elevation of 10 deg to excite purely luminance as well as chromatically opponent cells. All of the cells gave good responses to at least some of these stimuli.

Figure 13.8 shows the responses of four typical cells to these stimuli. In each panel, the abscissa indicates the stimulus azimuth, and the cells' responses are plotted on the ordinate. The horizontal dashed lines represent the cells' spontaneous firing rates to a uniform gray field having the same mean luminance as the stimuli. The solid horizontal lines show the cells' responses to the achromatic control stimulus.

The cell in Fig. 13.8A was located in the upper layers of V1, cells B and C were from V2, and cell D was from V3. As was previously shown (Lennie, Krauskopf, & Sclar, 1990), the responses of the V1 cell in 7A are well predicted by a model that assumes a linear combination of cone inputs. The predictions of this model are shown by the solid curve. The linear model allows a characterization of the chromatic properties of each cell with only two parameters, namely, the elevation and the azimuth of the color direction to which it best responds. Responses to all other colors can be derived from that vector because the shape of the tuning curve is assumed to be based on a linear mechanism and is identical for all cells. More precisely, the predicted response is given by the dot product of the color direction of the stimulus and the color direction of the cell's preferred direction:

$$R = b + A \sin(\mathit{elevation}) \sin\phi + \cos(\mathit{elevation}) \cos\phi \cos(\mathit{azimuth} - \psi), \quad (5)$$

where b is the response baseline, A is the response amplitude, ϕ is the cell's preferred elevation, and ψ is its preferred azimuth. For complex cells, we add a full-wave–rectification of the second term:

$$R = b + A\, |\sin(\mathit{elevation}) \sin\phi + \cos(\mathit{elevation}) \cos\phi \cos(\mathit{azimuth} - \psi)|. \quad (6)$$

The nonlinear function minimization program STEPIT (Chandler, 1969) was used to minimize the squared and normalized deviations of Eqn. (6) and the observed firing rates. As seen in the example V1 complex cell in Fig. 13.8A, the model fits the data quite well, and this is generally the case for cells in V1 (Lennie, Krauskopf, & Sclar, 1990). In that respect, V1 cells behave similarly to cells in the LGN (Derrington, Krauskopf, & Lennie, 1984).

The responses of the V2 complex cell illustrated in Fig. 13.8B and those of the V3 simple cell in Fig. 13.8D also follow the linear model's predictions closely. Like complex cells in V1 (Fig. 13.8A), complex cells in V2 (Fig. 13.8C) and V3 show a full-wave rectification along the direction opposite to their preferred azimuth. As the model predicts, a cell that prefers bright red and dark green gratings will also respond to dark red and bright green gratings. These can be characterized as linearly summing their cone inputs and full-wave–rectifying their output. Most cells in V2 (55 of 84, 65%) and the vast majority of V3 cells (94%) behaved in this simple, quasilinear way.

A minority of neurons in V2 showed systematic deviations from this linear model. For the cell shown

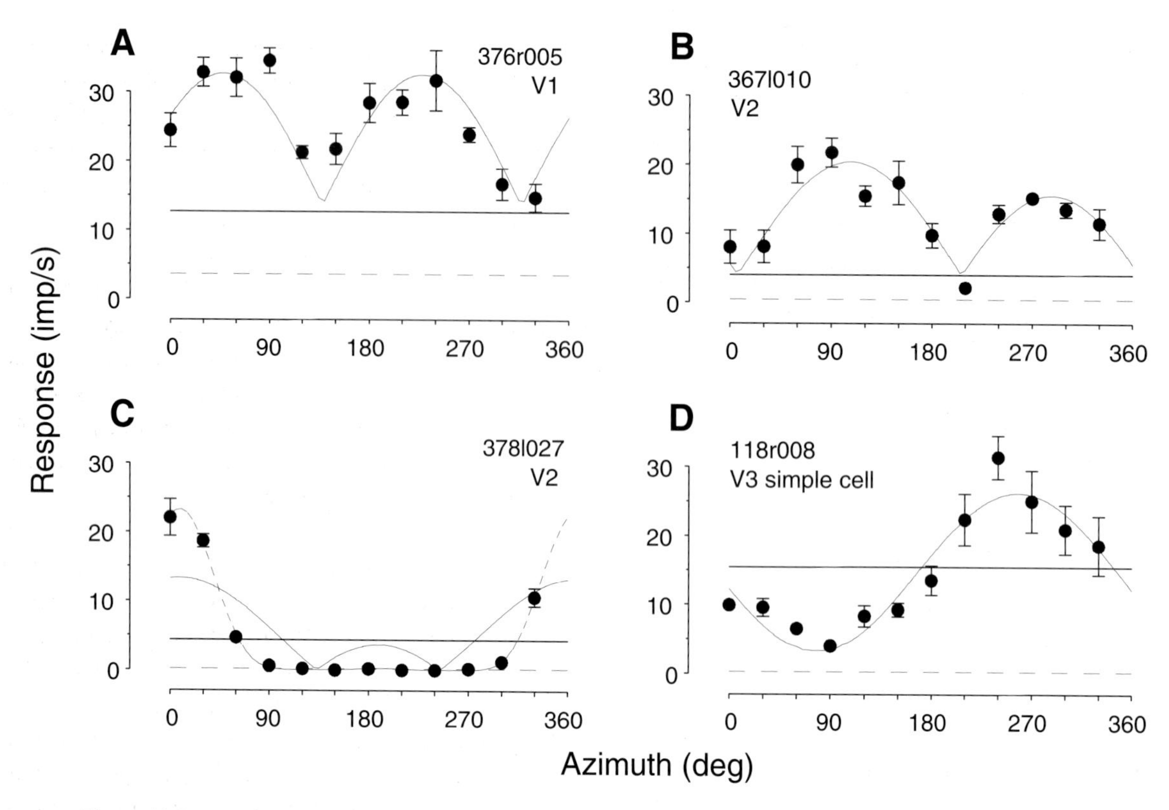

Figure 13.8: Responses of a typical V1 complex cell (A), two V2 complex cells (B,C), and a V3 simple cell (D) to stimuli of various azimuths and a constant elevation of 10 deg (as shown in Fig. 13.2). The solid curve fitted through the data represents the prediction of a model that assumes a linear summation of cone inputs. The horizontal dashed line (baseline) represents the response to a blank screen of space-averaged luminance equal to that of the stimuli. The horizontal full line represents the response to a grating modulated only in luminance, whose luminance contrast was the same as that of the chromatic stimuli. The dashed curves through the data in

in Fig. 13.8C, the tuning in color space is much narrower than the linear prediction (solid curve). About one-third of V2 cells (29 of 84, 35%) showed this narrow tuning, while hardly any could be found in V3. The cell in Fig. 13.8C, for example, responded only to stimuli consisting of bright red and dark green grating bars, and gave no response to stimuli of the opposite polarity (bright green and dark red) and no response to a bright orange stimulus (azimuth = 60 deg), even though this stimulus had a relatively high contrast along the color direction preferred by the cell. The narrowly tuned cell in Fig. 13.8C also shows a half-wave–rectification of its responses in its nonpreferred color direction.

To determine each cell's bandwidth of tuning in color space we used a nonlinear function to interpolate between the data points. The prediction of the nonlinear model is shown by the dashed curve in Fig. 13.8C. For linear cells, the predictions of both models are essentially identical, and they are not shown separately in Figs. 13.8A, B, and D. At a constant elevation of 10 deg, bandwidth was defined as the angular difference between the color vector that gave the best response and the color vector where the response had decreased to 50% of the difference between the firing rate at the peak and 90 deg away from it. This definition has the advantage that it predicts a constant bandwidth for linear cells, independently of their preferred azimuth and

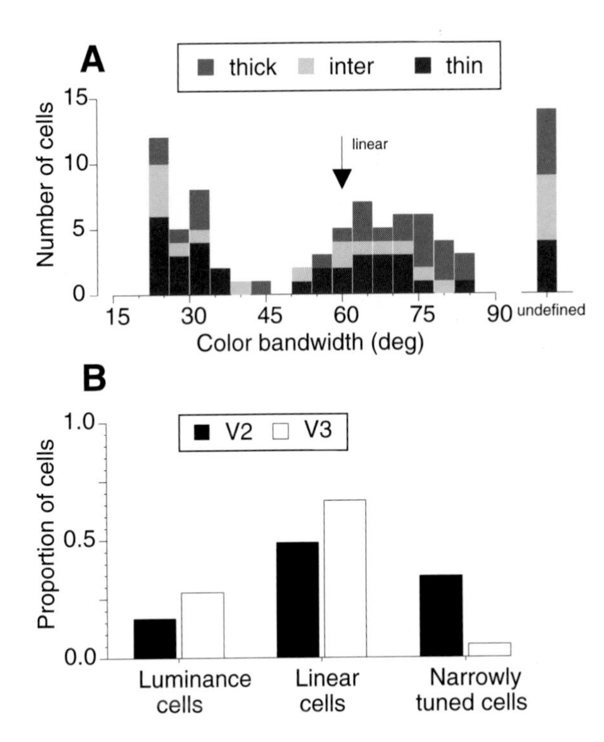

Figure 13.9: (A) Histogram of the bandwidth of tuning in color space for 84 V2 cells. To the right ("undefined") are cells that responded to the luminance component of the stimulus only. For each value of bandwidth, the proportion of cells located in thin, thick, or interstripe regions (as defined by CO staining) are indicated by different shades of gray. The bandwidth of 60 deg predicted by the linear model is indicated by the arrow. (B) Histogram for the distribution of luminance, chromatically linear, and narrowly tuned cells in our V2 and V3 samples. Narrowly tuned cells were defined as having a bandwidth of less than 45 deg.

elevation. Because of the additive term "sin (elevation) sin ϕ" in Eqn. (6), an actual response null does not have to occur in the direction orthogonal to the preferred color direction at the particular elevation of 10 deg that we used. A true null would occur only if the stimuli had an elevation of 0 deg, or for cells whose preferred elevation is 0 deg. But for all linear cells the bandwidth at half-height with respect to the orthogonal direction should be equal to 60 deg, the angle whose cosine is 0.5. The resulting histogram of bandwidths for our V2 neurons is shown in Fig. 13.9A.

The distribution is bimodal, showing distinct subpopulations of narrowly tuned cells (like the one in Fig. 13.8C) and of cells with an approximately linear tuning (like that of Fig. 13.8B), which are clustered around a bandwidth of 60 deg. To the right are cells for which we could not reliably determine any color-specific tuning because they responded only to the luminance component of the stimuli. It is important to remember that earlier in the visual system the responses are well fit by the linear model (Derrington, Krauskopf, & Lennie, 1984). Thus, most color selective LGN and V1 neurons have a bandwidth around 60 deg. Using this measure of bandwidth, our samples of V2 and V3 neurons could be classified into three categories: linear cells (bandwidth around 60 deg), narrowly tuned cells (bandwidth lower than 45 deg), and luminance cells (cells whose responses to the chromatic stimuli did not differ from that to the achromatic stimulus). The resulting classification for our samples of V2 and V3 neurons is shown in Fig. 13.9B. In V2 there is a significant subpopulation of cells whose bandwidths are much narrower than that predicted by the linear model. This subpopulation is almost entirely missing from the V3 sample. Only a handful (5 of 90) of V3 cells showed an indication of narrow tuning in color space. Accordingly, the proportions of luminance and linear cells were slightly higher in V3.

Preferred azimuths. One of the current mysteries of color processing concerns the mismatch between the early "cardinal directions" (Derrington, Krauskopf, & Lennie, 1984; Krauskopf, Williams, & Heeley, 1982) and the perceptually defined "unique hues," red, green, blue, and yellow (Hering, 1878; Hurvich & Jameson, 1955). While neurons in the LGN show clear preferences for the cardinal directions, it is being still debated whether neurons later in the pathway show such clustering around the unique hues (Zeki, 1978a; de Monasterio & Schein, 1982; Schein & Desimone, 1990). Using the nonlinear function that we fitted to the data of our first experiment, we derived the preferred azimuths and elevations for all of the cells in our samples. These distributions are shown in Fig. 13.10. In area V2 (Fig. 13.10A) we found a small preference for the cardinal directions, but the preferred azimuths were not restricted to them. Furthermore, there was no systematic clustering around the "unique hues." There was no systematic difference between the distributions of chromatically linear and narrowly tuned cells in that respect. However, Fig. 13.10B shows that there was an unexpectedly large proportion of V3 cells that showed a small but significant contribution from the short-wavelength–sensitive (S) cones (azimuth near 90 deg) to a luminance-based response. All of these neurons responded better to blue than to yellow.

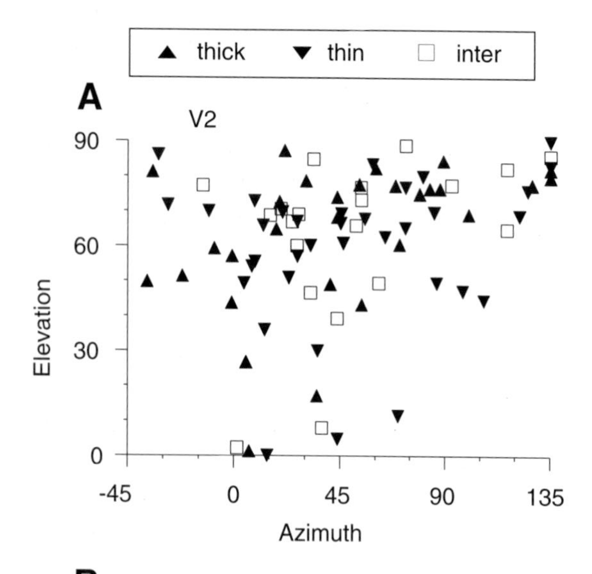

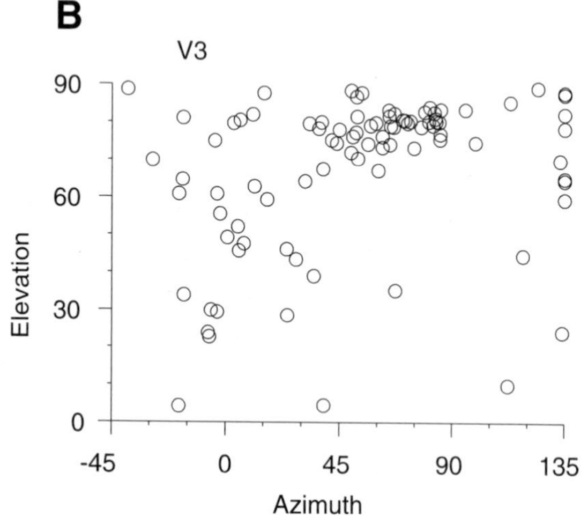

Figure 13.10: Scatterplot of the preferred azimuths and elevations of (A) 84 V2 cells and (B) 90 V3 cells. Different symbols and shades of gray indicate different CO compartments of V2. Azimuths and elevations were derived from the fit of the linear model proposed by Derrington, Krauskopf, and Lennie (1984).

Discussion

Segregation of stimulus attributes. Our results clearly show that for all stimulus attributes, selectivity of cells is distributed continuously and there are many cells in V2 and in V3 that are selective to more than a single stimulus attribute. Classifying cells in general, and classifying them based on one response property in particular, is therefore not appropriate.

In V2, there was a tendency toward more color selectivity in the thin stripes, but generally there was color selectivity in all CO compartments. The fact that statistical tests did not show significant differences in selectivities between different CO compartments might be explained by our relatively small sample size (119 cells), since the overall trends were in the directions where we expected differences. However, it should be clear that segregation was by no means absolute. Knowing the responses of a single cell does not allow one to classify that cell as belonging to a particular processing stream or CO compartment, nor does knowing the selectivity of a cell along one stimulus dimension allow predictions of its other properties.

In area V3, our results show that neurons display a larger variety of functional properties than previously appreciated. Almost all of the cells showed orientation selectivity, but directional selectivity, color selectivity, or endstopping were common as well. Furthermore, all combinations of tuning characteristics to these attributes were observed. This suggests that the processing of color, form, and motion does not follow segregated pathways within V3.

It is evident that within V2 there are no highly directional color cells. The cells that we found to be selective for color and direction of motion at the same time were either just marginally responsive to color or just marginally directional. It seems, therefore, that color selectivity in combination with directional selectivity, which can be observed psychophysically (see Chapter 15; Derrington & Henning, 1993; Metha, Vingrys, & Badcock, 1994; Gegenfurtner & Hawken, 1995; Stromeyer et al., 1995), is formed in another area of the cortex. Area V3, which also receives input from area V2 (see Felleman & Van Essen, 1991), might have that specific integration function (Gegenfurtner, Kiper, & Levitt, 1997) since we found many V3 cells that were both color-selective and directional. By comparison, directional color selective cells are almost absent in MT, another area associated with the motion pathway (Gegenfurtner et al., 1994). Interestingly, neurons in area V3 have a very broad temporal frequency tuning, and they respond briskly even at low temporal frequencies, unlike neurons in area MT. The pathway via areas V3 and V4 might therefore be a suitable candidate to underlie the perception of motion of slowly moving isoluminant stimuli.

The general picture that emerges from these studies of V2 and V3 is that the functional segregation of receptive field properties is much less striking than expected from the anatomical segregation of feedforward cortico–cortical connections of the CO compartments in areas V1 and V2. It appears that most cells in these areas can convey information about more than one stimulus attribute. Instead of treating various attributes separately to recombine them at a later stage, the visual system seems to allow for significant cross-talk at all stages of processing. These results have important implications for the elaboration of models of the visual system, and they seem to favor those that assume a simultaneous, integrated treatment of color, motion, and form information (see, for example, McClurkin & Optican, 1996; McClurkin, Zarbock, & Optican, 1996) over those that assume a parallel, segregated processing of various stimulus attributes.

These results suggest a relative uniformity of functional properties in areas V2 and V3 and the apparent lack of a separate pathway for color analysis. This would seem to contradict much anatomical evidence that there are distinct connectional pathways leading through V2 and V3 to the inferotemporal and posterior parietal visual cortex that seem to mediate different visual abilities (DeYoe & Van Essen, 1985; Shipp & Zeki, 1985; Morel & Bullier, 1990; Baizer, Ungerleider, & Desimone, 1991; Nakamura et al., 1993). This discordant result that the anatomically distinct pathways appear less functionally distinct is a puzzle. The resolution of this puzzle may lie in our not knowing which aspects of the multi-dimensional signal in each neuron are relevant to higher visual areas. Or perhaps those higher areas perform different analyses with the seemingly uniform information supplied to them by V2 and V3. This is clearly an important issue demanding further attention.

Cortical hierarchies. Most cortico–cortical anatomical connections are compatible with a processing hierarchy that proceeds from V1 through V2 and V3 and then splits into temporal and parietal streams. Some receptive field properties are compatible with this hierarchical organization. For example, receptive field sizes at a given eccentricity tend to increase at later stages, with V1 fields being the smallest and fields in MT or V4 being larger. This indicates summation processes over a large number of cells at earlier levels. V3 receptive fields are also larger on average than V2 fields (Felleman & Van Essen, 1987; Gattass, Sousa, & Gross, 1988), in accordance with a hierarchical model.

However, there are several aspects that seem to contradict a simple hierarchical processing scheme via V2 and V3. First, V3 cells are tuned to higher temporal frequencies and have higher temporal frequency cutoffs than V2 cells (Gegenfurtner, Kiper, & Levitt, 1997). Simple averaging of cells at earlier levels would predict lower temporal frequency preferences, for example, as is observed when going from the LGN to V1. A substantial proportion of cells in V3 is tuned to temporal frequencies to which no cells in V2 respond. They presumably receive their inputs directly from some other cortical area preceding it. Most likely this is layer 4B of V1, which has strong direct projections to V3 (Felleman, Burkhalter, & Van Essen, 1997). It is reasonable to suggest that there are functionally distinct V1 projections to V2 and V3. Levitt, Yoshioka, and Lund (1994) showed that the V1 layer 4B projection to V2 arises primarily from pyramidal neurons, while the projection to MT from the same layer is provided by the spiny stellate cell population (Shipp & Zeki, 1989b). Furthermore, V3 has few strongly directional-selective cells in the input layer 4, while V2 has many (Levitt, Kiper, & Movshon, 1994). This suggests that direction selectivity is created *de novo* within area V3, rather than from the same class of inputs as V2. Finally, the color properties of V3 cells are very much like what is observed in V1. Responses to chromatic stimuli are well described by a model linearly combining cone inputs. In V2, about one-third of the cells have a nonlinear tuning to specific color and luminance combinations. We did not observe that type of behavior in V3. Therefore it seems likely that V3 actually receives its largest input directly from V1, rather than from the preceding stage, area V2. This leaves open the intriguing question of the functional relevance of the V2 inputs to V3.

Our results show a substantial chromatic signal in area V3. This might seem puzzling given V3's place in a pathway nominally dominated by signals from the M-layers of the LGN, which show no color-opponent behavior. We are therefore left to explain the source of this signal. Felleman, Burkhalter, and Van Essen (1997) reported that the projection to V3 from V1 arose essentially exclusively from layer 4B, which is dominated by LGN M-inputs via layer 4Cα. However, it has long been suspected that 4B neurons could have access to P-signals via their apical dendrites in layer 3 (Lund, 1973). Sawatari and Callaway (1996) recently used laser photostimulation of cortical slices to show that there are indeed P-inputs to layer 4B neurons. Alternatively, the chromatic properties that we described here for cells in area V3 might derive from their inputs from the upper layers of V2, or from the less direct projection from the upper layers of V1. For example, layer 4A in V1 is known to contain cells with color-opponent behavior (Lennie, Krauskopf, & Sclar, 1990), and Levitt, Yoshioka, and Lund (1994) showed that V1 cells in layer 4A also project to the V2 thick stripes, which in turn project to V3. So there clearly are anatomical routes by which chromatic information, presumably derived from parvocellular LGN signals, could reach area V3.

Isoluminance in the extrastriate cortex. Our data show that color facilitates the response of neuronal populations in all CO compartments of V2 and in V3 in a similar way. No "isoluminant" stimulus can silence the entire population of neurons in these areas. This has a very strong implication for experiments that try to use color as a means of isolating neural pathways. There are good reasons to assume that isoluminant stimuli do indeed stimulate mainly parvocellular LGN neurons (Shapley, 1990; see Chapter 11), even though some magnocellular neurons might show a frequency-doubled response to low spatial frequency isoluminant stimuli (Schiller & Colby, 1983; Derrington, Krauskopf, & Lennie, 1984). However, given the large extent of mixing of parvo- and magnocellular signals in V1 and V2 (Lund & Boothe, 1975; Blasdel, Lund, & Fitzpatrick, 1985; Fitzpatrick, Lund, & Blasdel, 1985; Rockland, 1985; Lachica, Beck, & Casagrande, 1992; Nealey & Maunsell, 1994; Levitt, Yoshioka, & Lund, 1994; Yoshioka, Levitt, & Lund, 1994) it is not surprising to see only a little segregation of color processing at these and later stages.

Our results show that isoluminant stimuli cannot be used to isolate different functional streams at the level of cortical areas V2 and V3. This is consistent with quantitative psychophysical experiments, indicating that the primate color system is capable of supporting analysis of form and motion (for a review, see Gegenfurtner & Hawken, 1996b, and Chapter 15). What is typically found in quantitative psychophysical experimentation with isoluminant stimuli is a reduced sensitivity of the color system, when its performance is compared to performance for high-contrast luminance stimuli (Livingstone & Hubel, 1987). There are two major reasons that could explain this reduced sensitivity. First, because the spectral sensitivities of L- and

M-cones largely overlap, the input contrast to the visual system at the level of the cones is vastly reduced for isoluminant stimuli (Shapley, 1990; Gegenfurtner & Kiper, 1992). When stimuli are used that are equated for cone contrast, psychophysical performance at isoluminance was shown to equal or even exceed that for luminance in many psychophysical tasks (Chaparro et al., 1993; Derrington & Henning, 1993; Metha, Vingrys, & Badcock, 1994; Gegenfurtner & Hawken, 1995; Stromeyer et al., 1995). A second reason might lie in the continuing integration of signals along the various stages of processing in the visual cortex. In the LGN, about one-third of the parvocellular neurons respond exclusively to isoluminant stimulation (Derrington, Krauskopf, & Lennie, 1984), which has the advantage of decorrelating the signals from the L- and M-cones (Buchsbaum & Gottschalk, 1983). In the striate cortex less than 10% of all neurons show that behavior (Lennie, Krauskopf, & Sclar, 1990) and even fewer in the extrastriate cortex do (Schiller, Logothetis, & Charles, 1991; Gegenfurtner et al., 1994). Since there is little input to the visual system that is exactly isoluminant, and the magnitude of such input is naturally lower, it would be wasteful for the visual system to allocate extensive resources to process these stimuli. For the visual system, it seems to be more important to detect borders defined by luminance and color rather than isoluminant boundaries per se. Our data show that there are many cells that are selective for orientation and color, which can detect these combined luminance–color boundaries. Given these constraints, it is not surprising that in many cases performance at isoluminance equals that for luminance when the difference in sensitivities is taken into account (DeValois & Switkes, 1983; Switkes, Bradley, & DeValois, 1988; Webster, DeValois, & Switkes, 1990; Krauskopf & Farell, 1991; Gegenfurtner & Kiper, 1992; Würger & Landy, 1993; Reisbeck & Gegenfurtner, 1998). What is quite clear from our data is that isoluminant stimulation does not isolate different processing streams at the level of V2 or V3.

In summary, the results of our experiments suggest that the processing of color information within the brain is quite different from what has been described previously. First, the idea that color is processed within a specialized pathway independently of other stimulus attributes is not supported by our data. Indeed, virtually all of the color-sensitive cells that we recorded from were sensitive to luminance variations as well. Moreover, most of these cells were also sensitive to other stimulus attributes such as motion, form, or size and could be found anywhere within V2 and V3. Thus, these results indirectly argue against the existence of a color center in the brain. They suggest rather that color information, as any other stimulus attribute, is available to the cells encoding a particular region of the visual field at all stages of visual processing. However, it must be noted that this conclusion is difficult to reconcile with a variety of clinical results showing a specific loss of color vision in some brain-damaged patients, a condition known as cerebral achromatopsia (see Zeki, 1990, for a review). This discrepancy between clinical and experimental data remains a major puzzle in the study of color vision and certainly deserves particular attention in future research.

Finally, in light of these results, one can only wonder about the kind of treatment applied to color signals in the extrastriate cortex. Clearly, since color is a rather uniform object property, the narrowly tuned cells that we encountered within V2 could be extremely useful for image segmentation. In fact, the tuning of these cells resembles that of the higher order color mechanisms revealed in a variety of psychophysical experiments (Webster & Mollon, 1991; Gegenfurtner & Kiper, 1992; Krauskopf & Gegenfurtner, 1992; Zaidi & Halevy, 1993; Krauskopf, Wu, & Farell, 1996; see Chapter 16), and they might provide the main input to the V4 cells showing a very narrow wavelength tuning (Zeki, 1980; Schein & Desimone, 1990). One might thus speculate that within a V1, V2, and V4 pathway, a very specific color sensitivity is elaborated that allows one to attribute a particular color to a given surface and thus help to segregate objects from a background. Similarly, the frequent occurrence of color-sensitive cells showing direction selectivity in area V3 suggest that this area might be part of the "slow-motion" pathway that encodes the movement of colored targets (see Gegenfurtner & Hawken, 1996b).

However, the study of the chromatic properties of the extrastriate cortex leaves some important questions unresolved. In particular, the absence of a preferential representation of the unique hues or of the cardinal directions beyond the LGN seems difficult to reconcile with their perceptual salience. Clearly, more work is needed to comprehend the physiological basis of the mechanisms that are revealed psychophysically.

Acknowledgments

We would like to thank Doris Braun, Sue Fenstemaker, Mike Hawken, and Jenny Lund for comments on an earlier draft of this chapter. This work was supported in part by grants MRC G9203679N and NIH EY10021 to J.S. Lund, and by a grant from the ARC program of the DAAD and the BRC. KRG was supported by a Heisenberg fellowship from the Deutsche Forschungsgemeinschaft (DFG Ge 879/4-1).

14

Computational neuroimaging: color tuning in two human cortical areas measured using fMRI

Brian A. Wandell, Heidi A. Baseler, Allen B. Poirson, Geoffrey M. Boynton, and Stephen A. Engel

The neural representation of the visual world begins with the responses of four interleaved photoreceptor mosaics. These photoreceptor mosaics simultaneously encode information about pattern, motion, and many other aspects of the visual world. As early as the photoreceptors themselves, neural specializations exist that distinguish among the many kinds of information contained in the photoreceptor mosaics. For example, the photoreceptors are divided into rods and cones, which are specialized for different types of viewing conditions. At the next synapse, foveal cones are each contacted by bipolar cells with many different morphologies, interconnection patterns, and receptive field properties. These different types of cells appear to segregate information into streams that are specialized for different tasks.

The diversity of neural form and function observed within the retina is present thoughout the visual pathways. What is the relationship between this diversity and our perceptual experience of the world?

Theories of vision are dominated by the hypothesis that there is a direct correlation between the segregation of function at the neural level and the segregation of *perceptual attributes*. Namely, every day we make perceptual judgements that separate color from shape, motion, and other stimulus attributes. Most recent theories suppose that the cell diversity and perceptual experience can be related in a straightforward fashion.

For example, Zeki et al. (1991) have argued that, in the monkey brain, neurons within area V4 are specialized for representing color information. This view has been challenged by Cowey and Heywood (1995), who accept the basic conceptual framework but propose area TEO instead. Hubel and Livingstone (1987) have also argued in favor of the general hypothesis of functional segregation and proposed that the origins of neural specialization of color can be traced to the responses of neurons within specialized regions of V1 and V2. Retinal anatomists have suggested that the segregation of perceptual color information can be identified in the retina (e.g., Rodieck, 1991), and this view has obtained further support from Dacey and Lee's (1994) observations concerning the specialization of small-bistratified, retinal ganglion cells for carrying S-cone signals.

In the human cortex, two main types of evidence are used to support the hypothesis of a functional segregation and specialization for processing color appearance. Meadows (1974) analyzed a number of cases of brain damage in which several subjects' perception of color was disturbed, but acuity, motion, and other visual functions were not. This syndrome, which we will call cerebral dyschromatopsia, was reviewed thoroughly by Zeki (1990). While dyschromatopsia often is accompanied by several visual deficits, such as face blindness (prosopagnosia) and upper visual field loss, these associations are not always observed. Because there are some patients with color loss and no discernible prosopagnosia, and other patients with prosopagnosia but no discernible color loss (a double dissociation), Meadows, Zeki, and many others have accepted the view that color has a discrete representation in the cortex.

Neuroimaging studies provide a second source of evidence of the functional specialization of color representation in the human brain (Lueck et al., 1989; Zeki et al., 1991). These authors compared the size of the PET signal when subjects viewed a collection of full-color rectangular patches compared to monochrome images consisting of the same rectangles matched in luminance. The difference signal is particularly large in a region of the brain that is ventral to the calcarine cortex, near the upper field representation of primary and secondary visual cortical areas. Zeki and his colleagues take this observation to support the hypothesis that color processing is functionally specialized and segregated in the human cortex. The measurements themselves have been partially confirmed by Sakai et al. (1995), although other measurements by Gulyas et al. (1994) and Engel, Zhang, and Wandell (1997) suggest a different organization.

Neither the lesion data nor the neuroimaging data are conclusive. Let us first consider the case of cerebral dyschromatopsia. We might try to understand the implications for cerebral loss in this condition by comparing it with a fairly well understood color deficit: the loss of color vision caused by an inherited color deficiency resulting in the absence of at least one type of cone (dichromacy). The color confusions of dichromats are quite significant, although there is little or no loss of visual acuity or motion selectivity. Yet, we do not argue that color vision is functionally segregated and specialized in the photopigments. Nor do we argue that other types of information, such as motion and form, are not carried by the photoreceptors.

At the level of the photoreceptors, we already understand that color appearance reflects the coordinated action of a system of neurons and that the disruption of one type of photoreceptor leads to color anomalies; but no photoreceptor class is the unique site of color vision. We also understand that motion information is carried in the space–time response of the photoreceptors, so that the disruption of one photoreceptor class will not eliminate information needed to perform most motion discriminations or estimations. Beyond the receptors, functional specialization for perceptual attributes is a poor description of retinal pathways. Trichromacy is determined by the cone properties, but the cones are used for much more than color. The specialized rod pathway, via the AII amacrines, is important for rod vision, but the rod signals merge onto a cone pathway after only a few synaptic connections. There are few grounds to argue that retinal specializations are equated with perceptual experiences.

The neuroimaging data, based on the subtraction methodology (described by Posner & Raichle, 1994), are not decisive either. The experimental data reveal several cortical regions where activity is greater to the full-color image rather than to the monochrome image. Necessarily, there will be one region with the largest difference; it could not be otherwise. As used by Zeki, the subtraction methodology identifies color appearance within this single region. But the existence of a largest response does not imply that other differences should be ignored, or that color processing is localized rather than part of a distributed representation. Indeed, authors who favor distributed representations theoretically use neuroimaging data to emphasize the distributed nature of neural responses during color, motion, and stereo tasks (e.g., Gulyas et al., 1994).

Finally, let us return to the perceptual basis for the functional segregation hypothesis: some perceptual attributes, such as color, can be separated from other perceptual attributes, such as motion, depth, and form. As an empirical observation, this claim is only roughly true. The apparent velocity of an object varies depending on the object's contrast and color (Cavanagh & Anstis, 1991; see Chapter 15); the color of a pattern depends on its spatiotemporal properties: high spatial and temporal frequency patterns appear light-dark and very desaturated (Poirson & Wandell, 1993). It is difficult to find conditions in which we do not know both the location and the color, so that one might argue that color and spatial position (and motion and pattern) are tightly coupled, not dissociated. Hence, it is possible to use the perceptual literature to argue that there is a close coupling of color and other visual information, not a complete segregation.

We began this section with the claim that the arguments in favor of functional specialization are not conclusive. It is also true that the criticisms of the hypothesis presented here do not falsify it. Rather, the criticisms show that the evidence in favor of functional specialization and segregation is weak, and that it is a good moment to formulate alternatives and seek further empirical tests. To begin this process, we find it useful to consider the retina as a model system because we know much about the organization of information in the retina compared to the brain. What do the principles of information representation in the retina suggest that we might find in seeking to uncover the representation of information within cortex?

Computational neuroimaging. We are studying the hypothesis that the neural diversity observed using anatomical and electrophysiological methods represents a computational diversity, rather than functional specializations associated with perceptual attributes. Guided by this hypothesis, we are developing methods that we call *computational neuroimaging* to measure the neural computations applied to stimulus information. Our experiments are designed to answer questions about the nature of the computation performed along specialized pathways: How do neurons within the cortex transform visual information?

This approach flows from our current understanding of retinal processing. Within the retina, information about pattern, movement, and color often flow together in common pathways. The pathways themselves – such as the different types of horizontal cell pathways or the different ganglion cell pathways – appear to be specialized for certain kinds of computations. The HI- and HII-type horizontal cell mosaics perform separate computations that appear to regulate the gain of signals originating in different parts of the spectrum (Dacey et al., 1996). The midget and parasol cells appear specialized for encoding information about different spatiotemporal components of the image (Wandell, 1995).

If cortical diversity is also based on the need to perform different types of specialized computation, we would not ask: Where is color located? Rather, we would ask: How do neurons transform and relay the signals sent to them, and how do these transformations achieve the computational goals related to color (or motion, or depth) perception?

Adopting this computational view, we might also interpret lesion data differently. Rather than interpreting the lesion data as the destruction of a perceptual representation, we would view the lesions as a disruption of an information-processing task that is essential for color computations. Interference with certain types of processing routines may have serious consequences for color vision and perhaps other functions. We would not conclude that the lesion sites are also the locus of a unique color representation, any more than the wavelength transduction computed by the photopigment absorptions is the site of color vision.

To develop an empirical basis for this conceptual framework, we have been using functional magnetic resonance imaging (fMRI) to measure how color information is distributed and transformed within the human cortex. Our initial measurements of cortical color tuning in areas V1 and V2 were described in Engel, Zhang, and Wandell (1997); here we report on additional observations of color tuning in the extrastriate cortex located near the lingual and fusiform gyri. These measurements form part of a general project of measuring color responsivity across the visual cortex under a variety of stimulus conditions. In this way, we hope to understand the distribution of color information and to detect locations where computations specialized for color may reside within the cortex.

Methods

Because fMRI is a relatively new method for analyzing human cortical activity, we begin with some background on the method and our view on how it might be used to trace the distribution of cortical color information. Then we describe the specific methods that we used in the experiments reported here.

The fMRI signal: background. The magnetic resonance signal measures the rate at which dipoles,

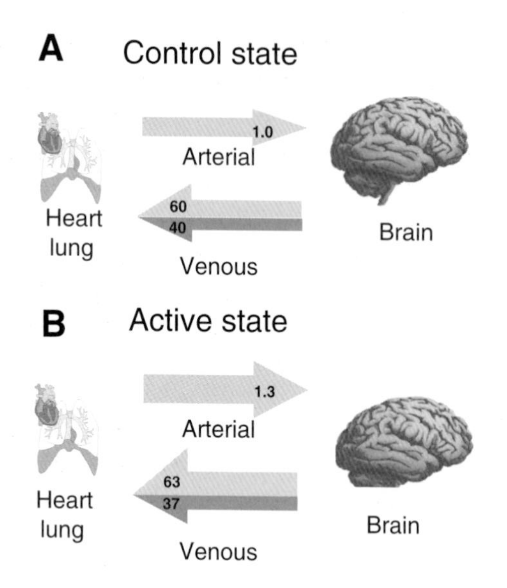

Figure 14.1: Activity-dependent changes in blood oxygenation. (A) In the control state, the arterial supply is fully oxygenated and 40% of the oxygen is used. (B) In an activated state, there is an increase in the total volume of blood oxygen delivered to the active region. The local oxygen consumption increases, but the oxygen supply exceeds the need, resulting in a relative increase in venous blood oxygen.

present within a uniform magnetic field, return to the equilibrium after being perturbed by a radio frequency (RF) pulse. The two rate parameters associated with the return of the dipole vector to equilibrium can be measured separately; these two rate parameters, T_1 and T_2, are influenced by different aspects of the local magnetic field. The physical significance of these rate parameters can be understood as follows. Suppose that we begin with a vector oriented in the vertical direction. When we rotate the vector off-axis, the vector difference between the initial position and the rotated position has a horizontal and a vertical component. The time constants describe the exponential relaxation back to equilibrium of the vertical (T_1) and horizontal (T_2) components.

In the case of the fMRI, the signal governed by the second rate parameter T_2 is measured. The fMRI signal depends on the observation that oxygenated (HbO) and deoxygenated (Hb) blood have different magnetic field properties (Thulborn et al., 1982). Oxygenated blood (HbO) is magnetically transparent (diamagnetic) while deoxygenated blood (Hb) is not (paramagnetic). After being perturbed by an RF pulse, dipoles near Hb return to their equilibrium state more rapidly than dipoles near HbO.

Fox and others (Fox & Raichle, 1986; Fox et al., 1988) showed that changes in neural activity cause a 30–50% increase in cerebral blood flow near active cortical regions (activity may be due to excitation, inhibition, or any metabolic process), but only a 5% increase in the oxygen metabolic rate. The situation is illustrated in Fig. 14.1. In a control condition, arteries supply nearly 100% of the oxygenated blood and roughly 40% of the oxygen is consumed locally, so that the blood returning in the veins comprises 60% oxygenated and 40% deoxygenated blood. During a stimulus condition that causes significant neural activity, an additional supply of oxygenated blood is delivered, but only a small fraction of this supply is metabolized. Hence, the fraction of extracted oxygen is smaller compared to the control state, and the proportion of oxygenated and deoxygenated blood becomes 63:37. The change in the ratio of oxygenated and deoxygenated blood changes the rate at which the dipoles return to steady state following an RF pulse. By comparing the rate of return one can make an inference about the local neural activity. This comparison is called the blood oxygen level dependent (**BOLD**) signal.[1]

The flow of information illustrated in Fig. 14.2 helps to remind us of the complexity of the signalling path in an fMRI experiment. The choice of signal and control stimuli result in two patterns of neural activity; the neural differences result in oxygenation differences. The relationship between the MR signal (output) and the stimulus level (input) is not well

[1]Fox and co-workers suggested that the large increase in blood flow compared to metabolized oxygen was a mismatch between two decoupled mechanisms. Buxton and Frank (1997) suggest that the difference is to be expected if we accept that (a) under rest conditions capillaries are perfused and (b) extracted oxygen is fully metabolized. Whether the two mechanisms are decoupled or intimately linked will be important for interpreting the neural significance of the fMRI signal.

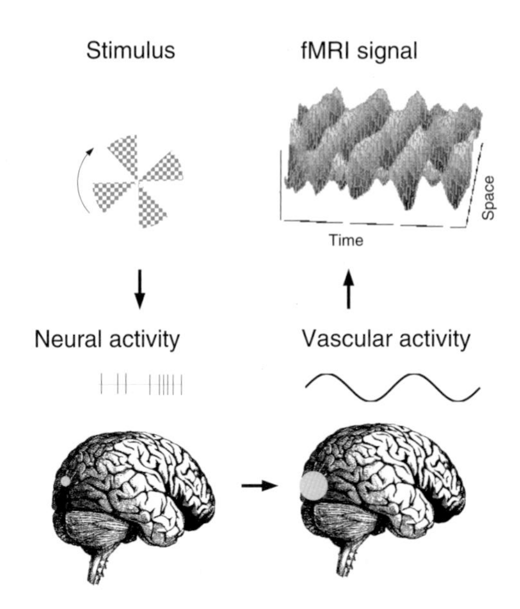

Figure 14.2: The fMRI imaging pathway. The stimulus causes neural activity. The neural activity changes the spatial distribution of blood oxygen. The blood oxygen distribution influences the local magnetic field, resulting in a spatial variation in the MR signal.

understood and depends on both of these intermediate steps (signal to neural activity, neural activity to oxygenation level), both of which may be nonlinear.

The stimulus sequence. Stimuli were 20-deg, contrast-reversing, circular, checkerboard patterns presented on a larger neutral mean field luminance, *Y*, of 72 cd/m^2 and an *xy* chromaticity of [0.30, 0.40]. We describe the stimuli using contrast levels based on the Smith–Pokorny cone fundamentals. Specifically, suppose that the values L_0, M_0, and S_0 represent the long-, medium-, and short-wavelength cone absorptions from the mean field, and ΔL, ΔM, and ΔS represent deviations from this mean value introduced by a specific color of the checkerboard pattern. Then, we describe the checkerboard contrast by the vector $\mathbf{s} = (\Delta L/L_0, \Delta M/M_0, \Delta S/S_0)$. The *color direction* of the stimulus is the unit length vector $\mathbf{s}/\|\mathbf{s}\|$.

The display system, which is based on a color LCD projector, was calibrated in situ using a PhotoResearch spectral radiometer and a hand-held Minolta photometer. We measured the spectral power distribution of each of the color channels at maximum intensity using a PhotoResearch spectroradiometer. We measured the gamma curve of the system (framebuffer to display intensity) for each of the channels and verified the independence of the three color channel outputs (cf. Brainard, 1989). Because of the constraints on placing equipment in the scanner room and inhomogeneities in the display equipment, these measurements are accurate to a factor of about 10%.

A spatial image of a black and white version of the checkerboard pattern, at a single moment in time, is shown in Fig. 14.3A. The contrast pattern consisted of a balanced set of positive and negative modulations about the mean field. The stimulus consisted of a fine pattern near its center that became increasingly coarse with eccentricity. Subjects fixated a mark at the center of the screen throughout the scan. For the data described here, the checkerboard contrast-reversed at 1 Hz.

During the MR scans, the contrast-reversing stimuli were presented as part of a stimulus-control pair. A contrast-reversing pattern (see Fig. 14.3) was presented for 15 s, followed by 27 s control consisting of a uniform field at the mean illumination level. In this way the response to the signal could always be compared with the control level presented in an adjacent time period.

Stimulus-control pairs were presented in triplets (see Fig. 14.3B). The three stimuli in a triplet shared a common color direction, but the stimulus contrast increased (or decreased), consecutive stimuli changing in contrast by a factor of two. Hence, a single triplet might consist of an (L,M,S) contrast series such as (0.1, 0.1, 0), (0.2, 0.2, 0), (0.4, 0.4, 0).

Six stimulus-control patterns were presented during each fMRI scan, representing two randomly chosen color directions within the (L,M) color plane (i.e., S = 0). Six fMRI scans were performed during a single experimental session. Hence, during a single session responses were measured to stimuli in six color directions, each consisting of three contrast levels. During each session every stimulus-control pair was presented twice, once as part of an increasing sequence and once as part of a decreasing sequence, for a total of 36 stimulus-control presentations.

Response amplitude. We measured the amplitude of the fMRI signal within a region of interest (ROI) using the following sequence of operations.

1. The ROIs were selected in two steps. First, two coarse ROIs were selected based on anatomical criteria. The first spanned area V1 as determined in separate studies of the retinotopic organization (Engel, Zhang, & Wandell, 1997). The second spanned a large region of the ventral occipital cortex and was selected to be distinct from the retinotopically mapped areas identified in earlier studies.

2. Second, the ROIs were refined by fMRI activity. Pixels in the coarse ROI were retained only if their response at the stimulus frequency rose to 30% of the signal variance for at least one of the color stimulus conditions. The reason for applying this second step is this: Voxels ($1.5 \times 1.5 \times 5$ mm) measured during the functional scans often include a mixture of gray matter and other unwanted substances, an effect called *partial voluming*. It is difficult to identify which voxels contain primarily gray matter and which do not. By restricting the ROI to responsive voxels, we reduce the noise introduced by partial voluming.

3. We measured the response using the average timeseries of the voxels in each ROI. First, we divided the time series into the 42-s stimulus-control periods and calculated the amplitude of the fMRI signal harmonic at the 42-s period (1/42 Hz). Second, we estimated the noise during a scan by calculating the amplitudes at other temporal frequencies. Based on earlier noise measurements, we determined that the noise power spectrum is well fit by a decreasing exponential function. Hence, we fit this function to the scan measurements and used the interpolated amplitude at 1/42 Hz as the noise amplitude estimate. Finally, we removed the estimated noise from the amplitude of the fMRI signal at the signal frequency, using the method described in Appendix 1 at the end of this chapter. We call the amplitude following removal of the noise the *response amplitude*, and this value, measured in percent modulation, is reported in our graphs.

Results

In a brief separate report, we described measurements of color tuning in areas V1 and V2 for contrast-reversing patterns at 1, 4, and 10 Hz. The color responses were strongest per unit cone contrast for signals in the L-M- (i.e., red-green) opponent channel (Engel, Zhang, & Wandell, 1997; see also Kleinschmidt et al., 1996). These measurements parallel, in some regards, the visual sensitivity to these same stimuli.

Here, we report color responses that we have measured in other positions within the extrastriate cortex. The next few paragraphs and figures describe some of this activity and color-tuning measurements from activity near the lingual and fusiform gyri of two observers.

The ventral occipital ROI (VO-ROI) spanned several imaging planes for both subjects. These planes are located roughly 2 cm anterior of the posterior pole and spaced by 8 mm. Figure 14.4 shows the location of VO-ROI in several planes located perpendicular to the calcarine. The locations of the measurement planes can be seen in the anatomical localizer shown at the left of the figure.

Figure 14.5 shows the spatial distribution of activity during one typical functional scan. The locations of brain regions in which the fMRI signal correlated with the stimulus presentation at a level of 0.3 or greater are shown. The methods used to calculate the correlations are described in detail in Engel, Zhang, and Wandell (1997). The stimulus caused significant activity outside of areas V1 and V2. The planes shown in Fig. 14.5 were chosen because they each contained significant activity in regions near the fusiform and lingual gyri, which have been correlated with prosopagnosia and achromatopsia (Meadows, 1974; Zeki et al., 1991). Naturally, we were interested to consider whether the color tuning in these areas differed significantly from the tuning reported in areas V1 and V2 (Engel, Zhang, & Wandell, 1997).

Figure 14.6 shows the fMRI time series measured in the two regions of interest during the presentation of 12 different-colored targets. Panels (A) and (B) show

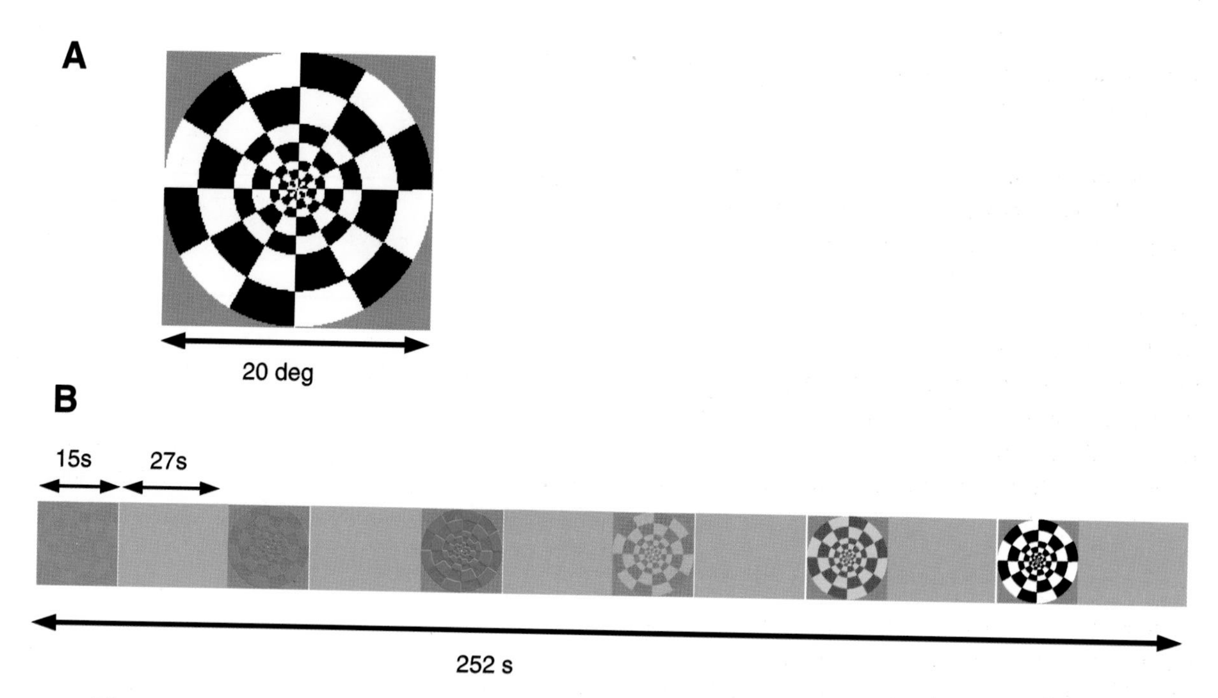

Figure 14.3: The stimulus and an example stimulus sequence shown during a single scan. (A) The spatial structure of a black and white version of the stimulus, at a single moment in time, is shown. (B) During each 252-s scan, six stimulus-control combinations were presented. The first three stimuli shared a common color (red-green in this example) but varied monotonically in contrast. The next group of three stimuli shared a common color (white-black in this example) that differed from the first three. In one-half of the scans the stimulus contrast increased by a factor of two between stimuli, as shown here. In the remaining scans, the stimulus contrast decreased by a factor of two between stimuli.

the fMRI signals as subject BW viewed four different-colored flickering patterns, each shown at three (increasing) contrast levels. Panels (C) and (D) show replications of this experiment for subject SE. The solid curves show the signals within area V1, and the dashed curves show the signals within the ventral occipital lobe.

The fMRI signal followed the time course of the flickering pattern, and the amplitude of the signal followed the contrast of the stimulus. For subject BW, the amplitude of the responses within area V1 and in the ventral occipital region were roughly equal; for subject SE, the amplitude in the ventral occipital region was approximately 1.5 times greater than the amplitude in the V1 region.

The absolute amplitudes measured in the two regions differed somewhat, but there was a close correlation between the response amplitudes measured within V1-ROI and VO-ROI. The response amplitudes in the two regions are compared in Fig. 14.7. The horizontal axis measures the amplitude of the response measured in V1-ROI, and the vertical axis measures the response in VO-ROI. The two panels show the results for the two observers. For BW, the response amplitudes in V1-ROI and VO-ROI are of similar magnitude; for SE, the ventral occipital response is roughly 1.5 times the response within area V1. The response amplitudes correlate at the 0.83 and 0.89 levels, suggesting that there is a common signalling path between these regions.

One puzzling aspect of the signals is the vertical offset: At low V1 signal levels, there is still a measurable signal in VO-ROI. We do not know if this is due to reduced noise levels in the VO-ROI region com-

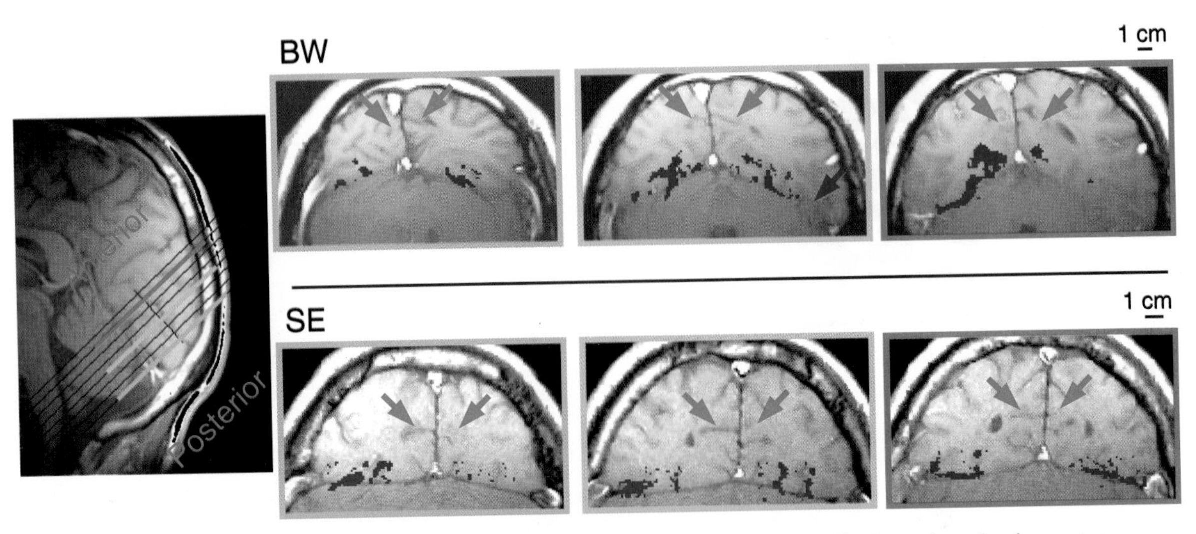

Figure 14.4: The location of the VO-ROI. The image planes were chosen perpendicular to the calcarine cortex. The positions of the planes are indicated by the colored lines on the sagittal image of subject SE on the left; image planes for subject BW were similar. The corresponding images for both subjects are shown on the right, with a bounding box in a color that matches the line shown on the left. The location of the VO-ROI within each plane is denoted by the blue overlay. The location of the calcarine sulcus is indicated by the magenta arrows. Area V1 is located several centimeters away from the VO-ROI, as measured along the surface of the cortex. The centimeter scale bar is slightly different for the two subjects, but it is consistent across the separate images for each subject.

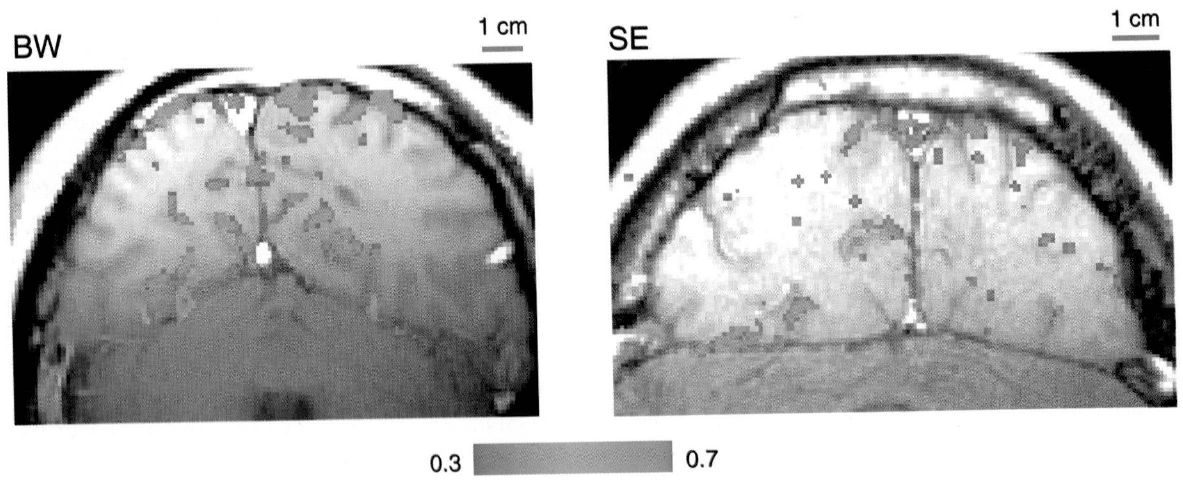

Figure 14.5: Spatial distributions of the active locations for a typical stimulus sequence. The image planes show activity in the planes shown as green (left image, BW) and blue (right image, SE) in Fig. 14.4. The color bar at the bottom indicates the correlation level. Only locations with correlation exceeding 0.3 are shown.

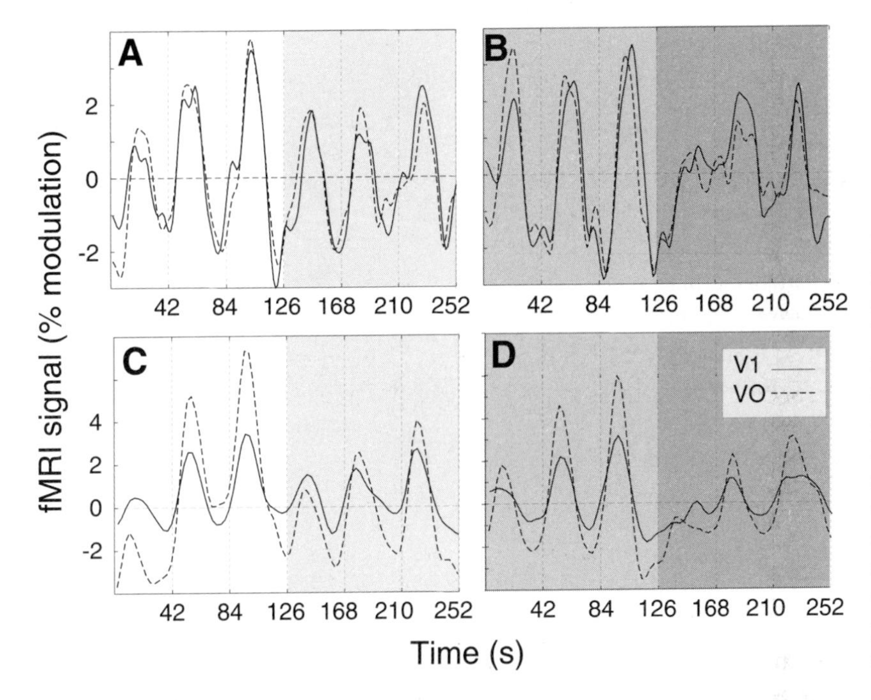

Figure 14.6: Time series of the fMRI signal within the V1 region of interest (V1-ROI) (solid lines) and the VO-ROI (dashed lines). (A,B) The fMRI signal during two separate experimental sessions, each lasting about 4 min, is shown for subject BW. Within each session, the observer viewed six separate targets that varied in color and contrast; the shading indicates targets with the same color, differing only by contrast (C,D). These are data for subject SE; the conditions and plotting conventions are the same.

pared to V1-ROI or whether the data represent a genuine difference in the brain activity.

A second method of comparing the responses in these two areas is to examine the isoresponse curves to different colored stimuli. Figure 14.8 shows the isoresponse curves in V1-ROI (A,B) and VO-ROI (C,D). Consistent with our previous report, the strongest response per unit cone contrast is in an opponent-color direction (LMS = 1,-1,0) (see Engel, Zhang, & Wandell, 1997). The color isoresponse curves for subject SE are nearly identical in the two regions. The isoresponse curves for BW are not precisely the same. Because of the uncertainty in the measurements (indicated by the large confidence interval), we cannot reject the hypothesis that the two data sets share a common color-tuning function under these measurement conditions. Methods for computing the confidence intervals are explained in Appendix 2 at the end of this chapter.

At these low spatiotemporal frequencies, and in the (L,M)-color plane, we are observing two copies of the same color signal. We plan to extend these test conditions using test and adapting stimuli that (a) strongly stimulate the S-cones, (b) have a variety of spatiotemporal properties, and (c) vary the adapting conditions. The systematic measurements of the large signals within these areas will allow us to determine whether the signal is copied across all conditions or whether certain aspects of the signal represented in V1 are transformed as the neural responses spread into other regions of the brain.

Related work

Neuroimaging. There have been roughly a dozen papers written on the use of functional neuroimaging methods to measure human cortical color representation. One set of papers, including both of those based on positron emission tomography (PET) and fMRI, use the subtraction methodology (Posner & Raichle, 1994) to identify a color center of the brain, an enterprise that is widely called *brain mapping*.

The studies by Zeki et al. (1991) and Sakai et al. (1995) set out to identify a focal area associated with color perception. Both groups used the subtraction methodology and compared the activity levels elicited by a pair of related stimuli. In both cases, one stimulus was a simple colored pattern and the comparison stimulus had the same shape and time course, but was an achromatic pattern. The achromatic pattern was matched in luminance (but not chromaticity) to the colored pattern. Both groups reported that, in comparing the difference between these two patterns, a high degree of statistical significance was observed near the VO-ROI in the lingual and fusiform gyri. Neither group reported the absolute size of the responses elicited by the individual stimuli, but they did report that the individual stimuli did generate considerable activity in calcarine sulcus.

Zeki et al. (1991) do not specify the color coordinates of their stimuli. Sakai et al. (1995) specify the color coordinates of the colored stimulus, but not of the matched achromatic stimuli. Our calculations, based on the assumption that achromatic patterns refer to a standard white, suggest that the largest contrast difference between the colored and achromatic stimuli is seen by the short-wavelength cone receptors. Hence, it is possible that the reason for the reported difference in activity between the colored and achromatic stimuli is due to differential signalling by a pathway specialized for carrying S-cone signals, a possibility that we have not examined in this chapter.

Gulyas and Roland (1994) reported using colored stimuli as part of their investigations of cortical processing. They set out to understand the distribution of activity throughout the brain as subjects engaged in tasks that involved color discrimination, form discrimination, and depth judgments. The use of color was only incidental to their main objective: to understand how different types of tasks evoke activity in different brain networks. For all experimental tasks, they report significant activity in a broad array of brain locations. The statistically significant regions activated during color detection were more numerous and widespread than those in form detection.

Kleinschmidt et al. (1996) and Engel, Zhang, and Wandell (1997) measured the fMRI signal obtained by alternating a colored contrast pattern and a neutral, uniform field. Kleinschmidt et al. (1996) compared the

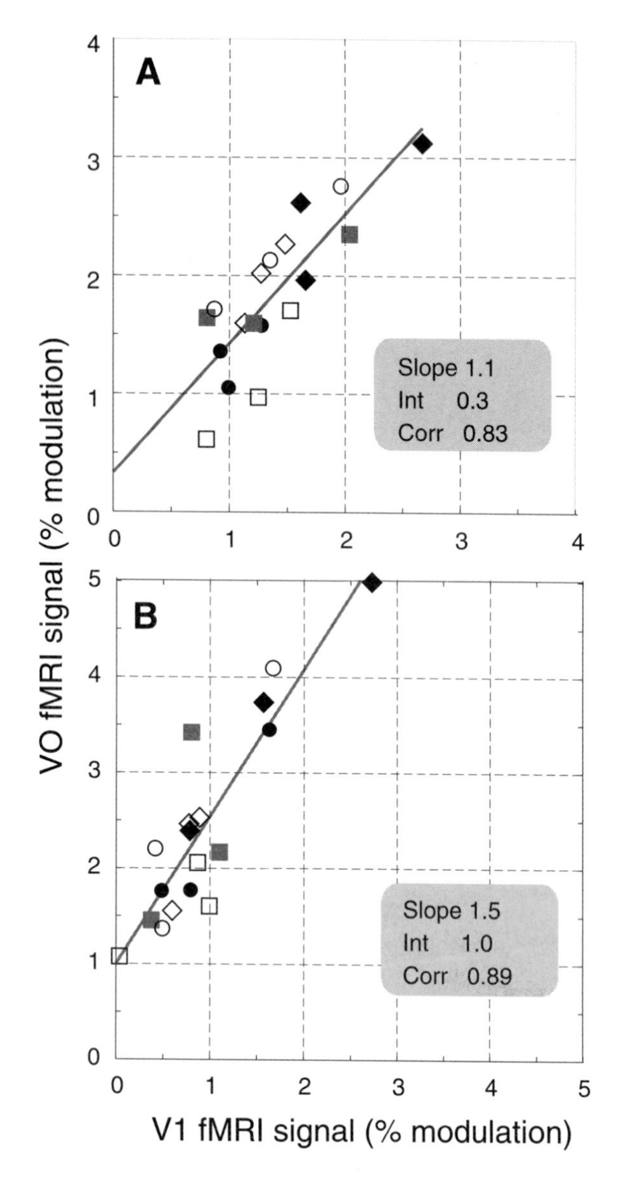

Figure 14.7: Response amplitudes in V1-ROI (horizontal axis) and VO-ROI (vertical axis) are compared. Different symbols refer to measurements with different colors; there are three symbols of each type, corresponding to the three contrast levels of each color. Panels (A) and (B) are for subjects BW and SE, respectively.

response to several colored stimuli, including L-M, S, and L+M stimuli. Engel, Zhang, and Wandell made similar measurements, but they: (a) concentrated on retinotopically organized areas V1 and V2, (b) mea-

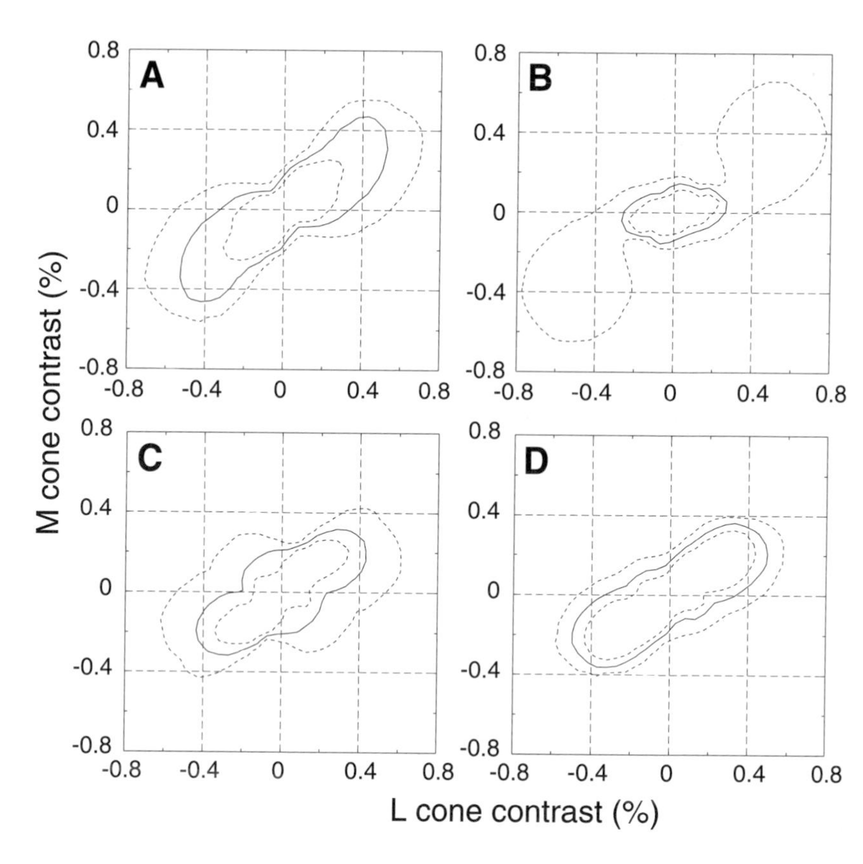

Figure 14.8: Isoresponse curves of fMRI signals measured within V1-ROI (A, C) and VO-ROI (B, D). The axes show the L- and M-cone contrast levels of the signal. The smooth curve is the isoresponse contour, and the dashed curves are 80% confidence intervals around the isoresponse contour. Panels A and B and C and D are for subjects BW and SE, respectively.

sured color tuning by using a larger variety of color stimuli, and (c) compared the fMRI color tuning with behavior. Figure 1C in the Kleinschmidt et al. paper and Fig. 14.5 here agree well, despite differences in MR methods and display instrumentation. Both groups report activity in the calcarine cortex and in the ventral occipital lobe. Both groups also report that, per unit cone contrast, the L-M direction produced a larger fMRI response. Here we confirm Kleinschmidt's observation that there is also significant signal in the ventral occipital lobe. We add the observation that the color tuning in this region is quite similar to the color tuning that we have reported on in area V1.

The approach taken by Kleinschmidt et al. and our group is based on establishing the nature of the color signal at different points in the cortex, rather than locating a single color center. Hence, the first set of questions posed by both groups is structured in terms of how signals initiated within the retina are distributed within the cortex. Subsequent experiments will address how these signals vary with spatial, temporal, and other properties of the input signal. Ultimately, we hope to understand how the set of transformations within the cortex explain our perceptions of color.

Single-unit work. The most detailed information we have about color information in the cortex comes from studies of single-unit recordings in monkey models. Because of the great technical challenges in making these measurements, there have been few attempts to make complete measurements of the spatio-temporal-chromatic response functions of individual units. Instead, the majority of the single-unit literature parallels the neuroimaging literature. Most authors seek partial characterizations, often with the goal of labeling a neuron to classify it as color-selective or not.

The papers containing the most complete characterizations to date are those from Lennie, Gegenfurtner, and their collaborators. In these studies, efficiency dictated that only a limited portion of the neuron's responsiveness be measured. For example, Lennie, Krauskopf, and Sclar (1990) describe their method of measuring color tuning as follows:[2]

> Since cortical neurons are often selective for several attributes of a stimulus, and since most cortical neurons respond to achromatic stimuli, our strategy was to characterize the receptive field with achromatic stimuli and then, with stimuli of the favored configuration, vary the chromatic properties for best response. (Lennie, Krauskopf, & Sclar, 1990, p. 651).

Because receptive fields are not likely to be pattern-color separable, the measured color tuning will depend on the selected spatial pattern (e.g., Poirson & Wandell, 1993, 1996). The use of an achromatic stimulus to define the optimal spatiotemporal parameters of a cell may have an influence on which aspect of the color tuning of the cell is reported.

Preliminary reports that promise to provide more complete summaries of the color responsivity of single cortical neurons have been announced. For example, Cottaris and De Valois (1998) have reported measurements of the spatiochromatic color tuning of single units in area V1. Their measurements are based on a white-noise method that has also been applied to measuring spatiochromatic tuning in the lateral geniculate nucleus (Reid & Shapley, 1992). Also, Gegenfurtner, Hawken, and their colleagues continue to make measurements in various visual areas (see Chapters 11 and 13).

[2]Lennie et al. (1990) performed some additional measurements using nonoptimal spatial patterns. These experiments were intended as a check on the generality of their main reported measurements.

Conclusions

The measurements described here and in Engel, Zhang, and Wandell (1997) offer us a view of cortical activity that complements the one obtained from single-unit recordings. Functional MRI offers a view of the brain that is somewhat analogous to reducing the magnification on our microscope: We see an overview of activity that is much harder to obtain from a sequence of single-unit measurements or even from optical imaging methods. This measurement resolution offers an opportunity for understanding the cortical architecture of color information processing. At this resolution – between behavior and single-unit recording – we may be able to learn new lessons about the neural computations that yield our experience of color.

It is too early to have a definitive view of whether cortical computations should be understood in terms of perceptual correlates or basic computational mechanisms. Over the next ten years, we may be able to combine behavioral, single-unit, neuroimaging experiments to clarify whether perception or computation is the organizing principal of the neural representation. By framing and carrying out empirical tests of these hypotheses, we may be able to learn more about the visual computations that determine our visual experience.

Appendix 1: response amplitude

We describe the method and rationale for estimating the response amplitude by subtracting the estimated noise at the stimulus presentation frequency from the fMRI signal amplitude at that frequency.

Suppose that the harmonic response at the stimulus presentation frequency has amplitude r and phase ϕ_r. We represent this quantity as a complex function $r\,exp(-i\,\phi_r)$. Further, we assume that this response is the sum of a signal and noise component,

$$r\exp(-i\phi_r) = s\exp(-i\phi_s) + \bar{n}\exp(-i\bar{\phi}_n),$$

where the amplitude, $\bar{n}$, is a Gaussian with nonzero mean and the phase, ϕ_n, is a uniformly distributed random variable. We measure this response at each voxel.

We compute the expected value, over a group of pixels defining an ROI, of the squared response amplitude as follows:

$$\|r\exp(-i\phi_r)\|^2 = \|s\exp(-i\phi_s) + \bar{n}\exp(-i\bar{\phi}_n)\|^2$$

$$= \|s\exp(-i\phi_s)\|^2 + \|\bar{n}\exp(-i\bar{\phi}_n)\|^2$$

$$+ 2\|s\exp(-i\phi_s)\bar{n}\exp(-i\bar{\phi}_n)\|^2.$$

The first and second terms are the squared amplitudes of the signal and the noise responses. Because the signal and the noise are uncorrelated, the third term has an expected value of zero. That is, because the phase term ϕ_n varies in all directions with equal likelihood, the expectation of the middle term is zero,

$$0 = E(\|2s\bar{n}\exp(-i\cdot(\bar{\phi}_n - \phi_s))\|^2).$$

Hence, the relationship between the squared magnitudes of the response, the signal, and the noise is given by the Pythagorean formula:

$$E(\|r\|^2) = E(\|s\|^2) + E(\|n\|^2).$$

We can estimate the squared magnitude of the signal from (a) the squared magnitude of the response harmonic at the stimulus frequency, and (b) an estimate of the squared magnitude of the noise at the stimulus frequency. Hence,

$$E(\|s\|^2) = E(\|r\|^2) - E(\|\bar{n}\|^2).$$

Finally, notice that $E(\|s\|^2)^{1/2}$ is not equal to $E(\|s\|)$ and that the isoresponse functions defined by equating with respect to $E(\|s\|^2)$ may be slightly different from those defined by equating with respect to $E(\|s\|)$.

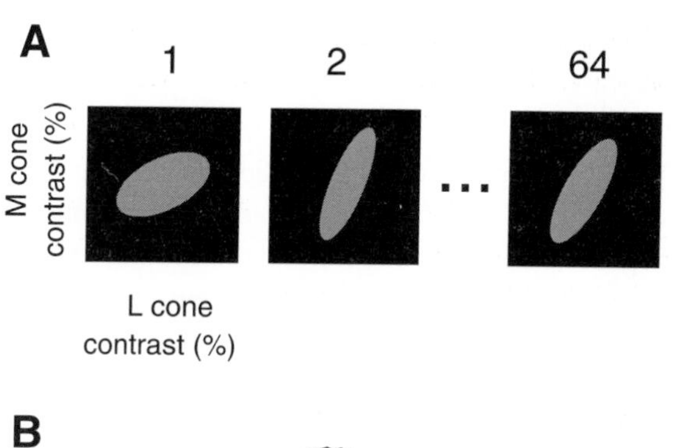

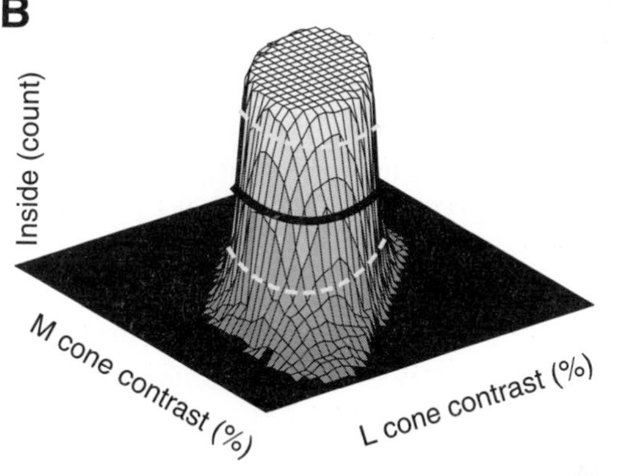

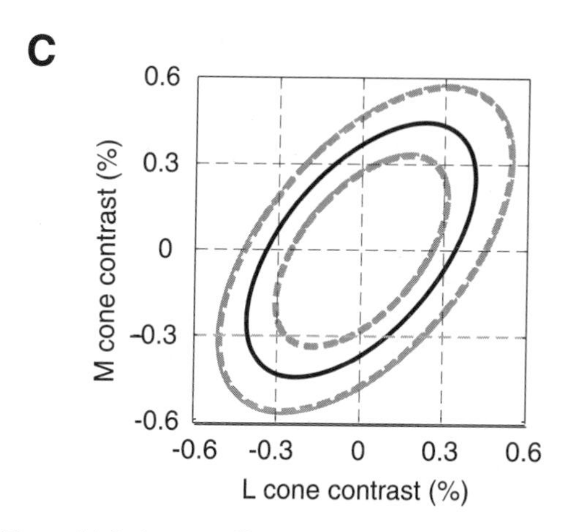

Figure 14.9: A resampling method for determining the isoresponse contour and confidence intervals. (A) Points on the interior of resampled contours are shown. (B) A surface plot showing the number of times each stimulus fell inside a simulated contour. The height of the mound shows the number of times each stimulus level falls within the isoresponse contour. (C) The solid curve shows the stimulus points that fall inside the resampled contour one-half of the time. The dashed curves show the stimulus points that fall within the resampled contours 90 (outer) and 10% (inner) of the time.

Appendix 2: Resampling

The isoresponse contours (solid lines) in Fig. 14.8 represent the set of stimuli that produce a common response amplitude. The isoresponse contour and the 80% confidence intervals (dashed lines) shown surrounding the contour are computed using a *resampling* procedure. The purpose of this appendix is to explain that procedure.

Each isoresponse contour is derived from 36 response amplitude measurements (see Fig. 14.9). To estimate the variability in this data set, we treat these 36 measurements as the entire population and we simulate the effects of repeat experiments by sampling from this population. Specifically, 36 random draws (with replacement) are made from the set of 36 measurements. A set of random draws is accepted for further analysis only if each color direction is represented by at least one data point. Each measurement represents a response level at a position within the stimulus plane. By gridding the sampled values onto the plane and using a standard two-dimensional contour algorithm (Matlab), we derive a single isoresponse contour. This is the *resampled* contour, and we determine those stimulus locations that fall inside and outside the resampled contour (Fig. 14.9A).

For the plots in this chapter, we performed 200 resamplings of the data to measure the chance that each stimulus point falls inside or outside of a resampled contour. Points near the origin of the stimulus plane, with low contrast in both the L- and M-coordinates, fall inside the resampled contours nearly every time. Points with high contrast fall outside the contour nearly every time (Fig. 14.9B). The isoresponse contour (Fig. 14.9C, solid line) is drawn though those stimulus levels that fall inside the resampled contour one-half of the time. The confidence interval shown by the outer dashed line shows stimulus points that fall inside a simulated response contour 90% of the time. The confidence interval shown by the outer dashed contour shows those points that fall within the resampled contour 10% of the time. Hence, the solid curve represents the typical isoresponse contour and the dashed curves define an 80% confidence interval about the isoresponse contour.

15

Interactions between color and motion in the primate visual system

Michael J. Hawken and Karl R. Gegenfurtner

Over the past decade it has become well established that there is motion perception for stimuli that are defined exclusively by chromatic contrast (equiluminant). Currently, the challenge is to determine the conditions for the perception of the motion of chromatic targets and to elucidate the nature of the underlying neural mechanisms. Much of the interest in color and motion over the past 15 to 20 years stems from the anatomical and physiological findings of Zeki and the well-known psychophysical experiment of Ramachandran and Gregory (1978). Dubner and Zeki (1971) found that the majority of cells in the middle temporal area of the primate extrastriate cortex (MT or V5) located on the posterior bank of the superior temporal sulcus were direction-selective. Neurons in MT gave a vigorous response to a bar of light moving in the preferred direction of motion and gave little or no response to a bar moving in the opposite direction. In a separate series of experiments Zeki (1978a) found color-selective neurons in a region of the extrastriate cortex adjacent to MT, called V4. From these and other anatomical and physiological studies Zeki (1978b) proposed that there were separate pathways and brain areas for the analysis of motion and color. In particular, he proposed that V4 was specialized for color processing and that V5 or MT was specialized for the analysis of motion.

Ramachandran and Gregory (1978) reported that the percept of apparent motion in human subjects was severely compromised at isoluminance. In their experiment, the motion of a central square region was defined by alternating the presentation of two random texture patterns. The texture elements (each about 3 deg across) were spatially offset in the central square region between the two stimuli. Hence, when viewing one pattern alone there was no cue for motion, but with an alternating presentation there was an oscillating central region. In this experiment the texture squares were red and green, and when the pattern was near to equiluminance the individual elements were clearly visible, but the apparent oscillatory motion that defined the central square was absent. Because they found that a motion percept was absent near to equiluminance, Ramachandran and Gregory proposed that the processing of motion was handled mainly by the luminance system. Their psychophysical result complemented the physiological findings of Zeki and his colleagues. Since then there have been a number of strong proposals claiming that one of the defining features of cortical organization is the segregation of processing for color, form, motion, and depth. In Livingstone and Hubel's (1987) formulation of the segregation of function scheme there are specific anatomical and functional streams that are concerned with processing each of the submodalities. In both Zeki's and Livingstone and Hubel's formulations of the segregation of function the separation of color and motion processing constitutes some of their strongest evidence in favor of their hypothesis.

Recently there have been many studies that have sought to clarify many of the issues surrounding color and motion processing. Psychophysical experiments have shown that, although motion perception is possible at isoluminance, the perceived speed is lower than for equivalent luminance stimuli (Cavanagh, Tyler, & Favreau, 1984). A number of studies have shown that chromatically opponent mechanisms most likely underlie both the detection and motion identification of chromatic targets, for both simple, one-dimensional patterns (Metha, Vingrys, & Badcock, 1994; Gegenfurtner & Hawken, 1995; Stromeyer et al.,1995) and for more complex pattern motion (Krauskopf & Farell, 1990). Here we review some of the studies. It has often been overlooked that motion, as a vector, consists of two parts: the direction and speed. One tentative conclusion from our review of the current studies is that the direction of motion is coded at isoluminance while the coding of speed is impaired.

Psychophysical studies

One of the most widely adopted psychophysical approaches to determining the effective contribution of chromatic targets to motion perception uses the fact that the luminance system [$V(\lambda)$] is, by definition, silenced at or near to photometric isoluminance (Lennie, Pokorny, & Smith, 1993; see Chapter 11). Therefore, any stimulus that is modulated purely in chromaticity, without any luminance component, can be thought of as functionally "lesioning" the luminance pathway. Many investigations have evaluated performance in psychophysical tasks using stimuli that are either achromatic or equiluminant. Similarly, there are now a number of physiological studies that investigate the properties of individual neurons at different levels of the visual pathway to achromatic and equiluminant stimuli to gain insight into how these two types of stimuli are represented in the visual pathways.

Following the work of Ramachandran and Gregory (1978), there have been a number of psychophysical studies suggesting that color and motion are not always processed independently, although, in many instances, motion perception is compromised at or near to isoluminance. At this point it is worth noting that there are a number of ways of approaching the question of independence. Some studies have endeavored to find whether there are conditions under which there are interactions between color and motion. Others have endeavored to minimize motion perception by using chromatic stimuli configured to silence the luminance motion pathway, and yet others have concentrated on defining the conditions that lead to a breakdown of motion perception.

Stimulus motion or velocity can be conveniently thought of as a vector quantity, because for a one-dimensional stimulus there is a direction and speed. Many studies have concentrated on one of these determinants of velocity, either measuring direction discrimination or estimating speed sensitivity. In the following sections we first discuss the results of studies that have concentrated mainly on directional selectivity. Then we discuss studies that have concentrated on studying the effects of color on speed judgments. At the end of the section on psychophysics we discuss some of the studies on higher order motion mechanisms.

Motion aftereffect. Some of the first studies that sought to determine the nature of the pathways involved in chromatic motion used the motion aftereffect as a tool. When a pattern moving in one direction is viewed for a relatively long period and then motion stops, the pattern is perceived to be moving in the opposite direction even though the stimulus is stationary. This is the motion aftereffect (MAE), sometimes called the waterfall illusion. One of the explanations for this effect is that the low-level motion detectors are adapted during the prolonged exposure (Barlow & Hill, 1963). In many theories of motion perception perceived motion is due to the relative balance of the activity in motion detectors that are sensitive to opposite directions of motion. It follows that when one of the pair is adapted then the balance of activity moves from the null toward the members of the group of detectors that were not adapted. It has been argued that if luminance and chromatic stimuli activate the same set of detectors, then adaptation to a luminance stimulus moving in one direction should result in a motion aftereffect for a chromatic stimulus. Similarly, adaptation to a moving chromatic stimulus should result in an aftereffect for a luminance stimulus. It has been demonstrated that adaptation to a high-contrast luminance target does result in a motion aftereffect for a chromatic target, and, correspondingly, adaptation to a chromatic target results in a measurable motion aftereffect for luminance targets (Cavanagh & Favreau, 1985; Derrington & Badcock, 1985; Mullen & Baker, 1985). However, the strength of these effects is quite asymmetric, in that luminance adaptation produces a stronger chromatic aftereffect than adaptation to an isoluminant target. Equating the degree of adaptation between achromatic and chromatic stimuli needs to be made with caution because the contrast scales are not equivalent; hence the degree of both activation and adaptation produced by a high-contrast luminance target may not be the same as that produced by a chromatic target. Unlike the stimuli used in more recent experiments, the stimuli employed in many of these experiments were not equated for threshold visibility. The results from these experiments should be interpreted qualitatively rather than quantitatively. Notwithstanding, results from the adaptation studies clearly show that there can be an interaction between chromatic and achromatic stimuli. However, the effects are relatively weak, so that, as pointed out by Derrington and Badcock (1985), it is difficult to use these results to distinguish between a system in which a single mechanism processes both luminance and chromatic stimuli and a system that has two separate motion mechanisms that are differentially sensitive to luminance and chromatic stimuli.

Motion-nulling studies. There have been a number of experiments that employ the fact that a counterphase modulating grating can be decomposed into two components (Levinson & Sekular, 1975; Kelly, 1979; Watson et al., 1980). Two vertically orientated gratings drifting in opposite directions, one drifting to the left and the other drifting to the right, result in a standing grating that is contrast-reversing at the same temporal frequency as the drifting components but with twice the contrast of the components. Cavanagh and Anstis (1991) found that the motion of either red/green or blue/yellow equiluminant gratings could be nulled by an achromatic luminance grating drifting in the opposite direction. From this result they argued that, since the luminance grating could null the motion of the chromatic grating, they must share a common mechanism; hence the motion of equiluminant chromatic gratings was signalled by the same directional mechanism that signalled the motion of the luminance targets. They used this method to determine the equivalent contrast of luminance gratings required to null a variety of equiluminant red/green and blue/yellow gratings. Because a low-contrast luminance grating is required to null a high-contrast chromatic grating, Cavanagh and Anstis argued that the net input signal for motion generated by equiluminant targets is equivalent to the input signal generated by low-contrast luminance targets. Smith (1994) points out that this result could also arise if there are two independent motion mechanisms that are relatively more sensitive to luminance and chrominance, respectively.

Chichilnisky, Heeger, and Wandell (1993) used the motion-nulling paradigm to determine whether there was a single monochromatic motion mechanism – a luminance-only mechanism. They tested many directions in color space measuring the cone contrast required to null a test grating. They argued that if there was a single mechanism, then they would have expected to find the nulling contours projected as a set of parallel lines that aligned in the orthogonal direction to the color direction of the monochromatic motion mechanism. Their experiments did not support the existence of a single monochromatic mechanism. They went on to test the more general hypothesis that motion perception depends on a univariate motion signal driven by all three color dimensions, and that the motion signal depends on the product of stimulus contrast and a term that depends only on the color direction (*Stiles invariance*). Their results support Stiles invariance for the conditions and color directions that they tested. Accordingly, contrast and color can be interchanged in their model with simple linear transformations to give equal nulling. A possible neural implementation is given through the summation of motion energy in the three color-opponent channels.

This is an attractive idea, but it is not clear whether the results generalize to all color directions and to other response measures such as contrast detection or speed judgments.

Contrast detection and identification thresholds. Levinson and Sekular (1975) showed that the contrast threshold for the detection of a moving luminance grating was the same as the contrast threshold for identifying its direction of motion. These results were confirmed and extended by Watson et al. (1980). Generally these results are interpreted to mean that the same set of neural mechanisms is used for detection and motion identification and that these mechanisms are direction-selective (Watson & Robson, 1981). The same logic can be applied to stimuli that are modulated in chromaticity. The question that is being posed here is whether the mechanism that detects stimuli that are modulated in chromaticity is the same mechanism that discriminates the direction of motion for the same set of stimuli. For chromatic stimuli there are a number of studies that indicate that there are is little difference between the thresholds for detection and motion identification (Cavanagh & Anstis, 1991; Mullen & Boulton, 1992; Derrington & Henning, 1993) while other studies suggest quite substantial differences in the detection and motion identification thresholds (Lindsey & Teller, 1990; Palmer, Mobley, & Teller, 1993), even when there are no differences between detection and orientation identification (Lindsey & Teller, 1990). More recent studies have resolved some of these differences by making parametric variations in a number of factors that seem to be important in determining performance. One rather useful measure of the difference between detection and motion identification is simply the ratio of their respective thresholds. This motion-to-detection ratio (M/D ratio) is 1 if detection and motion identification occur at the same contrast, as is the case for luminance under most conditions. A value greater than 1 means that the detection of the target occurs at a contrast that is lower than the contrast required for the identification of direction of motion, that is, the target is visible but its direction of motion cannot be clearly judged. A number of studies have shown that the M/D ratio depends on eccentricity (Derrington & Henning, 1993; Metha, Vingrys, & Badcock, 1994; Gegenfurtner & Hawken, 1995). In the fovea, the M/D ratio is closest to 1 and increases for more peripheral locations. These results go some way toward resolving the conflicting findings in the literature; most of the studies that found values of M/D close to 1 used foveal viewing while the studies showing the greatest effect used more eccentric targets.

It is also clear that the distinctive differences in the temporal contrast sensitivity function for luminance and equiluminant gratings contribute to the M/D ratio. For simple detection the luminance temporal contrast sensitivity function has a characteristic bandpass shape for low spatial frequencies (Robson, 1966; Kelly, 1971) while the red/green equiluminant temporal contrast sensitivity function is low-pass to about 0.2 Hz (Kelly, 1979; Burr & Morrone, 1993; Derrington & Henning, 1993; Gegenfurtner & Hawken, 1995), and at very low temporal frequencies it is also attenuated (Kelly, 1983). It does not matter if the temporal contrast-sensitivity function is determined with drifting or contrast-reversing gratings; the respective shapes remain the same (Kelly, 1979). The temporal contrast-sensitivity functions for *motion identification* of red/green chromatic equiluminant gratings and achromatic gratings are bandpass. Typical functions are shown in Fig. 15.1. This figure incorporates two effects that have lead to conflicting interpretations about the relative sensitivity of detection and motion identification. The first conflict concerns temporal frequency and the second concerns eccentricity. Thresholds for direction discrimination and simple detection are similar for achromatic gratings (Fig 15.1, bottom row) at all temporal frequencies and for all eccentricities within 5 deg of the fovea (Levinson & Sekular, 1975; Watson et al., 1980; Lindsey & Teller, 1990; Derrington & Henning, 1993; Palmer, Mobley, & Teller, 1993; Metha, Vingrys, & Badcock, 1994; Gegenfurtner & Hawken, 1995; Stromeyer et al., 1995). The situation is quite different for equiluminant stimuli (Fig. 15.1, top row). At low temporal frequencies, in the fovea, the sensitivity for detection is somewhat better than the sensitivity for motion identifica-

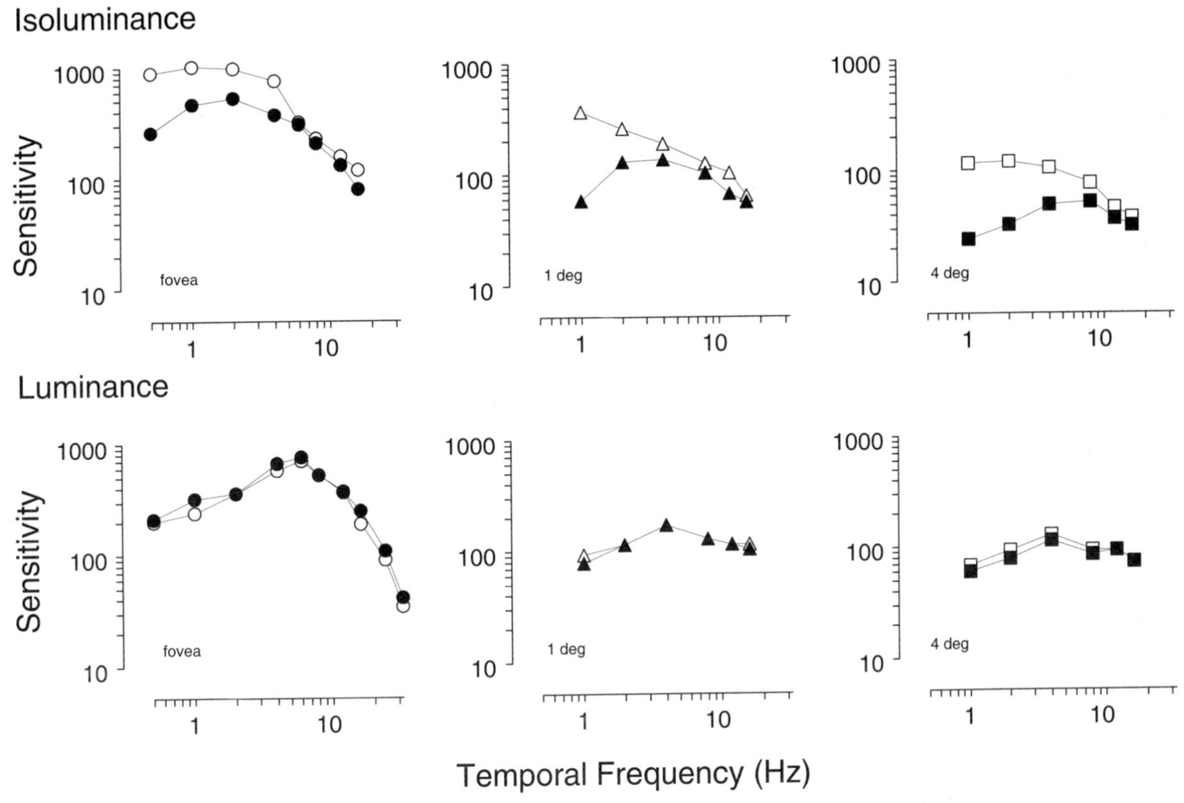

Figure 15.1: Temporal cone-contrast-sensitivity functions for isoluminant (top row) and luminance (bottom row) gratings for three eccentricities. Sensitivity for simple detection is shown by the open symbols and sensitivity for motion identification are shown by the filled symbols. At low temporal frequencies in the isoluminant condition, cone-contrast sensitivity for detection is considerably greater than the sensitivity for motion identification. At all temporal frequencies in the luminance condition there is no difference in the sensitivity for detection and motion identification. The graphs in the left column are for the foveal presentation of a large field (circular window diameter 10 deg); in the middle column for perifoveal (1 deg eccentric) presentation; in the right column for parafoveal (4 deg eccentric) presentation. In all conditions the stimulus was a 1 c/deg grating; for the peri- and parafoveal conditions the grating was vignetted by a Gaussian (1 deg standard deviation). From Gegenfurtner and Hawken (1995).

tion, while at more eccentric locations this difference becomes greater in amplitude and extends to higher temporal frequencies. There is no differential sensitivity above 10 Hz at any eccentricity.

Chromatic threshold contours. Recent studies have measured the ability of human observers to make detection or discrimination judgments for different combinations of lights. The threshold contrasts are represented in a cone contrast space (Stromeyer, Cole, & Kronauer, 1985). These thresholds as a function of color direction are called chromatic threshold contours. Many experiments involve determining the contours in long-wavelength– (L) and middle-wavelength– (M) cone contrast space while keeping the S-cone contrast constant. The small but significant contribution of S-cones will be ignored here; it has been characterized in detail by Lee and Stromeyer (1989). A schematic of the expected contour for a single mechanism that is sensitive to the difference of L- and M-cones, an opponent mechanism, is shown in Fig. 15.2A as the positive diagonal. The contour with the negative slope is sensitive to the sum of the L- and M-cone contrast and is the contour that is typical of the luminance mechanism $V(\lambda)$. The rationale for such experiments is that if there is a red/green opponent mechanism that differences L- and M-cones in a ratio of 1:1, then the direction of the contour that represents the contrast required for detection by this mechanism should run parallel to the line with the positive slope shown in Fig 15.2A. It should be noted that the use of the straight lines to fit the data in Fig. 15.2 is controversial, and a strict interpretation implies completely independent mechanisms. However, most studies find equally good fits for ellipses and rectangles, and in most cases the major and minor axes of the ellipse closely correspond to the straight lines describing the rectangle (Chichilnisky, Heeger, & Wandell, 1993; Gegenfurtner & Hawken, 1995).

Several studies have used chromatic contours to represent the thresholds for detection and motion identification. There are three principal questions that can be addressed using this approach. First, do the slopes of the contours indicate that chromatically opponent mechanisms contribute to motion identification? Second, are there differences between detection and motion identification thresholds? A significant difference would indicate a nondirectional chromatic detection mechanism that is independent of a motion-sensitive mechanism. Third, does the comparison of thresholds for chromatic and luminance motion identification suggest independent motion mechanisms. At relatively low temporal frequencies (or speeds) there is good agreement that the threshold contour for motion identification lies parallel to the slope expected for a red/green opponent mechanism (Metha, Vingrys, & Badcock, 1994; Gegenfurtner & Hawken, 1995; Stromeyer et al., 1995). The slope of the contour for chromatic detection is also close to 1. Examples of detection and identification contours are shown in Figs. 15.2B and 15.3A. It is a consistent finding that the threshold contour for motion identification lies outside the contour for detection, suggesting two mechanisms. Stromeyer et al. (1995) have called these two

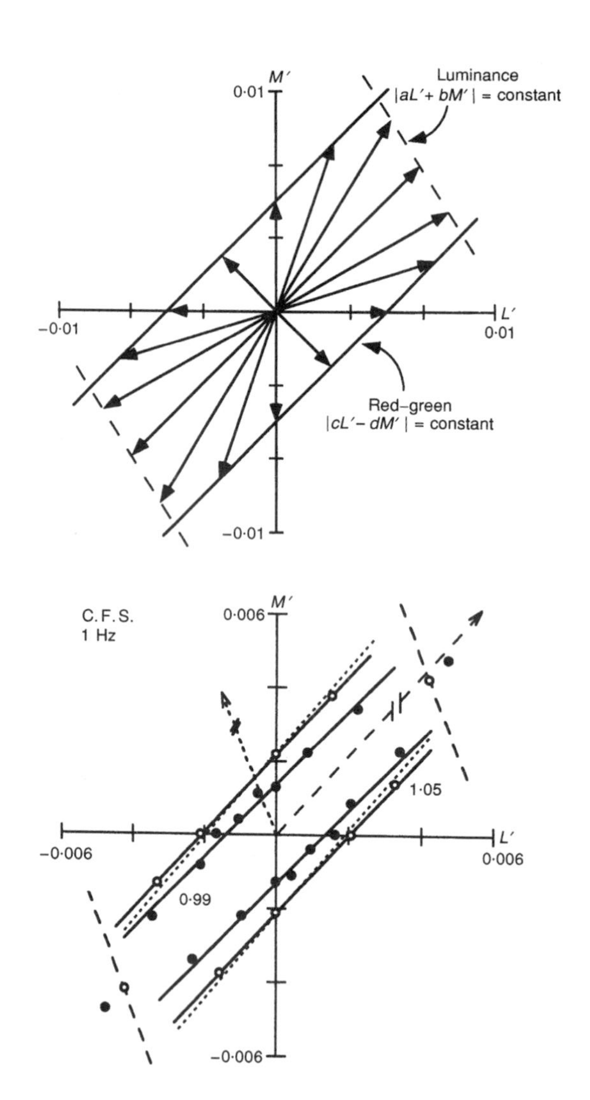

Figure 15.2: *Top*: The color space in cone contrast for the representation of grating stimuli in contour experiments. The x-axis represents modulation of L-cones with the M-cones "silenced" while the y-axis represents the modulation of M-cones alone. The top figure also shows the predicted contour of two putative mechanisms. The positive contour indicates the red/green chromatic opponent mechanism whose response is determined by the difference of modulations in L- and M-cones. The negative contour (dotted line) gives the response of the nonopponent luminance mechanism that sums the modulations in the L- and M-cones. *Bottom*: Thresholds for a single subject detecting a 1-Hz, 1-c/deg grating (solid symbols) or identifying the direction of motion (open symbols). From Stromeyer et al. (1995).

mechanisms the red-green hue mechanism (detection) and the spectrally opponent motion mechanism (motion identification). Note that, although the contours of the motion thresholds lie outside the contour for the detection thresholds, the major axes of the two contours are almost identical; they run parallel to the positive diagonal, which indicates an opponent process.

As the temporal frequency is increased the orientation of motion identification contours (Fig. 15.3B) in cone space tilts away from the positive slope of 1, which is characteristic for lower temporal frequencies. The direction of the tilt indicates that the motion mechanism is dominated by L-cones at high temporal frequencies (Gegenfurtner & Hawken, 1995; Stromeyer et al., 1995). However, the overall ratio of L- to M-cone input to the motion contour must depend on the relative contribution of the chromatically opponent and the luminance mechanisms. In addition, if there are changes in L- and M-cone contributions to these mechanisms that depend on temporal frequency and adapting background (Stromeyer et al., 1995), then teasing apart their contributions is complicated. At temporal frequencies greater than 10 Hz there are situations in which it seems as if the luminance mechanism can account for the detection and motion discrimination of gratings at all directions in L- and M-cone color space (Gegenfurtner & Hawken, 1995) while there are other conditions that favor a mixed contribution from an opponent motion mechanism and a luminance mechanism (Stromeyer et al., 1995). Certainly at suprathreshold levels it seems likely that there will be an interaction of a number of detection and motion mechanisms. Furthermore, there is a weak but measurable S-cone motion mechanism (Lee & Stromeyer, 1989; Gegenfurtner & Hawken, 1995).

The threshold cone contrast required for motion identification at equiluminance is lower than for luminance stimuli, which is seen clearly in the data from Stromeyer et al. (1995) presented in Fig. 15.2B. The main point here is that low-contrast chromatic stimuli give a significant percept of motion, and these are contrasts that are many times lower than those that would stimulate a luminance-based motion mechanism.

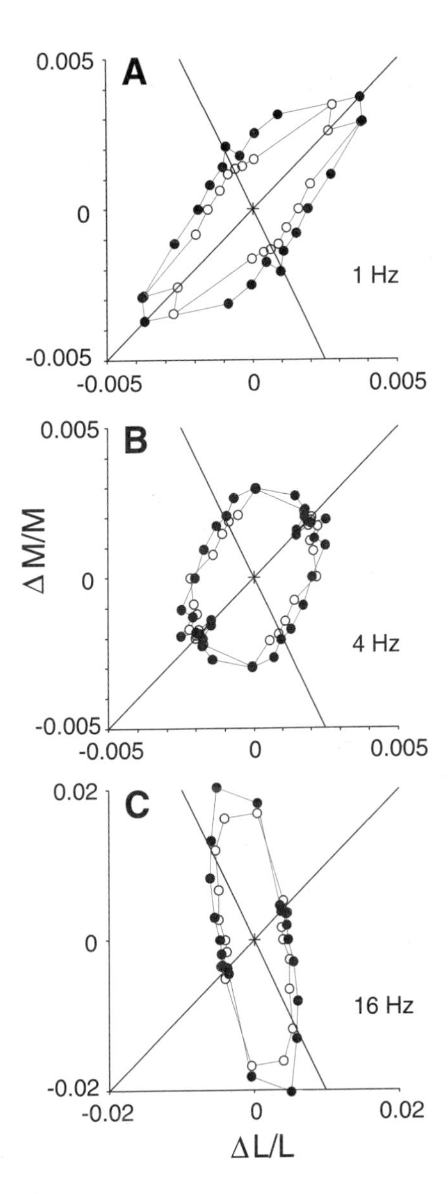

Figure 15.3: The effect of temporal frequency on the orientation of the chromatic threshold contours for foveal viewing. (A) At low temporal frequencies, 1 Hz, the principal orientation of the contour for both detection and motion identification is parallel to the positive diagonal, indicating the dominance of a red/green opponent mechanism in determining thresholds. (B) At 4 Hz, the contours for both detection and identification are not elongated in either the opponent or the luminance direction. (C) At 16 Hz, both contours are oriented toward an L-cone–dominated mechanism. From Gegenfurtner and Hawken (1995).

Timing. Cropper and Derrington (1994) asked whether the processing time required for subjects to accurately determine the motion of a stimulus was different for luminance and equiluminant first-order stimuli and second-order motion stimuli. Their premise was that if there are substantial differences in processing time, then the mechanisms responsible are also likely to be different. They used as their measure of motion perception the lowest velocity that was required for subjects to reliably report direction of motion, the lower threshold of motion (LTM). Stimuli were presented for periods from 15 to 960 ms at contrasts 0.5 and 1.5 log units above detection threshold. For the equiluminant chromatic gratings at the lower contrast (0.5 log unit above threshold) an LTM could not be measured until the stimulus had been presented for at least 120 ms, whereas for the higher contrast (1.5 log units above threshold) subjects could detect motion when the stimulus was presented for 15 ms. The integration time for the high-contrast equiluminant grating is the same as for either low- or high-contrast luminance gratings while the integration times for the lower contrast equiluminant gratings are similar to those for their second-order motion targets. Interestingly, they reported that the LTM for low-contrast equiluminant gratings is greater than 5 deg/s (or 5 Hz as they used 1-c/deg gratings) at the minimum duration for a measurable response to become apparent, yet at the higher contrast the LTM is 1 deg/s (1 Hz). Cropper and Derrington concluded that at low contrasts the motion of equiluminant targets is likely to be handled by a second-order motion system, whereas at higher equiluminant contrasts the mechanism that is used is equivalent to the one used for first-order luminance gratings.

In a more recent study Cropper and Derrington (1996) reported that when they used a two-frame sequence as a stimulus with an interframe interval of 8.33 ms (120 Hz) their subjects could identify the direction of motion of isoluminant stimuli at about ten times the detection threshold. Of significance is the fact that this motion identification performance was not masked by the addition of a stationary luminance grating at any spatial phase. There are two main points that can be taken from this study. First, the subjects cannot be using a "feature-tracking" mechanism to identify the motion of the chromatic targets because the duration (16.7 ms) is far below the limit for feature tracking to take place. Second, the motion in the chromatic target is detected by a chromatic mechanism because there is no influence of the luminance mask. The latter result gives independent support to the conclusions based on the chromatic contour studies (Metha, Vingrys, & Badcock, 1994; Gegenfurtner & Hawken, 1995; Stromeyer et al., 1995) that support the existence of a chromatic motion mechanism that signals the motion of chromatic targets and that is independent of the luminance-based motion mechanism.

Phase experiments. These experiments were designed to investigate the interaction between luminance and chromatic motion by changing the relative phase of the chromatic borders with time so that a chromatic mechanism would signal motion in one direction and luminance-based mechanisms would signal motion in the opposite direction. Dobkins and Albright (1993) found that human observers reported motion in the direction of a colored border where the motion was supplied by a relatively small change phase of the hue signal with time. They interpreted this to mean that the signal was carried by a phase-insensitive mechanism such as the M-cell pathway. All of their experiments were made with high-contrast chromatic targets, under conditions that would lead to a signal in a luminance-based pathway (Cropper & Derrington, 1994), but the experiments do not rule out a role for a chromatic motion mechanism that is required to account for the motion perception of lower contrast chromatic borders.

Perceived speed. Equiluminant chromatic targets appear to move more slowly than comparable luminance targets (Cavanagh, Tyler, & Favreau, 1984; Kooi & DeValois, 1992; Mullen & Boulton, 1992). In the experiments of Cavanagh, Tyler, and Favreau (1984) subjects had to adjust the speed of a luminance grating with 10% contrast to match the subjective speed of heterochromatic red/green gratings. The red/green gratings were displayed at a variety of luminance contrasts; in some conditions the red bars were brighter than the green bars, in others they were darker. Figure 15.4 shows the results of such a measurement (subject PC, 0.8 c/deg, 1.2 deg/s). It can be seen that the physical speed of the luminance grating was only at 40% of the physical speed of the equiluminant red/green grating when its perceived speeds matched. This implies that speed processing is dramatically impaired at isoluminance for slowly moving targets. No contrast dependence was found for fast-moving targets.

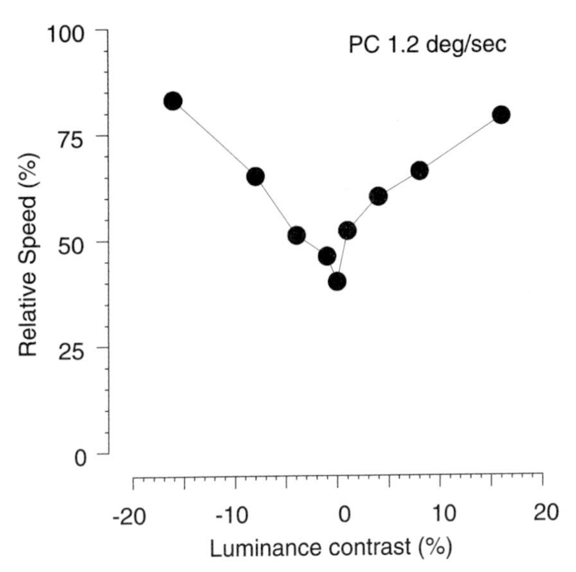

Figure 15.4: This figure shows a clear example of the slowing of chromatic grating at equiluminance. The relative speed of a red/green chromatic grating with different degrees of luminance modulation added to the chromatic grating. In this experiment there are two vertical gratings, one above and one below a fixation point. The top patch is always a luminance comparison grating that has 95% contrast. The lower patch is the test grating with mixed chromatic and luminance contrast. In the experiment shown here the observer's task was to adjust the speed of the comparison grating to match the speed of the test grating. The relative perceived speed scale on the ordinate is the speed of the high-contrast comparison grating when it matched the speed of the test grating divided by the physical speed. From Cavanagh, Tyler, and Favreau (1984).

Similar results were found for luminance gratings. The perceived speed of luminance gratings shows a modest but consistent dependence on contrast (Thompson, 1982; Stone & Thompson, 1992). The nature of the contrast dependency is related to the speed of the target; slow-moving gratings (1 deg/s) are slowed at low contrasts relative to a high-contrast standard while the perceived speed of fast-moving targets is relatively independent of contrast. Consistent with the relative independence of perceived speed with contrast are the findings that speed discrimination thresholds are relatively invariant with contrast over a range of speeds from 4 to 64 deg/s (McKee, Silverman, & Nakayama, 1986).

Various strategies of stimulus interaction have been used to tease apart the different types of motion processing. Cavanagh, Tyler, and Favreau (1984) found that a high-contrast red/green equiluminant grating added to a luminance grating produced a perceived slowing of the luminance grating. A similar, although reduced effect is seen when a high-contrast chromatic grating is added to one component of a luminance plaid (Kooi & DeValois, 1992). In both Cavanagh, Tyler, and Favreau's experiment and in Kooi and De Valois's experiment the addition of a chromatic grating is not equivalent to adding a low-luminance achromatic grating, which would tend to increase the speed, and therefore the color grating is not operating in a simple manner on the same channel as the achromatic grating or plaid.

Speed discrimination has been measured under different stimulus conditions to determine the similarity and differences in speed processing of chromatic and luminance targets. The Weber fraction for speed discrimination ($\Delta v/v$) for low-contrast moving chromatic gratings is considerably higher than for low-contrast luminance gratings (Cropper, 1994). In this study the range of base speeds was 0.25 – 4.0 deg/s and, in accord with previous studies, the Weber fraction for luminance-defined gratings, independent of contrast, was essentially constant for speeds greater than 1 deg/s but was elevated at the lowest speeds. The same pattern of Weber fractions was found for equiluminant gratings 1.5 log units above the detection threshold (Fig. 15.5); however, when the contrast was only 0.5

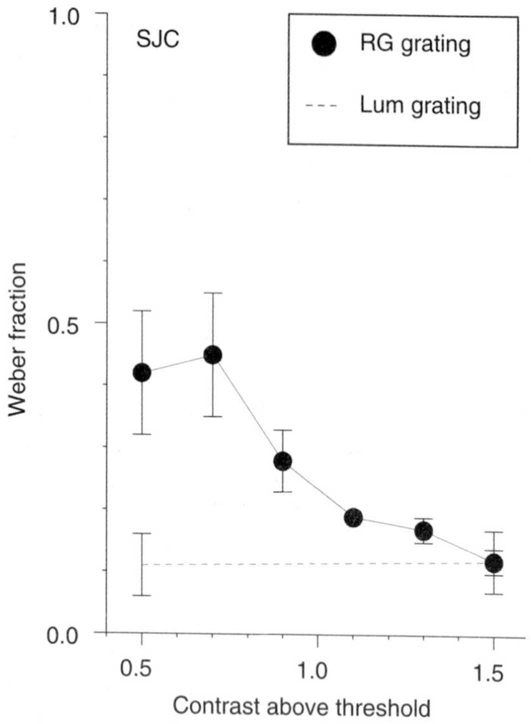

Figure 15.5: The Weber fraction for luminance (dotted line) and red/green isoluminant grating (solid line) as a function of contrast above the detection threshold. For both subjects the Weber fraction for luminance gratings is independent of contrast for contrasts that are 0.5 or 1.5 log units above detection. The low-contrast isoluminant gratings have elevated Weber fractions, indicating poor velocity discrimination. At high contrast isoluminance gratings show similar Weber fraction to luminance targets. Target speed 2 deg/s, spatial frequency 1 c/deg. Redrawn from Cropper (1994).

log unit above the detection threshold the Weber fractions were considerably higher than for the luminance or high-contrast equiluminant targets. It should be pointed out that, although the measurements were 0.5 log unit above the detection threshold, their direction of motion was "clearly discriminable" (Cropper, 1994). The results suggest that low-contrast equiluminant gratings are detected by a system that is independent of the first-order luminance motion mechanism. At higher equiluminant contrasts, where the performance is very similar to that for luminance stimuli, one mechanism could account for the performance with both sets of stimuli.

Relative velocity judgments for slow-moving equiluminant gratings show a pronounced dependence on contrast (Mullen & Boulton, 1992; Hawken, Gegenfurtner, & Tang, 1994; Gegenfurtner & Hawken, 1996a), and this dependence on contrast is much more pronounced for chromatic gratings than for luminance gratings (Fig 15.6). This is in accord with the results from Cropper (1994) and reinforces the suggestion of independent mechanisms for speed discrimination of slowly moving gratings. On the other hand, the relative perceived speed of fast-moving gratings, either luminance or chromatic, is almost unaffected by contrast (Fig. 15.6; Thompson, 1982; Hawken, Gegenfurtner, & Tang, 1994; Gegenfurtner & Hawken, 1996a). This means either that the same mechanism is providing the motion signals for the fast-moving chromatic and luminance stimuli or that there are independent mechanisms that have the same characteristics.

Second- and higher order mechanisms. Currently there are thought to be multiple stages of processing for motion stimuli. The first stage involves the extraction of a one-dimensional translational motion signal while the second stage provides a signal for complex motion. Examples of second-stage motion processing include the translational motion of a multidimensional spatial pattern, circular motion, expansion, or contraction. Each of these is thought to involve separate mechanisms. One of the most studied examples of a second-stage process concerns the analysis of plaid patterns (Adelson & Movshon, 1982). Plaid patterns are particularly attractive because there are two one-dimensional components; these are thought to be processed independently by a first-stage process, and the combined motion of the pattern is a result of the operation of a second-stage mechanism on inputs from the two independent first-stage processes. Two luminance gratings oriented 90 deg apart, with similar spatial and temporal frequencies, appear to combine and move as a plaid in a direction that is determined by the intersection of constraints (Adelson & Movshon, 1982).

Krauskopf and Farell (1990) found that if an equiluminant red/green chromatic grating is combined

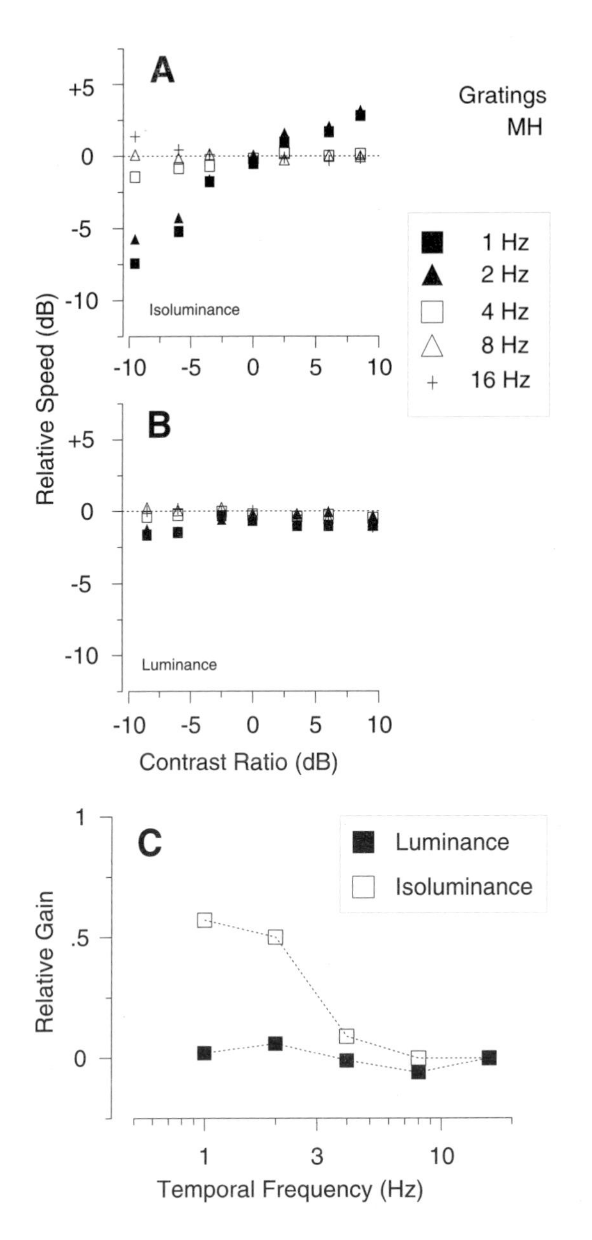

Figure 15.6: The effect of contrast on speed perception. At low temporal frequencies contrast effects speed perception of (A) isoluminant targets more than (B) luminance targets. At high speeds there is little effect of contrast on the speed perception of either luminance or isoluminant gratings. (C) Gain is a measure of the ratio of the change in perceived speed evoked by a change in contrast. A gain of 0 means that there is no change in the perceived speed with a change in contrast, and a gain of 1 would mean that a doubling of contrast would result in a doubling of the perceived speed. From Gegenfurtner and Hawken (1996a).

with a luminance grating, the two do not cohere; they appear to slide over each other. They found a similar result with the combination of luminance and equiluminant red/green or blue/yellow gratings (see Chapter 16, Fig. 16.15). In their experiments the spatial and temporal frequencies were 1 c/deg and 1 deg/s, respectively. They interpreted their results in terms of the cardinal directions of color space (Krauskopf, Williams, & Heeley, 1982), whereby stimulus components of a plaid that selectively stimulate orthogonal directions in color space, luminance vs. L–2M, for example, will not cohere. More recent experiments have called the strong cardinal direction hypothesis into question (Krauskopf, Wu, & Farell, 1996; see Chapter 16). In an extension of Krauskopf and Farell's experiments, Cropper, Mullen, and Badcock (1996) examined the effect of altering the angle between the plaid components while keeping the pattern speed constant (at 4 deg/s). They asked subjects to report the "most salient motion" direction of the stimulus pattern under conditions where the components differed in chromaticity. For angles of 90 deg between components, most of their subjects reported little coherence when the components were in different cardinal directions in color space and no coherence when the components were luminance and color. However, as the angular separation between components was reduced, the subjects reported a coherent percept for all orthogonal directions in the isoluminant plane. Their experiments, along with the more recent interpretations of Krauskopf, support the notion that chromatic processing is not limited to the cardinal directions and, furthermore, that motion processing may need to include a direct second-stage combination, independent of the component to pattern scheme of Adelson and Movshon (1982).

Kooi and DeValois (1992) and Kooi et al. (1992) employed a somewhat different strategy from that of Krauskopf and Farell (1990) to study plaid motion with chromatic stimuli. They did not ask the subjects for a binary decision ("coherence" or "no coherence") but rather sought to use a measure of the deviation from the expected direction of movement of a coherent equiluminant plaid when the components differ in

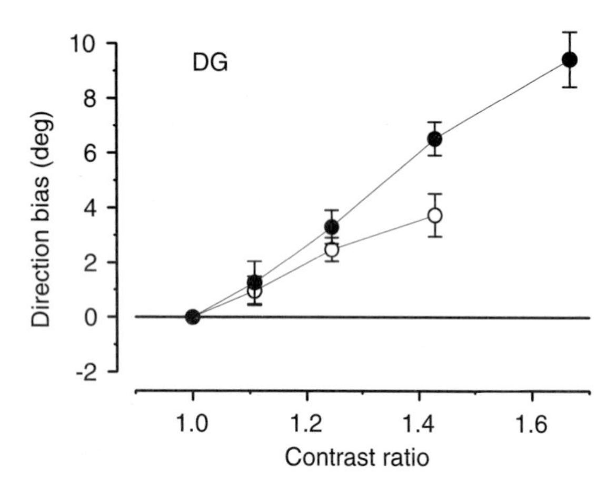

Figure 15.7: The perceived direction bias of motion of a plaid pattern when the components of the equiluminant plaid have different contrasts. The contrast is expressed as a ratio of the component contrasts. The direction bias away from the direction seen with equal component contrasts is plotted on the *y*-axis. The data are shown by the filled symbols. In a separate experiment, the relative perceived speed of each component was measured and then the direction bias was predicted based on the equation: direction bias = 45 − arctan (S_1/S_2), where S_1 and S_2 are the perceived speeds of each component.The predicted bias is shown by the open symbols. The component gratings were 1 c/deg, oriented 90 deg apart, with components moving at 3 deg/s. From Kooi and DeValois (1992).

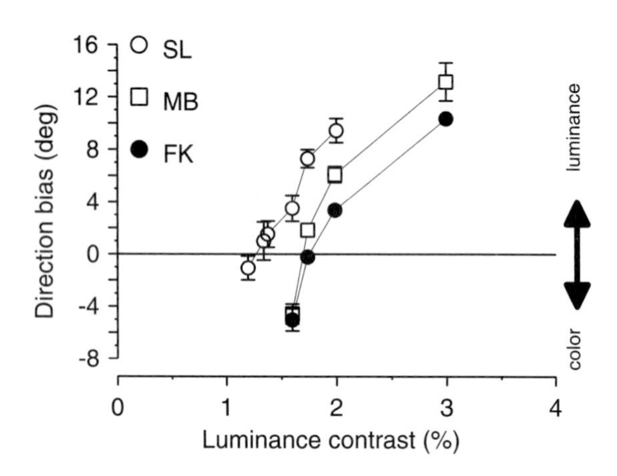

Figure 15.8: The equivalent luminance contrast of a 32-times threshold chromatic grating required to produce coherent pattern motion. Two grating components are red/green equiluminant and yellow/black luminance, moving in directions 90 deg apart. The red/green grating component has a fixed contrast of 32-times threshold for detection. In this experiment there are four or six low-luminance contrasts (plotted on the *x*-axis), chosen to span the zero-direction bias condition. The zero-direction bias point determines the equivalent luminance contrast required to match the strength of a color grating of 32-times threshold. From Kooi and DeValois (1992).

either spatial frequency or contrast. But the basic idea is to uncover what stimulus conditions determine the pattern direction of the plaid motion and thereby see if equiluminant plaid motion is equivalent to pattern motion at low-luminance contrast. In their first experiments Kooi and De Valois determined the effects of unequal contrast or spatial frequency of the components of equiluminant plaids on the perceived direction of the pattern. They called the deviation of the perceived direction from the physical direction of the intersections direction bias. There is a substantial bias introduced with unequal contrast or spatial frequency. The bias is quite well accounted for by the perceived speed of the components, as illustrated in the case of unequal contrast in Fig. 15.7.

Kooi and De Valois went on to determine the direction bias when the components of the plaid were a fixed, high color contrast red/green equiluminant grating and a variable, low-contrast yellow/black luminance grating. These experiments are somewhat different in concept from those of Krauskopf and Farell because, although the components do not appear to completely cohere, there is residual pattern motion. Kooi and De Valois asked their subjects to concentrate on the pattern component of the motion and then by using a range of luminance contrasts spanned the point of zero direction bias (Fig. 15.8). From this procedure they claimed to have obtained the effective or equivalent luminance grating that matched their high-contrast chromatic grating. The effective luminance contrast amounted to between 1 and 2%, which is only about 4 times the threshold for detection, whereas the chromatic grating was 32 times the detection threshold. They concluded that the chromatic gratings were about 8 times less effective than the luminance gratings. However, when they used luminance and equiluminant components or added a high-contrast chromatic grating to one of a pair of 4% luminance gratings, they obtained results that did not conform to the simple predictions that chromatic gratings act simply as low-contrast luminance gratings (Troscianko & Fahle, 1988). Their main conclusion was that a chromatic pathway provides a weak input to the motion system that is dominated by the magnocellular system but that there is further, as yet unexplained, motion at equiluminance.

Gegenfurtner and Hawken (1996a) found a similar pattern of the dependence of perceived speed on contrast for red/green equiluminant plaid patterns as for one-dimensional red/green equiluminant grating patterns. At low temporal frequencies there is a pronounced effect of contrast on the relative pattern speed, while at higher temporal frequencies this dependence is reduced until at temporal frequencies of 4 Hz or greater the perceived speed of either luminance or chromatically defined plaid patterns is essentially independent of contrast. It is not known whether fast-moving plaids whose components are luminance and red/green one-dimensional gratings appear to slide or cohere.

There is some question as to whether plaid pattern motion can be based simply on the combination of the independent detection of the components (Badcock & Derrington, 1992; Cox & Derrington, 1994) because at low speeds observers can identify the motion of the plaid pattern even when the motions of the components themselves are invisible. So some of the interactions that are seen with plaid patterns may not derive simply from the "perceived" speed and directions of the components.

With random square patterns made of red and green elements Morgan and Cleary (1992) found that substituting red for green between frames in a two-frame kinematograph disrupted motion perception. In an extension of this study Morgan and Ingle (1994) found that color could enhance luminance-based signals if the two sets of signals were consistent with a single motion direction. In situations where there was conflicting motion information, color could overcome luminance-based signals. They concluded that there is a distinct signed color motion signal, especially at low spatial frequencies, and the luminance-based and color-based motion signals are combined at a common site for the detection of motion. These conclusions concur with those reached by Gorea, Papathomas, and Kovacs (1993).

By using rotating grating stimuli, Cavanagh (1992) compared the thresholds for reporting the impression of global motion or attentive tracking of an annulus of luminance grating superimposed on an annulus of chromatic grating. Both of these annuli were undergoing circular motions in opposite directions. In global motion judgments, a task similar to the nulling task employed by Cavanagh and Anstis (1991), a low-luminance contrast grating (approximately 2–3%) nulled the motion of a high-contrast isoluminance grating. However, when the subjects had to attentively track one of the gratings, a much higher contrast (between 10 and 30%) was necessary before they could reliably follow the luminance grating in the presence of the colored grating. Furthermore, Cavanagh found that the relative global speed of a colored grating was slowed near to equiluminance when compared to a luminance grating. However, when observers were asked to make judgments using attentive tracking there was a markedly reduced effect at equiluminance. He argues that these experiments separate the contributions of low-level motion processes and those termed long-range motion processes (Braddick, 1980).

Motion capture. A moving luminance figure can induce the perception of motion in an adjacent chromatic figure even when the chromatic figure is physically stationary. This phenomenon is often termed motion capture. Ramachandran (1987) asked subjects to rate the motion of a red square on an equiluminant green background when the square was surrounded by an illusory border moving back and forth. His observers consistently reported substantial motion capture; the red square appeared to move with the surrounding border. The capture was most evident when the red square was nearly equiluminant with the background, and the effect diminished as a difference in luminance was introduced in addition to the chromaticity. Ramachandran concluded that the weak motion signal at equiluminance is captured by the luminance motion signal. In a different paradigm Carney, Shadlen, and Switkes (1987), using a dichoptic presentation of counterphase modulation of yellow/black and red/green gratings in spatiotemporal quadrature, found that the motion was integrated across the dichoptic gratings to give a coherent motion percept while the chromatic information exhibited rivalry. The color of the dominant percept was also captured by the motion signals.

Structure from motion. The visual system is able to use pure motion information to infer the structure of objects, structure-from-motion (SFM). The SFM task requires both the correct identification of the direction of motion and correct velocity discrimination. Würger and Landy (1993) performed a series of experiments to determine whether chromatic information could be used to generate structure-from-motion. They used stimuli that were modulated in luminance and in the two cardinal directions in the isoluminant plane. They found that, although the luminance mechanism dominates SFM performance, there is a significant, albeit weak, input from an L–M-opponent mechanism but no significant input from the S-cones. The weaker strength of the chromatic input was inferred because when performance was normalized for independently measured velocity discrimination and direction of motion thresholds the luminance performance remained superior to that obtained with chromatic targets.

Summary. There is overwhelming psychophysical evidence that we can make judgments about the direction of motion when targets are at or near isoluminance. Under many stimulus conditions we can make these direction judgments when the stimuli are close to the detection threshold. This is the case for luminance and chromatic targets. Direction-of-motion judgments are not made exclusively by the luminance mechanism; this is particularly evident at lower speeds, where chromatically opponent mechanisms provide directional information.

Speed judgments are profoundly affected when stimuli are at or near isoluminance, especially for lower speeds. Therefore, although direction judgments are possible at or near to threshold with a chromatic mechanism, there is a poor estimation of speed. These results suggest that the neural mechanisms for estimating the speed of chromatic targets have quite different characteristics from the mechanisms that estimate the speed of luminance targets.

Neural mechanisms

Retina and lateral geniculate nucleus. Second-stage chromatic and luminance opponent mechanisms are already established in the retina. The second stage here refers to the spatial and chromatic opponency and should not be confused with the second-stage motion mechanisms. Ganglion cells and lateral geniculate neurons can be divided into classes that are selectively sensitive to luminance and chromatic stimuli (DeValois, Abramov, & Jacobs, 1965; Wiesel & Hubel, 1966; Gouras, 1968; Kaplan & Shapley, 1982; Derrington & Lennie, 1984; Derrington, Krauskopf, & Lennie, 1984; Lee, Martin, & Valberg, 1988). What has emerged from studies on retinal ganglion cells and lateral geniculate neurons is a consensus for a number of functional classes of cells that, quite likely, have distinct morphological characteristics in the retina and layer locations in the geniculate. One functional class, consisting of phasic retinal ganglion cells (Gouras, 1968) and type IV cells in the LGN (Wiesel & Hubel, 1966) – more recently called M-cells (Shapley & Perry, 1986) – are most sensitive to modulations of luminance contrast. The M-cells have a concentric center-surround receptive field organization that combines L- and M-cone inputs of the same sign (ON or OFF) to the receptive field center and shows a variety of combinations of cone inputs to the surround. These cells probably account for 5 to 10% of retinal ganglion cells. They project to the two ventral layers of the LGN, the magnocellular layers. These M-cells have sufficient sensitivity to provide a signal in response to low spatial frequency, high temporal frequency targets at low contrasts. In contrast, another major class of neurons has receptive fields that are chromatically opponent, called tonic retinal ganglion cells (Gouras, 1968) and type I and type II cells in the LGN (Wiesel & Hubel, 1966); in the more recent classification these are called P-cells (Shapley & Perry, 1986). The P retinal ganglion cells project to the four dorsal layers of the LGN, the parvocellular layers. The P-cells have the requisite sensitivity to signal chromatic modulations at low contrast. They respond to achromatic stimuli at high contrast. A third functional class shows spatial opponency but not spectral opponency; in the LGN they were called type III cells by Wiesel and Hubel (1966). They are included in the tonic class of retinal ganglion cells and are generally located in the parvocellular layers of the LGN. Recently, the cells in the interlaminar spaces of the LGN, called koniocellular neurons, have been shown to be a distinct class in terms of anatomical projections (Hendry & Yoshioka, 1994). Functionally, they may include neurons that are S–(L+M) (blue/yellow) color-opponent (see Chapter 8).

The cells of the lateral geniculate are the principal functional thalamic inputs to the striate cortex (V1) of the Old-World primate. For the purposes of our discussion, V1 is the first site in the visual pathway where direction-selective neurons are found. In the primate V1 is necessary for detection and discrimination judgments of chromatic and luminance targets. Behavioral studies indicate that lesions of the M-cell division of the LGN result in a lower contrast sensitivity for achromatic luminance targets with low spatial frequency and high temporal frequencies (Merigan & Maunsell, 1990; Merigan, Byrne, & Maunsell, 1991). There is little or no deficit for either detection or direction discrimination at high contrasts or low temporal frequency after M-cell lesions (Merigan, Byrne, & Maunsell, 1991). Merigan, Byrne, and Maunsell (1991) conclude that "... magnocellular lesions reduce the visibility of stimuli used to test motion perception but they do not appear to alter motion perception otherwise." Other studies that have made selective lesions of the M- and P-layers of the geniculate (Schiller, Logothetis, & Charles, 1990) have used dot stimuli moving at different velocities, so the results cannot be easily interpreted in terms of those that have made threshold determinations for different combinations of spatial and temporal frequencies.

Striate cortex. The direct inputs to the striate cortex come from the three subdivisions of the LGN – magnocellular (M), parvocellular (P), and koniocellular (K). Initially, these inputs are kept somewhat separate in V1. M-cells have their main terminal arborizations in the upper part of layer 4C, called 4Cα, and in the lower part of layer 6 (Hubel & Wiesel, 1972; Hendrickson, Wilson, & Ogren, 1978). In addition, the cells in the lower half of layer 6 send descending connections to the ventral layers, the magnocellular, of the LGN (Lund et al., 1975; Fitzpatrick et al., 1994). The LGN P-cells terminate in the lower half of layer 4, called 4Cβ, layer 4a, and the upper part of layer 6, which also provides reciprocal connections to LGN P-cell layers (Lund et al., 1975; Fitzpatrick et al., 1994). The K-cells' terminations, which come from the inculcated layers of the LGN, terminate mainly in the cytochrome oxidase-rich (blob) regions of lower layer 3 in V1 (Hendry & Yoshioka, 1994).

In terms of functional properties it is quite well established that achromatic contrast sensitivity is relatively high in layers 4Cα and 4B (Blasdel & Fitzpatrick, 1984; Hawken & Parker, 1984; Livingstone & Hubel, 1984). Layer 4B gets a major input from layer 4Cα. In addition to high sensitivity, many of the cells in layers 4Cα, 4B, and 6 are direction-selective (Livingstone & Hubel, 1984; Orban, Kennedy, & Bullier, 1986; Hawken, Parker, & Lund, 1988). Of these V1 cells, there is a substantial proportion that are sensitive enough to account for behavioral detection and discrimination performance (Hawken & Parker, 1990), especially when we consider spatially restricted targets. Thus, there is considerable evidence that the M-cell pathway provides a sufficient direction-selective signal among a subpopulation of V1 cells to account for part of the achromatic behavioral performance. There are no published data on the temporal contrast sensitivity of direction-selective cells in V1, but response data suggest that most of the direction-selective cells are bandpass in temporal frequency (Hawken, Shapley, & Grosof, 1996). It has yet to be ascertained whether cells that are dominated by a magnocellular input are sensitive enough at low temporal frequencies to account for behavioral thresholds. Rather little has been reported concerning the chromatic sensitivity of direction-selective V1 neurons.

Extrastriate cortex. The striate cortex provides both direct and indirect input to the extrastriate visual areas. The outputs of regions of V1 that have the highest proportion of direction-selective cells (layers 4B and 6) project directly to area MT or V5 (Lund et al., 1975; Fries, Keiser, & Kuypers, 1985; Shipp & Zeki, 1989b; Movshon & Newsome, 1996). It is well established that area MT, which lies on the posterior bank of the middle temporal sulcus, is intimately related to the processing of visual motion information, including pursuit eye movements that require visual motion signals (Zeki, 1974; Movshon et al., 1985; Newsome et al., 1985). Up to 90% of the neurons in MT have direction-selective receptive fields, and they have a high sensitivity to achromatic contrast (Sclar, Maunsell, & Lennie, 1990). In the awake, behaving monkey it has been demonstrated that the performance of MT neurons is closely matched to behavioral judgments on the direction of motion (Newsome et al., 1991). Although lesions of area MT result in a quite dramatic reduction in direction discrimination thresholds for achromatic gratings immediately postlesion, there is quite a rapid recovery such that there is little long-term deficit in performance (Rudolph & Pasternak, 1996). Therefore, the role of MT as the pivotal extrastriate visual area involved in all motion performance is questionable. There is now strong evidence that other extrastriate areas have the requisite receptive fields to provide some of the neural signals underlying direction discrimination performance. In particular, recent physiological experiments show that more than 50% of the cells in area V3 are directionally selective and, of equal importance, that these direction-selective cells have contrast sensitivities that are at least as high as those found in area MT (Gegenfurtner, Kiper, & Levitt, 1997; see Chapter 13). Many of the neurons in V3 are well suited to provide a direction-selective signal that is sufficient to account for behavioral performance. As yet there have not been specific lesions that have included V3 alone or V3 and MT to determine if these areas together are pivotal for behavioral performance.

In the earliest studies of the properties of MT cells it was reported that the cells were not sensitive to color (Zeki, 1974). However, more recent studies have found that there are responses to isoluminant stimuli in MT (Charles & Logothetis, 1989; Saito et al., 1989; Dobkins & Albright, 1994; Gegenfurtner et al., 1994). The principle question that arises is whether the chromatic sensitivity of the cells in area MT is sufficient to account for behavioral detection and direction discrimination thresholds. Gegenfurtner et al. (1994) showed that the contrast sensitivity of MT cells to isoluminant targets was much poorer than the behavioral sensitivity. In their experiments they used the optimal spatial and temporal frequency for each cell and measured the cell's response as a function of contrast. In Fig.15.9 the average contrast response to isoluminant stimuli is shown in the open circles; the average response to achromatic gratings is shown by the filled circles. The vertical arrows in Fig. 15.9 indicate the behavioral thresholds for four different conditions. The main point here is that the behavioral thresholds for 1-Hz isoluminant stimuli are about 0.5 log unit lower than the neural thresholds of MT cells, while the neural thresholds for both luminance and isoluminant stimuli are about the same as those measured behaviorally for stimuli moving at 8 Hz. The conclusion drawn from these results is that the chromatic sensitivity of MT cells is not sufficient to account for behavioral performance at low temporal frequencies.

Even the neural substrate for the detection and motion discrimination of achromatic luminance gratings is unclear at present. The single-unit studies suggest that the neural sensitivity in MT is sufficient to provide the appropriate signal. The recent lesion studies in macaque monkeys do not lend support to the notion that MT alone is an essential component in the pathway. Lesions in a human patient (subject LM) that involve the analogue of MT give rise to selective motion deficits (cerebral akinetopsia) for fast-moving stimuli (Hess, Baker, & Zihl, 1989) and leave sensitivity to slow-moving stimuli intact. PET scans of patient LM indicate that there are significant responses in the analogue to V3 during the viewing of motion stimuli (Shipp et al., 1994). There are no studies that involve the selective lesions of extrastriate visual areas fol-

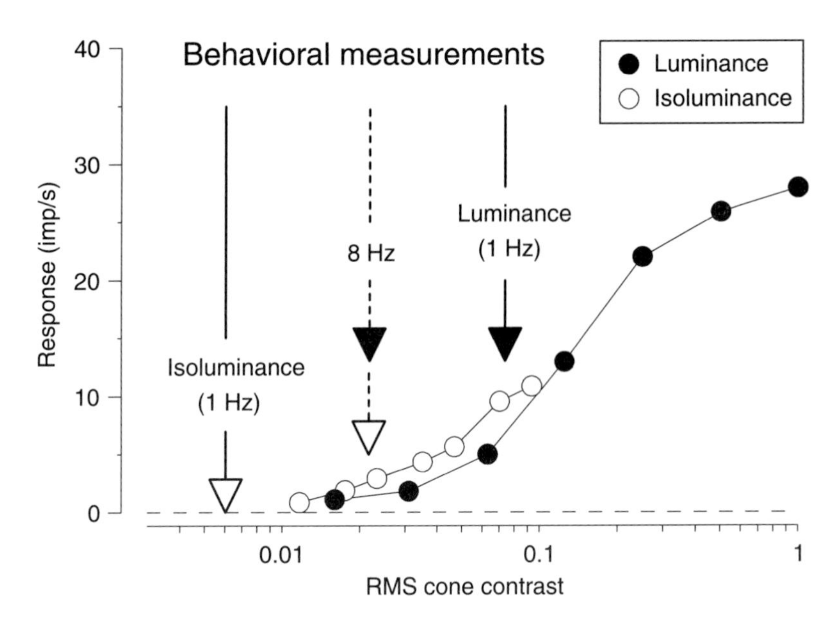

Figure 15.9: The average contrast response functions for MT neurons to red/green isoluminance (unfilled circles) and to black/white luminance (filled circles) grating stimuli of the optimal spatial and temporal frequencies. The arrows show the behavioral measurements of contrast thresholds for motion identification carried out in the same species of macaque monkey under similar conditions to those of the physiological experiments. From Gegenfurtner and Hawken (1996b).

lowed by testing for the sensitivity for detection and discrimination of chromatic targets.

Conclusions

The interaction between color and motion signals in visual processing critically depends on the experimental conditions. The two major hypotheses about motion processing at isoluminance both seem to apply over a certain range of conditions. For slowly moving stimuli (slower than 2-4 Hz) there are two distinct pathways for isoluminant and luminance stimuli. Isoluminant stimuli are processed by a chromatically opponent mechanism, which is highly sensitive to the direction of motion. However, its coding of speed is dramatically impaired. Perceived speed is lower than for comparable luminance gratings and highly dependent on contrast.

For fast-moving stimuli there seems to be a single mechanism that is mostly sensitive to luminance contrast and has a residual sensitivity to isoluminant stimuli. The physiological measurements in area MT make it quite likely that a mostly magnocellular pathway including area MT is the neural substrate for this mechanism. Its coding of direction of motion and speed is accurate, which makes it highly useful for functions that are typically associated with area MT and the dorsal stream in general, like the coordination of orienting behavior in space, and, in particular, the control of eye movements. It is known that area MT plays an important role in the generation of the visual signals that control smooth pursuit eye movements. Such signals need to be processed quickly to enable a fast and proper reaction to environmental stimuli. A fast neuronal pathway is advantageous, where there is little mixing of signals about different stimulus attributes, and therefore less synaptic integration is required. The magnocellularly dominated pathway via layers 4Cα and 4B in V1 that projects directly to area MT fulfills these requirements. The small variations of the isoluminant points of individual cells in MT may be an elegant and efficient implementation of cue-invariance.

Acknowledgments

The authors would like to thank John Krauskopf and Patrick Cavanagh for valuable comments on this chapter. This work was supported in part by NIH grant EY08300 to MH. KRG was supported by a Heisenberg fellowship from the German Research Council (DFG Ge 879/4-1).

Part IV: Perception

16

Higher order color mechanisms

John Krauskopf

By the end of the nineteenth century two major ideas about the nature of color vision were well established. Light is absorbed by three receptors with differing spectral responses, and the results of this process are transmitted by opponent processes, also assumed to be three in number. Approaching the end of the twentieth century we know, from a variety of physical, physiological, and psychophysical experiments, the intimate details of the former process. While we know a lot about the latter process, there is still a long way to go. In this chapter, I will review psychophysical evidence that bears on the nature of the second stage and higher order processes in color vision. I will briefly relate these findings to electrophysiological results.

Hering's notion of opponent processes was grounded in observations concerning the appearance of colors such as the impossibility of simultaneously seeing complimentary colors in the same place at the same time (Hering, 1905). Attempts to establish the precise nature of the second-stage processes were based on the assumption that stimulation of these processes results in the perception of "unique colors" (seen as "red," "yellow," "green," and "blue"). Wyszecki and Stiles (1982) discuss a number of "neural models," developed in this way, "in which the idea of trichromacy at the initial receptor stage is combined with schemes of coding the receptor signals into two signals carrying chromatic information and one signal carrying achromatic information." Included were the theories of Müller (1930a, 1930b), Judd (1949a), Adams (1923), Hurvich and Jameson (1955), Guth (1972), and Ingling (1977). All of these models are similar in assigning algebraic weights to the signals from the L-, M-, and S-cones so as to produce three mutually orthogonal second-, and in some models, higher, stage mechanisms.

More recently psychophysical, as opposed to phenomenological, evidence has been advanced in support of the existence of opponent mechanisms. Papers that argued for inhibition among chromatic mechanisms include those of Boynton, Ikeda, and Stiles (1964); Guth (1972); Sperling and Harwerth (1971); Krauskopf (1974); King-Smith and Carden (1976); Pugh (1976); Augenstein and Pugh (1977); Mollon and Polden (1977a); and Pugh and Mollon (1979). While these papers provided evidence for the interaction of cone signals on the basis of measurements of detection thresholds, they did not determine the contribution of the different cone classes to the second-stage mechanisms. Before considering an attempt to define the nature of the second-stage mechanisms by psychophysical means, it will be useful to define the color space used to represent the experiments.

Color space

The color space used for representing both psychophysical (Krauskopf, Williams, & Heeley, 1982) and electrophysiological (Derrington, Krauskopf, & Lennie, 1984) experiments is presented in Fig. 16.1. This

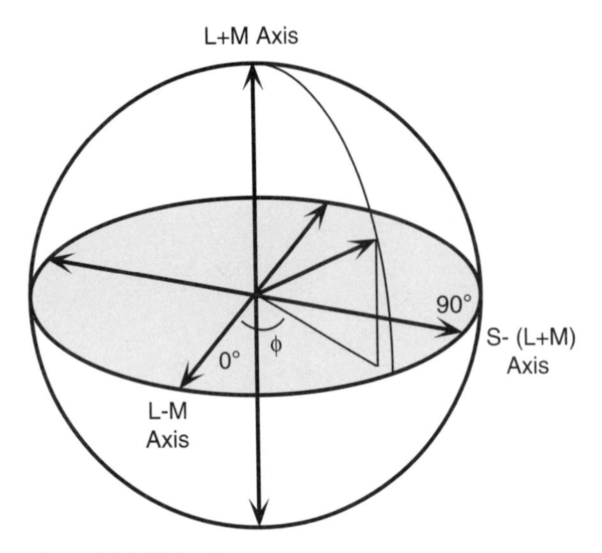

Figure 16.1: Color space.

diagram is an extension into three dimensions of the chromaticity diagram proposed by MacLeod and Boynton (1979) and as such is a linear transformation of a cone space. It is particularly suited to represent experiments in which the observer is adapted to an equal-energy white field, which is plotted at the origin.

There have been a number of labels applied to the axes. Two systems of labeling might be appropriate. One, which is theoretically neutral, would involve simply listing the variations in cone signals along each axis, in which case they might be called L–M, S, and S+L+M. Along one isoluminant axis the excitation of the S-cone excitation is held constant while the L- and M-cone excitations covary so as to keep luminance constant. Along the second isoluminant axis only the S-cone excitation varies, and along the third luminance direction all three cone excitations vary in proportion to their value at the equal-energy white point. Alternatively, the labeling might express the weights assigned to the cone receptors by the mechanisms that respond uniquely along each axis. In this case, the axes would be labeled L–M, S–(L+M), and L+M. The L+M-axis might also be called the "luminance axis" because a mechanism that reported only the change in luminance, assumed to be the sum of the sum of the L- and M-cone excitations, would respond exclusively to variation along this axis.

For completeness some complications must be faced with regard to axis labels. The Smith–Pokorny fundamentals that were used by MacLeod and Boynton (1979) give twice as much weight to the L-cones than to the M-cone in the definition of luminance. This means that the factor of two must appear in the labels. Then the axial cone signals would be L–M, S, and S+M+2L and the cone weights of the orthogonal mechanisms would become L–2M, S–(L+M), and L+M.

For theoretical reasons it is often desirable to represent cone signals as modulations, that is, as ΔL/L, ΔM/M, and ΔS/S (Stromeyer, Cole, & Kronauer, 1985). This would further complicate labeling.

These complications would miraculously go away if, in fact, equal weights should be assigned to the L- and M-cones in the definition of luminance. That possibility is considered elsewhere (Krauskopf, 1997).

Personal color spaces

There is an important difference in the way in which this color space has been used in psychophysical and electrophysiological experiments. In electrophysiological experiments the axes are scaled as a fraction of the maximum excursion from white available in the equipment. The spectral properties of an individual unit under study may be expressed in terms of its responsiveness to modulation in various directions in this space. If the unit linearly sums cone signals, the spectral properties can be summarized in terms of the preferred vector in color space or in the weight that the unit gives to the three classes of cone signals.

In psychophysical experiments it is preferred to scale the axes in units of the individual observer's detection thresholds along the cardinal axes because this scaling gives a common meaning to angles intermediate between the cardinal directions. Not only do thresholds along the cardinal axes vary reliably from observer to observer, but observers also exhibit reliable differences in the tilt of their isoluminant planes relative to that computed from radiometric calibrations

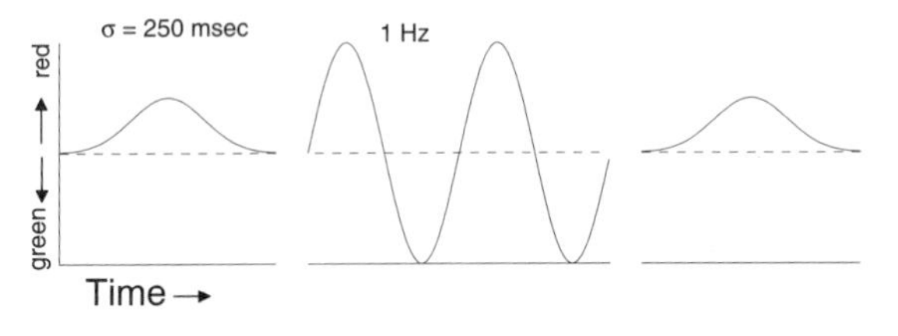

Figure 16.2: Habituation procedure. Thresholds measured for the detection of Gaussian pulses on steady equal-energy white field. "Preadaptation" measurements made with field steady between test pulses. "Postadaptation" measurements made with interspersed sinusoidal variation of field in habituating direction.

and the luminosity curve [$V(\lambda)$]. We have found that taking these factors into account can greatly reduce the variance of results (as in the plaid coherence experiments discussed later in this chapter).

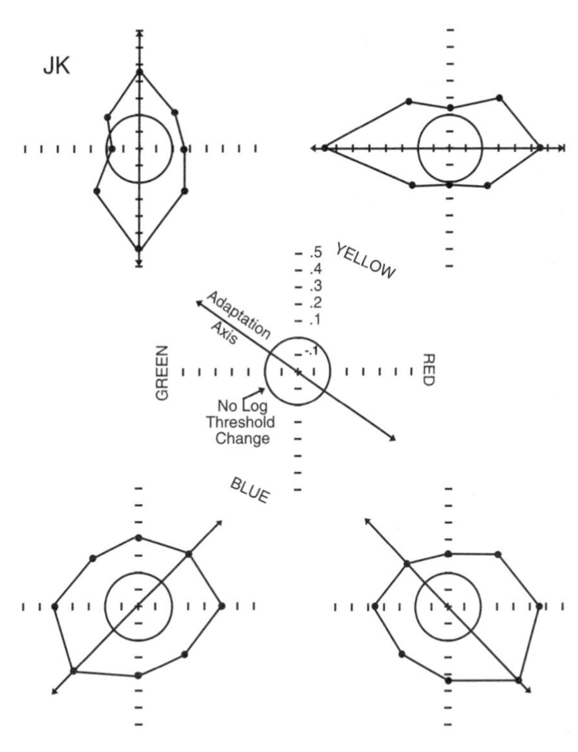

Figure 16.3: Polar plots representing changes in thresholds due to habituation. Vertical axis: S–(L+M), horizontal axis: L–M. Arrowheads indicate habituation direction. Angles in observer's personal color space.

Psychophysical evidence for mechanisms tuned along the cardinal directions

Habituation. Krauskopf, Williams, and Heeley (1982) set out to determine the contribution of the three classes of cone receptors to the three presumed second-stage mechanisms using psychophysical methods. One of the best methods for discovering independent sensory mechanisms is to find a procedure that selectively reduces the detectability of different classes of stimuli. Krauskopf, Williams, and Heeley (1982) tried to find directions in color space having unique properties by using a habituation procedure (Fig. 16.2). Thresholds for each observer were measured for detection of brief excursions in a number of directions away from the white point of the space illustrated in Fig. 16.1. Subsequently, they were repeatedly exposed to a field modulated sinusoidally in color along various lines through the white point between test exposures and then their postadaptation thresholds were measured.

Typical results for one observer for isoluminant stimuli are plotted in Fig. 16.3. The outstanding result was the selective effect of the direction of modulation of the habituation stimulus on the elevation of the detection thresholds. When the habituation stimulus was modulated along one isoluminant cardinal axes, thresholds for detection along that axis were raised substantially while thresholds along the other isoluminant cardinal axis were unchanged and thresholds along intermediates were modestly elevated. The effect of habituation along noncardinal directions was to raise thresholds more generally. However, there was at least a hint of selectivity in the data for nonaxial habituation, to which we will return shortly.

Similar selectivity was found when the habituation procedure was performed in the plane containing the luminance axis and the L–M-axis. Thresholds for detection of reddish and greenish pulses were raised following the viewing of fields modulated along the L–M-axis but not following the viewing of fields modulated along the luminance axis. Similar selectivity

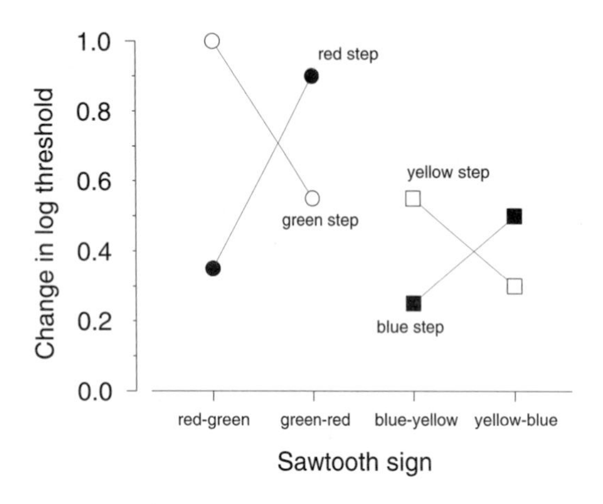

Figure 16.4: Changes in thresholds resulting from viewing fields varying in color in a sawtooth fashion as a function of time. Test is a chromaticity step.

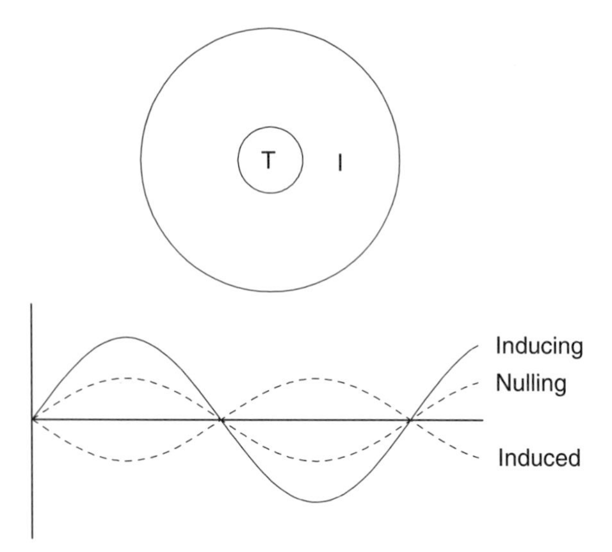

Figure 16.5: Measurement of dynamic induction. When the inducing annulus (I) is modulated sinusoidally in time the test disk (T) appears to be modulated in counterphase. Adding an in-phase modulation of the appropriate amount nulls the modulation of the disk.

was found in the plane defined by the S–(L+M)-axis and the luminance axis. These results are important because they place the site of the desensitization process beyond the point at which signals from the different classes of cones combine. This is because luminance modulation entails much greater variation in cone signals than modulation along either isoluminant axis and yet has no effect on the detection of isoluminant stimuli. There was no evidence of selectivity with nonaxial habituation in these nonisoluminant planes.

Sawtooth habituation. Another finding of interest concerns the effects of habituating stimuli modulated in a sawtooth fashion in time on detection of step changes in color along both isoluminant cardinal axes. The general procedure, illustrated in Fig. 16.2, was followed with the exception that the stimuli were steps that did not return to the adaptation level until the observer responded to assure that the detection was mediated by a mechanism responsive to changes in one direction. As shown in Fig. 16.4, thresholds were selectively raised for complementary colors depending on the sign of the sawtooth. Thus, if the habituation stimulus varied slowly from red to green and returned rapidly to red, thresholds for detecting step changes in the green direction were elevated more than thresholds for red changes, and vice versa. A similar result had been previously found for stimuli varying in luminance (Krauskopf, 1980). These results suggest there are separate mechanisms that are responsive to stimulus change in each direction along the three cardinal axes.

Spatial effects. Two more habituation experiments are worth mentioning. Flanagan, Cavanagh, and Favreau (1990) reported color- and orientation-specific tilt aftereffects when habituation and test gratings were modulated along the same cardinal direction. They also found smaller, apparently color-specific, effects for gratings modulated in different directions between the cardinal axes.

When an annulus is modulated in color about equal-energy white over time, an inscribed white disk is seen to vary in a complimentary fashion (Krauskopf, Zaidi, & Mandler, 1986) (Fig. 16.5). Krauskopf and Zaidi (1986) extended the habituation experiments using sawtooth modulation, comparing the effects of modulating an annulus with the direct effect of modulating the disk (Fig. 16.6A). As shown in Fig. 16.6B, modu-

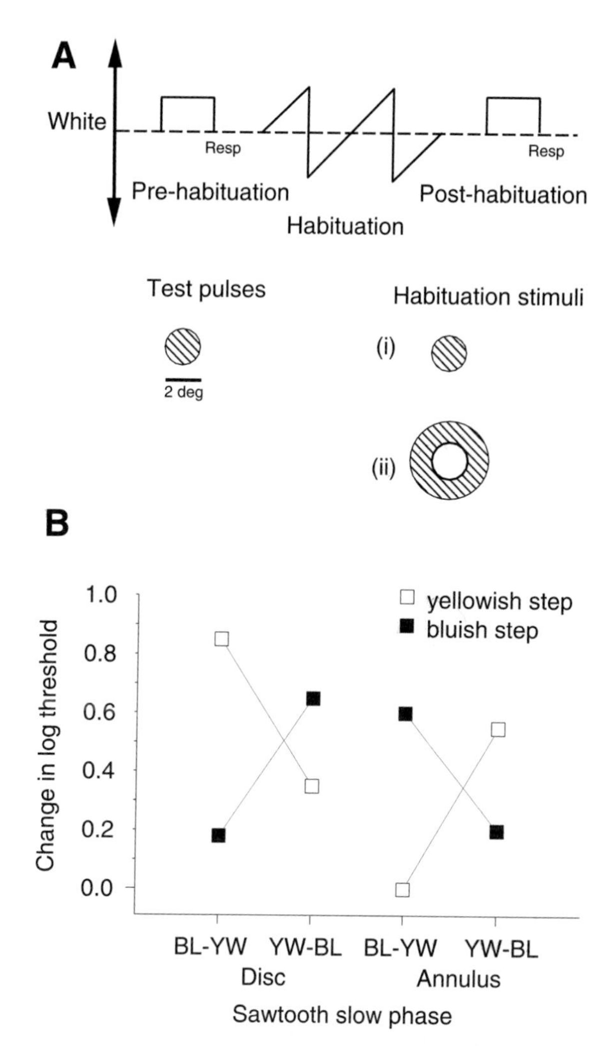

Figure 16.6: (A) Experimental design. (B) Changes in log thresholds due to viewing of modulated fields. Habituation stimulus: left – disk; right – annulus.

lating the annulus also resulted in differential threshold elevations, but the sign of the difference is opposite to that obtained when the disk is modulated. The reversal of the sign of the effect argues against a stray light interpretation. If stray light from the annulus falling in the test area was responsible for the habituation effect, we would not expect the sign reversal. The habituation effect appears to be related to the appearance of the test area. These two sets of experiments suggest that the habituation processes occur at a fairly high level in the visual pathway.

Evidence implying the existence of higher order mechanisms

Reanalysis of habituation experiments. Krauskopf, Williams, and Heeley (1982) noted, without quantitative analysis, that some of their results suggested selective effects when the habituating stimulus was modulated at noncardinal directions. As illustrated in Fig. 16.3, when the habituation stimuli are modulated along the cardinal axes thresholds are obviously raised more for tests along the habituation direction, but it is not clear whether the same is true for other habituation directions. What was needed was a method of averaging the data to produce an index of selectivity (Krauskopf et al., 1986).

If there were only two mechanisms in the isoluminant plane, habituation could act to reduce the sensitivity of one or the other, or both, of these mechanisms, resulting in a denormalization of threshold plots (Fig. 16.7). Predictions of the variation of test thresholds with azimuth (Θ) when habituation is along the 90-deg axis are illustrated in Fig. 16.7. The ratio of thresholds after habituation to those before habituation is represented by the length of the arrow divided by the radius of the circle. This ratio (d) is given in the following equation, where A is the horizontal semi-axis and B the vertical semi-axis:

$$d = AB \,/\, (A^2 \sin^2 \Theta + B^2 \cos^2 \Theta)^{0.5}. \quad (1)$$

As shown in Fig. 16.7B, this ratio is approximated by $\sin^2 \Theta$.

If the habituation stimulus is instead modulated along a direction between $\Theta = 45$ and 90 deg, both cardinal mechanisms will be desensitized, with a greater effect on the 90-deg mechanism. As shown in Fig. 16.8, thresholds are raised in all directions but the largest elevation occurs in the 90-deg direction. As seen in Fig. 16.8B, the ratio of post- to prehabituation thresholds is closely approximated by $A + B \sin^2 \Theta$. It is important to note that the phase remains the same as that for habituation at 90 deg.

Complementary effects are predicted for habituation in directions $\Theta = 0$ and 45 deg. In this region the

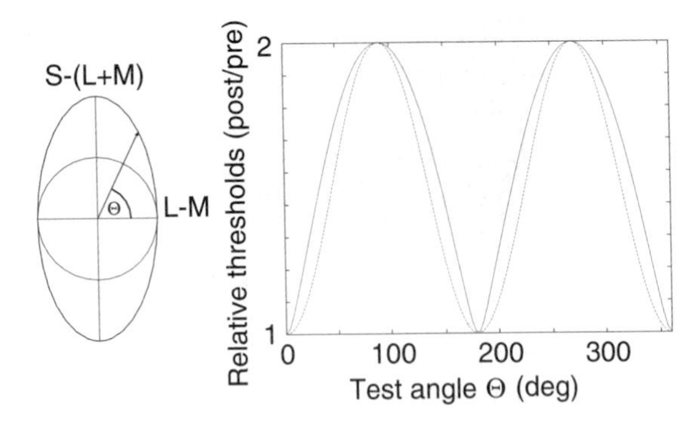

Figure 16.7: Theoretical thresholds assuming two cardinal mechanisms. (A) Habituation along vertical cardinal axis. Circle: normalized prehabituation thresholds. Ellipses: post-habituation thresholds. Θ: test angle. (B) Ratio of post- to prehabituation thresholds as a function of Θ and plot of $\sin^2 \Theta$.

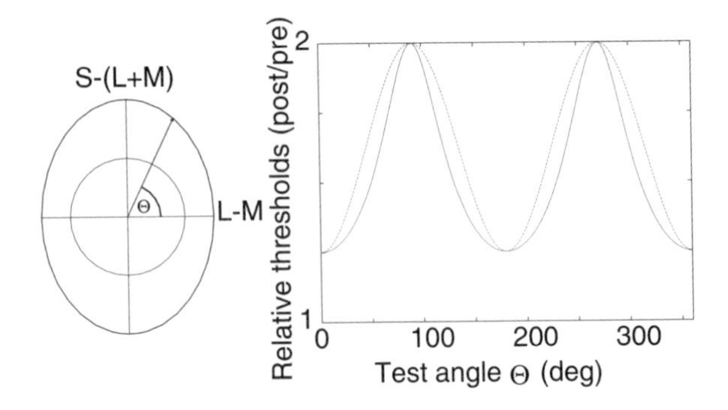

Figure 16.8: Same as Fig. 16.7, but for habituation along direction within 45 deg from vertical axis.

ratio of post- to prehabituation thresholds is approximated by $A + B\cos^2 \Theta$. That is, the phase differs by 90 deg from that for habituation directions between Θ = 45 and 90 deg. For habituation at exactly $\Theta = 45$ deg, thresholds would be elevated equally in all directions and phase would be indeterminant.

Changes in thresholds for 16 test directions following habituation at 45 and 90 deg are plotted in Fig. 16.9A (data from Krauskopf, Williams, & Heeley, 1982, as replotted in Krauskopf et al., 1986). These data conform to the predictions above. With habituation at $\Theta = 90$ deg, changes peak for tests at 90 and 270 deg and appear be nonexistent for tests at 0 and 180 deg. On the other hand, thresholds are fairly uniformly elevated with habituation at $\Theta = 45$ deg.

The modulations of the fundamental and second harmonics obtained by a Fourier analysis of the changes in threshold for all tests are plotted as a function of habituation direction in Fig. 16.9B. For both observers the modulation of the fundamental is small and not systematically related to the habituation direction while the second harmonic peaks as expected at habituation directions of 0, 90, 180, and 270 deg.

While the plots in Fig. 16.9B are consistent with the existence of independent mechanisms tuned along the isoluminant cardinal axes, the plots of the phase of the modulations as a function of habituation direction, plotted in Fig. 16.9C, strongly imply the insufficiency of that model. The phase of the fundamental bears no systematic relation to habituation direction, but the phase of the second harmonic is essentially equal to the habituation direction between 0 and 180 deg and is equal to the habituation angle minus 180 deg for habituation angles between 180 and 360 deg. This is the sort of result that we would expect if there were many mechanisms tuned to respond best at various directions about the isoluminant plane.

Krauskopf et al. (1986) used the same analytic approach on data from a "generalized transient tritanopia" experiment. This experiment was stimulated by that of Mollon and Polden (1977a). Thresholds were always measured on an equal-energy white background. Baseline measurements were made in eight equally spaced directions about white. Subsequently, thresholds were measured 200 ms after a 1.0-s step and return of the background in one of eight equally spaced directions. These data were also submitted to Fourier analysis, which revealed the phase following in both the fundamental and the second harmonic, supporting the ideas that there are multiple mechanisms and that, at least partially, independent mechanisms mediate responses to complementary colors.

Webster and Mollon (1991, 1994) studied the color matches made between targets presented to two retinal areas, one that was always a steady neutral field while the other was modulated in some color direction

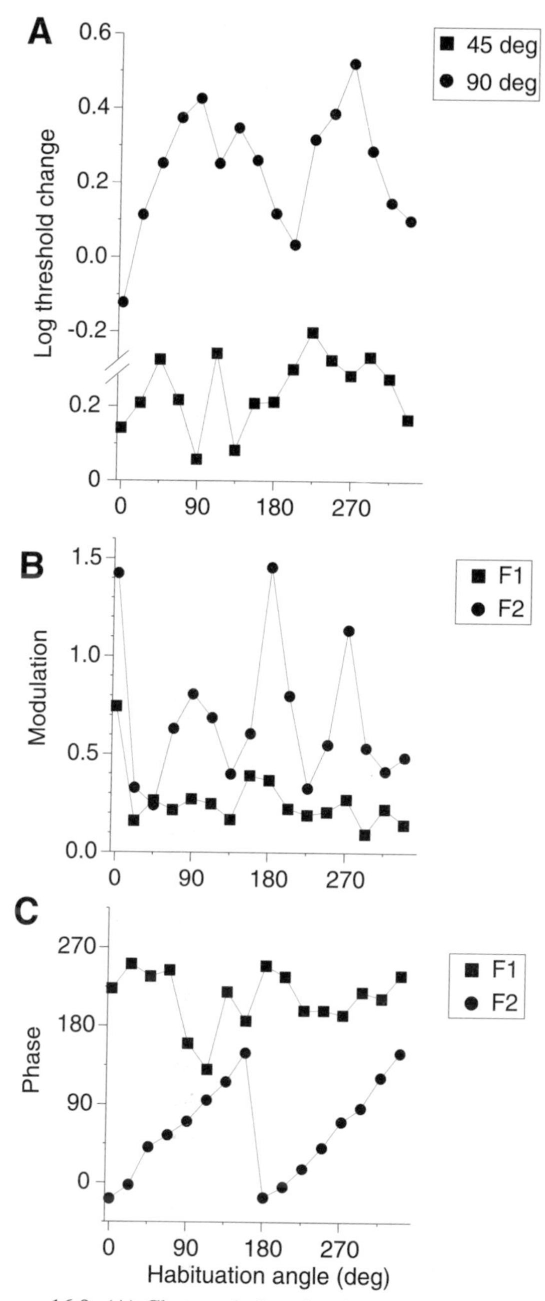

Figure 16.9: (A) Changes in log threshold as a function of test angle. Habituation vector: upper: 90–270 deg; lower: 45–225 deg. (B) Modulation (amplitude divided by mean threshold change) of first and second harmonics as a function of habituation angle. The data plotted in the upper and lower curves are represented by points plotted at the 90- and 45-deg loci on the abscissa. (C) Cosine phase of first and second harmonics as a function of habituation angle.

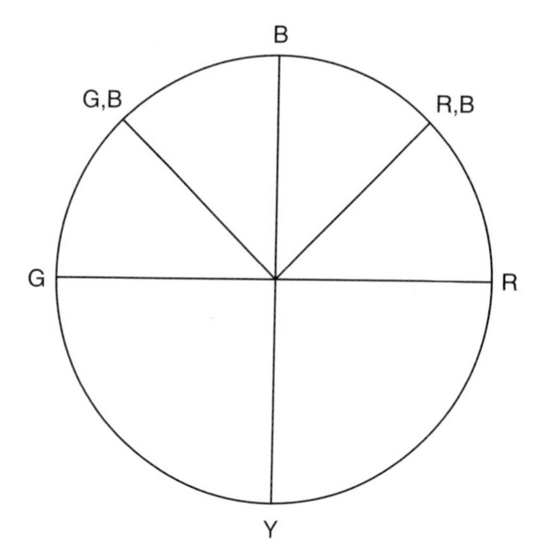

Figure 16.10: Stimuli used in detection and discrimination experiment.

between presentations of the test stimuli. They concluded that the changes in matches were inconsistent with sensitivity changes in only three independent channels. They also endorsed a model that included more than two channels lying with the isoluminant plane.

Zaidi and Shapiro (1993) have suggested that the effects of habituation could be explained by invoking adaptive orthogonalization among cardinal mechanisms. However, other evidence that is inconsistent with independent processing of stimuli within cardinal mechanisms comes from experiments that do not involve habituation.

Detection and discrimination of chromatic pulses. Evidence of a very different sort, requiring for its explanation more than independent axial isoluminant chromatic mechanisms, was presented by Krauskopf et al. (1986). In this experiment observers were required to tell in which of two temporally separated intervals a test pulse was presented and which of the two possible chromatic stimuli it was. In any session a particular pair of pulses was presented. There were four cardinal pairings with azimuths of 0, 90 deg; 90, 180 deg; 180, 270 deg; and 270, 0 deg and four noncar-

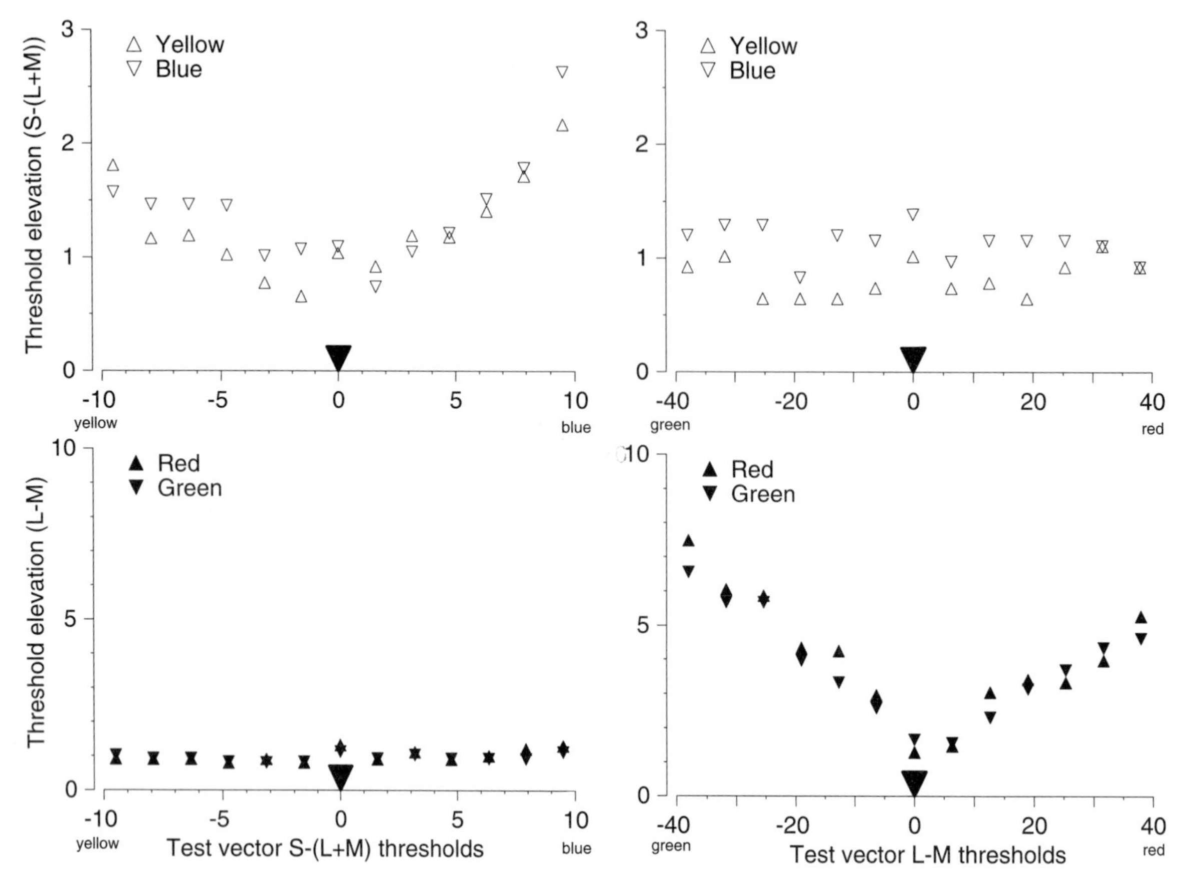

Figure 16.11: *Left*: Thresholds for detecting changes when delta and test lie on the same cardinal direction. Right: Delta and test on different axes. *Above*: S–(L+)M-axis; *below*: L–M-axis. *Above*: S–(L+)M-deltas, L–M-tests; *below*: L–M-deltas, S–(L+M)-tests.

dinal pairings with azimuths of 45, 135 deg; 135, 225 deg; 225, 315 deg; and 315, 45 deg.

The logic of this experiment is illustrated in Fig. 16.10. According to Müller's doctrine of specific energy of nerves, if stimuli along the cardinal directions were detected by different mechanisms, they would result in unique sensations. In Fig. 16.10 this is suggested by the "R," "B," "G," and "Y" associated with the axes. (The actual colors reported for pulses in the cardinal directions are "reddish," "reddish-blue," "greenish," and "greenish-yellow.") When appropriate corrections for guessing were made, cardinal stimuli actually detected could always be correctly identified. On the other hand, stimuli falling halfway between the cardinal axes should be detected by one or the other, and sometimes by both, of the neighboring cardinal mechanisms. This is indicated by the pairs of letters at the intermediate directions. Considering the 45–135-deg pair, we would expect the stimulus to be correctly identified when detected by the "R" or "G" mechanism, either alone or in concert with the "B" mechanism, but identification would be uncertain when only the "B" mechanism detected the stimulus. This theory predicts a measurable reduction in identifiability for intermediate pairs that was not observed in the experiment and thus once again cardinal mechanisms alone are insufficient.

Color discrimination and adaptation. Krauskopf and Gegenfurtner (1992), inspired by Craik's experi-

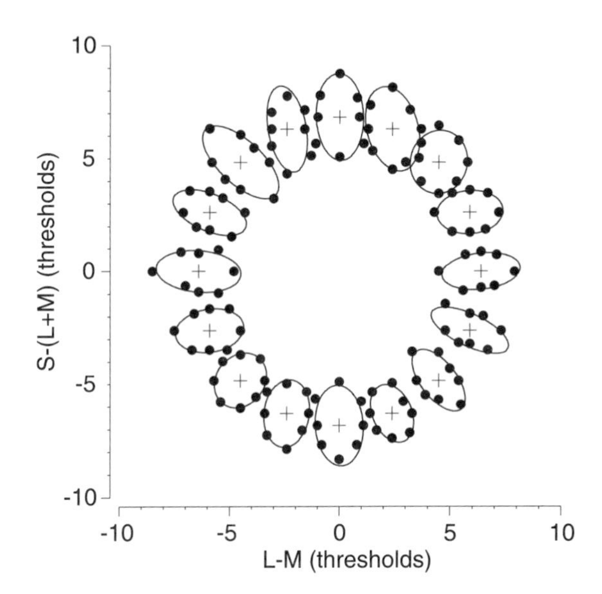

Figure 16.12: Discrimination ellipses for test vectors equally spaced in 16 directions about the white point. The adaptation point was white.

ments on the effect of the level of light adaptation on discrimination thresholds at various luminance levels, measured chromatic discrimination at loci removed from the adaptation point. The observer viewed a TV screen that had a luminance of about 35 cd/m^2, on which four disks were displayed simultaneously for 1 s. One of the disks was always different in some way from the other three test disks. The observers' task was to identify the odd disk. In one set of experiments the three test disks were excursions along one of the cardinal directions while the comparison disk was also an excursion along the same axis but differed in amplitude by the amount delta. This increment delta was varied in a staircase procedure to measure the threshold for discrimination. In a complementary set of experiments, the test disks were the same but the delta was in the direction perpendicular to the cardinal axis of the test. Thresholds were smallest when the test was the same as the adaptation field, whether it was an equal-energy white or a colored field. Thresholds were elevated linearly in proportion to the difference between the test and adaptation colors when delta was along the same cardinal direction but remained constant when the test and delta were along different cardinal directions (Fig. 16.11). Results having features in common with these have been reported by Loomis and Berger (1979); Zaidi, Shapiro, and Hood (1992); and Shapiro and Zaidi (1992).

In an additional experiment, discrimination thresholds were measured in 8 directions about test vectors in 16 directions around the equal-energy white point (Fig. 16.12). The theory that there are independent detection mechanisms along the cardinal axis predicts that the orientation of the major axes of the ellipses fitted to these data should be parallel to the cardinal axis nearer the test direction except for test vectors 45 deg removed from the cardinal axes. These latter should be circles. In the first and third quadrants the ellipses tend to behave as expected, but in the other quadrants the ellipses point toward the white point, in contradiction to the theory.

Chromatic induction. The method illustrated in Fig. 16.5 was used to measure chromatic induction along the three cardinal axes (Krauskopf, Zaidi, & Mandler, 1986). The functions relating the amplitude of the nulling modulation were different nonlinear functions of the inducing cardinal-axis amplitudes. These data were used to predict, by vector addition, the nulling amplitude and azimuth for inducing stimuli modulated along noncardinal directions. In general, the nulling modulations were predicted to be in different directions than the inducing stimuli. The experimental results deviated significantly from this expectation and thus implied, once again, the insufficiency of the cardinal mechanism model.

Coherence of plaid patterns. Observations on the role of color in the perception of coherent motion in plaid patterns initially appeared to demonstrate a special significance of the cardinal mechanisms. Adelson and Movshon (1982) found that pairs of moving luminance gratings drifting in different directions tended to exhibit coherent motion when they had similar spatial frequencies and contrasts. If the components differed, the observers were more likely to report that one grating appeared to move, or "slip," past the other.

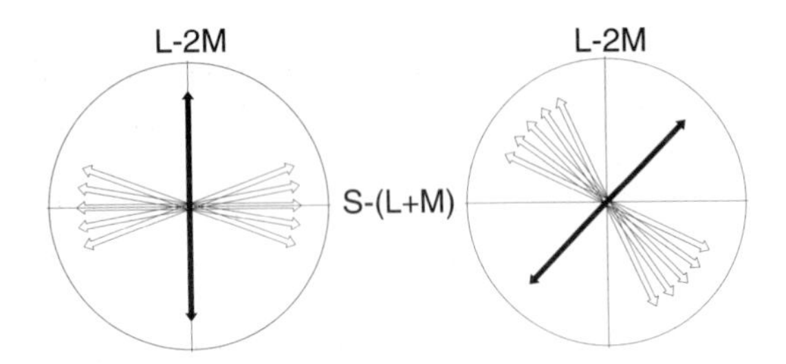

Figure 16.13: Stimuli used in paired comparison method. The black arrow depicts the fixed vector that was presented on all trials and the gray arrows the varied color directions. *Left:* Fixed direction along one cardinal axis, varied vectors around a second cardinal axis. *Right:* Fixed vector at 45 deg from the cardinal axis.

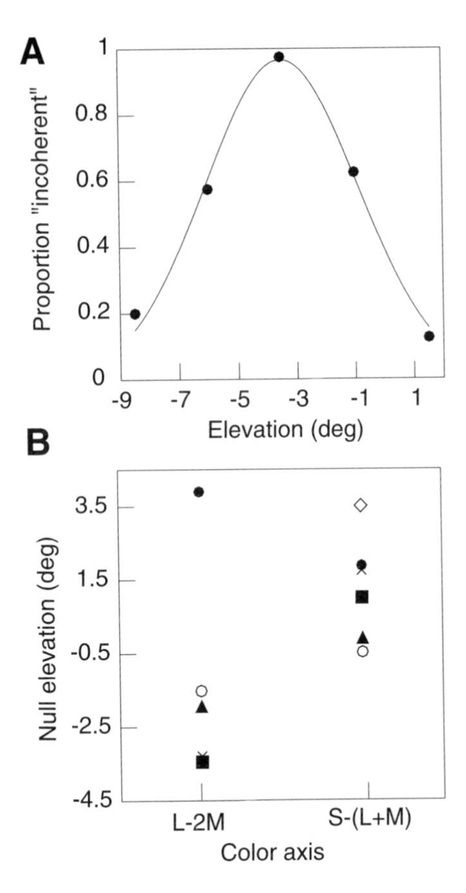

Figure 16.14: (A) Fitting the results with a Gaussian. Abscissa is the elevation of the variable component grating. Ordinate is the relative proportion with which each pairing was judged as less coherent compared with the other pairings. (B) Individual differences in isoluminant planes estimated from coherence judgments. Each observer is represented by the same symbol in the two columns. Instrumental coordinates are used in both plots.

Krauskopf and Farell (1990) asked whether the similarity rule applied to color. They found that observers judged the gratings to slip when they were composed of gratings separately modulated along two cardinal axes but to cohere when modulated along the two intermediate directions. Analytically, the intermediate gratings had equal vectorial components along the cardinal axes, and Krauskopf and Farell speculated that the similarity in the two gratings' effects on the cardinal mechanisms resulted in the appearance of coherence.

The observers had little difficulty making the judgments, and the results were very clearcut. However, further experiments in which the observers made judgments of *relative* coherence of pairs of patterns revealed that this explanation was insufficient (Krauskopf, Wu, & Farell, 1996). The basic design of their experiments is illustrated in Fig. 16.13. The stimuli were pairs of drifting gratings sinusoidally modulated in selected color directions about the equal-energy white background. In each of two 1-s intervals a grating modulated in a fixed direction in color space (Fig. 16.13, black arrow) was superimposed on a randomly selected grating modulated in one of five varied directions (gray arrows) whose orientation was orthogonal to that of the fixed grating. The observers' task was to choose which of the two pairs seemed to cohere less. Prior to the experiments detection thresholds were measured along the cardinal axes so that stimulus contrast could be expressed in individual threshold units.

The frequency with which the observers selected the pairing containing each of the variable directions as less coherent was plotted against the azimuth (or elevation) of the variable component (Fig. 16.14A). Such plots were well fit by Gaussians whose means provided measures of the least coherent pairs. The elevations of gratings that produced the least coherence when paired with a grating modulated in luminance are plotted in Fig. 16.14B. Individual observers show

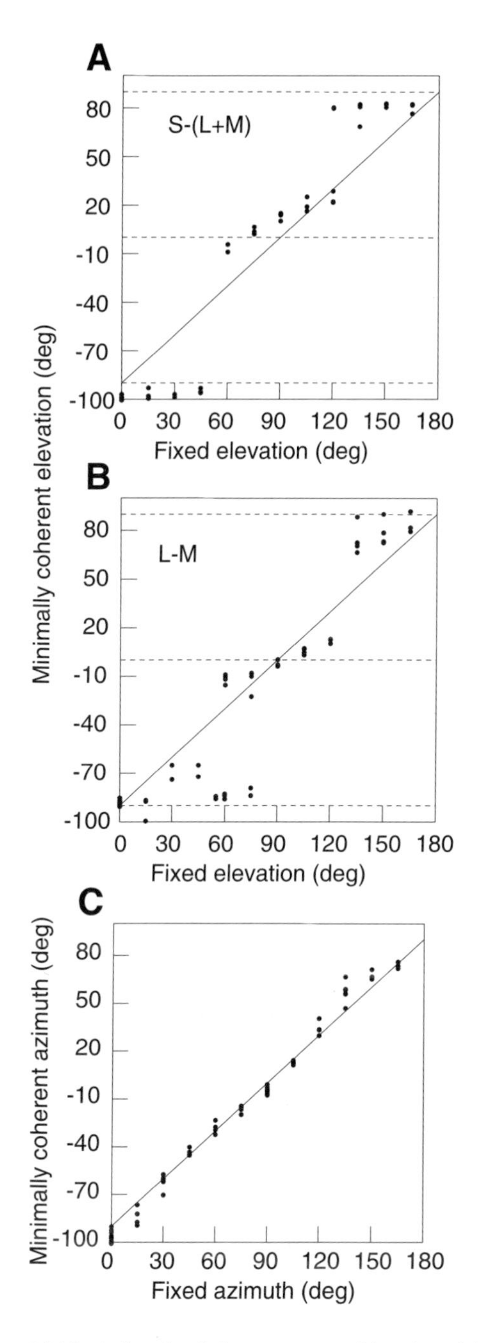

Figure 16.15: Azimuth of the vector resulting in minimal coherence (ordinate) when paired with fixed vectors (abscissa). The points are the results of individual runs, and their distribution gives an impression of the reliability of the measurements. Individual observer coordinates. Variations in (A) luminance and S–(L+M), (B) luminance and L–M, and (C) the isoluminant plane.

reliable differences in their isoluminant planes (the standard error of the means was 0.5 deg or less for all individual points). Therefore, in the main experiments the stimuli for each observer were computed in their own color space including a correction for the tilt of the isoluminant plane.

The results of performing this experiment for 12 fixed directions in the isoluminant plane are plotted in Fig. 16.15C. In this plane patterns are judged minimally coherent when the directions of modulation differ by approximately 90 deg from one another, independent of the absolute directions of the modulations. On the other hand, when the experiment was repeated in the planes through each of the isoluminant cardinal axes and the luminance axes (Figs. 16.15A & 16.15B) the results were those expected for two independent mechanisms.

In an auxiliary experiment, the same paired comparison procedure was applied to pairs of isoluminant gratings, separated by 90 deg, to determine which pair was least coherent. Observers chose ones that were close to the cardinal axes. Thus, just as in the habituation experiments, there appears to be a primacy of the cardinal mechanisms.

If the similarity rule held for color and there were only two cardinal mechanisms in the plane, then for fixed vectors lying between –45 and +45 deg of one cardinal axis the least coherent perception should be obtained when the variable vector is modulated along the other cardinal axis. However, this does not happen for stimuli lying within the isoluminant plane. Therefore, cardinal mechanisms are not sufficient to account for the results in this plane but do seem adequate for the other two major planes of color space. In the isoluminant plane patterns cohere to the degree that the component gratings share common higher order mechanisms, but in the vertical planes there seem to be only isoluminant mechanisms and a luminance mechanism. However, the question must be considered an open one in view of the evidence from noise-masking experiments (Gegenfurtner & Kiper, 1992), where stimuli modulated at an elevation of, say, 45 deg were masked more by a masker modulated at the same elevation than by one modulated at 135 deg.

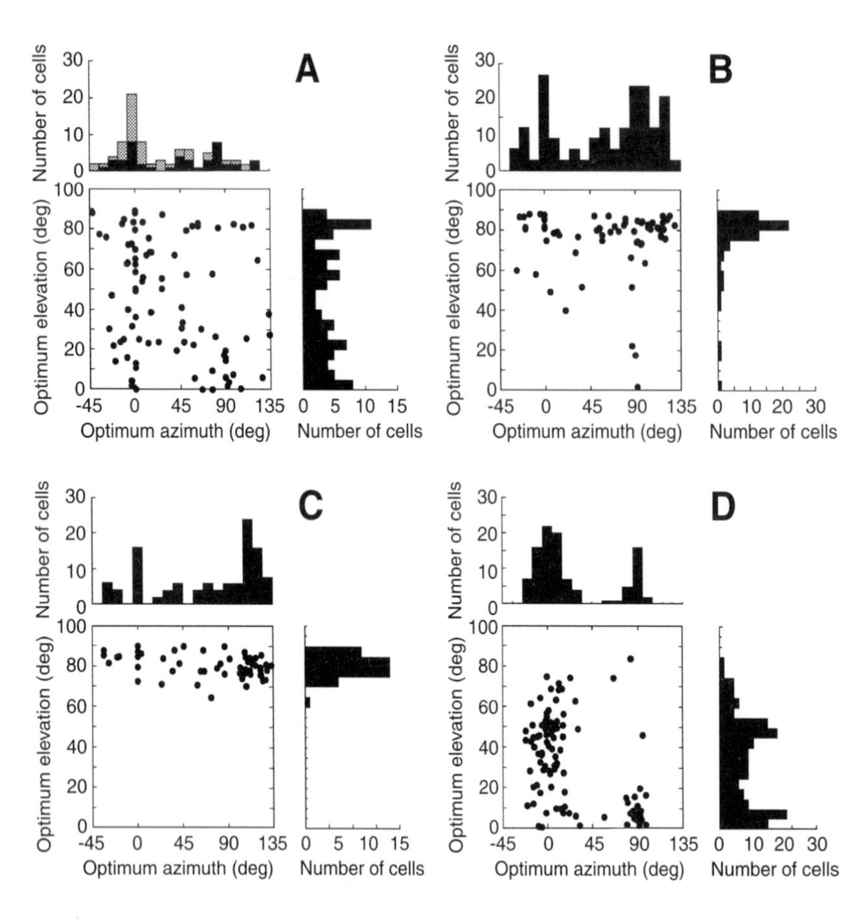

Figure 16.16: Distribution of preferred azimuths and elevations for units in V1: (A) nonoriented cells, (B) simple cells, (C) complex cells, and (D) in parvocellular layer of LGN. Marginal histograms above scattergrams provide distributions of azimuths and those to the right distributions of elevations. Reproduced with permission from Lennie, Krauskopf, and Sclar (1990).

Psychophysics and electrophysiology

Cardinal directions and LGN cells. Recordings of the responses of single units in the lateral geniculate nucleus (LGN) of macaque by DeValois, Abramov, and Jacobs (1966) and Wiesel and Hubel (1966) revealed chromatic opponency to be a property of many cells. Quantitative measurements of the responses of cells in the LGN to modulations of light in different directions in color space provided precise estimates of the preferred directions of individual units and thus the weights they assigned to the three classes of photoreceptors (Derrington, Krauskopf, & Lennie, 1984).

Initially there appeared to be a striking concordance between the results of the habituation experiments of Krauskopf, Williams, and Heeley (1982) and the preferred directions of single parvocellular units in the macaque lateral geniculate (LGN). The preferred directions fell into clusters that, while varying over a wide range of elevations, projected rather narrowly onto the two cardinal directions on the isoluminant plane (Fig. 16.16D). One difficulty is that LGN cells show no sign of fatigue no matter how strongly they are driven. Of the V1 cells illustrated in Figs. 16.16A–C, only the nonoriented units (Fig. 16.16A) show a significant response to chromatic stimuli. Their distribution of preferred color directions is not as concentrated as in the LGN (Lennie, Krauskopf, & Sclar, 1990), but there remains some residual concentration of units having preferred responses near the cardinal azimuths. Although habituation is seen in the cortical area V1 (Lennie, Lankheet, & Krauskopf, 1994), it would be premature to conclude that this is the substrate for the psychophysical habituation phenomena.

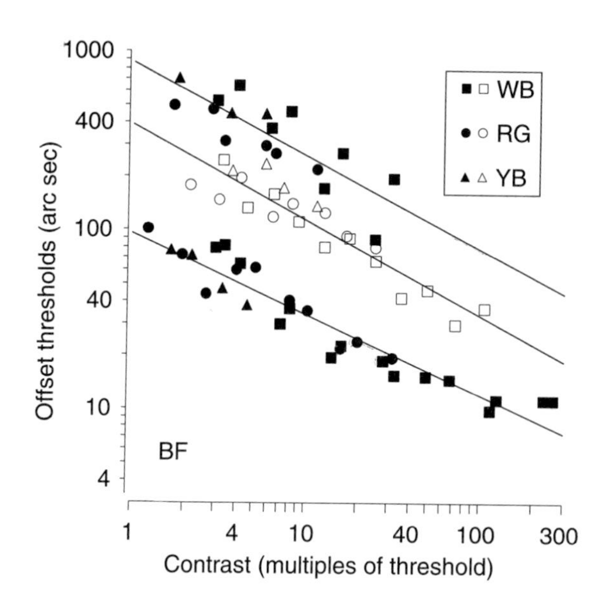

Figure 16.17: Log offset thresholds vs. log contrast for targets with Gabor profiles. The spatial frequency of the sinusoidal component of the Gabor pattern is the parameter.

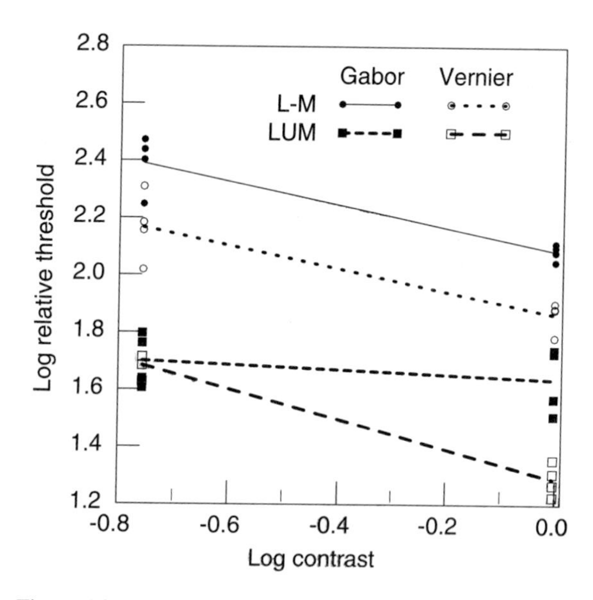

Figure 16.18: Log threshold amplitudes (re: 1 s of arc) for detection of 1-Hz sinusoidal motion of a one cycle per degree Gabor target as a function of contrast. Gabor: single 4 × 4 deg square target. Vernier: same but only lower half moves. LM: target modulated in L–M-direction. LUM: target modulated in luminance.

Chromatic and luminance mechanisms. It is clear that the processing of the chromatic aspects of stimuli must be mediated by the parvocellular neurons. No other pathway has been found. On the other hand, stimuli varying in luminance may be mediated by the magnocellular neurons, which are tuned to respond best to luminance variation or by parvocellular units. They may also provide input to higher level units that, by vector summation, respond best to luminance modulation (Lennie & D'Zmura, 1988). In the next section we consider evidence that different tasks may rely on different luminance mechanisms, as suggested by Lennie, Pokorny, and Smith (1993).

Vernier acuity. Krauskopf and Farell (1991) measured offset thresholds for Gaussian and Gabor targets. As shown in Fig. 16.17, the log offset threshold increased linearly with blur, decreased linearly with log contrast, and was essentially identical for targets modulated along the three cardinal directions when the targets were equated for detectability. The pattern of results is consistent with a model in which the photon catch is used with equal efficiency, regardless of color direction, to produce an estimate of the locus of the upper and lower distributions or of the tilt of the line segments connecting them. This makes it most likely that the luminance targets were signaled via a parvocellular pathway, because magnocellular neurons are very different in sensitivity and numerosity from the parvocellular units, which must be conveying the information about the chromatic targets.

Retinal- and object-relative motion. Movement can result in local changes in stimulation relative to the retina, which may be detected by a mechanism such as that proposed by Reichardt (1961). Motion may also be signaled by changes in form over time, a topic of considerable interest to the Gestalt psychologists (Brown, 1931). Here we consider how motion perception may be mediated by chromatic and luminance mechanisms when motion is defined in these two ways.

Krauskopf and Li (1996) measured amplitude thresholds for the detection of oscillatory motion of Gabor patches as a function of contrast for targets defined by luminance modulation and by isoluminant modulation along the L–M cardinal axis. Stimuli were presented on a TV monitor that generated an 8.5-deg square field maintained at a mean luminance of approximately 35 cd/m^2. A two-alternative forced-choice constant stimulus procedure was used to generate the data shown here. The targets occupied a central 4-deg square and were presented for 1 s with an inter-stimulus interval of 250 ms. In one interval, randomly chosen, the target oscillated horizontally at 1 Hz, while in the other interval it was stationary.

In one case a single Gabor target was moved relative to a uniform background, labeled "Gabor" in Fig. 16.18. In the other case, labeled "Vernier," only the lower half of the rectilinear Gabor patch moved, the upper half serving as a reference. The ordinate plots the log amplitude of the oscillation at threshold and the abscissa the log contrast relative to the maximum obtainable in the display.

The threshold amplitude for detecting motion of the single-luminance–modulated patch is independent of contrast, a result that was repeatedly obtained and confirms a conclusion of McKee, Silverman, and Nakayama (1986). Thresholds for isoluminant stimuli were generally much higher and improved markedly with increased contrast. No correction for detectability can make the luminance and chromatic curves superimpose. The difference in the form of the dependence on contrast strongly suggests that a different mechanism mediates the detection of motion for luminance and isoluminant stimuli. We speculate that motion of the luminance targets is mediated by the magnocellular motion mechanism. When only the lower half of the target moves, thresholds for both chromatic and luminance targets are improved and similarly dependent on contrast, suggesting that they are now detected by a similar, parvocellular, system.

Conclusions

While some evidence suggests the existence of three second-stage mechanisms mediating color vision, critical analysis of that evidence, reinforced by a variety of experiments, leads to the conclusion that there is a larger number of mechanisms tuned to a variety of directions within, at least, the isoluminant plane of color space.

Some of the experiments reviewed here not only reveal mechanisms selective to multiple directions in the isoluminant plane, but they also reveal no mechanisms outside the isoluminant plane other than those selective to the luminance cardinal direction. Conceivably, these experiments have sampled the luminance information carried by the magnocellular pathways. This holds open the possibility that the parvocellular representation of color space might provide an effectively continuous selectivity of hue, saturation, and brightness.

Whether this parvocellular representation is *isotropic* as well as continuous is unclear. The physiological data show anatomical variations, with the LGN neurons being more abundantly tuned to cardinal directions and cortical neurons being more uniform in their preferred directions. The psychophysical data also show task-dependence, with the cardinal directions being weighted somewhat more heavily than other directions in some, but not all, experiments. The anatomical variation and task dependence suggest a multistage decompression from the retina through the cortex of a few color space dimensions into a continuous isotropic representation.

Acknowledgments

The author wishes to thank Bart Farell for very useful comments on a draft version of this chapter. This work was supported by NIH grant EY06638.

17

Color and brightness induction: from Mach bands to three-dimensional configurations

Qasim Zaidi

The interplay of lights and objects in natural or man-made settings creates fascinating visual effects for the acute observer (Minnaert, 1993). Many of these effects can be explained satisfactorily by physical processes outside the observer or by the physics of light absorption by photopigments, but some are due to peculiarities of the visual system. To function effectively in a large variety of settings, the human visual system contains neural mechanisms that perform edge enhancement, spectral differencing, contrast induction, adaptation to steady and temporally varying lights, inference of motion and three-dimensional shape, and perceptual constancy. All of these mechanisms can be conceived of as performing particular transformations on the retinal image. These transformations generally fulfill functionally important roles, but they can also lead to blatantly nonveridical percepts, as shown by the literature on visual illusions (e.g., Luckiesh, 1965; Frisby, 1980; Griffiths & Zaidi, 1999).

A particularly important task for an observer is to segregate different objects from each other and from the background. A number of visual cues, including brightness, color, texture, motion, shape, stereo-disparity, and occlusion, facilitate this task. A number of studies have shown that the perceived magnitude of almost every visual modality is influenced by the magnitude of that modality in the surround. In fact, Chevreul's (1839) Law of Simultaneous Contrast of Colors can be safely generalized: "In the case where the observer sees at the same time two contiguous fields, they will appear as dissimilar as possible in almost every modality." Obviously, a visual process that enhances perceived differences between contiguous fields will facilitate object segregation. In a three-dimensional world, some segregations, like figure from ground, confer greater functional advantages than others.

A gray region appears brighter when viewed on a darker enclosing surround and darker when viewed on a brighter surround (Chevreul, 1839). When the surround is variegated in brightness, the perceived brightness of the test region can depend on a multitude of factors. For example, in Fig. 17.1 (Zaidi, Spehar, & Shy, 1997) the gray regions in the centers of the two configurations were made of identical materials and are of identical luminance (a purely physical measure), yet the gray region in the picture on the right appears considerably lighter than the gray region in the picture on the left. Figure 17.1 consists of a photograph of two figurally identical three-dimensional configurations, each consisting of an H shape in the foreground, with the gray square being the horizontal bar between the two vertical bars, and a larger square flap at an angle behind the H. If the perceived difference in the brightness of the two gray patches is due to induced contrast, then on the right the brightness induced by the dark background is of greater magnitude than the darkness induced by the light vertical bars. Similarly, the inducing effect of the light background prevails in the configuration on the left. In both pictures, each gray region has an equal perimeter

Figure 17.1: Two-dimensional pictures of a three-dimensional configuration. The two gray regions are of equal luminance, so any perceived difference in brightness is due to a combination of induced lightness from dark surrounds and induced darkness from light surrounds. The relative brightness of the two gray regions can be used to infer the strength of the induced effect from the background versus the effect from the flanking surrounds on the sides (Zaidi, Spehar, & Shy, 1997).

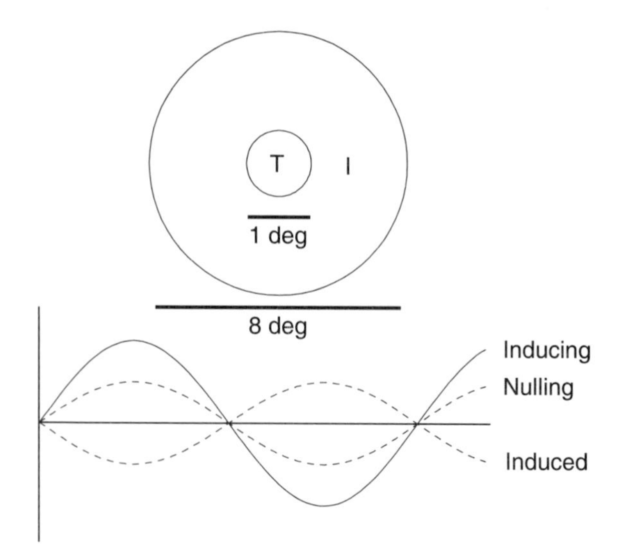

Figure 17.2: Nulling technique for measuring the magnitude of color induction. On the top, I indicates the inducing stimulus and T the test stimulus. On the bottom the solid line marked "Inducing" depicts the variation in time of the color of the inducing annulus. The dashed line marked "Induced" depicts the modulated appearance of the test disk when it is, in fact, not modulated. The dotted line marked "Nulling" depicts the real modulation of the disk required to make it appear steady (Krauskopf, Zaidi, & Mandler, 1986).

flanked by dark and light, and the visible areas of the background flaps are smaller than the visible areas of the vertical flanks. In spite of this, induced contrast from the flaps dominates that from the vertical flanks.

In this paper we will concern ourselves with the neural mechanisms that are responsible for color induction and the manner in which they affect color appearance. The main substrate for color induction is lateral connections between neural elements, most probably inhibitory. These connections range from mechanisms early in the visual stream that are responsible for edge enhancements, to higher level mechanisms that infer three-dimensional shape from two-dimensional cues in retinal images. Color appearance is also influenced by mechanisms of visual adaptation, and we will discuss models and experiments that isolate induction from adaptation effects, and explicitly examine their interaction.

Methods

Most studies on color induction have used measurements based on asymmetric matching or memory, despite Helmholtz's (1924, Vol. 2, pp. 264–271) detailed exposition of the drawbacks of these methods. To overcome these limitations, Krauskopf, Zaidi, and Mandler (1986) introduced a nulling method of measuring the magnitude of induction (Fig. 17.2). When an observer fixates on a white disk, the disk appears greenish when surrounded by a red annulus and reddish when surrounded by a green annulus. The effects of the annulus can be neutralized by adding some red light to the disk in the first case and some green light to the disk in the second. The amount of light needed to make the disk appear white may be taken as a measure of induction. Performing these operations in successive experiments requires the observer to keep a standard white in mind. On the other hand, if the color of the annulus is modulated from red to green, the disk appears to be modulated in counterphase to the annulus. If a modulated component is added to the disk, the disk can be made to appear steady. The nulling modulation can be used as a measure of induction. If induction is strictly complementary, the modulation added to the disk must be along the same line in color space as that applied to the annulus. Nulling modulations have been measured using methods of adjustment (Krauskopf, Zaidi, & Mandler, 1986; Chubb, Sperling, & Solomon, 1989), 2AFC staircases (Spehar, DeBonet, & Zaidi, 1996; DeBonet & Zaidi, 1997), and 2AFC methods of constant stimuli (Singer & D'Zmura, 1994, 1995). A spatial analogue of the method involving the nulling of induced sinusoidal gratings by superimposed physical gratings was devised by McCourt (1982).

Temporal and spatial nulling methods have the following advantages: (i) During each trial, the observer fixates on the central test. Successive contrast effects are kept to a minimum, because, except for small eye movements, the portion of the retina that receives the induced modulation is never exposed to the inducing field. (ii) The observer is required only to signal the absence or presence of modulation without extracting any subjective qualities of the test. At the null point, the induced effect of all phases of the surround modulation are equally canceled. No memory demands are made. The nulling method thus meets the requirements for Brindley's (1960) class A experiments. (iii) In every trial, the chromaticity and luminance of every point of the display, averaged over a cycle of modulation, can be kept the same for all conditions. The steady-state adaptation is therefore constant throughout the experiments. This enables induction effects on appearance to be separated from adaptation effects.

Neural locus of color induction

For a long time there was a debate concerning the neural locus of color induction. Induction implies that the effects of a stimulus falling on one part of the retina are modified by stimuli falling on another part of the retina at some level in the visual system. For example, it had been supposed that the stimulation of cones of one class in one part of the retina results in the desensitization of cones of the same class in other parts of the retina (e.g., Evans, 1948; Alpern, 1964). Alternatively, a similar form of lateral interaction had been proposed between second-stage opponent mechanisms (e.g., Jameson & Hurvich, 1964; Shevell, 1978; Guth, Massof, & Benzschawel, 1980; Ware & Cowan, 1982). The debate was resolved by Krauskopf, Zaidi, and Mandler (1986), who showed a cortical locus for color induction involving color mechanisms beyond the linear-opponent stage.

They first measured the amplitude of the induced modulation as a function of the amplitude of the inducing modulation for each of the three cardinal axes of color space (Krauskopf, Williams, & Heeley, 1982; Zaidi, 1992).[1] The amplitude of the nulling modulation versus the amplitude of the surround modulation is shown in Figs. 17.3A and B for two observers for each of the three cardinal directions. The lines are the best-fitting cubic polynomials of the form:

$$(1) \qquad N(A) = aA - bA^3,$$

where N is the required nulling modulation and A is the surround modulation amplitude. The curve for the luminance direction is the steepest, followed by that for the YV-direction, and that for the RG-direction is the shallowest. This experiment revealed that the nulling amplitudes are sufficiently different functions of inducing amplitudes in the three cardinal directions to allow a critical test of the locus of induction effects. This was carried out in a second experiment, in which the amplitude and direction of the modulation needed to null the effect of an inducing stimulus in various noncardinal directions was measured.

The second experiment was concerned with the question of whether induction takes place solely within independent opponent mechanisms. If this were true, induction would not, in general, result in changes of appearance in a strictly complementary direction. In the stimulus domain, modulation in a direction inter-

[1] A number of equivalent designations have been used for the three cardinal axes, including (Luminance, Constant B or S, Constant R&G or L&M), (L+M+S, L–M, S), amd (Light-Dark, Red-Green, Yellow-Violet). In the text we will use (LD, RG, YV), although some graphs taken from previous publications may use one of the others.

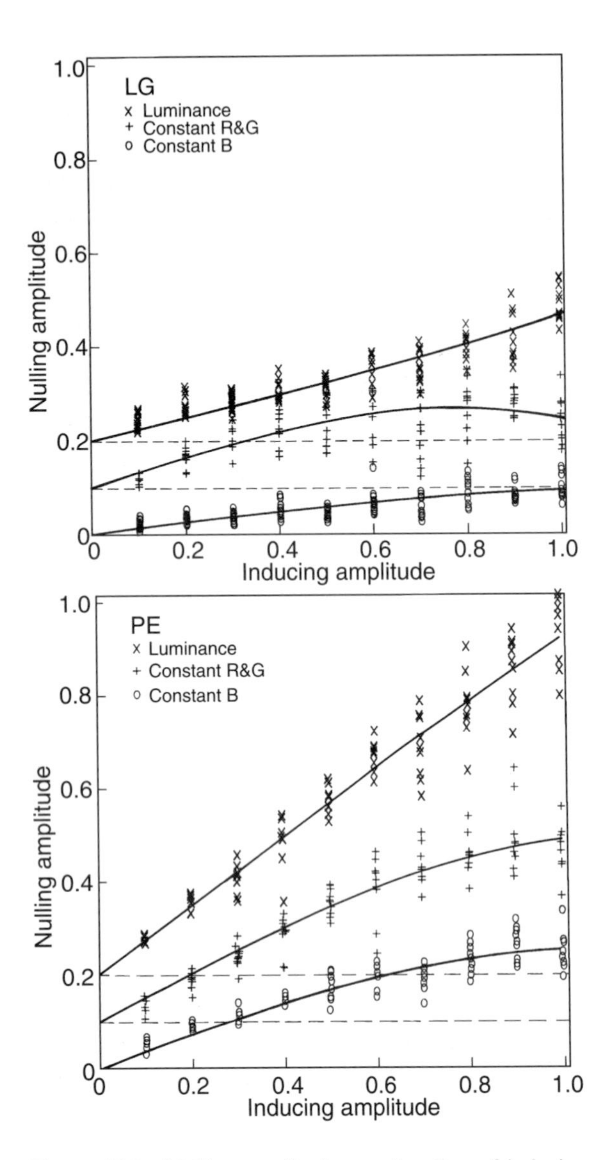

Figure 17.3: Nulling amplitude as a function of inducing amplitude for two observers. In each panel the results from top to bottom are for modulation along the LD-axis, along theYV-axis, and along the RG-axis. Note the displaced ordinates. Lines fitted to the data are best fits of cubic polynomials (Krauskopf, Zaidi, & Mandler, 1986).

mediate between two cardinal directions is equivalent to the vector sum of modulations of the appropriate amplitudes in the two cardinal directions. If induction took place solely within cardinal mechanisms, then the induced effect of this modulation could be nulled by a modulation that simultaneously nulled the induced effect in each of the cardinal directions. However, because the nulling functions are not identical for the cardinal directions, the induced effects in the two directions will be different multiples of the projected amplitudes of the inducing modulation. The predicted nulling modulation will therefore not be in the same color direction as the intermediate inducing direction. If induction took place in higher order color mechanisms that were tuned maximally to the inducing direction, then the best nulling modulation would be in the same direction as the inducing modulation. In forced-choice judgments for six intermediate directions in the isoluminant plane, the null in the inducing direction was consistently preferred to the null predicted from the cardinal directions. For isoluminant stimuli, the results were consistent with the complementary nature of chromatic induction.

The results ruled out interactions at both the cone and opponent mechanism levels as an explanation of induction. The observers preferred the settings in the direction of the inducing stimuli to those predicted from the cardinal-direction settings. Because modulation along the RG-axis varied the input only to the L- and M-cones, whereas modulation along the constant YV-axis only varied the input to the S-cones, explanations in terms of independent cone mechanisms also make the wrong prediction that the vector sum of cardinal-direction settings will be preferred. In fact, since the cardinal directions are linear combinations of cone absorptions, rejection of the vector-summation rule for chromatic induction also ruled out explanations in terms of any other linear combinations of cone absorptions. One possibility is that chromatic induction reflects lateral interactions beyond the stage of independent linear cardinal mechanisms within mechanisms preferentially tuned to many different directions of color space (Krauskopf et al., 1986). Recordings in the parvocellular layers of the macaque LGN have revealed two major classes of cells that are preferentially tuned to the chromatic cardinal axes (Derrington, Krauskopf, & Lennie, 1984), whereas similar experiments on the visual cortex reveal a diversity of cells. The cells that respond to pure chromatic modulation have maximal responses in many directions in color space, rather than only in the cardinal directions (Lennie, Krauskopf, & Sclar, 1990). It seems likely that lateral interactions between cortical cells cause chromatic induction.

Edge enhancement effects

In a stunning intellectual feat, from phenomenological observations of bright and dark bands at luminance edges where – based on physical considerations alone – none were expected, Mach (1865, 1866a, 1866b, 1868) inferred the retinal mechanisms of lateral inhibition and an excellent approximation of the mathematical form of center-surround receptive fields. These bands enhance the local lightness of the lighter side of the edge and the local darkness of the darker side. Mach bands have also been claimed to exist for chromatic stimuli at saturation edges (Pease, 1978). Consequently, it is worth examining the extent to which they contribute to brightness and chromatic induction (Bekesy, 1968).

The effect of Mach bands on appearance is illustrated nicely by the phenomenon called "grating induction." McCourt (1982) and Foley and McCourt (1985) showed that when a narrow uniform strip is inserted within a sinusoidal grating, an induced grating is perceived in the strip (Fig. 17.4A). Zaidi (1989) created a series of grating configurations to examine the role that proximal and distal elements of the inducing stimulus play in grating induction. Figures 17.4A–C show that gratings of different orientations can be induced by combining Mach bands created at the edges of inducing gratings of the same spatial frequencies and orientations, offset by different fractions of a cycle across the test region. In Fig. 17.4D there is no induced grating, and the Mach bands can be seen clearly. In Figs. 17.5A–D induced gratings of the same orientation and frequency have been created by offsetting inducing gratings of different orientations and spatial frequencies so that the row of pixels directly above the test field is identical in all four figures, as is the row directly below the test field. These local identities lead to similarity in the induced percept, demonstrating the primacy of edge effects in visual grating induction. However, the perceived contrast of the induced grating decreases progressively from Fig. 17.5A–D. This impression was confirmed by nulling measurements that showed that even though Mach bands are confined to local regions of the test, they are affected by more extended regions of the surround. For equiluminant chromatic stimuli, the amplitude of induced modulation was less than for luminance stimuli; otherwise, the effects were qualitatively similar.

In a grating induction display such as Fig. 17.4A, Zaidi and Sachtler (1991) noticed that if the real grating is drifted to the left, the induced grating also appears to drift to the left. If, after prolonged viewing of the drifting stimulus, the motion is abruptly stopped, the real grating appears to move to the right, as would be expected from the classical motion aftereffect (Wohlgemuth, 1911). Simultaneously, in the uniform central gap there appears a grating in counterphase to the surround that also appears to move to the right. It seemed as though motion adaptation could take place in areas of the retina that were not directly exposed to the physical adapting stimulus. To test whether the adaptation was caused by the induced grating per se, Zaidi and Sachtler measured the effect of prolonged viewing of aligned (Fig. 17.4A) and offset (Fig. 17.4D) inducing gratings on contrast thresholds for moving gratings confined to the physically uniform test region. The results showed that the desensitizing effect of offset surrounds was roughly equal to that of aligned surrounds. Therefore, the induced percept of cohesive gratings did not have a greater desensitization effect on test thresholds than did induced Mach bands. For both types of surrounds, the desensitizing effect was highest close to the edges of the adapting stimulus, and gradually decreased with distance from the edge.

Even though Mach bands can be combined to form induced gratings, whether edge enhancing mechanisms are important influences on the appearance of spatially extended tests is a different question. Zaidi, Yoshimi, and Flannigan (1991) presented data that bear on this question by separating the effects of area,

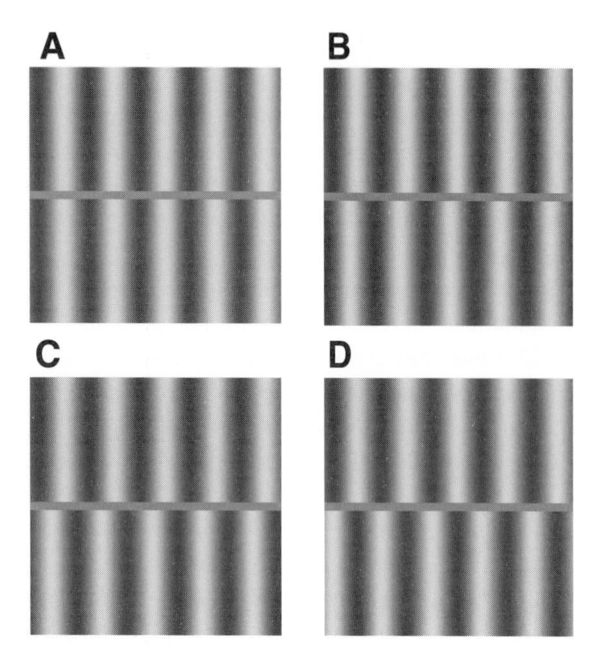

Figure 17.4: Displays showing sinusoidal grating induction in the central homogeneous test field. The orientation of the induced grating depends on the phase of the inducing grating above the test field relative to the phase of the grating below the test field. The relative shifts are (A) 0.0 cycles (in-phase), (B) 0.125 cycles, (C) 0.25 cycles, and (D) 0.5 cycles (counterphase). In (D) the induced percept consists of light and dark patches. Note: Mask inducing gratings to see actual homogeneity of test patch (Zaidi, 1989).

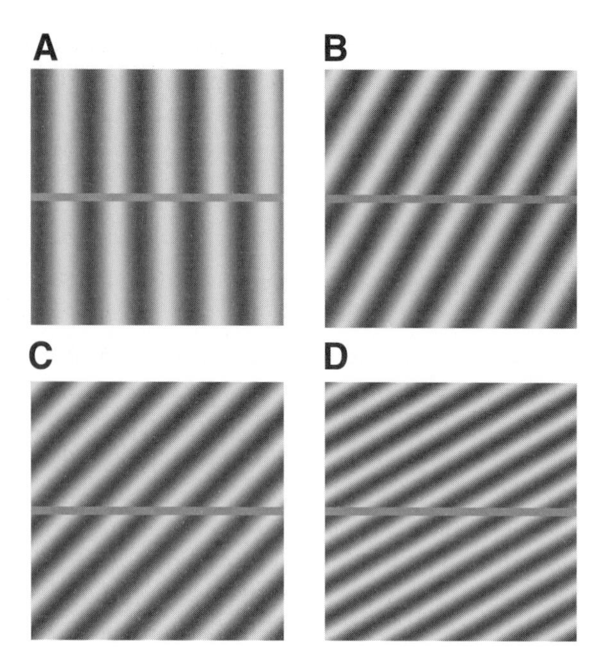

Figure 17.5: Displays showing induced gratings with the same orientation and spatial frequency. The inducing gratings have orientations of (A) 90 deg, (B) 60 deg, (C) 45 deg, and (D) 30 deg. The spatial frequency of the inducing gratings increases from (A) to (D) (Zaidi, 1989).

shape, and perimeter on the change in perceived color of a test region. The test regions were multiple-lobed shapes like those in Fig. 17.6A. The purposes of this experiment required tests whose shapes, areas, and perimeters could be varied independently. In addition, it was considered desirable to use shapes with smooth contours and without sharp corners. The analytic expression for Fourier descriptors provided by Zahn and Roskies (1972) proved to be suitable for generating the test shapes. The number of lobes was 3, 4, 5, 6, 7, 8, 9, 10, 20, or 40. The area of each test was equal to the area of a circle of radius 1 deg, that is, equal to π deg^2. The length of the perimeter of a test was 2.5π, 3.5π, 4.5π, or 5.5π deg. The test field was surrounded by a disk with an outer diameter of 5 deg. The nulling method described above was used to measure induction in the three cardinal directions. Since edge enhancements occur on the border between the test and the surround, for tests subtending equal areas and having comparable shapes, if the magnitude of induced contrast was proportional to the perimeter length, it would indicate that appearance was affected by the summation of Mach bands along the interior perimeter of the test. The results (Figs. 17.6B–D) showed that the magnitude of induced contrast does not depend on the length of the perimeter of the test. In all three panels the slope of the best-fitting regression line was not significantly different from zero. Hence, local edge enhancements do not have a significant effect on the appearance of spatially extended regions.

Spatially extended effects

The spatial extent of color induction has generally been studied with disk–annulus configurations of dif-

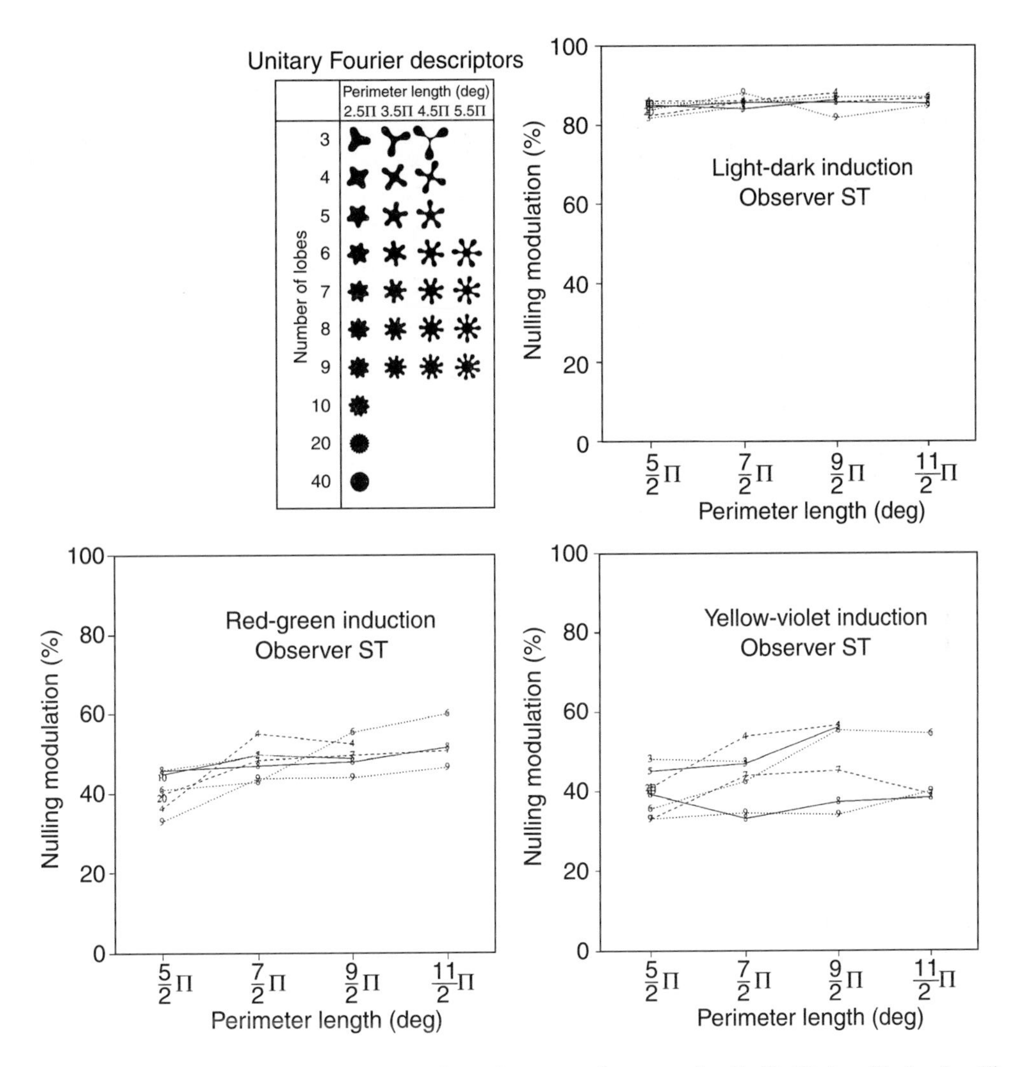

Figure 17.6: (A) Unitary Fourier descriptors of equal area, used as test stimuli. (B–D) Amplitude of nulling modulation in each cardinal direction expressed as a percentage of inducing modulation versus the length of the perimeter of the test. Each test stimulus is represented by the number of lobes that range from 3 to 40. Curves connect nulling amplitudes for tests of the same number of lobes (Zaidi, Yoshimi, & Flannigan, 1991).

ferent sizes (e.g., Yund & Armington, 1975; Reid & Shapley, 1988). These methods generally yield data where the amount of induction is a nonlinear function of parameters such as surround size. In these configurations, it is not possible to test whether the nonlinearity is due to nonlinear spatial interactions or just to a nonlinear decrease in effectiveness with distance from the test. To overcome these limitations, Zaidi et al. (1992) studied the differential effects of surrounding regions at different distances from the test by using stimuli similar to those shown in Fig. 17.7. The central disks are the test regions. In the surround, along every radial line through the center, the color of the surround varies as a spatial sine wave along a color line through mid-gray, resulting in a blurred bulls-eye target whose concentric rings vary sinusoidally in color. Each of the rows shows the color-varying surrounds of a single spatial frequency that were used in a single condition.

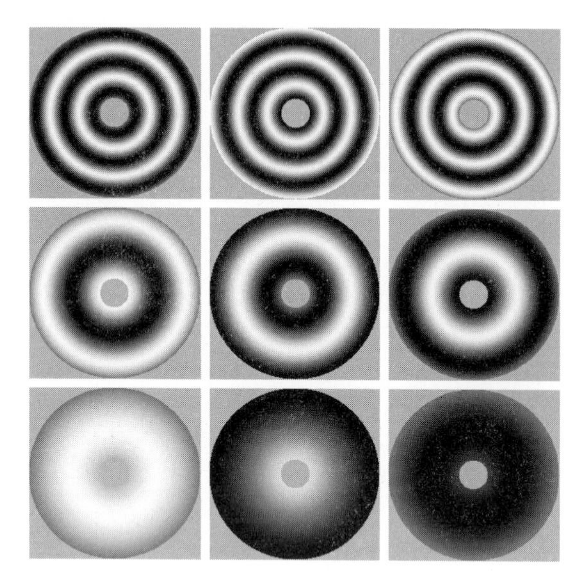

Figure 17.7: Single-frequency induction stimuli. Surrounds are annuli whose luminance varies sinusoidally along each radius. Each row depicts one spatial frequency condition. Across each row, the surround is shown at three different phases. The central disks are the tests, and in all nine pictures they are at the same mid-gray luminance; the differences in apparent brightnesses are due to different amounts of induced brightness from the surround (Zaidi et al., 1992).

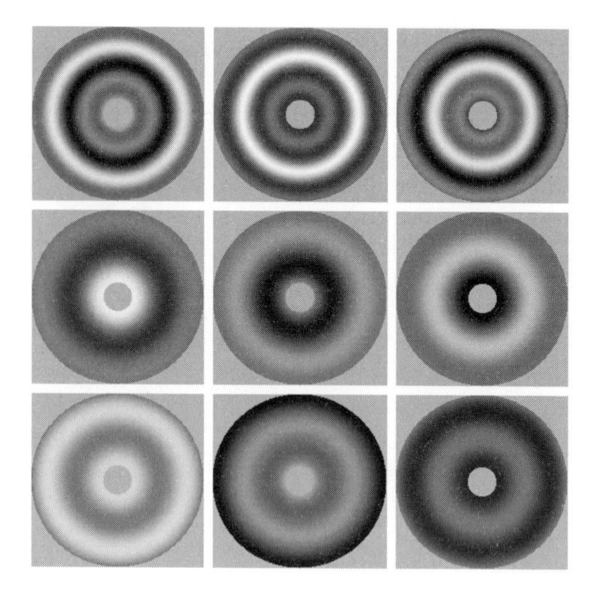

Figure 17.8: Compound-frequency induction stimuli. Surrounds are annuli whose luminance varies as the sum of two sine waves of different frequencies along each radius. The top row shows a high-frequency sine wave added to a medium-frequency sine wave. In the middle row, two different intermediate frequency sine waves are added. In the bottom row, a low-frequency sine wave is added to a medium-frequency sine wave. Across each row, the surround is shown at three different phases. The central disks are the tests, and in all nine pictures they are at the same mid-gray luminance; different apparent brightnesses are due to different amounts of induced brightness from the surround (Zaidi et al., 1992).

Each row shows three different phases of the sine wave with respect to the inner edge of the surround. The central disks in all nine pictures are physically the same mid-gray but appear to have different brightnesses. The surround consisted of a single sine wave of one of seven different spatial frequencies, and its color varied along one of the three cardinal directions of color space. Two aspects of the phenomenal appearance of the central test are directly relevant. First, within each row, as the phase of the surrounding sine-wave changes, the appearance of the test changes. The appearance of the test is roughly complementary to the appearance of the inner edge of the surround. In the experiments, as the phase of the surround (with respect to the inner edge) was changed uniformly in time, so that the sine wave appeared to drift toward the center at a constant velocity, the appearance of the center changed cyclically in time. The induced modulation was nulled by the addition of a real modulation to the test field, and the nulling modulation was used as the measure of the induced effect. Second, the magnitude of the change in the appearance is least in the top row, which has the surround with the highest spatial frequency, and largest in the bottom row, which has the surround with the lowest spatial frequency.

The measured amplitudes of the nulling modulation, when plotted against the spatial frequency of the surrounding sine waves, formed low-pass functions for all three color directions, LD, RG, and YV. For a variety of spatial models of induction, these low-pass functions indicate that the effects of elements of the surround decrease monotonically with distance from the test. Zaidi et al. (1992) and DeBonet and Zaidi (1997) showed that this spatial weighting function could be estimated from the Fourier transform of the nulling amplitude versus spatial frequency function for brightness induction but not for chromatic induction. The reason is that, for induced achromatic brightness, the total effect of the surround could be described as the sum of the induced effects of individual elements of the surround. The spatial summation inference was based on the results of superposition tests, that is, the brightness induced by every pair of surrounds presented simultaneously was equal to the sum of the brightnesses induced by each component presented singly. Chromatic induction failed the spatial superposition tests.

The spatial superposition assumption was tested by comparing the induced effect of surrounds composed of pairs of circularly symmetric sine waves to the sum of the induced effects of the constituent sine waves. The paired sine waves were set to be in identical phases at the inner edge of the surround. These compound stimuli are shown varying in luminance in Fig. 17.8. The central disks are the test regions set at the mean luminance level. Each row shows surrounds consisting of the sum of two spatial frequencies windowed by the edges of the surround. The three rows consist of the same medium frequency paired with a high (top row), medium (middle row), and low frequency (bottom row). Across each row, three phases (with respect to the inner edge) of the paired sine waves are shown. Although the central disks are all at the same luminance level, they appear to be different, depending on the frequencies and phases of the surrounding sine waves. Sine waves of each of the spatial frequencies used in the previous experiment were paired with each other. The amplitude of each constituent sine wave was 0.5 to give a maximum amplitude modulation of 1.0.

Figures 17.9A–C show the data for the paired sine waves. The ordinate of each point is the amplitude of the required nulling modulation. Each curve in the figures connects the data for a particular spatial frequency when paired with the spatial frequencies corresponding to the abscissa. Different line types have been used to distinguish the curves. The key to identifying the curves is to begin at the leftmost point where the curves are ordered from top to bottom in the same

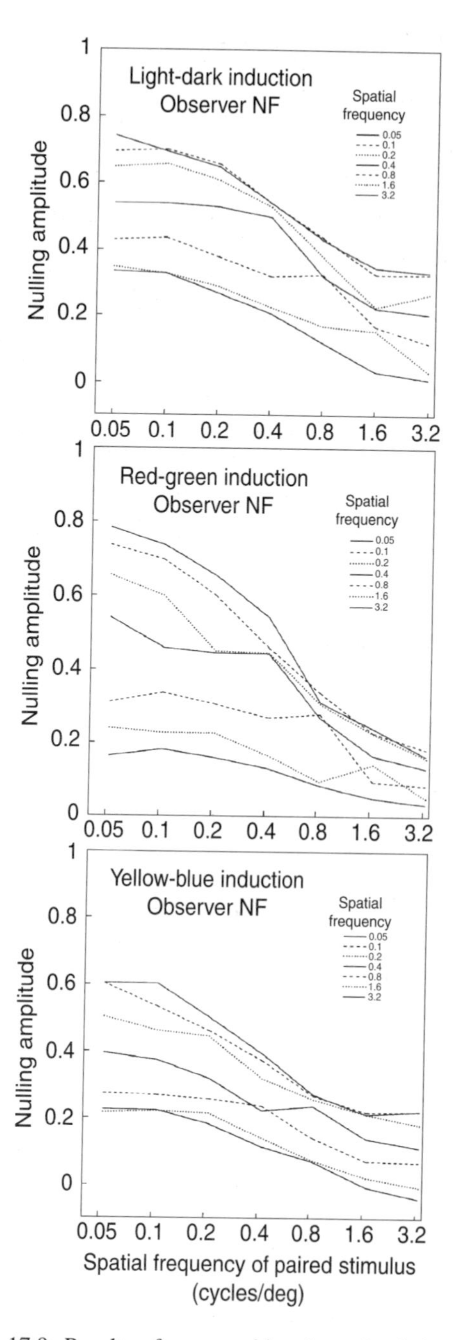

Figure 17.9: Results of superposition for color induction. The modulation required to null color induction is plotted against the spatial frequency of one sinusoidal component of the surrounding compound wave, with the other spatial frequency as the curve parameters indicated by the line types (Zaidi et al., 1992).

order as the spatial frequencies shown in the caption.

All of the curves for brightness induction are roughly parallel and have similar shapes, indicating that the amount of modulation required to null the induced effect decreases as each frequency is paired with progressively higher frequencies. Additionally, the incremental induced effect of adding sine waves of different spatial frequencies (represented by the curves) is fairly independent of the paired spatial frequencies (indicated on the abscissa). The main departure from parallelism is the noticeable upward protuberance in each curve where the two paired frequencies were equal, that is, the surround consisted of a single sine wave. At other pairings, the amplitude required to null the effect of a compound surround is a little less than the sum of the amplitudes required to null the effects of the two components. The slight subadditivity in nulling amplitude of the compound surround could be due to a phase mismatch in the modulation induced by the components. Because superposition holds, a punctate summation model can be used to fit the brightness induction data.

On the other hand, the curves for chromatic induction are not parallel. The range covered by the left-hand side of the curves (pairs that include the lowest spatial frequency) is considerably greater than the range covered by the right-hand side (pairs that include the highest spatial frequency). The effect of adding a high spatial frequency chromatic component severely attenuates the magnitude of induced contrast, especially for the RG condition.

Spatial variations in the region surrounding a central test disk, can be decomposed into variations along radial lines through the center of the disk (Figs. 17.7 and 17.8), and variations tangential to the radial lines. Zaidi and Zipser (1993) used radial patterns to examine the effect of surrounds whose color varied sinusoidally along concentric rings. The surrounds appeared to be composed of spokes, each spoke being of uniform color and the color of adjacent spokes varying continuously in a sinusoidal manner. The results showed that for both lightness and chromatic induction, on tests at the mean level, the induced effect of radially varying surrounds summed to zero, that is, to the sum of the induced effects of individual elements of the surround.

The results for brightness induction were interpreted in terms of a model that postulates a weighted spatial integration of induced effects. The perceived brightness at a point in visual space has two components, one due to the luminance of the light emanating from that point and the second due to the total induced effect of surrounding points. The model makes three assumptions about the induced effect. First, the induced effect of any surrounding point is in the complementary direction from the surround luminance relative to the test, with a magnitude proportional to the difference between the surround point and the mean level of the whole surround. Second, the induced effect of each surrounding point is weighted by a decreasing function of spatial distance from the test point. Third, the total induced effect is simply the sum of the induced effects of individual surrounding points. Algebraically, this model is defined by

$$(2) \qquad y = -\int_0^{2\pi} \frac{\int_0^{\infty} g(s)A(\Omega, s)s\,ds}{2\pi}\,d\Omega\,,$$

where y is the total induced effect at the test point, Ω is the angular orientation, s is the spatial distance between the test and induction point, $g(s)$ is the monotonically decreasing spatial weighting function of s, and $A(\Omega, s)$ is the signed magnitude of the luminance difference between the inducing point at (Ω, s) and the surround mean level.

The stimuli used in these experiments were circularly symmetric and varied only along radial lines; therefore, if the weighting function is assumed to be isotropic, Eqn. (2) can be reduced to a function of just the radial distance. For a surround consisting of a drifted single sinusoid of spatial frequency equal to ϕ cycles/deg, the induced effect at time t for the center point of the circular test can be expressed as:

$$(3) \quad y(t, \phi_i) = -A\int_L^X g(s)\cos[2\pi(\rho_0 t - \phi_i s + \phi_i L)]s\,ds,$$

where A is the amplitude of the surround sine wave, L

is the inner edge of the surround (i.e., the radius of the test disk), X is the outer edge of the surround, and ρ_0 is the temporal frequency of the drift (in cycles/s). For the present model, given that the test is uniform in contrast and luminance, the induced effect of all points inside the test on the test center is zero. Therefore, Eqn. (3) is expressed solely in terms of the effect of the surround. In the case of compound sine wave stimuli, with a second sinusoid of spatial frequency ϕ_i cycles/deg, the total induced effect is equal to:

$$y(t, \phi_i, \phi_j) = 0.5\, y(t, \phi_i) + 0.5\, y(t, \phi_j).$$

Since the induced contrast and brightness modulations could be suitably nulled with the addition of real sinusoidal modulation of the same temporal frequency as the inducing modulation, it is sufficient to describe the inducing, induced, and nulling modulations in terms of their amplitude and phase. For each condition, the amplitude and phase of the induced modulation were obtained from the Fourier transform of Eqn. (3) in the temporal frequency domain. By exploiting the fact that the drift was at a constant velocity given by ρ_0 divided by ϕ_i, the Fourier transform was simplified to

$$(4) \quad Y(\rho_0, \phi_i) = -\frac{A}{2}[\delta(\rho - \rho_0) + \delta(\rho + \rho_0)] \cdot e^{i2\pi\phi_i L} \cdot \int_L^X g(s) e^{i2\pi\phi_i s} sds,$$

where $Y(\rho_0, \phi_i)$ is the Fourier transform of induced modulation for a surround of spatial frequency ϕ_i drifted toward the test point at a temporal frequency of ρ_0, and δ is the Dirac delta function.

If the three assumptions of the model are satisfied, then, given the proper choice of $g(s)$, Eqn. (4) should fit the data. Since many smooth monotonic functions can be approximated by exponential functions, it was assumed that the spatial weighting function could be approximated by a negative exponential function of the form:

$$(5) \quad g(s) = \kappa e^{-\alpha s}.$$

Equation (5) was substituted into Eqn. (4), and solved to obtain the following expression:

$$(6) \quad Y(\rho_0, \phi_i) = \frac{-A\kappa}{(\alpha + i2\pi\phi_i)^2} \cdot [(1 + \alpha X + i2\pi\phi_i X) \cdot \exp(i2\pi\phi_i L - \alpha X - i2\pi\phi_i X) - (1 + \alpha L + i2\pi\phi_i L)e^{-aL}].$$

Expressions for the amplitude and the temporal phase of the induced modulation were then derived by transforming the right-hand side of Eqn. (6) into the polar form:

$$(7) \quad Y(\rho_0, \phi_i) = Amplitude \cdot e^{iPhase}.$$

This model provided a good fit to the contours of the brightness induction curves. The spatial weighting functions, $e^{-\alpha s}$, can be expressed in terms of the space constant $1/\alpha$ in degrees of visual angle, which gives the distance from the test at which the effectiveness of a surround point has fallen to $1/e$ of the maximum. The best estimates of the brightness induction space constants for two observers were 0.74 and 0.29 deg (DeBonet & Zaidi, 1997).

The failure of the superposition test for chromatic induction may not indicate a failure of weighted spatial summation if the nonlinearity in the data is due to punctate amplitude nonlinearities. Amplitude nonlinearities would also explain the nonlinear nulling functions in Fig. 17.3. To test this possibility, it was assumed that local chromatic signals from each point in the image pass through an amplitude compression in the visual system prior to the stage of lateral interactions responsible for chromatic induction. Such a nonlinearity will have two effects on the spatial summation model. Inside the surround, the nonlinearity will reduce the effective contrast of the surrounding wave. Inside the test, the nonlinearity will reduce the effectiveness of the nulling modulation. Mathematically, this is represented by

$$(8)\quad y=-\zeta[N]=-\int_0^{2\pi}\frac{\int_0^{\infty}g(\Omega,s)\zeta[A(\Omega,s)]sds}{2\pi}d\Omega,$$

where ζ is an odd-symmetric nonlinear compressive function, y is the actual induced modulation, and N is the measured nulling amplitude. The equation

$$(9)\quad \zeta[A]=A-cA^3$$

was used as the odd-symmetric compressive function of amplitude A.When applied to sinusoidal stimuli, ζ generates higher order harmonics. However, the optimal choice of the compressive nonlinearity for the fits to the present data generated higher harmonic energy that was less than 10% of the energy in the fundamental. Substituting $\zeta[A]$ and $g(s)$ into the one-dimensional form of Eqn. (8) yields an instantaneous induction level given by:

$$(10)\quad y(t) = -N(t)+cN^3$$
$$= -\int_X^L \kappa e^{-\alpha s}[A(s,t)-cA^3(s,t)]sds,$$

where $A(s, t)$ is the amplitude at radius s and time t of the surround. By using Fourier transforms similar to those used for the analysis of the linear model and removing higher order harmonics, an expression was derived for the amplitude of the inducing stimulus. However, numerical simulations showed that whereas adding a pointwise compressive nonlinearity to the model helped predict the compression in Fig. 17.3, it actually changed the predicted pattern of the results slightly in the direction opposite to the nonadditivity in the superposition measurements in Figs. 17.9B and C. The compressive nonlinearity had the effect of slightly attenuating predicted induction from low-frequency surrounds rather than from high-frequency surrounds. Consequently, the nonadditivity of chromatic induction is not due to a pointwise contrast compression. The causes of the subadditivity are discussed further in the section on habituation.

Adaptation mechanisms and brightness induction

The studies in the previous section isolated lateral effects from contamination by the effects of steady-state adaptation by keeping the time- and space-averaged mean luminance of all points in the stimulus equal. The results showed that brightness induction can be characterized as a summation process with a negative exponential spatial weighting function. Figure 17.10 from Zaidi, DeBonet, and Spehar (1995), however, demonstrates a failure of spatial summation. This figure consists of three vertical surround segments filled with a random binary texture. The space-averaged luminance of the three segments is equal, and the spatial contrast progressively decreases from left to right with values of 1.0, 0.3, and 0.0 Michelson contrast. Centered in each of the surround segments are five spatially uniform diamonds decreasing in luminance from top to bottom. The diamonds across each row have identical luminances, yet most observers see them as increasing in lightness from left to right in the top rows and from right to left in the bottom rows. This display shows that even from a nonfigural variegated surround, brightness induction depends in a complex manner on the relative luminance of the test and individual regions of the surround. A surround made of equal numbers of light and dark squares makes a test at higher than its average luminance appear darker than does a spatially uniform surround at the average luminance, and it makes a test at lower than its average luminance appear lighter than does the spatially uniform average luminance surround. The empirical ranked lightnesses of diamonds within each row are presented in Table 17.1, along with predicted rankings from the model described later in this section.

Other reports of additivity failure (Brown & MacLeod 1991; Schirillo & Shevell, 1996) also consisted of stimuli where the test was at a different level than the mean luminance and color of the surround, thus making it imperative to explicitly consider spatially local and extended adaptation mechanisms. In a steady display it is impossible to separate the effects of adaptation from induction. Spehar, DeBonet, and Zaidi

Figure 17.10: Brightness induction from random, binary textured surrounds. The three vertical surround segments have equal spatially averaged luminance, while the spatial contrast progressively decreases from left to right (1.0, 0.33, and 0.0). Centered in each of the surround columns are five spatially uniform diamonds with luminance decreasing from top to bottom. The luminance of the diamonds in the middle row is equal to the mean luminance of the surround segments. Diamonds across each row are of identical luminance, but their perceived lightnesses differ.

Observer JS			Observer BS		
3 (3)	2 (2)	1 (1)	3 (3)	2 (2)	1 (1)
1 (1)	2 (3)	3 (2)	3 (2)	2 (3)	1 (1)
1 (1)	2 (2)	3 (3)	1 (1)	2 (2)	3 (3)
1 (1)	2 (2)	3 (3)	1 (1)	2 (2)	3 (3)
1 (1)	2 (2)	3 (3)	1 (1)	2 (2)	3 (3)

Table 17.1: For two observers the empirical (and predicted) ranked lightness of diamonds within each row are presented in a similar configuration as the display shown in Fig. 17.10 (Zaidi, DeBonet, & Spehar, 1995).

(1996) used the time-varying nulling method to generate a general model for brightness induction from such variegated nonfigural surrounds to identify the conditions under which the induced effect can be described as spatially additive, and to delineate the processes that lead to failures of additivity.

The stimuli used were similar to those shown in Fig. 17.11A, in which a foveally fixated, spatially uniform disk was surrounded by an annulus filled with a binary random texture that was composed of equal numbers of two sets of randomly intermixed, equal-sized square elements. The luminance of each set of elements was modulated sinusoidally in time. The mean level, amplitude, and phase of temporal modulation were independently controlled for each set. Temporal modulation of the luminance of the surround elements resulted in an induced modulation of the brightness of the test. The amplitude of the nulling modulation was used as the measure of the overall induced effect.

In Experiments 1 and 2, the total induction on the test was measured when the luminance of both sets of texture elements was modulated sinusoidally at 0.5 Hz. The luminance modulation amplitudes of one set were 0.0, 0.2, 0.4, 0.6, 0.8, or 1.0, paired with modulation amplitudes of the other set of 1.0, 0.5, 0.0, –0.5, or –1.0, where a positive or negative amplitude denotes modulation in the same or opposite phase as the paired modulation. One cycle of each of these combinations of luminance modulations is shown schematically in Fig. 17.11B. In Experiment 1, to test for spatial additivity, the test and all elements of the surround had the same space- and time-averaged mean luminance. In Experiment 2, to test for local gain controls, the time-averaged luminance of the test disk was set at either 0.5 or 1.5 times this value. Experiment 3 tested for spatially extended gain controls by using the same spatial configuration to examine whether modulation in spatial contrast of the surround could produce brightness induction in the test. This was achieved by modulating the luminances of the two surround components sinusoidally at 0.5 Hz, with equal amplitudes, in opposite phase, around different mean luminance levels, as depicted in Fig. 17.11C. Experiment 4 measured induction from spatially uniform surrounds at three different mean levels on tests at the same three levels. This experiment tested for the effect of local gain controls in the surround, and whether spatially uniform

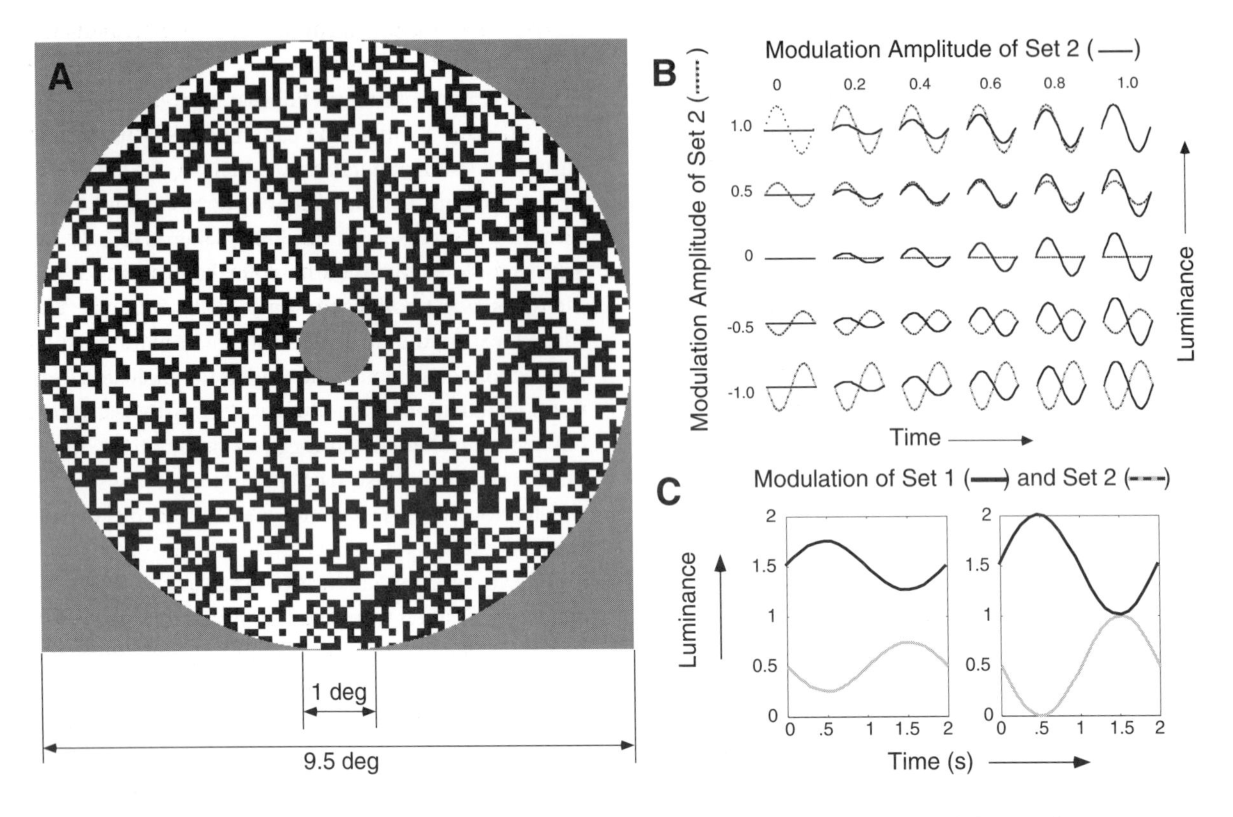

Figure 17.11: (A) Stimulus configuration: spatially uniform, 1-deg disk surrounded by 9.5-deg annulus composed of a binary random texture. (B) The luminance modulations of each set of the surround elements in Experiments 1 and 2. The luminances of both sets of texture elements were modulated sinusoidally at 0.5 Hz. The luminance modulation of one set was set at an amplitude of 0, 0.2, 0.4, 0.6, 0.8, or 1.0, paired with a modulation amplitude of the other set at 1.0, 0.5, 0, –0.5, or –1.0, where a negative sign denotes modulation in the opposite phase. (C) The luminance modulation of each set of the surround elements in time (sinusoidal modulation with frequency of 0.5 Hz) in Experiment 3. The two surround components were modulated with equal amplitudes in counterphase around different mean luminance levels. As a result, the space-averaged luminance of the surround was constant while the spatial contrast was modulated with amplitudes of 0.5 (left panel) or 1.0 (right panel) (Spehar, DeBonet, & Zaidi, 1996).

surrounds had similar brightness induction effects as textured surrounds.

Results from all four experiments, for one observer, are presented in Figs. 17.12A–E. The curves are the best fits from the model described below. For Experiments 1 and 2, the amplitude of nulling modulation is plotted as a function of the amplitude of the modulation of one set with the amplitude of modulation of the other set as a curve parameter. For each modulation level of one set, the magnitude of nulling modulation is a linear function of the amplitude of the paired set. In addition, the five curves for each observer are parallel and equally spaced, indicating that the amplitude of nulling modulation is a linear function of the amplitude of each of the surround sets. These results unambiguously support a model that postulates the spatial summation of lateral effects.

A comparison of Figs. 17.12A–C shows that the amount of required nulling modulation increases as the mean luminance of the test increases. It is important to note that the empirically measured nulling modulation amplitude is plotted, and not the amplitude of the induced modulation. It is well established (e.g., Watson, 1986) that the threshold for the detection of temporal modulation at 0.5 Hz is an increasing function of the mean luminance of the field. Craik (1938) conceptualized this fact in terms of a gain factor for the test modulation, set by the test mean luminance. In Experiments 1 and 2, even if the brightness induced from the surround were independent of the luminance level of the test, because of the gain set by the mean luminance of the test, the amount of real modulation needed to null the induced modulation should increase as a function of the test mean. The results of these experiments imply that local adaptation mechanisms in the test field should be incorporated into a general model of brightness induction.

For Experiment 3, the magnitude of nulling modulation is plotted as a function of the amplitude of contrast modulation. The three sets of symbols in each graph represent the data for the tests at the three different mean luminance levels. Contrast modulation of the surround does not produce any significant brightness induction for tests at the same mean luminance level as the surround (middle points). However, the results for test luminance levels at 0.5 and 1.5 show a nulling modulation of approximately equal amplitude but opposite sign. Phenomenally, this can be described in the following way: The test at a mean luminance of 0.5 appears lighter on the higher contrast surround and darker on the lower contrast surround; the opposite happens for the test at a mean luminance of 1.5. The change in sign of the required nulling modulation as a function of test mean level indicates that brightness induction is not a function of contrast modulation per se, and that the results may be better understood if the inducing effects of the two sets of surround elements are considered separately. A positive sign indicates that the nulling modulation was in the same phase as the modulation of the surround set with the higher mean luminance level in Fig. 17.11C, and a negative sign indicates that the nulling modulation was in phase with surround set of the lower mean luminance. The surround set whose mean luminance is closer to the mean level of the test seems to have a greater inducing effect. In the case where the test level is equidistant from the mean levels of the two surround sets, the induced effects cancel out and there is roughly zero induced modulation. These results suggest that the magnitude of the difference between the luminance level of the test and the mean luminance level of each surround element should be considered in modeling the total induced effect from complex surrounds. The model below postulates that there are pairwise lateral connections between points in the test and the surround, and that the magnitude of the induction signal between them is a decreasing function of the mean luminance difference between them.

For Experiment 4, the magnitude of nulling modulation is plotted as a function of the surround mean luminance level with the test luminance level as a curve parameter. For all three test levels the magnitude of the nulling modulation was the highest when the surround modulation was at the same mean luminance as the test. The magnitude of the nulling modulation decreased monotonically as a function of the difference between the test and the surround mean luminance levels. When the mean luminance levels of the test and the surround were equal (i.e., the highest points for each surround luminance) the magnitude of nulling modulation was approximately constant, thus indicating that the effect of the local gain control set by the surround mean luminance on the inducing signal roughly balances the effect on the nulling modulation of the gain set by the test mean.

To account for their results, Spehar, DeBonet, and Zaidi (1996) generalized the model in the previous section so that the induced effect from each point in the surround is proportional to its luminance attenuated by two gain controls and a spatial weighting function:

$$I(t) = -\int_0^{2\pi} \frac{\int_0^{\infty} W(s)\Gamma_D(\Omega, s)\Gamma_s(\Omega, s)L(\Omega, s, t)s\,ds}{2\pi}\,d\Omega. \quad (11)$$

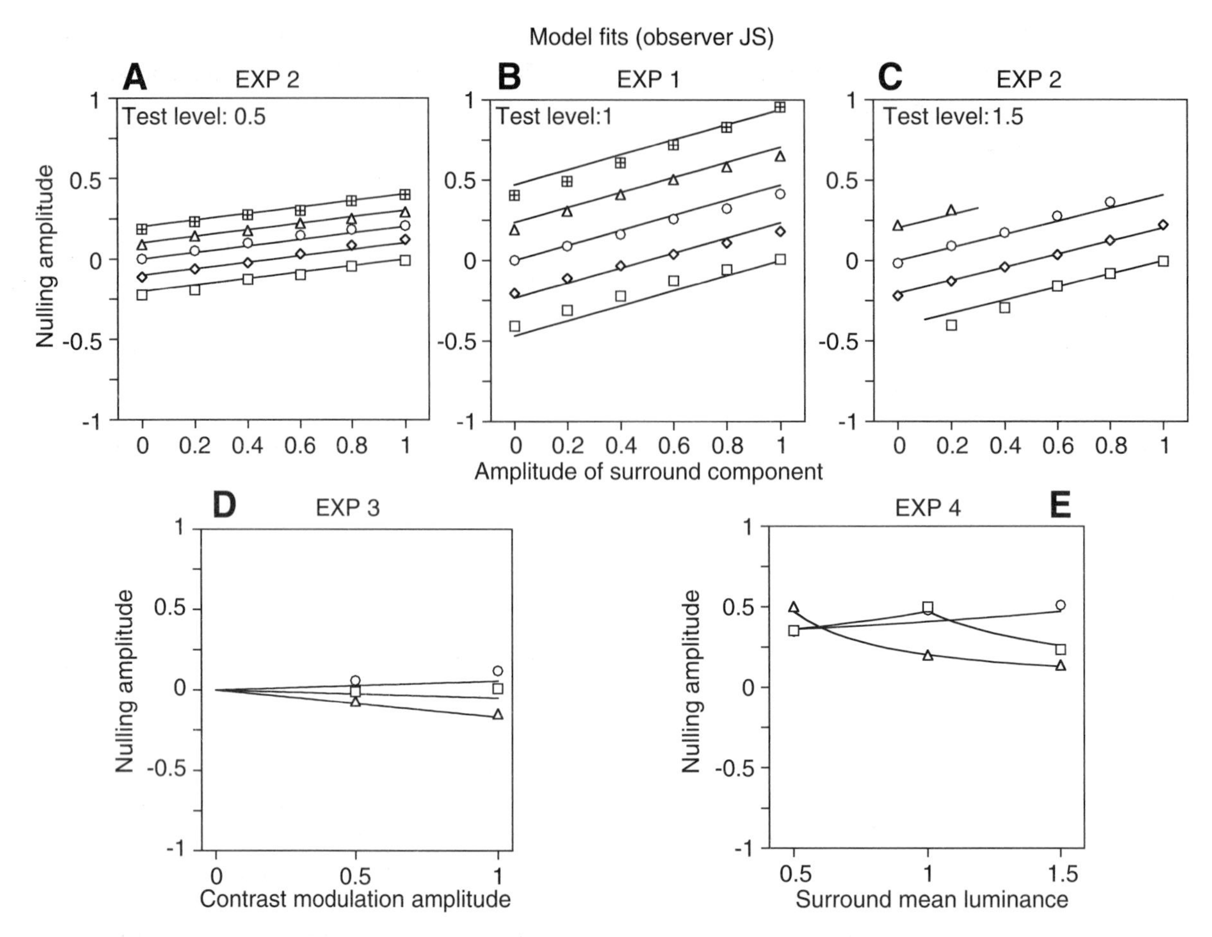

Figure 17.12: (A–C) Results of Experiments 1 and 2. The amplitude of nulling modulation (ordinate) is plotted as a function of the magnitude of the modulation of one set of the surround elements (abscissa) with the amplitude of modulation of the paired set as a curve parameter: (□: −1.0, ◊: -0.5, O: 0, Δ: 0.5, ⊞: 1.0). Each data point is the average of two 2AFC staircases (10 turns each). Test mean luminance equaled 0.5, 1.0, and 1.5 times 25 cd/m^2 for (A)–(C). (D) Results of Experiment 3. The magnitude of nulling modulation (ordinate) is plotted as a function of the amplitude of contrast modulation (abscissa). The curves represent data for tests at three different mean luminance levels. Triangles represent data for the test with a mean luminance of 0.5. Squares represent data for the test with a mean luminance of 1.0. Circles represent data for the test with a luminance level of 1.5. (E) Results of Experiment 4. The magnitude of nulling modulation is plotted as a function of the surround mean luminance level with the test mean luminance level as a curve parameter. Triangles represent data points for the test with a mean luminance equal to 0.5, squares test mean luminance 1.0, and circles test luminance level 1.5. Curves represent the best-fitting predictions of the model (Spehar, DeBonet, & Zaidi, 1996).

$I(t)$ is the total induced effect on the test patch at time t. (Ω, s) are the polar coordinates of a surround point, where Ω is the angular direction in radians and s the spatial distance from the test in degrees of visual angle. $L(\Omega, s, t)$ is the luminance at that point at time t. $\Gamma_s(\Omega, s)$ and $\Gamma_D(\Omega, s)$ are two gain control factors that affect the signals from (Ω, s). $W(s)$ is a negative exponential spatial weighting function of s (Zaidi et al., 1992).

In the model the response of the visual system to a luminance signal at every point is gain controlled by a factor that depends solely on the mean luminance level at that point. Following tradition (e.g., von Wiegand, Hood, & Graham, 1995), hyperbolic gain control functions were used to calculate the gain factor for each point in the surround:

$$(12) \quad \Gamma_S(\Omega, s) = \frac{\gamma_S}{\gamma_S + \int L(\Omega, s, t)dt},$$

where the parameter γ_S is a constant for each observer. By incorporating this gain control, the induction model is able to predict that surround modulation around a low mean luminance level generates more induction than the modulation of the same amplitude around a higher level.

To allow for the possibility that the local gain factor for the centered test field could be different from the surround gain, the factor Γ_C for the test was calculated by:

$$(13) \quad \Gamma_C = \frac{\gamma_C}{\gamma_C + \int L(0, 0, t)dt},$$

where $\int L(0, 0, t)dt$ is the time-averaged luminance of the center of the test and γ_C is a constant parameter for each observer.

The experimental results show that induced modulation depends on the pairwise differences between the mean levels of the test and individual surround points. This was modeled by attenuating the induction from each point by a gain factor set by the absolute difference between the time-averaged luminance at that point and the time-averaged luminance of the test:

$$(14) \quad \Gamma_D(\Omega, s) = \frac{\gamma_D}{\gamma_D + \left|\Gamma_S(\Omega, s)\int L(\Omega, s, t)dt - \Gamma_C\int L(0, 0, t)dt\right|},$$

where the parameter γ_D is constant for each observer. For conditions that include different mean levels, the local adaptation mechanism operating on the test field will influence the effectiveness of the added nulling modulation. Therefore, the true null will be achieved when the real modulation, after being gain-controlled by the test mean level, is equal and opposite to the induced modulation:

$$(15) \quad \Gamma_C N(t) = -I(t),$$

where $N(t)$ is the luminance modulation required to counteract the induction at time t. Thus the complete expression for the null is:

$$(16) \quad N(t) = \frac{1}{\Gamma_C} \cdot \int_0^{2\pi} \frac{\int_0^{\infty} W(s)\, \Gamma_D(\Omega, s)\, \Gamma_S(\Omega, s)\, L(\Omega, s, t) s ds}{2\pi} d\Omega .$$

For the spatial and temporal configurations used by Spehar, DeBonet, and Zaidi (1996), the general model can be simplified. Because the spatial composition of the binary texture in the surround was a uniform random distribution, it is sufficient to consider the effects of identical numbers and distribution of the two sets of surround elements, instead of considering each surround point individually. Therefore, instead of determining an observer's spatial weighting function, it is sufficient to estimate its aggregate effect on each of the two types of surround elements. In addition, the integrals in Eqn. (16) can be replaced by the sum of the independent effects of the two surround sets. Further simplification can be achieved because of the nature of the temporal modulation. The luminances of the two surround components were always modulated sinusoidally with the same frequency, either in phase or in the opposite phase. The model predicts that the induced modulation should also be sinusoidal with the same frequency and in the opposite phase with either one or both of the surround components. Therefore, for these conditions, it is sufficient to describe the inducing and nulling stimuli by just their signed amplitudes of modulation instead of considering instants of the modulating waveform.

As a result, for all of the conditions in the present study, Eqn. (16) can be simplified to predict the amplitude of the required nulling modulation N by the equation

$$(17) \quad N = \frac{w}{\Gamma_C} \sum_{i=1}^{2} \frac{\Gamma_{D_i} \cdot \Gamma_{S_i} \cdot A_i}{2},$$

where w incorporates the effect of the integrated spatial weighting function over the surround, A_i is the signed amplitude of luminance modulation of the ith component, and Γ_{D_i} and Γ_{S_i} are the gain controls that apply to the set i. This simplified model has only four free parameters: the three gain control constants γ_S, γ_D, and γ_C, the spatial weighting parameter w that scales the amplitude of induction for each observer.

The entire set of each observer's data was fit simultaneously with Eqn. (17). Figure 17.12 shows that the model's predictions fit the data extremely well. At all fixed mean levels of test and surround, the model predicts a linear relationship between the amplitude of the modulation of the surround components and the nulling amplitude. It also accounts for the changes in the nulling modulation amplitude due to variations in the mean luminance level of the test. The model correctly predicts the relative amplitude and phase of brightness induction from the contrast-modulated textured surround, and that the magnitude of brightness induction for different luminance levels of uniform surrounds and tests is a monotonically decreasing function of the difference between their luminance levels.

There is a large amount of psychophysical and physiological evidence for the spatially local gain controls used (Shapley & Enroth-Cugell, 1984; Chen, MacLeod, & Stockman, 1987). These adaptation mechanisms are known to occur relatively early in the visual system. The novel suggestion in this model is the pairwise spatially extended gain control on lateral interactions. Since the spatial weighting function for brightness induction falls off steeply as a function of distance from the test, these pairwise connections can be restricted to fairly short distances in retinal or cortical coordinates. A static compressive nonlinearity on these pairwise connections is not a viable alternative to this spatially extended gain control, because the predictions from a static nonlinearity depart significantly from the straight lines required to fit the data from Experiments 1 and 2. Another model for color induction that explicitly considers the role of adaptation mechanisms is by Courtney, Finkel, and Buchsbaum (1995).

This model decouples the inducing signal from a region and the signal induced into that region. The decoupling removes the need to make a recursive model, like Grossberg and Todorovic (1988), and results in computational simplicity. The assumption is that the incoming induced signals affect appearance but would not affect modulation thresholds or the nulling modulation. There is no psychophysical or physiological evidence that the outgoing inducing signal from a point is affected by in-coming induced signals. On the other hand, Spehar and Zaidi (1997a) have shown that the steady luminance level of the surround influences temporal contrast sensitivity only by presenting a contrast pedestal at the edge of the test and does not affect luminance modulation thresholds inside the test. Similar inferences were drawn on the basis of increment threshold measurements by Cornsweet and Teller (1965).

To judge how well the model performs for static displays, the effect of local adaptation on the appearance of the test was incorporated into the model by the equation:

$$(18) \quad P_C = \Gamma_C \cdot C + I_C,$$

where P_C, the predicted perceived gray level, is equal to C, the luminance of the test, multiplied by the gain factor for that luminance level, Γ_C [Eqn. (13)], plus I_C the total induced brightness on the test [Eqn. (11)]. Using the parameters estimated for the time-varying measurements, predictions were made for perceived brightness in Fig. 17.10. The predicted rankings, shown in Table 17.1, differed somewhat between observers, yet they agreed almost perfectly with the actual rankings made by each observer.

The success of the present model shows that perceived gray levels can be predicted in complex achromatic configurations by incorporating the effects of local and spatially extended adaptation mechanisms and linear summation of the induced effects of individual elements of the surround. The model consists of a simple nonrecursive integral equation with the only independent variables being the physical luminances of individual pixels, making it easy to implement for arbitrary, nonfigural, achromatic displays.

Habituation and chromatic induction

The failures of superposition tests for chromatic induction (Fig. 17.9) showed that high spatial frequency chromatic variations inhibit the inducing power of low-frequency surrounds. Since any nonlinear weighting function just multiplies the effectiveness of individual pixels, it is distributive over the addition of sine waves and is not the cause of the failure. Static response nonlinearities were ruled out earlier, and since the superposition tests were equated for the mean levels of all pixels in the display, gain control mechanisms cannot be the cause of the failure either. Even though it is difficult to conceive of hard-wired, nonlinear lateral connections that could lead to this particular failure of superposition, it is always a possibility. An alternative possibility is that high spatial frequency chromatic variations mask low-frequency variations. Some support for this alternative is provided by the study of Zaidi, Spehar, and DeBonet (1997), who showed that adapting to high-frequency chromatic texture will elevate thresholds for detecting low-frequency chromatic changes, but that this does not happen for luminance changes.

They used steady adapting fields subtending 14.14 deg horizontally and 10.63 deg vertically. The fields were covered with random binary and quaternary distributions of uniform-sized squares, 8.52 squares per deg^2 of visual angle. Three types of binary texture were used, which will be termed LD, RG, and YV for mnemonic purposes. Each type of texture consisted of equal numbers of randomly intermixed squares of two different colors, whose chromaticities and luminances were equal to points halfway between *W* and the extreme points on the corresponding cardinal axis. The three types of quaternary textures, LDRG, RGYV, and YVLD, were formed by adding the corresponding pairs of binary textures. For example, the LDRG texture consisted of light-red, dark-red, light-green, and dark green squares. In all six types of textures, the space-averaged chromaticity and luminance were equal to *W*. Thresholds for detecting full-field changes on the textured fields were compared to thresholds for detecting changes parallel to and toward one or the other end of the three cardinal axes, that is, the R, G, Y, V, L, and D color directions, from a uniform achromatic field whose chromaticity and luminance were equal to the space average of the texture. The chromaticity or luminance of all of the pixels of the screen was changed over 3 s as a half-cycle of a sinusoid. To control for criterion effects, each trial also included another interval in which the illuminant was not altered. The observers indicated the interval in which they perceived a color change. The observer adapted to the background for 2 min at the initiation of each session and readapted for 2 s after each trial.

The results for two observers are shown in Fig. 17.13. The chromatic content of the background texture is indicated on the abscissa. The log threshold elevation for detecting a change in each color direction as compared to the baseline threshold is plotted on the ordinate. The results are systematic and similar for the two observers. The presence of chromatic spatial variations makes it less likely that full-field chromaticity changes will be perceived, but thresholds for detection of full-field luminance changes are not affected by the presence of spatial variations. With one exception, changes toward a chromatic direction are affected only when there is spatial contrast along the same axis. There was no systematic effect of superimposing spatial contrast along a color axis orthogonal to the color direction of the simulated illumination change. The results indicate that the masking effect of spatial contrast is relatively independent within each of the two chromatic mechanisms.

The observers in these experiments were instructed to fixate on the center of the screen, but small eye movements are unavoidable when trying to maintain fixation (Carpenter, 1988). If the main effect of eye-movements were integration over space within recep-

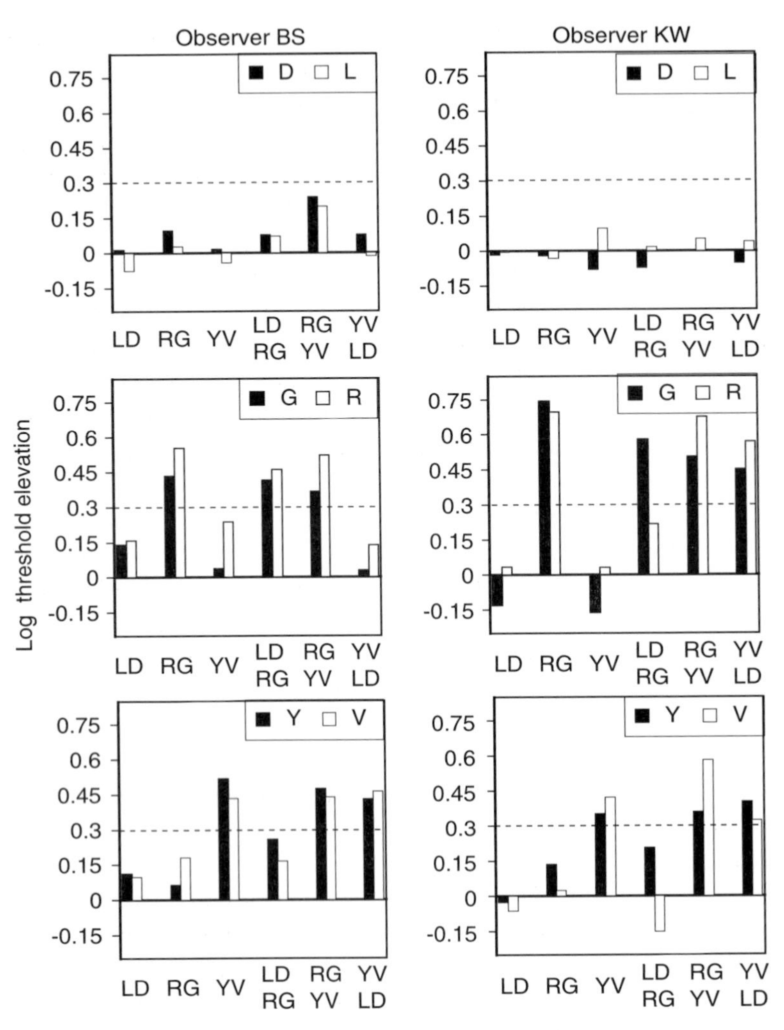

Figure 17.13: Results of adapting to high-frequency texture on full-field luminance and chromatic changes. The log of the threshold for detecting a change in each color direction minus the log of the baseline threshold for that color direction is plotted against the chromatic content of the background texture (see text). Symbols representing the color direction of the test are shown in the insets. A dashed horizontal line is drawn at 0.3 to indicate a doubling of threshold magnitude (Zaidi, Spehar, & DeBonet, 1997).

tive fields, as has been proposed by, for example, D'Zmura and Lennie (1986) and Fairchild and Lennie (1992), the adaptation level would be set by the mean chromaticity and luminance. Since the space-averaged colors of all of the backgrounds were identical to *W*, the presence of threshold elevations on the textured backgrounds rules out spatial integration as a major factor. In our view, however, eye movements lead to transient stimulation of receptive fields at the borders of the squares, thus creating a temporal modulation of stimulation to individual neurons, and prolonged temporal modulation has been shown to cause chromatically selective elevations of thresholds (Krauskopf, Williams, & Heeley, 1982; Zaidi & Shapiro, 1993).

The best clue for explaining why habituating to textures raises thresholds for large-field chromatic changes but not for luminance changes is provided by the study of Krauskopf and Zaidi (1985), which showed that habituating to modulation of a large disk raised thresholds for a concentric smaller disk in the chromatic case but not in the luminance case. Habituation to luminance modulation occurred only when the habituating stimulus shared the edge of the test stimulus. In the visual system, beginning from ganglion cells in the retina, neurons are spatially bandpass for luminance variations and hence are insensitive to variations that are uniform over their receptive fields (Shapley & Lennie, 1985). If detection of the large-field luminance changes on all of the backgrounds occurs at the edges of the display, then habituation to luminance modulations will not alter luminance thresholds. Neurons with receptive fields wholly within the boundary do not participate in the detection of luminance changes at the boundary, and habituation of neurons that are stimulated by eye movements across the boundary will be common to all conditions. On the other hand, since chromatically sensitive cortical neurons are responsive to chromatic variations that are uniform over their receptive field (Lennie, Krauskopf, & Sclar, 1990), large-field chromatic changes are detected inside the boundary and most probably near the fixation point. Habituation of neurons in the central field by eye movements across the internal edges in the texture will therefore raise chromatic thresholds. If the squares are too large, there will be few neurons whose receptive fields oscillate across boundaries; and if the squares are too small, there may be too much integration within receptive fields for there to be a substantial modulation of responses. Therefore, receptive field sizes and amplitudes of eye movements will jointly determine the sizes of the squares that elevate thresholds the most.

Using probe-flash measurements, Zaidi, Spehar, and DeBonet (1998) showed that the effects of adapting to textured fields were qualitatively similar to habituating to prolonged temporal modulations (Shapiro & Zaidi, 1992; Zaidi & Shapiro, 1993). Adapting to high spatial frequency textures, however, elevated thresholds considerably more than habituating to temporal modulations of spatially uniform fields.

There are a few differences between the texture adaptation conditions and the drifting compound sine-wave conditions, and a test of habituation by specific spatial frequencies should be done under drifting conditions. However, the most probable explanation for the spatial subadditivity in chromatic induction is that habituation to high-frequency chromatic variations reduces the total inducing signal from the surround to the test. A number of studies using other methods have shown that adding chromatically different dots or rings to spatially uniform surrounds reduces induction more than by adding stimuli differing in luminance (Shevell & Wesner, 1990; Jenness & Shevell, 1995). For example, Shevell and Wesner (1990) showed that the induced effect of a spatially uniform, red annulus on a central test is more severely attenuated by a thin equiluminant white ring embedded in the surround than by a dark ring. If their results are analyzed in terms of spatial frequency, a thin white ring equiluminant with the red surround adds high-frequency chromatic components, whereas a thin dark ring adds high-frequency luminance components. The results of Shevell and Wesner's study show that adding high spatial frequency chromatic components to a low-frequency chromatic surround attenuates the induced effect considerably more than adding high spatial frequency luminance components, which is consistent with the habituation results above.

Perceptual organization and object segmentation

Figure 17.1 is just one of many pictures in the literature showing a relationship between inferred three-dimensional organization and perceived brightness. Others can be found in Gilchrist (1977, 1980), Knill and Kersten (1991), Adelson (1993), and Sinha and Adelson (1993). In addition, Benary (1924), White (1979), and Zaidi (1990) presented brightness illusions that are incompatible with all extant low-level models of brightness induction, including the ones in this chapter. Inferred perceptual organizations have been suggested as explanations for these illusions (e.g., Benary, 1924; Spehar, Gilchrist, & Arend, 1992; Spehar, Gilchrist, & Arend, 1995; Taya, Ehrenstein, & Cavonius, 1995).

One important point of contention is whether the relevant perceptual organization is in terms of surface properties like appurtenance (Benary, 1924), belonging (Kanisza, 1979), transparency (Adelson, 1993; Taya, Ehrenstein, & Cavonius, 1995; Spehar, Gilchrist, & Arend, 1995), and so on, or whether the important factors are midlevel detections of junctions formed where edges meet in the retinal image. Junctions have been employed in recovering three-dimensional configurations from line drawings by Huffman (1971), Clowes (1971), Sugihara (1984), and Kanade (1980), and they have also been found to be useful in recovering shape from images, especially from raw range data. For example, in Sugihara's (1987) knowledge-guided system for range data analysis, a junction dictionary is used for the extraction and organization of edges and vertices, by consulting it to predict positions, orientations, and physical types of missing edges. These predictions guide the system as to where to search and what kinds of edges to search for, as well as how to label the extracted edges into an interpretation. An analysis of brightness illusions in terms of junctions affecting transparency and depth interpretations has previously been found useful by Adelson (1993) and Pessoa and Ross (1996).

Zaidi, Spehar, and Shy (1997) used three-dimensional configurations and two-dimensional pictures of the configurations to test whether there is a difference in induced contrast from surrounds that differ in perceived depth relationships, for example, background, coplanar, or occluding, but that are equal in retinal adjacency to the test. Each retinal image of a three-dimensional scene can be approximated by a two-dimensional projection, and retinal images on the two eyes are similar for objects at reasonably large distances from the observer. The use of pictures enables concentration on visual cues that are important in inferring three-dimensional organization from retinal images when stereoscopic disparity is not available. By using a variety of orthographic projections they were able to identify visual cues that enhance or inhibit induced contrast independent of the viewing angle and the presence or absence of other depth or perspective cues.

In the pictures in Fig. 17.1, to every observer, the gray test regions seemed to form part of the H, that is, to belong with the vertical bars. Domination of the induced effect by the background thus refutes Benary's notion that induced contrast is enhanced by appurtenance. Since the test regions were seen as parts of coplanar H's, these pictures also provide a counterexample to Gilchrist's (1977, 1980) coplanar ratio hypothesis. Another example where a larger inducing region has less effect despite appearing coplanar and belonging with the test region is shown in Fig. 17.14. The figure consists of a photograph of two figurally identical three-dimensional configurations, each consisting of an L shape in the foreground, a gray square embedded in the crook of the L, and a larger square flap at an angle behind the L. Induced contrast from the background flaps dominates that from the flanking L. In Figs. 17.1 and 17.14, the phenomenal effect was the same whether the three-dimensional configurations were viewed monocularly or binocularly. In terms of a junction-based analysis, both of these figures demonstrate that the largest effect on induced contrast is an attenuation across the edge of the test that is adjacent to the top of each T junction. Zaidi, Spehar, and Shy (1997) found that this rule held for all of the pictures that they tested, irrespective of any inferred surface properties or depth planes. They also showed that this rule could explain White's illusion and all of Benary's illusions.

Figure 17.14: This depiction of the Benary cross illusion refutes explanations that are based on "belonging" or coplanarity (e.g., Benary, 1924), because the L-shaped region has less induced effect despite appearing coplanar and "belonging" with the test region (Zaidi, Spehar, & Shy, 1997).

In a world composed of three-dimensional objects, the top edges of the T junctions are likely to signify occluding or more distant surrounds. In both of these cases, it would be advantageous to increase contrast from proximate backgrounds to facilitate figure-ground segregation. In fact, given the high utility of this segregation, especially for objects at a distance at which there are no other cues to the three-dimensional configuration, it may be useful for the visual system to always function as if T junctions separate the foreground from the background. The cost of illusory brightness differences in perceptually coplanar displays is probably much smaller than the benefits of a simple strategy that enhances figure-ground segregation.

Two pieces of evidence that were already discussed further support Chevreul's notion that the main functional role of color induction is to perceptually enhance marginal figure-ground differences. When the test and surround are at the same mean level, the amplitude functions in Fig. 17.3 show that fractional induced chromatic contrast is greatest for the smallest excursions in the color of the surround. In addition, the results in Fig. 17.12 show that induction accentuates differences most from those elements of the surround that are at luminance levels closest to the mean level of the test.

Figures 17.1 and 17.14 show that figural cues have to be factored into any adequate explanation of color induction. It is possible that incorporating the influence of midlevel perceptual cues like T junctions on spatial summation and lateral gain controls will eventually lead to mechanistic models that can explain brightness induction in natural settings.

Induction and color constancy

In most settings, even though changes in the spectral reflectances are possible (Nassau, 1983), particularly for living organisms (Zaidi & Halevy, 1993), in general a change in perceived colors in objects is going to be due to a change in illumination. A visual system can be said to possess the property of color constancy if the color percepts assigned to individual objects are invariant across illumination conditions. In terms of the responses of neurons or signal processing units, color constancy results if, at some stage of the visual system, the neurally transformed signals from objects in a scene vary by less than a discriminable difference across varying illuminations (Ives, 1912b). The logical first step in an analysis of color constancy is to specify the problem in terms of changes in neural signals caused by illumination changes.

In Fig. 17.15 are shown the excitations of the three cone classes and the cardinal mechanisms from each of 170 natural and man-made objects (Vrhel, Gershon, & Iwan, 1994) under zenith skylight (Z) and direct sunlight (T) (Taylor & Kerr, 1941). The spectrum labeled Z was measured by pointing the measuring instrument at the sky; of all of the phases of sunlight, it has the highest relative energy in the short wavelengths due to Rayleigh scattering. Direct sunlight at ground level (T) has the least relative energy in the short wavelengths and most in the long wavelengths. The plots in Fig. 17.15 represent the extreme case of comparing signals from an object under direct sunlight to that same object shadowed from the sun and reflecting pure skylight. In each panel, the solid line along the diagonal is the locus of equal signals under the two illuminants. Each point represents an individual object. The open circles at the top right corners in each of the L, M, and S panels represent direct cone absorptions of the two daylights. The most noticeable aspect of all three of the top panels is that the points lie close to straight lines, that is, there is a strong correlation ($r^2 > 0.998$) between the quanta absorbed by each cone type from different objects across daylight illumination changes. Comparable correlations between cone absorptions from other samples of objects and illuminants were found by Dannemiller (1993) and Foster and Nascimento (1994).

In the bottom left panel, the L/(L+M)-axes represent hues going from greenish to reddish (left to right and bottom to top). The points representing the objects all lie below the diagonal, indicating that the chromaticities of all of the objects have shifted toward green under illuminant Z as compared with under illuminant T. The S/(L+M)-axes represent hues going from yellowish to violet (left to right and bottom to top). The

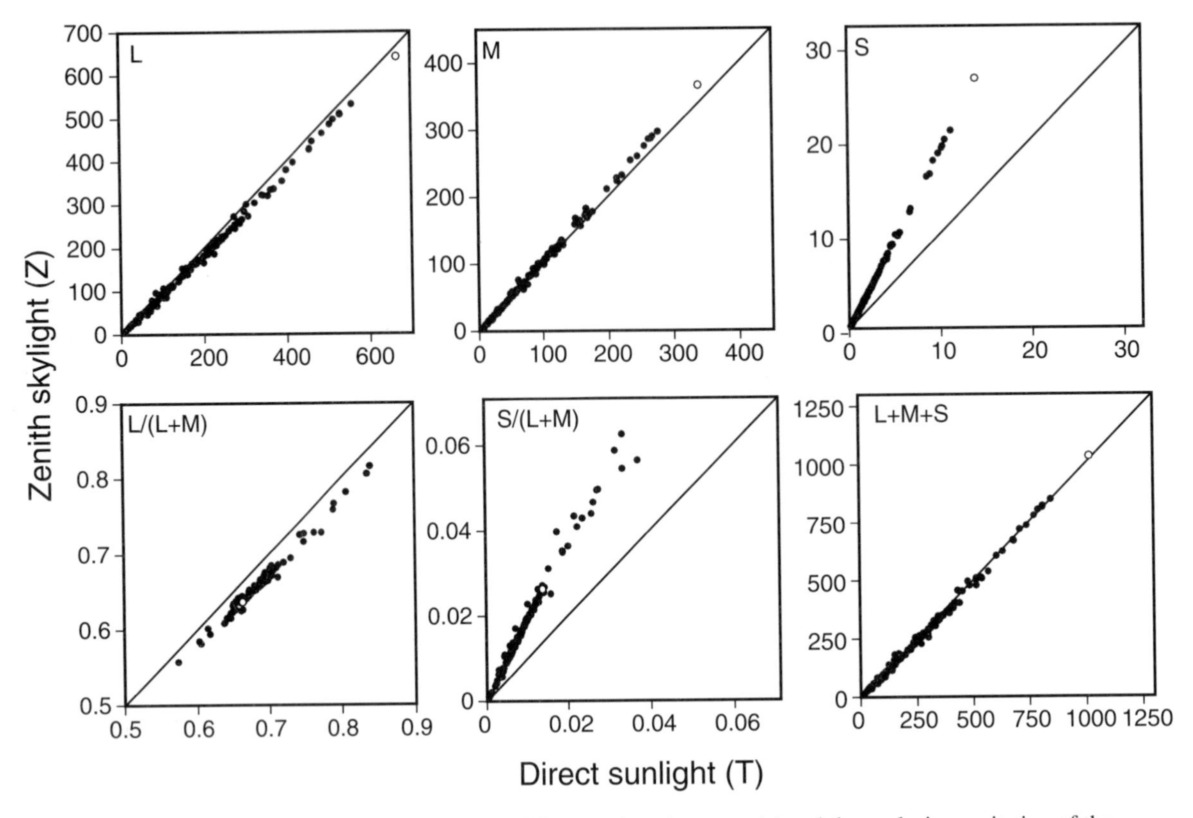

Figure 17.15: The excitation of the L-, M-, and S-cones (top three panels) and the exclusive excitation of the chromatic and luminance mechanisms along the RG, YV, and LD cardinal axes (bottom three panels) from each of the 170 objects from Vrhel, Gershon, and Iwan (1994) under illuminants T (abscissa) and Z (ordinate). The open circles represent the illuminants (Zaidi, Spehar, & DeBonet, 1997).

effect of an illuminant change is a shift in all of the chromaticities by approximately the same multiplicative factor, indicating that the chromaticities at the violet end are shifted the most. The L+M+S panel represents radiance changes at constant hue and saturation, and is dominated by L- and M-cone absorptions. Because the daylight spectra were equated for illuminance, a shift from T to Z causes almost no change in the total cone absorptions.

The illuminant-caused shifts in chromatic signals correlate well with everyday observations. Grass and bricks look appreciably bluer in shadows and yellower in sunshine. More systematic documentation is available in paintings made in the open air from the second half of the nineteenth century. A good example is Corot, who first used to sketch the scene and then painted it patch by patch, reproducing colors in each patch. Corot's painting, "Island of Saint Bartollemeo," shows walls of buildings in sunlight and shade. In cases where a wall is both lighted and shadowed, only on the shadowed bricks can be found traces of blue pigment. Da Vinci had remarked that a white dress appeared blindingly bright on the side turned toward the sun, but bluish on the other side. Similarly, Delacroix (1937) wrote: "The true colour of the flesh can be seen only in the sun and in the open air. If a man puts his head out of a window, its colouring is quite different from what it is indoors. Which shows the absurdity of studies done in a studio, where each one does his best to reproduce the wrong colour."

Given the above observations, the preoccupation with color constancy is perplexing. However, since the colors of objects depend not only on physical absorptions, but also on neural transformations (Monge, 1789; Chevreul, 1839; Ives, 1912b), it is worth asking whether any neural transformations help in discounting the effect of illumination changes. Because the shifts in Fig. 17.15 are systematic, there is a chance that the human visual system can attenuate their perceptual effects through the use of simple adaptation strategies without having to estimate reflectance or illumination spectra, as in the schemes proposed by Maloney and Wandell (1986) and D'Zmura and Iverson (1993a; b). An adaptation process that modifies cone signals so that they fall along the diagonals in the top panels of Fig. 17.15, or modifies second-stage signals so that they fall along the diagonals in the bottom panels, will lead to color constancy. As a corollary, any empirically measured color constancy will be consistent with models involving either type of these mechanisms, and to identify whether cone or second-stage adaptation mechanisms are responsible for constancy will require additional experiments like those done by Pugh and Mollon (1979) or Zaidi, Shapiro, and Hood (1992).

Historically, computational schemes for human color constancy have involved early adaptation mechanisms. In terms of our linking hypothesis, color constancy would be achieved at an early stage if neural processes equated first- or second-stage signals from each object in a scene across illumination conditions. As a result of this processing, the outputs of at least the second-stage mechanisms (and possibly even the first stage) should be transformed under each illuminant in a manner that, when plotted similar to Fig. 17.15, all of the points should fall on the diagonal of unit positive slope. The simplest mechanism that has been proposed for accomplishing this purpose is Von Kries's adaptation (Ives, 1912b; Brill, 1995), where each photoreceptor signal is gain-controlled by its own time-integrated signal, that is. for each object the signals (L, M, S) are transformed to:

$$(19) \quad \left(\frac{L}{\int Ldt/L_E}, \frac{M}{\int Mdt/M_E}, \frac{S}{\int Sdt/S_E}\right).$$

For a steady uniform field, the value of each integral is equal to the cone absorption from that field, and the ratio of cone absorptions for the integrated value is transformed to be equal to the ratios for an equal-energy light (L_E:M_E:S_E). This transformation could thus provide a simple explanation for the progressive desaturation of the perceived color of a continuously viewed, uniformly colored field (Vimal, Pokorny, & Smith, 1987). In his numerical simulations of color constancy, Ives (1912b) assumed that the integral for each photoreceptor was equal to the quantal absorptions by that class of receptors from the steady illuminant. Thus the result of the transformation was to make the illuminant appear achromatic. When Ives's assumption was applied separately for the two illuminants to signals from each object in Fig. 17.15 by Zaidi, Spehar, and DeBonet (1997), the result was rigid rotations of the lines between (0,0) and the open circles representing the illuminants in the S, M, and L panels, in a manner that the open circles were shifted to the unit diagonals. Since all of the points representing individual objects lie on or close to these lines, the transformed chromatic signals from individual objects were also fairly well equated across the illumination conditions, thus predicting color constancy. However, there are a number of conceptual problems in accepting this transformation as an explanation of human color constancy. First, the values of the integrals for free viewing of a variegated scene are difficult to predict a priori. In viewing a variegated scene, the ratios of time-integrated cone absorptions will only be equal to the absorption ratios of the illuminant spectrum if the integrated object reflectance spectrum for each photoreceptor is uniform, a condition that is unlikely for most natural scenes, even with the spatial averaging of reflectances due to active scanning (Brown, 1994). In reality, the spatially local values of the integrals will vary across the visual field; and to the extent that the gain for each photoreceptor is set by the spatially local signal that it receives from the particular region imaged on it, the transform in Eqn. (19) will shift the chromaticity of that object toward the achromatic point. A realistic version of this transform will thus not lead to color constancy. The second problem has to do with the stage in the visual system that is important for color constancy transformations. It is difficult to imagine why an equal-energy light should have a privileged status for an individual photoreceptor, that is. there is no theoretical justification for the L_E, M_E, and S_E terms in the denominators of Eqn. (19). On the other hand, in color-opponent cells, the achromatic signal can have a privileged position as the zero point toward which the response of the system is shifted by a high-pass temporal filter after opponent combination. However, in a variegated scene, discounting the integrated values of opponent signals creates problems similar to those discussed in the context of the integrals in Eqn. (19). Since it is unlikely that the integrals of the opponent signals will be proportional to the values from the illuminant, spatially local adaptation will shift all colors toward the achromatic point. This would be consistent with and an alternative explanation for the progressive desaturation of a colored scene that is stabilized on the retina, but similar to local photoreceptor adaptation it could only equate chromatic signals across illuminants at the cost of losing all perceived color differences in the scene. Third, and most importantly, the empirical results reviewed in this chapter show that when viewing a variegated field neither the appearance of colors nor the limen of discrimination are determined by the space-averaged level of stimulation. Consequently, models of color adaptation or constancy that rely on spatial- and/or temporal-integrated levels as the controlling parameters (Judd, 1940; West & Brill, 1982; Land, 1983; Worthey, 1985; Dannemiller, 1989; Brill, 1990; Valberg & Lange-Malecki, 1990; Brainard & Wandell, 1992; Finlayson, Drew, & Funt, 1993, 1994; Finlayson & Funt, 1996) may be consistent with some sets of empirical data, but are not sufficient to explain the results of experiments that isolate individual color mechanisms and adaptation processes (Zaidi, Spehar, & DeBonet, 1997).

It has often been proposed that color induction can lead to color constancy (e.g., Valberg & Lange-Malecki, 1990; Walraven et al., 1991). This assertion has usually been based on the results of studies that measure perceived shifts in colors of just one test patch rather than over the whole scene, and it seems irreconciliable with the finding that juxtaposing two patches shifts their appearances in complementary color directions (Chevreul, 1839; Krauskopf, Zaidi, & Mandler, 1986). Simultaneous color induction will shift the signals from juxtaposed objects in opposite directions, and therefore we cannot discount the effect of an illumination change by shifting signals from all objects in a scene in a correlated fashion, as, for example, toward the diagonals in Fig. 17.15. In some cases it is possible that induced contrast will counter a shift in the spectrum of the illuminant. This discounting is most likely to occur for unsaturated hues that are surrounded by more saturated hues. For saturated hues, the induced shift is more likely to be in a direction that exacerbates the effect of the illuminant change. Constancy of the appearance of an individual test patch could therefore be due to color induction but is unlikely to be a good measure of color constancy over the extent of a variegated scene. This objection applies particularly to methods that measure the achromatic loci of a test patch under different illuminants (e.g., Brainard & Speigle, 1994).

Zaidi (1998) argued that, since objects do appear to be of different colors under different illuminants, the nervous system could potentially derive information that aids in the recognition of both objects and illuminants by comparing the altered appearances of objects under different illuminants. Figure 17.15 shows that changes in the spectral composition of the illumination on a set of objects lead to affine transformations of the set of object chromaticities. Affine transformations have well-defined invariants, and these invariants can be used to derive the transformation parameters. This was accomplished by two shape-alignment types of algorithms that succeeded in identifying objects with identical reflectance functions, and also derived the relative chromaticities of the two illuminants. Because information about objects and illuminants is useful in many different tasks, it would be more advantageous for the visual system to use such algorithms to extract both sorts of information from retinal signals than to automatically discount either at an early neural stage.

Neural mechanisms of color induction

To paraphrase Koffka (1963), the central questions of visual perception are why things look the way they do, and why they look the way they are. Induced contrast and its contribution to appearance are broad enough topics, that this chapter presents only one of a number of essentially nonoverlapping treatments (compare, e.g., Gilchrist, 1994). The construction of appearance by the visual system involves memory, hard-wired perceptual priors, and cognitive inferences (Griffiths & Zaidi, 1999) as well as successive stages of neural transformations of photoreceptoral signals. We have concentrated on developments from the last decade that have led to progressively more general mechanistic models that delineate processes occurring at various stages of the visual system.

The result that color induction is due to lateral connections neither at the cone nor at the opponent mechanism level, but rather at the level of mechanisms tuned preferentially to many different color directions, implicates the visual cortex as the site of these interactions. It seems not to be generally recognized that human observers have exquisite memory for colors (Sachtler & Zaidi, 1992) and are intolerant of spurious hue changes (Li & Zaidi, 1997), and that such changes would occur frequently if appearance was determined by only two independently adapting color-opponent mechanisms.

The fact that Mach bands do not contribute substantially to the induced effect on spatially extended tests also points to a postretinal site, as do the estimated brightness induction space constants of 0.74 and 0.29 deg, centered on the fovea, for two observers. These estimates, however, are smaller than the space constants of 1.23 and 0.74 deg estimated for contrast–contrast induction for the same observers by DeBonet and Zaidi (1997).

The spatial summation of brightness induction is interesting because it indicates suprathreshold spatial additivity across a fairly large region of visual space. Early stages of the visual system are composed of neurons that have small receptive fields and fairly narrow bandpass sensitivity in the spatial frequency domain (Shapley & Lennie, 1985). In addition, psychophysical results are consistent with a small number of classes of spatial frequency selective mechanisms functioning independently at threshold (Graham, 1989). These mechanisms are generally thought to have spatial frequency bandwidths that are less than one octave wide. In brightness induction, the spatial additivity covered spatial scales over a range of six octaves, from 0.05 to 3.2 cycles/deg. Thus, at some stage of the visual system that is important for the computation of brightness contrast, the outputs of these mechanisms must be summed in a fairly simple manner to lead to the point-by-point additivity of lateral effects.

The spatial summation results may lead one to think that brightness induction could simply be the outcome of processing by neurons with large center-surround receptive fields that add or average the brightness of the surround. However, the results of induction from textured surrounds rule out this possibility and instead require gain controls on the connections between neurons covering spatially circumscribed areas. The effects of figural inferences on induced contrast also argue against simple lateral inhibition over extended cortical areas. In the present state of knowledge about visual neuro-physiology, it is not possible to even speculate about possible neural mechanisms for extracting T junctions or other figural features. It is clear, though, that perceived brightness and color are computed from the retinal image in conjunction with other perceptual attributes.

Acknowledgments

This work was supported by NEI grant EY07556. Michael Shy helped with the production of this manuscript.

18

Chromatic detection and discrimination

Rhea T. Eskew Jr., James S. McLellan, and Franco Giulianini

The modern history of the study of chromatic discrimination begins with the work of Yves LeGrand (1949/1994). Using MacAdam's (1942) chromatic discrimination ellipses, LeGrand showed that, for these constant-luminance conditions, a large fraction of the variation among the ellipses could be explained by considering two dimensions: one in which the short-wave–sensitive or S-cone signals varied alone, and another in which the long- and middle-wave–sensitive cone signals traded off against one another at constant sum. Rodieck (1973) performed a similar analysis, and Boynton, in a series of papers (Boynton & Kambe, 1980; Boynton, Nagy, & Olson, 1983; Boynton, Nagy, & Eskew, 1986) collected new data in support of ideas similar to LeGrand's. In a very influential paper, Krauskopf, Williams, and Heeley (1982) referred to these two dimensions, plus one along which luminance varied, as the "cardinal directions" of color space. In this chapter we continue down the path that LeGrand started, analyzing chromatic detection and discrimination in terms of cone signals. We will emphasize the representation of stimuli in cone-contrast space and focus on the roles of "first-site" or cone-specific adaptation and "second-site" or cone-nonspecific desensitization.

Color spaces

We begin by defining an important direction in color space: the direction in which all three cone signals are increased or decreased in the same proportion to their adapted baseline level of activity. One stimulus that lies along that direction is a "radiance modulation." Start with some light to which the observer is adapted, with a spectral distribution given by $E(\lambda)$. Alter its radiance by the same multiplier $(1 + k)$ at every wavelength $E'(\lambda) = (1 + k)\,E(\lambda)$. This modulation produces a shift up or down in the spectrum of the light without altering its shape on a log scale. A radiance modulation mimics the effect of shadowing by some object in a natural scene in which (to a crude first approximation) the shadowing reduces the radiance of incident light by the same factor across the visible spectrum. A radiance modulation is an example of what we will refer to as an "equichromatic" stimulus: It produces no change in chromaticity from the background (adapting) stimulus. We define a chromatic detection mechanism by exclusion, as a detector that does *not* respond to a pure equichromatic stimulus. Defined in this way, chromatic mechanisms are insensitive to modulations in overall light level (such as produced by shadows) but are *potentially* sensitive to any other stimulus. As will be discussed more fully below, in tristimulus color spaces such as the CIE or cone-excitation spaces, there are infinitely many equichromatic directions; in cone-contrast space there is only one.

Cone-excitation space. Lights may be represented as vectors in several different three-dimensional spaces. The first of these is the cone-excitation space

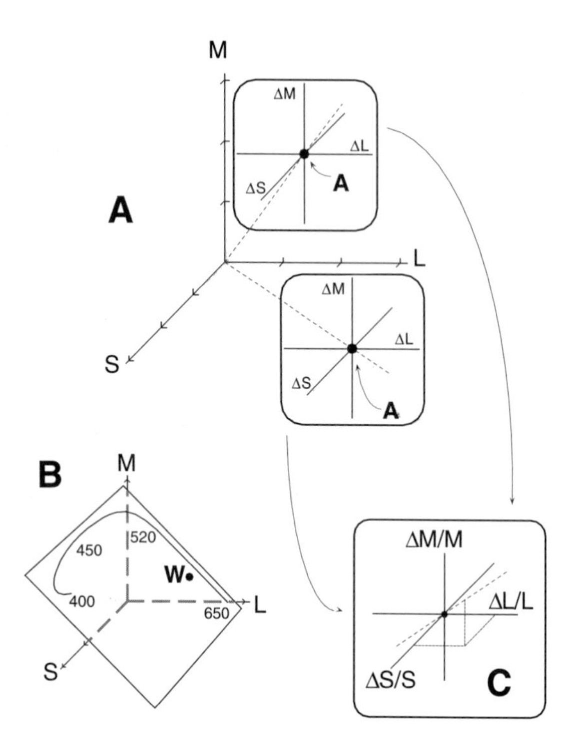

Figure 18.1: Color spaces. (A) Cone excitation space. The small coordinate axes represent local cone-excitation coordinates centered on the adapting condition (solid dot) at two different positions in cone space. The dashed lines connecting the origin with the adapting light are equichromatic directions. (B) The MacLeod–Boynton chromaticity plane located within cone space. Photopic luminance is constant in this plane and planes that are parallel to it. *W* represents an equal-energy white. (C) Cone-contrast space resulting from dividing the local cone-excitation coordinates (e.g., ΔL) by the corresponding cone space coordinate (e.g., L) of the adaping field. All adapting fields plot at the origin in cone-contrast space, and the equichromatic direction is $\pm\{1,1,1\}$ (dashed line) for all adapting chromaticities.

(Fig. 18.1A). The three axes of this space are proportional to the quantal catch rates in the three photoreceptor classes (L, M, S), and thus this space is a linear transformation of a color-matching space such as the CIE tristimulus space. The popular MacLeod–Boynton chromaticity diagram (MacLeod & Boynton, 1979) is a plane within this space (Fig. 18.1B). The cone-excitation space is most useful for specifying the adaptive state of the visual system, given by the vector of cone excitations produced by the adapting light: $\mathbf{A} = \{L_a, M_a, S_a\}$. The filled dots in the two insets to Fig. 18.1A represent the tips of two different adapting vectors. To represent test modulations around the adapting field, we subtract the adapting vector to make local cone-excitation coordinates ($\Delta L = L_{test} - L_a$, $\Delta M = M_{test} - M_a$, $\Delta S = S_{test} - S_a$). Here each of the three axes represents the change in cone excitation from the adapting stimulus (insets in Fig. 18.1A). A test stimulus that is *added* to an adapting field, such as is used in an increment threshold spectral sensitivity experiment, will be represented in the first octant of a local cone-excitation space, because none of the three cone signals is reduced below their adapted level by such a test light. The influential space used by Derrington, Krauskopf, and Lennie (1984) is a linear transformation of the axes of local cone-excitation space, so that they no longer lie along the cone-excitation coordinates but are along the cardinal directions mentioned above: ($\Delta L - \Delta M$, ΔS, and $\Delta L + \Delta M + \Delta S$).

The cone fundamentals used to define cone spaces are a special set of color matching functions (see Chapter 2). These color matching functions are arbitrarily scaled, so distances and angles in either of these cone-excitation spaces have no extrinsic meaning. However, it is conventional to define the units along the axes in luminance terms. The Smith–Pokorny (1975) cone fundamentals are designed such that the L- and M-cone fundamentals sum to a modified version of the CIE photopic luminosity function, $V(\lambda)$ (Judd, 1951; Vos, 1978). Boynton and Kambe (1980) partitioned the luminance of the light, in cd/m^2 or trolands (Td), into L- and M-components (e.g., L Td and M Td), such that L+M is the luminance of the light. Because the S-cone fundamental is not a component of $V(\lambda)$ in the Smith–Pokorny system, S-cone excitation must be defined differently: Boynton and Kambe chose a new unit that they called a "blue troland" (which will be referred to here as an "S troland") that they defined as the S-cone excitation produced by 1 Td of equal-energy white. A test presentation at constant luminance [as defined by $V(\lambda)$] requires that $\Delta L = -\Delta M$ in the local cone-excitation space; there is no such restriction on ΔS.

Cone-contrast space. The representation of most interest to us is the cone-contrast space (Fig. 18.1C). The first explicit use of cone-contrast space was by Noorlander and Koenderink (1983), although the basic ideas that underlie this representation are much older. In this space, the local cone-excitation coordinates of Fig. 18.1A (e.g., ΔL) are divided by the components of the adapting vector (L_a) to produce contrasts ($\Delta L/L_a$ or simply $\Delta L/L$) that mimick the sensitivity scaling effect of adaptation occurring in cone-specific pathways. Because contrasts are dimensionless, questions about the relative scaling of cone fundamentals are immaterial.

In cone-contrast space, like local cone-excitation space, the origin always represents the adapting condition. Note that, in general, cone-contrast space is *not* linearly related to cone-excitation space. The denominators of the cone contrasts are the coefficients of the transformation from the cone-excitation space to the cone-contrast space; and if the adapting state is not held constant, the coefficients are not constant. Thus a set of data collected under various adapting conditions will be nonlinearly mapped into cone-contrast space (there is a different, linear, mapping for each adapting condition, making the transformation of the entire set nonlinear). If the adapting state is constant, there is only a single, linear relationship.

In cone-contrast space, a radiance modulation by the factor $(1 + k)$ produces a cone-contrast triplet of $\{\Delta L/L, \Delta M/M, \Delta S/S\} = \{k, k, k\}$ (with a conventional luminance contrast of k) that is independent of the adapting condition. Any other means of producing equal cone contrasts (using stimuli that are metameric to a radiance modulation) also produces cone-contrast vectors along this "main diagonal" direction (dashed line in Fig. 18.1C). Thus, in cone-contrast space there is a fixed equichromatic direction. On the other hand, in cone-excitation space an equichromatic modulation moves along the line connecting the origin of the space to the adapting field [radiance modulation by $(1 + k)$ produces the vector $(1 + k)\ \{L_a, M_a, S_a\}$] and so is different for each adapting field (dashed lines in Fig. 18.1A). Just as each of the infinite number of adapting vectors in cone-excitation space maps to the origin of the cone-contrast space, each of the infinite number of equichromatic directions in cone-excitation space maps to the same $\pm\{1,1,1\}$ direction in cone-contrast space. As in the cone-excitation space, an incremental test light will be represented in the first octant of cone-contrast space. For calculating cone contrasts, see Cole and Hine (1992); Brainard (1996) provides a useful review and comparison of cone-contrast space and the Derrington, Krauskopf, and Lennie (1984) version of the cone-excitation space.

In the remainder of this chapter we use cone contrasts to represent stimuli. It is important to remember that most conclusions drawn in cone-contrast space apply to local cone-excitation space as well – they are simple linear transformations of one another – as long as the adapting state is held constant; however, across different adapting states the relationship is nonlinear.

Chromatic detection

Linear chromatic detection mechanisms. Classical opponent-color mechanisms are often modeled as approximately linear combinations of cone signals. For example, a red/green detection mechanism might be represented as:

$$(1) \qquad RG = \gamma_{RG}\,(W_{RG,L}\,\Delta L/L + W_{RG,M}\,\Delta M/M + W_{RG,S}\,\Delta S/S) + N(0, \sigma_{RG}) = \gamma_{RG}\,(\mathbf{W}_{RG} \cdot \mathbf{x}) + N(0, \sigma_{RG}),$$

where $\mathbf{W}_{RG} = \{W_{RG,L}, W_{RG,M}, W_{RG,S}\}$ and $\mathbf{x}^{\mathrm{T}} = \{\Delta L/L, \Delta M/M, \Delta S/S\}$.

RG is the number representing the mechanism's response (with $RG > 0$ meaning "green," arbitrarily); $\Delta L/L$, $\Delta M/M$, $\Delta S/S$ are the contrasts produced by the test stimulus in the three cone classes; $W_{RG,L}$, $W_{RG,M}$, and $W_{RG,S}$ are weights that may be positive or negative; and γ_{RG} is a positive number representing the gain of the opponent mechanism. N refers to the probability density of a stochastic process, with zero mean and standard deviation σ_{RG}, that is added to the mechanism to model all of the noise sources that corrupt its signal, viz., quantum light fluctuations, neural noises in cone-independent pathways, and neural noises at stages after cone signals have been combined. If we let $\overline{RG} = \gamma_{RG}\,(\mathbf{W}_{RG} \cdot \mathbf{x})$ be the deterministic portion of Eqn. (1), then $\overline{RG}$ is the mean of the mechanism response to a particular stimulus, and on any given presentation the expected response will differ from that mean by an amount determined by σ_{RG}. Defined thusly, $RG = N(\overline{RG}, \sigma_{RG})$.

The vector of weights $\mathbf{W}_{RG}$ will be referred to here as the "mechanism vector." It points in the direction of maximum (positive) responsivity of the mechanism in cone-contrast space (Fig. 18.2) and is required to have unit length (so only two of the three weights are independently determined). The stimulus is represented by the vector of cone contrasts that it produces, $\mathbf{x}$. The response of this hypothetical linear mechanism to any stimulus vector $\mathbf{x}$ is proportional to the projection of $\mathbf{x}$ onto the mechanism vector; because the mechanism response is the dot product of the stimulus and mechanism vectors, the strength of the response falls off as the cosine of the angle in color space away from the mechanism vector direction (Derrington, Krauskopf, & Lennie, 1984).

In the context of a linear model it is natural to define threshold as the length of a stimulus vector $|\mathbf{x}| = [(\Delta L/L)^2 + (\Delta M/M)^2 + (\Delta S/S)^2]^{0.5}$ producing a constant criterion response, θ. When the stimulus lies along the mechanism vector direction we define the threshold of the *mechanism*, T, to be

$$(2) \qquad |\mathbf{x}| \equiv T_{RG} = \frac{\theta_{RG}}{\gamma_{RG}}.$$

The set of all threshold vectors lies on two planes that are perpendicular to the mechanism vector, since these all produce the same projection onto the mechanism vector (Fig. 18.2); the planes represent the two response polarities. T is thus the smallest of the possible cone-contrast thresholds – the shortest Euclidean distance from the origin of the cone-contrast space to the mechanism's threshold plane. Similarly, we may define the yellow-blue detection mechanism as

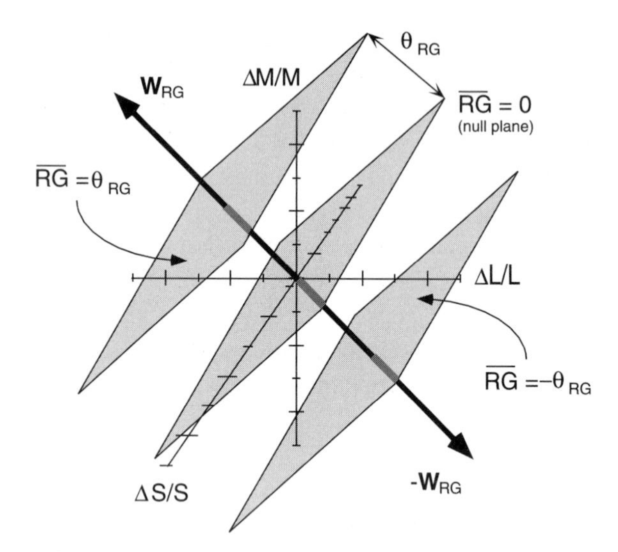

Figure 18.2: A mechanism vector $\mathbf{W}_{RG}$ and its negative in three-dimensional cone-contrast space. $\mathbf{W}_{RG}$ points in the direction of maximum responsivity of the *RG* detection mechanism. The three planes, all of which are at right angles to $\mathbf{W}_{RG}$, represent constant-response planes for this hypothetical mechanism. The central plane is the null plane of the mechanism; stimuli lying in this plane produce no response in *RG*. The two outer planes are threshold planes. T_{RG} is the shortest distance from the origin to a threshold plane, and thus it represents the stimulus that most efficiently stimulates the mechanism – the mechanism threshold.

$$(3) \qquad YB = \gamma_{YB}\,(\mathbf{W}_{YB} \cdot \mathbf{x}) + N(0, \sigma_{YB}) = N(\overline{YB}, \sigma_{YB}),$$

where $\mathbf{W}_{YB} = \{W_{YB,L}, W_{YB,M}, W_{YB,S}\}$ and $\mathbf{x}^T = \{\Delta L/L, \Delta M/M, \Delta S/S\}$, and we let "blue" be represented by $YB > 0$. For the mechanisms to span the cone-contrast space efficiently, the two mechanism vectors $\mathbf{W}_{RG}$ and $\mathbf{W}_{YB}$ will be at right angles, but there is no a priori restriction that this be so. Although we are using color names (e.g., *RG* and *YB*) to denote these mechanisms, we emphasize that this is a model of chromatic discrimination and detection, not of color appearance.

Additional mechanism vectors would be required to represent additional linear mechanisms in cone-contrast space. For instance, a third mechanism vector could be defined for the achromatic or luminance mechanism (*Lum*). The formalism would be identical for this nonopponent mechanism; we will let its mechanism vector be $\mathbf{W}_{Lum} = \{W_{Lum,L}, W_{Lum,M}, W_{Lum,S}\}$ and its gain be γ_{Lum}. If there are *broadband*, "higher order" chromatic mechanisms (see Chapters 13 and 16), these would be represented by still more linear mechanisms.

Assuming that the observer is unbiased and therefore sets detection criteria at $RG = 0$ and $YB = 0$, the detection probabilities in a 2AFC task are given by:

$$(4) \qquad P_{RG} = \int_0^{\infty} N(\overline{RG}, \sigma_{RG}/\sqrt{2})\, d(RG)$$

$$P_{YB} = \int_0^{\infty} N(\overline{YB}, \sigma_{YB}/\sqrt{2})\, d(YB)\,.$$

The standard deviations are divided by $\sqrt{2}$ because we are considering two alternative forced-choice (2AFC) tasks; in a single-interval procedure such as a Yes–No experiment, the unmodified standard deviations would be used (Green & Swets, 1974). We will define the 2AFC threshold to be where P_{RG} or P_{YB} equals 0.816 (because of its convenience when using the Weibull psychometric function). Intuitively, one may think of a univariate noise distribution N centered on the origin in cone-contrast space; the signal + noise distribution is a copy of N that has been moved outward, sliding along the mechanism vector direction, by the distance $\mathbf{W} \cdot \mathbf{x}$. Threshold is reached when 81.6% of the distribution has been moved past the origin. [Actually, N is a multivariate probability distribution with standard deviations ($\sigma_{RG}, \sigma_{YB}, \sigma_{Lum} \ldots$) taken in each of the mechanism vector directions.]

The weights $\{W_{RG,L}, W_{RG,M}, W_{RG,S}\}$ and $\{W_{YB,L}, W_{YB,M}, W_{YB,S}\}$ [Eqns. (1) and (3)] are interpretable as relative gains, in signal-to-noise or d′ units. For example, suppose that $\gamma_{RG} = 100$ and $W_{RG,M} = 0.2$ [Eqn. (1)]. This would indicate that an M-cone–isolating test stimulus with contrast of +5% would produce $100 \times 0.2 \times 0.05 = 1$ d′ unit of RG signal in the "greenish" direction. If we assume that the noise N is Gaussian, Eqn. (4) implies that at the 2AFC threshold,

$$(5) \qquad \frac{|\overline{RG}|}{\sigma_{RG}/\sqrt{2}} = \sqrt{2}\ \text{d}' = 0.90$$

(because the z-score corresponding to $P = 0.816$ is 0.90). In other words, at the 2AFC threshold the magnitude of the constant RG signal is

$$(6) \qquad |\overline{RG}| = \theta_{RG} = 0.636\ \sigma_{RG}\,.$$

Thus the mechanism threshold (the cone-contrast threshold vector along the mechanism direction) is

$$(7) \qquad T_{RG} = 0.636\,\frac{\sigma_{RG}}{\gamma_{RG}}$$

[cf. Eqn. 2]. As the postreceptoral gain γ decreases (at higher spatial or temporal frequencies, for example), the mechanism threshold T increases. If the noise is not Gaussian, the constant of proportionality between T and σ/γ is some number other than 0.636, depending on the form of the probability density function.

The formalism needs one additional component, a combination rule that specifies how the mechanisms interact. The usual assumption is that the mechanisms are stochastically independent and their outputs combine by probability summation. The detection contour – that is, the locus of threshold cone-contrast vectors – formed by the "inner envelope" of the *means* $\overline{RG}$, $\overline{YB}$, and $\overline{Lum}$ of three detection mechanisms is a parallelepiped, each face of which is an isoresponse surface for one polarity of one mechanism. The two outer planes in Fig. 18.2 represent two of the faces of such a parallelepiped. When the noise is considered, however, the shape of the contour becomes more rounded and smaller in the corners than the parallelepiped. If the noise in each mechanism is Gaussian, the probability sum of these three linear mechanisms creates an ellipsoidal contour, because an isoprobability surface of an n-dimensional multivariate Gaussian probability distribution is ellipsoidal (Silberstein & MacAdam, 1945).

We may approximate the operation of probability summation by use of the following combination rule: At threshold,

$$(8) \qquad |\overline{RG}|^{\beta} + |\overline{YB}|^{\beta} + |\overline{Lum}|^{\beta} = 1\,.$$

If the exponent β is unity, the mechanisms sum (a "city block" rule); when the exponent is large, then the largest signal dominates (a "winner takes all" rule) and a parallelepiped results; for intermediate values of β, the equation behaves like a probability summation. This sort of contour has frequently been used to model probability summation between chromatic mechanisms (Cole, Hine, & McIlhagga, 1993, 1994; Sankeralli & Mullen, 1996); a somewhat different approach is taken in the influential model of Guth and colleagues (Guth, Massof, & Benzschawel, 1980; Guth, 1991).

The Quick (1974) model of visual summation identifies the combination exponent β with the slope of the Weibull psychometric function (Graham, 1989). Estimates of the psychometric slope for isolated *RG* vary from about 1.6 to about 2.1 (Eskew et al., 1991; Eskew, Stromeyer, & Kronauer, 1994); *YB* probably also has a psychometric slope in this range (Watanabe, Smith, & Pokorny, 1997). However, *Lum* has a consistently higher slope, of about 2.2 (Stromeyer, Lee, & Eskew, 1992; Eskew, Stromeyer, & Kronauer, 1994; see also the foveal data of Vingrys & Metha, 1997).

Rather than assuming the Quick model and using psychometric slopes to estimate β, Eqn. (8) may be fit directly to detection data using β as a free parameter. When this is done, the estimates of β are often higher than the slopes of psychometric functions (e.g., 3–5; Cole, Hine, & McIlhagga, 1993, 1994). This discrepancy between psychometric slopes and fitted βs might indicate that some of the theoretical assumptions of the Quick approach, such as the high-threshold assumption (known to be wrong in detail, on other grounds), fail critically in these experiments. A value of $\beta = 4$ works well in many applications of this model.

Null planes and isolation of mechanisms. The plane through the origin that is orthogonal to the mechanism vector is the "null plane" of the mechanism (Fig. 18.2; Derrington, Krauskopf, & Lennie, 1984). The null plane comprises all of the stimuli (cone triplets) that do not affect the opponent mechanism, because their projection onto the mechanism vector is zero. One of the advantages of using vectors to represent these mechanisms is that doing so immediately reveals that these linear mechanisms are broadly tuned – they have some response to all stimuli except those in the null plane. To get narrowly tuned mechanisms – for example, a higher order mechanism that responds well to orange and hardly at all to red or yellow – the cone signals must be combined nonlinearly.

The null plane of the luminance mechanism is particularly important in the history of chromatic discrimination. This plane is called the *equiluminant plane*, and many studies of chromatic detection and discrimination restrict themselves to this plane; in practice, heterochromatic flicker photometry or minimally distinct borders are often used to find it. Some of the advantages and disadvantages of restricting attention to equiluminant stimuli are discussed in a subsequent section.

It is important to distinguish between mechanism directions and mechanism-isolating directions. To isolate one mechanism, stimuli are presented that lie within the null planes of the other mechanisms. For example, assume that there are three linear mechanisms, *RG*, *YB*, and *Lum*. To isolate one of them, for example, *RG*, one finds the intersection of the null planes of *YB* and *Lum*. These two null planes meet in a line going through the origin of the cone-contrast space; the only stimulus vectors that do not stimulate *YB* and *Lum*, and thus isolate the *RG* mechanism, lie along that line. The key point is that this *RG isolating direction* is the same as the *RG mechanism direction* only if the three mechanisms are mutually orthogonal; otherwise, the *RG* isolating direction stimulates the *RG* mechanism uniquely, but does so less efficiently than stimulation along the mechanism direction. If there are more than three linear mechanisms (e.g., if there are broadly tuned, higher order mechanisms; see Chapter 16), there may be no color direction that isolates any one of them, since more than two null planes need not meet in a single line. Knoblauch (1995) discusses relationships between mechanism directions and mechanism-isolating directions.

Sometimes the null plane is used to characterize a mechanism (Derrington, Krauskopf, & Lennie, 1984), but specifying the null plane alone gives no information about the sensitivity of the mechanism. Specifying the mechanism vector allows the gain of each mechanism to be defined. Of course, these gains [the γ's,

Eqns. (1) and (3)] will vary depending on the spatial and temporal characteristics of the stimuli, but for a fixed stimulus condition the specification of the gains, the mechanism vectors, and some combination rule (e.g., probability summation) allows a prediction of detection performance.

Estimation of cone weights and mechanism thresholds under neutral adaptation. Cole, Hine, and McIlhagga (1993) used a large blurred spot, presented as 200-ms flashes on a 1070-Td white background, and fitted a probability summation model like Eqn. (8) to measure the cone inputs to the three mechanisms. Sankeralli and Mullen (1996) used gratings on an ~500-Td white field and varied the spatial and temporal frequencies to attempt to better isolate the less-sensitive *YB* and *Lum* mechanisms. Both studies used two-interval forced-choice methods, and both had three observers. A summary of the results of these two studies is given in Table 18.1.

Values of the mechanism thresholds T_{RG} and T_{YB} [Eqn. (2)] are given in the table; these values might be expected to be fairly representative of most observers for reasonably low spatial frequency, long-duration (150-ms or more) flashed tests. Sankeralli and Mullen (1996) did not report thresholds directly; instead, these have been estimated from the two-dimensional fit parameters (thresholds were estimated from Table 1 of Sankeralli and Mullen as the mean length of the appropriate axes of their fitted superellipses).

No threshold is given for *Lum* since the stimulus used by Sankeralli and Mullen (1996) was a 24-Hz flicker, not a flash; however, for long-duration, foveally viewed, flashed spots, *RG* is generally about tenfold more sensitive than *Lum* (Cole, Stromeyer, & Kronauer, 1990; Chaparro et al., 1993; Eskew, Stromeyer, & Kronauer, 1994). This *RG* sensitivity advantage has at least three possible general causes: the noise could be relatively higher in *Lum* ($\sigma_{Lum} > \sigma_{RG}$); the gain could be relatively higher in *RG* ($\gamma_{RG} > \gamma_{Lum}$); or the detection uncertainty could be higher in *Lum*.

The less certain the observer is about the parameters of the test, the more irrelevant detection mechanisms the observer must monitor and the steeper the psychometric function (Tanner, 1961); the irrelevant mechanisms degrade performance at low test contrasts because, while they contribute noise, they never detect the test, being tuned to inappropriate values of the

	RG				*YB*				*Lum*		
Data from:	$W_{RG,L}$	$W_{RG,M}$	$W_{RG,S}$	T_{RG}	$W_{YB,L}$	$W_{YB,M}$	$W_{YB,S}$	T_{YB}	$W_{Lum,L}$	$W_{Lum,M}$	$W_{Lum,S}$
Cole et al. (1993)	-0.71	0.71	0.00	1.8×10^{-3}	-0.38	-0.59	0.71	1.9×10^{-2}	0.58	0.79	0.17
	-0.72	0.70	0.01	2.7×10^{-3}	-0.75	-0.06	0.66	1.5×10^{-2}	0.30	0.95	0.07
	-0.70	0.71	0.01	1.4×10^{-3}	-0.82	0.37	0.43	1.3×10^{-2}	0.94	-0.34	0.005
Sankeralli & Mullen (1996)	-0.72	0.70	0.02	2.3×10^{-3}	-0.40	-0.40	0.82	3.1×10^{-2}	0.95	0.30	-0.06
	-0.75	0.66	-0.01	2.6×10^{-3}	-0.34	-0.34	0.87	4.5×10^{-2}	0.98	0.21	-0.03
	-0.75	0.66	0.02	4.3×10^{-3}	-0.42	-0.42	0.80	7.2×10^{-2}	0.95	0.32	-0.05
Mean	-0.725	0.69	0.008	2.2×10^{-3}	-0.518	-0.24	0.715	2.2×10^{-2}	0.783	0.371	0.025
s.d.	0.021	0.024	0.012	5.9×10^{-3}	0.209	0.345	0.159	3.9×10^{-2}	0.281	0.458	0.089

Table 18.1: Cone-contrast weights estimated from data of six observers, from Cole, Hine, and McIlhagga (1993) and Sankeralli and Mullen (1996). Thresholds, in cone-contrast units, are shown for *RG* and *YB*, but not for *Lum* (see text).

stimulus parameters (such as test location, time, color, spatial frequency, etc.). Weibull slopes of 1.6 to 2.1, as measured for *RG* (preceding section), suggest that 3 to 10 total detection mechanisms contribute to the observer's performance, whereas the slightly steeper *Lum* psychometric functions suggest 10 to 30 mechanisms (Pelli, 1985, Table 1). This difference in uncertainty would cause *Lum* to be less sensitive than *RG* in a 2AFC procedure, but not by much: of the 1.0 log unit *RG* advantage, less than 0.2 log unit might reasonably be attributed to uncertainty differences.

The slope of the psychometric function is inversely related to the size of the standard deviation of the noise in the *relevant* mechanism [when the standard deviation is small, the detection probabilities in Eqn. (4) rise steeply with contrast]. The finding that the psychometric slope for luminance detection is steeper than the corresponding slope for chromatic detection suggests that the luminance pathway might be *less* noisy, implying in turn that the cause of the *RG* sensitivity advantage is higher effective gain. This might seem to contradict the finding that primate magnocellular (M) neurons, often claimed to subserve luminance detection, have higher contrast gain than parvocellular (P) neurons (e.g., Shapley, 1990). However, there is no contradiction. First, whereas M-cells do have higher contrast gain for *luminance* contrast, a comparison of color and luminance responsiveness using some form of cone-contrast metric shows that contrast gains for M-cells (for equichromatic stimuli) and P-cells (for red-green equiluminant stimuli) are similar (Lee, Martin, & Valberg, 1989b; Lee et al., 1993a), at least for low temporal frequencies. Second, P-cells as well as M-cells are likely to contribute to the detection of equichromatic stimuli (Lennie, Pokorny, & Smith, 1993). The cause of the higher effective gain for chromatic detection is likely to be a summation of the responses of many P-cells (Chaparro et al., 1993).

The L- and M-cone weights to *RG* are of opposite sign and nearly equal for all six observers in Table 18.1; the coefficients of variation (the standard deviation divided by the mean) are an order of magnitude or more smaller for $W_{RG,L}$ and $W_{RG,M}$ than for the other seven mean weights. The lack of individual differences in long-wavelength cone inputs to *RG* suggests that detection contours in the ($\Delta L/L$, $\Delta M/M$) plane of the cone-contrast space will always have slopes close to unity (as in Fig. 18.3, below), and this is the case for all studies of which we are aware. Similarly, Derrington, Krauskopf, and Lennie (1984) found that most macaque parvocellular neurons had approximately equal and opposite L- and M-cone inputs.

Sankeralli and Mullen (1996) were unable to determine the relative L and M weights for *YB*, so they set them equal to each other; Cole, Hine, and McIlhagga (1993) estimated weights separately, but the results were not well constrained by the data. The mean $W_{YB,L}$ and $W_{YB,M}$ values reported in Table 18.1 are nonetheless similar to the mean weights measured in a different experiment by Cole, Hine, and McIlhagga (1994), who attempted to isolate *YB* by using a spatial pattern to which the S-cones should be more sensitive (a Craik–Cornsweet parafoveal circular edge).

The mean value of $W_{RG,S}$ is very slightly positive. However, a reddish (negative) suprathreshold input of S-cones to the red-green *hue* mechanism is required to account for the violet appearance of short-wave lights (i.e., this S-cone input must have the same sign as the L-cones; Ingling, 1977; Werner & Wooten, 1979). Whether S-cones contribute similarly to the *RG detection* mechanism is controversial (Boynton, Nagy, & Olson, 1983; Mollon & Cavonius, 1987; Stromeyer & Lee, 1988). When measurements of the S-cone-contrast weight for *RG* detection and for red-green hue equilibria are made under identical conditions, the estimated weights have quite similar (very small) magnitudes and signs (Eskew & Kortick, 1994). Recently, Stromeyer et al. (1998) found that S-cone modulations facilitated and masked detection by *RG*, depending on whether the $+\Delta S/S$ signal was paired with the $+\Delta L/L$ signal (facilitation) or the $+\Delta M/M$ signal (masking). This and other evidence led them to suggest a small, "reddish" S-cone contribution to *RG*. In Table 18.2 we set $W_{RG,S} = -0.02$ in spite of the mean value of $W_{RG,S}$ in Table 18.1 being slightly above zero.

There is evidence that *RG*'s relative cone weights do not depend on temporal frequency, but for *Lum* the relative *L*/*M* inputs can vary widely with temporal frequency (Stromeyer et al., 1995). With the flashes used by Cole, Hine, and McIlhagga, the estimated L and M weights for *Lum* were essentially unconstrained by the data. The 24-Hz counterphase flicker used by Sankeralli and Mullen allowed more of the *Lum* mechanism to be revealed and permitted more certainty in the estimation of the weights – but these weights may not be appropriate for flashed stimuli. The mean weights shown in Table 18.1 for *Lum* are based on both studies and are therefore dubious, but they are given here for completeness. L-cones are generally believed to contribute more to *Lum* than M-cones, and the mean $W_{Lum,L}$ and $W_{Lum,M}$ are consistent with that belief; recall, however, that these are contrast gains (previous section) and do not necessarily imply anything about the scaling of the L- and M-cone fundamentals or cone numbers.

The "adjusted weights" shown in Table 18.2 are based on the mean weights from Table 18.1. Besides setting $W_{RG,S}$ at –0.02 as mentioned above, we also set the S-cone input to *Lum* to be zero (ignoring the small, negative contribution of the S-cones to *Lum*; Stockman, MacLeod, & DePriest, 1991), and we made changes to the other means in service of three desiderata: (1) that the vector of weights have unit length; (2) that the weights sum exactly to zero; (3) that the results have the same direction in cone-contrast space as the centroid vector formed by the mean weights. The first is a technical requirement of the model. The second, tiny adjustment makes the null planes of *RG* and *YB* contain the equichromatic vector, so they do not respond to radiance modulations and their metamers (see Color Spaces section). The adjusted weights represent an approximation to (3) given that we satisfy (1) and (2) (the reversal in the ratio of magnitudes of the adjusted $W_{RG,L}$ and $W_{RG,M}$ in Table 18.2, compared with the means in Table 18.1, is not an error, but is the result of requiring $W_{RG,S}$ to be negative and then applying these desiderata). Note that although $W_{RG,S}$ is 40 times smaller than $W_{YB,S}$, the sensitivity of *RG* is so much greater than the sensitivity of *YB* that the net result is that *YB* is only about four times as responsive to S-cone input as is *RG*.

For the observer represented by these adjusted weights, convenient basis vectors for the equiluminant plane are $\{\Delta L/L, \Delta M/M, \Delta S/S\} = \{0.43, -0.9, 0\}$ and $\{0, 0, 1\}$ (but the uncertainties expressed above about the luminance weights limit the usefulness of this plane). To stimulate the *YB* mechanism alone, one would find the intersection of the *RG* null plane and the equiluminant plane. This *YB* isolating direction is the vector that is orthogonal to both $\mathbf{W}_{RG}$ and $\mathbf{W}_{Lum}$; it is not exactly the S-cone direction, because of the small S-cone contribution to *RG*. Instead, $\{0.009, -0.019, 1\}$ is the direction that isolates *YB* (obviously, the S-cone direction is sufficient for most practical purposes). The *RG* isolating direction is $\{-0.43, 0.9, -0.015\}$. Desideratum (2) forces the equichromatic direction $\{1/\sqrt{3}, 1/\sqrt{3}, 1/\sqrt{3}\}$ to be the direction that isolates *Lum*.

Figure 18.3 shows an application of this model to a set of detection data collected in the $(\Delta L/L, \Delta M/M)$ plane (Giulianini & Eskew, 1998). The upper solid line drawn through the open symbols is given by:

(9) $$W_{RG,L}\,\Delta L/L + W_{RG,M}\,\Delta M/M = T_{RG}$$

(9a) $$-0.70\,\Delta L/L + 0.72\,\Delta M/M = 0.0078,$$

using the weights from Table 18.2 in Eqn. (1) and letting $\Delta S/S = 0$. The upper line represents the positive

	RG			*YB*			*Lum*		
Adjusted cone weigths	$W_{RG,L}$	$W_{RG,M}$	$W_{RG,S}$	$W_{YB,L}$	$W_{YB,M}$	$W_{YB,S}$	$W_{Lum,L}$	$W_{Lum,M}$	$W_{Lum,S}$
	-0.70	0.72	-0.02	-0.55	-0.25	0.8	0.90	0.43	0.00

Table 18.2: Adjusted cone-contrast weights, based on the means in Table 18.1. The vectors of adjusted weights have unit length, and for *RG* and *YB* they are exactly orthogonal to the equichromatic direction $\pm\{1,1,1\}$; $W_{RG,S}$ is set to –0.02 (see text).

("green") response polarity of *RG*; the lower solid line is obtained by changing the sign of T_{RG} (for the "red" polarity). The slope of both lines is $-W_{RG,L}/W_{RG,M} = 0.70/0.72 = 0.97$. The dashed lines are based on the analogous equation for *Lum*; their slope is $-W_{Lum,L}/W_{Lum,M} = -0.90/0.43 = -2.1$. The filled symbols in Fig. 18.3 result when red/green masking noise is added to the stimulus, raising all of the *RG* thresholds and exposing more of the *Lum* mechanism in the first and third quadrants (Giulianini & Eskew, 1998). The solid contour plotted near the filled symbols, which is a slice through a rounded parellelpiped, represents a fit of Eqn. (8), with $\beta = 4$ and with no *YB* contribution.

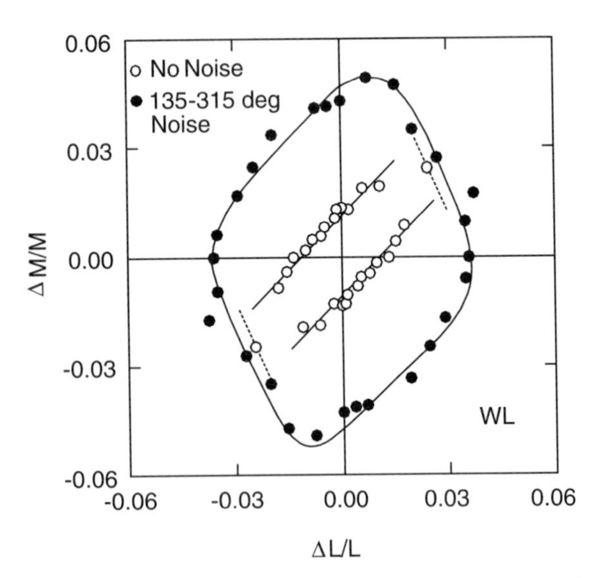

Figure 18.3: Detection thresholds from Giulianini and Eskew (1998, Fig. 5a), with (filled symbols) and without (open symbols) added masking noise. The test was a 1-cpd Gabor function; the noise consisted of flickering rings superposed on the test area, along the $\{\Delta L/L, \Delta M/M, \Delta S/S\} = \pm\{-1,1,0\}$ color direction (135 deg). Equation (9a) gives the equation of the solid line that describes the upper set of open symbols; it has slope $-W_{RG,L}/W_{RG,M} = 0.97$. The equation of the lower line is the negative of Eqn. (9a). The dotted lines have slope $-W_{Lum,L}/W_{Lum,M} = -2.1$. This *Lum* detection contour is positioned to pass through the threshold at 45 deg, since this polar direction lies approximately between the two *RG* detection contours and is presumably detected by *Lum*. The solid contour is a fit of Eqn. (8) to the masked data, with $YB = 0$ and $\beta = 4.0$, using the weights of Table 18.2, and with the mechanism thresholds allowed to vary freely.

Chromatic adaptation: first-site effects. The data reported in Fig. 18.3 do not constitute a test of the cone-contrast model, nor do the data of Cole, Hine, and McIlhagga (1993; 1994) or Sankeralli and Mullen (1996); all of these data could have been equally well described by a model based on the sums and differences of cone excitations rather than cone contrasts. Testing the contrast feature of the model requires changing the chromaticity of the adapting field, to see how well the first-site, cone-specific adaptation built into the model can account for changes in the data. We examine *RG* first.

Figure 18.4A shows spectral sensitivity data collected by Thornton and Pugh (1983) for a 3-deg Gaussian spot presented with a gradual time course, on adapting fields of 560, 580, and 600 nm. The Sloan notch (Sloan, 1928), generally taken to represent detection by *Lum* separating two regions of chromatic detection (Calkins, Thornton, & Pugh, 1992), shifts to the wavelength of the background field in each case. Figure 18.4B shows the same data, transformed to cone-contrast coordinates; only tests of 520 nm and above are shown, to eliminate S-cone contributions (Chaparro et al., 1995 performed a similar analysis). Since these flashes were all incremental monochromatic lights, they all plot in the first octant of the cone-contrast space; after eliminating the shorter wavelength tests that stimulate the S-cones, the remaining tests fall in the first quadrant of the ($\Delta L/L$, $\Delta M/M$) plane. The test light that has the same wavelength as the background field is a radiance modulation and therefore is represented at 45 deg in this plane of the cone-contrast space.

Note that most of the data fall along lines with near-unit slopes, just as in Fig. 18.3, and that *this is true regardless of the adapting field wavelength.* Increment thresholds of rhesus monkeys behave similarly when plotted in cone-contrast space: Tests that do not excite the S-cones fall along lines near slope 1 on various chromatic fields (Kalloniatis & Harwerth, 1991). This invariance across adapting chromaticity is the hallmark of cone-specific adaptation, such as that incorpo-

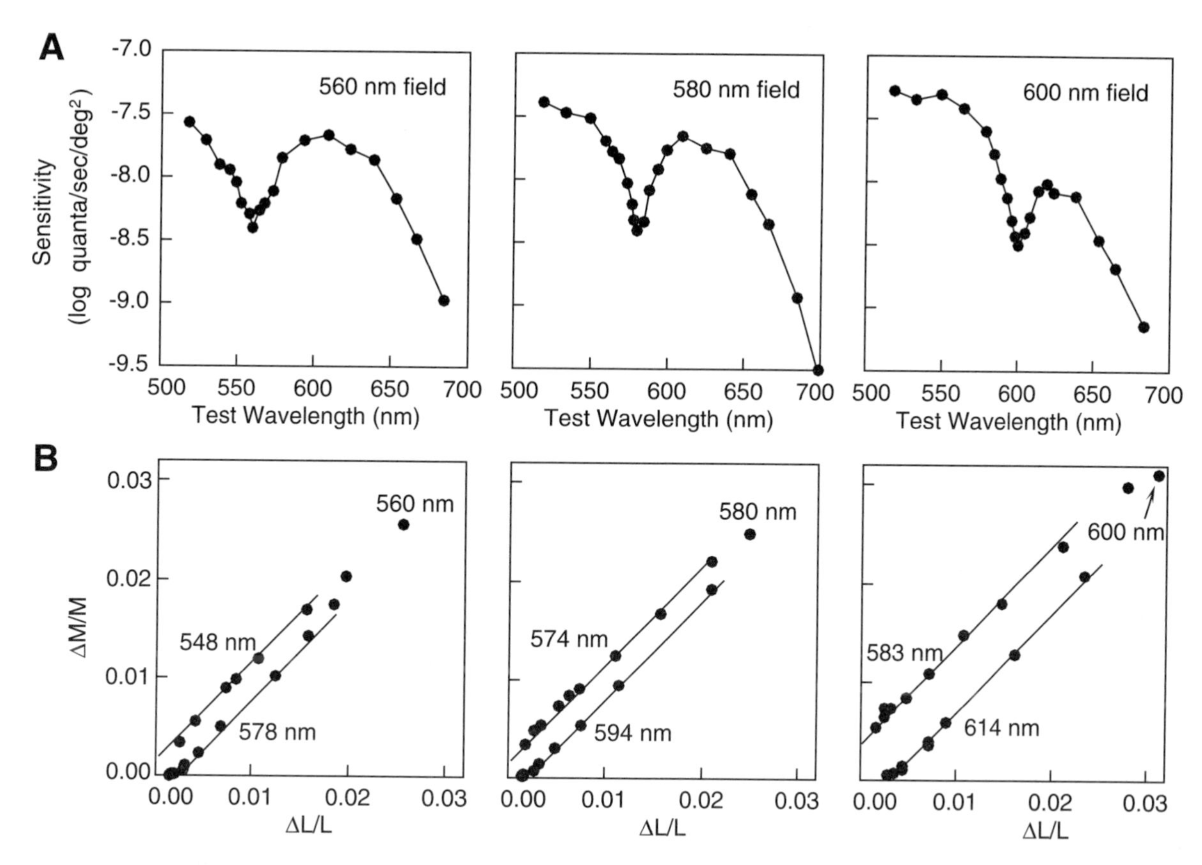

Figure 18.4: First-site adaptation. (A) Increment threshold spectral sensitivity curves, from Thornton and Pugh (1983, Fig. 1). Thresholds were measured for a Gaussian spot, presented with a gradual time course to reduce detectability by the luminance mechanism. Three background wavelengths were used; the notch in sensitivity occurs at the adapting wavelength. (B) Data from panel A transformed to cone-contrast space. Three of the test wavelengths are indicated in each panel, for illustration. When the test wavelength equals the field wavelength, the test is a radiance modulation and plots at 45 deg. The detection contours have a slope of ~1 on all adapting fields, indicating cone-independent adaptation; the radiance modulation lies in the null direction suggested by these detection contours.

rated in the cone-contrast representation. The constancy of the detection contour slopes implies that the relative L- and M-cone weights are unaffected by chromatic adaptation, whereas the small variation in the distance of the intercepts of the lines implies that the criterion signal [Eqn. (6)] must be larger on the 600-nm field (see next section).

Stromeyer, Cole, and Kronauer (1985) and Chaparro et al. (1995) measured detection thresholds in all four quadrants of the $(\Delta L/L, \Delta M/M)$ plane of the cone-contrast space and also found that the relative cone inputs to *RG* were unaffected by field color, in just this way (Fig. 18.5A); the slopes of the "red" and "green" detection contours are close to 1.0 when plotted in cone-contrast units regardless of the wavelength of the background field. In comparison, when the same contours are plotted in the $(\Delta L, \Delta M)$ plane of the local cone-excitation space (Fig. 18.5B), the slopes depend on the field wavelength; in particular, thresholds near the ΔL-axis are strongly elevated by the red field, just as expected from first-site adaptation (Eskew, Stromeyer, & Kronauer, 1992; Chaparro et al., 1995). Similarly, a reddish field reduces the relative L-cone input (in excitation units) to macaque P-retinal ganglion cells (Smith et al., 1992b). The 654-nm background in Fig. 18.5 is particularly important here, since it alone causes a large difference in the L and M adaptive states. Clearly, thresholds that are mediated by the L- and M-cones in *RG* are regulated by a first-site, cone-specific, approximately Weberian adaptation; there is no other simple way to account for the constant detection contour slope in cone-contrast space. Sensitivity regulation occurring at a later site, one receiving a pooled signal from multiple cone types, *cannot* produce a constant slope, since information about the individual cone's adaptive state has been lost.

LeGrand (1949/1994), in his analysis of MacAdam's ellipses, found that S-cone thresholds rose roughly in proportion to their adapting level for moderate and high levels of mean S-cone quantal catch. Others have also found results consistent with first-site, approximate Weber behavior in the S-cones (Boynton & Kambe, 1980; Nagy, Eskew, & Boynton, 1987; Krauskopf & Gegenfurtner, 1992; see Chromaticity Discrimination section). However, the highest S adapting level studied in any of these studies was about 400 S Td – except for one of the MacAdam PGN ellipses measured at about 4,000 S Td – and at an S-cone adapting level of about 6,000 S Td increment thresholds for violet lights begin to rise very steeply, showing saturation (Mollon & Polden, 1977b; Stromeyer, Kronauer, & Madsen, 1979).

Many measurements of the effects of adapting fields have been made for long-wave test lights, but these have employed incremental tests rather than silent substitutions and so they have not necessarily isolated a particular mechanism. These increment thresholds approximately obey Weber's law except at very low retinal illuminances (Stiles, 1978; Wyszecki & Stiles, 1982a). Of the studies that have isolated L- and M-cones and measured their inputs to *RG*, Chaparro et al. (1995) found that Weber-like adaptation occurred independently in the L- and M-cones down to the lowest level that they tested, about 60 M Td (400 Td total), across adapting chromaticities; Stromeyer, Lee, and Eskew (1992) showed that the *RG* detection

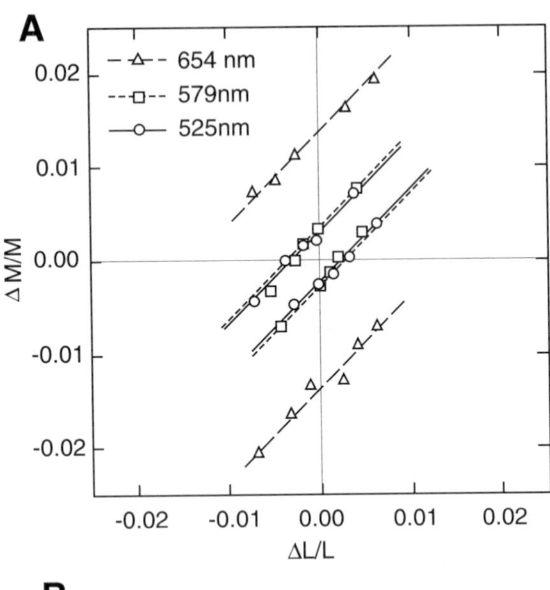

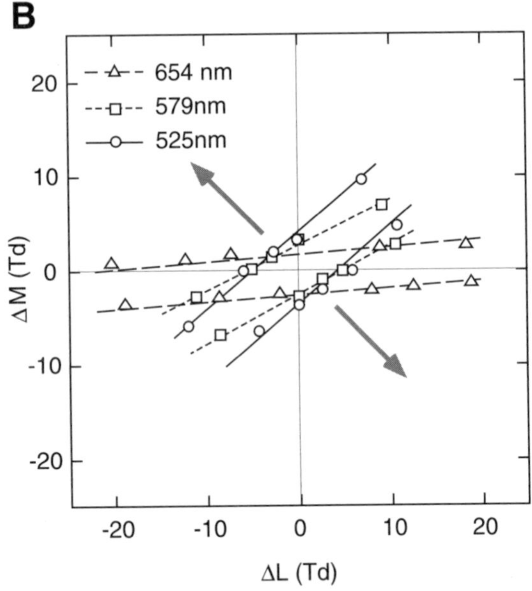

Figure 18.5: First-site adaptation. (A) Detection thresholds in the ($\Delta L/L$, $\Delta M/M$) plane of cone-contrast space, replotted from Fig. 4 of Chaparro et al., 1995 (3000-Td backgrounds). The equivalent wavelength of the background is indicated in the legend. As in Fig. 18.4, the slope of the detection contour is constant across the adapting chromaticities. (B) Data from panel A plotted in local cone space (as in Fig. 2 of Chaparro et al., 1995). In this space the slope of the contour varies across adapting chromaticities. Constant-luminance tests lie along the thick gray arrow, where $\Delta L = -\Delta M$.

contour had a slope near unity on a yellow field producing about 32 M Td (77 Td total); Stromeyer, Cole, and Kronauer (1985) showed that $\Delta L/L$ at threshold increased by only about 0.1 log unit as a 638-nm field was reduced from about 2970 to 39 L Td (right-most pair of points in Figs. 18.6A & B). All of these results are consistent with Weber behavior in the L- and M-cones, from low photopic levels and up.

Photocurrent recordings from isolated primate cones (Schnapf et al., 1990) suggest that there is little adaptation in cone outer segments until light levels are increased beyond about 1,000 Td. Since psychophysical cone thresholds begin to rise at 10–50 Td (Hood & Finkelstein, 1986), the photocurrent data might suggest that the "first site" is after the outer segment (at the cone-bipolar synapse, for example). The psychophysical results discussed in this section imply only that the adaptation takes place prior to the site(s) at which signals from different cone classes are combined. Psychophysical evidence suggests that the spatial integration area preceding the first nonlinearity in the visual system (presumably the nonlinearity representing adaptation) has the width of a single foveal cone (MacLeod, Williams, & Makous, 1992; MacLeod & He, 1993). Consistent with this psychophysical result, recent physiological recordings from both HI and HII monkey horizontal cells show cone-specific adaptation at this early level (Lee et al., 1997).

Chromatic adaptation: second-site desensitization. The evidence reviewed in the preceding section (and much more not discussed here) indicates that substantial sensitivity regulation occurs in cone-specific pathways, prior to any opponent combination. However, it is also abundantly clear that there are other sites of sensitivity regulation occurring at or after the point at which cone signals are combined. Cone-specific, first-site adaptation reduces the influence of the adapting chromaticity; once the first-site effect has been accounted for (at least approximately, by taking contrasts, for example), the remaining variation in sensitivity due to adapting condition is by definition a "second-site" effect. The term "adaptation" implies a loss of *absolute* sensitivity to retain *differential* or contrast sensitivity (Shapley & Enroth-Cugell, 1984). Because second-site effects are losses of both absolute and contrast sensitivity, perhaps due to a saturation of neural mechanisms at the opponent site, the second-site effects are often called desensitization rather than adaptation.

Chromatic fields raise *RG* thresholds above and beyond the first-site effect. This is illustrated in Fig. 18.5A, which shows detection contours in the $(\Delta L/L, \Delta M/M)$ plane of the cone-contrast space on red, yellow, and green adapting fields (Chaparro et al., 1995). As noted in the previous section, the constant slope of the *RG* detection contours shows first-site, cone-specific adaptation. However, the contours on the long-wave field are farther from the origin than on the green or yellow fields. The long-wave field raises the "red" (L increment and M decrement) and "green" (M increment and L decrement) test threshold contours *equally*, so the effect is not some masking effect that is specific to the color polarity of the test. We could modify Eqn. (9) to represent this result as:

$$(10)\quad \left|W_{RG,L}\Delta L/L + W_{RG,M}\Delta M/M\right| = T_{RG}(1 + J_{RG}),$$

in which T_{RG} is the mechanism threshold on a neutral field [Eqn. (2)], and the absolute value is taken so as to represent both response polarities. The scalar J_{RG} is a function of the adaptation conditions and represents the degree to which chromatic adaptation alters thresholds from those found on the neutral fields. Figure 18.5A requires J_{RG} to be about equal on the yellowish and greenish fields but higher on the long-wave field. Since the slope of the detection contour is $-W_{RG,L}/W_{RG,M}$, the data show that the relative cone weights are not altered by the second-site effect. Thus we may use the adjusted weights from Table 18.2 and, after substituting the definition of T_{RG} from Eqn. (2), Eqn. (10) becomes

$$(11)\quad \left|-0.70\ \Delta L/L + 0.72\ \Delta M/M\right| = \frac{\theta_{RG}}{\gamma_{RG}}(1 + J_{RG}).$$

The symbols in Fig. 18.6A and B show L-cone thresholds from Stromeyer, Cole, and Kronauer (1985) plotted in contrast units as a function of adapting

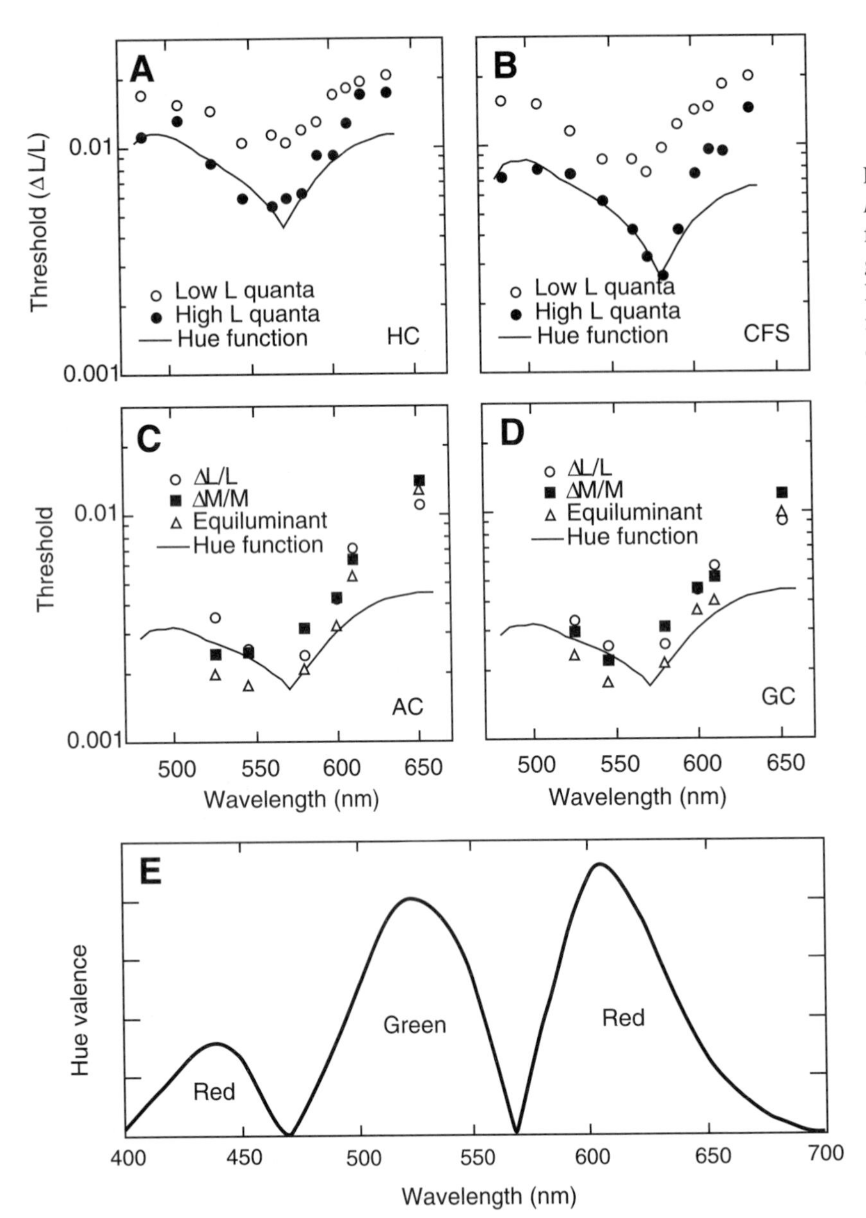

Figure 18.6: Second-site effects on *RG*. (A & B) $\Delta L/L$ thresholds as a function of the equivalent background wavelength, replotted from Fig. 7 of Stromeyer, Cole, and Kronauer (1985), for two observers. All of the fields were equated for quantal catch in the L-cones. Filled symbols: high intensity condition (3184 Td at 638 nm). Open symbols: low intensity condition (42 Td at 638 nm). Solid curves are red/green hue valence functions, plotted on a equal-L quantal basis. (C & D) Data from two observers, replotted from Fig. 3 of Chaparro et al. (1995). L, M, and equiluminant contrast thresholds are represented by different symbols. The abscissa is equivalent background wavelength; retinal illuminance was 3000 Td throughout. The solid curves are red/green valence functions, plotted on a constant-luminance basis. (E) The same red/green valence functions plotted over the whole spectrum on the familiar equal-energy basis. In all panels, the hue functions were computed as in Eqn. (14) (using the 2 deg from 10-deg cone fundamentals of Stockman, MacLeod, and Johnson, 1993) with $k_0 = 0.0004$, $k_2 = 1.244$, and $k_3 = 0.2$ (except for panel B, where $k_2 = 1.501$ and $k_3 = 0.3$ to make the minimum near 580 nm instead of near 570 nm).

wavelength for adapting fields that produce equal quantal catch in the L-cones (so first-site adaptation – the denominator of $\Delta L/L$ – is kept constant). L-cone thresholds are lowest near yellow and rise to either side; if there were no second-site desensitization, these thresholds should not vary with adapting condition since plotting cone contrast accounts for the multiplicative first-site effect. Given the log-unit sensitivity advantage of *RG* over *YB* or *Lum* (Table 18.1), we may assume that these L-cone tests are detected by *RG* (except perhaps for the most-desensitized cases, where another mechanism might intrude). Setting $\Delta M/M = 0$ in Eqn. (11) and rearranging,

$$(12)\quad J_{RG} = 0.70\,\frac{\gamma_{RG}}{\theta_{RG}}\left|\frac{\Delta L}{L}\right| - 1.$$

Thus the data in Figs. 18.6A and B provide a way to estimate the form of the second-site desensitization effect. J_{RG} is roughly U- or V-shaped in this spectral region, with a minimum in the yellow and a maximal elevating effect of about fivefold. The effect of adapting chromaticity is reduced at the lower radiance.

Figures 18.6C and D show similar data from Fig. 3 of Chaparro et al. (1995) but replotted in cone-contrast units, for L, M, and equiluminant tests. Whereas the adapting fields in panels A and B were equated for L-cone quantal catch, those in panels C and D were kept at 3000 Td; this accounts for part of the difference in shape of the data between the top two panels and the middle two panels. Again, thresholds are lowest for fields that should be near neutral for a red/green hue mechanism and rise elsewhere, particularly as the field wavelength is lengthened. The agreement between L, M, and equiluminant tests is support for the model since, according to Eqn. (10), the second-site effect alters all of the RG thresholds equally.

Recently, Yeh, Lee, and Kremers (1996) found analogous results in macaque P-retinal ganglion cells: two adapting fields that were equated for quantal catch in one cone class (e.g., L) were alternated, so that only the adaptive state of the other cone type (e.g., M) varied over time. Responsiveness to ΔL and ΔM tests was equally affected by the two fields, indicating strong second-site desensitization at the level of the ganglion cells.

Recall that the criterion response level θ_{RG} is proportional to the standard deviation of the noise within the mechanism [Eqn. (6)]. Equation (11) means that the higher threshold on the chromatic field could be the result of either an increase in the noise within the mechanism (σ_{RG}) or a decrease in its effective gain (γ_{RG}). Noise found at the level of the retinal ganglion cells of cat (Reich et al., 1994) is independent of mean illumination, and in both the cat and monkey it is apparently additive and generated within the ganglion cells themselves (Croner, Purpura, & Kaplan, 1993); if this is correct, then the noise cannot vary with adaptive state and thus could not explain second-site desensitization. The gain could be reduced via a compressive nonlinearity: A strongly chromatic field would raise the operating point on the nonlinear curve high enough that the tangent to the curve, which is the effective gain at the operating point, was reduced. However, if second-site desensitization is produced by means of a compressive nonlinearity, this must occur in a such a way that "reddish" and "greenish" test thresholds are raised approximately equally on, for example, a reddish field (Figs. 18.5, 18.6C & D).

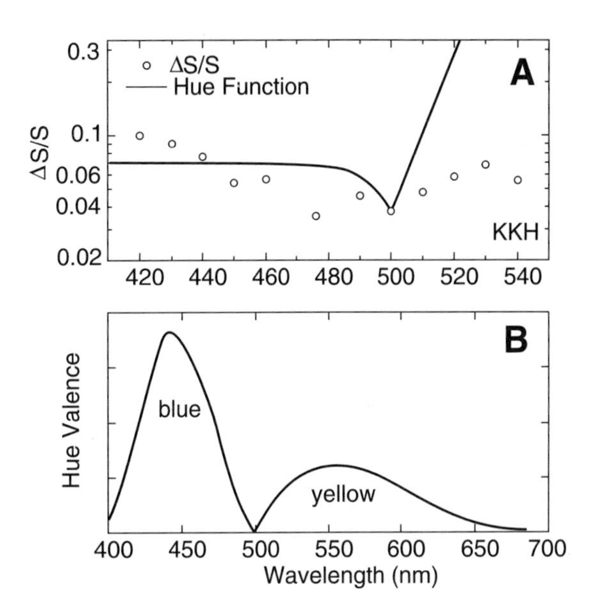

Figure 18.7: Second-site effects on *YB*. (A) S-cone-contrast thresholds as a function of the background wavelength for fields equated for S-cone quantal catch, from McLellan (1997). The 440-nm background was 3×10^8 quanta deg^{-2} s^{-1}. A 110-Td, "white" monitor field was always present. The solid line shows a *YB* hue valence function plotted on an equal S-cone quantal basis. The 540-nm field was $10^{4.7}$ Td, causing substantial bleaching of the L- and M-cones at this wavelength. (B) The same *YB* hue valence function as in (A), plotted over the whole spectrum on the familiar equal-energy basis. The hue functions were computed as in Eqn. (15) (using the 2 deg from 10-deg cone fundamentals of Stockman, MacLeod, & Johnson, 1993), with $k_1 = 0.0016$, $k_4 = 0.615$, and $k_5 = 4.808$. Note the different wavelength scale in (A) and (B).

The symbols in Fig. 18.7A are pure S-cone thresholds (McLellan, 1997), measured on fields that are equated for S-cone quantal catch (whereas the ratio of L- to M-cone quantal catch varies by roughly a factor of 2 across these conditions). Assuming that these are

YB-detected, then application of the same reasoning that was used for RG leads to:

$$(13)\quad J_{YB} = 0.80\,\frac{\gamma_{YB}}{\theta_{YB}}\left|\frac{\Delta S}{S}\right| - 1.$$

Thus, over this wavelength range, these S-cone thresholds may be used to estimate the second-site effect for YB. J_{RG} changes only very gradually with wavelength over most of this range, with most of the elevation occurring in the violet; the maximum second-site elevation here was about threefold.

Whatever J_{RG} and J_{YB} are, they cannot be functions of cone *contrasts* – they must depend on the chromaticity of the adapting field. There is a long line of work on color *appearance* that posits an additive contribution to the hue of a test light from the hue of the adapting field (e.g., Jameson & Hurvich, 1972; Shevell, 1978) above and beyond first-site adaptation; by applying this "two-process" approach to detection one might guess that J_{RG} and J_{YB} are similar to opponent-response functions derived from the color appearance of lights (Hurvich & Jameson, 1955). Thus, for example, we might let:

$$(14)\quad J_{RG} = k_0\,\left|L - k_2M + k_3S\right|,$$

$$(15)\quad J_{YB} = k_1\,\left|L + k_4M - k_5S\right|,$$

with positive constants k_2 through k_5 chosen so that the functions have zeros near the equilibrium hue wavelengths (unique yellow, green, and blue); see Figs. 18.6E and 18.7B. This choice would make J represent an increase in mechanism threshold T that is proportional to the hue valence of the adapting field. This version of J_{RG} is similar to the "J factor" used by Boynton to predict wavelength discrimination functions (Boynton, 1979).

The smooth curves in Figs. 18.6 and 18.7 show such hue functions plotted on the appropriate basis in each case. As shown in Fig. 18.6, J_{RG} is poorly approximated by a red-green valence function (one based on a linear combination of cone fundamentals, constrained to have a zero near "yellow"). The estimates of J_{YB} in Fig. 18.7 have almost no correspondence to a yellow-blue valence function. Thus, this version of the two-process model cannot account for second-site desensitization.

Although there are many other studies of chromatic detection under conditions of chromatic adaptation that might in principle be used to reveal second-site effects, most of them suffer from one or both of the following limitations: (1) the observer is allowed to adapt to the test as well as to the background light, making the actual adaptation state uncertain; (2) the postreceptoral mechanisms RG and YB are not isolated, so that it is not clear from the data which second-site effect could be estimated. Thus the form taken by J_{RG} and J_{YB} cannot yet be fully determined, nor is it known in detail how these second-site effects vary with field radiance and (perhaps) test parameters.

There is, however, a puzzle that should be addressed here: After complete first-site adaptation, all steady-state adapting fields would produce the same signal in each cone-specific pathway. No matter how signals in these pathways are then combined, there could be no second-site effects. Complete first-site adaptation means that all adapting fields should appear the same and produce the same detection thresholds, which is consistent with the mapping of all adapting fields to the origin in cone-contrast space (Figs. 18.1A & C). Given the large body of evidence favoring cone-specific adaptation, how can there also be second-site effects? There are at least two possible answers (which are not mutually exclusive): (1) *Complete* first-site adaptation never occurs. There is a residual, unadapted signal that is operated on at the second site. Recordings from individual turtle cones (Normann & Perlman, 1979; Burkhardt, 1993a) and extracellular recordings in primate retina (Valeton & van Norren, 1983) indicate that this is so: Although cones adapt, the response to a steady background does not stay constant, but instead increases with increasing quantal catch. (2) The modulation of cones near the edge of the adapting field, produced by eye movements, maintains the color appearance of the field and provides the signal that modulates sensitivity at the second site and raises threshold in the field center. Evidence from color matching experiments (Pokorny, Smith, & Starr,

1976; Vienot, 1983; Elsner, Burns, & Webb, 1993; Picotte, Stromeyer, & Eskew, 1994) suggests that, at least for relatively small fields, the edges determine field color under normal viewing, and when fields are stabilized on the retina they generally appear colorless regardless of their chromaticity (Sheppard, 1920; Hochberg, Triebel, & Seaman, 1951; Gur, 1986). However, there is some important evidence against this "filling-in" explanation of second-site threshold elevation. Nerger, Piantanida, and Larimer (1993) found that when a red disk was surrounded by a yellow annulus, stabilizing the edge between the two fields on the retina caused the yellow to fill in, making the disk appear yellow too. This filling-in affected the color *appearance* of small tests added to the disk, but it did not affect their increment threshold. If this result holds generally, then filling-in could still explain how the color appearance of an adapting field is maintained despite first-site adaptation, but that would not suffice to explain why chromatic adapting fields produce a second-site elevation of test thresholds. In case 1, cone contrasts may be regarded as approximations to the actual signal in the cone-specific pathways; in case 2, effects of field configuration and eye movements would need to be considered.

A note on equiluminance. Many studies of chromatic detection and discrimination restrict stimuli to the null plane of the luminance mechanism – the equiluminant plane. There are a number of advantages to doing so: If these mechanisms are really linear, and if equiluminance is accurately determined, then one is guaranteed to be studying chromatic mechanisms when in the equiluminant plane. The use of only one plane greatly reduces the need for data and simplifies the analysis of them. However, significant limitations may arise when only the equiluminant plane is studied.

First, it should be explicitly noted that, if first-site adaptation affects the cone signals supplied to *Lum* (Eisner & MacLeod, 1981; Stromeyer, Cole, & Kronauer, 1987; Pokorny, Jin, & Smith, 1993; Swanson, 1993), there can be no single plane in the cone-*excitation* space that is a null plane for a luminance detection mechanism across different adapting conditions. For example, a plane in which the photopic luminosity function $V(\lambda)$ is constant (Fig. 18.1B) cannot be a null plane for a luminance mechanism if the cones adapt independently, because the cone-specific rescaling would cause the mechanism's null plane to tilt differently in cone-excitation space for different adaptation states. $V(\lambda)$ is useful for specifying the photometric "effective intensity" of an adapting light, but since $V(\lambda)$ is a fixed function of cone fundamentals, it cannot be used to represent neural events that occur subsequent to first-site adaptation. We use the terms "constant luminance" or "constant-$V(\lambda)$" to refer to that fixed plane in cone-excitation space where $\Delta L = -\Delta M$ (see Color Spaces section) and reserve the term "equiluminant" for the null plane *in the cone-contrast space* of a putative nonopponent, linear (luminance) detection mechanism.

A drawback to using tests of *constant luminance* has to do with detecting first-site effects on *RG*. Cone-specific adaptation is most readily measured by using cone-specific tests. Presenting tests within the constant-$V(\lambda)$ plane makes it more difficult to detect first-site changes (Eskew, Stromeyer, & Kronauer, 1992; Chaparro et al., 1995). For example, consider the detection contours shown in Fig. 18.5B: Compared with the 525-nm adapting field, the 654-nm field suppresses L-cone sensitivity and relatively enhances M-cone sensitivity (in excitation units). A test presented along the 135-deg/315-deg, $\Delta L = -\Delta M$ direction (gray arrows) would miss the dramatic changes in sensitivity along the cone-specific axes, since the two first-site effects approximately trade off against one another for that test direction. First-site adaptation in S-cones can be easily measured in this constant-$V(\lambda)$ plane, since the S-cone axis lies in or near it.

A drawback to using *equiluminant* tests has to do with *YB*. Equiluminant lights produce contrasts of opposite signs in the L- and M-cones. Therefore they cannot produce very much differential signal in the long-wave side of *YB*, if, in fact, $W_{YB,L}$ and $W_{YB,M}$ have the same signs as indicated in Table 18.2. In the most extreme case, if the long-wave side of *YB* has the same spectral sensitivity as *Lum* (so that $\mathbf{W}_{YB}$ would be a weighted difference of $\{0,0,1\}$ and $\mathbf{W}_{Lum}$), then only the S-cone signal in *YB* could vary in the equiluminant plane. Thus, restricting test modulations to the equiluminant plane might lead one to believe that *YB* consists of S-cones alone, unopposed by a long-wave signal. Yet studies that modulate the luminance of test lights clearly show inhibitory interactions between long- and short-wave increment tests at threshold (e.g., Boynton, Ikeda, & Stiles, 1964; Thornton & Pugh, 1983; Kalloniatis & Harwerth, 1991), a notch in the increment threshold curve near unique green indicating opponency in the mechanism detecting short-wave tests (e.g., King-Smith & Carden, 1976), as well as inhibition at threshold between $\Delta S/S$ versus stimuli that modulate L and M equally for both increments and decrements in cone-contrast space (Cole, Hine, & McIlhagga, 1993, 1994; Sankeralli & Mullen, 1996).

In fact, chromatic detection is (at least) a three-dimensional problem, and it is unlikely that a full understanding of it can be reached by using stimuli lying in only one plane. Most cone modulations (<~15 Hz) are detected by *RG*, and it is difficult to isolate either *YB* or *Lum* over any substantial region of color space. This point is illustrated in Table 18.1 by the log-unit sensitivity advantage that *RG* has over *YB* and *Lum*, and by the unmasked data in Fig. 18.3 (this is for foveal stimulation; outside of the fovea the relative sensitivity of *Lum* increases: Mullen, 1991; Stromeyer, Lee, & Eskew, 1992). This sensitivity difference is why Sankeralli and Mullen (1996) and Cole, Hine, and McIlhagga (1994) varied spatiotemporal parameters to attempt to isolate the *YB* and *Lum* mechanisms. Equiluminance is not necessary, and often not sufficient, to fully reveal the properties of chromatic detection mechanisms.

Chromatic discrimination

Wavelength discrimination. In the typical wavelength discrimination study, subjects view a continuously present bipartite field, often of 2-deg diameter or so. In the two halves of the field, monochromatic lights of wavelength λ and $(\lambda + \Delta\lambda)$ appear; the observer turns a knob that alters $\Delta\lambda$ until the two halves appear discriminably different, and the threshold $\Delta\lambda$ is plotted against λ. In some cases the method of adjustment is abandoned for other psychophysical procedures. Typically, the experimenter attempts to keep photopic luminosity constant (usually at some rather low value such as 100 Td), but this is difficult and likely it is often not achieved. From our perspective, the classic wavelength discrimination experiment is a poor one: The adaptation state of the observer is not kept constant but rather varies with the wavelength under study, and it is not clear how much of the variation in $\Delta\lambda$ is due to changes in adaptation across wavelengths – changes in the *state* of the system – and how much is due to the direct "test" effects of wavelength difference.

The line in Fig. 18.8 shows a "typical" wavelength discrimination curve for a color-normal observer based on several published studies (see, e.g., Wyszecki & Stiles, 1982a, Sect. 7.10.2), but individual observers may differ substantially (e.g., Wright & Pitt, 1934), and the shape of the curve as well as its height depends on field size, luminance, and other factors (Bedford & Wyszecki, 1958; McCree, 1960). The curve often has a fairly broad minimum near 570–600 nm, where under optimal conditions $\Delta\lambda$ can be much less than 1 nm (e.g., 0.2 nm at 560 nm; Hilz, Huppmann, & Cavonius, 1974).

The points in Fig. 18.8 show predicted wavelength discrimination thresholds based on the cone weights in Table 18.2 and a 10:1 ratio of *RG* to *YB* sensitivity without including second-site desensitization. The poor fit illustrates the importance of second-site effects (see Boynton, 1979). Without second-site desensitization, the model of Eqn. (8) produces the best discriminability in the long-wave portion of the spectrum (Fig. 18.8). Incorporating second-site desensitization would lessen sensitivity, especially at the spectral extremes, and push the minimum $\Delta\lambda$ to shorter wavelengths (nearer 570 nm, cf. Fig. 18.6).

The importance of second-site desensitization in wavelength discrimination has also been emphasized by Mollon and colleagues (Mollon, Estévez, & Cavonius, 1990), who have studied the local maximum near 460 nm (the "short-wave pessimum"). Because

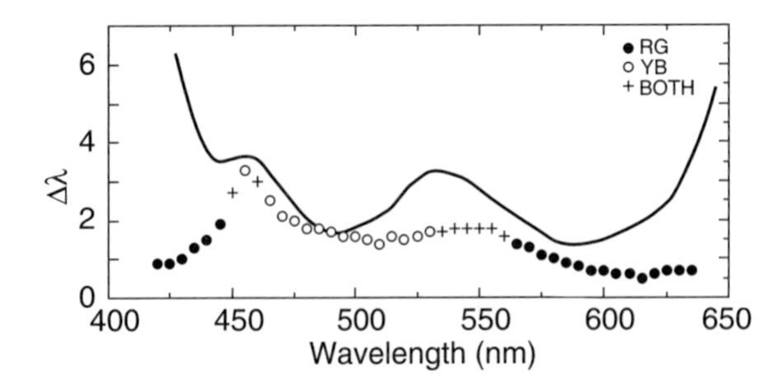

Figure 18.8: The solid line represents a "typical," W-shaped wavelength discrimination function. The points are the prediction of the linear model [Table 18.2, Eqn. (8)], assuming that the observer is always adapted to the mean of the λ and λ+Δλ fields and that there is no second-site desensitization. The summation exponent β was set to 40 to eliminate probability summation; filled symbols represent Δλ's (in nm) that are exclusively determined by *RG*, open ones by *YB*, and the "+" symbols by both mechanisms equally. The ratio of *RG* and *YB* thresholds was fixed as in Table 18.1, and the absolute thresholds were raised until the points matched the curve about 490 nm (due to the nonlinear relationship between Δλ and the cone contrasts, changing the thresholds also changes the shape of the prediction).

the L- and M-cone fundamentals change very slowly relative to each other in this wavelength region (see Chapter 2), discrimination here must depend on S-cones, consistent with the great exaggeration of the pessimum found in tritanopes, which lack functioning S-cones (Fischer, Bouman, & ten Doesschate, 1951; Wright, 1952; see Chapter 2). In color normals, discrimination at 460 nm becomes *better* when field luminance, duration, and saturation are reduced (Mollon & Estévez, 1988; Mollon, Estévez, & Cavonius, 1990), presumably because all three operations reduce the effects of second-site desensitization.

Chromaticity discrimination. MacAdam (1942) generated chromaticity discrimination ellipses by having observers set repeated color matches in a bipartite field at various reference chromaticities; the standard deviation of the match along each color direction was taken as the measure of discriminability of the two half-fields and was therefore used to determine the semiaxis length of the ellipse in that direction. Later studies by Brown and MacAdam (1949), who varied test luminance as well as chromaticity, and Wyszecki and Fielder (1971), who did not, also used the repeated color match technique whereas, more recently, Romero et al. (1993) used a constant-stimulus Yes–No method. As in the wavelength discrimination procedure, the observer's adaptation state covaries with the chromaticity under test.

Because a linear transformation converts an ellipsoid into a sphere (an ellipse into a circle), mechanism directions cannot be inferred a single elliptical contour – the short axis of an ellipse does not itself indicate the mechanism direction of a chromatic mechanism (Poirson et al., 1990; Knoblauch & Maloney, 1996) – because the sphere or circle has no privileged directions. However, this limitation does not hold when there is systematic variation in a *set* of ellipses, and, as noted in the Introduction, LeGrand (1949/1994) analyzed the variation of MacAdam's (1942) ellipses into ($\Delta L - \Delta M$) and ΔS directions, anticipating the "cardinal axes" of Krauskopf, Williams, and Heeley (1982). From the perspective outlined above, the ellipse should be regarded as an approximation to the rounded parallelepiped generated by the probability sum of linear or quasilinear detection mechanisms [Eqn. (8)].

Boynton and Kambe (1980) measured the effect of changing reference chromaticity on cardinal axis thresholds. They found that S-cone thresholds, presumably mediated via *YB*, varied with the excitation level of the adapting field as $\Delta S = k(S + S_0)$, where ΔS represents the threshold, S represents the excitation produced by the steady background, and S_0 is an "eigengrau" term that varied across subjects; all of these are in cone-excitation (S Td, not contrast) units. At high S levels, $\Delta S \approx kS$, which is Weber-like behavior (see also Nagy, Eskew, & Boynton, 1987; Krauskopf & Gegenfurtner, 1992; and Romero et al., 1993). Boynton and Kambe also tested the effect of the L/M adaptive ratio on ΔS thresholds, with the reference S quantal catch held constant, starting at a chromaticity metameric to about 478 nm and then increasing L/M by adding long-wave light. There was little effect of the L/M ratio, which is not inconsistent with the rightmost portion of Fig. 18.7A. Thus the main effect on S-cone thresholds observed by Boynton and Kambe seems to have been first-site adaptation, due (at least in part) to their not using deep blue and violet adapting conditions that would more strongly polarize the second site (as in the left portion of Fig. 18.7A). This is not to say that second-site desensitization played no role in elevating thresholds on Boynton and Kambe's yellow fields; but on yellow fields, where the *S* Td value was negligible, the only effect of second-site desensitization would be to raise the value of S_0 compared with its (unknown) value in the absence of second-site desensitization.

Boynton and Kambe also reported that constant luminance thresholds along the "red-green" ΔL–ΔM-axis (measured in ΔL Td) were lowest for chromatically neutral (yellow and white) conditions and rose with the change in reference *L/M* (Td) value away from neutral in either direction [at constant $V(\lambda)$]. Nagy, Eskew, and Boynton (1987) and Romero et al. (1993) found a similar pattern, but both Krauskopf and Gegenfurtner (1992) and Chaparro et al. (1995) found substantially less elevation of these thresholds, especially for long-wave–adapting conditions. As was discussed previously, however, testing only along the $\Delta L = -\Delta M$ direction can obscure first-site adaptive effects, and when Chaparro et al.'s data are plotted in cone-contrast terms a clear, second-site effect of adapting chromaticity is observed (Fig. 18.6).

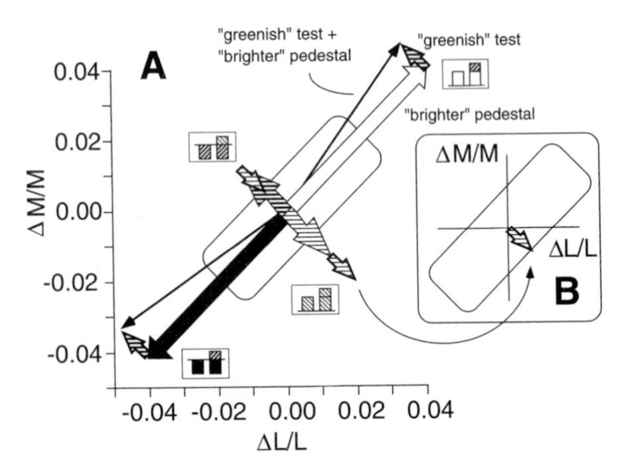

Figure 18.9: Pedestals and the vector difference hypothesis, illustrated in the ($\Delta L/L$, $\Delta M/M$) plane. The rounded rectangle symbolizes a detection contour. (A) The largest arrows represent four different pedestals, selected from the conditions shown in Fig. 18.10. The one at upper right is along the $\{\Delta L/L, \Delta M/M, \Delta S/S\} = \{1,1,0\}$ direction, and it appears "brighter" than the field (in Cole, Stromeyer, & Kronauer, 1990, the field was yellow and this pedestal appeared brighter yellow). The $\{-1,1,0\}$ direction test added to that pedestal appears "greenish." The pedestal alone is presented in one interval of a trial, and the vector sum of test and pedestal is presented in the other interval; the observer must discriminate the two intervals. The four small insets correspond to four of the conditions in Fig. 18.10. (B) A test vector translated to the origin. The vector difference hypothesis is that the pedestal and pedestal+test vectors in panel A will be discriminable when the test vector in panel B would be detectable.

Pedestal facilitation and masking. A "pedestal," as the term is used here, is a reference stimulus with the same shape and time course as the test that is to be detected, but of possibly different spectral composition (according to Tanner, 1961, the term "pedestal" was coined by psychoacoustican J. C. R. Licklider, who, incidentally, was one of the founders of the Internet). In a two-interval forced-choice experiment, for example, the pedestal would be presented in both temporal intervals and the test would be added in one interval. The task of the observer is to discriminate test + pedestal from pedestal alone. Thus, detection reduces to the special case of discrimination with a zero pedestal. A pedestal may also be used in a spatial-alternative task (e.g., Krauskopf & Gegenfurtner, 1992). Pedestal discrimination has been extensively studied in the spatial vision literature (e.g., Graham, 1989; Foley, 1994).

Many discrimination experiments may be represented as detection on a pedestal. Whereas second-site desensitization refers to changes in sensitivity due to changes in the adapting state, the pedestal effects include changes in sensitivity due to momentary shifts away from the adaptation state (shifts to which the observer presumably does not have time to adapt).

Like test stimuli, pedestals may be represented as vectors in cone-contrast space or cone-excitation space (Fig. 18.9A). One vector represents the pedestal alone; the other represents the pedestal plus the test. The vector difference between the two is the test signal that the observer must detect to make the discrimination. The utility of a vector representation becomes most apparent when pedestals are considered: The separate projections of the pedestal and test vectors onto the various mechanism vectors may be used to decompose the effects of quite complex stimuli into their relevant, simpler components.

Wandell (1982, 1985) examined the "vector difference hypothesis," which supposes that two vectors are discriminable when the vector difference between them is such that the difference would be detectable by itself (Fig. 18.9B). This hypothesis is equivalent to supposing that the only effect of a pedestal is to shift the origin of the representation to a new point. The hypothesis fails, in general, since strong pedestals generally raise thresholds, and weak pedestals may lower them. Were the vector difference hypothesis to hold, chromatic discrimination would, in essence, be identical to chromatic detection and we could ignore pedestal effects; it is because the vector difference hypothesis fails that pedestal effects are important to understanding chromatic discrimination.

When the test and the pedestal are detected by the same mechanism, weak pedestals generally facilitate detection, so that discrimination on near-threshold pedestals can be two or three times more acute than detection, whereas intense pedestals produce masking (Nachmias & Kocher, 1970; Nachmias & Sansbury, 1974; Foley & Legge, 1981; DeValois & Switkes, 1983; Mullen & Losada, 1994). Cole, Stromeyer, and Kronauer (1990) measured the pedestal discrimination of 1-deg spots in the *(ΔL/L, ΔM/M)* plane of cone-contrast space (on a yellow field); some of their data are replotted in threshold units in Fig. 18.10. The insets in each quadrant of the figure schematically represent the two intervals of a trial, with a test presented in one interval and the pedestal in both. The upper right-hand quadrant of Fig. 18.10A shows a typical pattern of results when the pedestal and test stimulate the same postreceptoral mechanism with the same polarity, in this case the "red" side of *RG*; both the pedestal and the test were along the $\{\Delta L/L, \Delta M/M, \Delta S/S\} = \{1,-1,0\}$ direction. When the pedestal is weak, thresholds fall almost linearly with pedestal contrast, decreasing by a factor of roughly 2 and then increasing again as pedestal contrast increases, so that pedestals of more than three to four times their own thresholds begin to mask detection of the test. A line of slope −1 through the weak-pedestal data gives the prediction of

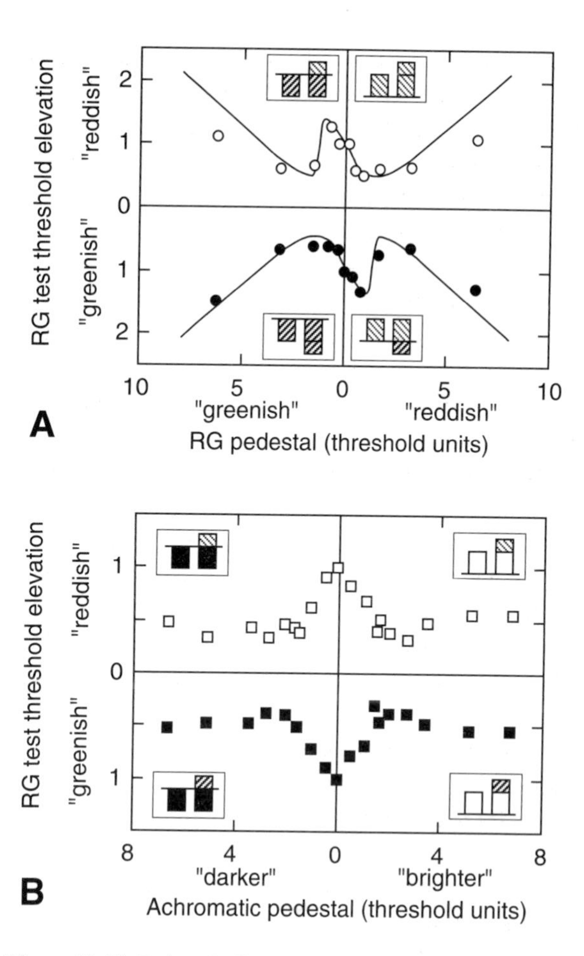

Figure 18.10: Pedestal effects. In each panel, the inset represents the presentation of two pedestals with a test added to one. (A) "Uncrossed" conditions. A greenish (solid symbols) or reddish (open symbols) test was presented with a greenish (left half of graph) or reddish (right half of graph) pedestal. Data are replotted, in units of "no pedestal" threshold, from Cole, Stromeyer, and Kronauer (1990, Fig. 4). (B) "Crossed" conditions. A greenish (solid symbols) or reddish (open symbols) test was presented with a brighter (right half of graph) or darker (left half of graph) pedestal (detected by *Lum*). Data are replotted, in units of "no pedestal" threshold, from Cole, Stromeyer, and Kronauer (1990, Fig. 5).

subthreshold summation, which approximately accounts for the facilitatory effect [however, the subthreshold summation idea is based on a repeatedly discredited psychophysical model, high-threshold theory (Graham, 1989); uncertainty reduction is another possibility (Pelli, 1985)]. That line also predicts the rise in threshold in the second quadrant that occurs when the pedestal polarity is opposite to the test ("green" instead of "red," a test along the {−1,1,0} direction), and this rise has been taken as evidence against the uncertainty reduction account of the weak pedestal effect (Bowen, 1997). The rise in the second quadrant continues until the point at which the pedestal again approaches its own threshold, where the test threshold drops abruptly. Cole, Stromeyer, and Kronauer (1990) attributed this abrupt drop in threshold to a change in the observer's criterion: The observer could choose the "less red" test interval ("green" pedestal + "red" test) by comparing it to the red pedestal, a strategy that can only be used when the pedestal itself is visible. For both pedestal polarities, the test is masked by high-contrast pedestals (little masking is shown in this particular data set, but substantial masking was reported in the same paper for another observer, and in the papers cited above). The curve shown in the upper right quadrant of Fig. 18.12A is sometimes called the "dipper" function.

In contrast to Fig. 18.10A, Fig. 18.10B shows that when the test and pedestal are crossed – that is, they stimulate different postreceptoral mechanisms – there is little masking by the pedestal. Here the test is green or red and the pedestal is equichromatic. Although this is not clear in Fig. 18.10B, in other crossed experiments the pedestal has little effect until it approaches or exceeds its own threshold, when the test threshold drops abruptly to roughly one-half of its zero-pedestal value (Switkes, Bradley, & DeValois, 1988; Chaparro et al., 1994). Cole, Stromeyer, and Kronauer found almost no masking using crossed pedestals; Switkes, Bradley, and DeValois (1988) and Mullen and Losada (1994) found slightly more, but only at very high pedestal contrasts (more than ~30× pedestal threshold). For Cole, Stromeyer, and Kronauer's (1990) disk stimuli, the red-green mechanism and luminance mechanism are approximately interchangeable: They behave similarly in uncrossed conditions, and in crossed conditions it does not much matter which one detects the test and which the pedestal. Whereas Switkes, Bradley, and DeValois (1988) report an asymmetry, in that luminance test gratings were masked but never facilitated by red-green pedestal gratings, Mullen and Losada (1994) find the two types of gratings to be interchangeable, as Cole, Stromeyer, and Kronauer did with spots.

Much evidence suggests that the mechanisms producing facilitation are different from those producing masking (Pelli, 1985; Eskew et al., 1991; Foley, 1994; Mullen & Losada, 1994). The largest effect of a pedestal is masking in the uncrossed conditions, with facilitation never reducing the threshold by much more than a factor of 2 or 3. The masking, which is often attributed to divisive inhibition (e.g., Foley, 1994), produces threshold elevations that approximate a power law, with reported exponents varying from about 0.5 to about 1.0, depending in part on the spatial and temporal frequencies of the stimuli. Similar or even larger masking, but not facilitation, can be obtained when the entire field is flashed instead of presenting a pedestal (e.g., Zaidi, Shapiro, & Hood, 1992) or when a noise-masking pattern is presented instead of a pedestal (Pelli, 1990).

A few studies have examined the effects of mixed pedestals, for example, a pedestal that stimulates both *RG* and *YB* (Krauskopf & Gegenfurtner, 1992) or both *RG* and *Lum* (Wandell, 1985; Cole, Stromeyer, & Kronauer, 1990). Mullen and Losada (1994) measured luminance grating thresholds in the presence of a mask grating that had both *RG* and *Lum* components (because the phases of the mask and test were independently randomized, the mask is not a "pedestal" as we have defined it). The chromatic contrast in the mask was always kept fixed at a level that would by itself slightly raise the threshold for the luminance test; the luminance contrast in the mask was varied in different runs. The resulting function more closely resembled an uncrossed pedestal function (Fig. 18.10A) than the crossed function (Fig. 18.10B) – there was facilitation at low mask contrasts despite the presence of the fixed masking *RG* component, suggesting that in some sense the uncrossed effect takes precedence over the crossed effect, perhaps because the uncrossed effect takes place at an earlier stage in the visual system.

Krauskopf and Gegenfurtner (1992) presented four 36-min-diameter disks around a central fixation spot, all on a 10-deg steady adapting field. Three of the four disks had the same chromaticity, one of them was different, and the observers' task was to select that one (a four-alternative forced-choice task). In the terminology used here, three pedestal stimuli were presented along with a fourth stimulus that consisted of the vector sum of the pedestal plus a test; the discrimination of the odd disk is equivalent to the detection of the test in the presence of a pedestal. The results were plotted in the equiluminant plane of a local cone-excitation space. Krauskopf and Gegenfurtner varied the chromaticity of the pedestal, with all stimuli restricted to the constant-$V(\lambda)$ plane (~370 Td in the main experiments). The results are shown in Fig. 16.11 (see Chapter 16), with the term "test" used for what we call a "pedestal." Red ($+\ \Delta(L\text{–}M)$) and green ($-\ \Delta(L\text{–}M)$) tests were equally masked by either red or green pedestals with contrasts ranging from about 6× threshold up to almost 40×, consistent with Fig. 18.10A (weak pedestals were not used, so no subthreshold facilitation was observed). Similarly, yellow ($-\Delta S$) and blue ($+\Delta S$) tests were approximately equally masked by yellow and blue pedestals, for pedestals up to about 10× threshold – here, weak pedestals produced some facilitation (Fig. 16.11, upper left). These data were collected on a white adapting field. When the background was made red instead of white, the minimum of the pedestal function shifted over to the adapting chromaticity, suggesting that adaptation occurs prior to the site of the pedestal effect (consistent with the first site being early in the visual system). When the crossed functions were measured (a red or green test on a yellow or blue pedestal, and vice versa), the pedestals produced a small facilitation or no effect (upper right and lower left of Fig. 16.11), which is reasonably consistent with the crossed luminance and red-green functions shown in Fig. 18.10B. This last result is important because it implies that *YB* and *RG* behave as *Lum* and *RG* have been shown many times to do (e.g., Fig. 18.10B), with pedestals detected by one mechanism producing a small facilitation and no masking of tests detected by the other mechanism, at least up to very high-contrast pedestals.

In another important experiment, Krauskopf and Gegenfurtner (1992) used tests of various chromaticities on pedestals of various chromaticities. The vector difference hypothesis requires that each set of thresholds be a simple translation of the roughly circular no-pedestal data. This was not the case: The pedestal does more than move the origin, as shown in Fig. 16.12.

The data of Fig. 16.12 were used by Krauskopf and Gegenfurtner to test the following two predictions: (1) Performance is determined by mechanisms tuned to the two chromatic dimensions of the graph, the so-called cardinal directions (Krauskopf, Williams, & Heeley, 1982). In this case, these contours should be oriented with their long axis parallel to the closest cardinal axis, because the pedestal masking would occur primarily in that direction, and for the pedestals that are equidistant from the two axes (45, 135, 225, and 315 deg) the contours should be roughly circular. (2) Performance is determined by many mechanisms tuned to various directions in color space ("higher order" mechanisms; Krauskopf et al., 1986; see Chapter 16). In this case, all of the contours should be of similar shapes with their long axis pointed at the origin, since each pedestal would mask maximally and equally in its own color direction. As Fig. 16.12 shows, neither hypothesis fully accounts for the data: Most of the sets of thresholds would be well described by hypothesis (1), but those on pedestals at 135 and 315 deg are consistent with hypothesis (2).

There is, however, a third possibility that was not discussed by Krauskopf and Gegenfurtner, which is that although there are only two mechanisms, they are not tuned to the "cardinal axes" but to some other, possibly nonorthogonal, directions in this plane of color space. For example, suppose that the red-green mechanism receives a small S-cone input, as we have argued above (Table 18.2). The inclusion of a small S-cone input to *RG* will cause the contours to tilt to the left, as many of them in Fig. 16.12 seem to do, because the negative signal generated by an increase in S will sum with the negative signals generated by an increase in L, and this will interact with the pedestal chromaticity. An example will illustrate this point.

Note that the 135- and 315-deg pedestal data (Fig. 16.12) are narrower than the 45- and 225-deg pedestal data along their $\Delta(L\text{–}M)$, horizontal direction. This asymmetry could be due to the S input to *RG*: At 45 deg, the incremental S-component produces a negative signal that *adds* to the negative produced by being on the reddish side of $\Delta(L\text{–}M)$, and this additional reddish signal could cause the *RG* component of the 45-deg pedestal to mask the *RG* mechanism. In comparison, at a pedestal angle of 135 deg the reddish S-cone signal *subtracts* from the greenish pedestal component produced by being on the other side of $\Delta(L\text{–}M)$, such that the *RG* component of the 45-deg pedestal produces some facilitation of *RG*. The net result would be that the 45-deg pedestal should produce a wider contour than the 135-deg pedestal contour.

Summary and conclusions

To understand chromatic detection and discrimination, we must understand the processes that regulate visual sensitivity. These can be categorized into first-site, cone-specific adaptation and second-site, opponent-level desensitization. The constant slope of detection contours across chromatic adapting conditions (Fig. 18.5A) strongly implies cone-specific, first-site adaptation. Representing data as cone contrasts approximately accounts for the first-site effect, dramatically revealing the second-site effect, but we do not currently have enough information to estimate the second-site effects except in limited regions of the spectrum (Figs. 18.6 and 18.7), nor do we yet understand the mechanisms underlying them.

In the future, the algebraic representations expressed here need to be generalized to time-varying models, to account for the dynamic nature of these pathways. In particular, there are phase shifts and cone-contrast weight changes as a function of temporal frequency between L and M inputs to *Lum*, but not *RG*, except at certain near-neutral adapting chromaticities (Stromeyer et al., 1997). These phase shifts make the linear detection contours into segments of ellipses. In the long run, the strictly linear model will not do; for example, *RG* and *YB* may each need to be half-wave rectified into two opposite polarity subsystems, to account for phenomena such as the decrease in sensitivity to green ($+\Delta M/M$ or $-\Delta L/L$) relative to red ($-\Delta M/M$ or $+\Delta L/L$) in the periphery observed by Stromeyer, Lee, and Eskew (1992).

Acknowledgments

Preparation of this chapter was supported by EY09712. We are grateful for comments on a draft manuscript and other assistance provided by Alex Chaparro, Mike D'Zmura, John Foley, Peter Gowdy, Ken Knoblauch, Barry Lee, Adam Reeves, Marcel Sankeralli, Andrew Stockman, and Charles Stromeyer.

19

Contrast gain control

Michael D'Zmura and Benjamin Singer

Electrophysiological work by Shapley and Victor (1979) initiated a number of investigations of the response gain of neurons in the retino-cortical pathways during the 1980s. Albrecht and Geisler (1991) and Heeger (1992) proposed influential models to account for the findings. Model neurons are tuned to varying spatial frequencies and orientations and possess their own gain control that acts through divisive normalization. The models normalize neuronal responses by the amount of contrast present within the visual field. Such a normalization can help to match neuronal channel capacity to the visual signal, and to provide for contrast constancy under changing conditions of viewing, for instance, to compensate for fog or haze (Robson, 1988; Brown & MacLeod, 1991).

At about this time, Sperling (1989) proposed a psychophysical model of human spatial vision with a contrast gain control, and work on contrast induction by Chubb, Sperling, and Solomon (1989) drew a direct parallel between physiological contrast gain control and the perceptual phenomenon. Contrast induction is demonstrated in Fig. 19.1. Although the two disks, at left and right, have physically identical contrasts, most observers agree that the disk on the left has a lower apparent contrast than the disk on the right. The visual system apparently has a contrast gain control that works automatically: In response to the high contrast on the left the visual system reduces contrast gain, while in response to the low contrast on the right it increases contrast gain. The result is the apparent difference in disk contrast.

Our own work with contrast induction started with the questions of whether contrast induction was a feature of color vision and whether there existed contrast gain controls within color-sensitive pathways. We created isoluminant versions of Fig. 19.1. The red and blue-green colors of the first version were drawn from the L&M-cone axis (MacLeod & Boynton, 1979; Krauskopf, Williams, & Heeley, 1982; Derrington, Krauskopf, & Lennie, 1984). Although the two disks had identical color contrasts, the disk that was surrounded by an area of low contrast typically appeared to possess a higher color contrast than the disk surrounded by an area of high color contrast. The yellow-green and purple colors of the second version were drawn from the S-cone axis. The two disks, again, had identical physical contrasts but different apparent contrasts.

We wished to characterize and understand further these perceptual effects of contrast induction, and turned to the work on contrast gain control to help provide physiological underpinnings. This chapter describes first our methods and initial results, and then turns to the topics of selectivity and spatial pooling in contrast induction. It continues with a discussion of multiresolution models for contrast gain control and ends with a look at the effects of contrast induction on color appearance.

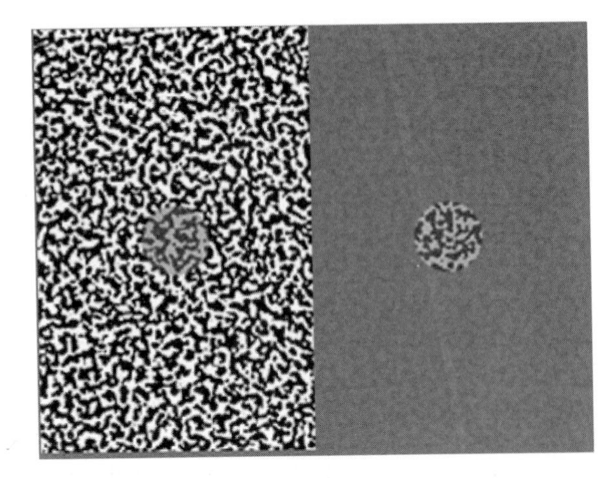

Figure 19.1: Contrast induction. Two disks with contrast levels that are physically identical appear to have different contrasts. The disk at left, surrounded by an area of relatively high contrast, appears to have lower contrast than the disk at right, which is surrounded by an area of relatively low contrast.

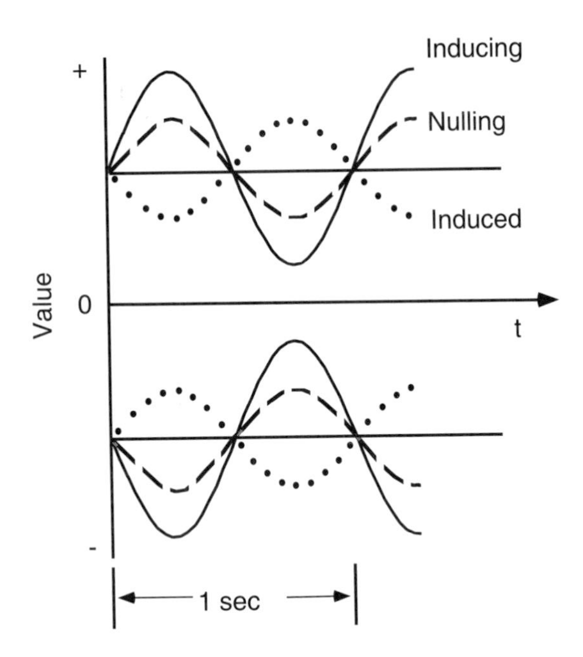

Figure 19.2: Nulling technique. A sinusoidal modulation of annulus contrast at 1 Hz ("inducing") causes an apparent modulation of disk contrast ("induced"). The disk can be made to appear steady by modulating the contrast of the disk physically ("nulling") at just the amplitude that is required to cancel the induced modulation.

Methods and initial results

Chubb, Sperling, and Solomon (1989) used a nulling technique in their work on contrast induction, and we followed suit. The technique is illustrated in Fig. 19.2 for an achromatic stimulus. The physical contrast of elements in an annular surround is made to wax and wane with a sinusoidal time course, typically at a rate of 1 Hz in our experiments. This inducing stimulus generally causes the apparent contrast of a central disk to change in counterphase: When the annulus contrast is high, the disk contrast appears low, and when the annulus contrast is low, the disk contrast appears high. With the nulling technique, one attempts to cancel this induced effect by adding a physical modulation of contrast to the central disk, in counterphase to the inducing modulation. A nulling modulation of the proper amplitude cancels the induced effect and the appearance of the central disk remains steady.

This nulling technique may be applied along arbitrary axes in color space. We describe stimulus color properties using the space of Derrington, Krauskopf, and Lennie (1984). This three-dimensional color space is centered on a neutral gray point, through which pass three axes (see Fig. 19.3): (1) the achromatic axis, along which light intensity varies; (2) the L&M-cone axis, lights along which are equal in luminance to the neutral gray point and that excite short-wavelength–sensitive S-cones equally, and (3) the S-cone axis, lights along which are also equal in luminance to the neutral gray point and that excite long- and medium-wavelength–sensitive L&M-cones equally (Smith & Pokorny, 1975). We typically used binary noise patterns in central and surrounding regions, as in Fig. 19.1, and varied the contrasts of these carrier patterns, presented on a computer monitor, by modulating values in color lookup tables.

In our first experiments, we assessed the basic intensive and spatiotemporal properties of color contrast induction (Singer & D'Zmura, 1994). We varied the amplitude of the inducing modulation and found that nulling modulations increase fairly linearly at low and moderate inducing modulations and then level off

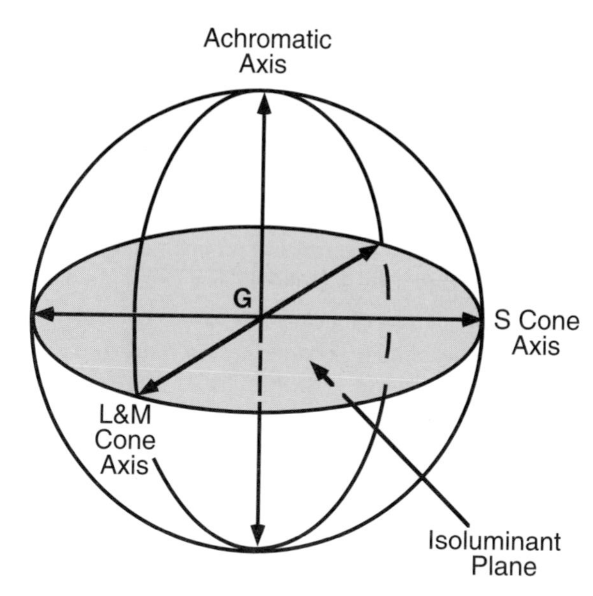

Figure 19.3: Color space of Derrington, Krauskopf, and Lennie (1984) with its three cardinal axes.

at some maximum value. These saturating contrast response functions were found for stimuli lying along achromatic, L&M-cone, and S-cone axes. We also varied the contrast of the central disk while keeping the amplitude of the inducing modulation constant. As the disk contrast increased, the nulling modulations increased fairly linearly up to a point of saturation for achromatic, L&M-cone, and S-cone axis stimuli.

We next varied the size of the inducing annular area systematically in an attempt to replicate and extend the earlier results of Cannon and Fullenkamp (1991). These researchers used sinusoidal carrier patterns and a contrast matching technique to measure the strength of contrast induction. They varied the annulus size and showed that induction increases with increasing annulus size with a roughly exponential saturation. We replicated this result with binary noise carrier patterns and the nulling technique for stimuli along the three color axes. The increase in the strength of induction was found to fall off in an approximately exponential fashion as the annulus size was increased. We then performed an experiment to pin down the conclusion that the modulation of contrast closest to the point of measurement (the central disk) is most efficacious in contrast induction. We used rings rather than annuli and varied the distance of the ring's inducing modulation from the central disk while simultaneously keeping the area of the ring constant. As expected, contrast induction decreases in an approximately exponential fashion as the ring of modulating contrast moves farther away from the central disk.

The space constants of the best-fit exponential curves to the data have values that depend very little on stimulus color, suggesting that the spatial pooling of contrast in contrast induction is independent of stimulus color. The space constants, which were found with binary noise carriers that had peak energy at about 1.8 c/deg of visual angle, were similar to those suggested by the data of Cannon and Fullenkamp (1991) for 2-c/deg sinusoidal carriers. These constants suggest that contrast pooling for foveal viewing occurs primarily over an area with radius 2–3 deg of visual angle centered on the point of measurement.

We next manipulated the temporal frequency of the inducing modulation and found that contrast induction is strongest at low temporal frequencies (as suggested by the static demonstrations of the effect in Fig. 19.1). The induction falls off sharply above 2 Hz and cuts off completely by about 8 Hz. Similar temporal frequency sensitivities were found along the three standard color axes. Although little phase lag in contrast induction is evident at 1 Hz, phase lag becomes more evident as the temporal frequency increases, and we measured the lag by varying the phase of the nulling modulation systematically. The measured phase lag at 1 Hz did not differ significantly from zero. Appreciable lag was measured at 2 and 4 Hz, although in no case were these difficult measurements highly reliable.

Our results on the large area within which contrast is pooled and on the sluggishness of contrast induction led us to suspect that contrast induction has a cortical locus. We tested this by using an interocular transfer experiment (Blakemore & Campbell, 1969). In the interocular transfer condition, the inducing annulus is presented to one eye of the observer and the central disk to the other eye. Contrast induction has about the same amplitude in this condition as in a monocular control condition, in which both the disk and the annulus are presented to a single eye for achromatic, L&M-cone, and S-cone axis stimuli. This result is a strong argument in favor of a cortical locus for contrast induction, because binocular neurons, which are sensitive to inputs from both eyes and presumably needed to mediate the result in the interocular transfer condition, are first found in cortex.

Selectivity

The question arises naturally as to whether contrast modulation of colors along one axis in color space, for instance red and green, affects the appearance of colors along some other axis, such as yellow and blue. In studying this question, we chose to examine first contrast induction among colors along the achromatic, L&M-cone, and S-cone axes (Singer & D'Zmura, 1994).

Figure 19.4 demonstrates the results. In the top row, three annuli with colors along the achromatic (left), L&M-cone (middle), and S-cone (right) axes surround identical achromatic disks. Most observers find that the apparent contrast of the achromatic disk at the left is lower than the apparent contrasts at the middle and right, which suggests that achromatic contrast produces the greatest contrast induction among achromatic colors. In the middle row of Fig. 19.4, the three annuli surround a nominally isoluminant disk with colors drawn from the L&M-cone axis. The apparent color contrast of the central disk is weakest when it is surrounded by an annulus of identical color properties (middle), and this suggests that contrast along the L&M-cone axis causes the greatest contrast induction among colors drawn from the same axis. Finally, most observers report that the apparent contrast of the S-cone disk, when surrounded by an S-cone annulus (bottom right), is lower than when surrounded by annuli of differing color.

We characterized this chromatic selectivity in contrast induction in several sets of experiments. In the first (Singer & D'Zmura, 1994), we used the nulling technique to measure induction for the nine possible combinations of annulus and disk color axes, as in Fig. 19.4. The results, which were found with both binary noise and sinusoidal patterns, showed that modulating achromatic contrast has a strong effect on apparent achromatic contrast but that modulating color contrast at isoluminance has a negligible effect on apparent achromatic contrast. The results also showed that modulating achromatic contrast has an appreciable effect on color appearance at isoluminance. Increasing the achromatic contrast of an annulus causes color saturation within a central disk to decrease, while decreasing achromatic contrast causes color saturation to increase. Such modulations of color appearance are most marked when they are caused by modulations of color contrast along the same chromatic axis. Modulating contrast along the L&M-cone axis is most effective in causing the apparent saturation of a disk with colors along the same axis to vary; modulating contrast along the achromatic or S-cone axes causes smaller changes in the apparent color contrast of the red and blue-green disk. Likewise, S-cone axis modulation induces apparent contrast modulation best among S-cone axis stimuli and less well among L&M-cone axis stimuli.

The simplest model for such chromatic selectivity has three color channels, each of which has its own contrast gain control that is influenced selectively by activity in all three channels. This style of model was first proposed by Chubb, Sperling, and Solomon (1989), who suggested that neurons tuned in the spatial frequency domain have contrast gain controls that normalize response according to the responses of neurons with similar spatial frequency tuning. We discuss spatial frequency selectivity below. Our immediate goal was to develop a chromatically selective model with parameters specified by the results of psychophysical experiments, in which the color properties of the disk and annulus were varied systematically (Singer & D'Zmura, 1995).

The earlier results on the intensive properties of contrast induction suggested that nulling modulations increase linearly as contrast modulation increases and as disk contrast increases. These two linearities suggest the plausibility of a bilinear model (Brainard & Wandell, 1991) for contrast gain control. The gain con-

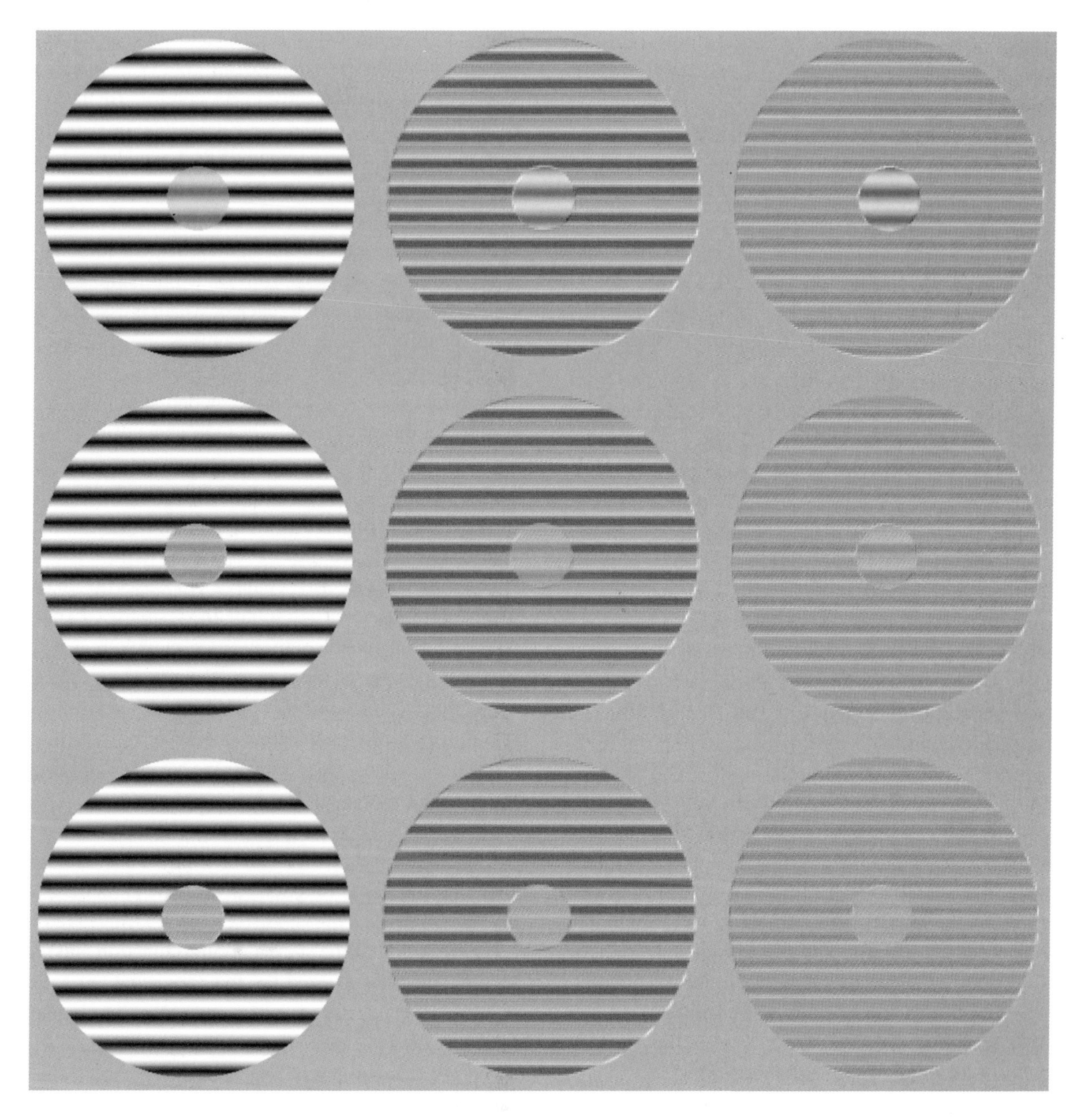

Figure 19.4: Chromatic selectivity. Demonstration for achromatic disks (top) with physically identical contrasts, L&M-cone axis disks (middle), and S-cone axis disks (bottom).

trol in such a model causes linear responses of second-stage achromatic and color-opponent channels to be multiplied by signals that depend linearly on contrast.

To validate such a model, we tested first whether nulling modulations increase linearly as the annulus contrast modulation is increased for each of the nine possible combinations (Fig. 19.4) of color axes for disk and annulus. The results showed that, at small and moderate contrast modulation levels, nulling modulations increase in an approximately linear fashion for

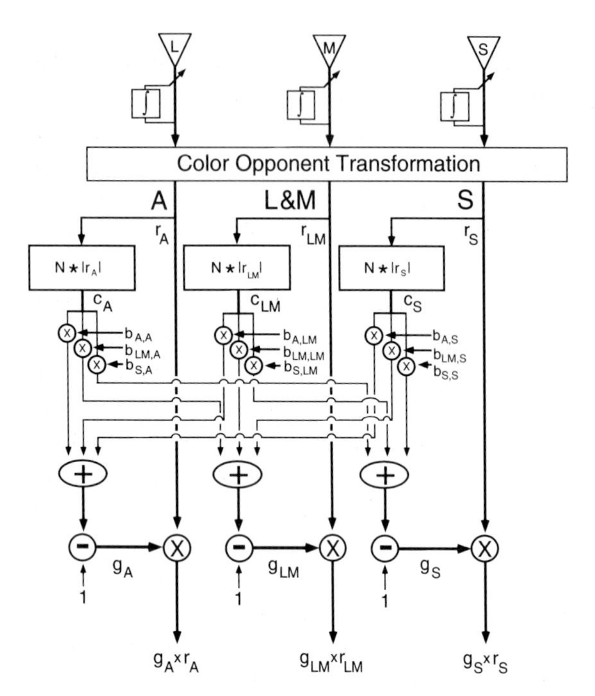

Figure 19.5: Feedforward, matrix-multiplicative model of contrast gain control that is chromatically selective. After Singer and D'Zmura (1995).

each combination of color axes. Saturation was found at higher levels of contrast. We then tested whether nulling modulations increase linearly as disk contrast is increased while holding the annulus contrast modulation constant for each of the nine possible color combinations. Small signal linearity with eventual saturation was again found. Finally, we varied the annulus *mean* contrast. The bilinear model predicts that this has no effect on the sizes of nulling modulations, and this was confirmed to be true at low and moderate contrast levels. Saturation in nulling modulations occurred at the highest contrasts.

We tried to fit the data using a model with three color channels, each with a contrast gain control that works through divisive normalization. Such a model produces linear increases in induction at low contrasts and saturation at higher contrasts, in agreement with our data. Yet functions suggested in earlier work with divisive normalization (Sperling, 1989; Heeger, 1992) led to poor fits of our data, particularly those concerning the effects of varying the annulus mean contrast.

The model that most easily accounts for the data on chromatic selectivity is shown in Fig. 19.5. It uses a feed-forward, multiplicative gain control. At the top are indicated the L-, M-, and S-cone photoreceptoral systems, which are provided with a von Kries-type gain control (von Kries, 1905b), and it is assumed that this gain control acts to scale cone signals by the average light. These scaled cone signals are then combined linearly in a second stage to form achromatic, L&M-cone and S-cone channels, each of which is assumed to signal positive and negative deviations about the average light.

Contrast in each of the three second-stage channels is determined by (1) rectifying the channel response $r(\mathbf{x})$ to create a nonnegative contrast signal $|r(\mathbf{x})|$ at each spatial position, and (2) determining local contrast by pooling the nonnegative contrast signal over an appropriate area (see Fig. 19.5).

We full-wave–rectify channel responses to determine contrast, rather than square the responses, to provide a linear increase in nulling modulation with increasing contrast modulation. The spatial pooling is accomplished formally by convolving a spatial pooling function $N(\mathbf{x})$ with the space-varying, rectified response $|r(\mathbf{x})|$ of each channel. We have in mind two-dimensional Gaussians or decaying exponentials for the spatial pooling functions, as suggested by our data and those of Cannon and Fullenkamp (1991), and we shall consider these functions in more detail in the following.

Convolution of the spatial pooling function and the rectified responses provides the (nonnegative) local contrast levels c_A, c_{LM}, and c_S in each color channel (see Fig. 19.5). These local contrasts are then combined in a manner that is consistent with the empirical results on chromatic selectivity (Singer & D'Zmura, 1995). The nine coefficients b determine the effects that the contrasts in each channel have on contrast gain. The selectively combined contrasts are used to determine the gains g_A, g_{LM}, and g_S, which act multiplicatively on the channel responses to produce the three outputs $g_A\, r_A$, $g_{LM}\, r_{LM}$, and $g_S\, r_S$. The three local contrasts c_A, c_{LM}, and c_S are combined linearly and then subtracted from 1 to determine these gains.

To specify more formally the action of the selective contrast gain control, consider first the responses of the second-stage mechanisms. By hypothesis, each mechanism responds to a space-varying visual stimulus by combining linearly the space-varying cone mechanism signals $L(\mathbf{x})$, $M(\mathbf{x})$, and $S(\mathbf{x})$:

$$(1) \quad r_i(\mathbf{x}) = a_{i,L}\, L(\mathbf{x}) + a_{i,M}\, M(x) + a_{i,S}\, S(\mathbf{x}) \,,$$

for mechanisms $i = A$, $L\&M$, and S. The average response of the achromatic mechanism, a nonnegative value, is subtracted to provide an achromatic response that takes on both positive and negative values. The local contrast within each second-stage mechanism, as a function of spatial position, is given in terms of the mechanisms' responses by:

$$(2) \quad c_j(\mathbf{x}) = N(\mathbf{x}) \cdot |r_j(\mathbf{x})| \,,$$

for $j = A$, $L\&M$, and S, in which $N(\mathbf{x})$ describes the spatial pooling of contrast. Finally, the contrast gains are determined by a selective linear combination of local contrasts:

$$(3) \quad g_i(\mathbf{x}) = 1 - \sum_j b_{ij} c_j(\mathbf{x}) \,,$$

for $i = A$, $L\&M$, and S and $j = A$, $L\&M$, and S.

The effect of increasing stimulus contrast is to reduce gain in an inhibitory fashion: The multiplicative gain g is equal to 1 if local contrast levels are 0 and is less than 1 otherwise. Note that if contrast levels are high enough, then it is possible for the weighted sum of contrasts to exceed 1, in which event the contrast gain is a negative number. Multiplying a second-stage channel's response by a negative number reverses its contrast. Two things prevent this undesirable consequence from happening in practice. First, the empirically determined coefficients b that describe the combination of contrasts are sufficiently small, relative to the contrasts encountered in natural images, and, second, the observed saturating nonlinearities limit the magnitude of contrast induction (Singer & D'Zmura, 1995).

We fit the saturation of contrast induction at higher contrasts by placing nonlinear saturation functions in the model of Fig. 19.5. We first tried to fit the experimental data by placing a saturating nonlinearity in each channel's gain control, acting after the summation of contributing contrasts from all channels, for a total of one nonlinearity per channel. Better fits were found by providing a saturating nonlinearity for each channel interaction, allowing the nonlinearities to act just prior to the summation of contributing contrasts, for a total of nine nonlinearities. The choice of spectral sensitivities for the second-stage, color-opponent mechanisms in this model is dictated primarily by convention. We tried to find privileged axes by searching for a pair of axes in the isoluminant plane for which contrast modulation along one axis produces no induction along the other. No such privileged axes were found.

We were puzzled that contrast induction alters color saturation but not hue. One might expect such a hue shift when viewing, for instance, an orange and blue disk surrounded by a contrast-modulated red and blue-green annulus. The amount of red within the orange, but not the yellow, could well vary in time, leading to an induced change in apparent hue. Our attempts to measure such hue shifts were fruitless. Apparently, the influence of red-green contrast on yellow-blue gain is sufficient to minimize such a hue shift. Channel interaction within the model is determined by the coefficients b that determine the influence of each channel's contrast on each channel's gain. While it is possible to choose coefficients that lead the model to predict substantial changes in apparent hue, the empirically determined coefficients lead to predicted changes almost exclusively in chromatic saturation (Singer & D'Zmura, 1995).

The chromatic selectivity exhibited by this model can be extended to account for other forms of selectivity. Chubb, Sperling, and Solomon (1989) used band-pass-filtered noise patterns to show that the change in apparent contrast of a central disk, found when contrast in a surrounding annulus is varied, depends on the relative spatial frequencies of the center and surround. Contrast induction is halved if the peak spatial frequency of the surround is moved to either an octave

Figure 19.6: Demonstrations of selectivity in spatial frequency (*top*), orientation (*middle*), and phase (*bottom*). A figure that demonstrates spatial frequency selectivity but uses noise carriers rather than sinusoids is presented by Chubb, Sperling, and Solomon (1989).

lower or an octave higher than that of the center. This selectivity can be captured by a model of visual processing that has several channels with different spatial frequency sensitivities. Activity within a spatial frequency channel affects most strongly the gain on that channel and more weakly the gain on channels with different sensitivities (Chubb, Sperling, & Solomon, 1989).

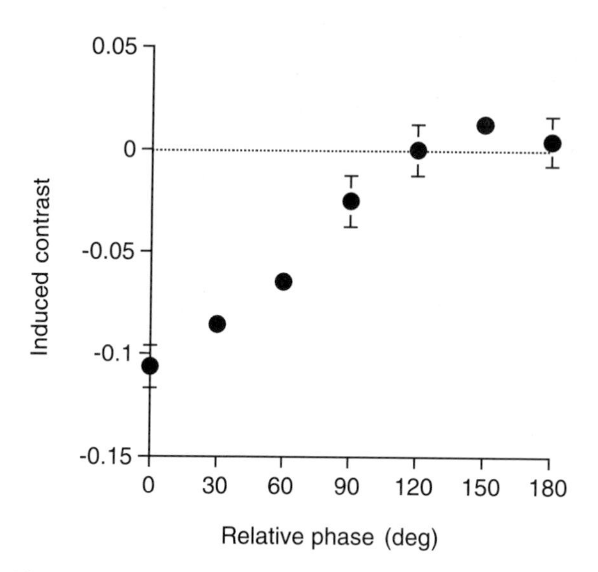

Figure 19.7: Measurement of phase selectivity. Data from a single observer (author MD), who used a contrast matching technique to match the apparent contrast of a disk of contrast 0.4, surrounded by an annulus of contrast 0.5, as relative phase was varied. Sinusoidal carrier patterns at 2 c/deg were used.

Spatial frequency selectivity is demonstrated along the top row of Fig. 19.6 using sinusoidal carrier patterns. The apparent contrast of the central disk is weakest in the case where the spatial frequencies of the center and surround are identical (middle).

The middle row of Fig. 19.6 demonstrates orientation selectivity, which was first investigated by Solomon, Sperling, and Chubb (1993). They used sinusoidal carriers as in Fig. 19.6 to show that contrast induction is strongest when the center and surround have identical orientations, and that this orientation selectivity is most marked at higher spatial frequencies. In our own work (Singer & D'Zmura, 1994), we used binary noise with energy limited to narrow orientation bands to investigate orientation selectivity. We found no evidence for orientation selectivity using achromatic, L&M-cone, and S-cone axis stimuli. However, the peak spatial frequency of our stimuli in these experiments, 1.8 c/deg, was low. The results of Solomon, Sperling, and Chubb (1993) suggest that orientation selectivity is first significant beyond 3–4 c/deg. One can capture orientation selectivity by a model of visual processing that has several channels with different orientation sensitivities. Activity within a channel of specific orientation tuning affects the gain within that channel most strongly.

The bottom row of Fig. 19.6 demonstrates phase selectivity. The relative phase of the central and surrounding sinusoids is 0 deg at left, 90 deg in the middle, and 180 deg at right. Contrast induction is strongest when the relative phase is 0 deg and weakest when it is 180 deg, although there is little difference between the effects found at 90 and 180 deg. Such phase selectivity was reported by Heeger and Robison (1994), who used plaid carrier patterns.

We have begun a more extensive study of phase selectivity (Za, Iverson, & D'Zmura, 1997), and a sample result with sinusoids at 2 c/deg is shown in Fig. 19.7. An annulus of contrast 0.5 reduces significantly the apparent contrast of a disk of contrast 0.4 for relative phases in the range 0–60 deg, but has small or negligible effects at greater relative phases. Preliminary results suggest that the variation of induction with relative phase depends on the choice of disk and annulus contrasts.

Pooling of contrast

We wished to extend the chromatically selective model to a more complete, multiresolution model with channels tuned to spatial frequency, orientation, and color. Such a model, specified using psychophysical data, could be applied to color images to quantify and visualize the effects of contrast induction on visual processing. Toward this end, we conducted further measurements of the spatial pooling of contrast to help specify the weighting functions for each channel (D'Zmura & Singer, 1996).

We modulated the contrast modulation of the inducing annular region with a spatial sinusoid (see Fig. 19.8). By varying the spatial frequency of the sinusoidal contrast modulation, we could determine the dependence of contrast induction on inducing spatial frequency (see also Zaidi et al., 1992; see Chapter 17). Transforming such a dependence from the spatial fre-

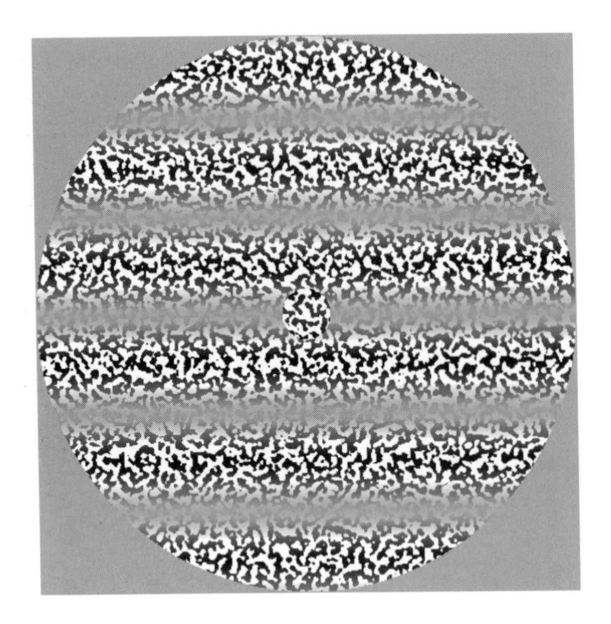

Figure 19.8: Spatially sinusoidal modulation of contrast of variable spatial frequency. After D'Zmura and Singer (1996).

quency domain back into the space domain provides the profile of a spatial pooling function. The spatial frequency sensitivity data for contrast pooling were fit well by Gaussian functions. The inverse Fourier transform of a Gaussian is again a Gaussian (Bracewell, 1978), so that spatial pooling functions for contrast induction can be modeled by Gaussians. The frequency-domain data were also fit with decaying exponential functions, but these accounted for the data less well than did the Gaussians.

We measured spatial-frequency sensitivity functions for coarse, medium, and fine carriers with peak spatial wavelengths of 0.69, 0.34, and 0.17 deg, respectively. The results showed that the linear extent of spatial pooling increases linearly, but not proportionally, with carrier wavelength. The linear extent of pooling, taken as two standard deviations of the best-fit Gaussian, was found to be approximately eight times the carrier wavelength, plus about 2 deg of visual angle. Cannon and Fullenkamp (1991) found earlier with their sinusoidal stimuli that spatial pooling is proportional to carrier wavelength. Our results differ in suggesting that the pooling does not go to zero as spatial frequency increases; rather, it tends toward a minimum extent of 2 deg. We also varied stimulus color properties and found only small differences in spatial pooling for achromatic, L&M-cone, and S-cone axis stimuli, in agreement with our earlier results.

The role that eye movements play in such results is uncertain. Although observers fixated on the central disk in the experiments, they undoubtedly moved their eyes during the course of single observations. These movements presumably cause the measured spatial pooling functions to be more extensive than they would be were eye movements absent.

Multiresolution model

With the spatial pooling functions in hand, we created a multiresolution model of contrast gain control that could be applied to color image data (D'Zmura, 1998). The natural extension to the chromatically selective model of contrast gain control (Fig. 19.5) includes channels that are selective not only for color, but also for spatial frequency and orientation. The response of each such channel is multiplied by a gain that depends on the local contrasts of the channels. These local contrasts depend, in turn, on contrast pooling functions that have sizes that depend on channel spatial frequency properties.

We used four octave-wide spatial frequency sensitivities, six orientation sensitivities of width 30 deg, and three chromatic sensitivities (A, L&M, and S) to create the model. These sensitivities were combined factorially to produce a model with 72 channels. The spatial frequency sensitivities of the channels are described by nonoverlapping regions of the spatial frequency plane (see Fig. 19.9). These regions are described by functions that take on the value 1 within the range of frequencies ω and orientations ϕ to which a mechanism is sensitive and take on the value 0 elsewhere.

A disadvantage of the complete localization of channel sensitivity in the spatial frequency domain is that the spatial receptive fields of the channels have unlimited extent. Yet the frequency sensitivities do not

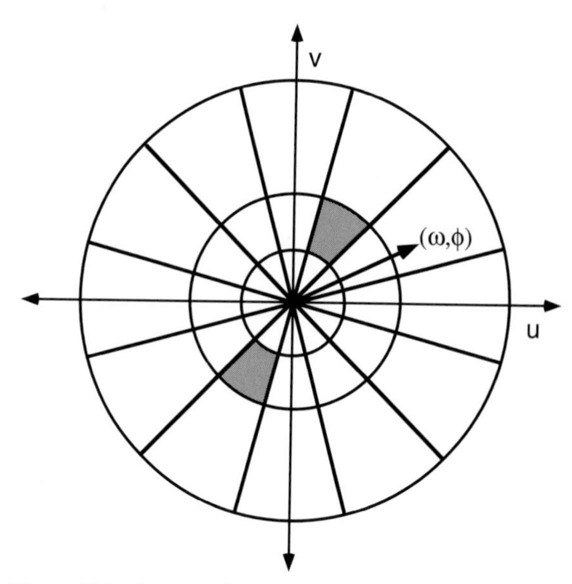

Figure 19.9: Sensitivities of model channels described by regions in the spatial frequency plane. The radial frequency coordinate ω and the azimuthal orientation coordinate φ can be used to describe a point in the spatial frequency domain. An exemplary channel sensitivity is shown by the shaded regions.

overlap, so that the channels are orthogonal to one another, and this orthogonality lets one reconstruct easily the output image that results from applying the model's contrast gain control to some input image. To ensure that the chromatic mechanisms are orthogonal, we treat lights along the three axes A, L&M, and S as contravariant quantities with coordinates that are related to those of three covariant, orthogonal mechanisms by an appropriate linear transformation (Knoblauch, 1995).

Model computation starts with a color decomposition of an input image into achromatic, L&M-cone axis, and S-cone axis components. These three color components r_A, r_{LM}, and r_S are determined by appropriate linear combinations of an input image's *RGB* values after the image has been stripped of its mean(R_{avg}, G_{avg}, B_{avg}):

$$(4) \quad r_i(\mathbf{x}) = a_{i,R}\,(R(\mathbf{x}) - R_{avg}) + a_{i,G}\,(G(\mathbf{x}) - G_{avg}) + a_{i,B}\,(B(\mathbf{x}) - B_{avg}),$$

for mechanisms $i = A$, $L\&M$, and S. The model then determines the responses of the channels tuned to spatial frequency and orientation. This is done by transforming the color responses [Eqn. (4)] into the spatial frequency domain, where it is easiest to parcel up the input image among the various channels. Suppose that the functions $G_j(\omega)$, for $j = 1, 2, 3, 4$, describe the four spatial frequency sensitivities and that the functions $H_k(\phi)$, for $k = 1, ..., 6$, describe the six orientation sensitivities. As was mentioned above, these functions have value 1 within their sensitivity range and are 0 elsewhere. The response $r_{ijk}(\mathbf{x})$ of the channel with the *i*th color, *j*th spatial frequency, and *k*th orientation sensitivities is then given by:

$$(5) \quad r_{ijk}(\mathbf{x}) = \mathcal{F}^{-1}\,[\,\mathcal{F}\,[\,r_i(\mathbf{x})\,]\,G_j(\omega)\,H_k(\phi)\,]\,,$$

for $i = A$, $L\&M$, S, $j = 1, ..., 4$, and $k = 1, ..., 6$. The original chromatic response $r_i(\mathbf{x})$ to an input image [Eqn. (4)] is subject to a Fourier transform $\mathcal{F}$ and is then filtered using G_j and H_k to produce the appropriate response in the spatial frequency domain. This response is then transformed back to the space domain to provide the desired response $r_{ijk}(\mathbf{x})$.

The model's contrast gain control works by multiplying each of the 72 channel responses at each position by channel-specific gains. The gains are determined by taking appropriate linear combinations of local contrasts found within each of the channels. Figure 19.10 shows a diagram of the model's contrast gain control for a single, spatiochromatically opponent channel with initial response r_{ijk}.

The local contrast within a channel is computed by taking the absolute value of the channel's response and convolving the result with a Gaussian contrast pooling function of appropriate size. The formula for computing the local contrast $c_{ijk}(\mathbf{x})$ within the *ijk*th channel, as a function of spatial position, is

$$(6) \quad c_{ijk}(\mathbf{x}) = N_j(\mathbf{x}) \cdot |r_{ijk}(\mathbf{x})|\,,$$

for $i = A$, $L\&M$, and S, $j = 1, ..., 4$, and $k = 1, ..., 6$. The Gaussian functions N_j are spatially isotropic and have standard deviations σ_j that increase linearly with chan-

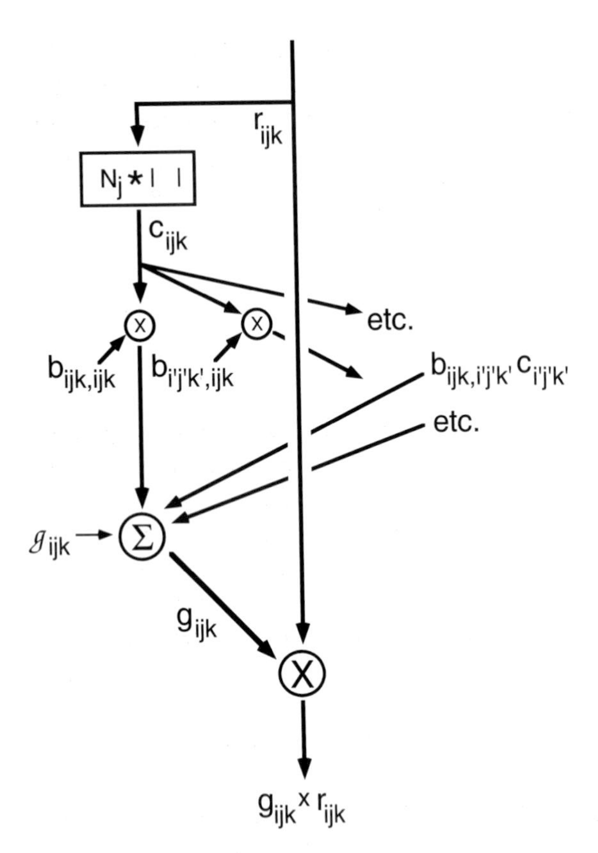

Figure 19.10: Single channel within a multichannel model of contrast gain control. See text for discussion.

nel peak wavelength (D'Zmura & Singer, 1996).

Note that if one omits the full-wave rectification and simply convolves a channel's response with a Gaussian, all that would be achieved would be a blurring that would not correspond at all to a contrast computation. The rectification causes the result of the local contrast computation to correspond to our intuition of contrast as a difference between the maximum and minimum values.

The gain control works by forming, for each of the 72 channels, an appropriate linear combination of the contrast responses, which it then uses to inhibit channel signals.

The coefficients b that describe channel interaction are set to values that incorporate the psychophysical results on selectivity in spatial frequency, orientation and color (Chubb, Sperling, & Solomon, 1989; Solomon, Sperling, & Chubb, 1993; Singer & D'Zmura, 1994). The values suggested by psychophysical results have always corresponded to inhibitory effects: Increasing contrast leads to decreases in gain and channel inhibition. Yet this is not a necessary feature of the model, which can readily accommodate the facilitation of channel response through the use of a value of opposite sign for one or more channel interaction coefficients.

It is sensible to suppose that a channel interaction coefficient b has a negative value if one channel's contrast inhibits another's response through the contrast gain control. This differs from the model in Singer and D'Zmura (1995), shown in Fig. 19.5, where the coefficients were positive and the linear combination of contrasts was subtracted from 1. In the present model, inhibitory interactions correspond to negative coefficients, and possible facilitatory interactions correspond to positive coefficients.

All weighted contrasts are summed within each channel's contrast gain control. The gain $g_{ijk}(\mathbf{x})$ on the ijkth channel as a function of spatial position is given formally by:

$$(7) \qquad g_{ijk}(\mathbf{x}) = \mathscr{g}_{ijk} + b_{ijk,abc}\, c_{abc}\, (\mathbf{x}),$$

for $i = A$, $L\&M$, and S, $j = 1, ..., 4$, and $k = 1, ..., 6$. The gain on the ijkth channel depends on the local contrasts in all of the other channels, to the extent that coefficients $b_{ijk,abc}$, which describe the influence of contrast within some abcth channel on the gain of the ijkth channel, are nonzero. Note the role of the term $\mathscr{g}_{ijk}$, which is the gain on the channel when the pooled contrast is 0. This we have set to 1 in model computations, so that increasing contrast leads to inhibition when the coefficients b are negative. Yet one may bias this term by setting it to a value other than 1. If this gain is greater than 1, then low image contrasts lead to facilitation through multiplication by a gain greater than 1, while higher contrasts inhibit channel activity through multiplication by a gain less than 1.

The model that we use is a simple, separable model: Selectivities in spatial frequency, orientation, and color are mutually independent. The coefficient that describes the contribution made by one channel's local contrast to the gain of a second channel is equal to the product of three independent factors: (1) a factor that depends on the spatial frequencies of the two channels; (2) a factor that depends on the two channels' relative orientations; and (3) a factor that depends on the two channels' chromatic properties.

In simulations described below, we have used for spatial frequency a factor that is maximum for channels of identical spatial frequency, falls to half-maximum for channels that differ by an octave, and falls to 0 for channels that differ by more than an octave. For orientation, the factor is maximal for channels of identical orientation, at half-maximum for channels that differ by 90 deg, and varies sinusoidally between maximum and half-maximum values for the intermediate relative orientations. For color, the factors are identical to those found by Singer and D'Zmura (1994, Fig. 12), with the exception that the contributions of isoluminant contrast to achromatic gain are taken to be 0.

A benefit of choosing a separable model of selectivity is that the 72×72 matrix of channel interactions is specified by far fewer quantities: 3×3 numbers to describe chromatic selectivity, plus 4×4 numbers for spatial frequency (of which we specify only three), plus 6×6 numbers for orientation (of which we specify only six) for a total of 61 channel interaction parameters for a separable model, a sum that we reduce further to 18. A second advantage of this model is that its effects on an image are roughly independent of image scale, because the model works nearly identically in all spatial frequency bands.

The receptive field model that is suggested by the results has two components. First is the "standard" linear receptive field, which is selective for stimulus spatial frequency, orientation, and color properties (e.g., Lennie, Krauskopf, & Sclar, 1990; Gegenfurtner, Kiper, & Fenstemaker, 1996). The standard receptive fields of neurons in a single channel determine the immediate response $r(\mathbf{x})$ of the channel. The second component is a large inhibitory surround that is somewhat but by no means completely matched to the properties of the center. This surround has a Gaussian profile, within which contrast is pooled and is centered on the standard receptive field. Activity within this inhibitory surround determines channel gain $g(\mathbf{x})$, which acts multiplicatively to determine the attenuated channel response $g(\mathbf{x})\ r(\mathbf{x})$. One may place a temporal filter in the pathway that computes local contrast to incorporate the temporal frequency sensitivity and phase lag results of Singer and D'Zmura (1994).

Appearance

The primary effect of contrast gain control processing is to equalize contrast levels across space and time. As shown in Fig. 19.1, low levels of contrast lead to relatively higher contrast gains in neighboring areas while high levels of contrast lead to lower gains in nearby areas. The local pooling of contrast by the gain control accounts for this spatial equalization. A gain control that pooled contrast over the entire visual field would wash out differences in contrast levels across regions, providing a gain that is identical at all points. Indeed, model simulations show that the spatial equalization of contrast at low frequencies is relatively poorer than at high frequencies, simply because the pooling functions at low frequencies are much broader in spatial extent (D'Zmura, 1998).

The center-surround structure of receptive fields that is imparted by the contrast gain control leads to a variety of visual illusions with "second-order," contrast-modulated stimuli. These illusions are analogous to classical visual illusions that depend on lateral inhibition.

Lu and Sperling (1996) measured the contrast induction found with achromatic, second-order Chevreul stripe, Mach band, and Craik–O'Brien–Cornsweet stimuli. They created these using achromatic contrast gradients rather than the standard lightness gradients. Lu and Sperling observed illusory variations in apparent contrast levels that are consistent with the operation of mechanisms with excitatory center and inhibitory surround. The model that we presented in the previous section generates Lu and Sperling's effects readily. Simulations of the visual processing of contrast-modulated Chevreul stimuli, for instance, produce output images that have reduced contrast in regions close to areas of higher physical contrast and that bear increased contrast in regions close to areas of lower physical contrast.

Moreover, the model suggests that these effects of lateral inhibitory processing should exist at isoluminance. Our own observations (Singer, 1994; Singer, Montag, & D'Zmura, 1995) of Chevreul stripes and Mach bands formed of isoluminant contrast modulations agree with these model predictions.

What is perhaps less obvious is that a contrast gain control that is selective for spatial frequency and orientation will also equalize contrast levels across spatial frequencies and orientations. These effects are illustrated by the image sequences in Fig. 19.11. Each sequence ("Lenna" in row one and "Space Shuttle" in row three) results from the repeated application of the contrast gain control model specified in the previous section to the input image in the leftmost column. Immediately below the successive model output images are shown the differences between the model output image and the previous image in the sequence (rows two and four). The first difference, shown in column two, is that between the original image and the immediate output of the contrast gain control. The first difference images show clearly the spatial equalization of contrast. Contrast in the space shuttle image, for example, is reduced strongly in areas of high contrast, like the shuttle and the rocket, but is little affected in areas of low contrast, like the sky.

The sequences of differences shown in rows two and four show that repeated application of model processing to an image produces an image that is eventually little changed by the contrast gain control: the differences fall off to zero. The resulting images pass through the contrast gain control unscathed, with the possible exception of a single scale factor for overall contrast. Such images are stable points of convergence that characterize the nonlinear processing of the model.

Inspecting these characteristic images (Fig. 19.11, rows one and three, rightmost column) reveals increased energy at orientations that are not present in the original image. Likewise, the characteristic images possess energy at high spatial frequencies that was not present in the original image. Energy is now spread more equally throughout all spatial frequency and orientation bands.

While the center-surround processing and the equalization of contrast by the model of contrast gain control help to account for many perceptual phenomena, there is a class of visual stimuli that give rise to changes in apparent contrast that elude the model. These stimuli, typified by White's (1979) illusion, have been of great interest in studies of lightness perception, because they show convincing effects of context that elude bottom-up, spatially homogeneous models of processing. Spehar and Zaidi (1997b) have created and studied achromatic, second-order versions of these stimuli. We show a second-order White's illusion stimulus in Fig. 19.12 (left), along with two versions at isoluminance, along the L&M-cone axis (middle) and along the S-cone axis (right). In all three cases, the apparent contrast of the four rectangles in the right column is less than that of the four rectangles in the left, even though the physical contrasts of the rectangles are identical.

An intuitive explanation for this illusion is that the rectangles along the right are seen somehow to belong to the horizontal stripes, while those on the left are not. As a result, the contrast of the rectangles on the right is compared to the contrast of the horizontal stripes and is found wanting, while the contrast of the rectangles on the left is compared to the low contrast of the background region.

Our model of contrast gain control, and others like it, cannot account for these perceptual phenomena. The rectangles at right have a longer border with regions of low contrast than they do with regions of high contrast, while the rectangles at left have a longer border with regions of high contrast than of low. The model predicts an effect in the direction opposite to that perceived.

Prominent in accounts of the standard White's illusion and related visual effects are mechanisms that sense the presence of a continuous border despite the presence of a change in lightness. One can segment an image in this way through an analysis of T-junctions

Figure 19.11: Equalization of contrast across space, spatial frequency, orientation, and color demonstrated through repeated application of the algorithm to input images. *Top row*: Lenna image (left) operated on repeatedly by the algorithm. *Second row*: Sequence of difference images showing change between input and output images. *Third row*: Space shuttle image (left) and its sequence. *Fourth row*: Difference images for the space shuttle.

(Todorovic, 1996). The most obvious construction to help account for the contrast-modulated White's illusion (Fig. 19.12) involves T-junctions among areas of different contrast levels and mechanisms that sense these.

Discussion

The model of contrast gain control that we have presented is designed to exhibit selectivity in color, spatial frequency, and orientation, like that detailed in psychophysical experiments. Its spatial pooling of contrast endows the receptive fields of model units with a secondary center-surround structure, and this helps to account further for the apparent lateral inhibition in the processing of contrast-modulated patterns. Model processing of color images suggests that the main effect of contrast gain control is to equalize contrast levels across spatial position, spatial frequency, orientation, and color.

The model bears two important simplifications. First, its spatial processing is largely scale-invariant: Processing does not vary among spatial frequency channels. Second, the model uses channel interactions that are separable in spatial frequency, orientation, and

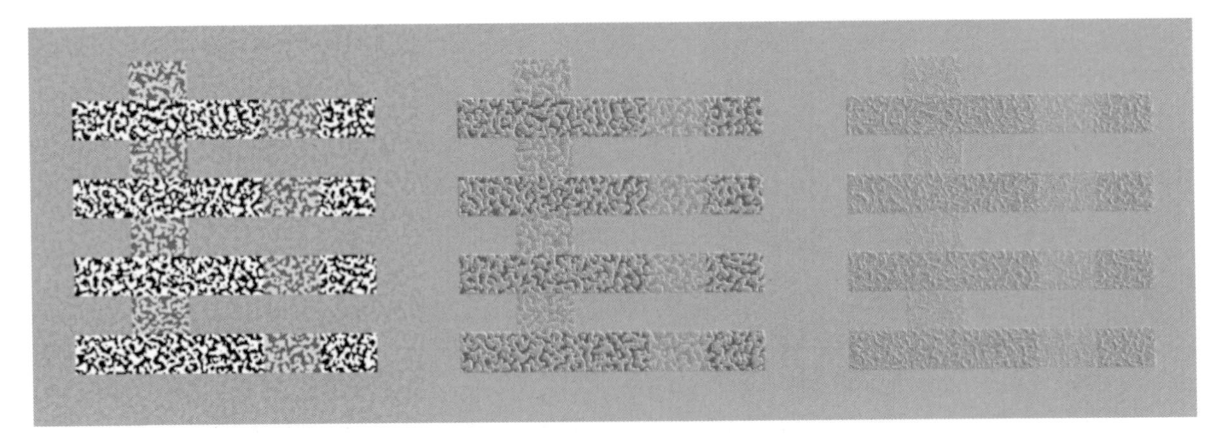

Figure 19.12: Second-order White's illusion for achromatic (left), L&M-cone axis (middle), and S-cone axis (right) stimuli.

color; yet scale invariance and separable interaction are known not to hold for human visual processing, and the question arises how one may amend the model to provide a closer account of human visual processing.

Color-sensitive channels have spatial frequency sensitivities that cut off at a frequency lower than that for the achromatic channel (Mullen, 1985). A model with reduced color sensitivity at the highest frequencies would thus agree better with human visual processing, yet it would no longer exhibit scale invariance. Likewise, the results on orientation selectivity of Solomon, Sperling, and Chubb (1993) and Singer and D'Zmura (1994) show that a lack of selectivity at low spatial frequencies gives way to substantial selectivity at high frequencies. Channel interaction is evidently not separable in spatial frequency and orientation, and changing the model to include this result would cause processing to vary with scale. Pooling areas that increase linearly (D'Zmura & Singer, 1996) but not in proportion (Cannon & Fullenkamp, 1991) to carrier spatial frequency provide a third assault on scale invariance. The size of a visual image on the retina must be known to determine the effects of scale-varying processing. Indeed, one suspects that the position of the image on the retina is also critical in contrast processing. Although we know of no studies that have investigated the dependence of contrast induction on eccentricity, human processing is likely to vary significantly as a function of retinal position.

The present model is thus a very simple one, and emendations to provide a closer match to human visual processing will lead to a more complex model that is inseparable, space-varying, scale-varying, and has a greater number of channels. However, a more complex model may provide a better account of the effects of low-level visual processing on appearance. Such a model may also help in understanding visual pattern detection; Foley (1994) has shown that divisive normalization of channel responses in pattern detection tasks can help to account for a variety of psychophysical masking data.

Perhaps the greatest weakness of the model is that it provides no role for higher level, perceptual mechanisms in contrast induction. The contrast-modulated White's illusions of Fig. 19.12 show that these higher level mechanisms alter apparent contrast, and one must wonder just how much of the processing that has been assigned to low-level mechanisms is best placed there. In viewing Figs. 19.4 and 19.6, for instance, one notes that contrast induction is accompanied by the perception of transparency. The central disks that have the weakest apparent contrast appear as though a Metelli-type, contrast-reducing transparent filter has been placed atop a much larger disk (Metelli, 1974). Are spatial frequency, orientation, and color selectivities in fact properties of transparency-sensitive mechanisms that are able to separate an image into two layers for transparent overlay and underlying surfaces? This question suggests that color induction may have a closer relationship to the faculty of color constancy than one might think at first. A critical problem in understanding both transparency (D'Zmura et al., 1997) and color constancy (Hurlbert, 1998) is the segmentation of images into representations of various physical processes that differ in their color and depth properties (e.g., light sources, transparent filters, surfaces) (Mausfeld, 1998), and we hope that studies of contrast induction will shed some light on how this is accomplished by our visual systems.

Acknowledgments

We thank Charlie Chubb, Carol Cicerone, Geoffrey Iverson, Kenneth Knoblauch, George Sperling, Jack Yellott, Stefania Za, and Qasim Zaidi for their helpful comments. This work was supported by National Eye Institute grant EY10014 to M. D'Zmura.

20

Physics-based approaches to modeling surface color perception

Laurence T. Maloney

The diversity of color results from the diversity of surfaces which absorb the light
– Ulrich von Strassburg (1262)

The study of surface color perception is a proper subset of the study of color perception, and one way to highlight the difference between them is to consider the effective stimulus appropriate to each. The effective stimulus for the study of color perception, broadly construed, is the spectral power distribution of light arriving at each point of the left and right retinas. There is no assumption that the patterns of light correspond to any possible arrangement of surfaces, objects, and illuminants in a three-dimensional scene.

In contrast, the study of surface color perception presupposes that the light reaching the retinas has a history. The effective stimulus is the result of the interaction of certain light sources (*the illuminant*) with the surfaces of objects in an environment. It is clear that any stimulus appropriate for the study of surface color perception is also appropriate for the study of color perception, but not vice versa.

Once we assume that the stimulus results from the interactions of lights and surfaces in an environment, it is natural to ask what information about the illuminants and surfaces in the environment is visually available to the observer. In particular, we can ask, to what extent does the color appearance assigned to a surface provide information concerning the physical, spectral properties of the surface? Visual systems whose estimates of surface color are determined by the spectral properties of surfaces exhibit a constancy, *color constancy*. Visual systems with color constancy have an objective capability: They remotely sense surface spectral information and represent it through color. This objective capability can be assessed in other species (Neumeyer, 1981; Ingle, 1985; Jacobs, 1990; Werner, 1990; see also Jacobs, 1981, 1993) as well as in humans.

Accordingly, while surface color perception is only a part of the study of color perception, it has long been recognized as an important part: "Colours have their greatest significance for us in so far as they are properties of bodies and can be used as marks of identification of bodies." (von Helmholtz, 1896/1962, Vol. II, p. 286).

Surface color perception is intimately linked with the precise physical, spectral properties of illuminants and surfaces in a scene. Indeed, without restrictions on illuminants, no degree of color constancy is possible: "The … problem … of constant color appearance is met by just one condition, namely restriction to one light source of constant spectral character " (Ives, 1912b, p. 70). In studying surface color perception, we cannot ignore the physical constraints that make it possible.

Finally, scene layout in three dimensions can profoundly affect lightness perception (Gilchrist, 1977, 1980; Gilchrist, Delman, & Jacobsen, 1983; see also Gilchrist, 1994). Consideration of surface color perception leads us to consider possible connections between depth and shape perception and color perception.

Scope of the review. This chapter is primarily concerned with recent algorithms for surface color perception based on explicit physical models of the environment. The first, and largest, part of the chapter describes these models and algorithms. Many of these algorithms are drawn from the computer vision literature. Even when there is no explicit claim by the authors that an algorithm could serve as a model of any aspect of human vision (or, more generally, biological vision), I have included it if it has implications for human vision.

The description of the algorithms is followed by a discussion of the relation between them and traditional models of color vision based on hypothesized color channels and the transformation of color information through successive stages (see Hurvich, 1981; Wyszecki & Stiles, 1982a; Kaiser & Boynton, 1996). I briefly discuss models of human surface perception that are not based on explicit physical models, such as Land and McCann's retinex theory (Land & McCann, 1971; Land, 1983, 1986) in the section on von Kries Algorithms.

The algorithms presented differ primarily in how they obtain information about the illuminant in a given scene. The last part of the chapter proposes that the problem of *illuminant estimation* can be formally treated as a cue combination or fusion problem, analogous to cue combination in depth or shape vision (Landy et al., 1995).

Difficulties. The study of surface color perception is beset by three difficulties, each of which I will return to below. The first is methodological, and it will be discussed in the section on Methodological Issues.

The second is our current lack of knowledge concerning the physics of light and surface interactions in three-dimensional scenes. Research in the last 30 years has led to a better representation of light–surface interaction. As a consequence, a number of new approaches to recovering surface properties have arisen. The bulk of this chapter concerns these algorithms and the models of light–surface interaction that underly them. A central theme in this chapter is the link between the recovery of surface color information and information concerning object shape and scene layout. It will become clear, however, that our understanding of the interactions of light and surface in real environments is far from complete.

The last difficulty, for lack of a better term, might be referred to as "conceptual clutter." The study of color constancy is beset by certain imprecisions in terminology. Over the course of the chapter, beginning with the section on Terminology, I will attempt to clarify some of them.

Related work. Previous reviews of the material discussed here include Hurlbert (1986, 1998) and Maloney (1992). Wandell (1995) is a useful introduction to both the empirical issues surrounding surface color perception and the necessary mathematical tools. Kaiser and Boynton (1996, pp. 570ff) contains a partial review of some of the material described here. Healey, Shafer, and Wolfe (1992) provide an overview of work in physics-based vision, and books by Hilbert (1987) and Thompson (1995) describe work in philosophy that is closely related to the material reviewed here. The two volumes edited by Byrne and Hilbert (1997a, 1997b) provide an interesting introduction to both color science and philosophy.

Terminology

Color constancy. The term "color constancy" is employed in different ways in the literature. For some authors (Jameson & Hurvich, 1989; Kaiser & Boynton, 1996; Hurlbert, 1998), the term describes human perceptual *performance*: "the tendency to see colors as unchanging even under changing illumination conditions" (Hurlbert, 1998). The emphasis here is on whatever it is that observers do achieve in any particular circumstance.

A second use of the term "color constancy" is to treat it as a synonym for discounting changes in the illumination of a scene (a sampling of authors: Beck,

1972; Arend, 1993; Foster & Nascimento, 1994; Bäuml, 1995). Brown and MacLeod (1997) have correctly criticized this use of the term "color constancy" as neglecting other factors that might affect surface color perception: the presence of other surfaces, atmospheric haze, and so on. We might reasonably refer to stability of color appearance despite changes in the illuminant as "illuminant color constancy," defining as many new constancies as there are factors that can potentially influence perceived color: "haze color constancy," and so on.

Nevertheless, it is awkward to define something by listing the many things on which it does not depend. Consequently, I adopt the following definition: *An observer has (perfect) color constancy precisely when the color appearance assigned to a small surface patch by the visual system is completely determined by that surface's local spectral properties.*[1] It should be clear that, if the color appearance of a surface patch is determined by its surface spectral reflectance, then it is not affected by changes in the illuminant, surrounding patches, haze, and so on.

To summarize: The term "color constancy" will be used in this chapter to describe a *goal*, one that is not necessarily achieved by any observer. Of central interest is the degree to which, and the circumstances under which, a human observer approximates "perfect color constancy" in his or her judgments of surface color.

Various authors (Brill, 1978, 1979; Craven & Foster, 1992; Foster & Nascimento, 1994; Foster et al., 1997) have considered alternative color invariances that are strictly weaker than color constancy, notably *relational color constancy* (Craven & Foster, 1992; Foster & Nascimento, 1994; Foster et al., 1997). Many of the issues raised here with respect to color constancy could also be raised with respect to these alternative invariances.

[1]By "local spectral properties" is meant the bidirectional reflectance density function at a point of the surface, defined later in the chapter.

The environment. It is often said that human color vision is "approximately color constant" (Hurvich, 1981, p. 199; Brainard, Brunt, & Speigle, 1997). This claim is misleading, if not further qualified, for the degree of color constancy exhibited can be very slight: "If changes in illumination are sufficiently great, surface colors may become radically altered ... weakly or moderately selective illuminants with respect to wavelength leave surface colors relatively unchanged ... but a highly selective illuminant may make two surfaces which appear different in daylight indistinguishable, and surfaces of the same daylight color widely different" (Helson & Judd, 1936). If one does not restrict the range of lights and surfaces used in an experiment, then human color constancy can be close to nonexistent.

What is usually meant by the claim that human color vision is approximately color constant is that there are circumstances, such as a range of possible illuminants and surfaces, similar to those encountered in everyday life, where human color vision approximates perfect color constancy: "With moderate departures from daylight in the spectral distribution of energy in the illuminant, external objects are seen ... nearly in their natural daylight colors" (Judd, 1940). The term "environment" will be used throughout to specify a set of assumptions concerning possible illuminants, surfaces, spatial layouts, and so on. The central issues of surface color perception include: (1) the determination of the degree of color constancy exhibited by a visual system for any given environment; and (2) the determination of the environments under which human color vision exhibits a given degree of color constancy.

Intrinsic colors. The estimates of surface color appearance produced by a visual system that is even approximately color constant must depend on certain physical spectral properties of the surface. These properties will be referred to as *intrinsic colors* (Shepard, 1992) for convenience in describing the algorithms below. The term intrinsic colors will typically refer to whatever it is that we are trying to estimate by means of a particular algorithm.

Methodological issues

Experimental methods. There are several methods that are commonly used to operationalize and measure the degree of color constancy exhibited by human observers: First, there is *simultaneous asymmetric color matching* (Arend & Reeves, 1986; Arend et al., 1991; Brainard, Brunt, & Speigle, 1997), in which the observer sees two regions of a scene comprising colored patches that are illuminated with two different illuminants. The observer adjusts a colored patch in one region (under one illuminant) to be the same color as a fixed test patch in the second. Brainard, Brunt, and Speigle (1997) argue that, if we wish to study the performance to be expected of an observer in a natural setting, he should be free to move his eyes around in the stimulus display. Nevertheless, the use of two illuminants in a single scene raises questions concerning the observer's adaptational state if he is allowed to look back and forth freely from one region to the other.

Second, there is *successive asymmetric color matching* (Brainard & Wandell, 1991, 1992; Bäuml, 1995), in which the observer views a single scene under first one illuminant and then a second. The observer adjusts a colored patch under the second illuminant to be the same color as a test patch seen under the first. The adjustable patch is typically in the same location as the test patch. Obviously, the observer's match depends on the observer's ability to remember colors accurately.

There is a potential confound with this method if the scene remains unchanged while the illuminant changes and the observer is aware that it does. Suppose, for example, that the observer noted that the fixed test patch is identical in appearance to a specific patch π under the first illuminant. He could then set the adjustable patch to match π in the second scene. This strategy could result in a good approximation to color constancy for visual systems that, in reality, have none. The same confound is present in simultaneous asymmetric color matching if the two scenes presented are the same and the observer knows that they are. In general, the scene containing the test patch and the scene containing the adjustable patch should be different.

Third, there is *achromatic matching* (Helson & Michels, 1948; Werner & Walraven, 1982; Fairchild & Lennie, 1992; Arend, 1993; Bäuml, 1994; Brainard, 1998), where an observer adjusts a specified patch to appear achromatic. This method provides less information concerning the remapping of colors induced by changes in illumination than the previous two methods. We know only that the observer's setting corresponds to some point in his *achromatic locus*, but not which point. Nevertheless, the task is apparently very easy to explain to naive observers and very easy to carry out.

Andres and Mausfeld (described in Mausfeld, 1998) require observers to set a test patch to be in an alternative *color locus*, comprising those colors that are neither reddish nor greenish. It is plausible that certain colors or color loci such as the achromatic are easier to remember, or to communicate to observers, and that color appearance measures based on these loci will be more stable.

Task and instruction dependence. Arend and Reeves (1986; Arend, 1993) report that observers are capable of reliably performing two different tasks in asymmetric color matching of surfaces under reference illuminants and can be instructed to perform either. They may equate the chromaticity of the *light* (*color signal* in the terminology developed below) radiating from the two patches, or they may choose the setting consistent with the same surface viewed under two different illuminants. Troost and DeWeert (1991) report that the effect of illuminant changes depends on the task that the observer undertakes. Such task dependence and sensitivity to instructions only complicate the interpretation of experiments on surface color perception. Speigle and Brainard (1996, p. 171), however, consider three tasks involving surface color judgment (simultaneous asymmetric matches, achromatic matches, and color naming). They report that "… all three tasks reveal similar and perhaps identical affects of the illumination … ." Speigle (1998) also examined performance in color naming and color scaling tasks and reached similar conclusions. While this issue is far from resolved, we may tentatively conclude that the same color percept mediates performance in many tasks.

Simulations and reality. In recent work, the stimulus arrangement (scene) is most often simulated on a CRT display. The settings of the CRT's guns are chosen so as to produce the precise stimulation of the observer's photoreceptors that would result from a particular scene composed of specified physical surfaces under specified illuminants.

Researchers working with physical surfaces and lights in real scenes (Berns & Gorzynski, 1991; Brainard, 1998) typically report greater degrees of color constancy than do researchers working with simulated scenes displayed on CRTs. Most recently, Brainard, Rutherford, and Kraft (1997) reported that they were unable to reproduce on a CRT display the degree of color constancy exhibited by observers in a real scene despite every attempt to make the simulated and real scenes identical. The degree of color constancy exhibited depends on the realism of the simulation in ways that we do not yet understand, suggesting that there are cues present in real scenes that we also do not understand.

We know that it possible to choose viewing conditions (an environment) where human observers exhibit little or no color constancy (Helson & Judd, 1936). However, it is difficult, given the methodological problems outlined above, to draw any firm conclusions about the upper limit of performance in realistic environments. The results of Brainard and colleagues suggest that the *upper limit* to color constancy performance is quite high. We do not yet know how high, or what sorts of environments produce optimal color constancy performance.

Environment I: flat world

In this section I describe a model (an environment) for surface color perception that abstracts away the three-dimensional layout of surfaces in a normal scene. The observer, in effect, views a scene painted on a planar surface, or perhaps the inside of a large sphere centered on him or her. The scene is illuminated uniformly by a single illuminant. There is no interreflection (mutual illumination) among surfaces nor any specularity. I will refer to the collection of assumptions that make up this environment as *flat world*. It is an idealization of typical experimental arrangements that have been used to measure human surface color perception, and it differs in many respects from the realistic scenes discussed in the previous section. Its importance stems from both the close relation between flat world and previous experimental work and also its importance to many of the models of color-constant color perception or models of observed human performance that we will review.

Illuminant. Light from a single, distant, punctate light source (the *illuminant*) is absorbed by surfaces within a scene and reemitted. $E(\lambda)$ will be used to denote the spectral power distribution of the incident illuminant at each wavelength λ in the electromagnetic spectrum. The reemitted light that reaches the observer will be referred to as the *color signal*. Its spectral power distribution is denoted $L(\lambda)$. It contains the information about illuminant and surface at each point in the scene that is available to the observer.

The color signals reaching the observer (Fig. 20.1) are imaged onto a two-dimensional *sensor array*. The sensor array can be thought of as simplified model of a retina. We assign coordinates xy to each point in the array. The exact choice of a coordinate frame is not very important so long as each point in the sensor array has a unique coordinate xy. The color signal arriving at point xy on the sensor array is then denoted $L^{xy}(\lambda)$.

Surface. Consider now a small patch of the surface plane shown in Fig. 20.1 that is imaged at location xy in the sensor array. Part of the light that is absorbed by the surface patch is reemitted and radiates in various directions. We characterize the effect of the surface patch on the resulting color signal by defining its *surface spectral reflectance*, which is denoted $S^{xy}(\lambda)$:

(1) $$L^{xy}(\lambda) = E(\lambda)S^{xy}(\lambda).$$

The above equation is assumed to hold for all λ.

In environments more complex than flat world, the function $S(\lambda)$ for a surface patch depends on the *viewing geometry*: the location in three dimensions of the surface patch, the locations of other surfaces, the location of the observer, and the location of the illuminant. We will return to this point below when we consider a second environment, *shape world.*

Photoreceptor classes. The sensor array contains multiple classes of sensors (photoreceptors). Let $R_K(\lambda)$, $k = 1, 2, \ldots, P$ denote the spectral sensitivity of P distinct classes of photoreceptors. For a trichromatic human observer, P is taken to be 3.

We assume that the initial information available to a color vision system at a single retinal location is the excitation of each of the P classes of receptor,

(2) $$\rho_1^{xy} = \int L^{xy}(\lambda)R_1(\lambda)d\lambda = \int E(\lambda)S^{xy}(\lambda)R_1(\lambda)d\lambda$$

$$\rho_2^{xy} = \int L^{xy}(\lambda)R_2(\lambda)d\lambda = \int E(\lambda)S^{xy}(\lambda)R_2(\lambda)d\lambda$$

$$\ldots$$

$$\rho_P^{xy} = \int L^{xy}(\lambda)R_P(\lambda)d\lambda = \int E(\lambda)S^{xy}(\lambda)R_P(\lambda)d\lambda.$$

The P numbers at each location form a vector $\rho^{xy} = [\rho_1^{xy}, \ldots, \rho_P^{xy}]$. A glance at Eqn. (2) discloses that the entries of the vector ρ^{xy} at each location xy depend on both the light $E(\lambda)$ and the surface reflectance $S^{xy}(\lambda)$.

The RGB heuristic. There is a persistent belief that crops up from time to time in the study of surface color perception that might be called the "RGB heuristic." Define the "color of the illuminant" $E(\lambda)$ to be the vector ρ^E whose components, $k = 1, 2, \ldots, P$, are given by,

(3) $$\rho_k^E = \int E(\lambda)R_k(\lambda)d\lambda.$$

These would be the excitations of the photoreceptors while looking directly at the illuminant $E(\lambda)$, or when viewing a perfectly reflecting surface, $S(\lambda) = 1$, under $E(\lambda)$. Define the "color of the surface" $S(\lambda)$ to be ρ^S with components

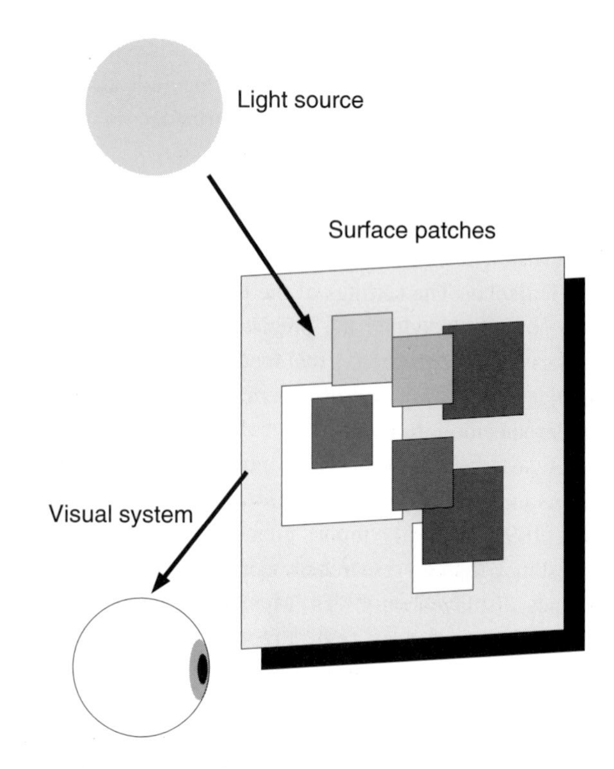

Figure 20.1: Flat world. A model environment for surface color perception that abstracts away the three-dimensional layout of surfaces in a normal scene. The observer views a scene painted on a planar surface. The scene is illuminated uniformly by a single, distant, punctate illuminant. There are no shadows and no interreflection (mutual illumination) among surfaces, nor any specularity.

(4) $$\rho_k^S = \int S(\lambda)R_k(\lambda)d\lambda.$$

These photoreceptor excitations correspond to those for the surface $S(\lambda)$ viewed under a light with constant, unit spectral power density.

Now, to what extent do the color of the surface ρ^S and the color of the illuminant ρ^E determine the actual excitations corresponding to the surface viewed under the illuminant (denoted ρ^{ES})? It might seem plausible that:

(5) $\rho_k^{ES} \approx \rho_k^E \rho_k^S$,

and this sort of approximation is widely used in computer graphics for modeling the interaction of light and surface. However, there is no *mathematical* reason to expect Eqn. (5) to hold, even approximately.

Evans (1948) gives a delightful counterexample in the form of two color plates showing objects illuminated under two lights of identical color appearance (ρ^E) but decidedly different spectral power distributions. Ives (1912b, p. 70) noted that an artificial daylight that matches daylight in color appearance need not render colors correctly: "A white surface under this [light] would look as it does under daylight but hardly a single other color would."

The accuracy (or inaccuracy) of the approximation in Eqn. (5) depends on the possible surfaces and illuminants present in the environment. It is natural, then, to ask what kinds of constraints on lights and surfaces are needed for the approximation of Eqn. (5) to achieve a specified degree of accuracy, and also to ask whether physical surfaces in our environment satisfy those constraints. We will continue this discussion in the von Kries Algorithms section below.

Linear models: mathematical notation

The geometry of $\mathbf{L}^2$. The functions $E(\lambda)$, $S^{xy}(\lambda)$, and $R_k(\lambda)$, introduced above, are assumed to be square-integrable functions, which we will define next. This assumption imposes no empirically significant constraint on the possible illuminants, surfaces, and receptors that are present in a scene.

The space of square-integrable functions $\mathbf{L}^2$ is defined as follows. The magnitude or norm of a function $f(\lambda)$ is defined to be

$$(6) \qquad \|f\| = \left[\int_{-\infty}^{\infty} f(\lambda)^2 d\lambda\right]^{1/2}.$$

The set of square-integrable real functions are those real functions that have finite magnitude, that is, the integral in Eqn. (6) converges to a finite limit. The set of square-integrable functions form a vector space $\mathbf{L}^2$, which has an inner product. The inner product of two square-integrable functions $f(\lambda)$, $g(\lambda)$ is defined to be

$$(7) \qquad \langle f, g\rangle = \int_{-\infty}^{\infty} f(\lambda)g(\lambda)d\lambda.$$

The two functions are orthogonal precisely when their inner product is 0.

$\mathbf{L}^2$ is an example of a linear function space, a vector space whose elements are functions. Apostol (1969, Chaps. 1 and 2) is an elementary introduction to linear function subspaces. Strang (1988) is a standard introduction to finite-dimensional linear algebra. See Maddox (1970) and Young (1988) for more advanced treatments. Wandell (1995) contains an introduction to the use of linear function subspaces in vision.

Basis functions. It is possible to show (Young, 1988) that we can find square-integrable functions $E_i(\lambda)$, $i = 1, 2, 3, \ldots$, such that for any light $E(\lambda)$ there are unique real numbers ε_i such that:

$$(8) \qquad E(\lambda) = \sum_{i=1}^{\infty} \varepsilon_i E_i(\lambda).$$

The functions $E_i(\lambda)$ form a *basis* of the linear function space $\mathbf{L}^2$. Just as in the finite dimensional case, there are infinitely many possible choices of a basis for $\mathbf{L}^2$. Our choice of basis for the lights $E(\lambda)$ will be guided by the empirical considerations described below.

Coordinates. The real numbers ε_i are the *coordinates* corresponding to the light $E(\lambda)$ in the space $\mathbf{L}^2$. The infinite vector $\varepsilon = [\varepsilon_1, \ldots, \varepsilon_i, \ldots]$ determines $E(\lambda)$. Not every function in $\mathbf{L}^2$ corresponds to a physically possible light $E(\lambda)$; some represent spectral power densities with negative power at some point of the visible spectrum. The physically realizable lights form a convex set in $\mathbf{L}^2$ that is analogous to the region within the spectral locus in CIE coordinates (Wyszecki & Stiles, 1982a).

The basis $E_i(\lambda)$, $i = 1, 2, 3, \ldots$ could also serve to express the coordinates of the surface reflectances. I will instead choose a second basis, $S_j(\lambda)$, $i = 1, 2, 3, \ldots$, to express the coordinates of the surface reflectances $S^{xy}(\lambda)$:

$$(9) \quad S^{xy}(\lambda) = \sum_{j=1}^{\infty} \sigma_j^{xy} S_j(\lambda).$$

The infinite vector $\sigma = [\sigma_1^{xy}, \ldots, \sigma_j^{xy}, \ldots]$ determines $S^{xy}(\lambda)$. The choice of this second basis will also be guided by empirical considerations. Note that while the coordinates σ_j^{xy} of $S^{xy}(\lambda)$ vary with location, the fixed basis elements do not. The same basis is used to model surface reflectance at every location in the scene.

History. Yilmaz (1962) first used truncated Fourier series expansions to model illuminants and surface reflectances across the visible spectrum, and Sällström (1973) first framed the problem in terms of truncated expansions using arbitrary bases. Brill (1978, 1979), Buchsbaum (1980), and Maloney and Wandell (1986; Maloney, 1984) independently developed this same representation of illuminants and surface spectral reflectance and their interactions. Maloney and Wandell (1986) termed these constraints on light and surface *linear models*.

Light–surface interaction. Substituting Eqs. (8) and (9) into Eqn. (2), we get

$$(10) \quad \rho_k^{xy} = \sum_{i=1}^{\infty} \sum_{j=1}^{\infty} \varepsilon_i \sigma_j^{xy} \int E_i(\lambda) S_j(\lambda) R_k(\lambda) d\lambda.$$

The integrals $\int E_i(\lambda) S_j(\lambda) R_k(\lambda) d\lambda$ contain only fixed elements that are independent of the particular scene viewed. Setting $\gamma_{ijk} = \int E_i(\lambda) S_j(\lambda) R_k(\lambda) d\lambda$, Eqn. (10) becomes

$$(11) \quad \rho_k^{xy} = \sum_{i=1}^{\infty} \sum_{j=1}^{\infty} \varepsilon_i \sigma_j^{xy} \gamma_{ijk}.$$

The values γ_{ijk} are fixed, known, and independent of any particular scene once the bases have been selected. It should be clear that the specific illuminant enters into the visual process only through its coordinates ε_i, $i = 1, 2, \ldots$, and the specific spectral reflectance functions only through their coordinates at each location, σ_j^{xy}, j=1, 2, ... Equation (11) is exact within the framework of assumptions adopted so far. It merely restates Eqn. (2) with respect to two infinite-dimensional coordinate systems; it can no more be solved for information about the surface reflectance (about the σ^{xy}) independent of the light than could Eqn. (2). Any of the coordinates σ_j^{xy} could serve as an intrinsic color (see the Terminology section above) – if we could reliably estimate it despite changes in the illuminant. But, in Eqn. (11), information about light and surface is irreversibly tangled.

We next approximate the infinite summations above (that perfectly capture light and surface reflectance) by truncated, finite summations:

$$(12) \quad E_\varepsilon^{xy}(\lambda) = \sum_{i=1}^{M} \varepsilon_i E_i(\lambda)$$

$$(13) \quad S_\sigma^{xy}(\lambda) = \sum_{j=1}^{N} \sigma_j^{xy} S_j(\lambda).$$

The class of lights that can be represented in this way for a fixed value of M and fixed basis elements $E_1(\lambda)$, $E_2(\lambda)$, ... $E_M(\lambda)$ is a finite-dimensional linear function subspace (linear model) of $\mathbf{L}^2$ that has dimension M. A linear function subspace of surface reflectances (linear model) with dimension N is defined analogously.

The *finite*-dimensional vectors $\varepsilon = [\varepsilon_1, \ldots, \varepsilon_M]$ and $\sigma^{xy} = [\sigma_1^{xy}, \ldots, \sigma_N^{xy}]$ will be referred to as the *coordinates* of light and surface within their respective linear subspaces. The subscripted variables $E_\varepsilon(\lambda)$ and $S_\sigma^{xy}(\lambda)$ will be used throughout to denote lights and surfaces constrained to lie in finite-dimensional linear models.

Linear models: fits to empirical data

Surfaces. The algorithms described in the next section depend crucially on the assumption that we can approximate real illuminants and surfaces by linear models with low values of M and N. In this section we review work concerning the realism of such models as descriptions of empirical surfaces and illuminants. An expanded version of this section may be found in Maloney (1998).

Fitting methods. I describe here how to fit an optimal least-squares linear model to the set of empirical surface reflectances described in Vrhel, Gershon, and Iwan (1994). More sophisticated methods permit the simultaneous choice of optimal linear models for any sets of empirical illuminants and surfaces (Vrhel & Trussel, 1992; Marimont & Wandell, 1992). I will refer to the data set of Vrhel, Gershon, and Iwan (1994) as the *Kodak data.*[2]

Suppose that we have a set of empirically measured surface spectral reflectances, $S^{\nu}(\lambda)$, $\nu = 1, 2, \ldots, V$. We will treat a sampled surface reflectance function, sampled at wavelengths λ_i, as a step function that is constant between λ_i and λ_{i+1} and has as its constant value the sampled value at λ_i. These step functions have values defined at all λ, and their use allows us to use the same notation and conventions for empirical and theoretical surface spectral reflectance functions.

For any fixed value of N and any choice of basis functions $S_1(\lambda), \ldots, S_N(\lambda)$ we can compute, by linear regression (Maloney, 1986), the weights $\hat{\sigma}_j^{\nu}$ in the truncated series of Eqn. (13) that minimize the least-squares error,

$$(14) \quad \Xi_{\nu} = \left\| S_{\nu} - S_{\hat{\sigma}}^{\nu} \right\|^2 = \int (S_{\nu}(\lambda) - S_{\hat{\sigma}}^{\nu}(\lambda))^2 d\lambda .$$

We now wish to choose the *basis functions* $S_j(\lambda)$, $j = 1, 2, \ldots, N$ to minimize the overall error,

$$(15) \quad \Xi = \Xi^1 + \ldots + \Xi^V .$$

Each of the optimal basis functions will be orthogonal to all of the others. It is convenient to assume that they are scaled to be of unit magnitude. Several authors fit empirical data using an alternative linear model to that of Eqn. (13):

$$(16) \quad \hat{S}^{\nu}(\lambda) = \bar{S}(\lambda) + \sum_{j=1}^{N} \sigma_j^{\nu} S_j(\lambda) ,$$

where $\bar{S}(\lambda)$ is the mean of the $S^{\nu}(\lambda)$. This model is presupposed by *principle component analysis* (Mardia, Kent, & Bibby, 1979), and authors who report using principle components analysis to fit their data will likely assume the model of Eqn. (16), not that of Eqn. (13).

The models are different. Note in particular that the mean in Eqn. (16) will not, in general, be orthogonal to the remaining basis elements or independent of them. It is even possible that $\bar{S}(\lambda)$ will be identical to $S_1(\lambda)$, an undesirable outcome. Further, with Eqn. (13), the scaled copies of any model surface reflectance $aS_{\sigma}(\lambda)$ are automatically in the space of model surface reflectances, a scaling property that does not, in general, hold for Eqn. (16). Many authors (e.g., Judd, MacAdam, & Wyszecki, 1964) normalize the vectors in their data set before fitting them. They may normalize the data vectors to have $\mathbf{L}^2$-norm 1 or to have value 1 at a specified wavelength. They then fit the normalized data, using principle components and, in reconstructing the original data, scale the mean by a value σ_0,

$$(17) \quad \hat{S}^{\nu}(\lambda) = \sigma_0 \bar{S}(\lambda) + \sum_{j=1}^{N} \sigma_j^{\nu} S_j(\lambda) ,$$

which is simply the inverse of the constant by which the original measured surface reflectance was scaled in normalizing it. Equations (17) and (13) appear to be identical (except for a change in the index), but, again, the vectors in Eqn. (17) need not be orthogonal. Accordingly, I use the models of Eqns. (12) and (13) in the sequel.

For empirically measured sets of surface reflectance functions, the optimal basis consistent with Eqns. (12) or (13) can be readily computed by standard linear algebra methods using the singular value

[2]The Kodak data and other sets of surface spectral reflectance functions collected by various authors are available from ftp.cns.nyu.edu:pub/ltm/SSR via anonymous ftp. There is currently no large set of illuminants publically available.

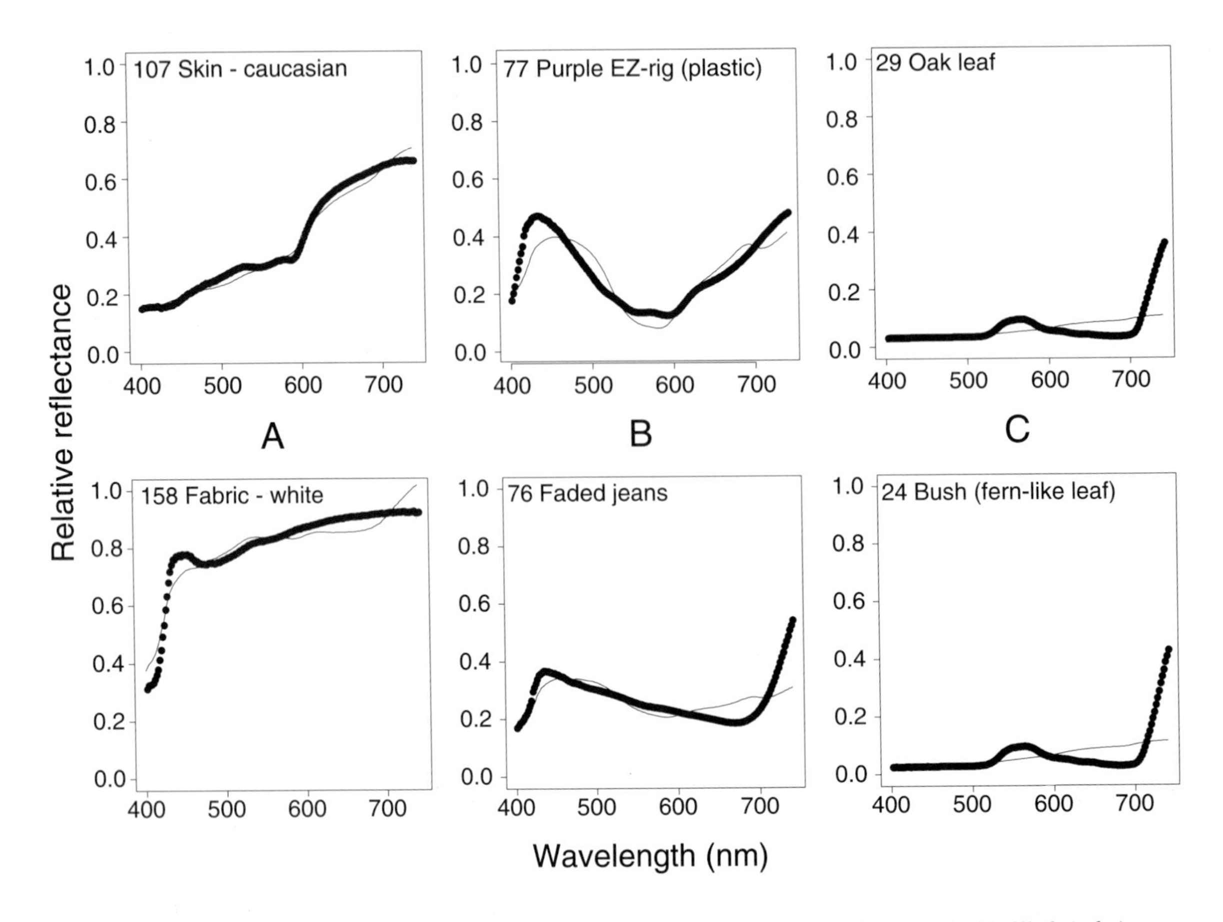

Figure 20.2: Surface spectral reflectance data from Vrhel, Gershon, and Iwan (1994) (plotted with *filled circles*) and reconstructions with three basis elements. The horizontal axes are wavelength, in nanometers, and the vertical are relative reflectance. (A) The two fits at the first quartile of Ξ^{ν}. (B) The two fits at the third quartile. (C) The worst two fits. See text for details.

decomposition (Mardia, Kent, & Bibby, 1979). Once we have computed the optimal basis $S_1(\lambda), \ldots, S_N(\lambda)$, we can compute the optimal approximation $\hat{S}^{\nu}(\lambda)$ by linear regression for each of the empirical functions $S^{\nu}(\lambda)$ and compare it to the original.

The Kodak data set. Figure 20.2 contains plots of six of the surfaces in a collection of 170 surfaces measured by Vrhel, Gershon, and Iwan (1994) and approximations[3] to those surfaces with $N = 3$. The two plots in Fig. 20.2A correspond to the two surfaces whose Ξ^{ν}'s fell at the first quartile of the 170 values of Ξ^{ν}. That is, about 25% of the surfaces in the sample have smaller (better) values of Ξ^{ν}. The two plots in Fig. 20.2B correspond to the two surfaces whose Ξ^{ν}'s fell at the third quartile of the 170 values of Ξ^{ν}. About 75% of the surfaces in the sample were better fit. The two plots in Fig. 20.2C correspond to the two surfaces with the largest values of Ξ^{ν} (the worst fits).

Figure 20.3 contains analogous plots for $N = 8$. These results illustrate the conclusion of Vrhel, Gershon, and Iwan (1994) that linear models with $N = 3$ provide poor approximations to the measured surface

[3] Vrhel, Gershon, and Iwan (1994) used principle component analysis to analyze these data. I refit their data by the least-squares model for reasons described in the main text and so that they could be more readily compared with earlier work.

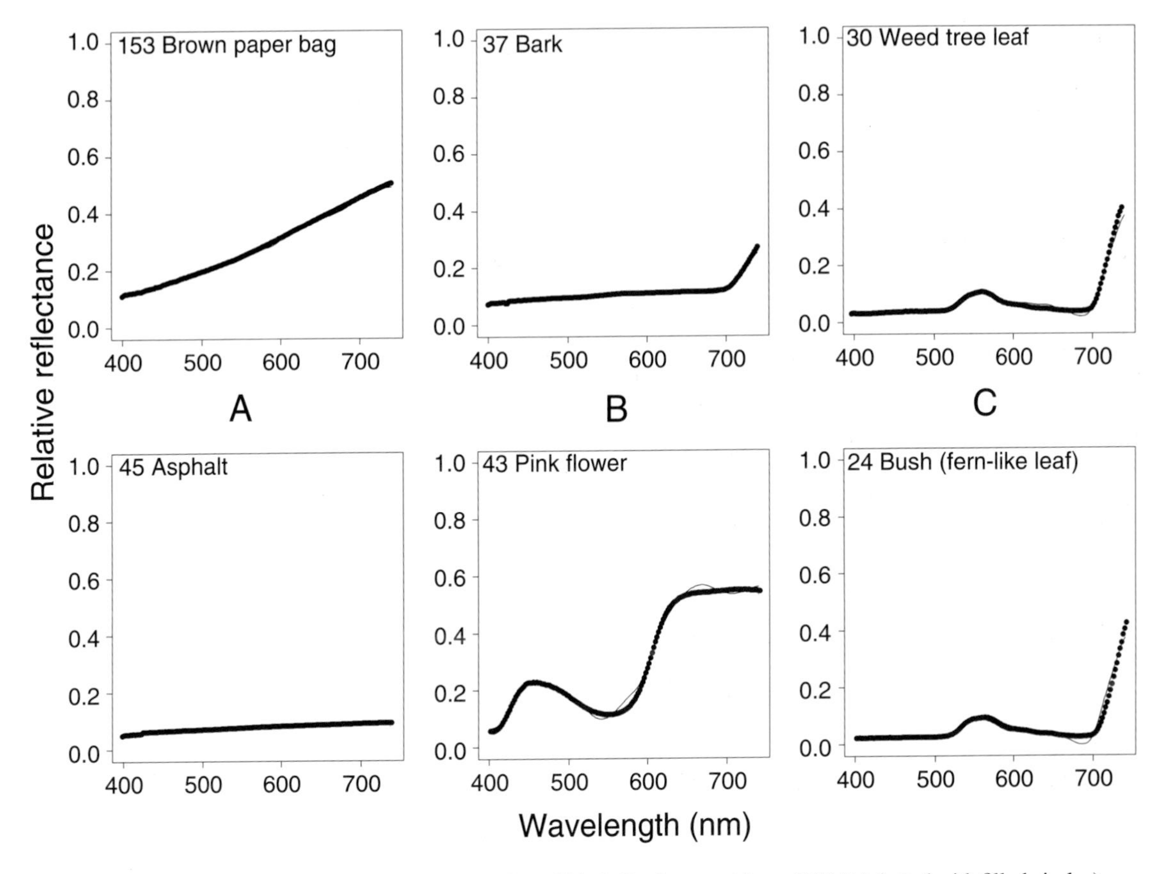

Figure 20.3: Surface spectral reflectance data from Vrhel, Gershon, and Iwan (1994) (plotted with filled circles) and reconstructions with eight basis elements. The horizontal axes are wavelength, in nanometers, and the vertical are relative reflectance. (A) The two fits at the first quartile of Ξ^V. (B) The two fits at the third quartile. (C) The worst two fits. See text for details.

reflectances in their data set while, with $N = 8$, the approximations are good.[4]

Figure 20.4 shows the variance accounted for (the Ξ term normalized) as a function of N for the data of Vrhel, Gershon, and Iwan (1994). The results of Maloney (1986) for the Krinov (1947) data set are also plotted in Fig. 20.4. It is clear that the fits to the Krinov data set reported by Maloney understate the difficulty of modeling empirical surface reflectances with low-dimensional linear models (if we take the Vrhel, Gershon, and Iwan data as representative of empirical surface reflectances). The results, however, do confirm the conclusions drawn in Maloney (1986): "... the number of parameters required to model ... spectral reflectances is five to seven, not three" (Vrhel, Gershon, & Iwan, 1994, p. 1674). The failure to find highly accurate approximations to real surfaces with $N = 3$ will have implications for the models and algorithms reviewed below, which we will return to in the section on Model Failure and Approximate Color Constancy.

History. Cohen (1964) used principle component analysis (Mardia, Kent, & Bibby, 1979) to fit the surface spectral reflectances of a subset of the Munsell

[4]Vrhel, Gershon, and Iwan (1994) report that roughly seven principle components suffice. The choice of seven or eight or nine is somewhat arbitrary. See Fig. 20.4.

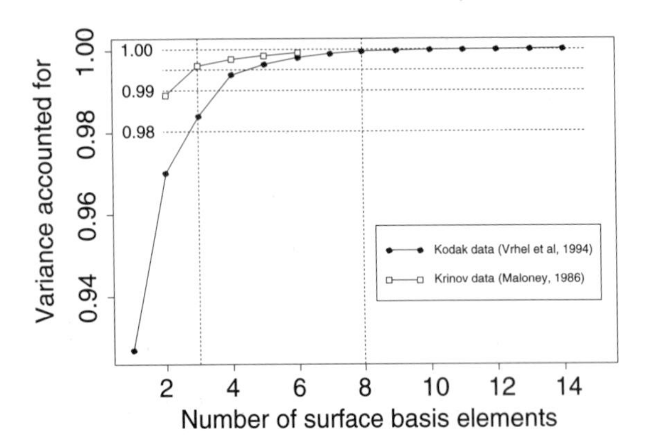

Figure 20.4: Variance accounted for versus number of basis elements for fits of the Vrhel, Gershon, and Iwan (1994) data (filled circles) and for fits of the Krinov data from (Maloney, 1986) plotted as open squares. The dashed horizontal and vertical lines mark cutoffs of interest.

color chips. He concluded that the mean surface reflectance and as few as two additional components provided good fits to the Munsell surface reflectances. Maloney (1986) refit the same data set using Eqn. (13) and the criterion in Eqns. (14) and (15) and measured how well the Munsell basis fit a a large set of surface reflectance functions collected by Krinov (1947/1953). The results of this work were described above. Parkkinen, Hallikainen, and Jaaskelainen (1989) measured the surface spectral reflectance of a large collection of Munsell chips and fit them using principle component analysis. As many as eight basis elements were needed to permit accurate fits to the surfaces.

Illuminants. The fitting methods and models for illuminants are identical to those for surface reflectance discussed above. Judd, MacAdam, and Wyszecki (1964) report summary results for a large set of measured spectral power distributions of daylight (principle components fit). Their results and later, more extensive work by other researchers (Das & Sastri, 1965; Sastri & Das, 1966, 1968; Dixon, 1978) indicate that sampled daylight may be well described by a small number of basis elements (possibly as small as $M = 3$). Romero, Garciá-Beltrán, and Hernández-Andrés (1997) sampled the spectral power distributions of daylight from 400 to 700 nm over a period of four days in Granada, Spain. They performed a principle component analysis on the resulting 99 spectral power distributions, each normalized to magnitude 1 (in $\mathbf{L}^2$). They found that approximations to the measured spectral power distributions using three basis elements accounted for 0.9997 of the variance. The number of samples collected was not large, the time period over which they were collected was short, and it is not clear how representative the climate of Granada is of climates in other regions of the world. Nevertheless, the fit is remarkable, and their results, together with the results of earlier research, indicate that low-dimensional linear models provide very good approximations to daylight spectral power distributions.

Issues in fitting empirical data. The results of this section suggest that surface reflectance functions and illuminants in natural environments are constrained. This idea is not new. Several authors (Stiles, Wyszecki, & Ohta, 1977; Lythgoe, 1979; MacAdam, 1981) have expressed the opinion that empirical surface reflectances are smooth, constrained curves. Land (1959/1961) asserts that “Pigments in our world have broad reflection characteristics.” Still, there are many open questions concerning the nature and importance of the constraints on “natural” surfaces and light and how they might best be modeled. In the remainder of this section I raise some of them.

Nonlinear models. Only linear models are considered here as candidate representations for natural surfaces and reflectances. It is certainly possible that a nonlinear model with N parameters such as

$$(18) \quad S_{\sigma}^{xy}(\lambda) = \prod_{j=1}^{N} S_j^{\sigma_j^{xy}}(\lambda)$$

might provide better approximations to empirical surface reflectance functions than any linear model with N parameters. There seems to have been no systematic attempt to find nonlinear models that provide bet-

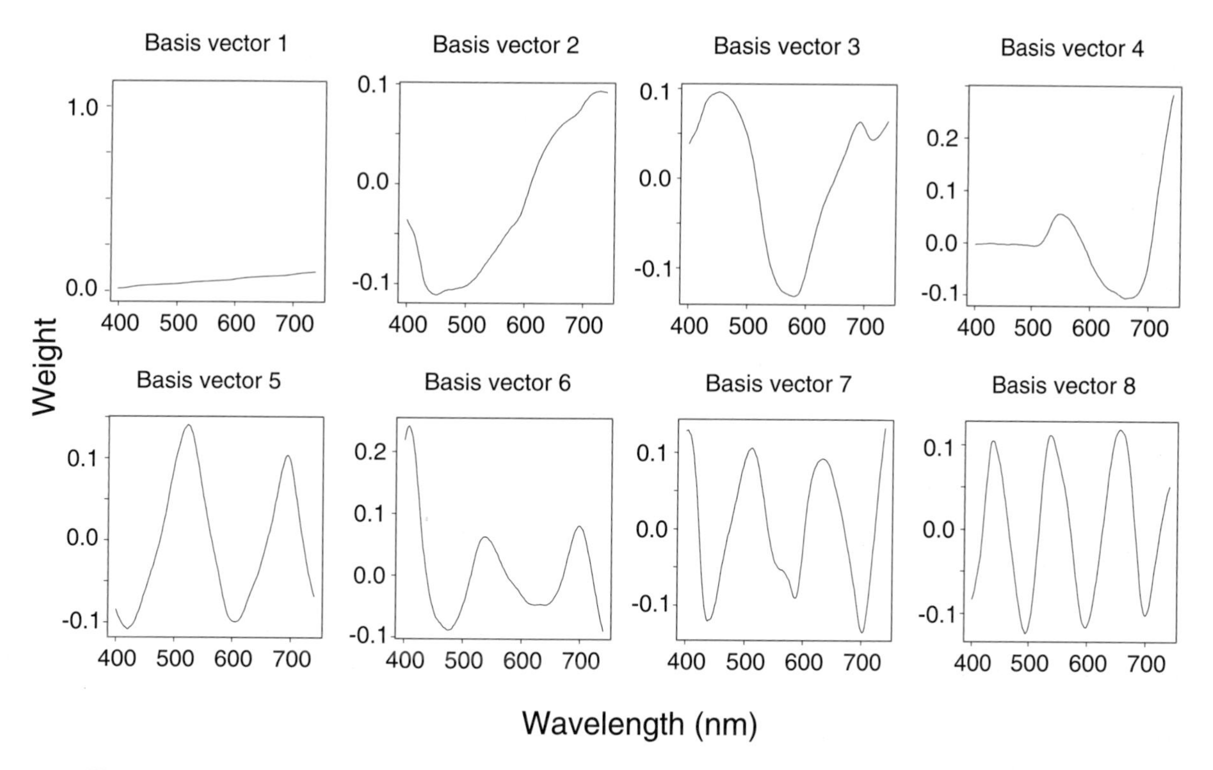

Figure 20.5: The first eight basis functions for the Vrhel, Gershon, and Iwan (1994) data. The units of the horizontal axes are wavelength in nanometers, and those of the the vertical axes are relative reflectance.

ter fits than linear models. Many of the algorithms below could be readily altered to take advantage of a nonlinear constraint such as that embodied in Eqn. (18).

Loss functions. The use of the least-squares error measure Ξ in fitting models to data is questionable. An advantage of the least-squares error measure is that it is independent of the properties of any particular visual system. Any conclusions drawn are statements about the empirically measured surface spectral reflectances themselves and, in attempting to understand the physical bases for the empirically observed constraints on surfaces, this is desirable (Maloney, 1986).

Yet it is also desirable to employ error measures that reflect the sensitivity of specific visual systems. Human vision, for example, is likely to be very insensitive to failures in approximation near the extremes of the (human) visible spectrum. Maloney (1986) refit the Krinov data with error weighted by human photopic visual sensitivity $V(\lambda)$ (see Wyszecki & Stiles, 1982a). He compared the variance accounted for by the weighted fit and the unweighted fit and concluded that the weighted fit accounted for markedly more of the variance. That is, the least-squares approximations to the Krinov data were enhanced by the spectral properties of the human visual system. A better approach (Dannemiller, 1992) is to measure the ability of an ideal observer to discriminate approximations from real surface functions. Marimont and Wandell (1992) developed fitting techniques that took into account human visual sensitivity and derived basis functions that, as expected, approached zero at the ends of the visible spectrum.

It is important to consider both the nature of the physical constraints on surfaces (independent of any visual system) and also the impact of the constraints on visual performance for particular visual systems.

Theoretical approaches. It is not clear what constitutes a "natural environment" for human vision or how to sample it, what surfaces should be included, and what weight each should be given. When we consider other biological visual systems, the problem is scarcely less difficult. It is, therefore, desirable to consider why at least some classes of physical surfaces exhibit physical constraints and what these constraints might be. If we understand the theoretical bases for these constraints, we need not wonder whether they might vanish with the next collection of empirical surface reflectance functions.

Stiles, Wyszecki, and Ohta (1977) and later Buchsbaum and Gottschalk (1984) suggested that many surface reflectance functions are low-pass: Their surface spectral reflectance functions, as a function of wavelength, are approximately low-pass. Maloney (1984, 1986) tested this low-pass hypothesis for the Krinov data and concluded that the Krinov reflectances contained little spectral energy above a band limit corresponding to three samples. He suggested specific physical processes responsible for this observed low-pass constraint for organic colorants.

Mollon, Estévez, and Cavonius (1990) argue that the low-pass constraint observed in the Krinov data is simply an artifact of the measurement process employed by Krinov. Krinov measured not isolated, homogeneous surfaces, but natural formations, each comprising many distinct materials. The spatial mixing of distinct surface spectral reflectances, they argue, leads to measured surface spectral reflectances that have been effectively passed through a low-pass filter and, in contrast, "[i]f the measurement is confined to part of an individual leaf or individual fruit, then fine detail is readily apparent in the spectra of the world of plants" (p. 129). This issue remains to be resolved. However, the first eight basis elements of the Vrhel, Gershon, and Iwan (1994) data (shown in Fig. 20.5) exhibit the same band-pass form with increasing peak frequencies as found by Maloney (1984, 1986) with the Krinov data. Further discussion of the low-pass hypothesis may be found in Maloney (1998).

Flat world algorithms

This section describes algorithms that presuppose a flat world environment with surfaces and illuminants perfectly described by low-dimensional linear models ($M = N = 3$).

We substitute $E_{\varepsilon}(\lambda)$ from Eqn. (12) for $E(\lambda)$ in Eqn. (2), and $S_{\sigma}(\lambda)$ from Eqn. (13) for $S(\lambda)$ in the same Eqn. (2). We then rewrite Eqn. (2), expressing the basic relations among the light, surface reflectances, and receptor responses, as the matrix equation:

$$(19) \quad \rho^{xy} = \Lambda_{\varepsilon}\sigma^{xy},$$

where $\rho^{xy} = [\rho_1^{xy}, \ldots, \rho_P^{xy}]$ is, as above, the vector formed from the receptor excitations of the P receptors at location xy. The matrix Λ_{ε} is P by N, and its kjth entry is of the form $\int E_{\varepsilon}(\lambda)S_j(\lambda)R_k(\lambda)d\lambda$. Note that the matrix Λ_{ε} depends only on the light (as the basis elements S_j and receptor spectral sensitivities R_k are fixed, independent of any particular scene). This matrix captures the role of the light in transforming surface reflectances at each location xy into receptor excitations.

The coordinates $\sigma = [\sigma_1^{xy}, \ldots, \sigma_N^{xy}]$ (or any convenient transform of them) could serve as intrinsic colors, that is, the quantities we seek to estimate given the ρ^{xy} at each location. Various limits on recovery of the σ^{xy} are dictated by Eqn. (19). We consider the limits on recovery when the light on the scene is assumed to be known and when the light on the scene is unknown.

In the simple case in which the ambient light and (therefore) the lighting matrix Λ_{ε} are known, we see that to recover the N weights that determine the surface reflectance we need merely solve a set of simultaneous linear equations. The recovery procedure reduces to matrix inversion when $P = N$. That is,

$$(20) \quad \Lambda_{\varepsilon}^{-1}\rho^{xy} = \sigma^{xy},$$

where the quantities on the left-hand side are all known or computable from known quantities. Recovery is also possible when $P > N$ whenever Λ_{ε} corresponds to an injective (1–1) linear transformation. If P

is less than N, Eqn. (19) is underdetermined and there is no (unique) solution see [Maloney (1984) for a discussion of the invertibility of the various matrices].

If the ambient light is unknown, then it is easy to show that we cannot do as well: We cannot, in general, recover the ambient light vector ε or the spectral reflectances even when $P = N$, so that matrix Λ_ε is square. Given any collection of photoreceptor excitations ρ^{xy} across a scene, there are, in general, many possible choices of surface reflectances σ^{xy} and illuminants ε that satisfy Eqn. (19). Since any such choice of a light vector ε and corresponding surface reflectances σ^{xy} could have produced the observed receptor excitations, we cannot determine which in fact did. Even if we restrict attention to the convex subset of "physically realizable" lights, such confusions are still possible for many choices of lights and surfaces.

Each of the linear models algorithms that will be described next takes a distinct approach to estimating the σ^{xy}.

The reference surface algorithms of Brill and Buchsbaum. Brill (1978, 1979) considers the case where $P = N = 3$ and there are three reference surfaces available in the scene at known locations xy_1, xy_2, and xy_3. (Brill's algorithm requires no restriction on the illuminant: $M = \infty$.) Then the receptor excitations ρ^{xy} at these locations (among others) are known and we have the simultaneous matrix equations:

$$(21) \quad \rho^{xy_1} = \Lambda_\varepsilon \sigma^{xy_1}$$

$$\rho^{xy_2} = \Lambda_\varepsilon \sigma^{xy_2}$$

$$\rho^{xy_3} = \Lambda_\varepsilon \sigma^{xy_3}.$$

Both sides of each equation are known. If the reference surfaces σ^{xy_l}, $l = 1, 2, 3$, are linearly independent, then it is possible to solve for Λ_ε (if the σ^{xy_i} are taken as the basis of the space of surfaces, then the matrix Ω whose columns are ρ^{xy_i}, $i = 1, 2, 3$, is the desired matrix Λ_ε). If another basis is used, then Λ_ε is simply the inverse of the matrix Ω premultiplied by the matrix that changes from the σ^{xy_i} basis to the second basis (see Strang, 1988). Once Λ_ε is known, we invert Eqn. (19) and solve for the coordinates (intrinsic colors) of all of the sources in the scene.

Buchsbaum (1980) assumes that $M = N = 3$ and requires that the location of one reference surface σ^{xy_0} be known. To explain his algorithm, we first define

$$(22) \quad \Lambda_1 = \Lambda_{[1,0,0]}$$

$$\Lambda_2 = \Lambda_{[0,1,0]}$$

$$\Lambda_3 = \Lambda_{[0,0,1]},$$

the light matrix Λ_ε for each of the known basis lights. Then the quantities

$$(23) \quad \rho^{xy_1} = \Lambda_1 \sigma^{xy_0}$$

$$\rho^{xy_2} = \Lambda_2 \sigma^{xy_0}$$

$$\rho^{xy_3} = \Lambda_3 \sigma^{xy_0}$$

are all known once the reference surface σ^{xy_0} is given. These are the receptor excitations corresponding to the reference surface under each of the basis lights in turn. It can be shown that (see Maloney, 1984, Chap. 3),

$$(24) \quad \Lambda_\varepsilon = \varepsilon_1 \Lambda_1 + \varepsilon_2 \Lambda_2 + \varepsilon_3 \Lambda_3,$$

giving the following expression for the receptor excitation of the reference surface under an unknown light ε:

$$(25) \quad \rho^{xy_0} = \Lambda_\varepsilon \sigma^{xy_0} = \varepsilon_1 \rho^{xy_1} + \varepsilon_2 \rho^{xy_2} + \varepsilon_3 \rho^{xy_3}.$$

If the fixed vectors ρ^{xy_l}, $l = 1, 2, 3$, are linearly independent, then the above equation can be solved for ε given ρ^{xy_0}. (The coordinates of the light ε are precisely the coordinates of the reference surfaces' receptor excitations ρ^{xy_0} with respect to the basis $\{\rho^{xy_1}, \rho^{xy_2}, \rho^{xy_3}\}$.) In summary, a single reference surface permits estimating the light when $P = N =$

$M = 3$. Once ε is known, Eqn. (19) may be solved for the intrinsic colors of surfaces.

The algorithm of Brill can be generalized to the case where $P = N$ takes on any arbitrary value; N linearly independent reference surfaces are then required. Buchsbaum's algorithm can be generalized to arbitrary values $M = N = P$, and it still requires only a single reference surface. Both algorithms can be applied to a visual system with any number of types of receptors P.

Note that Brill's algorithm makes no assumptions about the dimensionality of the linear model describing the illuminant (M). It can be used to estimate the matrix Λ_ε in environments where no restrictions are placed on the illuminant $E(\lambda)$. Brill's algorithm can, therefore, be used in environments where little is known about the illuminant. Of course, if the estimated Λ_ε is not invertible, then the surface parameters σ cannot be recovered by any method.

In Buchsbaum's algorithm, the three fixed receptor excitations ρ^{xy_l}, $l = 1, 2, 3$, generated by the single reference surface under the known basis lights serve much the same role as Brill's three reference surfaces: they "pin down" the light matrix.

Buchsbaum's gray world algorithm. In either Brill's or Buchsbaum's algorithm, the receptor excitation corresponding to one reference surface may be replaced by the average of the receptor excitations catches ρ^{xy} across a portion of the scene W, provided that we know the true mean of the intrinsic colors σ^{xy} across that part of the scene.

This substitution is an obvious consequence of the linearity of Eqn. (19): If, for all $xy \in W$,

(26) $\rho^{xy} = \Lambda_\varepsilon \sigma^{xy}$, then

(27) $$\sum_{xy \in W} \rho^{xy} = \Lambda_\varepsilon \sum_{xy \in W} \sigma^{xy}$$

and, dividing both sides by the number of locations in the set W,

(28) $$\rho^{mean} = \Lambda_\varepsilon \sigma^{mean}.$$

In particular, we can use the mean across the entire scene. If, for example, we assume that the mean intrinsic color of a scene is a specific gray, then this gray σ^{xy_0}, paired with ρ^{xy_0}, the mean of the observed receptor excitations in the scene, permit estimation of the illuminant using Buchbaum's algorithm. The assumption concerning the mean of the intrinsic colors in a scene is sometimes termed the *gray world* assumption (Buchsbaum, 1980, p. 24): "It seems that arbitrary, natural everyday scenes composed of dozens of colour subfields, usually none highly saturated, will have a certain, almost fixed spatial reflectance average. It is reasonable that this average will be some medium gray...." D'Zmura and Lennie (1986, p. 1667) make a similar claim: "... we expect that the space-averaged light from most natural scenes will bear a chromaticity that closely approximates that of the illuminant." The gray world assumption is close in spirit to Helson's "adaptation level" (Helson, 1934).

The term "gray world assumption" is misleading in two senses. First, for Buchsbaum's algorithm to work, the known average of the intrinsic surfaces need only be *known*; it need not be gray. If the average of the intrinsic colors across the scene corresponds to a green surface, the algorithm of Buchsbaum can as easily make use of a green world assumption. Second, the average need not include the entire scene: Any portion of the scene (e.g., the ground plane) can serve as an average reference. From this point on, I will refer to the gray world assumption in this wider sense as the "stable mean assumption."

The stable mean assumption is, fundamentally, a claim about the physical environment as seen through a particular set of photoreceptors, and such a claim is testable. It is obvious that a stable mean algorithm such as Buchsbaum's will erroneously "correct" any deviation of the mean intrinsic color away from the known reference color as readily as it corrects an imbalance induced by the illuminant. The algorithm is of value to the extent that typical excursions of the mean intrinsic color away from the reference are small and/or infrequent compared to the magnitude of changes in the illuminant. Empirical tests of these claims are lacking.

It should be noted that the stable mean assumption is specific to a particular set of photoreceptor types. It is entirely possible that a stable mean algorithm may fail dramatically for one species and succeed for another in the same environment. Discussion of the status of the stable mean assumption in human vision will be postponed until the von Kries Algorithms section.

The use of reference surfaces in two of the algorithms above makes them implausible candidates for a model of human color vision. The remaining algorithms described in this section illustrate ways of dispensing with reference surfaces altogether. Some of the algorithms first estimate the coordinates of the illuminant ε and then employ Eqn. (20) to solve for intrinsic colors; others estimate ε and the σ^{xy} simultaneously. Only Brill's algorithm avoids estimating ε and instead estimates Λ_ε directly. The remaining algorithms differ mainly in how they go about determining the illuminant.

The subspace algorithm of Maloney and Wandell. Maloney and Wandell (1986) assumed that there are more classes of receptors than dimensions in the linear model of surface reflectances: $P > N$. Suppose that there are $P = N + 1$ linearly independent receptors available to spectrally sample the image at each location. They proposed a method for computing the light coordinates ε and the N-dimensional surface reflectance vectors σ^{xy} given the $N + 1$-dimensional receptor response vector ρ^{xy} at each location. The matrix Λ_ε is then a linear transformation from the N-dimensional space of surface reflectances σ^{xy} into the $N + 1$-dimensional space of receptor excitations ρ^{xy}. As Λ_ε is a linear transformation, the receptor responses must fall in a proper subspace of the receptor space. In the case $P = 3$, $N = 2$, the vectors ρ^{xy} must lie on a plane in the three-dimensional receptor space. In the case $P = 4$, $N = 3$, the vectors ρ^{xy} must lie in a three-dimensional subspace (a "3-space") of the four-dimensional receptor space. The particular subspace is determined by Λ_ε and therefore by the lighting parameter ε. That is, as ε is varied for a fixed set of surfaces, the subspace spanned by ρ^{xy} moves about in the receptor space.

Maloney and Wandell (1986) proposed a two-step procedure to estimate normalized light and surface reflectance properties. First, they determine the plane spanning the receptor excitations. Second, knowledge of the plane permits recovery of the normalized ambient light vector $\hat{\varepsilon}$ by computations described by Maloney (1984). These computations use the vector at the origin normal to the subspace spanned by the receptor receptor excitations. This vector plays a role that is analogous to the single reference surface in Buchbaum's algorithm: It is a recoverable "landmark" that moves around as a function of the illuminant. By requiring $P > N$, Maloney and Wandell were able to replace Buchsbaum's reference surface with a geometrical landmark that serves the same purpose.

Note that the outcome of the algorithm consists of estimates of the intrinsic colors of surfaces known up to a single common lightness scaling factor C. That is, if ε is the true light and $\sigma^{xy_1}, \sigma^{xy_2}, \ldots, \sigma^{xy_Z}$ are the true intrinsic colors at Z locations, the algorithm returns a normalized estimate of the light $\bar{\varepsilon} = (1/C) \cdot \varepsilon$ and corresponding estimates of the surface properties $\bar{\sigma}^{xy_1} = C\sigma^{xy_1}, \ldots, \bar{\sigma}^{xy_Z} = C\sigma^{xy_Z}$, where C is an unknown scaling factor common to all of the estimates. (In contrast, the "reference surface" algorithms above can use the reference surface to estimate the absolute power output of the illuminant.)

The chromatic aberration algorithm of Funt and Ho. Ho (1988; Funt & Ho, 1989; Ho, Funt, & Drew, 1990) used the chromatic aberration inherent in lens systems to derive an estimate of the illuminant parameters ε up to an unknown scale factor, and showed that a visual system with only one receptor class can estimate the difference $\Delta(\lambda)$ between the color signals radiating from two adjoining surfaces separated by a sharp edge. Let S_{σ^1} and S_{σ^2} denote the two surface reflectance functions. Then,

$$(29) \quad \Delta(\lambda) = E_\varepsilon(\lambda)S_{\sigma^1}(\lambda) - E_\varepsilon(\lambda)S_{\sigma^2}(\lambda),$$

and the estimated color signal difference can be used to solve for ε up to an unknown scale factor.

Bayesian algorithms. D'Zmura, Iverson, and Singer (1995) and Freeman and Brainard (1995; Brainard & Freeman, 1997) reformulated the problem of estimating illuminant and surface within the linear models framework as a problem in Bayesian statistical decision theory. This sort of approach presupposes, first of all, that the *prior distribution* of possible scenes composed of illuminants and surfaces, and the *likelihood function*, the likelihood of any possible retinal excitation patterns conditional on a particular scene, are both known.

We need not review the details of their work except to note, first, that essentially any algorithm that can be formulated as a maximum likelihood estimate can be trivially reformulated as a Bayesian estimator that can take advantage of knowledge of the prior distribution (Blackwell & Girshick, 1954; Ferguson, 1967). That is, each of the algorithms reviewed here has a Bayesian counterpart that makes use of the known prior distribution. Second, the Bayesian estimator can do no worse than its maximum likelihood counterpart (Blackwell & Girshick, 1954; Ferguson, 1967). That is, there is no point in evaluating them computationally except to determine *how much* better they perform. Last, the expected advantage in performance depends on the prior distribution of illuminants and surfaces. The true prior distribution of illuminants and surfaces in realistic environments is currently unknown. Both D'Zmura, Iverson, and Singer and Freeman and Brainard simply assumed prior distributions of lights and surfaces. Both make the unrealistic assumption that surface patches at different locations in a scene are statistically independent.

Although there is little to be learned from their computational results, the *approach* that they propose is sound and their computational experiments are potentially important if they are redone with accurate estimates of prior distributions in realistic environments.

Multiple-view algorithms. Tsukada and Ohta (1990) and, independently, D'Zmura (1992) examined the information made available by viewing the same scene under multiple successive illuminants. Let the coordinates of the two illuminants be ε and ε'. D'Zmura first sets up the equations $\Lambda_{\varepsilon}\sigma^{xy} = \rho^{xy}$ and $\Lambda_{\varepsilon'}\sigma^{xy} = \rho'^{xy}$, describing the photoreceptor excitations corresponding to the surfaces at each location illuminated by each of the two illuminants. He then shows how to solve these simultaneous equations for the σ^{xy}. In particular, for $P = N = M = 3$, two views permit recovery of the intrinsic colors σ^{xy} up to an overall unknown scaling factor. Thus, if a trichromatic visual system can arrange to view a scene under two successive illuminants, it can recover the surface descriptors up to an overall scaling. In a later section, The Shadow Algorithm of D'Zmura, I describe a second application of these results.

The method of D'Zmura (1992) is a generalization of the subspace method of Maloney and Wandell. In an impressive series of articles, D'Zmura and Iverson (1993a, 1993b, 1994; Iverson & D'Zmura, 1995a, 1995b) study visual systems with P photoreceptors in the flat world environment with surface reflectance modeled by an N-dimensional linear model and illuminants modeled by an M-dimensional linear model. They derive necessary and sufficient conditions for the recovery of surface and light parameters when the same scene is seen under Q distinct illuminants, including the case $Q = 1$ treated by Maloney and Wandell.

Illuminant estimation. As has been discussed previously, several of the algorithms above so far share a common form. Each seeks an explicit estimate of the illuminant parameters ε using the constraints imposed by flat world and linear models. The results of Gilchrist and colleagues (Gilchrist, 1977, 1980; Gilchrist, Delman, & Jacobsen, 1983; see also Gilchrist, 1994) demonstrate that lightness perception depends on the visual system's representation of the three-dimensional layout of scenes. In the next section, we will examine more complex, three-dimensional environments where there are additional cues available concerning the illuminant.

Environment II: shape world

At this point, we must expand the mathematical models of illuminant and surface reflectance developed in the flat world environment to include more realistic models of light–surface interaction in a three-dimensional scene. The coordinate xy in $S^{xy}(\lambda)$ now specifies both a location xy on the sensor array (retina) and the surface point imaged on the sensor array at that location. We allow only one surface point corresponding to each retinal location (ignoring the possibility of transparency). In most of this section, it is assumed that there is a single, punctate illuminant distant from the observer and the scene (for some of the discussion there will be a small number of illuminants).

The algorithms in this section share the same common goal as the algorithms described in the previous section. All of them describe a particular computation designed to capture information about the illuminant ε. Some of the algorithms use this information to compute Λ_ε and invert it, just as in the algorithms described above, while others simultaneously estimate the illuminant and surface descriptors. The algorithms share a second common feature: Each goes beyond the assumptions of a simple flat world environment, basing its estimate of the illuminant on information that is available only from surfaces arranged in three dimensions.

Figure 20.6 indicates some of the additional structures introduced into the environment: shading, specularity, mutual illumination, and so on. The shape-from-shading literature is a source of models for illuminant–surface interaction in three-dimensional scenes (see Horn & Brooks, 1989; in particular, Horn & Sjoberg, 1989). This new environment will be referred to as *shape world.*

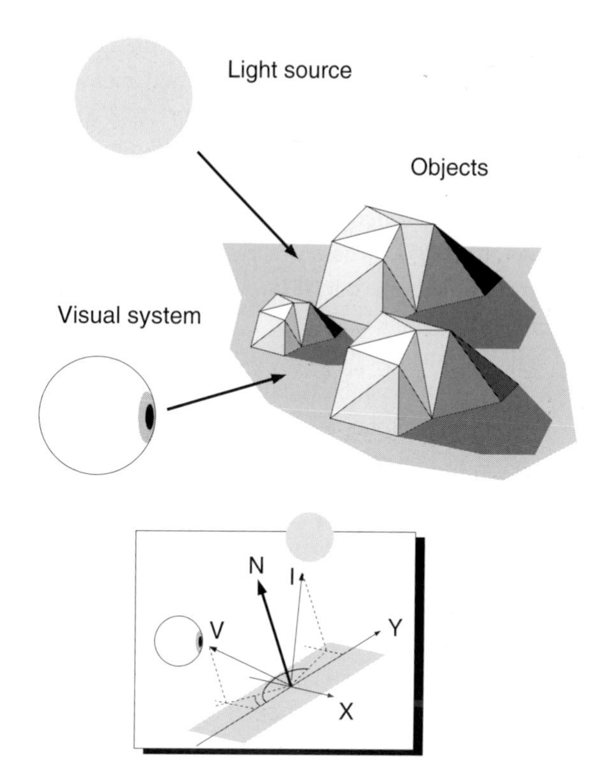

Figure 20.6: Shape world. A model environment for surface color perception that includes explicit representation of light–surface interaction in a three-dimensional scene, including shadows, interreflection (mutual illumination) among surfaces, and specularities. See text for an explanation. The inset illustrates the coordinate system used in expressing the relation between the illuminant, surface, and visual system.

Bidirectional reflectance density functions. The surface spectral reflectance function of flat world $S^{xy}(\lambda)$ is replaced by a *bidirectional reflectance density function* (BRDF), $S^{xy}(\lambda, V^{xy}, N^{xy}, I^{xy})$. N^{xy} is a unit normal vector to the surface at location xy. V^{xy} is a unit vector from the surface at xy in the direction of the visual system (the sensor array), and I^{xy} is the unit vector from the surface toward the (punctate) illuminant. The inset to Fig. 20.6 illustrates the interrelations among these vectors. Horn and Sjoberg (1989) provide a more detailed description of the BRDF and further references.

The possible interactions between light and surface are complex (Nassau, 1983; Weisskopf, 1968). Lee, Breneman, and Schulte (1990) summarize some of the common models of BRDFs. Hurlbert (1998) discusses many of the models of BRDFs in use in computer vision and computer graphics. Oren and Nayar (1995; Nayar & Oren, 1995) review more recent work and propose a new model of the BRDF that takes into account the roughness of the surface at fine scales. Lee, Breneman, and Schulte (1990) report measurements of BRDFs of a small number of surfaces, which we will return to below. Researchers at Columbia University and the University of Utrecht have measured BRDFs for 60 textured natural and artificial surfaces (Dana et al., 1997).[5] There is currently no larger collection of natural or artificial surfaces available to test existing or new models.

I will not attempt to summarize the literature on models of BRDFs, which is extensive. Instead, I will describe only the properties of BRDFs that are needed to understand the shape world algorithms reviewed below.

Geometry-reflectance separability. The first property we consider is *geometry-reflectance separability*:

$$(30) \quad S^{xy}(\lambda, V^{xy}, N^{xy}, I^{xy}) = S^{xy}(\lambda)G(V^{xy}, N^{xy}, I^{xy}).$$

It states that the effect of a change in viewing conditions V^{xy}, N^{xy}, I^{xy} is simply to scale the surface reflectance function (Shafer, 1985). $S^{xy}(\lambda)$ will be referred to as the *spectral component* of the BRDF. The surface spectral reflectance $S^{xy}(\lambda)$ of flat world is, of course, the spectral component $S^{xy}(\lambda)$ in shape world, scaled by the geometric factors. It will cause no confusion to use the same notation for the flat world spectral sensitivity function and the spectral component of a BDRF in shape world (when geometric-reflectance separability holds).

If there are multiple illuminants or nonpunctate illuminants, the color signal $L^{xy}(\lambda)$ is the superposition of the color signals corresponding to each illuminant point. It is possible that light from an illuminant can be successively absorbed and reemitted by more than one surface before reaching the sensor array. In such cases, the light emitted from one surface acts as an illuminant to a second and is treated accordingly.

[5]Their measurements and related discussion can be found at http://www.cs.columbia.edu/CAVE/curet.

Diffuse-specular superposition. Surfaces often do not satisfy geometry-reflectance separability. Shafer (1985; Klinker, Shafer, & Kanade, 1988) suggested that many surface BRDFs [corresponding to dielectric (nonconducting) surfaces such as plastics] may be represented as the sum of two surface BRDFs, each of which satisfies geometric-reflectance separability:

$$(31) \quad S^{xy}(\lambda, V^{xy}, N^{xy}, I^{xy}) = m_d S^{xy}(\lambda)G(V^{xy}, N^{xy}, I^{xy}) + m_s Spec(\lambda)G'(V^{xy}, N^{xy}, I^{xy}).$$

The first term in the summation is termed the *diffuse component*, and the second, the *specular component*. $Spec(\lambda) = 1$ is the surface spectral sensitivity function of a perfect reflector, $G(\cdot, \cdot, \cdot)$ is the geometric function for a diffuse surface, and $G'(\cdot, \cdot, \cdot)$ is the geometric function for a specular (mirrorlike) surface. The weights m_d and m_s control the diffuse-specular balance. Note that $Spec(\lambda)$ is the same at every location while the $S^{xy}(\lambda)$ may vary. The constraint on surfaces embodied in Eqn. (31) will be referred to as the *diffuse-specular superposition* property. The neutral interface model of Lee, Breneman, and Schulte (1990) exhibits this property. I will refer to $S^{xy}(\lambda)$ in Eqn. (31) as the spectral component of the BDRF.

Lee, Breneman, and Schulte (1990) tested whether surfaces satisfied the diffuse-specular superposition property. They measured the spectral reflectance functions of nine surface materials for different viewing geometries. They found that the property was satisfied for some of the surface materials (including yellow plastic cylinder, green leaf, and orange peel) but not all (e.g., blue paper, maroon bowl). Tominaga and Wandell (1989, 1990) report empirical tests of the property as well.

In the description of the algorithms, I will assume that $N = M = P = 3$. That is, the possible illuminants are drawn from a linear model with three parameters $(\varepsilon_1, \varepsilon_2, \varepsilon_3)$, the visual system has three classes of photoreceptor, and the possible spectral components of surfaces are also drawn from a linear model with three parameters $(\sigma_1, \sigma_2, \sigma_3)$. I will generally assume that either geometric-reflectance separability or diffuse-specular condition holds. The term *viewing geometry* refers to the relative positions of surfaces, illuminants, and the observer in shape world.

Viewing the light/white surface. The first model we consider is one where the observer is able to look around and somehow identify the illuminant itself. Viewing the illuminant is equivalent to having a single white reference surface in Buchsbaum's algorithm. We can solve directly for the light. Alternatively, the observer may attempt to identify a perfectly reflective (white) patch in the scene and use it just as a glimpse of the illuminant. These approaches fail, of course, when the scene contains no white patch or directly observable illuminant, or when the observer cannot identify one or the other.

Specularity algorithms. Lee (1986) and D'Zmura and Lennie (1986) proposed algorithms based on the diffuse-specular superposition property. Figure 20.7 illustrates the key idea. Suppose that we have two *objects*, each with a uniform diffuse-specular surface and with sufficient variation in the viewing geometry so that the color signals from the object are not uniform. Further, assume that the objects have distinct diffuse components $S(\lambda)$ and $S'(\lambda)$.

The receptor excitations from the first object will be weighted mixtures of a diffuse color signal,

$$(32) \quad D_k = \int E(\lambda)S(\lambda)R_k(\lambda)d\lambda, \quad k = 1, 2, 3,$$

and a specular color signal

$$(33) \quad Spec_k = \int E(\lambda)Spec(\lambda)R_k(\lambda)d\lambda, \quad k = 1, 2, 3,$$

where the weights are determined by the surface mixture parameters m_d and m_s and the viewing geometry. The receptor excitations for the second object will also be a weighted mixture of a diffuse color signal,

$$(34) \quad D'_k = \int E(\lambda)S'(\lambda)R_k(\lambda)d\lambda, \quad k = 1, 2, 3,$$

and the same specular color signal

$$(35) \quad Spec_k = \int E(\lambda)Spec(\lambda)R_k(\lambda)d\lambda, \quad k = 1, 2, 3.$$

Then the photoreceptor excitation ρ^{xy} for the first object at location xy is a weighted mixture,

$$(36) \quad \rho^{xy} = \alpha^{xy}D + \beta^{xy}Spec,$$

where $D = [D_1, D_2, D_3]'$ and $Spec = [Spec_1, Spec_2, Spec_3]$. $Spec$ is the color of the illuminant (as defined in the *RGB Heuristic*). As shown in Fig. 20.7, all of the photoreceptor excitations must lie in a plane through the origin spanned by D and $Spec$.

By a similar argument, all of the photoreceptor excitations for the second object must lie in the plane through the origin spanned by D' and $Spec$:

$$(37) \quad \rho'^{xy} = \alpha'^{xy}D' + \beta'^{xy}Spec.$$

If there are enough different points from each of the objects to permit detection and estimation of the two planes, and if the planes are distinct, the intersection of the planes determines $Spec$ up to an unknown scaling. From this we can learn ε up to an overall scaling.

One strength of the method is that it does not assume that bright, specular highlights visible on the surface reflect precisely the spectral power distribution of the illuminant. Instead, the light from each point on the surface is modeled as a weighted mixture of a specular component and a diffuse component. The specular component has the spectral power distribution of the illuminant. The spectral power distribution of the diffuse component is characteristic of the object and is assumed to be independent of the viewing angle.

Klinker, Shafer, and Kanade (1988) analyze the diffuse-specular superposition constraint further and demonstrate that the diffuse-specular mixtures are not simply confined to a plane in Fig. 20.6 but will typically form a characteristic skewed T shape within the plane. The skewed T is a color space feature that can be extracted and used in estimating the illuminant. Healey (1991) shows that it is possible to recover the illuminant by using one diffuse specular object if the viewing geometry guarantees that some point on the object will exhibit a pure diffuse reflectance, and

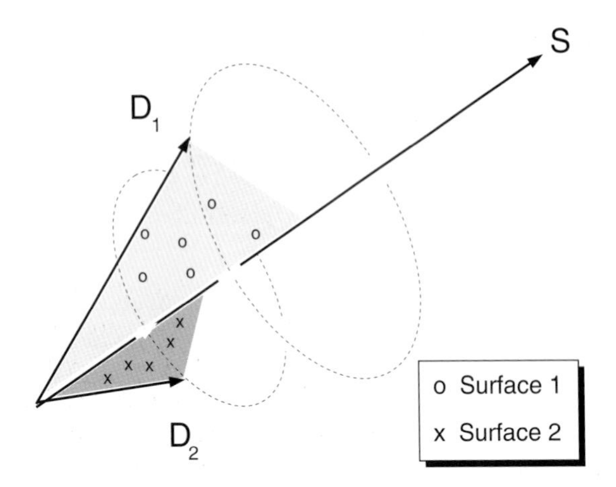

Figure 20.7: The specular-diffuse constraint. The photoreceptor excitations for an object with a homogeneous surface that satisfies the specular-diffuse constraint (see text) must lie in a plane. The intersection of two such planes, corresponding to two different specular-diffuse objects, marks the chromaticity of the illuminant.

another point a pure specular reflectance with 100% reflectance.

The mutual reflection algorithm of Drew and Funt. Drew and Funt (1990) describe how it is possible to estimate the illuminant parameters ε using the mutual reflection between adjacent surfaces in shape world. Light emitted from one surface with spectral component $S^1(\lambda)$ may reach a second nearby surface with spectral component $S^2(\lambda)$ and be absorbed and reemitted in the direction of the observer. The color signal reaching the observer is proportional to $E(\lambda)S^1(\lambda)S^2(\lambda)$, where the constant of proportionality is determined by the viewing geometry.

Drew and Funt (1990) restrict attention to the case where each ray of light from the illuminant encounters either one or two surfaces in its transit to the sensor array (a single-bounce or one-bounce model of mutual reflection). They express the color signal as a sum of zero-bounce and one-bounce components for each surface, substitute Eqns. (12) and (13) into their expressions, and show that the resulting third-degree equations in ε and σ^1, σ^2 can be solved by iterative methods. The method is "reasonably robust" (p. 399) and can be used even when the two mutually illuminating surfaces have the same spectral components, for example, at a corner in a room.

The shadow algorithm of D'Zmura. D'Zmura (1992) points out that his multiple illuminant algorithm can be applied across shadow boundaries. At a shadow boundary that does not coincide with the boundary between two different surfaces, the same surface is illuminated by two different illuminants. If the shadow boundary intersects three or more surfaces, the multiple illuminant algorithm of D'Zmura (1992) can be applied to recover estimates of three-dimensional illuminants and surface parameters with only three photoreceptor classes.

Other work. We have, so far, not made use of an evident constraint on $S(\lambda)$, the surface spectral reflectance function, that $0 \leq S(\lambda) \leq 1$. Forsyth (1990) develops an approach to color constancy that makes use of such physical-realizability constraints. Rubner and Schultern (1989) apply a regularization approach that is similar in spirit to the Bayesian. Other work, not further discussed here, includes that of Gershon and Jepson (1988, 1989).

Other environments. The shape world environment is not a realistic model of natural scenes. One glaring deficiency is the restriction to one punctate illuminant or a small number of punctate illuminants that illuminate the scene successively or illuminate nonoverlapping parts of the scene. Research is needed to develop realistic models of light–surface interaction in scenes in a true *real world.*

Model failure and approximate constancy

The consequences of model failure. As noted above, linear models of illuminant and surface reflectance with three dimensions do not provide perfect fits to empirical data sets. Accordingly, if any of the algorithms described above for flat world or shape world are to be applied in an environment composed of illuminants and surfaces drawn from such empirical collections, the consequences of discrepancies between idealization and application must be considered.

We can model the effect of linear model failure within the linear model framework as follows (Maloney, 1984). We represent the true spectral power distribution of the illuminant as the sum of the linear model approximation to it and a residual illumination term $e(\lambda)$:

$$(38) \quad E(\lambda) = E_{\varepsilon}(\lambda) + e(\lambda) .$$

A similar decomposition is adopted for the surface reflectance at location xy:

$$(39) \quad S^{xy}(\lambda) = S_{\sigma}^{xy}(\lambda) + s^{xy}(\lambda) ,$$

where $s^{xy}(\lambda)$ is the residual surface spectral reflectance.

The color signal $L^{xy}(\lambda)$ at location xy is then,

$$(40) \quad L^{xy}(\lambda) = E_{\varepsilon}(\lambda)S_{\sigma}^{xy}(\lambda) + e(\lambda)S_{\sigma}^{xy}(\lambda) + E_{\varepsilon}(\lambda)s^{xy}(\lambda) + e(\lambda)s^{xy}(\lambda).$$

Let

$$(41) \quad L_{\varepsilon\sigma}^{xy}(\lambda) = E_{\varepsilon}(\lambda)S_{\sigma}^{xy}(\lambda) ,$$

the color signal from location xy if the linear models had accurately captured the illuminant spectral power distribution and the surface spectral reflectance. The discrepancy between $L^{xy}(\lambda)$ and $L_{\varepsilon\sigma}^{xy}(\lambda)$ is the sum of three terms,

$$(42) \quad e(\lambda)S_{\sigma}(\lambda) + E_{\varepsilon}(\lambda)s(\lambda) + e(\lambda)s(\lambda) ,$$

corresponding to (a) the model light $E_{\varepsilon}(\lambda)$ shining on the residual surface $s^{xy}(\lambda)$, (b) the residual light $e(\lambda)$ shining on the model surface $S_{\sigma}^{xy}(\lambda)$, and (c) the residual light $e(\lambda)$ shining on the residual surface $s^{xy}(\lambda)$. We can view the scene, then, as the superposition of two scenes: the *model scene* $L_{\varepsilon\sigma}^{xy}(\lambda)$, whose parameters we wish to estimate, and the *error scene* in Eqn. (42), whose presence perturbs our estimates. The magnitude of the perturbation at location xy is $\Delta\rho^{xy} = [\Delta\rho_1^{xy}, \Delta\rho_2^{xy}, \Delta\rho_3^{xy}]$, where:

$$\Delta\rho_k^{xy} = \int e(\lambda)S_{\sigma}(\lambda)R_k(\lambda)d\lambda + \int E_{\varepsilon}(\lambda)s(\lambda)R_k(\lambda)d\lambda + \int e(\lambda)s(\lambda)R_k(\lambda)d\lambda.$$

The error term $\Delta\rho^{xy}$ depends, first of all, on the bases E_i and S_j, and, of course, we would like to choose bases to minimize the impact of the error scene. Marimont and Wandell (1992) describe how to do this for any empirical collection of lights, surfaces, and photoreceptor sensitivities. Brill (1978, 1979), Buchsbaum (1980), Maloney (1984), and Wandell (1987) describe perturbation analyses of some of the flat world algorithms. A few results can be found in Maloney (1984, Chap. 4).

Linear model algorithms and human vision

In this section, we consider the relation between the work reviewed above and classical models of color vision framed in terms of hypothesized color channels and transformations of color information through successive stages. Earlier forms of this discussion may be found in Maloney (1984, 1992).

Adaptational state and adaptational control. Let ρ^{xy} denote, as above, the excitations of the three photoreceptor classes in a small retinal patch near xy. It is well known that ρ^{xy} alone does not determine the perceived color appearance of the patch at xy that may be profoundly influenced by photoreceptor excitations in other retinal areas, in one or both eyes. I assume that there are *mechanisms of color appearance* whose excitations, written in vector form $\mu^{xy} = [\mu_1^{xy}, \ldots, \mu_W^{xy}]$, determine performance in tasks measuring color appearance in a small patch at location xy.

The number W of mechanisms of color appearance is unknown, but, for the normal trichromatic observer, it is plausible that there are three independent mechanisms, and possibly others. For simplicity, I assume that $W = 3$ in the following. The relationship between

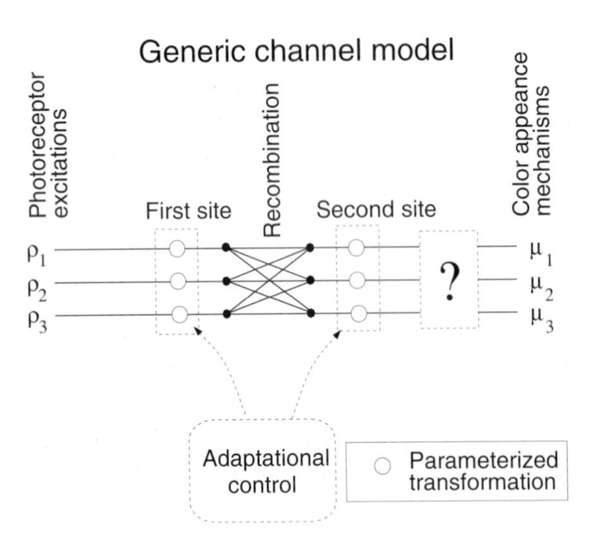

Figure 20.8: The generic channels framework. The open circles mark points in channels where variable gain controls or additive biases may be applied. See text for an explanation.

μ^{xy} and ρ^{xy} can be written as

$$\mu^{xy} = \tau_{\Sigma}(\rho^{xy}), \quad (43)$$

where $\tau_{\Sigma}: \mathbf{R}^3 \rightarrow \mathbf{R}^3$ is a transformation from one three-dimensional vector space to a second, and where Σ denotes the influence of the surround: the photoreceptor excitations of previous retinal stimulations, and current retinal stimulations at locations other than *xy*.

The *color transformations*, denoted $\tau_{\Sigma}(\cdot)$, represent the net effect of all of the transformations (at various stages) performed by the visual system on the initial retinal information corresponding to location *xy*. The precise form of $\tau_{\Sigma}(\cdot)$ has received considerable attention. There is widespread agreement that initial information in three color pathways passes through a "first-site" (Fig. 20.8), where it it scaled multiplicatively (Chichilnisky & Wandell, 1995) and shifted additively (Burnham, Evans, & Newhall, 1957; Jameson & Hurvich, 1964; Walraven, 1976; Shevell, 1978), followed by an opponent recombination (Hurvich & Jameson, 1957; see also Hurvich, 1981; Wandell, 1995; Kaiser & Boynton, 1996) and by "second site" multiplicative attenuation (Hurvich & Jameson, 1957; Webster & Mollon, 1995; Brown & MacLeod, 1997).

The transformation $\tau_{\Sigma}(\cdot)$ is likely nonlinear if it is examined across a wide range of overall scene intensities, but "[f]or modest signals under a constant adaptation state, single-cell responses and psychophysical sensitivity are consistent with mechanisms that respond to simple sums or differences of the cone contrasts" (Webster, 1996, p. 595). Under such circumstances, we can assume that the $\tau_{\Sigma}(\cdot)$ are approximately affine transformations of the form

$$\tau_{\Sigma}(\rho) = D_2 H D_1(\rho - O_1), \quad (44)$$

where D_1 and D_2 are 3×3 diagonal matrices corresponding to the multiplicative attenuations at first and second sites, respectively, O_1 is an offset 3-vector corresponding to a first site additive shift, and H is a 3×3 opponent transformation matrix.

Of interest to us here are the components of $\tau_{\Sigma}(\cdot)$ that may change in response to Σ: the multiplicative attenuations at first and second sites, and the additive shifts. I refer to these variable components of models of $\tau_{\Sigma}(\cdot)$ as *parameters*. These parameters depend on the surround Σ. If we regard the opponent transformation as fixed (nonparameterized), the parameters above include the first-site offsets and attenuations, $O = [\pi_1, \pi_2, \pi_3]$ and $D_1 = Diag[\pi_4, \pi_5, \pi_6]$, and the second-site attenuations $D_2 = Diag[\pi_7, \pi_8, \pi_9]$. The state vector of nine parameters $\pi = [\pi_1, \ldots, \pi_9]$ determines $\tau_{\Sigma}(\cdot)$, which may be written $\tau_{\pi(\Sigma)}(\cdot)$ or $\tau_{\pi}(\cdot)$ to emphasize that the influence of the surround/scene reduces to selecting the settings of a handful of parameters, π. This formulation of the study of color appearance is essentially that employed by Krantz (1968) in modeling the effect of "context" on color appearance.

Stiles (1961, p. 264) proposed that the study of color vision be considered as the study of two processes, *local retinal adaptational state* and the process that selects the local state, *the control of adaptation*: "we anticipate that a small number of variables – adaptation variables – will define the condition of a particular visual area at a given time, instead of the indefinitely many that would be required to specify the conditioning stimuli. The adaptation concept – if it works – divides the original problem into two: What are the values of the adaptation variables corresponding to different sets of conditioning stimuli, and how does adaptation, so defined, modify the visual response to given test stimuli?"

The exact number and nature of the local retinal adaptational state parameters π is not important to the discussion below. Various authors have suggested models with from three to nine parameters. See the discussion in Brainard, Brunt, and Speigle (1997). What is important is that π determines the affine transformation $\tau_{\pi}(\Sigma)$ and that the surround/scene Σ determines π:

$$(45) \quad \Sigma \rightarrow \pi(\Sigma) \rightarrow \tau_{\pi(\Sigma)}.$$

Again, the advantage of this approach, "if it works," is that is reduces the study of color vision to two questions:

1. *What are the possible local retinal adaptation states?*[6] That is, what are the local retinal state parameters and what possible settings can they take on? Let the set of all possible parameter settings π be denoted Π.

2. *How are possible conditioning stimuli mapped to a choice of retinal adaptational state?* In our terminology, the set Π of possible parameter settings corresponds to the possible retinal adaptational states and the control of adaptation is the selection process $\pi(\Sigma)$ that sets the retinal state, selecting the color transformation as a function of the surround Σ.

As was discussed above, we know something about the structure of the $\tau_{\pi(\Sigma)}$, the affine transformations of Eqn. (44). Until recently, the second question has received considerably less attention than the first. I will argue below in the Linear Models and Adaptational Control section that the linear model algorithms make their most original and significant contributions as candidate theories of adaptational control, addressing the second of the two questions. In the intervening sections, I describe a small number of earlier theories of adaptational control that are necessary to the argument. The following is neither a comprehensive nor a representative discussion of models of adaptational control, omitting important work from Judd (1940) to Hunt (1991).

Recent work of Zaidi, Spehar, and DeBonet (1997, 1998) is not directly relevant to this review, but certainly deserves mention. Zaidi and colleagues model the effect of illuminant changes at both first and second sites in the color channels. They then derive simple transformations that are intended to cancel the illumination-induced changes at the second site. They propose mechanisms of adaptational control that are sensitive to spatial variation in the scene viewed. The resulting low-level (their terminology) approach to color constancy is an elegant alternative to the sort of physics-based approach described here. It would be particularly interesting to analyze the behavior of their model in response to specularity, shadow, and so on. See Poirson and Maloney (1998) for additional discussion. Poirson and Maloney (1998) and Webster (1996) propose contrasting models of chromatic adaptation and adaptational control.

von Kries algorithms. Von Kries (1902/1970, 1905/1970) proposed that retinal adaptation states be identified with scalings of photoreceptor excitations. Translated into the notation above, this is equivalent to restricting the $\tau(\cdot)$ to be diagonal matrices,

$$(46) \quad \mu^{xy} = D\rho^{xy},$$

where $D = Diag[\pi_1, \pi_2, \pi_3]$. Von Kries's discussion of the Coefficient Law for simple center-surround configurations of stimuli includes the assumption that $\pi_k^{xy} = 1/\rho_k^{\Sigma}$, where ρ_k^{Σ} is the excitation of the kth class of receptor in response to the large conditioning surround field. It is not made clear how the visual system might choose the coefficients π_k in more complex scenes, lacking a center-surround structure.

[6] The term "local retinal adaptational state" will be used to refer to the state of the pathways that carry chromatic information corresponding to a small region of the visual scene. I do not mean to imply that chromatic adaptation is purely retinal, but rather to emphasize that the local retinal color information is transformed at successive stages of the visual system in ways that are not fixed.

Helson (1934, 1938) proposed a model of color adaptation in which the coefficient π_k was the inverse of the average of photoreceptor excitations within the kth class across a region of the retina. This von Kries–Helson model coincides with von Kries's proposal for the simple center-surround configuration and is one possible extension of a von Kries adaptation to arbitrary scenes. The von Kries–Helson model of adaptation is closely related to gray world (Buchsbaum, 1980, p. 24), and Buchsbaum's algorithm is intended as a natural generalization of von Kries–Helson that takes into account the linear model constraints on light and surface, making explicit the dependence of the algorithm's performance on environmental constraints.

The retinex algorithm of Land and McCann (1971) assumes Eqn. (46) and propose a complicated computation of the parameters (π_1, π_2, π_3). In particular, the value of the parameter π_k depends only on the photoreceptor excitations of the kth class of photoreceptors, $k = 1, 2, 3$. For the purposes of adaptational control, then, the visual system is divided into three, isolated *retinexes*, that is, pathways driven by photoreceptor excitations from a single class of photoreceptor but from widely separated areas in the retina.

Brainard and Wandell (1986) analyzed this retinex algorithm and concluded that the computation, in effect, selects, as π_k, the inverse of the largest photoreceptor excitation in the kth class of photoreceptors across a region of the retina. A heuristic motivation for this choice of π is that, if there is a perfectly reflecting white surface visible in the scene, the retinex (Land & McCann, 1971) algorithm will "lock on" to it and, in effect, scale by coefficients inversely proportional to the color of the illuminant (see The RGB Heuristic).

Brainard and Wandell (1986) analyzed the later retinex algorithms of Land (1983, 1986) and concluded that they, in effect, tend to choose the inverse of the spatially weighted *geometric mean* of the photoreceptors in the kth class of photoreceptors as the π_k. In contrast, the von Kries–Helson algorithm described above used the inverse of the spatially weighted arithmetic mean.

These algorithms are all examples of so-called lightness algorithms (see Hurlbert, 1986, 1998; Maloney, 1992, for reviews). The term "lightness algorithm" is unfortunate, for there is no special link between such algorithms and the study of lightness (Gilchrist, 1994). I refer to them as *von Kries algorithms.*

Brill and West (1981; West & Brill, 1982) study necessary and sufficient conditions under which von Kries adaptation can result in color constancy in response to changes in the illuminant.

How well does von Kries describe human vision? It is believed by many that von Kries adaptation, possibly with an additive bias, is an accurate model for human observers, at least in simple scenes. See Brainard (1998) for a discussion and Chichilnisky and Wandell (1995) for representative results. Foster and his colleagues (Craven and Foster, 1992; Foster & Nascimento, 1994) found that simulated von Kries transformations on images were judged illuminant changes more often than simulated illuminant changes. Note, however, that these results imply that a von Kries transform accounts for much of visual adaptation, not all. The analyses typically performed do not look for small systematic patterns of deviation that are consistent with a contribution from second-site adaptation.

Recent experimental work is consistent with von Kries adaptation but rejects von Kries–Helson adaptation in even modestly complex scenes (Jenness & Shevell, 1995; Brown & MacLeod, 1997; Mausfeld, 1998; Brainard, 1998). For example, Brainard (1998; discussed in Kaiser & Boynton, 1996, p. 519) draped part of a room in red vinyl cloth. This change in average chromaticity of a large part of the scene should have resulted, according to von Kries–Helson, in a substantial change in the von Kries adaptational coefficients and a marked change in the color appearances of objects in the scene. It did not. In more recent experiments, reported at the Optical Society Meetings, Kraft and Brainard (1997) described experiments in which the mean chromaticity of the color signals across the scene was held fixed while the illuminant was varied. The observer looked into a three-foot cubical chamber. He viewed one of two scenes, one created with a neutral illuminant on a variety of mostly gray surfaces and one created with a red illuminant on a variety of mostly blue surfaces. The mean color signal across the observer's field of view was the same. Even with this mean held fixed, the visual system still adjusted substantially to the illuminant. Their results suggest that, under these experimental circumstances, the mean color signal is influential but not decisive (D. H. Brainard, personal communication).

von Kries–Ives adaptation. Clearly, the von Kries adaptation is not coextensive with the von Kries–Helson adaptation. The retinex algorithms assume the same local retinal adaptation states (von Kries) but propose different rules for controlling the choice of adaptational state.

We can distinguish one particular class of von Kries algorithm in which the coefficients π_k are set equal to the chromaticity of the illuminant defined above in The RGB Heuristic. I will refer to this form of adaptation as the *on Kries–Ives* adaptation. Again, without constraints on illuminants and surfaces in the scene, such an algorithm will not, in general, discount the illuminant or result in approximate color constancy.[7]

A crucial difference between the von Kries–Ives adaptation and the other von Kries algorithms, described above, is that it is not obvious how to compute the chromaticity of the illuminant from retinal excitations. We can, however, view this limitation in another way. Given any source of information about the chromaticity of the illumination, we can implement a von Kries–Ives algorithm based on it. Knowledge of the illumination parameters ε is enough to determine the chromaticity of the illuminant. Hence, any of the linear model algorithms that estimate ε can be used to implement a von Kries–Ives algorithm. We next consider the relation between linear model algorithms and adaptational control in biological vision.

[7]It is sometimes claimed that von Kries algorithms do not require strong assumptions about possible illuminants and surface reflectances such as are presupposed by "linear model algorithm." On the contrary, they do. The relation between performance and environment is often left unstated, disguised as some form of the *RGB Heuristic* and Eqn. (5).

Linear models and adaptational control. Taken as models of chromatic adaptation the linear models algorithms highlight the control mechanisms of adaptation, the second of Stiles's processes. Each of the linear model algorithms described above can be taken as a theory of the sources of information in a scene that affect the local adaptational state (Maloney & Varner, 1986).

Many but not all of the linear models algorithms derive an estimate $\hat{\varepsilon}$ of the illuminant vector ε and then compute estimates of the intrinsic surface colors σ^{xy} by applying the inverse matrix to the photoreceptor receptor excitations,

$$(47) \quad \hat{\sigma}^{xy} = \Lambda_{\hat{\varepsilon}}^{-1}\rho^{xy}.$$

Any one-to-one transformation of the $\hat{\sigma}^{xy}$ would serve equally well to determine color appearance, at least in the absence of further assumptions concerning the behavior of the mechanisms. If we let, $F: \mathbf{R}^3 \to \mathbf{R}^3$ denote any one-to-one function, we can hypothesize a link between the linear models algorithms and Eq. (45) by writing

$$(48) \quad \mu^{xy} = F(\hat{\sigma}^{xy}) = F(\Lambda_{\hat{\varepsilon}}^{-1}\rho^{xy}).$$

Again, comparing the two equations, we see that the linear model algorithms reduce the information present in the surround (the remainder of the scene) to the handful of numbers $\hat{\varepsilon}$. That is, the control processes implicit in each of the algorithms above may use information based on specularity cues, mutual illumination cues, shading cues, and so forth, but their contribution to the equation is reduced to their effect on the handful of parameters $\hat{\varepsilon}$.

The algorithms described above puts the possible *local retinal adaptational states* into a one-to-one correspondence with the matrices $\Lambda_{\varepsilon}^{-1}$, the matrices that compensate for each of the possible illuminants ε. Recall that these matrices form a three-parameter family,

$$(49) \quad \Lambda_{\varepsilon}^{-1} = [\varepsilon_1\Lambda_1 + \varepsilon_2\Lambda_2 + \varepsilon_3\Lambda_3]^{-1},$$

where the Λ_i are fixed, determined by the linear models. Within the linear model framework, this family of matrices, parameterized by ε, corresponds to the family of transformations $\tau_\pi(\cdot)$ parameterized by π. It is interesting to ask, How close are the two matrix families to one another for realistic models of illuminant and surface and various channel-stage models of early vision, including the von Kries model and the full affine model of Eqn. (44)? This is an open question.

If the two sets of linear transformations are not identical, then we could propose a constrained form of any of the linear model algorithms where the illuminant estimate is computed as prescribed by the algorithm, but the matrix $\tau_\pi(\cdot)$ chosen is not the correct matrix $\Lambda_{\hat{\varepsilon}}^{-1}$ but the best approximation to it among the possible choices of $\tau_\pi(\cdot)$. Linear model algorithms constrained to use the transformations available in human vision are, in effect, models of adaptation control based on a particular cue to the illuminant.

Privileged environments. Many of the algorithms described above, if used across a range where their environmental assumptions are satisfied, would provide accurate estimates of surface properties. This range, described by low-dimensional linear models, may include illuminants that markedly differ in color and surface spectral reflectances that also span a wide chromatic gamut and still permit perfect color constancy. *Large* changes in the physical light can be consistent with essentially perfect color constancy.

Yet once lights or surfaces are drawn from outside the linear models to which the algorithm is attuned, the estimates of surface properties will fail to be constant for almost all changes in the light. Consequently, if one were to select lights and surfaces haphazardly, one would almost certainly conclude that a visual system that, in fact, embodied one of the algorithms above was at best approximately color-constant, failing less on some occasions than on others. It would be easy to overlook the class of illuminants and surfaces where the algorithms operate flawlessly. The existence of such *privileged environments* of lights and surfaces is perhaps the most significant prediction for the study of human color vision.

The preceding discussion leads to the following considerations. What, precisely, are the linear models for surface reflectances and for illuminants that lead to optimal human performance in surface color perception experiments? This question remains open. Researchers typically pick plausible simulated illuminants and surface reflectances drawn from Judd, MacAdam, and Wyszecki's (1964) linear model of daylight and a linear model derived from Munsell sample measurements (e.g., Brainard & Wandell, 1991). The resulting measures of performance likely understate optimal human performance, as noted in the section on methodology.

Light estimation as cue combination

Consideration of the algorithms above suggests that there are several possible cues to the illuminant. It is natural to consider the estimation of the illuminant as a *cue combination problem*, analogous to cue combination in depth/shape vision. Kaiser and Boynton (1996, p. 521) previously suggested that illuminant estimation is best thought of as the combination of information from multiple cues.

Consider the following simple model of illuminant estimation: Each of several cues (specularity, etc.) is used to estimate the illuminant parameters ε. One cue could correspond to the stable mean assumption (also known as gray world), deriving an estimate of the illuminant parameters ε from a weighted average of the photoreceptor excitations across the scene. This estimate is denoted $\hat{\varepsilon}_{SM}$. One or more cues might be based on specular information, leading to the estimate $\hat{\varepsilon}_{Spec}$. The weighted average of estimates based on these cues is taken to be a single illuminant estimate $\hat{\varepsilon}$ that controls the color transformation as discussed in the previous section

$$(50)\quad \hat{\varepsilon} = \alpha_{SM}\hat{\varepsilon}_{SM} + \alpha_{Spec}\hat{\varepsilon}_{Spec} + \cdots .$$

The weights $\alpha_.$ are nonnegative and sum to 1, and each value $\alpha_.$ reflects the relative importance assigned to the corresponding cue.

Landy et al. (1995) report empirical tests that imply that depth cue weights change. The implication for surface color perception is that the relative weight assigned to different estimates of the illuminant from different cue types may also change. In particular, consider the sort of experiment where almost all cues to the illuminant are missing: The observer views a large, uniform surround with a small number of test regions superimposed. It is plausible that the only cue to the illuminant remaining is the stable mean cue, the chromaticity of the surround. Then the visual system might set α_{SM} to essentially 1, effectively basing its light estimate on this single cue. The result would be the sort of von Kries–Helson behavior observed experimentally. In more complex scenes, the weight assigned to the stable mean cue might be reduced as other cues become available. This would explain the puzzling observation discussed previously, that the von Kries–Helson pattern of response found in many laboratories does not generalize to the outside world (or even David Brainard's laboratory).

A second and surprising analogy between depth cue combination and illuminant estimation is that not all cues to the illuminant provide full information about the illuminant parameters ε. Some of the methods (such as the subspace algorithm) lead to estimates of ε that include an unknown multiplicative scale factor. It would be meaningless to average such an estimate of "relative" ε together with a full estimate of ε.

An analogous problem arises in depth cue combination since certain depth cues (such as relative size) provide depth information up to an unknown multiplicative scale factor. The problem of combining depth estimates, some of which are in meters and others in only relative units, is termed *cue promotion* by Maloney and Landy (1989) and is treated further in Landy et al. (1995).

In summary, I advance the proposal that a central step in the control of adaptation is the computation of illuminant estimates based on illuminant cues. In scenes that are rich in accurate illuminant cues, the estimate will be not too far from correct, and approximate color constancy results. In simple center-surround scenes, the single cue would seem to be the large surround, the stable mean cue, resulting in approximate von Kries–Helson behavior. What the cues to the illuminant employed in human vision are, and how they are combined, remain open questions. Many of the algorithms above can be identified with potential cues to the illuminant, but it is likely that some of them will prove to be irrelevant to human vision. It is also plausible that better models of illuminant and surface interactions in complex scenes will lead to the discovery of other candidate cues to the illuminant.

Summary

Physics-based models of surface color perception have implications for the study of color vision as a whole.These include, first of all, the importance of studying adaptational control as well as the structure of single color channels. A second implication is that we need to know more about how light and surface interact in scenes if we are to understand how color vision proceeds in complex, natural environments.

In addition, the investigation of explicit models of light and surface in scenes has led us to a number of candidate cues to the illumination in scenes and has permitted the development of precise models of how this information may be recovered. It remains to be seen whether any of these cues to the illuminant are used by biological vision systems.

Last, once the possibility that there are many possible cues to the illuminant is considered, it is natural to ask how a visual system, ideal or biological, might pick and choose among them, and this leads to consideration of the similarities between surface color perception and depth/shape perception.

Not addressed in this chapter is the issue of the relationship between shape perception and surface color perception. If the spatial layout of the scene is known, including the precise shapes and positions of objects, it is entirely plausible that more accurate estimates of the illuminant parameters can be computed. Also not addressed is the issue of complex illuminant environments with multiple, nonpunctate sources of light.

Acknowledgments

This work was supported by the National Eye Institute under grant EY08266 and a fellowship from the Deutscher Akademischer Austauschdienst. The data from Vrhel, Gershon, and Iwan (1994) were provided by Ron Gershon. I thank Ron Gershon, his co-authors, and the Kodak company for making them available. Several people were kind enough to read and comment on earlier drafts: David Brainard, Michael D'Zmura, Karl Gegenfurtner, Susan Hodge, Michael Landy, and Joong Nam Yang. I am very grateful to them all. Special thanks go to Rainer Mausfeld for sharing the quote from Ulrich von Strassburg, and to Michael Brill for pointing out the enduring importance of Ives (1912b).

References

Abney, W. de W. (1913) *Researches in colour vision and the trichromatic theory*. Longmans, London.

Adam, A. (1969) Foveal red-green ratios of normals, colorblinds and heterozygotes. *Proceedings of the Tel-Hashomer Hospital, Tel-Aviv, Israel*, 8, 2–6.

Adams, E.Q. (1923) A theory of color vision. *Psychological Review*, 30, 56.

Adams, A.J., Hine, K.E., Schefrin, B.E., Bresnik, G.H., & Zisman, F. (1987) A simple clinical test of blue cone sensitivity in early eye disease. *Documenta Ophthalmological Proceedings Series*, 46, 237–242.

Adelson, E.H. (1993) Perceptual organization and the judgment of brightness. *Science,* 262, 2042–2044.

Adelson, E.H. & Movshon, J.A. (1982) Phenomenal coherence of moving visual patterns. *Nature*, 300, 523–525.

Ahnelt, P. & Kolb, H. (1994a) Horizontal cells and cone photoreceptors in primate retina: A golgi-light microscopic study of spectral connectivity. *Journal of Comparative Neurology*, 343, 387–405.

Ahnelt, P. & Kolb, H. (1994b) Horizontal cells and cone photoreceptors in human retina: A golgi-electron microscopic study of spectral connectivity. *Journal of Comparative Neurology,* 343, 406–427.

Ahnelt, P.K., Keri, C., & Kolb, H. (1990) Identification of pedicles of putative blue-sensitive cones in the human retina. *Journal of Comparative Neurology*, 293, 39–53.

Ahnelt, P.K., Kolb, H., & Pflug, R. (1987) Identification of a subtype of cone photoreceptor, likely to be blue sensitive, in the human retina. *Journal of Comparative Neurology*, 255, 18–34.

Albrecht, D.G. & Geisler, W.S. (1991) Motion sensitivity and the contrast-response function of simple cells in the visual cortex. *Visual Neuroscience,* 7, 531–546.

Alpern, M. (1964) Relation between brightness and color contrast. *Journal of the Optical Society of America A*, 54, 1491–1492.

Alpern, M.L. (1974) What is it that confines in a world without color? *Investigative Ophthalmology*, 13, 648–674.

Alpern, M. (1976) Tritanopia. *American Journal of Optometry and Physiological Optics,* 53, 340–349.

Alpern, M. (1979) Lack of uniformity in colour matching. *Journal of Physiology,* 288, 85–105.

Alpern, M. (1987) Variation in the visual pigments of human dichromats and normal human trichromats. In *Frontiers of Visual Science: Proceedings of the 1985 Symposium* (ed. N.R.C. Committee on Vision), pp. 169–193. National Academy Press, Washington, DC.

Alpern, M., Kitahara, H., & Fielder, G.H. (1987) The change in color matches with retinal angle of incidence of the colorimeter beams. *Vision Research,* 27, 1763–1778.

Alpern, M., Kitahara, K., & Krantz, D.H. (1983) Perception of colour in unilateral tritanopia. *Journal of Physiology*, 335, 683–697.

Alpern, M., Kitahara, K., & Tamaki, R. (1983) The dependence of the colour and brightness of a monochromatic light upon its angle of incidence on the retina. *Journal of Physiology,* 338, 651–668.

Alpern, M., Lee, G.B., & Spivey, B.E. (1965) π_1-cone monochromatism. *Archives for Ophthalmology*, 74, 334–337.

Alpern, M., Lee, G.B., Maaseidvaag, F., & Miller, S.S. (1971) Colour vision in blue-cone "monochromacy". *Journal of Physiology,* 212, 211–233.

Alpern, M. & Moeller, J. (1977) The red and green cone visual pigments of deuteranomalous trichromacy. *Journal of Physiology*, 266, 647–675.

Alpern, M. & Pugh, E.N. (1977) Variation in the action spectrum of erythrolabe among deuteranopes. *Journal of Physiology,* 266, 613–646.

Alpern, M. & Wake, T. (1977) Cone pigments in human deutan color vision defects. *Journal of Physiology,* 266, 595–612.

Apostol, T.M. (1969) *Calculus, 2nd Edition, Volume II.* Xerox, Waltham, MA.

Arbour, N.C., Zlotogora, J., Knowlton, R., Merin, S., Rosenmann, A., Kanis, A.B., Rokhlina, T., Stone, E.M., & Sheffield, V.C. (1997) Homozygosity mapping of achromatopsia to chromosome 2 using DNA pooling. *Human Molecular Genetics*, 6, 689–694.

Arden, G., Wolf, J., Beringer, T., Hogg, C.R., Tzekov, R., & Holder, G.E. (1999) S-cone ERGs elicited by a simple technique in normals and in tritanopes. *Vision Research*, 39, 641–650.

Arend, L.E. Jr. (1993) How much does illuminant color affect unattributed colors? *Journal of the Optical Society of America A*, 10, 2134–2147.

Arend, L.E. & Reeves, A. (1986) Simultaneous color constancy. *Journal of the Optical Society of America A*, 3, 1743–1751.

Arend, L.E., Reeves, A., Schirillo, J., & Goldstein, R. (1991) Simultaneous color constancy: Papers with diverse Munsell values. *Journal of the Optical Society of America A*, 8, 661–672.

Asenjo, A.B., Rim, J., & Oprian, D.D. (1994) Molecular determinants of human red/green color discrimination. *Neuron,* 12, 1131–1138.

Augenstein, E.J. & Pugh, E.N. Jr. (1977) The dynamics of the π_1 colour mechanism: Further evidence for two sites of adaptation. *Journal of Physiology*, 272, 247–281.

Ayyagari, R., Kaluk, L.E., Szczesny, J., Bingham, E.L., Coats, C.L., Toda, Y., & Sieving, P.A. (1998) Variation in the molecular rearrangements of color genes and phenotype in blue cone monochromacy. *Investigative Ophthalmology & Visual Science*, 39, S296.

Babcock, H.W. (1953) The possibility of compensating astronomical seeing. *Publications of the Astronomical Society of the Pacific*, 65, 229–236.

Badcock, D.R. & Derrington, A.M. (1992) Two stage analysis of the motion of 2-dimensional patterns. *Vision Research,* 32, 691–698.

Bader, C.R., MacLeish, P.R., & Schwartz, E.A. (1979) A voltage-clamp study of the light response in solitary rods of the tiger salamander. *Journal of Physiology*, 296, 1–26.

Baizer, J.S., Ungerleider, L.G., & Desimone, R. (1991) Organization of visual inputs to the inferior temporal and posterior parietal cortex in macaques. *Journal of Neuroscience,* 11, 168–190.

Balding, S.D., Sjoberg, S.A., Neitz, J., & Neitz, M. (1998) Pigment gene expression in protan color vision defects. *Vision Research*, 38, 3359–3364.

Barbur, J.L., Harlow, A.J., & Plant, G.T. (1994) Insights into the different exploits of colour in the visual cortex. *Proceedings of the Royal Society, London, Series B*, 258, 327–334.

Barlow, H.B. (1972) Dark and light adaptation: psychophysics. In *Handbook of Sensory Physiology*, Vol. VII/4 (eds. Jameson, D. & Hurvich, L.M.), pp. 1-28. Springer, Berlin.

Barlow, H.B. (1982) What causes trichromacy? A theoretical analysis using comb-filtered spectra. *Vision Research,* 22, 635–643.

Barlow, H.B. & Hill, R.M. (1963) Evidence for a physiological explanation of the waterfall phenomenon and figural after-effects. *Nature,* 200, 1345–1246.

Bastian, B.L. (1976) *Individual differences among the photopigments of protan observers.* Ph.D. thesis, University of Michigan, Ann Arbor.

Bäuml, K.H. (1994) Color appearance: effects of illuminant changes under different surface collections. *Journal of the Optical Society of America A*, 12, 531–542.

Bäuml, K.H. (1995) Illuminant changes under different surface collections: Examining some principles of color appearance. *Journal of the Optical Society of America A*, 12, 261–271.

Baylor, D.A. (1996) How photons start vision. *Proceedings of the National Academy of Sciences, USA,* 93, 560–565.

Baylor, D.A., Fuortes, M.G.F., & O'Bryan, P.M. (1971) Receptive fields of cones in the retina of the turtle. *Journal of Physiology*, 214, 265–294.

Baylor, D.A. & Hodgkin, A.L. (1974) Changes in time scale and sensitivity in turtle photoreceptors. *Journal of Physiology*, 242, 729–758.

Baylor, D.A., Lamb, T.D., & Yau, K.-W. (1979) The membrane current of single rod outer segments. *Journal of Physiology,* 288, 589–611.

Baylor, D.A., Nunn, B.J., & Schnapf, J.L. (1984) The photocurrent, noise and spectral sensitivity of rods of the monkey *Macaca fascicularis. Journal of Physiology,* 357, 575–607.

Baylor, D.A., Nunn, B.J., & Schnapf, J.L. (1987) Spectral sensitivity of cones of the monkey *macaca fascicularis. Journal of Physiology,* 390, 145–160.

Beck, J. (1972) *Surface Color Perception.* Cornell University Press, Ithaca, NY.

Bedford, R.E. & Wyszecki, G.W. (1958) Wavelength discrimination for point sources. *Journal of the Optical Society of America*, 48, 129–135.

Bekesy, von G. (1968) Mach- and Hering-type lateral inhibition in vision. *Vision Research,* 8, 1483–1499.

Benary, W. (1924) Beobachtungen zu einem Experiment über Helligkeitskontrast. *Psychologische Forschung*, 5, 131–142.

Bender, B.G., Ruddock, K.H., De Vries-De Mol, E.C., & Went, L.N. (1972) The colour vision characteristics of an observer with unilateral defective colour vision. *Vision Research*, 12, 2035–2057.

Berendschot, T.T.J.M., van de Kraats, J., & van Norren, D. (1996) Foveal cone mosaic and visual pigment density in dichromats. *Journal of Physiology,* 492, 307–314.

Berns, R.S. & Gorzynski, M.E. (1991) Simulating surface colors on CRT displays: The importance of cognitive clues. *AIC Conference: Colour and Light*, 21–24.

Berson, E.L., Sandberg, M.A., Rosner, B., & Sullivan, P.L. (1983) Color plates to help identify patients with blue cone monochromatism. *American Journal of Ophthalmology*, 95, 741–747.

Birch, J. (1993) *Diagnosis of Defective Colour Vision.* Oxford, Oxford University Press.

Birch, J.B., Chisholm, I.A., Kinnear, P., Marré, M., Pinckers, A.J.L.G., Pokorny, J., Smith, V.C., & Verriest, G. (1979) Acquired color vision defects. In *Congenital and Acquired Color Vision Defects* (eds. Pokorny, J., Smith, V.C., Verriest, G., & Pinckers, A.J.L.G.), pp. 243–348. Grune & Stratton, New York.

Birren, F. (1963) *The Natural System of Colours by Moses Harris (1766). A facsimile edition of what is perhaps the rarest known book in the literature of color* (with historical notes and commentary by Birren, F.). Whitney Library of Design, New York.

Birren, F. (1980) *A facsimile edition of Coloritto by J.C. Le Blon* (with an introduction by Birren, F.). Van Nostrand Reinhold, New York.

Blackwell, D. & Girshick, M.A. (1954) *Theory of Games and Statistical Decisions*. Wiley, New York.

Blackwell, H.R. & Blackwell, O.M. (1957) Blue mono-cone monochromacy: A new color vision defect. *Journal of the Optical Society of America,* 47, 338.

Blackwell, H.R. & Blackwell, O.M. (1961) Rod and cone receptor mechanisms in typical and atypical congenital achromatopsia. *Vision Research,* 1, 62–107.

Blakemore, C. & Campbell, F.W. (1969) On the existence of neurones in the human visual system selectively sensitive to the orientation and size of retinal images. *Journal of Physiology,* 203, 237–260.

Blakemore, C. & Vital-Durand, F. (1986) Organization and post-natal development of the monkey's lateral geniculate nucleus. *Journal of Physiology*, 380, 453–491.

Blasdel, G.G. & Fitzpatrick, D. (1984) Physiological organization of layer 4 in macaque striate cortex. *Journal of Neuroscience*, 4, 880–895.

Blasdel, G.G., Lund, J.S., & Fitzpatrick, D. (1985) Intrinsic connections of macaque striate cortex: Axonal projections of cells outside lamina 4C. *Journal of Neuroscience*, 5, 3350–3369.

Bone, R.A. & Sparrock, J.M.B. (1971) Comparison of macular pigment densities in the human eye. *Vision Research,* 11, 1057–1064.

Bone, R.A., Landrum, J.T., & Cains, A. (1992) Optical density spectra of the macular pigment *in vivo* and *in vitro*. *Vision Research,* 32, 105–110.

Bone, R.A., Landrum, J.T., Fernandez, L., & Tarsis, S.L. (1988) Analysis of the macular pigment by HPLC: Retinal distribution and age study. *Investigative Ophthalmology & Visual Science,* 29, 843–849.

Bongard, M.M. & Smirnov, M.S. (1954) Determination of the eye spectral sensitivity curves from spectral mixture curves. *Doklady Akademiia nauk S.S.S.R.,* 102, 1111–1114.

Bongard, M.M., Smirnov, M.S., & Friedrich, L. (1958) The four-dimensional colour space of the extra-foveal retinal area of the human eye. *Visual Problems of Colour,* Vol. I, p. 325. NPL Symposium No. 8. London: Her Majesty's Stationery Office.

Bonneau, D., Souied, E., Gerber, S., Rozet, J.M., Daens, E., Journel, H., Plessis, G., Weissenbach, J., Munnich, A., & Kaplan, J. (1995) No evidence for genetic heterogeneity in dominant optic atrophy. *Journal of Medical Genetics*, 32, 951–953.

Boring, E.G. (1942) *Sensation and Perception in the History of Experimental Psychology.* Appleton-Century-Crofts, New York.

Born, G., Grützner, P., & Hemminger, H. (1976) Evidenz für eine Mosaikstruktur der Netzhaut bei Konduktorinnen für Dichromasie. *Human Genetics*, 32, 189–196.

Bouma, P.J. (1942) Mathematical relationship between the colour vision system of trichromats and dichromats. *Physica,* 9, 773–784.

Bowen, R.W. (1997) Isolation and interaction of ON and OFF pathways in human vision: Contrast discrimination at pattern offset. *Vision Research*, 37, 185–198.

Bowmaker, J.K. (1991a) Visual pigments and colour vision in primates. In *From Pigments to Perception. Advances in Understanding Visual Processes* (eds. Valberg, A. & Lee, B.B.), pp. 1–9. Plenum Press, New York, London.

Bowmaker, J.K. (1991b) The evolution of vertebrate visual pigments and photoreceptors. In *Evolution of the Eye and Visual System* (eds. Cronly-Dillon, J. & Gregory, R.L.), pp. 63–81. The Macmillan Press, Houndsmills, Basingstoke, Hampshire, UK.

Bowmaker, J.K. (1991c) Visual pigments, oil droplets and photoreceptors. In *The Perception of Colour* (ed. Gouras, P.), pp. 108–127. The Macmillan Press, Houndsmills, Basingstoke, Hampshire, UK.

Bowmaker, J.K., Astell, S., Hunt, D.M., & Mollon, J.D. (1991) Photosensitive and photostable pigments in the retinae of old world monkeys. *Journal of Experimental Biology,* 156, 1–19.

Bowmaker, J.K., Dartnall, H.J.A., Lythgoe, J.N., & Mollon, J.D. (1978) The visual pigments of rods and cones in the rhesus monkey *Macaca mulatta*. *Journal of Physiology*, 274, 329–348.

Bowmaker, J.K. & Dartnall, H.J.A. (1980) Visual pigment of rods and cones in a human retina. *Journal of Physiology*, 298, 501–511.

Bowmaker, J.K., Dartnall, H.J.A., & Mollon, J.D. (1980) Microspectrophotometric demonstration of four classes of photoreceptors in the Old World primate, *Macaca fascicularis*. *Journal of Physiology*, 298, 131–143.

Boycott, B.B. & Dowling, J.E. (1969) Organization of the primate retina: Light microscopy. *Philosophical transactions of the Royal Society, London, Series B*, 255, 109–184.

Boycott, B.B., Hopkins, J.M., & Sperling, H.G. (1987) Cone connections of the horizontal cells of the rhesus monkey's retina. *Proceedings of the Royal Society, London, Series B*, 229, 345–379.

Boycott, B.B. & Hopkins, J.M. (1991) Cone bipolar cells and cone synapses in the primate retina. *Visual Neuroscience*, 7, 49–60.

Boycott, B.B. & Hopkins, J.M. (1993) Cone synapses of a flat diffuse cone bipolar cell in the primate retina. *Journal of Neurocytology*, 22, 765–778.

Boycott, B.B. & Kolb, H. (1973) The connections between bipolar cells and photoreceptors in the retina of the domestic cat. *Journal of Comparative Neurology*, 148, 91–114

Boycott, B.B. & Wässle, H. (1991) Morphological classification of bipolar cells of the primate retina. *European Journal of Neuroscience*, 3, 1069–1088.

Boyle, R. (1688) *Some Uncommon Observations about Vitiated Sight*. J. Taylor, London.

Boynton, R.M. (1979) *Human Color Vision*. Holt, Rinehart, & Winston, New York.

Boynton, R.M., Das, S.R., & Gardiner, J. (1966) Interactions between photopic visual mechanisms revealed by mixing conditioning fields. *Journal of the Optical Society of America*, 56, 1775–1780.

Boynton, R.M., Ikeda, M., & Stiles, W.S. (1964) Interactions among chromatic mechanisms as inferred from positive and negative increment thresholds. *Vision Research*, 4, 87–117.

Boynton, R.M. & Kambe, N. (1980) Chromatic difference steps of moderate size measured along theoretically critical axes. *Color Research and Application*, 5, 13–23.

Boynton, R.M., Nagy, A.L., & Eskew, R.T. Jr. (1986) Similarity of normalized discrimination ellipses in the constant-luminance chromaticity plane. *Perception*, 15, 755–763.

Boynton, R.M., Nagy, A.L., & Olson, C.X. (1983) A flaw in equations for predicting chromatic differences. *Color Research and Application*, 8, 69–74.

Bracewell, R.N. (1978) *The Fourier Transform and its Applications, 2nd Edition*. McGraw-Hill, New York.

Braddick, O.J. (1980) Low-level and high-level processes in apparent motion. *Philosophical Transactions of the Royal Society of London, Series B*, 290, 131–151.

Bradley, A., Switkes, E., & DeValois, K. (1988) Orientation and spatial frequency selectivity of adaptation to colour and luminance gratings. *Vision Research*, 28, 841–856.

Brainard, D. (1989) Calibration of a computer controlled color monitor. *Color Research and Applications*, 14, 23–34.

Brainard, D.H. (1996) Cone contrast and opponent modulation color spaces. In *Human Color Vision, 2nd Edition* (eds. Kaiser, P.K. & Boynton, R.M.), pp. 563–579. Optical Society of America, Washington, DC.

Brainard, D.H. (1998) Color constancy in the nearly natural image. 2. Achromatic loci. *Journal of the Optical Society of America A*, 15, 307–325.

Brainard, D.H. & Williams, D.R. (1993) Spatial reconstruction of signals from short-wavelength cones. *Vision Research*, 33, 105–116.

Brainard, D.H., Brunt, W.A., & Speigle, J.M. (1997) Color constancy in the nearly natural image. 1. Asymmetric matches. *Journal of the Optical Society of America A*, 14, 2091–2110.

Brainard, D.H. & Freeman, W.T. (1997) Bayesian color constancy. *Journal of the Optical Society A*, 14, 1393–1411.

Brainard, D.H., Rutherford, M.D., & Kraft, J.M. (1997) Color constancy compared: Experiments with real images and color monitors. *Investigative Ophthalmology & Visual Science* (Suppl.), 38, 476.

Brainard, D.H. & Speigle, J.M. (1994) Achromatic loci measured under realistic viewing conditions. *Investigative Ophthalmology & Visual Science (Suppl.)*, 35, 1328.

Brainard, D.H. & Wandell, B.A. (1986) An analysis of the retinex theory of color vision. *Journal of the Optical Society of America A*, 3, 1651–1661.

Brainard, D.H. & Wandell, B.A. (1991) A bilinear model of the illuminant's effect on color appearance. In *Computational Models of Visual Processing* (eds. Landy, M.S. & Movshon, J.A.), pp. 171–186. MIT, Cambridge, MA.

Brainard, D.H. & Wandell, B.A. (1992) Asymmetric color matching: How color appearance depends on the illuminant. *Journal of the Optical Society of America A*, 9, 1433–1448.

Brandstätter, J.H., Greferath, U., Euler, T., & Wässle, H. (1995) Co-stratification of GABA A receptors with the directionally selective circuitry of the rat retina. *Visual Neuroscience*, 12, 345–358.

Breton, M.E. & Cowan, W.B. (1981) Deuteranomalous color matching in the deuteranopic eye. *Journal of the Optical Society of America*, 71, 1220–1223.

Breton, M.E., Schueller, A.W., Lamb, T.D., & Pugh, E.N. Jr. (1994) Analysis of ERG a-wave amplification and kinetics in terms of the G-protein cascade of phototransduction. *Investigative Ophthalmology & Visual Science,* 35, 295-309.

Brettel, H., Viénot, F., & Mollon, J.D. (1997) Computerized simulation of color appearance for dichromats. *Journal of the Optical Society of America A*, 14, 2647–2655.

Brewster, D. (1832) On the undulations excited in the retina by the action of luminous points and lines. *London and Edinburgh Philosophical Magazine and Journal of Science*, 1, 169–174.

Brill, M.H. (1978) A device performing illuminant-invariant assessment of chromatic relations. *Journal of Theoretical Biology*, 71, 473.

Brill, M.H. (1979) Further features of the illuminant-invariant trichromatic photosensor. *Journal of Theoretical Biology*, 78, 305.

Brill, M.H. (1990) Image segmentation by object color: a unifying framework and connection to color constancy. *Journal of the Optical Society of America A*, 7, 2041–2049.

Brill, M.H. (1995) Commentary on Ives "The relation between the color of the illuminant and the color of the illuminated object." *Color Research and Application*, 20, 70–71.

Brill, M.H. & West, G. (1981) Contributions to the theory of invariance of color under the condition of varying illumination. *Journal of Mathematical Biology*, 11, 337–350.

Brindley, G.S. (1954a) The order of coincidence required for visual threshold. *Proceedings of the Physical Society London,* 67, 673–676.

Brindley, G.S. (1954b) The summation areas of human colour-receptive mechanisms at increment threshold. *Journal of Physiology,* 124, 400–408.

Brindley, G.S. (1955) The colour of light of very long wavelength. *Journal of Physiology,* 130, 35–44.

Brindley, G.S. (1960) *Physiology of the Retina and Visual Pathway.* Williams & Wilkins, Baltimore, MD.

Brindley, G.S., Du Croz, J.J., & Rushton, W.A.H. (1966) The flicker fusion frequency of the blue-sensitive mechanism of colour vision. *Journal of Physiology,* 183, 497–500.

Brody, J.A., Hussels, I., Brink, E., & Torres, J. (1979) Herditary blindness among Pingelapese people of eastern Caroline Islands. *Lancet*, 1, 1253–1257.

Brown, J.F. (1931) The thresholds for visual movement. *Psychologische Forschung*, 1931, 14, 249–268.

Brown, R.O. (1994) The world is not grey. *Investigative Ophthalmology & Visual Science (Suppl.),* 35, 2165.

Brown, R.O. & MacLeod, D.I.A. (1991) Induction and constancy for color saturation and achromatic contrast variance. *Investigative Ophthalmology & Visual Science (Suppl.)*, 32, 1214.

Brown, R. & MacLeod, D.I.A. (1997) Color appearance depends on the variance of surround colors. *Current Biology*, 7, 844–849.

Brown, W.R.J. & MacAdam, D.L. (1949) Visual sensitivities to combined chromaticity and luminance differences. *Journal of the Optical Society of America*, 39, 808–834.

Brunner, W. (1932) Über den Vererbungsmodus der verschiedenen Typen der angeborenen Rotgrünblindheit. *Albrecht von Graefes Archiv für Ophthalmologie,* 124, 1–52.

Buchsbaum, G. (1980) A spatial processor model for object colour perception. *Journal of the Franklin Institute*, 310, 1–26.

Buchsbaum, G. & Gottschalk, A. (1983) Trichromacy, opponent colours coding and optimum colour information transmission in the retina. *Proceedings of the Royal Society, London, Series B*, 220, 80–113.

Buchsbaum, G. & Gottschalk, A. (1984) Chromaticity coordinates of frequency-limited functions. *Journal of the Optical Society of America*, 1, 885–887.

Bumsted, K. & Hendrickson, A.E. (1999) Distribution and development of short-wavelength cones differ between Macaca monkey and human fovea. *Journal of Comparative Neurology*, 403, 502–516.

Burkhardt, D.A. (1993a) Light adaptation and photopigment bleaching in cone photoreceptors in situ in the retina of the turtle. *Journal of Neuroscience*, 14, 1091–1105.

Burkhardt, D.A. (1993b) Synaptic feedback, depolarization, and color opponency in cone photoreceptors. *Visual Neuroscience,* 10, 981–989.

Burkhardt, D.A. (1994) Light adaptation and photopigment bleaching in cone photoreceptors in-situ in the retina of the turtle. *Journal of Neuroscience*, 14, 1091–1105.

Burnham, R.W., Evans, R.M., & Newhall, S.M. (1957) Prediction of color appearance with different adaptation illuminations. *Journal of the Optical Society of America*, 47, 35–42.

Burns, S.A. & Elsner, A.E. (1993) Color matching at high luminances: photopigment optical density and pupil entry. *Journal of the Optical Society of America A,* 10, 221–230.

Burr, D.C. & Morrone, M.C. (1993) Impulse-response functions for chromatic and achromatic stimuli. *Journal of the Optical Society of America A*, 10, 1706–1713.

Buxton, R.B. & Frank, L.R. (1997) A model for the coupling between cerebral blood flow and oxygen metabolism during neural stimulation. *Journal Of Cerebral Blood Flow And Metabolism*, 17, 64–72.

Byrne, A. & Hilbert, D.R. (1997a) *Readings on Color; Volume 1: The Philosophy of Color.* MIT Press, Cambridge, MA.

Byrne, A. & Hilbert, D.R. (1997b) *Readings on Color; Volume 2: The Science of Color.* MIT Press, Cambridge, MA.

Cajal, S.R. (1893) La rétine des vertébrès. *La cellule,* 9, 119–257.

Calkins, D.J. (1994) Microcircuitry of M and L cone midget ganglion cell pathways in the primate fovea. Ph.D. Dissertation, University of Pennsylvania, Dept. of Neuroscience.

Calkins, D.J., Schein, S.J., & Sterling, P. (1995) Cone inputs to three types of non-midget ganglion cell in macaque fovea. *Investigative Ophthalmology & Visual Science (Suppl.),* 36, 15.

Calkins, D.J., Schein, S.J., Tsukamoto, Y., & Sterling, P. (1994) M and L cones in macaque fovea connect to midget ganglion cells by different numbers of excitatory synapses. *Nature,* 371, 70–72.

Calkins, D.J. & Sterling, P. (1996) Absence of spectrally specific lateral inputs to midget ganglion cells in primate retina. *Nature*, 381, 613–615.

Calkins, D.J., Thornton, J.E., & Pugh, E.N. Jr. (1992) Monochromatism determined at a long-wavelength/middle-wavelength cone-antagonistic locus. *Vision Research*, 32, 2349–2367.

Calkins, D.J., Tsukamoto, Y., & Sterling, P. (1996) Fovea l cones form basal as well as invaginating junctions with diffuse ON bipolar cells. *Vision Research*, 36, 3373–3381.

Campbell, F.W. & Rushton, W.A.H. (1955) Measurement of the scotopic pigment in the living human eye. *Journal of Physiology*, 130, 131–147.

Cannon, M.W. & Fullenkamp, S.C. (1991) Spatial interactions in apparent contrast: Inhibitory effects among grating patterns of different spatial frequencies, spatial positions and orientations. *Vision Research,* 31, 1985–1998

Carney, T., Shadlen, M., & Switkes, E. (1987) Parallel processing of motion and colour information. *Nature*, 328, 647–649.

Carpenter, R.H.S. (1988) *Movements of the eyes*. Pion, London.

Carr, R.E., Morton, N.E., & Siegel, I.M. (1970) Pingelap eye disease. (Letter). *Lancet*, I, 667.

Casagrande, V.A. (1994) A third parallel visual pathway to primate area V1. *Trends in Neuroscience,* 17, 305–311.

Castano, J.A. & Sperling, H. (1982) Sensitivity of the blue-sensitive cones across the central retina. *Vision Research*, 22, 661–673.

Cavanagh, P. (1991) The contribution of colour to motion. In *From Pigments to Perception* (eds. Valberg, A. & Lee, B.B.), pp. 151–164. Plenum, New York.

Cavanagh, P. (1992) Attention based motion perception. *Science*, 257, 1563–1565.

Cavanagh, P. & Anstis, S. (1991) The contribution of color and motion in normal and color-deficient observers. *Vision Research*, 31, 2109–2148.

Cavanagh, P. & Favreau, O.E. (1985) Color and luminance share a common motion pathway. *Vision Research*, 25, 1595–1601.

Cavanagh, P., MacLeod, D.I.A., & Anstis, S.M. (1987) Equiluminance: spatial and temporal factors and the contribution of blue-sensitive cones. *Journal of the Optical Society of America A,* 4, 1428–1438.

Cavanagh, P., Saida, S., & Rivest, J. (1995) The contribution of color to depth perceived from motion parallax. *Vision Research*, 35, 1871–1878.

Cavanagh, P., Tyler, C.W., & Favreau O.E. (1984) Perceived velocity of moving chromatic gratings. *Journal of the Optical Society of America A*, 1, 893–899.

Cavonius, C.R. & Robbins, D.O. (1973) Relationship between luminance and visual acuity of the rhesus monkey. *Journal of Physiology*, 232, 501–511.

Chan, T., Lee, M., & Sakmar, T.P. (1992) Introduction of hydroxyl-bearing amino acids causes bathochromic spectral shifts in rhodopsin. *Journal of Biological Chemistry*, 267, 9478–9480.

Chandler, J.P. (1969) STEPIT: Finds local minima of a smooth function of several parameters. *Behavioral Science*, 14, 81–82.

Chaparro, A., Stromeyer, C.F. III, Chen, G., & Kronauer, R.E. (1995) Human cones appear to adapt at low light levels: Measurements on the red-green detection mechanism. *Vision Research*, 35, 3103–3118.

Chaparro, A., Stromeyer, C.F. III, Huang, E.P., Kronauer, R.E., & Eskew, R.T. Jr. (1993) Colour is what the eye sees best. *Nature*, 361, 348–350.

Chaparro, A., Stromeyer, C.F. III, Kronauer, R.E., & Eskew, R.T. Jr. (1994) Separable red-green and luminance detectors for small flashes. *Vision Research*, 34, 751–762.

Charles, E.R. & Logothetis, N.K. (1989) The responses of middle temporal (MT) neurons to isoluminant stimuli. *Investigative Ophthalmology & Visual Science (Suppl.)*, 30, 427.

Chen, B., MacLeod, D.I.A., & Stockman, A. (1987) Improvement in human vision under bright light: Grain or gain? *Journal of Physiology,* 394, 41–46.

Chevreul, M.E. (1839) *De la loi du contraste simultane des couleurs*. Pitois Levreault, Paris.

Chichilnisky, E.J., Heeger, D.J., & Wandell, B.A. (1993) Functional segregation of color and motion perception examined in motion nulling. *Vision Research*, 33, 2113–2125.

Chichilnisky, E.J. & Wandell, B.A. (1995) Photoreceptor sensitivity changes explain color appearance shifts induced by large uniform backgrounds in dichoptic matching. *Vision Research*, 35, 239–254.

Chiu, M.I. & Nathans, J. (1994) Blue cones and cone bipolar cells share transcriptional specificity as determined by expression of human blue visual pigment-derived transgene. *Journal of Neuroscience*, 14, 3246–3436.

Chiu, M.I., Zack, D.J., Wang, Y., & Nathans, J. (1994) Murine & bovine blue cone pigment genes: Cloning and characterization of two new members of the S family of visual pigments. *Genomics*, 21, 440–443.

Chubb, C., Sperling, G., & Solomon, J.A. (1989) Texture interactions determine perceived contrast. *Proceedings of the National Academy of Sciences, USA*, 86, 9631–9635.

Chun, M.-H., Grünert, U., Martin, P.R., & Wässle, H. (1996) The synaptic complex of cones in the fovea and in the periphery of the macaque monkey retina. *Vision Research*, 36, 3383–3395.

Cicerone, C.M. (1990) Color appearance and the cone mosaic in trichromacy and dichromacy. *Proceedings of the symposium of the International Research Group on Colour Vision Deficiencies, Tokyo, Japan* (ed. Ohta, Y.), pp. 1–12. Kugler & Ghedini, Amstelveen, The Netherlands.

Cicerone, C.M., Gowdy, P.D., & Otake, S. (1994) Composition and arrangement of the cone mosaic in the living human eye. *Investigative Ophthalmology & Visual Science (Suppl.)*, 35, 1571.

Cicerone, C.M. & Nerger, J.L. (1989) The relative numbers of long-wavelength-sensitive and middle-wavelength-sensitive cones in the human fovea centralis. *Vision Research*, 19, 115–128.

Cideciyan, A.V. & Jacobson, S.G. (1996) An alternative phototransduction model for the human rod and cone ERG a-waves: Normal parameters and variation with age. *Vision Research*, 36, 2609–2621.

CIE (1926) *Commission Internationale de l'Éclairage Proceedings, 1924*. Cambridge University Press, Cambridge, UK.

CIE (1932) *Commission Internationale de l' Éclairage Proceedings, 1931*. Cambridge University Press, Cambridge, UK.

Clarke, S. & Miklossy, J. (1990) Occipital cortex in man: Organization of callosal connections, related myeloarchiture, and putative boundaries of functional visual areas. *Journal of Comparative Neurology*, 298, 188–214.

Cleland, B.G. & Levick, W.R. (1974) Properties of rarely encountered types of ganglion cells in the cat's retina and an overall classification. *Journal of Physiology*, 240, 457–492.

Clowes, M.B. (1971) On seeing things. *Artificial Intelligence*, 2, 79–116.

Coblentz, W.W. & Emerson, W.B. (1917) Relative sensibility of the average eye to light of different colors and some practical applications to radiation problems. *Bulletin of Bureau of Standards*, 14, 167–236.

Cohen, J. (1964) Dependency of the spectral reflectance curves of the Munsell color chips. *Psychonomic Science*, 1, 369.

Cohen, J. (1965) Some electron microscopic observations on interreceptor contacts in the human and macaque retinae. *Journal of Anatomy*, 99, 595–610.

Cohen, E. & Sterling, P. (1990a) Demonstration of cell types among cone bipolar neurons of cat retina. *Philosophical Transactionsof the Royal Society, London, Series B*, 330, 305–321.

Cohen, E. & Sterling, P. (1990b) Convergence and divergence of cones onto bipolar cells in the central area of the cat retina. *Philosophical Transactions of the Royal Society, London, Series B*, 330, 323–328.

Cohn, S.A., Emmerich, D.S., & Carlson, E.A. (1989) Differences in the responses of heterozygous carriers of color-blindness and normal controls to briefly presented stimuli. *Vision Research*, 29, 255–262.

Cole, B.L., Henry, G.H., & Nathans, J. (1965) Phenotypical variations of tritanopia. *Vision Research*, 6, 301–313.

Cole, G.R. & Hine, T. (1992) Computation of cone contrasts for color vision research. *Behavior Research Methods, Instruments, & Computers*, 24, 22–27.

Cole, G.R., Hine, T., & McIlhagga, W. (1993) Detection mechanisms in L-, M-, and S-cone contrast space. *Journal of the Optical Society of America A*, 10, 38–51.

Cole, G.R., Hine, T.J., & McIlhagga, W. (1994) Estimation of linear detection mechanisms for stimuli of medium spatial frequency. *Vision Research*, 34, 1267–1278.

Cole, G.R., Stromeyer, C.F. III, & Kronauer, R.E. (1990) Visual interactions with luminance and chromatic stimuli. *Journal of the Optical Society of America A*, 7, 128–140.

Coletta, N.J. & Williams, D.R. (1987) Psychophysical estimation of extrafoveal cone spacing. *Journal of the Optical Society of America A*, 4, 1503–1513.

Cornsweet, T.N. & Teller, D.Y. (1965) Relation of increment thresholds to brightness and luminance. *Journal of the Optical Society of America*, 55, 1303–1308.

Cornwall, M.C., MacNichol, E.F., & Fein, A. (1984) Absorptance and spectral sensitivity measurements of rod photoreceptors of the tiger salamander, *Ambystoma tigrinum*. *Vision Research*, 24, 1651–1659.

Cottaris, N.P. & DeValois, R.L. (1998) Temporal dynamics of chromatic tuning in macaque primary visual cortex. Nature, 395, 896–900.

Courtney, S.M., Finkel, L.H., & Buchsbaum, G. (1995) Network simulations of retinal and cortical contributions to color constancy. *Vision Research*, 35, 413–434.

Cowey, A. & Heywood, C.A. (1995) There's more to colour than meets the eye. *Behavioural Brain Research*, 71, 89–100.

Cox, J. (1961) Unilateral color deficiency, congenital and acquired. *Journal of the Optical Society of America*, 51, 992–999.

Cox, M.J. & Derrington, A.M. (1994) The analysis of motion of two-dimensional patterns: do Fourier components provide the first stage? *Vision Research*, 34, 59–72.

Craik, K.J.W. (1938) The effect of adaptation on differential brightness discrimination. *Journal of Physiology,* 92, 406–421.

Craven, B.J. & Foster, D.H. (1992) An operational approach to color constancy. *Vision Research*, 32, 1359–1366.

Crawford, B.H. (1949) The scotopic visibility function. *Proceedings of the Physical Society, London, Series B*, 62, 321–334.

Critchley, M. (1965) Acquired anomalies of colour perception of central origin. *Brain*, 88, 711–724.

Crognale, M.A., Teller, D.Y., Yamaguchi, T., Motulsky, A.G., & Deeb, S.S. (1999) Analysis of red/green color discrimination in subjects with a single X-linked photopigment gene. *Vision Research*, 39, 707–719.

Crone, R.A. (1956) Combined forms of congenital color defects: a pedigree with atypical total color blindness. *British Journal of Ophthalmology*, 40, 462.

Crone, R.A. (1959) Spectral sensitivity in colour-defective subjects and heterozygous carriers. *American Journal of Ophthamology*, 48, 231–238.

Crone, R.A. (1968) Incidence of known and unknown colour vision defects. *Ophthalmologica, Basel*, 155, 37–55.

Croner, L.J. & Kaplan, E. (1995) Receptive fields of P and M ganglion cells across the primate retina. *Vision Research,* 35, 7–24.

Croner, L.J., Purpura, K., & Kaplan, E. (1993) Response variability in retinal ganglion cells of primates. *Proceedings of the National Academy of Sciences, USA*, 90, 8128–8130.

Crook, J.M., Lange-Malecki, B., Lee, B.B., & Valberg, A. (1988) Visual resolution of macaque retinal ganglion cells. *Journal of Physiology,* 396, 205–224.

Crook, J.M., Lee, B.B., Tigwell, D.A., & Valberg, A. (1987) Thresholds to chromatic spots of cells in the macaque geniculate nucleus as compared to detection sensitivity in man. *Journal of Physiology*, 392, 193–211.

Crooks, J. & Kolb, H. (1992) Localization of GABA, glycine, glutamate and tyrosine hydroxylase in the human retina. *Journal of Comparative Neurology*, 315, 287–302.

Cropper, S.J. (1994) Velocity discrimination in chromatic gratings and beats. *Vision Research*, 34, 41–48.

Cropper, S.J. & Derrington, A.M. (1994) Motion of chromatic stimuli: first-order or second-order? *Vision Research,* 34, 49–58.

Cropper, S.J. & Derrington, A.M. (1996) Rapid colour specific detection of motion in human vision. *Nature,* 379, 72–74.

Cropper, S.J., Mullen, K.T., & Badcock, D.R. (1996) Motion coherence across different chromatic axes. *Vision Research*, 36, 2475–2488.

Cruz-Coke, R. (1964) *Colour blindness – an evolutionary approach.* Thomas, Springfield, Illinois.

Curcio, C.A., Allen, K.A., Sloan, K.R., Lerea, C.L., Hurley, J.B., Klock, I.B., & Milam, A.H. (1991) Distribution and morphology of human cone photoreceptors stained with anti-blue opsin. *Journal of Comparative Neurology,* 312, 610–624.

Curcio, C.A. & Sloan, K.R. (1992) Packing geometry of human cone photoreceptors: variation with eccentricity and evidence for local anisotropy. *Visual Neuroscience*, 9, 169–180.

Curcio, C.A., Sloan, K.R., Kalina, R.E., & Hendrickson, A.E. (1990) Human photoreceptor topography. *Journal of Comparative Neurology*, 292, 497–523.

D'Zmura, M. (1992) Color constancy: Surface color from changing illumination. *Journal of the Optical Society of America A*, 9, 490–493.

D'Zmura, M. (1998) Color contrast gain control. In *Color Vision: Perspectives from Different Disciplines* (eds. Backhaus, W.G.K., Kliegl, R., & Werner, J.S.). Walter de Gruyter, Berlin, New York.

D'Zmura, M. & Iverson, G. (1993a) Color constancy. I. Basic theory of two-stage linear recovery of spectral descriptions for lights and surfaces. *Journal of the Optical Society of America A*, 10, 2148–2165.

D'Zmura, M. & Iverson, G. (1993b) Color constancy II. Results for two-stage linear recovery of spectral descriptions for lights and surfaces. *Journal of the Optical Society of America A*, 10, 2166–2180.

D'Zmura, M. & Iverson, G. (1994) Color Constancy: III. General linear recovery of spectral descriptions for lights and surfaces. *Journal of the Optical Society of America A*, 11, 2389–2400.

D'Zmura, M., Iverson, G., & Singer, B. (1995) Probabilistic color constancy. In *Geometric Representations of Perceptual Phenomena; Papers in Honor of Tarow Indow on His 70th Birthday* (eds. Luce, R.D., D'Zmura, M., Hoffman,

D., Iverson, G.J., & Romney, A.K.), pp. 187–202. Lawrence Erlbaum Associates, Mahwah, NJ.

D'Zmura, M., Colantoni, P., Knoblauch, K., & Laget, B. (1997) Color transparency. *Perception,* 26, 471–492.

D'Zmura, M. & Lennie, P. (1986) Mechanisms of color constancy. *Journal of the Optical Society of America A*, 3, 1662–1672.

D'Zmura, M. & Singer, B. (1996) The spatial pooling of contrast in contrast gain control. *Journal of the Optical Society of America A,* 13, 2135–2140.

Dacey, D.M. (1989) Axon-bearing amacrine cells of the macaque monkey retina. *Journal of Comparative Neurology,* 284, 275–293.

Dacey, D.M. (1993a) Morphology of a small-field bistratified ganglion cell type in the macaque and human retina. *Visual Neurosci*ence, 10, 1081–1098.

Dacey, D.M. (1993b) The mosaic of midget ganglion cells in the human retina. *Journal of Neuroscience,* 13, 5334–5355.

Dacey, D.M. (1994) Physiology, morphology and spatial densities of identified ganglion cell types in primate retina. In *Higher-Order Processing in the Visual System,* Ciba Foundation Symposium 184, pp. 12–34. Wiley, Chichester, UK.

Dacey, D.M. (1996a) Circuitry for color coding in the primate retina. *Proceedings of the National Academy of Sciences, USA,* 93, 582–588.

Dacey, D.M. (1996b) Morphology and physiology of the AII amacrine cell network in the macaque monkey retina. *Society for Neurosciences, Abstracts*, 22, 492.

Dacey, D.M. & Lee, B.B. (1994a) Physiology of identified ganglion cell types in an in vitro preparation of macaque retina. *Investigative Ophthalmology & Visual Science (Suppl.),* 35, 2001.

Dacey, D.M. & Lee, B.B. (1994b) The "blue-on" opponent pathway in primate retina originates from a distinct bistratified ganglion cell type. *Nature,* 367, 731–735.

Dacey, D.M. & Lee, B.B. (1995) Physiological identification of cone inputs to HI and HII horizontal cells in macaque retina. *Investigative Ophthalmology & Visual Science (Suppl.)*, 36, S3.

Dacey, D.M. & Lee, B.B. (1997) Cone inputs to the receptive field of midget ganglion cells in the periphery of macaque retina. *Investigative Ophthalmology & Visual Science (Suppl.)*, 38, 708.

Dacey, D.M. & Petersen, M.R. (1992) Dendritic field size and morphology of midget and parasol ganglion cells of the human retina. *Proceedings of the National Academy of Sciences, USA,* 89, 9666–9670.

Dacey, D.M., Lee, B.B., Stafford, D.K., Pokorny, J., & Smith, V.C. (1996) Horizontal cells of the primate retina: Cone specificity without spectral opponency. *Science,* 271, 656–659.

Dacheux, R.F. & Raviola, E. (1986) The rod pathway in the rabbit retina. A depolarizing bipolar and amacrine cell. *Journal of Neuroscience*, 6, 331–345.

Dacheux, R.F. & Raviola, E. (1990) Physiology of H1 horizontal cells in the primate retina. *Proceedings of the Royal Society, London, Series B,* 239, 213–230.

Dalton, J. (1798) Extraordinary facts relating to the vision of colours. *Memoirs of the Literary and Philosophical Society of Manchester*, 5, 28–45.

Damasio, A., Yamada, M.D.T., Damasio, H., Corbett, J., & McKee, J. (1980) Central achromatopsia: Behavioral, anatomic, and physiologic aspects. *Neurology,* 30, 1064–1071.

Dana, K.J., van Ginneken, B., Nayar, S.K., & Koenderink, J.J. (1997) Reflectance and texture of real-world surfaces. *Proceedings of the IEEE Computer Society: Conference on Computer Vision and Pattern Recognition, June 1997.*

Dannemiller, J.L. (1989) Computational approaches to color constancy: Adaptive and ontogenetic considerations. *Psychological Review,* 96, 255–266.

Dannemiller, J.L. (1992) Spectral reflectance of natural objects: how many basis functions are necessary? *Journal of the Optical Society of America A*, 9, 507–515.

Dannemiller, J.L. (1993) Rank ordering of photoreceptors catches from objects are nearly illumination invariant. *Vision Research,* 33, 131–137.

Dartnall, H.J.A. (1953) The interpretation of spectral sensitivity curves. *British Medical Bulletin,* 9, 24-30.

Dartnall, H.J.A. (1962) Extraction, measurement and analysis of visual photopigment. In *The Eye* (ed. Davson, H.), pp. 323–365. Academic Press, London.

Dartnall, H.J.A., Arden, G.B., , G.B., Luck, C.P., Rosenberg, M.E., Pedler, C.M.H., & Tansley, K. (1965) Anatomical, electrophysiological and pigmentary aspects of vision in the bush baby: An interpretative study. *Vision Research,* 5, 399–424.

Dartnall, H.J.A., Bowmaker, J.K., & Mollon, J.D. (1983) Human visual pigments: microspectrophotometric results from the eyes of seven persons. *Proceedings of the Royal Society, London, Series B,* 220, 115–130.

Das, S.R. & Sastri, V.D.P. (1965) Spectral distribution and color of tropical daylight. *Journal of the Optical Society of America*, 55, 319.

Davanger, S., Ottersen, O.P., & Storm-Mathisen, J. (1991) Glutamate, GABA, and glycine in the human retina: An immunocytochemical investigation. *Journal of Comparative Neurology,* 311, 483–494.

Daw, N.W. & Enoch, J.M. (1973) Contrast sensitivity, Westheimer function and Stiles-Crawford effect in a blue cone monochromat. *Vision Research,* 13, 1669–1680.

Daw, N.W. & Pearlman, A.L. (1970) Cat colour vision: Evidence for more than one cone process. *Journal of Physiology*, 211, 125–137.

Deeb, S.S. & Motulsky, A.G. (1998) Disorders of color vision. In *Genetic Diseases of the Eye* (ed., Traboulsi, E.I.), pp. 303-325. Oxford University Press, Oxford.

de la Villa, P., Kurahashi, T., & Kaneko, A. (1995) L-glutamate-induced responses and cGMP-activated channels in three subtypes of retinal bipolar cells dissociated from the cat. *Journal of Neuroscience,* 15, 3571–3582.

de Monasterio, F.M. (1978) Properties of concentrically organized X and Y ganglion cells of macaque retina. *Journal of Neurophysiology,* 41, 1394–1417.

de Monasterio, F.M. (1979) Asymmetry of ON- and OFF-pathways of blue-sensitive cones of the retina of macaques. *Brain Research,* 166, 39–48.

de Monasterio, F.M. & Gouras, P. (1975) Functional properties of ganglion cells of the rhesus monkey retina. *Journal of Physiology*, 251, 167–195.

de Monasterio, F.M. & Schein, S.J. (1982) Spectral bandwidths of color-opponent cells of geniculocortical pathway of macaque monkeys. *Journal of Neurophysiology*, 47, 214–224.

de Monasterio, F.M., McCrane, E.E.P., Newlander, J.K., & Schein, S.J. (1985) Density profile of blue-sensitive cones along the horizontal meridian of macaque retina. *Investigative Opthalmology & Visual Science*, 26, 289–302.

DeValois, K.K. & Switkes, E. (1983) Simultaneous masking interactions between chromatic and luminance gratings. *Journal of the Optical Society of America*, 73, 11–18.

DeValois, R.L. (1965) Analysis and coding of color vision in the primate visual system. *Cold Spring Harbor Symposium of Quantitative Biology,* 30, 567–580.

DeValois, R.L., Abramov, I., & Jacobs, G.H. (1966) Analysis of Response Patterns of LGN Cells. *Journal of the Optical Society of America*, 56, 966–977.

DeValois, R.L., Albrecht, D.G., & Thorell, L.G. (1982) Spatial frequency selectivity of cells in macaque visual cortex. *Vision Research*, 22, 549–559.

DeValois, R.L. & DeValois, K.K. (1993) A multi-stage color model. *Vision Research,* 33, 1053–1065.

DeValois, R.L., Morgan, H.C., Polson, M.C., Mead, W.R., & Hull, E.M. (1974) Psychophysical studies of monkey vision I. Macaque luminosity and color vision tests. *Vision Research*, 14, 53–67.

DeValois, R.L., Morgan, H., & Snodderly, D.M. (1974) Psychophysical studies of monkey vision – III. Spatial luminance contrast sensitivity tests of macaque and human observers. *Vision Research,* 14, 75–81.

DeVries, H. (1948a) The heredity of the relative numbers of red and green receptors in the human eye. *Genetica*, 24, 199–212.

DeVries, H.L. (1948b) The luminosity curve of the eye as determined by measurements with the flicker photometer. *Physica,* 14, 319–348.

De Vries-De Mol, E.C., Went, L.N., van Norren, D., & Pols, L.C.W. (1978) Increment spectral sensitivity of hemizygotes and heterozygotes for different classes of colour vision. *Modern Problems in Ophthalmology*, 19, 224–228.

DeBonet, J.S. & Zaidi, Q. (1997) Comparison between spatial interactions in perceived contrast and perceived brightness. *Vision Research*, 37, 1141–1155.

Deeb, S.S., Lindsey, D.T., Sanocki, E., Winderickx, J., Teller, D.Y., & Motulsky, A.G. (1992) Genotype-phenotype relationships in human red/green color vision defects: Molecular and psychophysical studies. *American Journal of Human Genetics*, 51, 687–700.

Delacroix, E. (1937) *The Journal of Eugene Delacroix* (trans. from French by Walter Pach). Covici, Friede, New York.

DeMarco, P., Pokorny, J., & Smith, V.C. (1992) Full spectrum cone sensitivity functions for X-chromosome linked anomalous trichromats. *Journal of the Optical Society of America A,* 9, 1465–1476.

Derrington, A.M. & Badcock, D.R. (1985) The low level motion system has both chromatic and luminance inputs. *Vision Research*, 25, 1879–1884.

Derrington, A.M. & Henning, G.B. (1993) Detecting and discriminating the direction of motion of luminance and colour gratings. *Vision Research*, 33, 799–811.

Derrington, A.M. & Lennie, P. (1984) Spatial and temporal contrast senstivities of neurones in lateral geniculate nucleus of macaque. *Journal of Physiology*, 357, 219–240.

Derrington, A.M., Krauskopf, J., & Lennie, P. (1984) Chromatic mechanisms in lateral geniculate nucleus of macaque. *Journal of Physiology,* 357, 241–265.

DeYoe, E.A., Hockfield, S., Garren, H., & Van Essen, D.C. (1990) Antibody labeling of functional subdivisions in visual cortex: Cat-301 immunoreactivity in striate and extrastriate cortex of the macaque monkey. *Visual Neuroscience*, 5, 67–81.

DeYoe, E.A. & Van Essen, D.C. (1985) Segregation of efferent connections and receptive field properties in visual area V2 of the macaque. *Nature*, 317, 58–61.

DeYoe, E.A. & Van Essen, D.C. (1988) Concurrent processing streams in monkey visual cortex. *Trends in Neuroscience,* 11, 219–226.

Diaz-Araya, C.M., Provis, J.M., & Billson, F.A. (1993) NADPH-diaphorase histochemistry reveals cone distributions in adult human retinae. *Australian and New Zealand Journal of Ophthalmology*, 21, 171–179.

Diggle, P.J. (1993) *Statistical analysis of spatial point patterns*. Academic Press, London.

Dixon, E.R. (1978) Spectral distribution of Autralian daylight. *Journal of the Optical Society of America*, 68, 437–450.

Dobkins, K.R. & Albright, T.D. (1993) What happens if it changes color when it moves? Psychophysical experiments on the nature of chromatic input to motion detectors. *Vision Research*, 33, 1019–1036.

Dobkins, K.R. & Albright, T.D. (1994) What happens if it changes color when it moves? The nature of chromatic input to macaque visual areas MT. *Journal of Neuroscience*, 14, 4854–4870.

Dobkins, K.R. & Albright, T.D. (1995) Behavioral and neural effects of chromatic isoluminance in the primate visual motion system. *Visual Neuroscience*, 12, 321–332.

Dogiel, A.S. (1891) Über die nervösen Elemente in der Retina des Menschen. *Archiv für Mikroskopie und Anatomische Entwicklungsmechanik,* 38, 317–344.

Dow, B.M. & Gouras, P. (1973) Color and spatial specificity of single units in rhesus monkey foveal striate cortex. *Journal of Neurophysiology*, 36, 79–100.

Dowling, J.E. & Boycott, B.B. (1966) Organization of the primate retina: electron microscopy. *Proceedings of the Royal Society, London, Series B,* 166, 80–111.

Drew, M.S. & Funt, B.V. (1990) Calculating surface reflectance using a single-bounce model of mutual reflection. *Proceedings of the Third International Conference on Computer Vision, Osaka, Japan,* December 4-7, 1990. IEEE Computer Society, Washington, DC.

Drummond-Borg, M., Deeb, S.S., & Motulsky, A.G. (1989) Molecular patterns of X chromosome-linked color vision genes among 134 men of European ancestry. *Proceedings of the National Academy of Science, USA*, 86, 983–987.

Dryja, T.P., Berson, E.L., Rao, V.R., & Oprian, D.D. (1993) Heterozygous missense mutation in the rhodopsin gene as a cause of congenital stationary night blindness. *Nature Genetics*, 4, 280–283.

Dryja, T.P., Finn, J.T., Peng, Y.W., McGee, T.L., Berson, E.L., & Yau, K.W. (1995) Mutations in the gene encoding the α-subunit of the rod cGMP-gated channel in autosomal recessive retinitis pigmentosa. *Proceedings of the National Academy of Science, USA*, 92, 10177–10181.

Dubner, R. & Zeki, S.M. (1971) Response properties and receptive fields of cells in an anatomically defined region of the superior temporal sulcus in the monkey. *Brain Research*, 35, 528–532.

Dulai, K.S., Bowmaker, J.K., Mollon, J.D., & Hunt, D.M. (1994) Sequence divergence, polymorphism and evolution of the middle-wave and long-wave visual pigment genes of great apes and old world monkeys. *Vision Research,* 34, 2483–2491.

Earle, P. (1845) On the inability to distinguish colors. *American Journal of Medical Science*, 9, 346–354.

Ebrey, T.G. & Honig, B. (1977) New wavelength dependent visual pigment nomograms. *Vision Research,* 17, 147–151.

Eiberg, H., Kjer, P., & Rosenberg, T. (1994) Dominant optic atrophy (OPA1) mapped to chromosome 3q region. Linkage analysis. *Human Molecular Genetics*, 3, 977–980.

Eisner, A. & MacLeod, D.I.A. (1981) Flicker photometric study of chromatic adaptation: Selective suppression of cone inputs by colored backgrounds. *Journal of the Optical Society of America,* 71, 705–718.

Eisner, A. & MacLeod, D.I.A. (1980) Blue sensitive cones do not contribute to luminance. *Journal of the Optical Society of America*, 70, 121–123.

Elenbaas, W. (1951) *The High Pressure Mercury Vapour Discharge*. North-Holland Publishers, Amsterdam.

Elsner, A.E., Burns, S.A., & Webb, R.H. (1993) Mapping cone photopigment optical density. *Journal of the Optical Society of America A*, 10, 52–58.

Engel, S.A., Zhang, X., & Wandell, B.A. (1997) Color tuning in human visual cortex measured using functional magnetic resonance imaging. *Nature*, 388, 68–71.

Engelking, E. (1925) Die Tritanomalie, ein bisher unbekannter Typus anomaler Trichromasie. *Graefes Archiv für Ophthalmologie*, 116, 196–244.

Enoch, J.M. (1961) Nature of the transmission of energy in the retinal receptors. *Journal of the Optical Society of America,* 51, 1122–1126.

Enoch, J.M. (1963) Optical properties of the retinal receptors. *Journal of the Optical Society of America,* 53, 71–85.

Enoch, J.M. & Stiles, W.S. (1961) The colour change of monochromatic light with retinal angle of incidence. *Optica Acta,* 8, 329–358.

Enroth-Cugell, C. & Robson, J.G. (1966) The cones of retinal ganglion cells of the cat. *Journal of Physiology*, 517–552.

Esfahani, P., Schein, S., Klug, K., Tsukamoto, Y., & Sterling, P. (1993) Characterization of L, M, and S cone pedicles in primate fovea. *Society for Neuroscience Abstracts,* 19, 1201.

Eskew, R.T. Jr. & Kortick, P.M. (1994) Hue equilibria compared with chromatic detection in 3D cone contrast space. *Investigative Ophthalmology & Visual Science (Suppl.)*, 35, 1555.

Eskew, R.T. Jr., Stromeyer, C.F. III, & Kronauer, R.E. (1992) The constancy of equiluminant red-green thresholds examined in two color spaces. *Advances in Color Vision*

Technical Digest Series (Optical Society of America), 4, 195–197.

Eskew, R.T. Jr., Stromeyer, C.F. III, & Kronauer, R.E. (1994) Temporal properties of the red-green chromatic mechanism. *Vision Research*, 34, 3127–3137.

Eskew, R.T. Jr., Stromeyer, C.F. III, Picotte, C.J., & Kronauer, R.E. (1991) Detection uncertainty and the facilitation of chromatic detection by luminance contours. *Journal of the Optical Society of America A*, 8, 394–403.

Estévez, O. (1979) *On the Fundamental Database of Normal and Dichromatic Color Vision*. Ph.D. thesis, Amsterdam University.

Estévez, O. & Spekreijse, H. (1974) A spectral compensation method for determining the flicker characteristics of the human colour mechanism. *Vision Research*, 14, 823–830.

Estévez, O. & Spekreijse, H. (1982) The "Silent Substitution" method in visual research. *Vision Research*, 22, 681–691.

Euler, T., Schneider, H., & Wässle, H. (1996) Glutamate responses of bipolar cells in a slice preparation of the rat retina. *Journal of Neuroscience*, 16, 2934–2944.

Euler, T. & Wässle, H. (1995) Immunocytochemical identification of cone bipolar cells in the rat retina. *Journal of Comparative Neurology*, 361, 461–478.

Evans, R.M. (1948) *An Introduction to Color*. John Wiley, New York.

Fairchild, M.D. & Lennie, P. (1992) Chromatic adaptation to natural and incandescent illuminants. *Vision Research*, 32, 2077–2085.

Falls, H.F., Wolter, J.R., & Alpern, M. (1965) Typical total monochromacy. *Archives of Ophthalmology*, 74, 610–616.

Famiglietti, E.V. (1981) Functional architecture of cone bipolar cells in mammalian retina. *Vision Research*, 21, 1559–1563.

Famiglietti, E.V. (1990) A new type of wide-field horizontal cell, presumably linked to blue cones, in rabbit retina. *Brain Research*, 535, 174–179.

Famiglietti, E.V. & Kolb, H. (1976) Structural basis for ON- and OFF-center responses in retinal ganglion cells. *Science*, 194, 193–195.

Feig, K. & Ropers, H.-J. (1978) On the incidence of unilateral and bilateral colour blindness in heterozygous females. *Human Genetics*, 41, 313–323.

Feil, R., Aubourg, P., Helig, R., & Mandel, J.L. (1990) A 195-kb cosmid walk encompassing the human Xq28 color vision pigment genes. *Genomics*, 6, 367–373.

Felleman, D.J., Burkhalter, A., & Van Essen, D.C. (1997) Cortical connections of areas V3 and VP in macaque monkey extrastriate visual cortex. *Journal of Comparative Neurology*, 379, 21–47.

Felleman, D.J. & Van Essen, D.C. (1987) Receptive field properties of neurons in area V3 of macaque monkey extrastriate cortex. *Journal of Neurophysiology*, 57, 889–920.

Felleman, D.J. & Van Essen, D.C. (1991) Distributed hierarchical processing in the primate cerebral cortex. *Cerebral Cortex*, 1, 1–47.

Ferguson, T.S. (1967) *Mathematical Statistics: A Decision Theoretic Approach*. Academic Press, New York.

Ferrera, V.P., Nealey, T.A., & Maunsell, J.H. (1992) Mixed parvocellular and magnocellular geniculate signals in visual area V4. *Nature*, 358, 756–761.

Ferrera, V.P., Nealey, T.A., & Maunsell, J.H. (1994) Responses in macaque visual area V4 following inactivation of the parvocellular and magnocellular LGN pathways. *Journal of Neuroscience*, 14, 2080–2088.

Fincham, E.F. (1953) Defects of the colour-sense mechanism as indicated by the accommodation reflex. *Journal of Physiology*, 121, 570–580.

Finlayson, G.D., Drew, M.S., & Funt, B.V. (1993) Color constancy: Enhancing Von Kries adaptation via sensor transformations. In *Human Vision, Visual Processing, and Digital Display IV, Proceedings SPIE* (eds. Allebach, J.P. & Rogowitz, B.E.), pp. 473–484.

Finlayson, G.D., Drew, M.S., & Funt, B.V. (1994) Color constancy: Generalized diagonal transforms suffice. *Journal of the Optical Society of America A*, 11, 3011–3019.

Finlayson, G.D. & Funt, B.V. (1996) Coefficient channels: Derivation and relationship to other theoretical studies. *Color Research and Application*, 21, 87–96.

Fischer, F.P., Bouman, M.A., & ten Doesschate, J. (1951) A case of tritanopy. *Documenta Ophthalmologica*, 5/6, 73–87.

Fitzgibbon, J., Appukuttan, B., Gayther, S., Wells, D., Delhanty, J., & Hunt, D.M. (1994) Localisation of the human blue cone pigment gene to chromosome band 7q31.3–32. *Human Genetics*, 93, 79–80.

Fitzpatrick, D., Lund, J.S., & Blasdel, G.G. (1985) Intrinsic connections of macaque striate cortex: Afferent and efferent connections of lamina 4C. *Journal of Neuroscience*, 5, 3329–3349.

Fitzpatrick, D., Usrey, W.M., Schofield, B.R., & Einstein, G. (1994) The sublaminar organization of corticogeniculate neurons in layer 6 of macaque striate cortex. *Visual Neuroscience*, 11, 307–315.

Flanagan, P., Cavanagh, P., & Favreau, O.E. (1990) Independent orientation-selective mechanisms for the cardinal directions of colour space. *Vision Research*, 30, 769–778.

Fleagle, J.G. (1988) *Primate Adaptation and Evolution*. Academic Press, San Diego, London.

Fleischman, J.A. & O'Donnell, F.E. (1981) Congenital X-linked incomplete achromatopsia: Evidence for slow progression, carrier fundus findings, and possible genetic linkage with glucose-6-phosphate dehydrogenase locus. *Archives of Ophthalmology*, 99, 468–472.

Fletcher, R. & Voke, J. (1985) *Defective colour vision: fundamentals, diagnosis and management*. Adam Hilger, Bristol.

Foley, J.M. (1994) Human luminance pattern-vision mechanisms: Masking experiments require a new model. *Journal of the Optical Society of America A*, 11, 1710–1719.

Foley, J.M. & Legge, G.E. (1981) Contrast detection and near-threshold discrimination in human vision. *Vision Research*, 21, 1041–1053.

Foley, J.M. & McCourt, M.E. (1985) Visual grating induction. *Journal of the Optical Society of America A*, 2, 1220–1230.

Forsyth, D. (1990) A novel algorithm for color constancy. *International Journal of Computer Vision*, 5, 5–36.

Foster, D.H. & Nascimento, S.M.C. (1994) Relational colour constancy from invariant cone-excitation ratios. *Proceedings of the Royal Society, London, Series B*, 257, 115–121.

Foster, D.H., Nascimento, S.M.C., Craven, B.J., Linnell, K.J., Cornelissen, F.W., & Brenner, E. (1997) Four issues concerning colour constancy and relational colour constancy. *Vision Research*, 37, 1341–1345.

Fox, P.T. & Raichle, M.E. (1986) Focal physiological uncoupling of cerebral blood flow and oxidative metabolism during somatosensory stimulation in human subjects. *Proceedings of the National Academy of Sciences, USA*, 83, 1140–1144.

Fox, P.T., Raichle, M.E., Mintum, M.A., & Dence, C. (1988) Nonoxidative glucose consumption during focal physiologic neural activity. *Science*, 241, 462–464.

Franceschetti, A. (1928) Die Bedeutung der Einstellungsbreite am Anomaloskop für die Diagnose der einzelnen Typen der Farbsinnstörungen, nebst Bemerkungen über ihren Vererbungsmodus. *Schweizer medizinisches Wochenschrift*, 58, 1273–1279.

François, J., Verriest, G., Matton-Van Leuven, M.T., de Rouck, A., & Manavian, D. (1966) Atypical achromatopsia of sex-linked recessive inheritance. *American Journal of Ophthalmology*, 61, 1101–1108.

François, J., Verriest, G., Mortier, V., & Vanderdonck, R. (1957) De la fréquence des dyschromatopsies congénitales chez l'homme. *Annales Oculist, Paris*, 190, 5–16.

Freeman, W.T. & Brainard, D.H. (1995) Bayesian decision theory, the maximum local mass estimate, and color constancy. *Proceedings of the 5th International Conference on Computer Vision*, pp. 210–217. Cambridge University Press, Cambridge, UK.

Fries, W., Keiser, K., & Kuypers, H.G.J.M. (1985) Layer IV cells in macaque striate cortex project to both superior colliculus and prestriate visual area 5. *Experimental Brain Research*, 58, 613–616.

Frisby, J.P. (1980) *Seeing: Illusion, Brain and Mind*. Oxford University Press, New York.

Funt, B.V. & Ho, J. (1989) Color from black and white. *International Journal of Computer Vision*, 3, 109–117.

Furukawa, T., Morrow, E.M., & Cepko, C.L. (1997) *Crx*, a novel *otx*-like homeobox gene, shows photoreceptor-specific expression and regulates photoreceptor differentiation. *Cell*, 91, 531–541.

Galezowski, X. (1868) *Du diagnostic des Maladies des Yeux par la Chromatoscopie rétinienne: Précéde d'une Etude sur les Lois physiques et physiologiques des Couleurs.* J.B. Baillière et Fils, Paris.

Garth, T. (1933) The incidence of color-blindness among races. *Science*, 77, 333.

Gattass, R., Sousa, A.P., & Gross, C.G. (1988) Visuotopic organization and extent of V3 and V4 of the macaque. *Journal of Neuroscience*, 8, 1831–1845.

Gegenfurtner, K.R. & Hawken, M.J. (1995) Temporal and chromatic properties of motion mechanisms. *Vision Research*, 35, 1547–1563.

Gegenfurtner, K.R. & Hawken, M.J. (1996a) Perceived velocity of luminance, chromatic and non-Fourier stimuli: Influence of contrast and temporal frequency. *Vision Research*, 36, 1281–1290.

Gegenfurtner, K.R. & Hawken, M.J. (1996b) Interaction of motion and color in the visual pathways. *Trends in Neuroscience*, 19, 394–401.

Gegenfurtner, K.R. & Kiper, D.C. (1992) Contrast detection in luminance and chromatic noise. *Journal of the Optical Society of America A*, 9, 1880–1888.

Gegenfurtner, K.R., Kiper, D.C., Beusmans, J.M.H., Carandini, M., Zaidi, Q., & Movshon, J.A. (1994) Chromatic properties of neurons in macaque MT. *Visual Neuroscience*, 11, 455–466.

Gegenfurtner, K.R., Kiper, D.C., & Fenstemaker, S.B. (1996) Processing of color, form and motion in macaque area V2. *Visual Neuroscience*,13, 161–172.

Gegenfurtner, K.R., Kiper, D.C., & Levitt, J. (1997) Functional properties of neurons in macaque area V3. *Journal of Neurophysiology*, 77, 1906–1923.

Geisler, W.S. & Albrecht, D.G. (1997) Visual cortex neurons in monkeys and cats: detection, discrimination, and identification. *Visual Neuroscience*, 14, 897–919.

Gershon, R. & Jepson, A.D. (1988) Discounting illuminants beyond the sensor level. *Proceedings of the SPIE Conference on Intelligent Robots and Computer Vision VII*, 1002, 250–257.

Gershon, R. & Jepson, A.D. (1989) The computation of color constant descriptors in chromatic images. *Color Research and Application*, 14, 325–334.

Ghosh, K.K., Goodchild, A.K., Sefton, A.E., & Martin, P.R. (1996) The morphology of retinal ganglion cells in the new world marmoset monkey *Callithrix jacchus*. *Journal of Comparative Neurology*, 366, 76–92.

Ghosh, K.K., Martin, P.R., & Grünert, U. (1997) Morphological analysis of the blue cone pathway in the retina of a new world monkey, the marmoset *Callithrix jacchus*. *Journal of Comparative Neurology*, 379, 211–225.

Gibson, I.M. (1962) Visual mechanisms in a cone-monochromat. *Journal of Physiology*, 161, 10P–11P.

Gibson, K.S. & Tyndall, E.P.T. (1923) Visibility of radiant energy. *Scientific Papers of the Bureau of Standards*, 19, 131–191.

Gielen, C.C.A.M., van Gisbergen, J.A.M., & Vendrik, A.J.H. (1982) Reconstruction of cone system contributions to responses of colour-opponent neurones in monkey lateral geniculate. *Biological Cybernetics*, 44, 211–221.

Gilchrist, A.L. (1977) Perceived lightness depends on perceived spatial arrangement. *Science*, 195, 185–187.

Gilchrist, A.L. (1980) When does perceived lightness depend on perceived spatial arrangement? *Perception & Psychophysics*, 28, 527–538.

Gilchrist, A.L. (1994) *Lightness, Brightness, & Transparency*. Lawrence Erlbaum Associates, Hillsdale, NJ.

Gilchrist, A.L., Delman, S., & Jacobsen, A. (1983) The classification and integration of edges as critical to the perception of reflectance and illumination. *Perception & Psychophysics*, 33, 425–436.

Giulianini, F. & Eskew, R.T. Jr. (1998) Chromatic masking in the (ΔL/L, ΔM/M) plane of cone-contrast space reveals only two detection mechanisms. *Vision Research*, 38, 3913–3926.

Glickstein, M. & Heath, G.G. (1975) Receptors in the monochromat eye. *Vision Research*, 15, 633–636.

Goethe, J. von (1810) *Zur Farbenlehre*. I. Band (Didaktischer und polemischer Teil), Tafel I. J.G. Kotta, Stuttgart, Berlin. (A colored reproduction is available in the 1928 edition from Verlags Diederichs, Jena, edition.)

Goldsmith, T.H. (1991) The evolution of visual pigments and colour vision. In *The Perception of Colour, Vol. 6, Vision and visual dysfunction* (ed. Gouras, P.), Chapter 5, pp. 62–89. Macmillan, Houndmills, UK.

Goodchild, A.K., Chan, T.L., & Grünert, U. (1996) Horizontal cell connections with short wavelength sensitive cones in macaque monkey retina. *Visual Neuroscience*, 13, 833–845.

Goodchild, A.K., Ghosh, K.K., & Martin, P.R. (1996) A comparison of photoreceptor spatial density and ganglion cell morphology in the retina of human, macaque monkey, cat, and the marmoset *Callithrix jacchus*. *Journal of Comparative Neurology*, 366, 55–75.

Goodeve, C.F. (1936) Relative luminosity in the extreme red. *Proceedings of the Royal Society A*, 155, 664-683.

Gorea, A., Papathomas, T.V., & Kovacs, I. (1993) Motion perception with spatiotemporally matched chromatic and achromatic information reveals a "slow" and "fast" motion system. *Vision Research*, 33, 2515–2534.

Göthlin, G.F. (1924) Congenital red-green abnormality in colour vision and congenital total colour blindness from the point of view of heredity. *Acta Ophthalmologica (Kopenhagen)*, 2, 15–34.

Gouras, P. (1968) Identification of cone mechanisms in monkey ganglion cells. *Journal of Physiology*, 199, 533–547.

Gouras, P. (1974) Opponent-colour cells in different layers of foveal striate cortex. *Journal of Physiology*, 238, 583–602.

Gouras, P. & Link, K. (1966) Rod and cone interaction in dark-adapted monkey ganglion cells. *Journal of Physiology*, 184, 499–510.

Gouras, P. & Zrenner, E. (1979) Enhancement of luminance flicker by color-opponent mechanisms. *Science*, 205, 587–589.

Gowdy, P. & Cicerone, C. M. (1998) The spatial arrangement of the L and M cones in the central fovea of the living human eye. *Vision Research*, 38, 2575–2589.

Graham, N. (1989) *Visual Pattern Analyzers*. Oxford University Press, New York.

Graham, C.H. & Hsia, Y. (1958) Color defect and color theory. *Science*, 127, 675–682.

Graham, C.H. & Hsia, Y. (1967) Visual discriminations of a subject with acquired unilateral tritanopia. *Vision Research*, 6, 469–479.

Graham, C.H., Sperling, H.G., Hsia, Y., & Coulson, A.H. (1961) The determination of some visual functions of a unilaterally color-blind subject: Methods and results. *Journal of Psychology*, 51, 3–32.

Grassman, H. (1853) Zur Theorie der Farbenmischung. *Poggendorf's, Annalen der Physik, Leipzig*, 89, 69–84.

Green, D.G. (1972) Visual acuity in the blue cone monochromat. *Journal of Physiology*, 222, 419–426.

Green, D.M. & Swets, J.A. (1974) *Signal Detection Theory and Psychophysics*. Krieger, Huntington, NY.

Greferath, U., Grünert, U., Fritschy, J.M., Stephenson, A., Möhler, H., & Wässle, H. (1995) GABA A receptor subunits have differential distributions in the rat retina: *In situ* hybridization and immunohistochemistry. *Journal of Comparative Neurology*, 353, 553–571.

Greferath, U., Grünert, U., & Wässle, H. (1990) Rod bipolar cells in the mammalian retina show protein kinase C-like immunoreactivity. *Journal of Comparative Neurology*, 301, 433–442.

Griffin, D.R., Hubbard, R., & Wald, G. (1947) The sensitivity of the human eye to infra-red radiation. *Journal of the Optical Society of America,* 37, 546–554.

Griffiths, A.F. & Zaidi, Q. (1999) A three-dimensional tilt illusion: the roles of perspective distortions and perceptual priors. *Perception* (submitted).

Grossberg, S. & Todorovic, D. (1988) Neural dynamics of 1-D and 2-D brightness perception: A unified model of classical and recent phenomena. *Perception & Psychophysics*, 43, 241-277.

Grünert, U., Martin, P.R., & Wässle, H. (1994) Immunocytochemical analysis of bipolar cells in the macaque monkey retina. *Journal of Comparative Neurology,* 348, 607–627.

Grünert, U. & Wässle, H. (1990) GABA-like immunoreactivity in the monkey retina: A light and electron microscopic study. *Journal of Comparative Neurology,* 297, 509–524.

Grünert, U. & Wässle, H. (1996) Glycine receptors in the rod pathway of the macaque monkey retina. *Visual Neuroscience*, 13, 101–115.

Grüsser, O.-J. & Landes, T. (1991) The world turns grey: Achromatopsia, colour agnosia and other impairments of colour vision caused by cerebral lesions. In *Vision and Visual Dysfunction* (ed. Cronly-Dillon, J.), pp. 385–410. MacMillan, Hammersmith, Hampshire, UK.

Grützner, P. (1964) Der normale Farbensinn und seine Abweichungen. *Bericht der Deutschen ophthalmologischen Gesellschaft,* 66, 161–172.

Grützner, P., Born, G., & Hemminger, H.J. (1976) Coloured stimuli within the central visual fields of carriers of dichromatism. *Modern Problems of Ophthalmology*, 17, 147–150.

Guild, J. (1931) The colorimetric properties of the spectrum. *Philosophical Transactions of the Royal Society, London, A*, 230, 149–187.

Gulyas, B., Heywood, C.A., Popplewell, D.A., Roland, P.E., & Cowey, A. (1994) Visual form discrimination from color or motion cues: Functional anatomy by positron emission tomography. *Proceedings of the National Academy of Sciences, USA*, 91, 9965–9969.

Gulyas, B. & Roland, P.E. (1994) Processing and analysis of form, colour and binocular disparity in the human brain: functional anatomy by positron emission tomography. *European Journal of Neuroscience*, 6, 1811–28.

Gur, M. (1986) The physiological basis of wavelength discrimination: Evidence from dichoptic and Ganzfeld viewing. *Vision Research*, 26, 1257–1262.

Gur, M. & Akri, V. (1992) Isoluminant stimuli may not expose the full contribution of color to visual functioning: Spatial contrast sensitivity measurements indicate interaction between color and luminance processing. *Vision Research*, 32, 1253–1262.

Guth, S.L. (1972) A new vector. In *Color Metrics* (eds. Vos, J.J., Friele, L.F.C., & Walraven, P.L.), pp. 82–98. AIC/Holland Institute for Perception TNO, Soesterberg.

Guth, S.L. (1991) Model for color vision and light adaptation. *Journal of the Optical Society of America A*, 8, 976–993.

Guth, S.L., Massof, R.W., & Benzschawel, T. (1980) Vector model for normal and dichromatic color vision. *Journal of the Optical Society of America*, 70, 197–212.

Hagstrom, S.A., Neitz, J., & Neitz, M. (1997) Ratio of M/L pigment gene expression decreases with retinal eccentricity. In *Colour Vision Deficiencies XIII* (ed. Cavonius, C.R.), pp. 59–65. Kluwer, Dordrecht.

Hagstrom, S.A., Neitz, J., & Neitz, M. (1998) Variation in cone populations for red-green color vision examined by analysis of mRNA. *NeuroReport*, 9, 1963–1967.

Hanna, M.C., Platts, J.T., & Kirkness, E.F. (1997) Identification of a gene within the tandem array of red and green color pigment genes. *Genomics*, 43, 384–386.

Hargrave, P.A., McDowell, J.H., Feldmann, R.J., Atkinson, P.H., Rao, J.K.M., & Argos, P. (1984) Rhodopsin's protein and carbohydrate structure: selected aspects. *Vision Research*, 24, 1487–1499.

Harosi, F.I. (1987) Cynomolgus and rhesus monkey visual pigments. Application of Fourier transform smoothing and statistical techniques of the determination of spectral parameters. *Journal of General Physiology,* 89, 717–743.

Harrison, R., Hoefnagel, D., & Hayward, J.N. (1960) Congenital total color blindness, a clinico–pathological report. *Archives of Ophthalmology*, 64, 685–692.

Hartveit, E. (1996) Membrane currents evoked by ionotropic glutamate receptor agonists in rod bipolar cells in the rat retinal slice preparation. *Journal of Neurophysiology,* 76, 401–422.

Harwerth, R.S. & Smith, E.L. III (1985) Rhesus monkey as a model for normal vision of humans. *American Journal of Optometry and Physiological Optics,* 62, 633–641.

Harwerth, R.S. & Sperling, H.G. (1975) Effects of intense visible radiation on the increment threshold spectral sensitivity of the Rhesus monkey eye. *Vision Research,* 15, 1193–1204.

Hawken, M.J., Gegenfurtner, K.R., & Tang, C. (1994) Contrast dependence of color and luminance motion mechanisms in human vision, *Nature*, 367, 268–270.

Hawken, M.J. & Parker, A.J. (1984) Contrast sensitivity and orientation selectivity in lamina IV of the striate cortex of Old World monkeys. *Experimental Brain Research,* 54, 367–72.

Hawken, M.J., Parker, A.J., & Lund, J.S. (1988) Laminar organization and contrast sensitivity of direction-selective cells in the striate cortex of the old-world monkey. *Journal of Neuroscience*, 8, 3541–3548.

Hawken, M.J. & Parker, A.J. (1990) Detection and discrimination mechanisms in the striate cortex of the old-world monkey. In *Vision: Coding and Efficiency* (ed. Blakemore, C.), pp. 103–116, Cambridge University Press, Cambridge, UK.

Hawken, M.J., Shapley, R.M., & Grosof, D.H. (1996) Temporal frequency selectivity in monkey visual cortex. *Visual Neuroscience*, 13, 477–492.

Hayashi, T., Motulsky, A.G., & Deeb, S.S. (1999) The molecular basis of deuteranomaly: position of a green-red hybrid gene in the visual pigment array determines colour vision phenotype. *Nature genetics* (in press).

Hayhoe, M.M. & MacLeod, D.I.A. (1976) A single anomalous photopigment? *Journal of the Optical Society of America*, 66, 276–277.

Hays, W.L. (1981) *Statistics, third edition*. CBS College Publishing, New York.

Hays, T.R., Lin, S.H., & Eyring, H. (1980) Wavelength regulation in rhodopsin: Effects of dipoles and amino acid side chains. *Proceedings of the National Academy of Science, USA*, 77, 6314–6318.

He, J.C. & Shevell, S.K. (1994) Individual differences in cone photopigments of normal trichromats measured under dual Rayleigh-type color matches. *Vision Research*, 34, 367–376.

He, J.C. & Shevell, S.K. (1995) Variation in color matching and discrimination among deuteranomalous trichromats: Theoretical implications of small differences in photopigments. *Vision Research,* 18, 2579–2588.

Healey, G. (1991) Estimating spectral reflectances using highlights. *Image and Vision Computing*, 9, 333–337.

Healey, G., Shafer, S., & Wolfe, L. (1992) *Physics-Based Vision: Principles and Practice*. Jones & Bartlett, London.

Hecht, S. (1949) Brightness, visual acuity and color blindness. *Documenta Ophthalmologica,* 3, 289–306.

Hecht, S. & Hsia, Y. (1948) Colorblind vision. I. Luminosity losses in the spectrum for dichromats. *Journal of General Physiology*, 31, 141–153.

Heeger, D.J. (1992) Normalization of cell responses in cat striate cortex. *Visual Neuroscience,* 9, 181–197.

Heeger, D.J. & Robison, R.R. (1994) Is simultaneous contrast divisive? *Investigative Ophthalmology & Visual Science (Suppl.),* 35, 2006.

Helmholtz, H. von (1867) *Handbuch der Physiologischen Optik*. Voss, Hamburg and Leipzig.

Helmholtz, H.L.F. von (1852) Über die Theorie der zusammengesetzten Farben. *Annales de Physique, Leipzig*, 887, 45–66.

Helmholtz, H. von (1896/1962) *Helmholtz's Treatise on Physiological Optics* (ed. Southall, J.P.C.). Dover, New York.

Helmholtz, H. von (1924) *Physiological Optics 2*, pp. 264–271. Optical Society of America, Washington, DC.

Helmreich, E.J.M. & Hofmann, K.-P. (1996) Structure and function of proteins in G-protein-coupled signal transfer. *Biochimica et Biophysica Acta,* 1286, 285–332.

Helson, H. (1934) Some factors and implications of color constancy. *Journal of the Optical Society of America*, 33, 555–567.

Helson, H. (1938) Fundamental problems in color vision. I. The principle governing changes in hue saturation and lightness of non-selective samples in chromatic illumination. *Journal of Experimental Psychology*, 23, 439.

Helson, H. & Judd, D.B. (1936) An experimental and theoretical study of changes in surface colors under changing illuminations. *Psychological Bulletin*, 33, 740–741.

Helson, H. & Michels, W.C. (1948) The effect of chromatic adaptation on achromaticity. *Journal of the Optical Society of America*, 38, 1025–1032.

Hendley, C.D. & Hecht, S. (1949) The colors of natural objects and terrains, and their relation to visual color deficiency. *Journal of the Optical Society of America*, 39, 870–873.

Hendrickson, A.E., Koontz, M.A., Pourcho, R.G., Sarthy, P.V., & Goebel, D.J. (1988) Localization of glycine-containing neurons in the *Macaca* monkey retina. *Journal of Comparative Neurology,* 273, 473–487.

Hendrickson, A.E., Wilson, J.R., & Ogren, M.P. (1978) The neuroanatomical organization of pathways between the dorsal lateral geniculate nucleus and the visual cortex in old and new world primates. *Journal of Comparative Neurology*, 182, 123–136.

Hendry, S.H.C. & Yoshioka, T. (1994) A neurochemically distinct third channel in the macaque dorsal lateral geniculate nucleus. *Science*, 264, 575–578.

Henry, G.H. (1991) Afferent inputs, receptive field properties and morphological cell types in different laminae of the striate cortex. In *The Neural Basis of Visual Function* (ed. Leventhal, A.G.), pp. 223–245. Macmillan Press, Houndsmills, Basingstoke, Hampshire, UK.

Henry, G.H., Cole, B.L., & Nathans, J. (1964) The inheritance of congenital tritanopia with the report of an extensive pedigree. *Annals of Human Genetics*, 27, 219–231.

Henry, W.C. (1854) *Memoirs of the Life and Scientific Researches of John Dalton*. Cavendish Society, London.

Hering, E. (1878) *Zur Lehre vom Lichtsinne*. Carl Gerold's Sohn, Wien.

Hering, E. (1905) *Outline of a Theory of the Light Sense* (eds. and trans. Hurvich, L.& Jameson, D., 1964) Harvard University Press, Cambridge, MA.

Herr, S.S., Tiv, N., Klug, K., Schein, S.J., & Sterling, P. (1995) L and M cones in macaque fovea make different numbers of synaptic contacts with OFF (but not ON) midget bipolar cells. *Investigative Ophthalmology & Visual Science,* 36, 2368.

Herr, S.S., Tiv, N., Sterling, P., & Schein, S.J. (1996) S cones in macaque fovea are invaginated by one type of ON bipolar cell, but L and M cones are invaginated by midget and diffuse bipolar cells. *Investigative Ophthalmology & Visual Science,* 37, 4864.

Herschel, J.F.W. (1845) Light. *Encyclopedia Metropolitana,* IV, 343.

Hess, R.F., Baker, C.L., & Zihl, J. (1989) The "motion-blind" patient: Low level spatial and temporal filters. *Journal of Neuroscience*, 9, 1626–1640.

Hess, R.F., Mullen, K.T., Sharpe, L.T., & Zrenner, E. (1989) The photoreceptors in atypical achromatopsia. *Journal of Physiology,* 417, 123–149.

Hess, R.F. , Mullen, K.T., & Zrenner, E. (1989) Human photopic vision with only short wavelength cones: Post-receptoral properties. *Journal of Physiology,* 417, 150–169.

Heywood, C.A. & Cowey, A. (1987) On the role of cortical visual area V4 in the discrimination of hue and pattern in macaque monkeys. *Journal of Neuroscience*, 7, 2601–2616.

Heywood, C.A., Gadotti, A., & Cowey, A. (1992) Cortical area V4 and its role in the perception of color. *Journal of Neuroscience*, 12, 4056–4065.

Hicks, T.P., Lee, B.B., & Vidyasagar, T.R. (1983) The responses of cells in macaque lateral geniculate nucleus to sinusoidal gratings. *Journal of Physiology,* 337, 183–200.

Hilbert, D.R. (1987) *Color and color perception: A study in anthropocentric realism.* CSLI Lecture Notes Number 9. Center for the Study of Language and Information, Stanford, CA.

Hilz, R.L., Huppmann, G., & Cavonius, C.R. (1974) Influence of luminance contrast on hue discrimination. *Journal of the Optical Society of America*, 64, 763–766.

Hippel, A. von (1880) Ein Fall von einseitiger kongenitaler Rot-Grün-Blindheit bei normalem Farbensinn des anderen Auges. *Albrecht von Graefes Archiv für Ophthalmologie*, 26, 176–186.

Hippel, A. von (1881) Über einseitige Farbenblindheit. *Albrecht von Graefes Archiv für Ophthalmologie*, 27, 47–55.

Hisatomi, O., Kayada, S., Aoki, Y., Iwasa, T., & Tokunaga, F. (1994) Phylogenetic relationships among vertebrate visual pigments. *Vision Research,* 34, 3097–3102.

Ho, J. (1988) *Chromatic aberration: A New Tool for Colour Constancy.* Master's thesis, School of Computer Science, Simon Fraser University, Vancouver, Canada.

Ho, J., Funt, B.V., & Drew, M.S. (1990) Separating a color signal into illumination and surface reflectance components: Theory and applications. *IEEE Transactions on Pattern Analysis and Machine Intelligence*, 12, 966–977.

Hochberg, J.E., Triebel, W., & Seaman, G. (1951) Color adaptation under conditions of homogeneous visual stimulation (Ganzfeld). *Journal of Experimental Psychology,* 41, 153–159.

Hochstein, S. & Shapley, R. (1976) Quantitative analysis of retinal ganglion cell classifications. *Journal of Physiology,* 262, 237–264.

Hollmann, M. & Heinemann, S. (1994) Cloned glutamate receptors. *Annual Review of Neuroscicience,* 17, 31–108.

Holmgren, F. (1881) How do the colour-blind see the different colours? *Proceedings of the Royal Society, London,* 31, 302–306.

Holmgren, F. (1884) Über den Farbensinn. *Congres Périodique International des Sciences Medicales, 8th Session, Copenhagen, Compte rendu,* 1, 80–98 (1884).

Hood, D.C. & Birch, D.G. (1993) Human cone receptor activity: The leading edge of the *a*-wave and models of receptor activity. *Visual Neuroscience*, 10, 857–871.

Hood, D.C. & Birch, D.G. (1995) Phototransduction in human cones measured using the a-wave of the ERG. *Vision Research,* 35, 2801–2810.

Hood, D.C. & Finkelstein, M.A. (1986) Sensitivity to light. In *Handbook of Perception and Human Performance. Volume 1: Sensory Processes and Perception* (eds. Boff, K.R., Kaufman, L., & Thomas, J.P.). pp. 5-1 – 5-66. Wiley, New York.

Hopkins, J.M. & Boycott, B.B. (1992) Synaptic contacts of a two-cone flat bipolar cell in a primate retina. *Visual Neuroscience,* 8, 379–384.

Hopkins, J.M. & Boycott, B.B. (1995) Synapses between cones and diffuse bipolar cells of a primate retina. *Journal of Neurocytology,* 24, 680–694.

Hopkins, J.M. & Boycott, B.B. (1996) The cone synapses of DB1 diffuse, DB6 diffuse and invaginating midget bipolar cells of a primate retina. *Journal of Neurocytology,* 25, 391–403.

Hopkins, J.M. & Boycott, B.B. (1997) The cone synapses of cone bipolar cells of primate retina. *Journal of Neurocytology,* 26, 313–325.

Horn, B.K.P. & Sjoberg, R.W. (1989) Calculating the reflectance map. In *Shape from Shading* (eds. Horn, B.K.P. & Brooks, M.J.), pp. 215–244. The MIT Press, Cambridge, MA.

Horn, R. & Marty, A. (1988) Muscarinic activation of ionic currents measured by a new whole-cell recording method. *Journal of General Physiology*, 92, 145–159.

Horner, J.F. (1876) Die Erblichkeit des Daltonismus. In *Amtlicher Bericht über die Verwaltung des Medizinalwesens des Kantons Zürich vom Jahr 1876*, pp. 208–211. Druck der Genossenschaftsbuchdruckerei, Zürich.

Horowitz, B.R. (1981) Theoretical considerations of the retinal receptor as a waveguide. In *Vertebrate Photoreceptor Optics* (eds. Enoch, J.M. & Tobey, F.L.), pp. 217–300. Springer-Verlag, Berlin, New York.

Hsia, Y. & Graham, C.H. (1957) Spectral luminosity curves for protanopic, deuteranopic, and normal subjects. *Proceedings of the National Academy of Science, USA,* 43, 1011–1019.

Hubel, D.H. & Livingstone, M.S. (1987) Segregation of form, color, and stereopsis in primate area 18. *Journal of Neuroscience*, 7, 3378–3415.

Hubel, D.H. & Wiesel, T.N. (1968) Receptive fields and functional architecture of monkey striate cortex. *Journal of Physiology*, 195, 215–243.

Hubel, D.H. & Wiesel, T.N. (1972) Laminar and columnar distribution of geniculo-cortical fibres in macaque monkey. *Journal of Comparative Neurology*, 146, 421–450.

Huffman, D.A. (1971) Impossible objects as nonsense sentences. In *Machine Intelligence 6* (eds. Mettzer, B. & Michie, D.), pp. 295–323. Edinburgh University Press, Edinburgh.

Hunt, R.W.G. (1991) Revised colour-appearance model for related and unrelated colours. *Colour Research an Applications*, 16, 146–165.

Hunt, D.M., Dulai, K.S., Bowmaker, J.K., & Mollon, J.D. (1995) The chemistry of John Dalton's color blindness. *Science*, 267, 984–988.

Hunt, D.M., Williams, A.J., Bowmaker, J.K., & Mollon, J.D. (1993) Structure and evolution of polymorphic photopigment gene of the marmoset. *Vision Research,* 33, 147–154.

Hurlbert, A. (1986) Formal connections between lightness algorithms. *Journal of the Optical Society of America A*, 3, 1684–1693.

Hurlbert, A. (1998) Computational models of color constancy. In *Perceptual Constancies* (eds. Walsh, V. & Kulikowski, J.), pp. 283–322. Cambridge University Press, Cambridge, UK.

Hurvich, L.M. (1972) Color vision deficiencies. In *Visual Psychophyics, Vol. 7/4* in *The Handbook of Sensory Physiology* (eds. Jameson, D. & Hurvich, L.M.), pp. 582–624. Springer-Verlag, Berlin.

Hurvich, L.M. (1981) *Color Vision.* Sinauer, Sunderland, MA.

Hurvich, L.M. & Jameson, D. (1955) Some quantitative aspects of an opponent-colors theory. II. Brightness, saturation, and hue in normal and dichromatic vision. *Journal of the Optical Society of America*, 45, 602–616.

Hurvich, L.M. & Jameson, D. (1957) An opponent process theory of color vision. *Psychological Review*, 64, 384–404.

Hurvich, L.M. & Jameson, D. (1964) Does anomalous color vision imply color weakness? *Psychological Science*, 1, 11–12.

Hyde, E.P., Forsythe, W.E., & Cady, F.E. (1918) The visibility of radiation. *Astrophysics Journal,* 48, 65–83.

Ibbotson, R.E., Hunt, D.M., Bowmaker, J.K., & Mollon, J.D. (1992) Sequence divergence and copy number of the middle- and long-wave photopigment genes in Old World monkeys. *Proceedings of the Royal Society, London, Series B*, 247, 145–154.

Iinuma, I. & Handa, Y. (1976) A consideration of the racial incidence of congenital dyschromats in males and females. *Modern Problems in Ophthalmology*, 17, 151–157.

Ikeda, M., Hukami, K., & Urakubo, M. (1972) Flicker photometry with chromatic adaptation and defective colour vision. *American Journal of Ophthalmology*, 73, 270–277.

Ikeda, H. & Ripps, H. (1966) The electroretinogram of a cone monochromat. *Archives of Ophthalmology,* 75, 513–517.

Ingle, D. (1985) The goldfish as a retinex animal. *Science*, 227, 651–654.

Ingling, C.R. Jr (1977) The spectral sensitivity of the opponent-color channels. *Vision Research*, 17, 1083–1089.

Ingling, C.R. Jr. & Martinez-Uriegas, E. (1983a) The relationship between spectral sensitivity and spatial sensitivity for the primate r-g X-channel. *Vision Research,* 23, 1495–1500.

Ingling, C.R. Jr. & Martinez-Uriegas, E. (1983b) The spatiochromatic signal of the r-g channel. In *Colour Vision: Physiology and Psychophysics* (eds. Mollon, J.D. & Sharpe, L.T.), pp. 433–444. Academic Press, London.

Ishida, A.T., Stell, W.K., & Lightfoot, D.O. (1980) Rod and cone inputs to bipolar cells in goldfish retina. *Journal of Comparative Neurology,* 191, 315–335.

Iverson, G. & D'Zmura, M. (1995a) Criteria for color constancy in trichromatic bilinear models. *Journal of the Optical Society of America A*, 11, 1970–1975.

Iverson, G. & D'Zmura, M. (1995b) Color constancy: Spectral recovery using trichromatic bilinear models. In *Geometric Representations of Perceptual Phenomena, Papers in Honor of Tarow Indow on His 70th Birthday* (eds. Luce, R.D., D'Zmura, M., Hoffman, D., Iverson, G.J., &

Romney, A.K.), pp. 169–185. Lawrence Erlbaum Associates, Mahwah, NJ.

Ives, H.E. (1912a) Studies in the photometry of lights of different colours. I. Spectral luminosity curves obtained by the equality of brightness photometer and flicker photometer under similar conditions. *Philosophical Magazine Series 6,* 24, 149–188.

Ives, H.E. (1912b) the relation between the color of the illuminant and the color of the illuminated object. *Transactions of the Illuminating Engineering Society,* 62–72. (Reprinted in *Color Research and Application*, 20, 70–75.)

Jacobs, G.H. (1981) *Comparative color vision.* Academic Press, New York, London.

Jacobs, G.H. (1983) Differences in spectral response properties of LGN cells in male and female squirrel monkeys. *Vision Research,* 23, 461–468.

Jacobs, G.H. (1984) Within-species variations in visual capacity among squirrel monkeys (*Saimiri Sciureus*): Color vision. *Vision Research,* 24, 1267–1277.

Jacobs, G.H. (1990) Evolution of mechanisms for color vision. *Proceedings of the SPIE,* 1250, 287–292.

Jacobs, G.H. (1993) The distribution and nature of colour vision among the mammals. *Biological Reviews,* 68, 413–471.

Jacobs, G.H. (1996) Primate photopigments and primate color vision. *Proceedings of the National Academy of Sciences, USA,* 93, 577–581.

Jacobs, G.H. & Deegan, J.F. II (1993a) Photopigments underlying color vision in ringtail lemurs (*Lemur catta*) and brown lemurs (*Eulemur fulvus*). *American Journal of Primatology,* 30, 243–256.

Jacobs, G.H. & Deegan, J.F. (1993b) Polymorphism of cone photopigments in new world monkeys: Is the spider monkey unique? *Investigative Ophthalmology & Visual Science (Suppl.)* 34, 749.

Jacobs, G.H. & Deegan, J.F. (1997) Spectral sensitivity of macaque monkeys measured with ERG flicker photometry. *Visual Neurosciences* 14, 921–928.

Jacobs, G.H., Deegan, J.F. II, Neitz, J., Crognale, M.A., & Neitz, M. (1993) Photopigments and color vision in the nocturnal monkey, *Aotus. Vision Research,* 33, 1773–1783.

Jacobs, G.H., Deegan, J.F. II, Neitz, M., & Neitz, J. (1996a) Presence of routine trichromatic color vision in New World monkeys. *Investigative Ophthalmology & Visual Science (Suppl.),* 37, 346.

Jacobs, G.H. & DeValois, R.L. (1965) Chromatic opponent cells in squirrel monkey lateral geniculate nucleus. *Nature,* 206, 487–489.

Jacobs, G.H. & Harwerth, R.S. (1989) Color vision variations in Old and New World primates. *American Journal of Primatology,* 18, 35–44.

Jacobs, G.H. & Neitz, J. (1985) Color vision in squirrel monkeys: sex-related differences suggest the mode of inheritance. *Vision Research,* 25, 141–143.

Jacobs, G.H. & Neitz, J. (1987a) Inheritance of color vision in a New World monkey (*Saimiri Sciureus*). *Proceedings of the National Academy of Sciences,USA,* 84, 2545–2549.

Jacobs, G.H. & Neitz, J. (1987b) Polymorphism of the middle wavelength cone in two species of South American monkey: *Cebus apella* and *Callicebus moloch. Vision Research,* 27, 1263–1268.

Jacobs, G.H. & Neitz, J. (1993) Electrophysiological estimates of individual variation in the L/M cone ratio. In *Color Vision Deficiencies XI* (ed. Drum, B.), pp. 107–112. Kluwer Academic Publishers, The Netherlands.

Jacobs, G.H., Neitz, J., & Crognale, M. (1987) Color vision polymorphism and its photopigment basis in a callitrichid monkey (*saguinus fuscicollis*). *Vision Research,* 27, 2089–2100.

Jacobs, G.H., Neitz, M., Deegan, J.F., & Neitz, J. (1996b) Trichromatic colour vision in New World monkeys. *Nature,* 382, 156–158.

Jacobs, G.H., Neitz, J., & Krogh, K. (1996) Electroretinogram flicker photometry and its applications. *Journal of the Optical Society of America A,* 13, 641–648.

Jacobs, G.H., Neitz, J., & Neitz, M. (1993) Genetic basis of polymorphism in the color vision of platyrrhine monkeys. *Vision Research,* 33, 269–274.

Jacoby, R.A. & Marshak, D.W. (1995) Diffuse bipolar cell inputs to parasol ganglion cells in macaque retina. *Investigative Ophthalmology & Visual Science (Suppl.),* 36, 16.

Jacoby, R.A. & Marshak, D.W. (1996) Inputs to parasol ganglion cells in macaque retina. *Investigative Ophthalmology & Visual Science (Suppl.),* 37, 4349.

Jacoby, R., Stafford, D., Kouyama, N., & Marshak, D. (1996) Synaptic inputs to ON parasol ganglion cells in the primate retina. *Journal of Neuroscience,* 16, 8041–8056.

Jaeger, W. (1972) Genetics of congenital colour deficiencies. In *Handbook of Sensory Physiology, Vol. 7/4, Visual Psychophysics* (eds. Jameson, D. & Hurvich, L.M.), pp. 625–642. Springer-Verlag, Berlin.

Jaeger, W. & Kroker, K. (1952) Über das Verhalten der Protanopen und Deuteranopen bei großen Reißflächen. *Klinische Monatsblätter für Augenheilkunde*, 121, 445–449.

Jägle, H., Sharpe, L.T., & Nathans, J. (1999) Rayleigh matches and X-chromosome-linked pigment genes. *Vision Research* (in preparation).

Jameson, D. & Hurvich, L.M. (1964) Theory of brightness and color contrast in human vision. *Vision Research*, 4, 135–154.

Jameson, D. & Hurvich, L.M. (1972) Color adaptation: Sensitivity, contrast, after-images. In *Handbook of Sensory Physiology. VII/4: Visual Psychophysics* (eds. Jameson, D. & Hurvich, L.M.), pp. 568–581. Springer-Verlag, Berlin.

Jameson, D. & Hurvich, L.M. (1989) Essay concerning color constancy. *Annual Review of Psychology*, 40, 1–22.

Jenness, J.W. & Shevell, S.K. (1995) Color appearance with sparse chromatic context. *Vision Research*, 35, 797–805.

Jolly, A. (1985) *The Evolution of Primate Behavior.* Macmillan Publishing Company, New York.

Jonasdottir, A., Eiberg, H., Kjer, B., Kjer, P., & Rosenberg, T. (1997) Refinement of the dominant optic atrophy locus (OPA1) to a 1.4-cM interval on chromosome 3q28-3q29, within a 3-Mb YAC contig. *Human Genetics*, 99, 115–120.

Jordan, G. & Mollon, J.D. (1988) Two kinds of men? *Investigative Ophthalmology & Visual Science (Suppl.)*, 32, 1212.

Jordan, G. & Mollon, J.D. (1993) A study of women heterozygous for colour deficiencies. *Vision Research*, 33, 1495–1508.

Jørgensen, A.L., Deeb, S., & Motulsky, A.G. (1990) Molecular genetics of X-chromosome-linked color vision among populations of African and Japanese ancestry: High frequency of a shortened red pigment gene among Afro-Americans. *Proceedings of the National Academy of Sciences, USA*, 87, 6512–6516.

Jørgensen, A.L., Philip, J., Raskind, W.H., Matsushita, M., Christensen, B., Dreyer, V., & Motulsky, A.G. (1992) Different patterns of X inactivation in MZ twins discordant for red-green color-vision deficiency. *American Journal of Human Genetics*, 51, 291–298.

Judd, D.B. (1940) Hue saturation and lightness of surface colors with chromatic illumination. *Journal of the Optical Society of America*, 30, 2–32.

Judd, D.B. (1943) Facts of color-blindness. *Journal of the Optical Society of America,* 33, 294–307.

Judd, D.B. (1945) Standard response functions for protanopic and deuteranopic vision. *Journal of the Optical Society of America*, 35, 199–121.

Judd, D.B. (1948) Color perceptions of deuteranopic and protanopic observers. *Journal of Research of the National Bureau of Standards*, 41, 247–271.

Judd, D.B. (1949a) Response functions for types of vision according to the Müller theory. *Journal of Research of the National Bureau of Standards*, 42.

Judd, D.B. (1949b) Standard response functions for protanopic and deuteranopic vision. *Journal of the Optical Society of America*, 39, 505.

Judd, D.B. (1951) Report of U.S. Secretariat Committee on Colorimetry and Artificial Daylight. *Proceedings of the Twelfth Session of the CIE, Stockholm,* p. 11. Bureau Central de la CIE, Paris.

Judd, D.B., MacAdam, D.L., & Wyszecki, G.W. (1964) Spectral distribution of typical daylight as a function of correlated color temperature. *Journal of the Optical Society of America*, 54, 1031–1040.

Kaas, J.H. & Krubitzer, L.A. (1991) The organization of extrastriate visual cortex. In *Neuroanatomy of the Visual Pathways and their Development* (eds. Dreher, B. & Robinson, S.R.), pp. 302–323. The Macmillan Press, Houndmills, Basingstoke, Hampshire, UK.

Kainz, P.M., Neitz, M., & Neitz, J. (1998) Molecular genetic detection of female carriers of protan defects. *Vision Research*, 38, 3365–3369.

Kaiser, P.K. & Boynton, R.M. (1996) *Human Color Vision, 2nd Edition*. Optical Society of America, Washington, DC.

Kaiser, P.K., Lee, B.B., Martin, P.R., & Valberg, A. (1990) The physiological basis of the minimally distinct border demonstrated in the ganglion cells of the macaque retina. *Journal of Physiology,* 422, 153–183.

Kalloniatis, M. & Harwerth, R.S. (1991) Effects of chromatic adaptation on opponent interactions in monkey increment-threshold spectral-sensitivity functions. *Journal of the Optical Society of America A*, 8, 1818–1831.

Kalmus, H. (1955) The familial distribution of congenital tritanopia with some remarks on some similar conditions. *Annals of Human Genetics*, 20, 39–56.

Kalmus, H. (1965) *Diagnosis and Genetics of Defective Colour Vision*. Pergamon Press, Oxford.

Kamermans, M. & Spekreijse, H. (1995) Spectral behavior of cone-driven horizontal cells in teleost retina. *Progress in Retinal and Eye Research,* 14, 313–360.

Kanade, T. (1980) A theory of origami world. *Artificial Intelligence,* 13, 279–311.

Kaneko, A. (1973) Receptive field organization of bipolar and amacrine cells in the goldfish retina. *Journal of Physiology*, 235, 133–153.

Kaneko, A. & Tachibana, M. (1983) Double color-opponent receptive fields of carp bipolar cells. *Vision Research,* 23, 381–388.

Kanisza, G. (1979) *Organization in Vision*. Praeger, New York.

Kaplan, E., Lee, B.B., & Shapley, R. (1990) New views of primate retinal function. In *Progress in Retinal Research, Vol. 9* (eds. Osborne, N. & Chader, G.), pp. 273–336. Pergamon Press, Oxford, NY.

Kaplan, E., Mukherjee, P., & Shapley, R. (1993) Information filtering in the lateral geniculate nucleus. In *Contrast Sensitivity* (eds. Shapley, R. & Lam, D.M.-K.), pp. 183–200. MIT Press, Cambridge, MA.

Kaplan, E., Purpura, K., & Shapley, R.M. (1987) Contrast affects the transmission of visual information through the mammalian lateral geniculate nucleus. *Journal of Physiology,* 391, 267–288.

Kaplan, E. & Shapley, R. (1982) X and Y cells in the lateral geniculate nucleus of macaque monkeys. *Journal of Physiology,* 330, 125–143.

Kaplan, E. & Shapley, R. (1986) The primate retina contains two types of ganglion cells, with high and low contrast sensitivity. *Procreedings of the National Academy of Sciences,USA,* 83, 2755–2757.

Kay, R.F. (1984) On the use of anatomical features to infer foraging behavior in extinct primates. In *Adaptations for Foraging in Nonhuman Primates.* (eds. Rodman, P.S. & Cant, J.G.H.), pp. 21–53. Columbia University Press, New York.

Kay, R.F., Ross, C., & Williams, B.A. (1997) Anthropoid origins. *Science,* 275, 797–804.

Kelly, D.H. (1971) Theory of flicker and transient responses. I. Uniform fields. *Journal of the Optical Society of America,* 61, 537–546.

Kelly, D.H. (1974) Spatio-temporal frequency characteristics of color-vision mechanisms. *Journal of Physiology,* 228, 55.

Kelly, D.H. (1979) Motion and vision. II. Stabilized spatio-temporal threshold surface. *Journal of the Optical Society of America,* 69, 1340–1349.

Kelly, D.H. (1983) Spatiotemporal variation of chromatic and achromatic contrast thresholds. *Journal of the Optical Society of America,* 73, 742–750.

Kennard, C., Lawden, M., Morland, A.B., & Ruddock, K.H. (1995) Colour identification and colour constancy are impaired in a patient with incomplete achromatopsia associated with prestriate cortical lesions. *Proceedings of the Royal Society, London, Series B,* 260, 169–175.

King-Smith, P.E. (1973a) The optical density of erythrolabe determined by a new method. *Journal of Physiology,* 230, 551–560.

King-Smith, P.E. (1973b) The optical density of erythrolabe determined by retinal densitometry using the self-screening method. *Journal of Physiology,* 230, 535–549.

King-Smith, P.E. & Carden, D. (1976) Luminance and opponent-color contributions to visual detection and adaptation and to temporal and spatial integration. *Journal of the Optical Society of America,* 66, 709–717.

King-Smith, P.E. & Webb, J.R. (1974) The use of photopic saturation in determining the fundamental spectral sensitivity curves. *Vision Research,* 14, 421–429.

Kingdom, F.A.A., Moulden, B., & Collyer, S. (1992) A comparison between colour and luminance contrast in a spatial linking task. *Vision Research,* 32, 709–717.

Kiorpes, L. (1992) Development of vernier acuity and grating acuity in normally reared monkeys. *Visual Neuroscience,* 9, 243–251.

Kiper, D.C., Fenstemaker, S.B., & Gegenfurtner, K.R. (1997) Chromatic properties of neurons in macaque area V2. *Visual Neuroscience,* 14, 1061–1072.

Kjer, P. (1959) Infantile optic atrophy with dominant mode of inheritance. *Acta Opthalmological Supplement,* 54, 1–146.

Kleinschmidt, A., Lee, B.B., Requart, M., & Frahm, J. (1996) Functional mapping of color processing by magnetic resonance imaging of responses to selective p-and m-pathway stimulation. *Experimental Brain Research,* 110, 279–288.

Klinker, G.J., Shafer, S.A., & Kanade, T. (1988) The measurement of highlight in color images. *International Journal of Computer Vision,* 2, 7–32.

Klug, K., Tiv, N., Tsukamoto, Y., Sterling, P., & Schein, S.J. (1992) Blue cones contact OFF-midget bipolar cells. *Society for Neuroscience Abstracts,* 18, 838.

Klug, K., Tsukamoto, Y., Sterling, P., & Schein, S.J. (1993) Blue cone off–midget ganglion cells in Macaque. *Investigative Ophthalmology & Visual Science,* 34, 1398.

Knau, H. & Sharpe, L.T. (1998) Psychophysical estimates of cone pigment densities in dichromats & trichromats. *Investigative Ophthalmology & Visual Science,* 38, S120

Knight, R. & Buck, S.L. (1993) Cone pathways and the pi 0 and pi 0' rod mechanisms. *Vision Research,* 33, 2203–2213.

Knight, J.D., Li, R., & Botchan, M. (1991) The activation domain of the bovine papillomavirus E2 protein mediates association of DNA-bound dimers to form DNA loops. *Proceedings of the National Academy of Science, USA,* 88, 3204–3208.

Knill, D.C. & Kersten, D. (1991) Apparent surface curvature affects lightness perception. *Nature,* 351, 228–230.

Knoblauch, K. (1995) Dual bases in dichromatic color space. In *Colour Vision Deficiencies XII* (ed. Drum, B.), pp. 165–176. Kluwer Academic Publishers, Dordrecht.

Knoblauch, K. & Maloney, L.T. (1996) Testing the indeterminacy of linear color mechanisms from color discrimination data. *Vision Research,* 36, 295–306.

Knowles, A. & Dartnall, H.J.A. (1977) The photobiology of vision. In *The Eye, Vol 2B* (ed. Davson, H.). Academic, London, New York.

Koffka, K. (1963) *Principles of Gestalt Psychology.* Harcourt, Brace & World, New York.

Kohl, S., Marx, T., Giddings, I., Jägle, H., Jacobson, S.G., Apfelstedt-Sylla, E., Zrenner, E., Sharpe, L.T., & Wissinger, B. (1998) Total colorblindness is caused by mutations in the gene encoding the α-subunit of the cone photoreceptor cGMP-gated cation channel. *Nature Genetics*, 19, 257–259.

Kolb, H. (1970) Organization of the outer plexiform layer of the primate retina: electron microscopy of Golgi-impregnated cells. *Philosophical Transactions of the Royal Society, London, Series B,* 258, 261–283.

Kolb, H. (1991) Anatomical pathways for color vision in the human retina. *Visual Neuroscience,* 7, 61–74.

Kolb, H. (1994) The architecture of functional neural circuits in the vertebrate retina. *Investigative Ophthalmology & Visual Science,* 35, 2385–2404.

Kolb, H., Boycott, B.B., & Dowling, J.E. (1969) A second type of midget bipolar cell in the primate retina. Appendix. *Philosophical Transactions of the Royal Society, London, Series B,* 255, 177–184.

Kolb, H. & Dekorver, L. (1991) Midget ganglion cells of the parafovea of the human retina: A study by electron microscopy and serial section reconstructions. *Journal of Comparative Neurology,* 303, 617–636.

Kolb, H. & Famiglietti, E.V. (1974) Rod and cone pathways in the inner plexiform layer of cat retina. *Science,* 186, 47–49.

Kolb, H. & Lipetz, L.E. (1991) The anatomical basis for colour vision in the vertebrate retina. In *The Perception of Colour* (ed. Gouras, P.), pp. 128–145. The Macmillan Press, Houndsmills, Basingstoke, Hampshire, UK.

Kolb, H., Linberg, K., & Fisher, S.K. (1992) Neurons of the human retina: A Golgi study. *Journal of Comparative Neurology,* 318, 147–187.

Kolb, H. & Nelson, R. (1995) The organization of photoreceptor to bipolar synapses in the outer plexiform layer. In *Neurobiology and Clinical Aspects of the Outer Retina* (eds. Djamgoz, M.B.A., Archer, S.N., & Vallerga, S.). Chapman & Hall, London.

Kolb, H., Nelson, R., & Mariani, A. (1981) Amacrine cells, bipolar cells and ganglion cells of the cat retina: A Golgi study. *Vision Research,* 21, 1081–1114.

Koliopoulos, J., Iordanides, P., Palimeris, G., & Chimonidou, E. (1976) Data concerning colour vision deficiencies amongst young Greeks. *Modern Problems in Ophthalmology,* 17, 161–164.

Komatsu, K., Ideura, Y., Kaji, S., & Yamane, S. (1992) Color selectivity of neurons in the inferior temporal cortex of the awake macaque monkey. *Journal of Neuroscience,* 12, 408–424.

Kondo, T. (1941) Untersuchungen bei angeborenen Farbensinn-Anomalien. Über das Zustandekommen und Wesen der angeborenen Farbensinn-Anomalien. *Acta of the Society of Ophthalmology of Japan,* 45, 659.

König, A. (1894) Über den menschlichen Sehpurpur und seine Bedeutung für das Sehen. *Sitzungsberichte der Akademie der Wissenschaften, Berlin,* 1894, 577–598.

König, A. & Dieterici, C. (1886) Die Grundempfindungen und ihre Intensitäts-Vertheilung im Spektrum. *Sitzungsberichte Akademie der Wissenschaften, Berlin,* 1886, 805–829.

Kooi, F.L. & DeValois, K.K. (1992) The role of color in the motion system. *Vision Research,* 32, 657–688.

Kooi, F.L., DeValois, K.K., Grosof, D.H., & DeValois, R.L. (1992) Properties of the recombination of one-dimensional motion signals into a pattern motion signal. *Perception & Psychophysics,* 52, 415–424.

Koontz, M.A. & Hendrickson, A.E. (1990) Distribution of GABA-immunoreactive amacrine cell synapses in the inner plexiform layer of macaque monkey retina. *Visual Neuroscience,* 5, 17–28.

Koontz, M.A., Hendrickson, A.E., Brace, S.T., & Hendrickson, A.E. (1993) Immunocytochemical localization of GABA and glycine in amacrine and displaced amacrine cells of Macaque monkey retina. *Vision Research,* 33, 2617–2628.

Koulischer, L., Zanen, J., & Meunier, A. (1968) La théorie Lyon peut-elle expliquer la disparité exceptionellement observéede la perception colorée chez des jumelles unvitellines? *Journale de Genetique Humaine, Supplemente,* 15, 242–254.

Koutalos, Y. & Yau, K.-W. (1996) Regulation of sensitivity in vertebrate rod photoreceptors by calcium. *Trends in Neurosciences,* 19, 73–81.

Kouyama, N. & Marshak, D.W. (1992) Bipolar cells specific for blue cones in the macaque retina. *Journal of Neuroscience,* 12, 1233–1252.

Kraft, J.M. & Brainard, D.H. (1997) An analysis of cues contributing to color constancy. *Program of the Optical Society of Americal Annual Meeting,* Long Beach, CA, October 12–17, 1997, p. 110.

Kraft, T.W. (1988) Photocurrents of cone photoreceptors of the golden-mantled ground squirrel. *Journal of Physiology,* 404, 199–213.

Kraft, T.W., Neitz, J., & Neitz, M. (1998) Spectra of human L cones. *Vision Research,* 38, 3663–3670.

Kraft, T.W., Schneeweis, D.M., & Schnapf, J.L. (1993) Visual transduction in human rod photoreceptors. *Journal of Physiology,* 464, 747–765.

Krantz, D. (1968) A theory of context effects based on cross-context matching. *Journal of Mathematical Psychology,* 5, 1–48.

Krantz, D.H. (1975) Color measurement and color theory: II. Opponent-colors theory. *Journal of Mathematical Psachology,* 12, 304–327.

Krauskopf, J. (1964) Color appearance of small stimuli and the spatial distribution of color receptors. *Journal of the Optical Society of America*, 54, 1171.

Krauskopf, J. (1974) Interaction of chromatic mechanisms in detection. In *Colour Vision Deficiencies II. International Symposium, Edinburgh* 1973, *Modern Problems of Ophthalmology*, 13, 92–97.

Krauskopf, J. (1980) Discrimination and detection of changes in luminance. *Vision Research*, 20, 671–677.

Krauskopf, J. (1997) On the relative effectiveness of L- and M-cones. *Investigative Ophthalmology & Visual Science (Suppl.)*, 38, 14.

Krauskopf, J. & Farell, B. (1990) Influence of colour on the perception of coherent motion. *Nature*, 348, 328–331.

Krauskopf, J. & Farell, B. (1991) Vernier acuity: effects of chromatic content, blur and contrast. *Vision Research*, 31, 735–749.

Krauskopf, J. & Gegenfurtner, K. (1992) Color discrimination and adaptation. *Vision Research*, 32, 2165–2175.

Krauskopf, J. & Li, X. (1996) Retinal- and object-relative cues to motion are used differently by luminance and chromatic mechanisms. *Investigative Ophthalmology & Visual Science (Suppl.)*, 37, 2.

Krauskopf, J., Williams, D.R., & Heeley, D.W. (1982) The cardinal directions of color space. *Vision Research,* 22, 1123–1131.

Krauskopf, J., Williams, D.R., Mandler, M.B., & Brown, A.M. (1986) Higher order color mechanisms. *Vision Research*, 26, 23–32.

Krauskopf, J., Wu, H.-J., & Farell, B. (1996) Coherence, cardinal directions and higher-order mechanisms. *Vision Research*, 36, 1235–1245.

Krauskopf, J. & Zaidi, Q. (1985) Spatial factors in desensitization along cardinal directions of color space. *Investigative Ophthalmology & Visual Science (Suppl.)*, 26, 206.

Krauskopf, J. & Zaidi, Q. (1986) Induced desensitization. *Vision Research*, 26, 759–762.

Krauskopf, J., Zaidi, Q., & Mandler, M.B. (1986) Mechanisms of simultaneous color induction. *Journal of the Optical Society of America A*, 3, 1752–1757.

Kremers, J. (1996) Responses of marmoset lateral geniculate cells to rotating stimuli. *Perception,* 25, 117–118.

Kremers, J., Lee, B.B., & Kaiser, P.K. (1992) Sensitivity of macaque retinal ganglion cells and human observers to combined luminance and chromatic modulation. *Journal of the Optical Society of America A,* 9, 1477–1485.

Kremers, J., Lee, B.B., & Yeh, T. (1995) Receptive field dimensions of primate retinal ganglion cells. In *Color Vision Deficiencies XII* (ed. Drum, B.), pp. 399–406. Kluver Academic Publishers, Dordrecht.

Kremers, J. & Weiss, S. (1997) Receptive field dimensions of lateral geniculate cells in the common marmoset (*Callithrix jacchus*). *Vision Research,* 37, 2171–2181.

Kremers, J., Weiss, S., & Zrenner, E. (1997) Temporal properties of marmoset lateral geniculate cells. *Vision Research,* 37, 2649–2660.

Kremers, J., Weiss, S., Zrenner, E., & Maurer, J. (1997) Spectral responsivity of lateral geniculate cells in the dichromatic common marmoset (*Callithrix jacchus*). In *Colour Vision Deficiencies XIII* (ed. Drum, B.), pp. 87–97. Kluwer Academic Publishers, Dordrecht, Boston, London.

Kremers, J., Zrenner, E., Wiess, S., & Meierkord, S. (1998) Chromatic processing in the lateral geniculate nucleus of the common marmoset (*Callithrix jacchus*). In *Color Vision: Perspectives from Different Disciplines* (eds. Backhaus, W.G.K., Kliegl, R., & Werner, J.S.), pp. 89–99. Walter de Gruyter, Berlin, New York.

Krill, A.E. (1964) A technique for evaluating photopic and scotopic flicker function with one light intensity. *Documental Ophthalmologica*, 18, 452.

Krill, A.E. (1969) X-chromosomal-linked diseases affecting the eye: status of the heterozygote female. *Transactions of the American Ophthalmological Society,* 67, 535.

Krill, A. & Beutler, E. (1965) Red-light thresholds in heterozygote carriers of protanopia: genetic implications. *Science*, 149, 186–188.

Krill, A.E. & Schneidermann, A. (1964) A hue discrimination defect in so called normal carriers of color vision defects. *Investigative Ophthalmology,* 3, 445–450.

Krill, A.E., Smith, V.C., & Pokorny, J. (1971) Further studies supporting the identity of congenital tritanopia and hereditary dominant optic atrophy. *Investigative Ophthalmology,* 10, 457–465.

Krinov, E.L. (1947/1953) *Spectral'naye otrazhatel'naya sposobnost'prirodnykh obrazovanii.* Izd. Akad. Nauk USSR (Proc. Acad. Sci. USSR); translated by G. Belkov, *Spectral reflectance properties of natural formations*; Technical translation: TT-439. Ottawa, Canada: National Research Council of Canada, 1953.

Kropf, A. & Hubbard, R. (1958) The mechanism of bleaching rhodopsin. *Annals of the New York Academy of Science*, 74, 266–280.

Lachica, E.A., Beck, P.D., & Casagrande, V.A. (1992) Parallel pathways in macaque monkey striate cortex: Anatomically defined columns in layer III. *Proceedings of the National Academy of Sciences, USA*, 89, 3566–3570.

Ladd-Franklin, C. (1932) *Colour and Colour Theories.* Kegan Paul, London.

Lakowski, R. (1969a) Theory and practice of colour vision testing. A review. Part I. *British Journal of Industrial Medicine*, 26, 173–189.

Lakowski, R. (1969b) Theory and practice of colour vision testing. A review. Part 2. *British Journal of Industrial Medicine*, 26, 265–288.

Lamb, T.D. (1994) Stochastic simulation of activation in the G-protein cascade of phototransduction. *Biophysical Journal,* 67, 1439–1454.

Lamb, T.D. (1995) Photoreceptor spectral sensitivities: common shape in the long-wavelength region. *Vision Research,* 35, 3083–3091.

Lamb, T.D. (1996) Gain and kinetics of activation in the G-protein cascade of phototransduction. *Proceedings of the National Academy of Sciences, USA,* 93, 566–570.

Lamb, T.D. & Pugh, E.N. Jr. (1992) A quantitative account of the activation steps involved in phototransduction in amphibian photoreceptors. *Journal of Physiology,* 449, 719–757.

Lamb, T.D. & Simon, E.J. (1977) Analysis of electrical noise in turtle cones. *Journal of Physiology,* 272, 435–468.

Lambrecht, H.-G. & Koch, K.-W. (1991) A 26 kd calcium binding protein from bovine rod outer segments as modulator of photoreceptor guanylate cyclase. *EMBO Journal,* 10, 793–798.

Land, E.H. (1959/1961) Experiments in color vision. *Scientific American*, 201, 84–99; reprinted in *Color Vision; An Enduring Problem in Psychology* (eds. Teevan, R.C. & Birney, R.C.). Van Nostrand, Toronto.

Land, E.H. (1983) Recent advances in retinex theory and some implications for cortical computations: Color vision and the natural image. *Proceedings of the National Academy of Sciences, USA*, 80, 5163–5169.

Land, E.H. (1986) Recent advances in retinex theory. *Vision Research*, 26, 7–22.

Land, E.H. & McCann, J.J. (1971) Lightness and retinex theory. *Journal of the Optical Society of America*, 61, 1–11.

Landy, M.S., Maloney, L.T., Johnston, E.B., & Young, M. (1995) Measurement and modeling of depth cue combination: In defense of weak fusion. *Vision Research*, 35, 389–412.

Larsen, H. (1921) Demonstration mikroskopischer Präparate von einem monochromatischen Auge. *Klinische Monatsblätter der Augenheilkunde*, 67, 301–302.

Le Blon, J.C. (1722) *Il Coloritto, or the harmony of colouring in painting reduced to mechanical practice*. London.

Le Grand, Y. (1968) *Light, Colour and Vision, 2nd ed.* Chapman and Hall, London.

Lee, B.B. (1991) On the relation between cellular sensitivity and psychophysical detection. In *From Pigments to Perception* (eds. Valberg, A. & Lee, B.B.), pp. 105–115. Plenum Press, London.

Lee, B.B. (1996) Receptive field structure in the primate retina. *Vision Research,* 36, 631–644.

Lee, B.B., Dacey, D.M., Smith, V.C., & Pokorny, J. (1997a) Time course and cone specificity of adaptation in primate outer retina. *Investigative Ophthalmology & Visual Science (Suppl.)*, 38, 1163.

Lee, B.B., Kremers, J., & Yeh, T. (1998) Receptive fields of primate retinal ganglion cells studied with a novel technique. *Visual Neuroscience*, 15, 161-175.

Lee, B.B., Martin, P.R., & Valberg, A. (1988) The physiological basis of heterochromatic flicker photometry demonstrated in the ganglion cells of the macaque retina. *Journal of Physiology*, 404, 323–347.

Lee, B.B., Martin, P.R., & Valberg, A. (1989a) Nonlinear summation of M- and L-cone inputs to phasic retinal ganglion cells of the macaque. *Journal of Neuroscience*, 9, 1433–1442.

Lee, B.B., Martin, P.R., & Valberg, A. (1989b) Sensitivity of macaque retinal ganglion cells to chromatic and luminance flicker. *Journal of Physiology*, 414, 223–243.

Lee, B.B., Martin, P.R., Valberg, A., & Kremers, J. (1993a) Physiological mechanisms underlying psychophysical sensitivity to combined luminance and chromatic modulation. *Journal of the Optical Society off America A*, 10, 1403–1412.

Lee, B.B., Pokorny, J., Smith, V.C., & Kremers, J. (1994) Responses to pulses and sinusoids in macaque ganglion cells. *Vision Research,* 34, 3081–3096.

Lee, B.B., Pokorny, J., Smith, V.C., Martin, P.R., & Valberg, A. (1990) Luminance and chromatic modulation sensitivity of macaque ganglion cells and human observers. *Journal of the Optical Society of America A,* 7, 2223–2236.

Lee, B.B., Silveira, L.C.L., Yamada, E., & Kremers, J. (1996a) Parallel pathways in the retina of old and new world primates. *Revista Brasileira de Biologia,* 56, 323–338.

Lee, B.B., Smith, V.C., Pokorny, J., & Kremers, J. (1996b) Rod inputs to macaque ganglion cells and their temporal dynamics. *Investigative Ophthalmology & Visual Science* (Suppl.), 36, 689.

Lee, B.B., Smith, V.C., Pokorny, J., & Kremers, J. (1997b) Rod inputs to macaque ganglion cells. *Vision Research*, 37, 2813–2828.

Lee, B.B., Valberg, A., Tigwell, D.A., & Tryti, J. (1987) An account of responses of spectrally opponent neurons in macaque lateral geniculate nucleus to successive contrast. *Proceedings of the Royal Society, London, Series B*, 230, 293–314.

Lee, B.B., Wehrhahn, C., Westheimer, G., & Kremers, J. (1993b) Macaque ganglion cell responses to stimuli that elicit hyperacuity in man: Detection of small displacements. *Journal of Neuroscience,* 13, 1001–1009.

Lee, B.B., Wehrhahn, C., Westheimer, G., & Kremers, J. (1995) The spatial precision of macaque ganglion cell responses in relation to Vernier acuity of human observers. *Vision Research,* 35, 2743–2758.

Lee, B.B. & Yeh, T. (1995) Tritan pairs estimated by modulation photometry of red, green and blue lights. In *Color Vision Deficiencies XII* (ed. Drum, B.), pp. 177–184. Kluwer Academic Publishers, Dordrecht.

Lee, H.-C. (1986) Method for computing the scene-illuminant chromaticity from specular highlights. *Journal of the Optical Society of America A*, 3, 1694–1699.

Lee, H.-C., Breneman, E.J., & Schulte, C.P. (1990) Modeling light reflection for computer color vision. *IEEE Transactions on Pattern Analysis and Machine Intelligence*, 12, 402–409.

Lee, J. & Stromeyer, C.F. III (1989) Contribution of human short-wave cones to luminance and motion detection. *Journal of Physiology*, 413, 563–595.

LeGrand, Y. (1949/1994) (Les seuils différentiels de couleurs dans la théorie de Young, *Revue d'Optique*, 28, 261–278.) Color difference thresholds in Young's theory (trans. by Knoblauch, K.). *Color Research and Application*, 19, 296–309.

Lennie, P. (1984) Recent developments in the physiology of color vision. *Trends in Neuroscience,* 7, 243–248.

Lennie, P. (1991) Color vision. *Optics and Photonics News*, 1, 10–16.

Lennie, P. (1993) Roles of M and P pathways. In *Contrast Sensitivity* (eds. Shapley, R. & Lam, D.M.K.), pp. 201–214. MIT Press, Cambridge, MA.

Lennie, P. (1998) Single units and visual cortical organization. *Perception*, 27, 889–935.

Lennie, P. & D'Zmura, M. (1988) Mechanisms of color vision. *CRC Critical Reviews in Neurobiology*, 3, 333–400.

Lennie, P. & Fairchild, M.D. (1994) Ganglion cell pathways for rod vision. *Vision Research*, 34, 477–482.

Lennie, P., Haake, P.W., & Williams, D.R. (1991) The design of chromatically opponent receptive fields. In *Computational Models of Visual Processing* (eds. Landy, M.S. & Movshon, J.A.), pp. 71–82. MIT Press, Cambridge, MA.

Lennie, P., Krauskopf, J., & Sclar, G. (1990) Chromatic mechanisms in striate cortex of macaque. *Journal of Neuroscience,* 10, 649–669.

Lennie, P., Lankheet, M.J.M., & Krauskopf, J. (1994) Chromatically-selective habituation in monkey striate cortex. *Investigative Ophthalmology & Visual Science (Suppl.)*, 35, 1662.

Lennie, P., Pokorny, J., & Smith, V.C. (1993) Luminance. *Journal of the Optical Society of America A*, 10, 1283–1293.

Lennie, P., Trevarthen, C., Van Essen, D., & Wässle, H. (1990) Parallel processing of visual information. In *Visual Perception: The Neurophysiological Foundations* (eds. Spillmann, L. & Werner, J.), pp.103–128. Academic Press, New York.

Lettvin, J.Y., Maturana, H.R., McCulloch, W.S., & Pitts, W.H. (1959) What the frog's eye tells the frog's brain. *Proceedings of the IRE*, 47, 1940–1951.

Leventhal, A.G., Ault, S.J., Vitek, D.J., & Shou, T. (1989) Extrinsic determinants of retinal ganglion cell development in primates. *Journal of Comparative Neurology,* 286, 170–189.

Leventhal, A.G., Rodieck, R.W., & Dreher, B. (1981) Retinal ganglion cell classes in the Old World monkey: Morphology and central projections. *Science,* 213, 1139–1142.

Leventhal, A.G., Thompson, K.G., & Liu, D. (1993) Retinal ganglion cells within the foveola of New World (*saimiri sciureus*) and Old World (*macaca fascicularis*) monkeys. *Journal of Comparative Neurology,* 338, 242–254.

Leventhal, A.G., Thompson, K.G., Liu, D., Zhou, Y., & Ault, S.J. (1995) Concomitant sensitivity to orientation, direction, and color of cells in layers 2, 3, and 4 of monkey striate cortex. *Journal of Neuroscience*, 15, 1808–1818.

Levinson, E. & Sekular, R. (1975) The independence of channels in human vision selective for direction of movement. *Journal of Physiology,* 250, 347–366.

Levitt, J.B., Kiper, D.C., & Movshon, J.A. (1994) Receptive fields and functional architecture of macaque V2. *Journal of Neurophysiology,* 71, 2517–2542.

Levitt, J.B., Yoshioka, T., & Lund, J.S. (1994) Intrinsic cortical connections in macaque visual area V2: Evidence for interaction between different functional streams. *Journal of Comparative Neurology,* 342, 551–570.

Lewis, P.R. (1955) A theoretical interpretation of spectral sensitivity curves at long wavelengths. *Journal of Physiology,* 130, 45–52.

Li, A. & Lennie, P. (1997) Mechanisms underlying segmentation of colored textures. *Vision Research*, 37, 83–97.

Li, A. & Lennie, P. (1999) Color and brightness in segmenting textured surfaces (in preparation).

Li, A. & Zaidi, Q. (1997) Image enhancement of colored images. *Investigative Ophthalmology & Visual Science*, 38, S254.

Liang, J., Grimm, B., Goelz, S., & Bille, J.F. (1994) Objective measurement of wave aberrations of the human eye with the use of a Hartmann-Shack wavefront sensor. *Journal of the Optical Society of America A*, 11, 1949–1957.

Liang, J. & Williams, D.R. (1997) Aberrations and retinal image quality of the normal human eye. *Journal of the Optical Society of America A*, 14, 2873–2883.

Liang, J., Williams, D.R., & Miller, D.T. (1997) Supernormal vision and high-resolution retinal imaging through adaptive optics. *Journal of the Optical Society of America A*, 14, 2884–2892.

Lima, S.M.A., Silveira, L.C.L., & Perry, V.H. (1993) The M-ganglion cell density gradient in new world monkeys. *Brazilian Journal of Medical and Biological Research,* 26, 961–964.

Lima, S.M.A., Silveira, L.C.L., & Perry, V.H. (1996) Distribution of M retinal ganglion cells in diurnal and nocturnal new world monkeys. *Journal of Comparative Neurology,* 368, 538–552.

Lindsey, D.T. & Teller, D.Y. (1990) Motion at isoluminance: Discrimination/detection ratios for moving isoluminant gratings. *Vision Research*, 30, 1751–1761.

Livingstone, M.S. & Hubel, D.H. (1983) Specificity of cortico-cortical connections in monkey visual system. *Nature*, 304, 531–534.

Livingstone, M.S. & Hubel, D.H. (1984) Anatomy and physiology of a color system in primate visual cortex. *Journal of Neuroscience*, 4, 309–356.

Livingstone, M.S. & Hubel, D.H. (1987) Psychophysical evidence for separate channels for the perception of form, color, movement and depth. *Journal of Neuroscience*, 7, 3416–3468.

Livingstone, M. & Hubel, D. (1988) Segregation of form, color, movement and depth: Anatomy, physiology and perception. *Science*, 240, 740–750.

Logothetis, N.K., Schiller, P.H., Charles, E.R., & Hurlbert, A.C. (1990) Perceptual deficits and the role of color opponent and broad band channels in vision. *Science,* 247, 214–217.

Loomis, J.M. & Berger, T. (1979) Effects of chromatic adaptation on color discrimination and color appearance. *Vision Research*, 19, 891–901.

Lu, Z. & Sperling, G. (1996) Second-order illusions: Mach bands, Chevreul, and Craik-O'Brien-Cornsweet. *Vision Research,* 36, 559–572.

Luckiesh, M. (1965) *Visual Illusions*. Dover, New York.

Lueck, C.J., Zeki, S., Friston, K.J., Deiber, M.-P., Cope, P., Cunningham, V.J., Lammertsma, A.A., Kennard, C., & Frackowiak, R.S.J. (1989) The color center in the cerebral cortex of man. *Nature*, 340, 386–389.

Lund, J.S. (1973) Organisation of neurons in the visual cortex, area 17, of the monkey (*Macaca mulatta*). *Journal of Comparative Neurology,* 147, 455–495.

Lund, J.S. (1988) Anatomical organization of macaque monkey striate visual cortex. *Annual Review of Neurosciences,* 11, 253–288.

Lund, J.S. & Boothe, R.G. (1975) Interlaminar connections and pyramidal neuron organization in the visual cortex, area 17, of the macaque monkey. *Journal of Comparative Neurology*, 159, 305–334.

Lund, J.S., Lund, R.D., Hendrickson, A.E., Bunt, A.H., & Fuchs, A.F. (1975) The origin of efferent pathways from the primary visual cortex, area 17, of the macaque monkey shown by retrograde transport of horseradish peroxidase. *Journal of Comparative Neurology*, 164, 287–304.

Lunkes, A., Hartung, U., Magarino, C., Rodriguez, M., Palmero, A., Rodriguez, L., Heredero, L., Weissenbach, J., Weber, J., & Auburger, G. (1995) Refinement of the OPA 1 gene locus on chromosome 3q28-q29 to a region of 2-8 cM, in one Cuban pedigree with autosomal dominant optic atrophy type Kjer. *American Journal of Human Genetics,* 57, 968–970.

Luther, R. (1927) Aus dem Gebiet der Farbreizmetrik. *Zeitschrift für technische Physik,* 8, 540–558.

Lutze, M., Cox, N.J., Smith, V.C., & Pokorny, J. (1990) Genetic studies of variation in Rayleigh and photometric matches in normal trichromats. *Vision Research*, 30, 149–162.

Lynch, D.K. & Soffer, B.H. (1999) On the solar spectrum and the color sensitivity of the eye. *Optics & Photonics News*, 10, 28-30.

Lyon, M.F. (1972) X-chromosome inactivation and developmental patterns in mammals. *Biological Review of the Cambridge Philosophical Society*, 47, 1–35.

Lyon, M.F. (1961) Gene action in the X-chromosome of the mouse (*mus Musculus L*). *Nature*, 190, 372.

Lythgoe, J.N. (1972) The adaptation of visual pigments to their photic environment. In *Handbook of Sensory Physiology, volume VII/I, Photochemistry of Vision* (ed. Dartnall, H.J.A.), pp. 566–603. Springer-Verlag, Berlin.

Lythgoe, J.N. (1979) *The Ecology of Vision*. Clarendon, Oxford.

MacAdam, D.L. (1942) Visual sensitivities to color differences in daylight. *Journal of the Optical Society of America*, 32, 247–274.

MacAdam, D.L. (1981) *Color Measurement. Theme and Variations*. Springer-Verlag, Berlin.

Mach, E. (1865) Über die Wirkung der räumlichen Vertheilung des Lichtreizes auf die Netzhaut. *Sitzungsberichte der mathematisch-naturwissenschaf tlichen Classe der kaizerlichen Akademie der Wissenschaften*, 52, 303–322.

Mach, E. (1866a) Über den physiologischen Effect räumlich vertheilter Lichtreize (zweite Abhandlung). *Sitzungsberichte der mathematisch-naturwissenschaftlichen Classe der kaiserlichen Akademie der Wissenschaften,* 54, 131–144.

Mach, E. (1866b) Über die physiologische Wirkung räumlich vertheilter Lichtreize (dritte Abhandlung). *Sitzungsberichte der mathematisch-naturwissenschaftlichen Classe der kaiserlichen Akademie der Wissenschaften,* 54, 393–408.

Mach, E. (1868) Über die physiologische Wirkung räumlich vertheilter Lichtreize (vierte Abhandlung). *Sitzungsberichte der mathematisch-naturwissenschaftlichen Classe der kaiserlichen Akademie der Wissenschaften,* 57, 11–19.

Macke, J.P. & Nathans, J. (1997) Individual variation in the size of the human red and green visual pigment gene array. *Investigative Ophthalmology & Visual Science*, 38, 1040–1043.

MacLeod, D.I.A. & Boynton, R.M. (1979) Chromaticity diagram showing cone excitation by stimuli of equal luminance. *Journal of the Optical Society of America,* 69, 1183–1186.

MacLeod, D.I.A. & Hayhoe, M. (1974) Three pigments in normal and anomalous color vision. *Journal of the Optical Society of America*, 64, 92–96.

MacLeod, D.I.A. & He, S. (1993) Visible flicker from invisible patterns. *Nature*, 361, 256–258.

MacLeod, D.I.A. & Lennie, P. (1976) Red-green blindness confined to one eye. *Vision Research*, 16, 691–702.

MacLeod, D.I.A. & Webster, M.A. (1983) Factors influencing the color matches of normal observers. In *Colour Vision: Physiology and Psychophysics* (eds. Mollon, J.D. & Sharpe, L.T.), pp. 81–92. Academic Press, London.

MacLeod, D.I.A., Williams, D.R., & Makous, W.A. (1992) A visual nonlinearity fed by single cones. *Vision Research*, 32, 347–363.

MacNichol, E.F. (1986) A unifying presentation of photopigment spectra. *Vision Research,* 26, 1543–1556.

MacNichol, E.F., Levine, J.S., Mansfield, R.J.W., Lipetz, L.E., & Collins, B.A. (1983) Microspectrophotometry of visual pigments in primate photoreceptors. In *Colour vision: Physiology and Psychophysics* (eds. Mollon, J.D. & Sharpe, L.T.), pp. 13–38. Academic Press, London.

Maddox, J.J. (1970) *Elements of Functional Analysis*. Cambridge University Press, Cambridge.

Maloney, L.T. (1984) Computational approaches to color constancy. Dissertation; Stanford University. Reprinted (1985) as *Stanford Applied Psychology Laboratory Report* 1985–01.

Maloney, L.T. (1986) Evaluation of linear models of surface spectral reflectance with small numbers of parameters. *Journal of the Optical Society of America A*, 3, 1673–1683.

Maloney, L.T. (1992) Color constancy and color perception: The linear-models framework. In *Attention and Performance XIV: Synergies in Experimental Psychology, Artificial Intelligence, and Cognitive Neuroscience – A Silver Jubilee* (eds. Meyer, D.E. & Kornblum, S.), pp. 59–78. MIT Press, Cambridge, MA.

Maloney, L.T. (1998) Surface spectral reflectance: Models and evaluation. In *Colour Vision: From Light to Object* (eds. Mausfeld, R. & Heyer, D.) (in preparation).

Maloney, L.T. & Landy, M.S. (1989) A statistical framework for robust fusion of depth information. In *Visual Communications and Image Processing, IV Proceedings of the SPIE* (ed. Pearlman, W.A.), p. 1199, 1154–1163.

Maloney, L.T. & Varner, D.C. (1986) Chromatic adaptation, the control of chromatic adaptation, and color constancy (abstract). *Optics News*, 12, 134.

Maloney, L.T. & Wandell, B.A. (1986) Color constancy: A method for recovering surface spectral reflectance. *Journal of the Optical Society of America A*, 3, 29–33.

Mansfield, R.J.W. (1985) Primate photopigments and cone mechanisms. In *The Visual System* (eds. Fein, A. & Levine, J.S.), pp. 89–106. Alan R. Liss, New York.

Marc, R. & Sperling, H.G. (1997) Chromatic organization of primate cones. *Science*, 196, 454–456.

Mardia, K.V., Kent, J.T., & Bibby, J.M. (1979) *Multivariate Analysis*. Academic Press, London.

Mariani, A.P. (1983) The neuronal organization of the outer plexiform layer of the primate retina. *International Review of Cytolology,* 86, 285–320.

Mariani, A.P. (1984) Bipolar cells in monkey retina selective for the cone likely to be blue-sensitive. *Nature,* 308, 184–186.

Marimont, D. & Wandell, B.A. (1992) Linear models of surface and illuminant spectra. *Journal of the Optical Society of America A*, 9, 1905–1913.

Marshak, D.W., Aldrich, L.B., Del Valle, J., & Yamada, T. (1990) Localization of immunoreactive cholecystokinin precursor to amacrine cells and bipolar cells of the macaque monkey retina. *Journal of Neuroscience,* 10, 3045–3055.

Martin, R.D. (1990) *Primate Origins and Evolution. A Phylogenetic Reconstruction.* Chapman and Hall Ltd., London.

Martin, R.D. (1993) Primate origins: plugging the gaps. *Nature,* 363, 223–234.

Martin, P.R. & Grünert, U. (1992) Spatial density and immunoreactivity of bipolar cells in the macaque monkey retina. *Journal of Comparative Neurology,* 323, 269–287.

Martin, P.R., White, A.J.R., Goodchild, A.K., Wilder, H.D., & Sefton, A.E. (1997) Evidence that blue-on cells are part of the third genicolocortical pathway in primates. *European Journal of Neuroscience,* 9, 1536–1541.

Masland, R.H. (1988) Amacrine cells. *Trends in Neurosciences,* 11, 405–410.

Masland, R.H. (1996) Processing and encoding of visual information in the retina. *Current Opinion in Neurobiology*, 6, 467–474.

Massey, S.C. (1990) Cell types using glutamate as a neurotransmitter in the vertebrate retina. In *Progress in Retinal Research Volume 9* (eds. Osborne, N.N. & Chader, G.), pp. 399–425. Pergamon Press, London.

Massof, R.W. & Bailey, J.E. (1976) Achromatic points in protanopes and deuteranopes. *Vision Research*, 16, 53–57.

Mastrangelo, I.A., Courey, A.J., Wall, J.S., Jackson, S.P., & Hough, P.V. (1991) Looping and Sp1 multimer links: A mechanism for transcriptional synergism and enhancement. *Proceedings of the National Academy of Science USA*, 88, 5670–5674.

Mathies, R. & Stryer, L. (1976) Retinal has a highly dipolar vertically excited singlet state: implications for vision. *Proceedings of the National Academy of Science, USA,* 73, 2169–2173.

Maumenee, I.H., Li, Y., Hurd, J.R., Mitchell, T.N., & Zhu, D. (1998). Achromatopsia in the Pinelapese maps to chromosome 8. *Investigative Ophthalmology & Visual Science (Suppl.)*, 39, S297 (withdrawn at the meeting).

Maunsell, J.H., Nealey, T.A., & DePriest, D.D. (1990) Magnocellular and parvocellular contributions to responses in the middle temporal visual area (MT) of the macaque monkey. *Journal of Neuroscience*, 10, 3323–3334.

Mausfeld, R. (1998) Colour perception: From Grassman codes to a dual code for object and illuminant colours. In *Color Vision: Perspectives from different Disciplines* (eds. Backhaus, W.G.K., Kliegl, R., & Werner, J.S.). Walter de Gruyter, Berlin, New York.

Maxwell, J.C. (1855) Experiments on colours, as perceived by the eye, with remarks on colour-blindness. *Transactions of the Royal Society of Edinburgh,* 21, 275–298.

Maxwell, J.C. (1856) On the theory of colours in relation to colour-blindness. A letter to Dr. G. Wilson. *Transactions of the Royal Scottish Society of Arts,* 4, 394–400.

Maxwell, J.C. (1860) On the theory of compound colours and the relations of the colours of the spectrum. *Philosophical Transactions of the Royal Society, London,* 150, 57–84.

McClurkin, J.W. & Optican, L.M. (1996) Primate striate and prestriate cortical neurons during discrimination. I. Simultaneous temporal encoding of information about color and pattern. *Journal of Neurophysiology*, 75, 481–495.

McClurkin, J.W., Zarbock, J.A., & Optican, L.M. (1996) Primate striate and prestriate cortical neurons during discrimination. II. Separable temporal codes for color and pattern. *Journal of Neurophysiology*, 75, 496–507.

McCollough, C. (1965) Color adaptation of edge detectors in the human visual system. *Science*, 149, 1115–1116.

McCourt, M.E. (1982) A spatial frequency dependent grating-induction effect. *Vision Research,* 72, 119–134.

McCree, K.J. (1960) Small-field tritanopia and the effects of voluntary fixation. *Optica Acta*, 7, 317–323.

McGuire, B.A., Stevens, J.K., & Sterling, P. (1984) Microcircuitry of bipolar cells in the cat retina. *Journal of Neuroscience,* 4, 2920–2938.

McIlhagga, W., Hine, T., Cole, G.R., & Snyder, A.W. (1990) Texture segregation with luminance and chromatic contrast. *Vision Research*, 30, 489–495.

McKee, S.P., Silverman, G.H., & Nakayama, K. (1986) Precise velocity discrimination despite random variation in temporal frequency and contrast. *Vision Research*, 26, 609–619.

McLellan, J.S. (1997) *Cone-opponent effects on S-cone increment and decrement detection*. Ph.D. Dissertation, Northeastern University, unpublished.

McMahon, M.J., Lankheet, M.J.M., Lennie, P., & Williams, D.R. (1995) Fine strucuture of P-cell receptive fields in the fovea, revealed by laser interferometry. *Investigative Ophthalmology & Visual Science (Suppl.)*, 36, 4.

McMahon, M.J. & MacLeod, D.I.A. (1998) Dichromatic color vision at high light levels: red/green discrimination using the blue-sensitive mechanism. *Vision Research*, 38, 973-83.

McNaughton, P.A. (1990) Light response of vertebrate photoreceptors. *Physiological Reviews,* 70, 847-883.

Meadows, J.C. (1974) Disturbed perception of colours associated with localized cerebral lesions. *Brain*, 97, 615–632.

Meagher, M.J., Jørgensen, A.L., & Deeb, S.S. (1996) Sequence and evolutionary history of the length polymorphism in intron 1 of the human red photopigment gene. *Journal of Molecular Evolution*, 43, 622–630.

Merbs, S.L. & Nathans, J. (1992a) Absorption spectra of human cone pigments. *Nature*, 356, 433–435.

Merbs, S.L. & Nathans, J. (1992b) Absorption spectra of the hybrid pigments responsible for anomalous color vision. *Science*, 258, 464–466.

Merbs, S.L. & Nathans, J. (1993) Role of hydroxyl-bearing amino acids in differentially tuning the absorption spectra of the human red and green cone pigments. *Photochemistry and Photobiology*, 58, 706–710.

Merigan, W.H. (1989) Chromatic and achromatic vision of macaques: role of the P pathway. *Journal of Neuroscience,* 9, 776–783.

Merigan, W.H., Byrne, C., & Maunsell, J.H.R. (1991) Does primate motion vision depend on the magnocellular pathway? *Journal of Neuroscience*, 11, 3422–3429.

Merigan, W.H. & Maunsell, J.H.R. (1990) Macaque vision after magnocellular lateral geniculate lesions. *Visual Neuroscience*, 5, 347–352.

Merigan, W.H. & Maunsell, J.H. (1993) How parallel are the primate visual pathways? *Annual Review of Neuroscience*, 16, 369–402.

Merighi, A., Raviola, E., & Dacheux, R.F. (1996) Connections of two types of flat cone bipolars in the rabbit retina. *Journal of Comparative Neurology,* 371, 164–178.

Metelli, F. (1974) The perception of transparency. *Scientific American,* 230, 91–98.

Metha, A.B., Vingrys, A.J., & Badcock, D.R. (1994) Detection and discrimination of moving stimuli: The effects of color, luminance and eccentricity. *Journal of the Optical Society of America*, 11, 1697–1709

Michael, C.R. (1969) Retinal processing of visual images. *Scientific American,* 220, 104–114.

Michael, C.R. (1978a) Color vision mechanisms in monkey striate cortex: Dual-opponent cells with concentric receptive fields. *Journal of Neurophysiology*, 41, 572–588.

Michael, C.R. (1978b) Color vision mechanisms in monkey striate cortex: Simple cells with dual opponent-color concentric receptive fields. *Journal of Neurophysiology*, 41, 1233–1249.

Michael, C.R. (1978c) Color-sensitive complex cells in monkey striate cortex. *Journal of Neurophysiology*, 41, 1250–1266.

Michael, C.R. (1979) Color-sensitive hypercomplex cells in monkey striate cortex. *Journal of Neurophysiology*, 42, 726–744.

Milam, A.H., Dacey, D.M., & Dizhoor, A.M. (1993) Recoverin immunoreactivity in mammalian cone bipolar cells. *Visual Neuroscience,* 10, 1–12.

Miller, D.T., Williams, D.R., Morris, G.M., & Liang, J. (1996) Images of cone photoreceptors in the living human eye. *Vision Research*, 36, 1067–1079.

Miller, S.S. (1972) Psychophysical estimates of visual pigment densities in red-green dichromats. *Journal of Physiology,* 223, 89–107.

Mills, S.L. & Massey, S.C. (1992) Morphology of bipolar cells labeled by DAPI in the rabbit retina. *Journal of Comparative Neurology,* 321, 133–149.

Milunsky, A., Huang, X., Milunsky, J., DeStefano, A., & Baldwin, C. (1998) A second genetic locus for autosomal recessive achromatopsia (ARA). *American Journal of Human Genetics*, 63, supplement, A1740.

Minnaert, M.G.J. (1993) *Light and Color in the Outdoors* (trans. L. Seymour). Springer-Verlag, New York.

Missotten, L. (1965) *The Ultrastructure of the Human Retina*. Editions Arscia S.A, Bruxelles.

Mitchell, D.E. & Rushton, W.A.H. (1971) Visual pigments in dichromats. *Vision Research,* 11, 1033–1043.

Miyahara, E., Pokorny, J., Smith, V.C., Baron, R., & Baron, E. (1998) Color vision in two observers with highly based LWS/MWS cone ratios. *Vision Research*, 38, 601–612.

Miyake, Y., Yagasaki, K., & Ichikawa, H. (1985) Differential diagnosis of congenital tritanopia and dominantly inherited juvenile optic atrophy. *Archives of Ophthalmology,* 103, 1496–1501.

Modarres, M., Mirsamadi, M., & Peyman, G.A. (1996–1997) Prevalence of congenital color deficiences in secondary-school students in Tehran. *International Ophthalmology,* 20, 221–222.

Möller-Ladekarl, P. (1934) Über Farbendistinktion bei Normalen und Farbenblinden. *Acta Ophthalmogica (supplement)*, 3, 1–128.

Mollon, J.D. (1982) Color Vision. *Annual Review of Psychology,* 33, 41–85.

Mollon, J.D. (1987) On the origins of polymorphisms. In *Frontiers of visual science*, pp. 160–168. National Academy Press, Washington, DC.

Mollon, J.D. (1989) "Tho' she kneel'd in that place where they grew..." the uses and origins of primate colour vision. *Journal of Experimental Biology,* 146, 21–38.

Mollon, J.D. (1991) Uses and evolutionary origins of primate colour vision. In *Evolution of the Eye and Visual System* (eds. Cronly-Dillon, J. & Gregory, R.L.), pp. 306–319. The Macmillan Press, Houndsmills, Basingstoke, Hampshire, UK.

Mollon, J.D. (1996) The evolution of trichromacy: an essay to mark the bicentennial of Thomas Young's graduation in Göttingen. In *Brain and Evolution. Proceedings of the 24th Göttingen Neurobiology Conference 1996, Volume I* (eds. Elsner, N. & Schnitzler, H.-U.), pp. 124–139. Georg Thieme Verlag, Stuttgart.

Mollon, J.D. (1997) "...aus dreyerley Arten von Membranen oder Mokekulen": George Palmer's legacy. In *Colour Vision Deficiencies XIII* (ed. Cavonius, C.R.), pp. 3–20. Kluwer, Dordrecht.

Mollon, J.D. & Bowmaker, J.K. (1992) The spatial arrangement of cones in the primate fovea. *Nature*, 360, 677–679.

Mollon, J.D., Bowmaker, J.K., & Jacobs, G.H. (1984) Variations of colour vision in a new world primate can be explained by polymorphism of retinal photopigments. *Proceedings of the Royal Society, London, Series B,* 222, 373–399.

Mollon, J.D. & Cavonius, C.R. (1987) The chromatic antagonisms of opponent process theory are not the same as those revealed in studies of detection and discrimination. In *Colour Vision Deficiencies VIII* (ed. Verriest, G.), pp. 473–483. Junk, Dordrecht.

Mollon, J.D. & Estévez, O. (1988) Tyndall's paradox of hue discrimination. *Journal of the Optical Society of America A*, 5, 151–159.

Mollon, J.D., Estévez, O., & Cavonius, C.R. (1990) The two subsystems of colour vision and their rôles in wavelength discrimination. In *Vision: Coding and Efficiency* (ed.

Blakemore, C.), pp. 119–131. Cambridge University Press, Cambridge.

Mollon, J.D. & Jordan, G. (1988) Eine evolutionäre Interpretation des menschlichen Farbensehens. *Die Farbe,* 35/36, 139–170.

Mollon, J.D., Newcombe, F., Polden, P.G., & Ratcliff, G. (1980) On the presence of three cone mechanisms in a case of total achromatopsia. In *Color vision deficiencies V,* Chapter 3 (ed. Verriest, G.), pp. 130–135. Adam Hilger, Bristol.

Mollon, J.D. & Polden, P.G. (1977a) An anomaly in the response of the eye to light of short wavelengths. *Philosophical Transactions of the Royal Society, London, Series B*, 278, 207–240.

Mollon, J.D. & Polden, P.G. (1977b) Saturation of a retinal cone mechanism. *Nature*, 265, 243–246.

Monge, G. (1789) Memoire sur quelques phenomenes de la vision. *Annales de Chimie*, 3, 131–147.

Montag, E.D. (1994) Surface color naming in dichromats. *Vision Research*, 16, 2137–2151.

Morel, A. & Bullier, J. (1990) Anatomical segregation of two cortical visual pathways in the macaque monkey. *Visual Neuroscience*, 4, 555–578.

Moreland, J.D. (1984) Analysis of variance in anomaloscope equations. In *Colour Vision Deficiencies VII* (ed. Verriest, G.), Documental Ophthalmologica Proceedings Series 39, pp. 111–119. Dr. W. Junk, The Hague.

Moreland, J.D. & Kerr, J. (1979) Optimization of a Rayleigh-type equation for the detection of tritanomaly. *Vision Research,* 19, 1369–1375.

Moreland, J.D. & Roth, A. (1987) Validation trials on an optimum blue-green equation. In *Colour Vision Deficiencies VIII* (ed. Verriest, G.), Documenta Ophthalmologica Proceedings Series 46, pp. 233–236. Nijhoff-Junk, Dordrecht.

Morgan, M.J. & Cleary, R.F. (1992) Effects of colour substitutions upon motion detection in spatially random patterns. *Vision Research,* 32, 815–821.

Morgan, M.J. & Ingle, G. (1994) What direction of motion do we see if luminance but not colour contrast is reversed during displacement? Psychophysical evidence for a signed-colour input to motion detection. *Vision Research,* 34, 2527–2535.

Movshon, J.A. & Newsome, W.T. (1996) Visual response properties of striate cortical neurons projecting to area MT in macaque monkeys. *Journal of Neuroscience*, 16, 7733–7741.

Movshon, J.A., Thompson, I.D., & Tolhurst, D.J. (1978) Receptive field organization of complex cells in the cat's striate cortex. *Journal of Physiology*, 283, 79–99.

Movshon, J.A., Adelson, E.H., Gizzi, M.S., & Newsome, W.T. (1985) The analysis of moving visual patterns. In *Pattern Recognition Mechanisms* (eds. Chagas, C., Gattass, R., & Gross, C.), pp. 117–151, Vatican Press, Rome.

Mullen, K.T. (1985) The contrast sensitivity of human colour vision to red-green and blue-yellow chromatic gratings. *Journal of Physiology,* 359, 381–400.

Mullen, K.T. (1991) Colour vision as a post-receptoral specialization of the central visual field. *Vision Research*, 31, 119–130.

Mullen, K.T. & Baker, C.L. (1985) A motion aftereffect form an isoluminant stimulus. *Vision Research*, 25, 685–688.

Mullen, K.T. & Boulton, J.C. (1992) Absence of smooth motion perception in color vision. *Vision Research*, 32, 483–488.

Mullen, K.T., Cropper, S.J., & Losada, M.A. (1997) Absence of linear subthreshold summation between red-green and luminance mechanisms over a wide range of spatio-temporal conditions. *Vision Research*, 37, 1157–1165.

Mullen, K.T. & Kingdom, F.A.A. (1996) Losses in peripheral colour sensitivity predicted from "hit and miss" postreceptoral cone connections. *Vision Research*, 36, 1995–2000.

Mullen, K.T. & Losada, M.A. (1994) Evidence for separate pathways for color and luminance detection mechanisms. *Journal of the Optical Society of America A*, 11, 3136–3151.

Müller, G.E. (1924) *Darstellung und Erklärung der verschiedenen Typen der Farbenblindheit.* Vandenhoed & Ruprecht, Göttingen.

Müller, G.E. (1930a) Über die Farbenempfindungen. *Zeitschrift für Psychologie und Physiologie der Sinnesorgane, Ergänzungsband*, 17, 1–430.

Müller, G.E. (1930b) Über die Farbenempfindungen. *Zeitschrift für Psychologie und Physiologie der Sinnesorgane, Ergänzungsband*, 18, 435–647.

Müller, J.R. & Lennie, P. (1995) A unified model of chromatic induction and habituation. *Investigative Ophthalmology & Visual Science (Suppl.)*, 36, 1832.

Nachmias, J. & Kocher, E.C. (1970) Visual detection and discrimination of luminance increments. *Journal of the Optical Society of America*, 60, 382–389.

Nachmias, J. & Sansbury, R.V. (1974) Grating contrast: discrimination may be better than detection. *Vision Research*, 14, 1039–1042.

Nagel, W.A. (1905) Dichromatische Fovea, trichromatische Peripherie. *Zeitschrift für Psychologie und Physiologie des Sinnesorgane*, 39, 93–101.

Nagel, W.A. (1907) Neue Erfahrungen über das Farbensehen der Dichromaten auf großem Felde. *Zeitschrift für Sinnesphysiologie*, 41, 319–337.

Nagy, A.L. (1980) Large-field substitution Rayleigh matches of dichromats. *Journal of the Optical Society of America*, 70, 778–784.

Nagy, A.L. & Boynton, R.M. (1979) Large-field color naming of dichromats with rods bleached. *Journal of the Optical Society of America*, 69, 1259–1265.

Nagy, A.L., Eskew, R.T. Jr., & Boynton, R.M. (1987) Analysis of color-matching ellipses in a cone-excitation space. *Journal of the Optical Society of America A*, 4, 756–768.

Nagy, A.L., MacLeod, D.I.A., Heynemann, N.E., & Eisner, A. (1981) Four cone pigments in women heterozygous for color deficiency. *Journal of the Optical Society of America*, 71, 719–722.

Nagy, A.L. & Purl, K.F. (1987) Color discrimination and neural coding in color deficients. *Vision Research*, 27, 483–489.

Naka, K.I. & Rushton, W.A.H. (1966) S-potentials from colour units in the retina of fish (*Cyprinidae*). *Journal of Physiology*, 185, 536–555.

Nakamura, H., Gattass, R., Desimone, R., & Ungerleider, L.G. (1993) The modular organization of projections from areas V1 and V2 to areas V4 and TEO in macaques. *Journal of Neuroscience*, 13, 3681–3691.

Napier, J.R. & Napier, P.H. (1985) *The Natural History of the Primates*. British Museum (Natural History), London.

Nassau, K. (1983) *The Physics and Chemistry of Color: The Fifteen Causes of Color*. Wiley, New York.

Nathans, J. (1987) Molecular biology of visual pigments. *Annual Review of Neuroscience*, 10, 163–194.

Nathans, J., Davenport, C.M., Maumenee, I.H., Lewis, R.A., Hejtmancik, J.F., Litt, M., Lovrien, E., Weleber, R., Bachynski, B., Zwas, F., Klingaman, R., & Fishman, G. (1989) Molecular genetics of human blue cone monochromacy. *Science*, 245, 831–838.

Nathans, J., Maumenee, I.H., Zrenner, E., Sadowski, B., Sharpe, L.T., Lewis, R.A., Hansen, E., Rosenberg, T., Schwartz, M., Heckenlively, J.R., Traboulsi, E., Klingaman, R., Bech-Hansen, N.T., LaRoche, G.R., Pagon, R.A., Murphey, W.H., & Weleber, R.G. (1993) Genetic heterogeneity among blue-cone monochromats. *American Journal of Human Genetics*, 53, 987–1000.

Nathans, J., Piantanida, T.P., Eddy, R.L., Shows, T.B., & Hogness, D.S. (1986) Molecular genetics of inherited variation in human color vision. *Science*, 232, 203–210.

Nathans, J., Thomas, D., & Hogness, D.S. (1986) Molecular genetics of human color vision: The genes encoding blue, green and red pigments. *Science*, 232, 193–202.

Navarro, R., Artal, P., & Williams, D.R. (1993) Modulation transfer of the human eye as a function of retinal eccentricity. *Journal of the Optical Society of America A*, 10, 201–212.

Nawy, S. & Copenhagen, D.R. (1987) Multiple classes of glutamate receptor on depolarizing bipolar cells in retina. *Nature*, 325, 56–58.

Nawy, S. & Copenhagen, D.R. (1990) Intracellular cesium separates two glutamate conductances in retinal bipolar cells of goldfish. *Vision Research*, 30, 967–972.

Nayar, S.K. & Oren, M. (1995) Visual appearance of matte surfaces. *Science*, 267, 1153–1156.

Nealey, T.A. & Maunsell, J.H.R. (1994) Magnocellular and parvocellular contributions to the responses of neurons in macaque striate cortex. *Journal of Neuroscience*, 14, 2069–2079.

Neitz, J. & Jacobs, G.H. (1984) Electroretinogram measurements of cone spectral sensitivity in dichromatic monkeys. *Journal of the Optical Society of America*, 1, 1175–1180.

Neitz, J. & Jacobs, G.H. (1986) Polymorphism of the long-wavelength cone in normal human colour vision. *Nature*, 323, 623–625.

Neitz, J. & Jacobs, G.H. (1989) Polymorphism of cone pigments among color normals: Evidence from color matching. In *Colour Vision Deficiencies IX* (eds. Drum, B. & Verriest, G.), 27–34. Kluwer Academic, Dordrecht.

Neitz, J. & Jacobs, G.H. (1990) Polymorphism in normal human color vision and its mechanisms. *Vision Research*, 30, 621–636.

Neitz, J. & Neitz, M. (1994) Color vision defects. In *Molecular Genetics of Inherited Eye Disorders* (eds. Wright, A.F. & Jay, B.), pp. 217–257. Harwood Academic, Chur.

Neitz, M., Kraft, T.W., & Neitz, J. (1998) Expression of L cone pigment gene subtypes in females. *Vision Research*, 38, 3221–3255.

Neitz, M. & Neitz, J. (1995) Numbers and ratios of visual pigment genes for normal red-green color vision. *Science*, 267, 1013–1016.

Neitz, M. & Neitz, J. (1998) Molecular genetics and the biological basis of color vision. In *Color vision: perspectives from different disciplines* (eds. Backhaus, W.G.K., Kliegl, R., & Werner, J.S.) pp. 101–119. Walter de Gruyter, Berlin.

Neitz, M., Neitz, J., & Grishok, A. (1995) Polymorphism in the number of genes encoding long-wavelength-sensitive cone pigments among males with normal color vision. *Vision Research*, 35, 2395–2407.

Neitz, M., Neitz, J., & Jacobs, G.H. (1989) Analysis of fusion gene and encoded photopigment of colour-blind humans. *Nature*, 342, 679–682.

Neitz, M., Neitz, J., & Jacobs, G.H. (1991) Spectral tuning of pigments underlying red-green color vision. *Science*, 252, 971–974.

Neitz, M., Neitz, J., & Jacobs, G.H. (1993) More than three different cone pigments among people with normal color vision. *Vision Research*, 33, 117–122.

Neitz, M., Neitz, J., & Jacobs, G.H. (1995) Genetic basis of photopigment variations in human dichromats. *Vision Research*, 35, 2095–2103.

Neitz, M., Neitz, J., & Kainz, P.M. (1996) Visual pigment gene structure and the severity of color vision defects. *Science*, 274, 801–804.

Nelson, J.H. (1938) Anomalous trichromatism and its relation to normal trichromatisim. *Proceedings of the Physical Society*, 49, 332–356.

Nelson, R. (1977) Cat cones have rod input: A comparison of the response properties of cones and horizontal cell bodies in the retina of the cat. *Journal of Comparative Neurology*, 172, 109–135.

Nelson, R., Famiglietti, E.V. Jr., & Kolb, H. (1978) Intracellular staining reveals different levels of stratification for on- and off-center ganglion cells in cat retina. *Journal of Neurophysiology*, 41, 472–483.

Nelson, R. & Kolb, H. (1983) Synaptic patterns and response properties of bipolar and ganglion cells in the cat retina. *Vision Research*, 23, 1183–1195.

Nerger, J.L., Piantanida, T.P., & Larimer, J. (1993) Color appearance of filled-in backgrounds affects hue cancellation, but not detection thresholds. *Vision Research*, 33, 165–172.

Neuhann, T., Kalmus, H., & Jaeger, W. (1976) Ophthamological findings in the tritans, described by Wright and Kalmus. *Modern Problems in Ophthalmology*, 17, 135–142.

Neumeyer, C. (1981) Chromatic adaptation in the honey bee: Successive color contrast and color constancy. *Journal of Comparative Physiology*, 144, 543–553.

Newsome, W.T., Britten, K.H., Salzman, C.D., & Movshon, J.A. (1991) Neuronal mechanisms of motion perception. *Cold Spring Harbor Symposia on Quantitative Biology*, LV, 697–705.

Newsome, W.T., Wurtz, R.H., Dürsteler, M.R., & Mikami, A. (1985) The deficits in visual motion processing following ibotenic acid lesions of the middle temporal visual area of the macaque monkey. *Journal of Neuroscience*, 5, 825–840.

Noorlander, C. & Koenderink, J.J. (1983) Spatial and temporal discrimination ellipsoids in color space. *Journal of the Optical Society of America*, 73, 1533–1543.

Nordby, K. (1990) Vision in a complete achromat: a personal account. In *Night Vision: Basic, Clinical and Applied Aspects* (eds. Hess, R.F., Sharpe, L.T., & Nordby, K.), Chapter 7, pp. 290–315. Cambridge University Press, Cambridge.

Nordström, S. & Polland, W. (1980) Different expressions of one gene for congenital achromatopsia with amblyopia in northern Sweden. *Human Heredity*, 39, 122–128.

Normann, R.A. & Perlman, I. (1979) The effects of background illumination on the photoresponses of red and green cones. *Journal of Physiology*, 286, 491–507.

Norton, T.T. & Casagrande, V.A. (1982) Laminar organization of receptive-field properties in Lateral Geniculate Nucleus of bush baby (*Galago crassicaudatus*). *Journal of Neurophysiology*, 47, 715–741.

Nunn, B.J., Schnapf, J.L., & Baylor, D.A. (1984) Spectral sensitivity of single cones in the retina of *Macaca fascicularis*. *Nature*, 309, 264–266.

Ogden, T.E. (1974) The morphology of retinal neurons of the owl monkey, *Aotes*. *Journal of Comparative Neurology*, 153, 399–428.

Okano, T., Kojima, D., Fukada, Y., Shichida, Y., & Yoshizawa, T. (1992) Primary structures of chicken cone visual pigments: vertebrate rhodopsins have evolved out of cone visual pigments. *Proceedings of the National Academy of Sciences, USA*, 89, 5932–5936.

Oloff, H. (1935) Über angeborene blauanomale Trichromasie. *Klinische Monatsblätter für Augenheilkunde*, 94, 11–20.

Oprian, D.D., Asenjo, A.B., Lee, N., & Pelletier, S.L. (1991) Design, chemical synthesis, expression of genes for the three human color vision pigments. *Biochemistry*, 30, 11367–11372.

Orban, G.A., Kennedy, H., & Bullier, J. (1986) Velocity sensitivity and direction selectivity of neurons in areas V1 and V2 of the monkey: influence of eccentricity. *Journal of Neurophysiology*, 56, 462–480.

Oren, M. & Nayar, S.K. (1995) Generalization of the Lambertian model and implications for machine vision. *International Journal of Computer Vision*, 14, 227–251.

Osorio, D. & Vorobyev, M. (1996) Colour vision as an adaptation to frugivory in primates. *Proceedings of the Royal Society, London, Series B*, 263, 593–599.

Packer, O., Hendrickson, A.E., & Curcio, C.A. (1989) Photoreceptor topography of the retina in the adult pigtail macaque (*macaca nemestrina*). *Journal of Comparative Neurology*, 288, 165–183.

Packer, O.S., Williams, D.R., & Bensinger, D.G. (1996) Photopigment transmittance imaging of the primate photoreceptor mosaic. *Journal of Neuroscience*, 16, 2251–2260.

Palmer, G. (1777) *Theory of Colours and Vision*. S. Leacroft, London.

Palmer, G. (1786) *Théorie de la Lumière, applicable aux arts, et principalement à la peinture*. Hardouin et Gattey, Paris.

Palmer, J., Mobley, L.A., & Teller, D.Y. (1993) Motion at isoluminance: Discrimination/detection ratios and the summation of luminance and chromatic signals. *Journal of the Optical Society of America A*, 10, 1353–1362.

Parkkinen, J.P.S., Hallikainen, J., & Jaaskelainen, T. (1989) Characteristic spectra of Munsell colors. *Journal of the Optical Society of America A*, 6, 318–322.

Parra, I. & Windle, B. (1993) High resolution visual mapping of stretched DNA by fluorescent hybridization. *Nature Genetics*, 5, 17–21.

Parsons, J.H. (1924) *An Introduction to Colour Vision, 2nd ed.* Cambridge University Press, Cambridge.

Partridge, J.C. & De Grip, W.J. (1991) A new template for rhodopsin (vitamin A_1 based) visual pigments. *Vision Research,* 31, 619-630.

Paulus, W. & Kröger-Paulus, A. (1983) A new concept of retinal colour coding. *Vision Research*, 23, 529–540.

Pease, P.L. (1978) On color Mach bands. *Vision Research*, 18, 751–755.

Pease, P.L., Adams, A.J., & Nuccio, E. (1987) Optical density of human macular pigment. *Vision Research,* 27, 705–710.

Peichl, L. & Wässle, H. (1979) Size, scatter and coverage of ganglion cell receptive field centres in the cat retina. *Journal of Physiology*, 291, 117–141.

Peichl, L. & Wässle, H. (1981) Morphological identification of on- and off-centre brisk transient (Y) cells in the cat retina. *Proceedings of the Royal Society, London, Series B,* 212, 139–156.

Pelli, D.G. (1985) Uncertainty explains many aspects of visual contrast detection and discrimination. *Journal of the Optical Society of America A*, 2, 1508–1532.

Pelli, D.G. (1990) The quantum efficiency of vision. In *Vision: Coding and Efficiency* (ed. Blakemore, C.), pp. 3–24. Cambridge University Press, Cambridge.

Pentao, L., Lewis, R.A., Lebetter, D.H., Patel, P.I., & Lupski, J.R. (1992) Maternal uniparental isodisomy of chromosome 14: Association with autosomal recessive rod monochromasy. *American Journal of Human Genetics*, 50, 690–699.

Perry, V.H. & Cowey, A. (1981) The morphological correlates of X- and Y-like retinal ganglion cells in the retina of the monkey. *Experimental Brain Research,* 43, 226–228.

Perry, V.H., Oehler, R., & Cowey, A. (1984) Retinal ganglion cells that project to the dorsal lateral geniculate nucleus in the macaque monkey. *Neuroscience,* 12, 1101–1123.

Pessoa, L. & Ross, W.D. (1996) A contrast/filling in model of 3-D lightness perception: Benary cross, White's effect and coplanarity. *Investigative Ophthalmology & Visual Science (Suppl.)*, 37, 1066.

Peter, L. (1926) Zur Kenntniss der Vererbung der totalen Farbenblindheit mit besonderer Berücksichtigung der in der Schweiz bis jetzt nachgewiesen Fälle. *Archives der Julius Klaus-Stiftung für Vererbungsforschung*, 2, 143–180.

Peterhans, E. & von der Heydt, R. (1993) Functional Organization of area V2 in the alert macaque. *European Journal of Neuroscience*, 5, 509–524.

Petry, H.M. & Harosi, F.I. (1990) Visual pigments of the tree shrew (*tupaia belangeri*) and greater galago (*galago crassicaudatus*): A microspectrophotometric investigation. *Vision Research,* 30, 839–851.

Philip, J., Andersen, C.H., Dreyer, V., Freiesleben, E., Gurtler, H., Hauge, M., Kissmeyer-Nielsen, F., Nielsen, L.S., Pers, M., Robson, E.B., Svejgaard, A., & Sorensen, B. (1969) Colour vision deficiency in one of two presumably monozygotic twins with secondary amenorrhoea. *Annals of Human Genetics*, 33, 185–195.

Piantanida, T.P. & Gille, J. (1992) Methodology-specific Rayleigh-match distributions. *Vision Research*, 32, 2375–2377.

Pickford, R.W. (1944) Women with colour-blind relatives. *Nature*, 153, 409.

Pickford, R.W. (1947) Sex differences in colour vision. *Nature*, 159, 606–607.

Pickford, R.W. (1949) Colour vision of heterozygotes for sex-linked red-green defects. *Nature*, 163, 804–805.

Pickford, R.W. (1957) Colour vision of achromats' parents. *Nature*, 180, 926–927.

Pickford, R.W. (1959) Some heterozygous manifestations of colour blindness. *British Journal of Physiological Optics*, 16, 83–95.

Pickford, R.W. (1967) Variability and consistency in the manifestation of red-green colour vision defects. *Vision Research,* 7, 65–77.

Picotte, C.J., Stromeyer, C.F. III, & Eskew, R.T. Jr. (1994) The foveal color-match-area effect. *Vision Research*, 34, 1605–1608.

Pitt, F.H.G. (1935) Characteristics of dichromatic vision. Committee on the Physiology of Vision, Report No. 14. Medical Research Council, Special Report Series, No. 200. His Majesty's Stationery Office, London.

Pitt, F.H.G. (1944) Monochromatism. *Nature,* 154, 466–468.

Piçanco-Diniz, C.W., Silveira, L.C.L., Yamada, E.S., & Martin, K.A.C. (1992) Biocytin as a retrograde tracer in the mammalian visual system. *Brazilian Journal of Medical and Biological Research,* 25, 57–62.

Planta, P. von (1928) Die Häufigkeit der angeborenen Farbensinnstörungen bei Knaben und Mädchen und ihre Feststellung durch die üblichen klinischen Proben. *Albrecht von Graefes Archiv für Klinische und Experimentelle Ophthalmologie*, 120, 253–281.

Poirson, A.B. & Maloney, L.T. (1998) Surface color appearance in simple and complex scenes. In *Colour Vision: From Light to Object* (eds. Mausfeld, R. & Heyer, D.) (in preparation).

Poirson, A.B. & Wandell, B.A. (1993) The appearance of colored patterns: Pattern–color separability. *Journal of the Optical Society of America A*, 10, 2458–2471.

Poirson, A.B. & Wandell, B.A. (1996) Pattern-color separable pathways predict sensitivity to simple colored patterns. *Vision Research*, 36, 515–526.

Poirson, A.B., Wandell, B.A., Varner, D.C., & Brainard, D.H. (1990) Surface characterizations of color thresholds. *Journal of the Optical Society of America A*, 7, 783–789.

Pokorny, J., Jin, Q., & Smith, V.C. (1993) Spectral luminosity functions, scalar linearity, and chromatic adaptation. *Journal of the Optical Society of America A*, 10, 1304–1313.

Pokorny, J., Moreland, J.D., & Smith, V.C. (1975) Photopigments in anomalous trichromacy. *Journal of the Optical Society of America*, 65, 1522–1524.

Pokorny, J. & Smith, V.C. (1977) Evaluation of a single-pigment shift model of anomalous trichromacy. *Journal of the Optical Society of America*, 67, 1196–1209.

Pokorny, J. & Smith, V.C. (1987) L/M cone ratios and the null point of the perceptual red/green opponent system. *Die Farbe*, 34, 53–57.

Pokorny, J. & Smith, V.C. (1993) Monochromatic tritan metamers. *Optical Society of America Technical Digest*, 16, 84.

Pokorny, J., Smith, V.C., & Lutze, M. (1988) Aging of the human lens. *Applied Optics*, 26, 1437–1440.

Pokorny, J., Smith, V.C., & Lutze, M. (1989) Heterochromatic modulation photometry. *Journal of the Optical Society of America A*, 6, 1618–1623.

Pokorny, J., Smith, V.C., & Starr, S.J. (1976) Variability of color mixture data – II. The effect of viewing field size on the unit coordinates. *Vision Research*, 16, 1095–1098.

Pokorny, J., Smith, V.C., & Swartley, R. (1970) Threshold measurements of spectral sensitivity in a blue cone monochromat. *Investigative Ophthalmology & Visual Science*, 9, 807–813.

Pokorny, J., Smith, V.C., & Verriest, G. (1979) Congenital color defects. In *Congenital and Acquired Color Vision Defects* (eds. Pokorny, J., Smith, V.C., Verriest, G., & Pinckers, A.J.L.G.), pp. 183–241. Grune & Stratton, New York.

Pokorny, J., Smith, V.C., & Went, L.N. (1981) Color matching in autosomal dominant tritan defect. *Journal of the Optical Society of America*, 71, 1327–1334.

Pokorny, J., Smith, V.C., & Wesner, M. (1991) Variability in cone populations and implications. In *From Pigments to Perception* (eds. Valberg, A. & Lee, B.B.), pp. 23–34. Plenum, New York.

Polyak, S.L. (1941) *The Retina*. University of Chicago Press, Chicago.

Posner, M.I. & Raichle, M.E. (1994) *Images of Mind*. W. H. Freeman, New York.

Post, R.H. (1962) Population differences in red and green color vision deficiency: A review and a query on selection relaxation. *Eugenics Quarterly*, 9, 131–146.

Post, R.H. (1963) "Color-blindness" distribution in Britain, France and Japan. A review with notes on selection relaxation. *Eugenics Quarterly*, 10, 110–118.

Pu, M., Berson, D.M., & Pan, T. (1994) Structure and function of retinal ganglion cells innervating the cat's geniculate wing: An in vitro study. *Journal of Neuroscience*, 14, 4338–4358.

Pugh, E.N. Jr. (1976) The nature of the π_1 mechanism of W. S. Stiles. *Journal of Physiology*, 257, 713–747.

Pugh, E.N. Jr. & Lamb, T.D. (1993) Amplification and kinetics of the activation steps in phototransduction. *Biochimica et Biophysica Acta*, 1141, 111–149.

Pugh, E.N. Jr. & Mollon, J.D. (1979) A theory of the π_1 and π_3 color mechanisms of Stiles. *Vision Research*, 19, 293–312.

Pugh, E.N. & Sigel, C. (1978) Evaluation of the candidacy of the π-mechanisms of Stiles for color-matching fundamentals. *Vision Research*, 18, 317–330.

Purpura, K., Kaplan, E., & Shapley, R.M. (1988) Background light and the contrast gain of primate P and M retinal ganglion cells. *Proceedings of the National Academy of Sciences, USA*, 85, 4534–4537.

Purpura, K., Tranchina, D., Kaplan, E., & Shapley, R.M. (1990) Light adaptation in the primate retina: Analysis of changes in gain and dynamics of monkey retinal ganglion cells. *Visual Neuroscience*, 4, 75–93.

Quick, R.F. Jr. (1974) A vector-magnitude model of contrast detection. *Kybernetik*, 16, 65–67.

Ramachandran, V.S. (1987) Interaction between colour and motion in human vision. *Nature*, 328, 645–647.

Ramachandran, V.S. & Gregory, R.L. (1978) Does colour provide an input to human motion perception? *Nature*, 275, 55–56.

Rao-Mirotznik, R., Harkins, A.B., Buchsbaum, G., & Sterling, P. (1995) Mammalian rod terminal: Architecture of a binary synapse. *Neuron*, 14, 561–569.

Raviola, E. & Gilula, N.B. (1973) Gap junctions between photoreceptor cells in the vertebrate retina. *Proceedings of the National Academy of Sciences, USA*, 70, 1677–1681.

Raviola, E. & Gilula, N.B. (1975) Intramembrane organization of specialized contacts in the outer plexiform layer of the retina. *Journal of Cell Biology*, 65, 192–222.

Rayleigh, Lord (Strutt, R.J.) (1881) Experiments on colour. *Nature*, 25, 64–66.

Rayleigh, Lord (Strutt, R.J.) (1890) *Report of Committee on Colour-Vision*. The Royal Society, London.

Regan, B.C., Viénot, F., Charles-Dominique, P.C., Pefferkorn, S., Simmen, B., Juillot, C., & Mollon, J.D. (1996) The colour signals that fruits present to primates. *Investigative Ophthalmology & Visual Science,* 37, 648.

Reich, D.S., Sanchez-Vives, M., Mukherjee, P., & Kaplan, E. (1994) Response variability of retinal ganglion cells is independent of the synaptic pathway activated and of retinal illumination. *Investigative Ophthalmology & Visual Science (Suppl.),* 35, 2124.

Reichardt, W. (1961) Autocorrelation, a principle for the evaluation of sensory information by the central nervous system. In *Sensory Communication* (ed. W.A. Rosenblith), pp. 303–317, John Wiley, New York.

Reichel, E., Bruce, A.M., Sandberg, M.A., & Berson, E.L. (1989) An electroretinographic and molecular genetic study of X-linked cone degeneration. *American Journal of Ophthalmology,* 108, 540–547.

Reid, R.C. & Shapley, R. (1988) Brightness induction by local contrast and the spatial dependence of assimilation. *Vision Research,* 28, 115–132.

Reid, R.C. & Shapley, R. (1992) Spatial structure of cone inputs to receptive fields in primate lateral geniculate nucleus, *Nature*, 356, 716–718.

Reisbeck, T.E. & Gegenfurtner, K.R. (1998) Effects of contrast and temporal frequency on orientation discrimination for luminance and isoluminant stimuli. *Vision Research*, 38, 1105–1117.

Reitner, A., Sharpe, L.T., & Zrenner, E. (1991) Is colour vision possible with only rods and blue cones? *Nature*, 352, 798–800.

Reyniers, E., Van Thienen, M.-N., Meire, F., de Boulle, K., Devries, K., Kestelijn, P., & Willems, P.J. (1995) Gene conversion between red and defective green opsin gene in blue cone monochromacy. *Genomics*, 29, 323–328.

Robson, J.G. (1966) Spatial and temporal contrast sensitivity functions of the visual system. *Journal of the Optical Society of America,* 56, 1141–1142.

Robson, J.G. (1988) Linear and non-linear operations in the visual system. *Investigative Ophthalmology & Visual Science, (Supplement),* 29, 117.

Rockland, K.S. (1985) A reticular pattern of intrinsic connections in primate area V2 (area 18). *Journal of Comparative Neurology,* 235, 467–478.

Rodieck, R.W. (1973) *The Vertebrate Retina: Principles of Structure and Function.* W.H. Freeman, San Francisco.

Rodieck, R.W. (1988) The Primate Retina. In *Comparative Primate Biology, Vol. 4, Neuroscience,* (ed. Steklis, H.D.), pp. 203–278. Alan R. Liss, New York.

Rodieck, R.W. (1991) Which cells code for color? In *From Pigments to Perception. Advances in Understanding Visual Processes.* (eds. Valberg, A. & Lee, B.B.), pp. 83–93. Plenum Press, New York, London.

Rodieck, R.W., Binmoeller, K.F., & Dineen, J. (1985) Parasol and midget ganglion cells of the human retina. *Journal of Comparative Neurology,* 233, 115–132.

Rodieck, R.W., Brening, R.K., & Watanabe, M. (1993) The origin of parallel visual pathways. In *Contrast sensitivity* (eds. Shapley, R. & Lam, D.M.K.), pp. 117–144. MIT Press, Cambridge, MA.

Rodieck, R.W. & Watanabe, M. (1993) Survey of the morphology of macaque retinal ganglion cells that project to the pretectum, superior colliculus, and parvicellular laminae of the lateral geniculate nucleus. *Journal of Comparative Neurology,* 338, 289–303.

Romero, J., Garciá-Beltrán, A., & Hernández-Andrés, J. (1997) Linear bases for representation of natural and artificial illuminants. *Journal of the Optical Society of America A*, 14, 1007–1014.

Romero, J., García, J.A., Jiménez del Barco, L., & Hita, E. (1993) Evaluation of color-discrimination ellipsoids in two-color spaces. *Journal of the Optical Society of America A*, 10, 827–837.

Romeskie, M. (1978) Chromatic opponent-response functions of anomalous trichromats. *Vision Research*, 18, 1521–1532.

Roorda, A. & Williams, D.R. (1999) The arrangement of the three cone classes in the living human eye. *Nature*, 397, 520–522.

Rosa, M.G.P., Pettigrew, J.D., & Cooper, H.M. (1996) Unusual pattern of retinogeniculate projections in the controversial primate Tarsius. *Brain, Behaviour and Evolution,* 48, 121–129.

Ross, J., Sharpe, L.T., Johnstone, J.R., & Wiese, J. (1999) Rudimentary color vision in a cone monochromat (in preparation).

Rubin, M.L. (1961) Spectral hue loci of normal and anomalous trichromats. *American Journal of Ophthalmology,* 52, 166–172.

Rubin, J.M. & Richards, W.A. (1982) Color vision and image intensities: when are changes material? *Biological Cybernetics*, 45, 215–226.

Rubner, J. & Schultern, K. (1989) A regularized approach to color constancy. *Biological Cybernetics*, 61, 29–36.

Ruddock, K.H. (1965) The effect of age upon colour vision II. Changes with age in light transmission of the ocular media. *Vision Research,* 5, 47–58.

Rudolph, K.K. & Pasternak, T. (1996) Motion and form perception after lesions of areas MT/MST and V4 in macaque. *Investigative Ophthalmology & Visual Science*, 37, S486.

Rushton, W.A.H. (1963) The density of chlorolabe in the foveal cones of a protanope. *Journal of Physiology,* 168, 360–373.

Rushton, W.A.H. (1972) Visual pigments. In *Handbook of Sensory Physiology, Vol. VII/1, Photochemistry of Vision* (ed. Dartnall, H.J.), Chapter 9, pp. 364–394. Springer-Verlag, Berlin.

Rushton, W.A.H. & Baker, H.D. (1964) Red/green sensitivity in normal vision. *Vision Research*, 4, 75–85.

Rushton, W.A.H., Powell, D.S., & White, K.D. (1973) Pigments in anomalous trichromats. *Vision Research*, 13, 2017–2031.

Rüttiger, L., Braun, D.I., Gegenfurtner, K.R., Petersen, D., Schönle, P., & Sharpe, L.T. (1999) Selective color constancy deficits after circumscribed unilateral brain lesions. *Journal of Neuroscience*, 19, 3094–3106.

Sachtler, W. & Zaidi, Q. (1992) Chromatic and luminance signals in visual memory. *Journal of the Optical Society of America A*, 9, 877–894.

Sacks, O. (1997) *The Island of the Colorblind*. Knopf, New York.

Said, F.S. & Weale, R.A. (1959) The variation with age of the spectral transmissivity of the living human crystalline lens. *Gerontologia*, 3, 213–231.

Saito, H., Tanaka, K., Isono, H., Yasuda, M., & Mikami, A. (1989) Directionally selective response of cells in the middle temporal area (MT) of macaque monkey to movement of equiluminous opponent color stimuli. *Experimental Brain Research*, 75, 1–14.

Sakai, K., Watanabe, E., Onodera, Y., Uchida, I., Kato, H., Yamamoto, E., Koizumi, H., & Miyashita, Y. (1995) Functional mapping of the human colour centre with echo-planar magnetic resonance imaging. *Proceedings of the Royal Society, London, Series B*, 261, 89–98.

Salin, P.-A. & Bullier, J. (1995) Corticocortical connections in the visual system: Structure and function. *Physiological Review*, 75, 107–154.

Sällström, P. (1973) Colour and physics: Some remarks concerning the physical aspects of human colour vision. *University of Stockholm: Institute of Physics Report*, 73–09.

Sankeralli, M.J. & Mullen, K.T. (1996) Estimation of the L-, M-, and S-cone weights of the postreceptoral detection mechanisms. *Journal of the Optical Society of America A*, 13, 906–915.

Sanocki, E., Lindsey, D.T., Winderickx, J., Teller, D.Y., Deeb, S.S., & Motulsky, A.G. (1993) Serine/alanine amino acid polymorphism in the L and M cone pigments: Effects on Rayleigh matches among deuteranopes, protanopes and color normal observers. *Vision Research*, 33, 2139–2152.

Sanocki, E., Shevell, S.K., & Winderickx, J. (1994) Serine/Alanine amino acid polymorphism of the L-cone photopigment assessed by dual Rayleigh-type color matches. *Vision Research*, 34, 377–382.

Sastri, V.D.P. & Das, S.R. (1966) Spectral distribution and color of north sky at Delhi. *Journal of the Optical Society of America*, 56, 829.

Sastri, V.D.P. & Das, S.R. (1968) Typical spectra distributions and color for tropical daylight. *Journal of the Optical Society of America*, 58, 391.

Sawatari, A. & Callaway, E.M. (1996) Convergence of magno- and parvocellular pathways in layer 4B of macaque primary visual cortex. *Nature*, 380, 442–446.

Scharff, L.V. & Geisler, W.S. (1992) Stereopsis at isoluminance in the absence of chromatic aberrations. *Journal of the Optical Society of America A*, 9, 868–876.

Scheibner, H.M.O. & Boynton, R.M. (1968) Residual red-green discrimination in dichromats. *Journal of the Optical Society of America*, 58, 1151–1158.

Schein, S.J. & Desimone, R. (1990) Spectral properties of V4 neurons in the macaque. *Journal of Neuroscience*, 10, 3369–3389.

Schein, S.J., Marrocco, R.T., & de Monasterio, F.M. (1982) Is there a high concentration of color-selective cells in area V4 of monkey visual cortex? *Journal of Neurophysiology*, 47, 193–213.

Schiller, P.H. & Colby, C.L. (1983) The responses of single cells in the lateral geniculate nucleus of the rhesus monkey to color and luminance contrast. *Vision Research*, 23, 1631–1641.

Schiller, P., Logothetis, N.K., & Charles, E.R. (1990) Role of color-opponent and broad-band channels in vision. *Visual Neuroscience*, 5, 321–346.

Schiller, P.H., Logothetis, N.K., & Charles, E.R. (1991) Parallel pathways in the visual system: Their role in perception at isoluminance. *Neuropsychologia*, 29, 433–441.

Schirillo, J.A. & Shevell, S.K. (1996) Brightness contrast from inhomogeneous surrounds. *Vision Research*, 36, 1783–1796.

Schmidt, H.-J.A., Sharpe, L.T., Knau, H., & Wissinger, B. (1999) Gene copy number in the human photopigment gene array (in preparation).

Schmidt, I. (1934) Über manifeste Heterozygotie bei Konduktorinnen für Farbensinnstörungen. *Klinische Monatsblätter für Augenheilkunde und augenärztliche Fortbildung*, 92, 456–467.

Schmidt, I. (1936) Über einer Massenuntersuchung des Farbensinnes mit dem Anomaloskop. *Zeitschrift für Bahnärtz*, 31, 44–53.

Schmidt, I. (1955) A sign of manifest heterozygosity in carriers of color deficiency. *American Journal of Optometry*, 32, 404–408.

Schnapf, J.L., Kraft, T.W., & Baylor, D.A. (1987) Spectral sensitivity of human cone photoreceptors. *Nature*, 325, 439–441.

Schnapf, J.L., Kraft, T.W., Nunn, B.J., & Baylor, D.A. (1988) Spectral sensitivity of primate photoreceptors. *Visual Neuroscience*, 1, 255–261.

Schnapf, J.L. & McBurney, R.N. (1980) Light-induced changes in membrane current of cone outer segments of tiger salamander and turtle. *Nature*, 287, 239–241.

Schnapf, J.L., Nunn, B.J., Meister, M., & Baylor, D.A. (1990) Visual transduction in cones of the monkey *Macaca fascicularis*. *Journal of Physiology*, 427, 681–713.

Schneeweis, D.M. & Schnapf, J.L. (1995) Photovoltages of rods and cones in the macaque retina. *Science*, 268, 1053–1056.

Scholes, J.H. (1975) Colour receptors, and their synaptic connexions, in the retina of a cyprinid fish. *Philosophical Transactions of the Royal Society, London, Series B*, 270, 61–118.

Sclar, G., Lennie, P., & DePriest, D.D. (1989) Contrast adaptation in striate cortex of macaque. *Vision Research*, 29, 747–755.

Sclar, G., Maunsell, J.H.R., & Lennie, P. (1990) Coding of image contrast in central visual pathways of macaque monkeys. *Vision Research*, 30, 1–10.

Sekiguchi, N., Williams, D.R., & Brainard, D.H. (1993) Efficiency in detection of isoluminant and isochromatic interference fringes. *Journal of the Optical Society of America A*, 10, 2118–2133.

Sereno, M.I. & Allman, J.M. (1991) Cortical visual areas in mammals. In *The Neural Basis of Visual Function.* (ed. Leventhal, A.G.), pp. 160–172. The Macmillan Press, Houndmills, Basingstoke, Hampshire, UK.

Shaaban, S.A. & Deeb, S.S. (1998) Functional analysis of the promoters of the human red and green visual pigment genes. *Investigative Ophthalmology & Visual Science*, 39, 885–896.

Shafer, S.A. (1985) Using color to separate reflectance components. *Color Research and Applications*, 10, 210–218.

Shapiro, A.G. & Zaidi, Q. (1992) The effects of prolonged temporal modulation on the differential response of color mechanisms. *Vision Research*, 32, 2065–2075.

Shapley, R. (1990) Visual sensitivity and parallel retinocortical channels. *Annual Review of Psychology*, 41, 635–658.

Shapley, R.M. & Brodie, S. (1993) Responses of human ERG to rapid color exchange: implications for M/L cone ratios. *Investigative Ophthalmology & Visual Science*, 34, 911.

Shapley, R. & Enroth-Cugell, C. (1984) Visual adaptation and retinal gain controls. *Progress in Retinal Research*, 3, 263–346.

Shapley, R. & Kaplan, E. (1989) Responses of magnocellular LGN neurons and M retinal ganglion cells to drifting heterochromatic gratings. *Investigative Ophthalmology & Visual Science (Suppl.)*, 30, 323

Shapley, R., Kaplan, E., & Soodak, R. (1981) Spatial summation and contrast sensitivity of X and Y cells in the lateral geniculate nucleus of the macaque. *Nature*, 292, 543–545.

Shapley, R.M. & Lennie, P. (1985) Spatial frequency analysis in the visual system. *Annual Reviews of Neuroscience*, 8, 547–583.

Shapley, R.M. & Perry, V.H. (1986) Cat and monkey retinal ganglion cells and their visual functional roles. *Trends in Neuroscience*, 9, 229–235.

Shapley, R., Reid, R.C., & Kaplan, E. (1991) Receptive field structure of P and M cells in the monkey retina. In *From Pigments to Perception* (eds. Valberg, A. & Lee, B.B.), pp. 95–104. Plenum, New York.

Shapley, R.M. & Victor, J.D. (1978) The effect of contrast on the transfer properties of cat retinal ganglion cells. *Journal of Physiology*, 285, 299–310.

Shapley, R.M. & Victor, J.D. (1979) The contrast gain control of the cat retina. *Vision Research*, 19, 431–434.

Shapley, R.M. & Victor, J.D. (1981) How the contrast gain control modifies the frequency responses of cat retinal ganglion cells. *Journal of Physiology*, 318, 162–179.

Sharpe, L.T. & Nordby, K. (1990a) Total colour-blindness: an introduction. In *Night Vision: Basic, Clinical and Applied Aspects* (eds. Hess, R.F., Sharpe, L.T., & Nordby, K.), Chapter 7, pp. 253–289. Cambridge University Press, Cambridge.

Sharpe, L.T. & Nordby, K. (1990b) The photoreceptors in the achromat. In *Night Vision: Basic, Clinical and Applied Aspects* (eds. Hess, R.F., Sharpe, L.T., & Nordby, K.), Chapter 10, pp. 335–389. Cambridge University Press, Cambridge.

Sharpe, L.T., Stockman, A., Jägle, H., Knau, H., Klausen, G., Reitner, A., & Nathans, J. (1998) Red, green, and red-green hybrid pigments in the human retina: correlations between deduced protein sequences and psychophysically-measured spectral sensitivities. *Journal of Neuroscience*, 18, 10053–10069.

Sharpe, L.T., Stockman, A., Jägle, H., Knau, H., & Nathans, J. (1999) L, M and L–M hybrid cone photopigments in man: deriving λ_{max} from flicker photometric spectral sensitivities. *Vision Research*, 39, in press.

Shatz, C.J. (1996) Emergence of order in visual system development. *Proceedings of the National Academy of Sciences, USA*, 93, 602–608.

Shepard, R.N. (1992) The perceptual organization of colors: An adaptation to regularities of the terrestrial world? In *The Adapted Mind; Evolutionary Psychology and the Generation of Culture* (eds. Barkow, J.H., Cosmides, L.,

& Tooby, J.), pp. 495–531. Oxford University Press, New York.

Sheppard, H. (1920) Foveal adaptation to color. *American Journal of Psychology*, 31, 34–58.

Sherman, S.M. & Koch, C. (1986) The control of retinogeniculate transmission in the mammalian lateral geniculate nucleus. *Experimental Brain Research,* 63, 1–20.

Shevell, S.K. (1978) The dual role of chromatic backgrounds in color perception. *Vision Research*, 18, 1649–1661.

Shevell, S.K., He, J.C., Kainz, P., Neitz, J., & Neitz, M. (1998) Relating color discrimination to photopigment genes in deutan observers. *Vision Research*, 38, 3371–3376.

Shevell, S.K. & Wesner, M.F. (1990) Chromatic adapting effect of an achromatic light. *OSA Technical Digest Series*, 15, 149.

Shipp, S., de Jong, B.M., Zihl, J., Frackowiak, R.S.J., & Zeki, S. (1994) Brain activity related to residual motion vision in a patient with bilateral lesions of V5. *Brain*, 117, 1023–1038.

Shipp, S. & Zeki, S.M. (1985) Segregation of pathways leading from area V2 to areas V4 and V5 of macaque monkey visual cortex. *Nature*, 315, 322–325.

Shipp, S. & Zeki, S. (1989a) The organization of connections between areas V5 and V2 in macaque monkey visual cortex. *European Journal of Neuroscience*, 1, 333–354.

Shipp, S. & Zeki, S.M. (1989b) The organization of connections between areas V5 and V1 in macaque monkey visual cortex. *European Journal of Neuroscience*, 1, 309–322.

Shyue, S.-K., Hewett-Emmett, D., Sperling, H.G., Hunt, D.M., Bowmaker, J.K., Mollon, J.D., & Li, W.-H. (1995) Adaptive evolution of color vision genes in higher primates. *Science,* 269, 1265–1267.

Sigel, C. & Pugh, E.N. (1980) Stiles's π_5 color mechanism: Tests of field displacements and field additivity properties. *Journal of the Optical Society of America*, 70, 71–81.

Silberstein, L. & MacAdam, D.L. (1945) The distribution of color matchings around a color center. *Journal of the Optical Society of America*, 35, 32–39.

Silveira, L.C.L., Lee, B.B., Yamada, E.S., Kremers, J., & Hunt, D.M. (1998) Post-receptoral mechanisms of colour vision in new world primates. *Vision Research*, 38, 3329–3337.

Silveira, L.C.L., Lee, B.B., Yamada, E.S., & Kremers, J. (1997) Morphology and physiology of S-cone pathways in New-World primates. *Investigative Ophthalmology & Visual Science (Suppl.)* 38, 708.

Silveira, L.C.L., Perry, V.H., & Yamada, E.S. (1993) The retinal ganglion cell distribution and the representation of the visual field in area 17 of the owl monkey, *Aotus trivirgatus. Visual Neuroscience,* 10, 887–897.

Silveira, L.C.L., Piçanco-Diniz, C.W., Sampaio, L.F.S., & Oswaldo-Cruz, E. (1989) Retinal ganglion cell distribution in the *Cebus* monkey: A comparison with the cortical magnification factors. *Vision Research*, 29, 1471–1483.

Silveira, L.C.L., Yamada, E.S., Perry, V.H., & Piçanco-Diniz, C.W. (1994) M and P retinal ganglion cells of diurnal and nocturnal New-World monkeys. *NeuroReport,* 5, 2077–2081.

Simmons, D.R. & Kingdom, F.A.A. (1997) On the independence of chromatic and achromatic stereopsis mechanisms. *Vision Research*, 37, 1271–1280.

Singer, B. (1994) *Color Contrast Gain Control.* Ph.D. dissertation, University of California, Irvine.

Singer, B., Montag, E., & D'Zmura, M. (1995) Spatial opponency at isoluminance. *Investigative Ophthalmology & Visual Science (Suppl.),* 36, 664.

Singer, B. & D'Zmura, M. (1994) Color contrast induction. *Vision Research,* 34, 3111–3126.

Singer, B. & D'Zmura, M. (1995) Contrast gain control: A bilinear model for chromatic selectivity. *Journal of the Optical Society of America A,* 12, 667–685.

Sinha, P. & Adelson, E. (1993) Recovering reflectance and illumination in a world of painted polyhedra. In *Fourth International Conference on Computer Vision*, pp. 156–163. IEEE Computer Society Press, Los Alamitos, CA.

Siniscalco, M., Filippi, G., & Latte, B. (1964) Recombination between protan and deutan genes: Data on their relative positions in respect to the G6PD locus. *Nature*, 204, 1062–1064.

Sjoberg, S.A., Neitz, M., Balding, S.D., & Neitz, J. (1998) L-cone pigment genes expressed in normal colour vision. *Vision Research*, 18, 3213–3219.

Sloan, L.L. (1928) The effect of intensity of light, state of adaptation of the eye, and size of photometric field on the visibility curve: A study of the Purkinje phenomenon. *Psychological Monographs*, 38, 1–87.

Sloan, L.L. & Habel, A. (1955) Color signal systems for the red-green color blind. An experimental test of the three-color signal system proposed by Judd. *Journal of the Optical Society of America*, 45, 592–598.

Sloan, L.L. & Wallach, L. (1948) A case of unilateral deuteranopia. *Journal of the Optical Society of America*, 38, 502–509.

Smith, A.T. (1994) The detection of second-order motion. In *Visual Detection of Motion* (eds. Smith, A.T. & Snowden, R.J.), pp. 145–176, Academic Press, London.

Smith, D.P., Cole, B.L., & Isaacs, A. (1973) Congenital tritanopia without neuroretinal disease. *Investigative Ophthalmology*, 12, 608–617.

Smith, N.P. & Lamb, T.D. (1997) The a-wave of the human electroretinogram recorded with a minimally invasive technique. *Vision Research,* 37, 2943–2952.

Smith, R.G. (1987) Montage: a system for three-dimensional reconstruction by personal computer. *Journal of Neuroscience Methods,* 21, 55–69.

Smith, V.C., Lee, B.B., Pokorny, J., Martin, P.R., & Valberg, A. (1992a) Responses of macaque ganglion cells to the relative phase of heterochromatically modulated lights. *Journal of Physiology,* 458, 191–221.

Smith, V.C. & Pokorny, J. (1973) Psychophysical estimates of optical density in human cones. *Vision Research,* 13, 1199–1202.

Smith, V.C. & Pokorny, J. (1975) Spectral sensitivity of the foveal cone photopigments between 400 and 500 nm. *Vision Research,* 15, 161–171.

Smith, V.C. & Pokorny, J. (1977) Large-field trichromacy in protanopes and deuteranopes. *Journal of the Optical Society of America*, 67, 213–221.

Smith, V.C., Pokorny, J., & Swartley, R. (1973) Continuous hue estimation of brief flashes by deueranomalous observers. *American Journal of Psychology,* 86, 115–131.

Smith, V.C., Pokorny, J., & Zaidi, Q. (1983) How do sets of color-matching functions differ? In *Colour Vision: Physiology and Psychophysics* (eds. Mollon, J.D. & Sharpe, L.T.), pp. 93–105. Academic, London.

Smith, V.C., Pokorny, J., Lee, B.B., Kremers, J., & Yeh, T. (1992b) Chromatic adaptation in P-pathway cells. *Investigative Ophthalmology & Visual Science (Suppl.)*, 33, 907.

Smith, V.C., Pokorny, J., Delleman, J.W., Cozijnsen, M., Houtman, W.A., & Went, L.N. (1983) X-linked incomplete achromatopsia with more than one class of functional cones. *Investigative Ophthalmology & Visual Science,* 24, 451–457.

Snodderly, D.M., Brown, P.K., Delori, F.C., & Auran, J.D. (1984) The macular pigment. I. Absorbance spectra, localization, and discrimination form other yellow pigments in primate retina. *Investigative Ophthalmology & Visual Science,* 25, 660–673.

Snyder, A.W. (1975) Photoreceptor optics - theoretical principles. In *Photoreceptor Optics.* (eds. Snyder, A.W. & Menzel, R.), pp. 38–55. Springer-Verlag, Berlin, Heidelburg, New York.

Solomon, J.A., Sperling, G., & Chubb, C. (1993) The lateral inhibition of perceived contrast is indifferent to on-center/off-center segregation, but specific to orientation. *Vision Research,* 33, 2671–2683.

Spehar, B., DeBonet, J.S., & Zaidi, Q. (1996) Brightness Induction from Uniform and Complex Surrounds: A General Model. *Vision Research*, 36, 1893–1906.

Spehar, B., Gilchrist, A.L., & Arend, L.E. (1992) White's illusion: the role of intensity and figural relationships. *Perception (Suppl. 2),* 21, 81.

Spehar, B., Gilchrist, A.L., & Arend, L.E. (1995) The critical role of relative luminance relations in White's effect and grating induction. *Vision Research*, 35, 2603–2614.

Spehar, B. & Zaidi, Q. (1997a) Surround effects on the shape of the temporal contrast sensitivity function. *Journal of the Optical Society of America A*, 14, 2517–2525.

Spehar, B. & Zaidi, Q. (1997b) New configurational effects on perceived contrast and brightness: Second-order White's effects. *Perception*, 26, 409–417.

Speigle, J.M. (1998) Testing whether a common representation mediates the effects of viewing context on color appearance. Unpublished Ph.D. thesis, University of California, Santa Barbara.

Speigle, J.M. & Brainard, D.H. (1996) Is color constancy task independent? *Proceedings of the 4th IS&T/SID Color Imaging Conference*, 167–172.

Speranskaya, N.I. (1959) Determination of spectrum color co-ordinates for twenty-seven normal observers. *Optics and Spectroscopy,* 7, 424–428.

Sperling, G. (1989) Three stages and two systems of visual processing. *Spatial Vision,* 4, 183–207.

Sperling, H.G. (1958) An experimental investigation of the relationship between colour mixture and luminance efficiency. In *Visual Problems of Colour, Volume 1,* pp. 249–277. Her Majesty's Stationery Office, London.

Sperling, H.G. (1960) Case of congenital tritanopia with implications for a trichromatic model of color perception. *Journal of the Optical Society of America*, 50, 156–163.

Sperling, H.G. & Harwerth, R.S. (1971) Red-green cone interaction in the increment-threshold spectral sensitivity of primates. *Science*, 172, 180–184.

Spitzer, H. & Hochstein, S. (1985) Simple- and complex-cell response dependences on stimulation parameters. *Journal of Neurophysiology,* 53, 1244–1265.

Spivey, B.E. (1965) The X-linked recessive inheritance of atypical monochromatism. *Archives of Ophthalmology,* 74, 327–333.

Spivey, B.E., Pearlman, J.T., & Burian, H.M. (1964) Electroretinographic findings (including flicker) in carriers of congenital X-linked achromatopsia. *Documenta Ophthalmogica*, 18, 367–375.

Stafford, D.K. & Dacey, D.M. (1997) Physiology of the A1 amacrine cell: a spiking, axon bearing interneuron of the macaque monkey retina. *Visual Neuroscience*, 14, 507–522.

Stavenga, D.G., Smits, R.P., & Hoenders, B.J. (1993) Simple exponential functions describing the absorbance bands of visual pigment spectra. *Vision Research,* 33, 1011–1017.

Stell, W.K., Ishida, A.T., & Lightfoot, D.O. (1977) Structural basis for on- and off-center responses in retinal bipolar cells. *Science,* 198, 1269–1271.

Sterling, P. (1990) Retina. In *The Synaptic Organization of the Brain* (ed. Shepherd, G.M.), pp. 170–213. Oxford University Press, New York.

Sterling, P., Smith, R.G., Rao, R., & Vardi, N. (1995) Functional architecture of mammalian outer retina and bipolar cells. In *Neurobiology and Clinical Aspects of the Outer Retina* (eds. Djamgoz, M.B.A., Archer, S.N., & Vallerga, S.), pp. 323–348. Chapman & Hall, London.

Stiles, W.S. (1937) The luminous efficiency of monochromatic rays entering the eye pupil at different points and a new colour effect. *Proceedings of the Royal Society, London, Series B,* 123, 90–118.

Stiles, W.S. (1939) The directional sensitivity of the retina and the spectral sensitivity of the rods and cones. *Proceedings of the Royal Society, London, Series B,* 127, 64–105.

Stiles, W.S. (1948) The physical interpretation of the spectral sensitivity curve of the eye. In *Transactions of the Optical Convention of the Worshipful Company of Spectacle Makers*, pp. 97–107. Spectacle Makers' Co., London (Reprinted in Stiles, W.S., 1978, *Mechanisms of Colour Vision*. Academic, London).

Stiles, W.S. (1949) Incremental thresholds and the mechanisms of colour vision. *Documenta Ophthalmologica,* 3, 138–163.

Stiles, W.S. (1953) Further studies of visual mechanisms by the two-colour threshold technique. *Coloquio sobre problemas opticos de la vision,* 1, 65–103.

Stiles, W.S. (1959) Color vision: The approach through increment threshold sensitivity. *Proceedings of the National Academy of Science, USA,* 45, 100–114.

Stiles, W.S. (1961) Adaptation, chromatic adaptation, colour transformation. *Anales Real Sociedad Espanola Fisica e Quimie*, Series A, 57, 149–175.

Stiles, W.S. (1964) Foveal threshold sensitivity on fields of different colors. *Science,* 145, 1016–1018.

Stiles, W.S. (1978) *Mechanisms of Colour Vision*. Academic, London.

Stiles, W.S. & Burch, J.M. (1955) Interim report to the Commission Internationale de l'Éclairage Zurich, 1955, on the National Physical Laboratory's investigation of colour-matching (1955) with an appendix by Stiles, W.S.& Burch, J.M. *Optica Acta,* 2, 168–181.

Stiles, W.S. & Burch, J.M. (1959) NPL colour-matching investigation: Final report (1958). *Optica Acta,* 6, 1–26.

Stiles, W.S., Wyszecki, G., & Ohta, N. (1977) Counting metameric object-color stimuli using frequency-limited spectral reflectance functions. *Journal of the Optical Society of America*, 67, 779.

Stockman, A. & MacLeod, D.I.A. (1987) An inverted S-cone input to the luminance channel: evidence for two processes in S-cone flicker detection. *Investigative Ophthalmology & Visual Science (Suppl.),* 28, 92.

Stockman, A., MacLeod, D.I.A., & DePriest, D.D. (1991) The temporal properties of the human short-wave photoreceptors and their associated pathways. *Vision Research*, 31, 189–208.

Stockman, A., MacLeod, D.I.A., & Johnson, N.E. (1993) Spectral sensitivities of the human cones. *Journal of the Optical Society of America A*, 10, 2491–2521.

Stockman, A., MacLeod, D.I.A., & Vivien, J.A. (1993) Isolation of the middle- and long-wavelength sensitive cones in normal trichromats. *Journal of the Optical Society of America A,* 10, 2471–2490.

Stockman, A. & Mollon, J.D. (1986) The spectral sensitivities of the middle- and long-wavelength cones: An extension of the two-colour threshold technique of W. S. Stiles. *Perception,* 15, 729–754.

Stockman, A. & Sharpe, L.T. (1998) Human cone spectral sensitivities: a progress report. *Vision Research*, 18, 38, 3193–3206.

Stockman, A. & Sharpe, L.T. (2000a) The spectral sensitivities of the middle- and long-wavelength-sensitive cones derived from measurements in observers of known genotype. *Vision Research* (submitted).

Stockman, A. & Sharpe, L.T. (2000b) Tritanopic color matches and the long- and middle-wavelength-sensitive cone spectral sensitivities. *Vision Research* (submitted).

Stockman, A., Sharpe, L.T., & Fach, C.C. (1999) The spectral sensitivity of the human short-wavelength sensitive cones derived from thresholds and color matches. *Vision Research*, 39, 2901-2927.

Stockman, A., Sharpe, L.T., Merbs, S., & Nathans, J. (1999) Spectral sensitivities of human cone visual pigments determined *in vivo* and *in vitro*. *Methods in Enzymology*, in press.

Stone, J. & Johnston, E. (1981) The topography of primate retina: A study of the human, bushbaby, and New- and Old-World monkeys. *Journal of Comparative Neurology,* 196, 205–223.

Stone, L.S. & Thompson, P. (1992) Human speed perception is contrast dependent. *Vision Research*, 32, 1535–1549.

Strang, G. (1988) *Linear Algebra and its Applications*. Harcourt, Brace, Jovanovich, New York.

Strassburg, U. von (1262) De Pulchro, In Grabmann, M. (1926) *Des Ulrich Engelberti O. Pr. (1277) Abhandlung de Pulchro: Untersuchung und Texte.* München: Sitzungsberichte der Bayerischen Akademie der Wissenschaften, Philosophische-historische Klasse, Jg. 1925.

Strege, S., Neitz, M., Kainz, P.M., & Neitz, J. (1996) Cone opsin gene expression in the retinas of two men with two different L-cone pigment genes. *Investigative Ophthalmology & Visual Science*, 37, S338.

Stromeyer, C.F. III, Chaparro, A., Rodriguez, C., Chen, D., Hu, E., & Kronauer, R.E. (1998) Short-wave cone signal in the red-green detection mechanism. *Vision Research*, 38, 813–826.

Stromeyer, C.F. III, Chaparro, A., Tolias, A.S., & Kronauer, R.E. (1997) Colour adaptation modifies the long-wave versus middle-wave cone weights and temporal phases in human luminance (but not red-green) mechanism. *Journal of Physiology*, 499.1, 227–254.

Stromeyer, C.F. III, Cole, G.R., & Kronauer, R.E. (1985) Second-site adaptation in the red-green chromatic pathways. *Vision Research*, 25, 219–237.

Stromeyer, C.F. III, Cole, G.R., & Kronauer, R.E. (1987) Chromatic suppression of cone inputs to the luminance flicker mechanism. *Vision Research*, 27, 1113–1137.

Stromeyer, C.F., III & Lee, J. (1988) Adaptational effects of short wave cone signals on red-green chromatic detection. *Vision Research*, 28, 931–940.

Stromeyer, C.F. III, Lee, J., & Eskew, R.T. Jr. (1992) Peripheral chromatic sensitivity for flashes: A post-receptoral red-green asymmetry. *Vision Research*, 32, 1865–1873.

Stromeyer, C.F. III, Kronauer, R.E., & Madsen, J.C. (1979) Response saturation of short-wavelength cone pathways controlled by color-opponent mechanisms. *Vision Research*, 19, 1025–1040.

Stromeyer, C.F. III, Kronauer, R.E., Ryu, A., Chaparro, A., & Eskew, R.T. Jr. (1995) Contributions of human long-wave and middle-wave cones to motion detection. *Journal of Physiology*, 485.1, 221–243.

Sugihara, K. (1984) An algebraic approach to shape-from-image problems. *Artificial Intelligence*, 23, 59–95.

Sugihara, K. (1987) Use of vertex-type knowledge for range data analysis. In *Three-Dimensional Machine Vision* (ed. Kanade, T.), pp. 267–298. Kluwer, Boston.

Svaetichin, G. & MacNichol, E.F. Jr. (1958) Retinal mechanisms for chromatic and achromatic vision. *New York Academy of Sciences - Annals,* 74, 385–404.

Swain, M.J. & Ballard, D.H. (1991) Color indexing. *International Journal of Computer Vision*, 7, 11–32.

Swanson, W.H. & Fiedelman, M. (1997) Sensitivity and spectral tuning of the red-green pathway in heterozygous carriers of congenital colour vision defect. In *Colour Vision Deficiencies XIII, Documenta Ophthalmologica Proceedings Series (Vol. 59)* (ed. Cavonius, C.R.), pp. 77–86. Kluwer, Dordrecht.

Swanson, W.H. (1993) Chromatic adaptation alters spectral sensitivity at high temporal frequencies. *Journal of the Optical Society of America A*, 10, 1294–1303.

Swanson, W.H., Ueno, T., Smith, V.C., & Pokorny, J. (1987) Temporal modulation sensitivity and pulse-detection thresholds for chromatic and luminance perturbations. *Journal of the Optical Society of America A,* 4, 1992–2005.

Swindale, N.V. (1991) Coverage and design of striate cortex. *Biological Cybernetics*, 65, 415–424.

Switkes, E., Bradley, A., & DeValois, K.K. (1988) Contrast dependence and mechanisms of masking interactions among chromatic and luminance gratings. *Journal of the Optical Society of America A*, 5, 1149–1162.

Szél, A., Diamantstein, T., & Röhlich, P. (1988) Identification of the blue-sensitive cones in the mammalian retina by anti-visual pigment antibody. *Journal of Comparative Neurology,* 273, 593–602.

Szél, A., Röhlich, P., & van Veen, T. (1993) Short-wave sensitive cones in the rodent retinas. *Experimental Eye Research,* 57, 503–505.

Tamura, H., Sato, H., Katsuyama, N., Hata, Y., & Tsumoto, T. (1996) Less segregated processing of visual information in V2 than in V1 of the monkey visual cortex. *European Journal of Neuroscience*, 8, 300–309.

Tanner, W.P. (1961) Physiological implications of psychological data. *Annals of the New York Academy of Sciences*, 89, 752–765.

Tauchi, M. & Masland, R.H. (1984) The shape and arrangement of the cholinergic neurons in the rabbit retina. *Proceedings of the Royal Society, London, Series B,* 223, 101–119.

Taya, R., Ehrenstein, W.H., & Cavonius, C.R. (1995) Varying the strength of the Munker-White's effect by stereoscopic viewing. *Perception*, 24, 685–694.

Taylor, A.H. & Kerr, G.P. (1941) The distribution of energy in the visible spectrum of daylight. *Journal of the Optical Society of America*, 31, 3.

Taylor, W.O.G. (1975) Constructing your own P.I.C. test. *British Journal of Physiological Optics*, 30, 22–24.

Teller, D.Y. (1990) The domain of visual science. In *Visual Perception: The Neurophysiological Foundations* (eds. Spillman, L. & Werner, J.S.), pp. 11–19. Academic Press, San Diego.

Teller, D.Y. & Pugh, E.N. Jr. (1983) Linking propositions in color vision. In *Colour Vision: Physiology and Psychophysics* (eds. Mollon, J.D. & Sharpe, L.T.), pp. 577–589. Academic Press, New York.

Terborgh, J. (1983) *Five New World Primates.* Princeton University Press, New Jersey.

Terstiege, H. (1967) Untersuchungen zum Persistenz- und Koeffizientesatz. *Die Farbe,* 16, 1–120.

Thompson, E. (1995) *Colour Vision: A Study in Cognitive Science and the Philosophy of Perception.* Routledge, London.

Thompson, P. (1982) Perceived rate of movement depends on contrast. *Vision Research*, 22, 377–380.

Thorell, L.G., DeValois, R.L., & Albrecht, D.G. (1984) Spatial mapping of monkey V1 cells with pure color and luminance stimuli. *Vision Research*, 24, 751–769.

Thornton, J.E. & Pugh, E.N. Jr. (1983) Red/green opponency at detection threshold. *Science*, 219, 191–193.

Thulborn, K.R., Waterton, J.C., Matthews, P.M., & Radda, G.K. (1982) Oxygenation dependence of the transferse relation time of water protons in whole blood at high field. *Biochimica et Biophysica Acta*, 714, 265–270.

Toda, Y., Coats, C.L., Kakuk, L.E., Szczesny, J., Bingham, E.L., Ayyagari, R., & Sieving, P.A. (1998) Bilateral macular atrophy in X-linked family with intact red/green color genes but loss of the locus control region. *Investigative Ophthalmology & Visual Science*, 39, S968.

Todorovic, D. (1996) Lightness and junctions. Universität Bielefeld, Zentrum für interdisziplinäre Forschung, *Technical Report*, 26/96.

Tominaga, S. & Wandell, B.A. (1989) The standard surface reflectance model and illuminant estimation. *Journal of the Optical Society of America A*, 6, 576–584.

Tominaga, S. & Wandell, B.A. (1990) Component estimation of surface spectral reflectance. *Journal of the Optical Society of America A*, 7, 312–317.

Torre, V., Matthews, H.R., & Lamb, T.D. (1986) Role of calcium in regulating the cyclic nucleotide cascade of phototransduction in retinal rods. *Proceedings of the National Academy of Sciences, USA*, 83, 7109–7113.

Tovée, M.J. (1994) The molecular genetics and evolution of primate colour vision. *Trends in Neuroscience*, 17, 30–37.

Tovée, M.J., Bowmaker, J.K., & Mollon, J.D. (1992) The relationship between cone pigments and behavioural sensitivity in a new world monkey (*Callithrix jacchus jacchus*). *Vision Research*, 32, 867–878.

Trendelenburg, W. (1941) Ein Anomaloskop zur Untersuchung von Tritoformen der Farbenfehlsichtigkeit mit spektraler Blaugleichung. *Klinische Monatsblätter der Augenheilkunde*, 106, 537–546.

Trezona, P.W. (1973) The tetrachromatic colour match as a colorimetric technique. *Vision Research*, 13, 9–25.

Troilo, D., Howland, H.C., & Judge, S.J. (1993) Visual optics and retinal cone topography in the common marmoset (*Callithrix jacchus*). *Vision Research*, 33, 1301–1310.

Troost, J.M. & DeWeert, C.M. (1991) Naming versus matching in color constancy. *Perception & Psychophysics*, 50, 591–602.

Troscianko, T. & Fahle, M. (1988) Why do isoluminant stimuli appear slower. *Journal of the Optical Society of America A*, 5, 871–880.

Ts'o, D.Y. & Gilbert, C.D. (1988) The organization of chromatic and spatial interactions in the primate striate cortex. *Journal of Neuroscience*, 8, 1712–1727.

Tsukada, M. & Ohta, Y. (1990) An approach to color constancy using multiple images. *Proceedings of the Third International Conference on Computer Vision*, 3, 385–393.

Tsukamoto, Y., Masarachia, P., Schein, S.J., & Sterling, P. (1992) Gap junctions between the pedicles of macaque foveal cones. *Vision Research*, 32, 1809–1815.

Tyler, D.E. (1991) The evolutionary relationships of *Aotus*. *Folia Primatologica*, 56, 50–52.

Unger, V.M. & Schertler, G.F.X. (1995) Low resolution structure of bovine rhodopsin determined by electron cryo-microscopy. *Biophysical Journal*, 68, 1776–1786.

Ungerleider, L.G. & Mishkin, M. (1982) Two cortical visual systems. In *Analysis of Visual Behavior* (eds. Ingle, D.J., Goodale, M.A., & Mansfield, R.J.W.), pp. 549–586. The MIT Press, Cambridge, MA.

Usui, T., Kremers, J., Sharpe, L.T., & Zrenner, E. (1998) Flicker cone electroretinogram in dichromats and trichromats. *Vision Research*, 38, 18, 3391–3396.

Vajoczki, L. & Pease, P.L. (1997) Cone monochromacy: a case report. In *Colour Vision Deficiencies XIII* (ed. Cavonius, C.R.) pp. 283–290. Kluwer Academic, Dordrecht.

Valberg, A. & Lange-Malecki, B. (1990) "Colour constancy" in Mondrian patterns: a partial cancellation of physical chromaticity shifts by simultaneous contrast. *Vision Research*, 30, 371–380.

Valberg, A., Lee, B.B., & Tigwell, D.A. (1986) Neurons with strong inhibitory S-cone inputs in the macaque lateral geniculate nucleus. *Vision Research*, 26, 1061–1064.

Valeton, J.M. & van Norren, D. (1983) Light adaptation of primate cones: an analysis based on extracellular data. *Vision Research*, 23,1539–1547.

van de Merendonk, S. & Went, L.N. (1980) Two cases of inherited deutan and tritan disturbances in the same person, and a study of their family. In *Colour Vision Deficiencies V* (ed. Verriest, G.), pp. 268–272. Adam Hilger, Bristol.

Van Essen, D.C., Anderson, C.H., & Felleman, D.J. (1992) Information processing in the primate visual system: An integrated systems perspective. *Science*, 255, 419–423.

Van Essen, D.C. & Gallant, J.L. (1994) Neural mechanisms of form and motion processing in the primate visual system. *Neuron*, 13, 1–10.

van Heel, L., Went, L.N., & van Norren, D. (1980) Frequency of tritan disturbances in a population study. In *Colour Vision Deficiencies V* (ed. Verriest, G.), pp. 256–260. Adam Hilger, Bristol.

van Norren, D. & Vos, J.J. (1974) Spectral transmission of the human ocular media. *Vision Research,* 14, 1237–1244.

van Norren, D. & Went, L.N. (1981) New test for the detection of tritan defects evaluated in two surveys. *Vision Research*, 21, 1303–1306.

Vanderdonck, R. & Verriest, G. (1960) Femme protanomale et hétérozygote mixte (genes de la protanomalie et de la deuteranopie en position de repulsion) ayant deux fils deuteranopes, un fils protanomal et deux fils normaux. *Biotypologie*, 21, 110–120.

Vaney, D.I. (1990) The mosaic of amacrine cells in the mammalian retina. *Progress in Retinal Research,* 9, 49–100.

Vardi, N., Kaufman, D.L., & Sterling, P. (1994) Horizontal cells in cat and monkey retina express different isoforms of glutamic acid decarboxylase. *Visual Neuroscience*, 11, 135–142

Vardi, N. & Sterling, P. (1994) Subcellular localization of $GABA_A$ receptor on bipolar cells in macaque and human retina. *Vision Research*, 34, 1235–1246.

Vautin, R.G. & Dow, B.M. (1985) Color cell groups in foveal striate cortex of the behaving macaque. *Journal of Neurophysiology*, 54, 273–292.

Verdon, W. & Adams, A.J. (1987) Short-wavelength sensitive cones do not contribute to mesopic luminosity. *Journal of the Optical Society of America A,* 4, 91–95.

Verriest, G. (1972) Chromaticity discrimination in protan and deutan heterozygotes. *Die Farbe*, 21, 7–16.

Verriest, G. & Seki, R. (1965) Nouveaux cartons pseudoisochromatique destine à la reconnaisance d'une vision de type scotopique. In *Tagungsbericht der Internationale Farbtagung Luzern, Band I* (ed. Verriest, G.), pp. 229–239. Musterschmidt Verlag, Göttingen.

Victor, J.D., Maiese, K., Shapley, R., Sidtis, J., & Gazzaniga, M.S. (1989) Acquired central dyschromatopsia: Analysis of a case with preservation of color discrimination. *Clinical Vision Sciences*, 4, 183–196.

Viénot, F. (1983) Can variation in macular pigment account for the variation of colour matches with retinal position? In *Colour Vision: Physiology and Psychophysics* (eds. Mollon, J.D. & Sharpe, L.T.), pp. 107–116. Academic Press, London.

Viénot, F., Brettel, H., Ott, L., Ben M'Barek, A., & Mollon, J.D. (1995) What do colourblind people see? *Nature*, 376, 127–128.

Vierling, O. (1935) *Die Farbensinnprufung bei der deutschen Reichsbahn*. Verlag Bernecker, Melsungen.

Vimal, R.L.P., Pokorny, J., & Smith, V.C. (1987) Appearance of steadily viewed lights. *Vision Research,* 27, 1309–1318.

Vimal, R.L.P., Pokorny, J., Smith, V.C., & Shevell, S.K. (1991) Foveal cone detection statistics. *Vision Research*, 29, 61–78.

Vimal, R.L.P., Smith, V.C., Pokorny, J., & Shevell, S.K. (1989) Foveal cone thresholds. *Vision Research,* 29, 61–78.

Vingrys, A.J. & Metha, A.B. (1997) Psychometric functions in the central visual field. In *Colour Vision Deficiences XIII* (ed. Cavonius, C.R.), pp. 377–384. Kluwer Academic, Dordrecht.

Virsu, V. & Lee, B.B. (1983) Light adaptation in cells of macaque lateral geniculate nucleus and its relation to human light adaptation. *Journal of Neurophysiology*, 50, 864–877.

Voigt, J.H. (1781) Das herrn Giros von Gentilly Muthmassungen über die Gesichtsfehler bey Untersuchung der Farben. *Magazin für das Neueste aus der Physik und Naturgeschichte (Gotha)*, 1, 57–61.

Vollrath, D., Nathans, J., & Davis, R.W. (1988) Tandem array of human visual pigment genes at Xq28. *Science*, 240, 1669–1672.

von Kries, J. (1902/1970) Chromatic adaptation. Selection translated and reprinted in *Sources of Color Science* (ed. MacAdam, D.L.), pp. 109–119. The MIT Press, Cambridge, MA.

von Kries, J. (1905/1970) Influence of adaptation on the effects produced by luminous stimuli. Selection translated and reprinted in *Sources of Color Science* (ed. MacAdam, D.L.), pp. 120–126. The MIT Press, Cambridge, MA.

von Kries, J. (1919) Über einen Fall von einseitiger angeborener Deuteranomalie (Grünschwäche). *Zeitschrift für Sinnesphysiologie*, 50, 137–152.

von Kries, J. & Nagel, W. (1896) Über den Einfluss von Lichtstärke und Adaptation auf das Sehen des Dichromaten (Grünblinden). *Zeitschrift für die Psychologie und Physiologie des Sinnesorgane*, 12, 1–38.

von Wiegand, T.E., Hood, D.C., & Graham, N. (1995) Testing a computational model of light-adaptation dynamics. *Vision Research,* 35, 3037–3051.

Vos, J.J. (1972) *Literature Review of Human Macular Absorption in the Visible and its Consequences for the Cone Receptor Primaries*. Netherlands Organization for Applied Scientific Research, Institute for Perception, Soesterberg, The Netherlands.

Vos, J.J. (1978) Colorimetric and photometric properties of a 2° fundamental observer. *Color Research and Application*, 3, 125–128.

Vos, J.J. & Walraven, P.L. (1971) On the derivation of the foveal receptor primaries. *Vision Research,* 11, 799–818.

Vos, J.J., Estévez, O., & Walraven, P.L. (1990) Improved color fundamentals offer a new view on photometric additivity. *Vision Research,* 30, 936–943.

Votruba, M., Moore, A.T., & Bhattacharya, S.S. (1997) Genetic refinement of the dominant optic atrophy (OPA1) locus to within a 2cM interval of chromosome 3q. *Journal of Medical Genetics*, 43, 117–121.

Vrhel, M.J., Gershon, R., & Iwan, L.S. (1994) Measurement and analysis of object reflectance spectra. *Color Research and Application*, 19, 4–9.

Vrhel, M.J. & Trussel, H.J. (1992) Color correction using principal components. *Color Research and Applications*, 17, 328–338.

Waaler, G.H.M. (1927) Über die Erblichkeitsverhältnisse der verschiedenen Arten von angeborener Rotgrünblindheit. *Zeitschrift für induktive Abstammungs- und Vererbungslehre*, 45, 279–333.

Waardenburg, P.J. (1963a) Colour sense and dyschromatopsia. In *Genetics and Ophthalmology, Vol. 2* (eds. Waardenburg, P.J., Franceschetti, A. & Klein, D.), pp. 1425–1566. Blackwell, Oxford.

Waardenburg, P.J. (1963b) Achromatopsia congenita. In *Genetics and Ophthalmology, Vol. II* (eds. Franceschetti, A. & Klein, D.), pp. 1695–1718. Royal van Gorcum, Assen, The Netherlands.

Wagner, G. & Boynton, R.M. (1972) Comparison of four methods of heterochromatic photometry. *Journal of the Optical Society of America,* 62, 1508–1515.

Wald, G. (1945) Human vision and the spectrum. *Science,* 101, 653–658.

Wald, G. (1964) The receptors of human color vision. *Science,* 145, 1007–1016.

Wald, G. (1966) Defective color vision and its inheritance. *Proceedings of the National Academy of Sciences, USA*, 55, 1347–1363.

Wald, G. (1967) Blue-blindness in the normal fovea. *Journal of the Optical Society of America,* 57, 1289–1301.

Wald, G., Wooten, B.R., & Gilligan, K. (1974) Retinal mocaicism in women heterozygous for red and green color vision defects. *Annual Meeting Program, The Association for Research in Vision and Ophthalmology*, 27.

Walls, G. (1956) The G. Palmer story. *Journal of the Medicine and Allied Science*, XI, 66–96.

Walls, G.L. (1964) Notes on four tritanopes. *Vision Research*, 4, 3–16.

Walls, G.L. & Heath, G.G. (1956) Neutral points in 138 protanopes and deuteranopes. *Journal of the Optical Society of America*, 46, 640–649.

Walls, G.L. & Mathews, R.W. (1952) New means of studying color-blindness and normal foveal color vision. *University of California Publications in Psychology*, 7, 1–172.

Walraven, P.L. (1974) A closer look at the tritanopic confusion point. *Vision Research,* 14, 1339–1343.

Walraven, J. (1976) Discounting the background: The missing link in the explanation of chromatic induction. *Vision Research*, 16, 289–295.

Walraven, P.L. (1993) The Stiles-Crawford effects in normal and anomalous vision. *Optical Society of America Technical Digest,* 3, 118–121.

Walraven, P.L. & Bouman, M.A. (1960) Relation between directional sensitivity and spectral response curves in human cone vision. *Journal of the Optical Society of America,* 50, 780–784.

Walraven, J., Benzschawel, T.L., Rogowitz, B.E., & Lucassen, M.P. (1991) Testing the contrast explanation of color constancy. In *From Pigments to Perception* (eds. Valberg, A. & Lee, B.), pp. 369–378. Plenum, New York.

Wandell, B.A. (1982) Measurement of small color differences. *Psychological Review*, 89, 281–302.

Wandell, B.A. (1985) Color measurement and discrimination. *Journal of the Optical Society of America A*, 2, 62–71.

Wandell, B.A. (1987) The synthesis and analysis of color images. *IEEE Transactions on Pattern Analysis and Machine Intelligence*, PAMI-9, 2–13.

Wandell, B.A. (1995) *Foundations of Vision.* Sinauer & Associates, Sunderland, MA.

Wandell, B.A. & Pugh, E.N. (1980a) Detection of long-duration incremental flashes by a chromatically coded pathway. *Vision Research,* 20, 625–635.

Wandell, B.A. & Pugh, E.N. (1980b) A field additive pathway detects brief-duration, long-wavelength incremental flashes. *Vision Research,* 20, 613–624.

Wang, Y., Macke, J.P., Merbs, S.L., Zack, D.J., Klaunberg, B., Bennett, J., Gearhart, J., & Nathans, J. (1992) A locus control region adjacent to the human red and green visual pigment genes. *Neuron*, 9, 429–440.

Wang, Y., Smallwood, P., Cowan, M., Blesh, D., & Lawler, A. (1999) Mutually exclusive expression of human red and green visual pigment-reporter transgenes occurs at high frequency in murine cone photoreceptors. *Proceedings of the National Academy of Sciences, USA,* 96, in press.

Ware, C. & Cowan, W.B. (1982) Changes in perceived color due to chromatic interactions. *Vision Research*, 22, 1035–1062.

Wässle, H. & Boycott, B.B. (1991) Functional architecture of the mammalian retina. *Physiological Reviews*, 71, 447–480.

Wässle, H., Boycott, B.B., & Röhrenbeck, J. (1989) Horizontal cells in the monkey retina: Cone connections and dendritic network. *European Journal of Neuroscience,* 1, 421–435.

Wässle, H., Grünert, U., Chun, M.-H., & Boycott, B.B. (1995) The rod pathway of the macaque monkey retina: Identification of AII-amacrine cells with antibodies against calretinin. *Journal of Comparative Neurology,* 361, 537–551.

Wässle, H., Grünert, U., Martin, P.R., & Boycott, B.B. (1994) Immunocytochemical characterization and spatial distribution of midget bipolar cells in the macaque monkey retina. *Vision Research,* 34, 561–579.

Wässle, H., Grünert, U., Röhrenbeck, J., & Boycott, B.B. (1990) Retinal ganglion cell density and cortical magnification factor in the primate. *Vision Research,* 30, 1897–1911.

Wässle, H., Yamashita, M., Greferath, U., Grünert, U. & Muller, F. (1991) The rod bipolar cell of the mammalian retina. *Visual Neuroscience,* 7, 99–112.

Watanabe, M. & Rodieck, R.W. (1989) Parasol and midget ganglion cells of the primate retina. *Journal of Comparative Neurology,* 289, 434–454.

Watanabe, A., Smith, V.C., & Pokorny, J. (1997) Psychometric functions for chromatic discriminations. In *Colour Vision Deficiences XIII* (ed. Cavonius, C.R.), pp. 369–376. Kluwer Academic, Dordrecht.

Watson, A.B. (1986) Temporal sensitivity. In *Handbook of Perception and Human Performance* (eds. Boff, K.R., Kaufman, L., & Thomas, J.P.), pp. 6-1–6-43. Wiley, New York.

Watson, A.B. (1990) Algotecture of visual cortex. In *Vision: Coding and efficiency* (ed. Blakemore, C.), pp. 393–410. Cambridge University Press, Cambridge.

Watson, A.B. & Robson, J.G. (1981) Discrimination at threshold: labelled detectors in human vision. *Vision Research,* 21, 1115–1122.

Watson, A.B., Thompson, P.G., Murphy, B.J., & Nachmias, J. (1980) Summation and discrimination of gratings moving in opposite directions. *Vision Research*, 20, 341–347.

Weale, R.A. (1953) Cone-monochromatism. *Journal of Physiology,* 121, 548–569.

Weale, R.A. (1959) Photo-sensitive reactions in foveae of normal and cone-monochromatic observers. *Optica Acta*, 6, 158–174.

Weale, R.A. (1988) Age and the transmittance of the human crystalline lens. *Journal of Physiology,* 395, 577–587.

Webster, M.A. (1992) A reanalysis of λ_{max} variation in the Stiles and Burch 10 degree color matching functions. *Journal of the Optical Society of America A*, 9, 1419–1421.

Webster, M.A. (1996) Human colour perception and its adaptation: topical review. *Network: Computation in Neural Systems*, 7, 587–634.

Webster, M.A., DeValois, K.K., & Switkes, E. (1990) Orientation and spatial-frequency discrimination for luminance and chromatic gratings. *Journal of the Optical Society of America A*, 7, 1034–1049.

Webster, M.A. & MacLeod, D.I.A. (1988) Factors underlying individual differences in the color matches of normal observers. *Journal of the Optical Society of America A,* 5, 1722–1735.

Webster, M.A. & Mollon, J.D. (1991) Changes in colour appearance following post-receptoral adaptation. *Nature*, 349, 235–238.

Webster, M.A. & Mollon, J.D. (1994) The influence of contrast adaptation on color appearance. *Vision Research*, 34, 1993–2020.

Webster, M.A. & Mollon, J.D. (1995) Colour constancy influenced by contrast adaptation. *Nature*, 373, 694–698.

Weinhaus, R., Burke, J., Delori, F., & Snodderly, M. (1995) Comparison of Fluorescein angiography with microvascular anatomy of macaque retinas. *Experimental Eye Research*, 61, 1–16.

Weiss, S., Kremers, J., & Maurer, J. (1998) Interaction between rod and cone signals in responses of lateral geniculate neurons in dichromatic marmosets (*callithrix jacchus*). *Visual Neuroscience*, 15, 931–943.

Weiss, S., Kremers, J., & Zrenner, E. (1995) Cone and rod input of parvocellular lateral geniculate cells in the common marmoset (*Callithrix jacchus*). *Investigative Ophthalmology & Visual Science (Suppl.)* 36, 691.

Weisskopf, V.F. (1968) How light interacts with matter. *Scientific American*, 219, 59–71.

Weitz, C.J., Miyake, Y., Shinzato, K., Montag, E., Zrenner, E., Went, L.N., & Nathans, J. (1992) Human tritanopia associated with two amino acid substitutions in the blue-sensitive opsin. *American Journal of Human Genetics*, 50, 498–507.

Weitz, C.J., Went, L.N., & Nathans, J. (1992) Human tritanopia associated with a third amino acid substitution in the blue-sensitive visual pigment. *American Journal of Human Genetics*, 51, 444–446.

Went, L.N. & Pronk, N. (1985) The genetics of tritan disturbances. *Human Genetics*, 69, 255–262.

Went, L.N., Völker-Dieben, H., & De Vries-De Mol, E.C. (1974) Colour vision, ophthalmological and linkage studies in a pedigree with a tritan degree. *Modern Problems in Ophthalmology*, 13, 272–276.

Werner, A. (1990) *Farbkonstanz bei der Honigbiene, Apis Mellifera.* Doctoral dissertation, Fachbereich Biologie, Freie Universität Berlin.

Werner, J.S. & Walraven, J. (1982) effect of chromatic adaptation on the achromatic locus: The role of contrast, luminance, and background color. *Vision Research*, 22, 929–944.

Werner, J.S. & Wooten, B.R. (1979) Opponent chromatic mechanisms: Relation to photopigments and hue naming. *Journal of the Optical Society of America*, 69, 422–434.

Wesner, M.F., Pokorny, J., Shevell, S.K., & Smith, V.C. (1991) Foveal cone detection statistics in color normals and dichromats. *Vision Research*, 31, 1021–1037.

West, G. & Brill, M.H. (1982) Necessary and sufficient conditions for von Kries chromatic adaptation to give color constancy. *Journal of Mathematical Biology*, 15, 249–258.

White, M. (1979) A new effect of pattern on perceived lightness. *Perception*, 8, 413–416.

Wiesel, T.N. & Hubel, D.H. (1966) Spatial and chromatic interactions in the lateral geniculate body of the rhesus monkey. *Journal of Neurophysiology*, 29, 1115–1156.

Wikler, K.C. & Rakic, P. (1990) Distribution of photoreceptor subtypes in the retina of diurnal and nocturnal primates. *Journal of Neuroscience*, 10, 3390–3401.

Wilder, H.D., Grünert, U., Lee, B.B., & Martin, P.R. (1996) Topography of ganglion cells and photoreceptors in the retina of a New World monkey: the marmoset *Callithrix jacchus*. *Visual Neuroscience*, 13, 335–352.

Williams, D.R. (1985) Aliasing in human foveal vision. *Vision Research*, 25, 195–205.

Williams, D.R. (1990) The invisible cone mosaic. In *Advances in Photoreception: Proceedings of a Symposium on Frontiers of Visual Science*, pp. 135–148. National Academy Press, Washington, D.C.

Williams, A.J., Hunt, D.M., Bowmaker, J.K., & Mollon, J.D. (1992) The polymorphic photopigments of the marmoset: Spectral tuning and genetic basis. *EMBO Journal*, 11, 2039–2045.

Williams, D.R., MacLeod, D.I.A., & Hayhoe, M.M. (1981a) Foveal tritanopia. *Vision Research*, 21, 1341–1356.

Williams, D.R., MacLeod, D.I.A., & Hayhoe, M. (1981b) Punctate sensitivity of the blue sensitive mechanism. *Vision Research*, 21, 1357–1375.

Williams, D.R., Sekiguchi, N., & Brainard, D.H. (1993) Color, contrast sensitivity, and the cone mosaic. *Proceedings of the National Academy of Sciences, USA*, 90, 9770–9777.

Williams, D.R., Sekiguchi, N., Haake, P.W., Brainard, D.H., & Packer, O.S. (1991) The cost of trichromacy for spatial vision. In *From Pigments to Perception* (eds. Valberg, A. & Lee, B.B.) pp. 11–22. Plenum Press, New York.

Wilmer, E.N. (1944) Color of small objects. *Nature*, 153, 774–775.

Wilmer, E.N. & Wright, W.D. (1945) Colour sensitivity of the fovea centralis. *Nature*, 156, 774–775.

Wilson, E.B. (1911) The sex chromosomes. *Archiv für Mikroskopic und Anatomische Enwicklungsmechanik*, 77, 249–271.

Wilson, G. (1845) John Dalton's autopsy. *British Quarterly Review*, 1, 157–160.

Wilson, G. (1855) *Researches on Colour Blindness*. Sutherland-Knox, Edinburgh.

Winderickx, J., Battisti, L., Hibiya, Y., Motulsky, A.G., & Deeb, S.S. (1993) Haplotype diversity in the human red and green opsin genes: evidence for frequent sequence exchange in exon 3. *Human Molecular Genetics*, 2, 1413–1421.

Winderickx, J., Battisti, L., Motulsky, A.G., & Deeb, S.S. (1992a) Selective expression of human X chromosome-linked green opsin genes. *Proceedings of the National Academy of Sciences, USA*, 89, 9710–9714.

Winderickx, J., Lindsey, D.T., Sanocki, E., Teller, D.Y., Motulsky, A.G., & Deeb, S.S. (1992b) Polymorphism in red photopigment underlies variation in colour matching. *Nature*, 356, 431–433.

Winderickx, J., Sanocki, E., Lindsey, D.T., Teller, D.Y., Motulsky, A.G., & Deeb, S.S. (1992c) Defective colour vision associated with a missense mutation in the human green visual pigment gene. *Nature Genetics*, 1, 251–256.

Winick, J.D., Blundell, M.L., Galke, B.L., Salam, A.A., Leal, S.M., & Karayiorgou, M. (1999) Homozygosity mapping of the achromatopsia locus in the Pingelapese. *American Journal of Human Genetics* (in press).

Wissinger, B., Jägle, H., Kohl, S., Broghammer, M., Baumann, B., Hanna, D.B., Hedels, C., Apfelstedt-Sylla, E., Anastasi, M., Jacobson, S.G., Zrenner, E., & Sharpe, L.T. (1998) Human rod monochromacy: Linkage analysis and mapping of a candidate gene expressed in cone photoreceptors. *Genomics*, 51, 325–331.

Wohlgemuth, A. (1911) On the aftereffect of seen movement. *British Journal of Psychology Monographs (Suppl.)*, 1.

Wolf, S., Sharpe, L.T., Schmidt, H.-J.A., Knau, H., Weitz, S., Kioschis, P., Poustka, A., Zrenner, E., Lichter, P., & Wissinger, B. (1999) Direct visual resolution of gene copy number in the human photopigment gene array. *Investigative Ophthalmology & Visual Science*, 40 (in press).

Worthey, J. (1985) Limitations of color constancy. *Journal of the Optical Society of America A*, 2, 1014–1026.

Wright, W.D. (1928–29) A re-determination of the trichromatic coefficients of the spectral colours. *Transactions of the Optical Society*, 30, 141–164.

Wright, W.D. (1952) The characteristics of tritanopia. *Journal of the Optical Society of America*, 42, 509–521.

Wright, W.D. & Pitt, F.H.G. (1934) Hue-discrimination in normal colour-vision. *Proceedings of the Physical Society (London)*, 46, 459–473.

Würger, S.M. & Landy, M.S. (1993) Role of chromatic and luminance contrast in inferring structure from motion.

Journal of the Optical Society of America A, 10, 1363–1372.

Wyszecki, G. & Fielder, G.H. (1971) New color-matching ellipses. *Journal of the Optical Society of America*, 61, 1135–1152.

Wyszecki, G. & Stiles, W.S. (1967) *Color Science: Concepts and Methods, Quantitative Data and Formulae,1st Ed.* Wiley & Sons, New York.

Wyszecki, G. & Stiles, W.S. (1982a) *Color Science. Concepts and Methods, Quantitative Data and Formulae, 2nd Ed.* John Wiley & Sons, New York.

Wyszecki, G. & Stiles, W.S. (1982b) High-level trichromatic color matching and the pigment-bleaching hypothesis. *Vision Research,* 20, 23–37.

Yamada, E.S. (1995) *Organizacao Morfofuncional do Sistema Visual de Primates Platirrinos.* Thesis/Dissertation, Belem, Brazil.

Yamada, E.S., Silveira, L.C.L., Gomes, F.L., & Lee, B.B. (1996) The retinal ganglion cell classes of New World primates. *Revista Brasileira de Biologia (Suppl. 1),* 56, 381–396.

Yamada, E.S., Silveira, L.C.L., & Perry, V.H. (1996) Morphology, dendritic field size, somal size, density and coverage of M and P retinal ganglion cells of dichromatic *Cebus* monkeys. *Visual Neuroscience*, 13, 1011–1029.

Yamaguchi, T., Motulsky, A.G., & Deeb, S.S. (1997) Visual pigment gene structure and expression in human retinae. *Human Molecular Genetics*, 6, 981–990.

Yamashita, M. & Wässle, H. (1991) Responses of rod bipolar cells isolated from the rat retina to the glutamate agonist 2-amino-4-phosphonobutyric acid (APB). *Journal of Neuroscience*, 11, 2372–2382.

Yang, G. & Masland, R.H. (1994) Receptive fields and dendritic structure of directionally selective retinal ganglion cells. *Journal of Neuroscience,* 14, 5267–5280.

Yasuma, T., Tokuda, H., & Ichikawa, H. (1984) Abnormalities of cone photopigments in genetic carriers of protanomaly. *Archives of Ophthalmology*, 102, 897–900.

Yates, J.T. (1974) Chromatic information processing in the foveal projection (area striata) of unanesthetized primate. *Vision Research*, 14, 163–173.

Yau, K.-W. (1994) Phototransduction mechanism in retinal rods and cones – the Friedenwald lecture. *Investigative Ophthalmology & Visual Science,* 35, 9–32.

Yeh, T., Lee, B.B., & Kremers, J. (1996) The time course of adaptation in macaque ganglion cells. *Vision Research*, 36, 913–931.

Yeh, T., Lee, B.B., Kremers, J., Cowing, J.A., Hunt, D.M., Martin, P.R., & Troy, J.B. (1995) Visual responses in the lateral geniculate nucleus of dichromatic and trichromatic marmosets (*Callithrix jacchus*). *Journal of Neuroscience,* 15, 7892–7904.

Yilmaz, H. (1962) Color vision and a new approach to color perception. In *Biological Prototypes and Synthetic Systems, Vol. 1*. Plenum, New York.

Yokota, A., Shin, Y., Kimura, J., Senoo, T., Seki, R., & Tsubota, K. (1990) Congenital deuteranomaly in one of monozygotic triplets. In *Color Vision Deficiencies VIII* (ed. Ohta, Y.), pp. 199–203. Kugler & Ghedini, Amsterdam.

Yokoyama, S., Starmer, W.T., & Yokoyama, R. (1993) Paralogous origin of the red- and green-sensitive visual pigment genes in vertebrates. *Molecular Biology and Evolution*, 10, 527–538.

Yokoyama, S. & Yokoyama, R. (1989) Molecular evolution of human visual pigment genes. *Molecular Biology and Evolution*, 6, 186–197.

Yoshioka, T. & Hendry, S.H.C. (1994) Immunocytochemical and quantitative analyses of a third geniculocortical population in the macaque LGN. *Investigative Ophthalmology & Visual Science (Suppl.),* 35, 1975.

Yoshioka, T., Levitt, J.B., & Lund, J.S. (1994) Independence and merger of thalamocortical channels within macaque monkey primary visual cortex: anatomy of interlaminar projections. *Visual Neuroscience*, 11, 467–489.

Young, N. (1988) *An Introduction to Hilbert Space*. Cambridge University Press, Cambridge.

Young, R.S.L. & Price, J. (1985) Wavelength discrimination deteriorates with illumination in blue cone monochromats. *Investigative Ophthalmology & Visual Science*, 26, 1543–1549.

Young, T. (1802) The Bakerian lecture: On the theory of light and colours. *Philosophical Transactions of the Royal Society, London*, 92, 12–48.

Young, T. (1807) *A Course of Lectures on Natural Philosophy and the Mechanical Arts*. J. Johnson, London.

Yund, E.W. & Armington, J.C. (1975) Color and brightness contrast effects as a function of spatial variables. *Vision Research,* 15, 917–929.

Za, S., Iverson, G., & D'Zmura, M. (1997) Phase selectivity in contrast induction. *Investigative Ophthalmology & Visual Science,* 38, S632.

Zahn, C.T. & Roskies, R.Z. (1972) Fourier descriptors for plane closed curves. *IEEE Transactions on Computing,* 21, 269–281.

Zaidi, Q. (1989) Local and distal factors in visual grating induction. *Vision Research*, 29, 691–697.

Zaidi, Q. (1990) Apparent brightness in complex displays: A reply to Moulden and Kingdom. *Vision Research*, 30, 1253–1255.

Zaidi, Q. (1992) Parallel and serial connections between human color mechanisms. In *Applications of Parallel Processing in Vision* (ed. Brannan, J.R.), pp. 227–259. Elsevier Science Publishers, New York.

Zaidi, Q. (1998) Identification of illuminant and object colors: heuristic-based algorithms. *Journal of the Optical Society of America A*, 15, 1767–1776.

Zaidi, Q., DeBonet, J.S., & Spehar, B. (1995) Perceived grey-levels in complex configurations. *Proceedings of the Third Annual IS&T/SID Color Imaging Conference*, 14–17.

Zaidi, Q. & Halevy, D. (1993) Visual mechanisms that signal the direction of color changes. *Vision Research*, 33, 1037–1051.

Zaidi, Q. & Sachtler, W. (1991) Motion adaptation from surrounding stimuli. *Perception*, 20, 703–714.

Zaidi, Q. & Shapiro, A.G. (1993) Adaptive orthogonalization of opponent-color signals. *Biological Cybernetics*, 69, 415–428.

Zaidi, Q., Shapiro, A., & Hood, D. (1992) The effect of adaptation on the differential sensitivity of the S-cone color system. *Vision Research*, 32, 1297–1318.

Zaidi, Q., Spehar, B., & DeBonet, J. (1997) Color constancy in variegated scenes: Role of low-level mechanisms in discounting illumination changes. *Journal of the Optical Society of America A*, 14, 2608–2621.

Zaidi, Q., Spehar, B., & DeBonet, J. (1998) Adaptation to textured chromatic fields. *Journal of the Optical Society of America A*, 15, 23–31.

Zaidi, Q., Spehar, B., & Shy, M. (1997) Induced effects of backgrounds and foregrounds in pictures of three-dimensional configurations: The role of T junctions. *Perception*, 26, 395–408.

Zaidi, Q., Yoshimi, B., & Flannigan, N. (1991) The influence of shape and perimeter length on induced color contrast. *Journal of the Optical Society of America A*, 8, 1810–1817.

Zaidi, Q., Yoshimi, B., Flannigan, N., & Canova, A. (1992) Lateral interactions within color mechanisms in simultaneous induced contrast. *Vision Research*, 32, 1695–1707.

Zaidi, Q. & Zipser, N. (1993) Induced contrast from radial patterns. *Vision Research*, 33, 1281–1286.

Zanen, J. & Meunier, A. (1958a) Disparité de la perception chromatique chez des jumelles univitellines. *Bulletin Societé de Belge Ophthalmologie*, 118, 356–368.

Zanen, J. & Meunier, A. (1958b) Nouvelle observation de disparité de la perception chromatique chez des jumeiles univitellines. *Bulletin Societé de Belge Ophthalmologie*, 119, 444–450.

Zeki, S.M. (1973) Colour coding in rhesus monkey prestriate cortex. *Brain Research*, 53, 422–427.

Zeki, S.M. (1974) Functional organization of a visual area in the posterior bank of the superior temporal sulcus of the rhesus monkey. *Journal of Physiology*, 236, 549–573.

Zeki, S.M. (1977) Colour coding in the superior temporal sulcus of rhesus monkey visual cortex. *Proceedings of the Royal Society, London, Series B*, 197, 195–223.

Zeki, S.M. (1978a) Uniformity and diversity of structure and function in rhesus monkey prestriate visual cortex. *Journal of Physiology*, 277, 273–290.

Zeki, S.M. (1978b) Functional specialization in the visual cortex of the rhesus monkey. *Nature*, 274, 423–428.

Zeki, S.M. (1980) The representation of colours in the cerebral cortex. *Nature*, 284, 412–418.

Zeki, S.M. (1983a) Colour coding in the cerebral cortex: the reaction of cells in monkey visual cortex to wavelengths and colours. *Neuroscience*, 9, 741–765.

Zeki, S.M. (1983b) Colour coding in the cerebral cortex: the responses of wavelength-selective and colour-coded cells in monkey visual cortex to changes in wavelength composition. *Neuroscience*, 9, 767–781.

Zeki, S. (1990) A century of cerebral achromatopsia. *Brain*, 113, 1721–1777.

Zeki, S.M. & Shipp, S. (1988) The functional logic of cortical connections. *Nature*, 335, 311–317.

Zeki, S.M. & Shipp, S. (1989) Modular connections between area V2 and area V4 of macaque monkey visual cortex. *European Journal of Neuroscience*, 1, 494–506.

Zeki, S., Watson, J.D.G., Lueck, C.J., Friston, K.J., Kennard, C., & Frackowiak, R.S.J. (1991) A direct demonstration of functional specialization in human visual cortex. *Journal of Neuroscience*, 11, 641–649.

Zhou, Z.J., Marshak, D.W., & Fain, G.L. (1994) Amino acid receptors of midget and parasol ganglion cells in primate retina. *Proceedings of the National Academy of Sciences, USA*, 91, 4907–4911.

Zrenner, E. & Gouras, P. (1981) Characteristics of the blue sensitive cone mechanism in primate retinal ganglion cells. *Vision Research*, 21, 1605–1609.

Zrenner, E. & Gouras, P. (1983) Cone opponency in tonic ganglion cells and its variation with eccentricity in rhesus monkey retina. In: *Colour Vision: Physiology and Psychophysics* (eds. Mollon, J. & Sharpe, L.T.), pp. 211–223. Academic, New York.

Zrenner, E., Magnussen, S., & Lorenz, B. (1988) Blauzapfenmonochromasie: Diagnose, genetische Beratung und optische Hilfsmittel. *Klinische Monatsblätter für Augenheilkunde und augenärztliche Fortbildung*, 193, 510–517.

Zrenner, E., Nelson, R., & Mariani, A. (1983) Intracellular recordings from a biplexiform ganglion cell incë Macaque retina, stained with horseradish peroxidase. *Brain Research*, 262, 181–185.

Author index

A

B

C

D

E

F

G

H

I

L

M

N

O

P

Q

R

S

T

U

V

W

Y

Z

Subject index

A

B

C

D

E

F

G

H

I

K

L

M

N

O

P

R

S

T